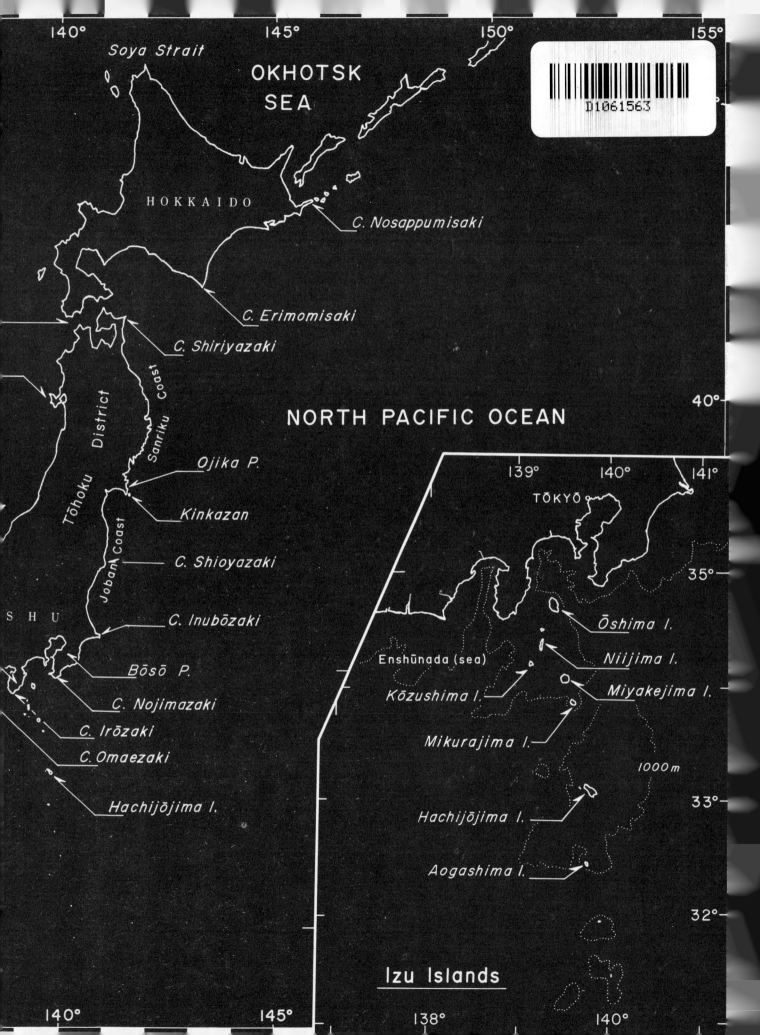

OKHOTSK SEA

Soya Strait

HOKKAIDO

C. Nosappumisaki

C. Erimomisaki

C. Shiriyazaki

NORTH PACIFIC OCEAN

Sanriku Coast

Tōhoku District

Ojika P.

Kinkazan

Jōban Coast

C. Shioyazaki

C. Inubōzaki

S H U

Bōsō P.

C. Nojimazaki

C. Irōzaki

C. Omaezaki

Hachijōjima I.

TŌKYŌ

Ōshima I.

Niijima I.

Enshūnada (sea)

Kōzushima I.

Miyakejima I.

Mikurajima I.

1000m

Hachijōjima I.

Aogashima I.

Izu Islands

KUROSHIO

PHYSICAL ASPECTS OF THE JAPAN CURRENT

"This is an International Book Year Publication"

Edited by HENRY STOMMEL and KOZO YOSHIDA

UNIVERSITY OF WASHINGTON PRESS

Seattle and London

The Japanese edition of this book is entitled
KUROSHIO: Its Physical Aspects.

Printed in Japan

UNIVERSITY OF WASHINGTON PRESS edition,
1972

Library of Congress Cataloging in Publication Data

Stommel, Henry   1920-
    Kuroshio: physical aspects of the Japan Current.

    Bibliography: p.
    1. Oceanography—Kuroshio. I. Yoshida, Kōzō,
1922-   joint author. II. Title.
GC296.K85S8   551.4′75′55   72-378
ISBN 0-295-95225-3

Contributors

Hideo Kawai

Department of Fisheries
Faculty of Agriculture
Kyoto University
Kyoto, Japan

Jotaro Masuzawa

Marine Division
Japan Meteorological Agency
Tokyo, Japan

Akio Mogi

The 7th Maritime Safety Headquarters
Maritime Safety Agency
Moji, Japan

Shigeo Moriyasu

Meteorological Research Institute
Koenji, Tokyo, Japan

Hideo Nitani

Hydrographic Department
Maritime Safety Agency
Tokyo, Japan

Daitaro Shoji

Hydrographic Department
Maritime Safety Agency
Tokyo, Japan

Henry Stommel

Department of Meteorology
Massachusetts Institute of Technology
Cambridge, Mass. 02139, USA

Bruce A. Taft

Scripps Institution of Oceanography
University of California, San Diego
La Jolla, Calif. 92037, USA

Toshihiko Teramoto

Ocean Research Institute
University of Tokyo
Tokyo, Japan

Thomas Winterfeld

National Oceanographic Data Center
National Oceanic & Atmospheric Administration
Rockville, Md.20852, USA

L. V. Worthington

Woods Hole Oceanographic Institution
Woods Hole
Mass. 02543, USA

Kozo Yoshida

Geophysical Institute
University of Tokyo
Tokyo, Japan

Yoko Shimizu

Geophysical Institute
University of Tokyo
Tokyo, Japan

Preface

ORIGIN OF THE TREATISE

In 1964 plans were being laid for the Cooperative Study of the Kuroshio, an international field study which was to go on for many years. One of the difficulties experienced at the time was the lack of a book about the Kuroshio that summarized the work done on the Kuroshio up until that time. We were forced, therefore, to make plans for future work without the benefit of a systematic review and assessment of the considerable material already at hand. One of the difficulties of preparing such a review was obviously that there really was a great deal of material to study; the region of the Kuroshio had more observations made in it than any other deep-sea region in the world. Even though much of this data was tabulated in oceanographic data centers, and susceptible to computer compilation, it was evident that the task of discussing it would be formidable. Certainly the Japanese observational material deserved a deeper and more thorough discussion than I had been able to afford it in observations in my book on the Gulf Stream and it seemed that a single author would be unequal to the task. We decided, therefore, to build the treatise with chapters written by different authors. There has been some exchange of manuscripts between various authors in the process of preparing them for the treatise, and in this way we have tried to avoid too much overlap and duplication. On the whole we think we have been able to provide a fairly balanced and complete account of work done prior to 1964, and, therefore, a foundation for future workers concerned with the new results of the Cooperative Study of the Kuroshio. We hope our treatise will also be useful to students who may find it a helpful guide through the intricacies of the literature of past studies listed in the Bibliography.

Throughout all stages of the preparation of this treatise we have enjoyed the support of the Japan Society for the Promotion of Science and the U.S. National Science Foundation, under the auspices of the U.S.-Japan Cooperative Science Program.

FUTURE STUDIES OF THE KUROSHIO

During the next five years we may hope to see many new results from the Cooperative Study of the

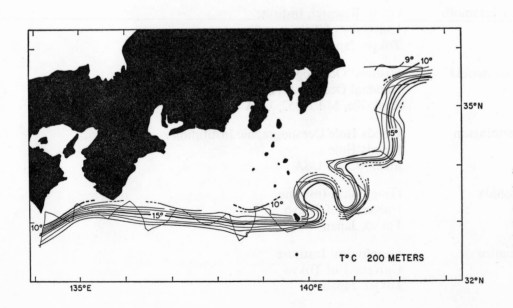

FIG. 1 A chart of temperature at 200 meters as determined by *Atlantis II,* September 12-15, 1965. The zig-zag line is the track of the ship, along which the temperature observations were made (courtesy of L. V. Worthington).

Kuroshio. Certainly the excellent hydrographic sections which are being accumulated should serve to clarify the general hydrographic conditions in the large oceanic region of the northwestern Pacific Ocean surrounding the Kuroshio. However, much of the C.S.K. program is carried out with station spacing and frequency of observation which are more suited to "climatology" of the whole Kuroshio region than to detailed synoptic mapping of such features as the development of rapidly moving meanders, shifting of the axis of the current, breaking off eddies, etc. These, of course, are very interesting features and various theoretical models are being developed in attempts to compute, and even to predict their behavior. We cite the work of Warren[1] (1963), of Robinson and Niiler[2] (1967) and of Niiler and Robinson[3] (1967) as the first stage in the development of predictive models. Given adequate synoptic observational material, one could hope for rapid progress in numerical modeling of these features. Consider, for example, the remarkable shape of the axis of the Kuroshio revealed by a simple zig-zag bathythermograph survey in 1965, shown in Fig. 1. The current seems to be tightly kinked, or "wrapped," around the island of Hachijo Jima. The studies of tide guage records at Hachijo Jima by D. Shoji[4,5,6]

(1957a,b, 1961) and S. Yosida[7] (1961) indicate that sometimes the current flows north and sometimes south of the island. The evolution of such a stream pattern in time could be determined by making weekly cruises of the zig-zag variety over several years. A series of 150–250 such weekly maps would provide a powerful stimulus to the development of theoretical and predictive models. If such knowledge could be successfully acquired, it seems it would constitute substantial increase in the understanding of the dynamics of the Kuroshio. There is no large study of the Gulf Stream in progress that can compare in scope with the C.S.K., nor does there appear to be any hope for one in the foreseeable future. It seems to be a matter of fact that oceanographic activity in the Kuroshio area is more intense than in the Gulf Stream area. If the momentum and enthusiasm evident in C.S.K. can be extended to finer scale features of the Kuroshio, then some very fundamental understanding of the dynamics of western boundary currents can be applied to other currents, such as the Gulf Stream, Brazil Current, Somali Current, etc. Thus the Japanese studies of the Kuroshio could be of great value to oceanographers all over the world.

Henry Stommel
Paris, 10 May 1970

[1] Warren, B. A. (1963): Topographic influences on the path of the Gulf Stream. *Tellus,* **15**(2): 167–183.

[2] Robinson, A. R. and P. P. Niiler (1967): The theory of free inertial currents. I. Path and structure. *Tellus,* **19**(2): 269–291.

[3] Niiler, P. P. and A. R. Robinson (1967): The theory of free inertial jets. II. A numerical experiment for the path of the Gulf Stream. *Tellus,* **19**(4): 601–619.

[4] Shoji, Daitaro (1957a): Kuroshio during 1955. *Rec. Oceanogr. Wks. Japan,* Spec. No. **1**: 21–26.

[5] ——— (1957b): On the variations of daily mean sea levels and the Kuroshio from 1954 to 1955. *Proc. UNESCO Symp. Phys. Oceanogr.*: 130–136.

[6] ——— (1961): On the variations of the daily mean sea levels along the Japanese Islands. *J. Oceanogr. Soc. Japan,* **17**(3): 141–152.

[7] Yosida, Shozo (1961): On the short period variation of the Kuroshio in the adjacent sea of the Izu Islands. *Hydrogr. Bull., New Ser.,* **65**: 1–18.

Contents

INDEX CHART FOR CHAPTERS

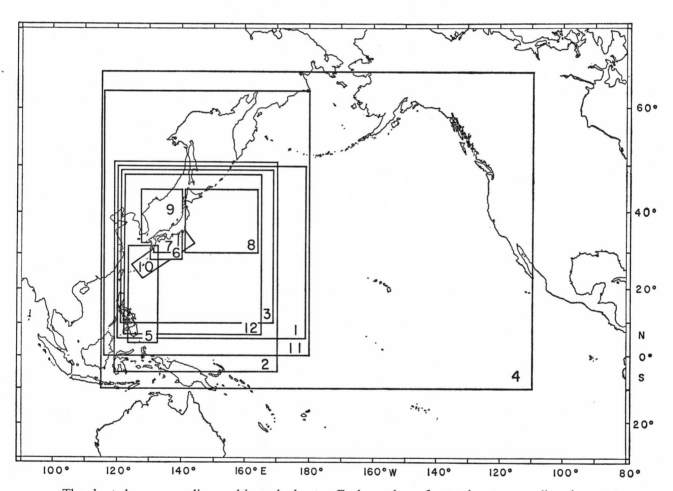

The chart shows areas discussed in each chapter. Each number refers to the corresponding chapter.

Chapter 1

HISTORY OF THE JAPANESE OBSERVATION PROGRAM OF THE KUROSHIO AND ADJACENT REGIONS

Toshihiko Teramoto

Ocean Research Institute, University of Tokyo, Tokyo, Japan.

1 INTRODUCTION

"It is said that in the sea between the Izu Peninsula and an uninhabited islet 130 to 140 *ri** south of the Peninsula there is a region called the Kuroshio, where a strong stream 30 to 40 *ri* wide is found. This stream is like a large river in the sea and has surges of water rolling back against the stream. It is further said that at some distance from the Boso Peninsula the water flows eastward and that ships, if caught in this stream, can never return." This is an approximate translation of a description of the Kuroshio found in the sequel to the book *Saiyuki,* written by Nankei Tachibana and published in the period from 1795 to 1799. Tachibana was a famous physician as well as a famous traveller who made a tour round Japan, and the *Saiyuki* was a journal of a part of the tour. According to Uda (1969), it may be the oldest description of the Kuroshio written by a Japanese. In a book titled *Hachijo-Hikki,* which was published a little later than Tachibana's work, a description of the Kuroshio is also found. One interesting section runs in part: "In the region of the Kuroshio, the sea is as dark as if an ink stick had been rubbed on its surface, and many hundreds of eddies can be seen floating downstream in succession." The characteristics of the Kuroshio are very well expressed in these descriptions.

The Kuroshio is very closely associated with the lives of the Japanese, and much information has been accumulated through the experiences of the people, especially fishermen and navigators, from ancient times, although few written records were kept. There was of course no systematic approach to the subject until modern scientific methods were introduced into

Japan in the latter half of the 19th century.

Scientific observations of the sea in Japan were initiated and maintained primarily by government agencies concerned with fisheries, hydrography and meteorology, and their approach was based largely on their own respective requirements. Since the Kuroshio has a great influence on the fishing, hydrography and meteorology in the vicinity of Japan, many of the observations were conducted in the Kuroshio region. Although there are differences in the ultimate purpose of their observations, the emphasis has generally been on the fundamental investigation of oceanographical conditions: namely, conditions of water movement and distribution of water properties. In the early stages, pioneering scientists established the foundations for future observations. Since then these observations have been routinely conducted, although there has been a rise and fall in activity due to the influence of national events such as wars. Thus, the synoptic features** of oceanographical conditions in the seas adjacent to Japan have been studied, and exchanges of methods of observation as well as of data have, in general, been actively supported by the agencies. After World War II, conferences were initiated among the agencies in order to work out effective observation plans for each and to eliminate overlaps and gaps in the various observation networks. But it was found impossible to establish a systematic national program planned to focus upon specified subjects, because each agency had its own purposes for observation. In the observation plan of each agency the emphasis was usually on forming as vast an observation network as possible to cover the seas adjacent to Japan. Therefore, little effort could be expended on digging structures or mechanisms of

---

* One *ri* was equivalent to about 4 km about 350 years ago but had been ordinarily used as equivalent to about 650 m until 1816.

** According to Hideo Kawai (1969), the term "synoptic" features should be used to represent oceanographical conditions as they exist simultaneously over a broad area. But in this chapter the term was used to represent oceanographical conditions obtained from observations of a broad area made with the expectation to give a certain joint space-time average of the conditions.

complicated processes. Thus, these questions are still largely unexplored.

This chapter will provide a brief historical review of the observation programs for the Kuroshio and its adjacent seas from the standpoint of physical oceanography. The description is based on the activities of the three agencies previously mentioned. It consists of two parts, the first describing the period before the end of World War II and the second the period after the war. These two periods are in turn divided into two stages, and for each an introductory summary will be made. A chronological table of the programs and cruise-track charts compiled by Masahito Sakurai are given in the appendix to indicate the type of observations made by any agency in a given area and at a given time. In the "Remarks" column of the table, publication data for the observations in question are provided.

2 OBSERVATIONS IN THE PERIOD UP TO THE END OF WORLD WAR II

2.1 Observation in the preliminary stage

In this stage efforts were mainly devoted to establishment of observation methods and to organization of observation systems.

Fisheries agencies

As was mentioned earlier, much information on oceanographical conditions was accumulated in Japan through the experiences of fishermen and navigators working at sea. In the field of fisheries, oceanographic activity commenced with the selection and subsequent compilation of the accumulated information. At the request of Daisuke Suzuki, director of Suisan-kyoku (the Fisheries Bureau), this work was accomplished by Shin'nosuke Matsubara and was published as a report titled "Suisan yosatsu chosa hokoku" (Fisheries preliminary investigation reports) in 1889. This work is pertinent as a first step in oceanographic investigation.

In 1893, the first governmental institute for oceanographical and fisheries inquiry, Suisan Chosa-Sho (the Fisheries Investigation Institute), was established. It had a consultative committee, called Suisan Chosa-Kai (Board of Fisheries Inquiry). At a meeting of this committte, Yuji Wada, a member of the committee and the staff of Chuo Kisho-Dai (the Central Meteorological Observatory), emphasized the pressing necessity for synoptic surveys of surface currents by means of drifting bottles. In spite of the strong objections of some members, his proposal was

adopted and put into practice. Unfortunately, however, the institute was abolished four years later because of financial difficulties caused by a marked increase in the use of funds for naval and military purposes, influenced by strained relations between Japan and Russia, and the survey plan had to be given up halfway. Through Wada's untiring effort, the survey was resumed later on an expanded scale as will be described in the following section. The abolishment of the Fisheries Investigation Institute was followed by the establishment of the considerably smaller Suisan Chosa-ka (the Fisheries Inquiry Section) of Fisheries Bureau.

Tasaku Kitahara, a young official working at the section, saw the necessity for conducting fundamental investigations of the motion and properties of sea water for developing fisheries. He made a proposal emphasizing the need for such investigations but his suggestion was not immediately accepted, and he retired from the bureau. In 1901 the first meeting of the International Council for the Exploration of the Sea was held, and the Japanese participant was the chief of the Fisheries Inquiry Section. The activities of northern European countries in oceanographic investigations of the North Sea, made in close association with fisheries, greatly impressed the Japanese participant. Subsequently, Kitahara's proposal was re-evaluated. He was called back to the bureau in 1904 and devoted himself to the fundamental investigation of fisheries. He invented observation instruments such as a sea-water sampler, a plankton net and a dredger. At about that time Tokuro Akanuma developed a new areometer, which has since been used widely for rough but rapid measurements of the specific gravity of sea water. Since 1910 hydrographic sections had gradually been made in the Tokyo, the Sagami and the Suruga Bays by Kintaro Kitahara, Okamura, Akanuma and other staff members of the section. The Kuroshio flows just south of these bays, and oceanographic conditions of the bays are closely related to changes of the Kuroshio. In 1912 chlorinity determination through titration of silver nitrate was introduced by Hisatoshi Marukawa, who was a professor at Suisan Koshu-Jo (the Fisheries Training Institute). The regions of oceanographic observation were gradually extended to the Yellow Sea, the East China Sea and the Okhotsk Sea.

In 1914, a minor change in organization was made and the fundamental investigations were transferred to the superintendence of the Fisheries Training Institute; at the same time the staff engaging in the investigations was also transferred. That same year, Hikotaro Asano of the institute attempted to make dynamic computations of the geostrophic current in the Kuroshio region and compared the computed current with that measured by Ekman-Merz current meters. These fundamental investigations were essentially oceanographic observations. It should be noted that these early workers placed great weight on fundamental oceanographic investigations, and stressed the need for working with physicists in planning and conducting oceanographic observations and in interpreting their results. A few years before at the suggestion of Marukawa, who was at that time studying oceanography in Berlin, Keisuke Shimo, the director of the institute, took steps to invite an active physicist to the institute as professor. As a result, Torahiko Terada, who was a professor of physics at the Imperial University of Tokyo, came to lecture on experimental physics on a part-time basis for subsequent several years. In addition, he took part in planning and carrying out oceanographic observations. An excellent physicist with inexhaustible scientific inspiration, he provided many useful suggestions for the oceanographers. The observation program which was established under his lead consisted of two parts: regular observations at specified coastal and off-shore stations, and synoptic oceanographic observations at specified sections in the sea adjacent to Japan. From the first study temporal variations in oceanographic conditions were expected to be revealed, and from the second, synoptic oceanographic conditions. The method for oceanographic observations was modelled after that used in the northern European countries.

The long-cherished desire of these first pioneers of modern fisheries was fulfilled with the allocation of sufficient expense-money for fundamental investigations of oceanographic conditions in the national budget in 1917. In the next year, Kaiyo Chosa-Bu (Oceanographical Inquiry Division) was established at the institute, and the *Ten'o-Maru* (a ship of 161 gross tons) was completed. The following year, publication of a monthly oceanographic chart and of a quarterly inquiry bulletin (Kaiyo Chosa-Yoho) was begun for quick reporting of observational results. In the charts and bulletins, synopses of fishing and oceanographic conditions in the sea adjacent to Japan were summarized. In 1925, the *Soyo-Maru* (202 gross tons) was constructed and the *Ten'o-Maru* was retired. For about 30 years the *Soyo-Maru* engaged in oceanographic observations and made a great contribution to observational researches.

In addition to these activities at the Fisheries Training institute, efforts in oceanographic observation were made at local Fisheries Experimental Stations in Korea, Formosa, Sakhalin and many prefectures, with emphasis on fisheries. The work achieved at the stations of the Government-General of Korea deserves special mention here. Since 1915, accurate observations had been made by Keizo Nishida and others at the station. This group developed a method for measuring velocity profiles by simultaneous use of two current meters, one at a specified deep layer as reference and the other at the desired layer of depth. The method was later used by the Hydrographic Office of the Japanese Navy.

In 1929, Norin-Sho Suisan Shiken-Jo (the Imperial Fisheries Experimental Station of the Ministry of Agriculture and Forestry) was established, and Fisheries Inquiry Division of the Fisheries Training Institute was transferred to it. Shin'ichi Kasuga, director of the station, persuaded the directors of the local fisheries stations to organize a committee (Suisan Renraku Shiken Uchiawase-Kai) at which arrangements for exchanges with respect to oceanographic observations could be made. Staffs from 57 stations engaging in observations around the country gathered to participate in the meeting. During the assembly, important subjects to be studied through future observations were selected, and the policies and techniques for carrying out these studies were discussed. Subsequently, the assembly became an annual event. The opening of the first assembly was epoch-making with regard to oceanographic observations in the field of fisheries. The event is considered to mark the second stage of observations, in which studies by all the Experimental Fisheries Stations in Japan were conducted systematically under preplanned programs.

In the preliminary stage, most of the observations were carried out in the region of the Kuroshio and its branch currents, but they were not always conducted systematically. The observations at this stage, however, can still be considered significant in initiating observational investigations of the ocean on the one hand and in contributing to a partial synopsis of oceanographic conditions on the other. I regard Kitahara as the man responsible for initiating oceanographic observations in Japan, and Terada as the greatest contributor to the development of oceanographic observations.

Hydrographic Office of the Japanese Navy

In the initial stage of oceanic surveying, the Hydrographic Office made special efforts in the sounding of the coastal and adjacent seas of Japan. Although observations of oceanographic conditions were widely recognized as important to ensure safe navigation, the office was unable to undertake it because of work with the sounding. In connection with coastal sounding, measurements of tidal elevation were required.

Tidal measurements were first made at Shinagawa, Tokyo and continued for 40 days, from January to February 1882. After that temporal measurements of tidal elevation were made at various places. The establishment of a permanent tidal station was realized much later. The first station was established at Tsukiji, Tokyo in 1931, and thereafter the number of stations gradually increased. Although the initial reason for establishing the stations was to use them as a reference for soundings, records from tidal stations in regions adjacent to ocean currents later proved to reflect current fluctuations, and much effort was later expended in constructing tidal stations in the regions of the Kuroshio and its branch currents in order to examine the dynamic structure of the currents. Oceanographic observations from a survey ship of the office were first obtained in 1909. Until this time, the collection of oceanographic data such as surface water temperatures and dead reckoned velocity was carried out with the help of naval and merchant ships. In order to simplify data processing, data presentation was standardized in 1881 on the format of a ship's log-books, and all Japanese ships were asked to employ the standardized form. These data were to be routinely summarized for every ten days, published three times a month, and distributed. In 1911, the office conducted the first oceanographic observations in which tracking of a drift buoy was made by the survey ship (Sokuryo Kan) *Kuzuki* (1500 displacement tons) in the Straits of Osumi and in the Kuroshio region east of Kyushu. In the following year this current measurement was tested by the survey ship *Matsue* (2550 displacement tons) while drifting along the Kuroshio between Shionomisaki and Choshi. Until about 1929, the following three methods of current measurement by buoy tracking were employed. In the first, a buoy with surface drogue was thrown overboard from the moored ship; the wire was paid out as the buoy moved. By measuring the length of paid-out wire in relation to the elapsed time movement of the buoy and so of the surface-water was obtained. In the second and third methods, a drift buoy with a flag on top of a vertical rod was thrown from the ship, and by tracking the buoy the water movement could be followed. In order to explore the water movement in a rectangular area 30 to 40 miles in length and width, 20 to 30 buoys were released at intervals of 1 to 2 miles and were tracked throughout one day. Numerous measurements were made at various places in the western North Pacific from 1933 to 1938. From these measurements interesting features of small-scale eddies, convergences and divergences were often found (Kishi, 1967).

In addition, drift bottles were also released by the office. This method was first tried in 1894 and was extensively used in the period from 1920 to 1930. The number of bottles released every year during the period amounted to about 10,000, and 500 to 700 bottles were picked up. Konishi (1921a, 1921b, 1922a, 1922b, 1923) of the office processed the accumulated data from the drift bottle experiments starting in 1908 and published the results. The synoptic current-patterns in the sea adjacent to Japan were illustrated.

In 1917, Hydrographic Office finished a preliminary sounding in the coastal area of Japan. As the next step, the office decided to devote time to oceanographic surveys in off-shore regions in addition to more detailed soundings in the coastal area. The oceanographic surveys included sounding, sampling of bottom materials, current measurement and measurement of temperatures, densities and the chemical composition of sea water at specified subsurface layers. Ekman-Merz current meters were imported to carry out the current measurements. Prior to making the surveys, current charts of the North Pacific for four seasons were published in 1917. The charts were based on dead-reckoned velocities reported by Japanese naval and merchant ships during the 26-year period from 1891 to 1916. In 1918, trial surveys were conducted, and in the following year yearly survey programs were planned. The next five years were devoted to training for the oceanographic surveys. In 1923, the *Matsue* succeeded in making a hydrographic cast down to a depth of 400 m in the region between the Bonin Islands and the Palau Islands. The sampling bottles used were developed by Kitahara for water sampling in shallow and intermediate layers. Following this success, every survey ship began to make attempts at hydrographic casts.

In 1924, the survey ship *Yamato* (1330 displacement tons) made three hydrographic sections in the Japan Sea. Ryoichi Shigematsu, captain of the ship, analyzed the data and reported the results. From 1925 to 1930 the survey ship *Manshu* (3510 displacement tons) carried out extensive oceanographic surveys throughout the western North Pacific. The number of hydrographic stations reached more than 1000, and at many of the stations hydrographic casts down to layers deeper than 5000 m were successfully made using Nansen bottles. Shigematsu, the former captain of the *Manshu*, analyzed the observation data from these stations, and his results were in part reported and published in the *Hydrographic Report* (vol. 6, 1933). The report comprised 500 pages with additional charts, and summarized current patterns and distributions of water properties at selected layers. Shigematsu reported on the characteristics and accuracy of the instruments used in addition to the analytical results of observations, and pointed out the defects of certain instruments and facilities. On the basis of his advice and experiences at sea, Saburo Kishindo and Tamejiro Kondo, both of whom were members of Survey Section of the office, began the laborious work of overcoming some of these defects. Existing instruments and facilities were improved, and some technically superior instruments were imported. In 1933 the survey ship *Komahashi* (1230 displacement tons), with new instrumentation, greatly intensified her functions in oceanographic surveys, and began full surveys in the western North Pacific. Soon afterwards the survey ship *Koshu* (2080 displacement tons) and the *Katsuriki* (1540 displacement tons) also joined in the surveys with new and more powerful instruments and facilities. Prior to these surveys Kishindo had intended to introduce the method of dynamic computations for determining geostrophic current and had made a study of the method. The method had already been successfully put into practice in the Atlantic, and it was his intention to apply it to current estimates in the Pacific. After studying oceanographic conditions in seas adjacent to Japan for three years, he proposed a practical survey method including the computations. His proposal was not readily accepted by the head of the navy staff because it seemed much too academic. The importance of oceanographic surveys was still not adequately recognized at the time, and in spite of the acknowledged need of the office to increase the number of members who could lead surveys, the existing leaders numbered only five.

The Manchurian Incident in 1931 gave the navy an opportunity to intensify surveys which were needed only for naval movements and to neglect surveys for maritime safety. In conjunction with the intensification of protection of military secrets, most of current charts and data reports published by the office were limited for circulation within the navy from 1933.

From September to October 1936, the Japanese Combined Fleet held grand naval maneuvers off the Kii Peninsula. Around that time the path of the Kuroshio had meandered greatly to the south from its averaged position in connection with a large cold water mass that had appeared in the region. The fleet, however, had no previous information on the meandering and failed to accomplish satisfactory arrangement of two fleets temporarily organized for the maneuvers in the region. This failure in arrangement upset the maneuver plan. After the maneuvers, the Survey Section analyzed the hydrographic observa-

tion data obtained by the *Komahashi* in the region a little before the maneuvers and showed that the meander of the Kuroshio was revealed through dynamic computation of the data. This incident underlined for the navy the significance of the proposed method. The activities of Survey Section began to develop greatly after 1937, during which time Kishindo had returned to the office as chief of the section. Oceanographic surveys now entered the second stage in which oceanographic observations were made extensively in the North Pacific. By the end of the preliminary stage, the office's capabilities for oceanographic observations had grown to the point where its observations could be considered accurate even by present-day standards. This was chiefly due to the efforts of Kishindo and Shigematsu. Unfortunately, observation data from the early days of the preliminary stage were reduced to ashes by a fire at Hydrographic Office during the Great Earthquake of 1923. Subsequent data from the surveys of the *Koshu, Komahashi* and *Katsuriki* in the deep-sea region of the North Pacific during the period from 1933 to 1938 as well as data from the *Manshu* are, however, recognized as important contributions to oceanography.

Meteorological agencies

A close relationship between changes in climate in the Northeastern Districts of Japan and changes in the oceanographic conditions of the adjacent sea had been noted in connection with rice crops since comparatively early in the 20th century. Kichisaburo Endo, a specialist in seaweed research, pointed out that the poor crop of 1902 as well as the sudden decrease in seaweed in the Straits of Tsugaru might be the result of a cold water mass associated with an unusual intensification of the Oyashio. Nobuo Tsukiji suggested that there was probably a close relation between the poor crop of 1909 and the unusually low water temperatures in the sea off the Pacific side of the northern part of the Northeastern Districts (Sanriku) in the summer of the same year; Takematsu Okada made a similar suggestion with respect to the poor crop in 1912. The theory that crop failures are generally associated with low water temperatures in the sea off Sanriku having therefore achieved some support, the idea of conducting oceanographic observations to reveal atmospheric-oceanic interactions was naturally considered for submission to the meteorological agency. The main office of the meteorological agency in those days was called Chuo Kisho-Dai (the Central Meteorological Observatory). Interest in the subject gradually died, however, due to the lack of crop failures for about 20 years after 1914.

Oceanographic observations became part of the official program of the meteorological agency due to the following circumstances. During World War I the import of sextants and chronometers necessary for navigation ceased, and the shipping industry, which had become active owing to a sudden increase in marine transportation, suffered from a shortage of these instruments. Okada, who realized the necessity of advancing marine meteorology and oceanography in order to make navigation safer and more economical, endeavored to establish a marine observatory. According to his plan, supervision of the manufacture and calibration of navigational instruments would be the responsibility of the observatory in addition to oceanographic and meteorological observations. With the help of the Ministry of Education and the shipping industry, the establishment of the observatory at Kobe was realized in 1920. Okada as the first director laid the foundations for the oceanographic program of the observatory. Under his guidance many oceanographers who advanced oceanography in Japan emerged. In 1927, the research ship *Shumpu Maru I* (150 gross tons) was completed, and the Observatory started observations.

Before this time the Central Meteorological Observatory had already been engaged in some oceanographical activities, such as regular measurements of surface water temperatures and tidal elevations at coastal stations, although its activities were limited to those which did not require a ship. Regular measurements of surface water temperature and density started at Nemuro in 1907, and the measurements were extended to about 50 stations. Continuous measurements with respect to tidal elevations were first carried out at Toba in 1924, and by 1936 more than 10 stations had been established. Recently, the number of tidal stations maintained by the agency reached 60.

In addition to these activities, other oceanographic studies were undertaken by the agency. Extended measurements of current had been undertaken earlier, in a semi-official capacity, by Wada and others using drift bottles. These measurements were carried out in the sea adjacent to Japan, but unfortunately had to cease after a few years. Twenty years later the measurements were resumed, this time with the financial support of the Osaka Mainichi Press Co., whose head was aware both of the importance and the urgency of these measurements. Measurements were carried out extensively in the offshore region around Japan in the period from 1913 to 1917. The number of bottles used amounted to 13,357, and of these, 2990 bottles (22.4%) were retrieved. The results were later summarized by Kumata (1922).

Through this study the synoptic patterns of the Kuroshio, Oyashio and other currents in the sea adjacent to Japan became clearer. In parallel with current measurements, Wada initiated a program to complete a synoptic chart of the average monthly-mean surface-water temperature of the western North Pacific for every month of the year. The data used were collected over a 20-year period from 1882 to 1902 with the cooperation of Japanese ships. The charts were published in 1910. The study of ocean currents and the associated distribution of water properties is of the utmost importance in oceanography. In this respect, the programs conceived and put into practice by Wada can be considered of primary importance as the first steps in a study requiring minimum use of ships, even if the results obtained presented only a rough pattern of current and monthly synoptic features of surface temperature distribution. Moreover, these current measurements formed the first extensive survey systematically conducted for a specific purpose. The compilation of surface temperature data into charts since then was carried out at the observatory.

In 1927, Marine Observatory started oceanographic and marine-meteorological observations on board the *Shumpu Maru* in the sea adjacent to Japan. The observations were aimed at the fundamental investigation of oceanographical conditions, and special effort was exerted to establish accurate, standard observation techniques. In the following year, and for the four subsequent years, the observatory conducted observations in the Japan Sea. In addition to hydrographic casts with Nansen bottles and reversing thermometers, measurements of current by means of Ekman-Merz current meters were carried out from moored ships. Determinations of chlorinity through silver nitrate titration and pH measurements of sea water as well as analyses of chemical constituents such as dissolved oxygen, phosphates and silicates were made on board ship. Dynamic computations of the geostrophic current were carried out by Kanji Suda, Koji Hidaka and other members of the observatory staff. Suda studied the vertical stability of the stratification of sea water from an analysis of the observed data. Hidaka developed useful formulae to estimate in situ temperature and depth from temperatures observed by pairing protected and unprotected reversing thermometers. These formulae are still in use at present. From 1929, the observatory began to collect data on surface-water temperatures and meteorological data such as atmospheric pressure and temperature, wind speed and direction and humidity reported by ships at sea. Surface temperature data were compiled at the observatory in reports of yearly, five-yearly and ten-yearly averaged distributions of surface-water temperature.

From 1933 to 1936, the observatory strove to carry out oceanographic observations in the sea adjacent to the Nansei Islands and Kyushu to reveal synoptic features of current and water-property distributions associated with the Kuroshio in the region. In 1936, the standard method for observing oceanographic conditions was established at the observatory with the help of such oceanographers as Hidaka, Rikichi Sekiguchi, Yasuo Matsudaira, Yasuo Miyake and Tadami Yanagisawa. The method included new techniques of data sampling and processing. Although revised several times to keep up with advances in instrumentation, it has been widely adopted as the standard method by Japanese oceanographers up to the present, and probably represents one of the greatest achievements of the observatory prior to World War II.

In 1934, the Northeastern Districts were subjected to an exceptionally cold summer, which resulted in a poor crop. In autumn of the same year, the Japanese islands were hit by a huge typhoon which caused major disasters. These disasters provided the opportunity to re-arouse interest in an investigation of the relationship between atmospheric and oceanographic conditions. A multiple-ship survey conducted by the fisheries agencies for two years in the Pacific region off the Northeastern Districts had clearly shown that the water temperature in that year was very low compared with that of the preceding year. This fact prompted the meteorological agency to intensify its observational program, and to attempt a complete observational network covering the sea adjacent to Japan. Thus, the observation operations of the agency shifted into the second stage.

2.2   Observations in the second stage

In this stage, the observation method had already been established, and active and extensive observations were being conducted by every agency.

Fisheries agencies

At an assembly of the committee organized by Kasuga, the importance of conducting regular observations at selected coastal stations and at selected sections was recognized. Coastal stations to be occupied regularly six times per month were specified together with hydrographic sections to be made regularly once a month. The stations and sections are illustrated in Fig. 1. Observations were carried out

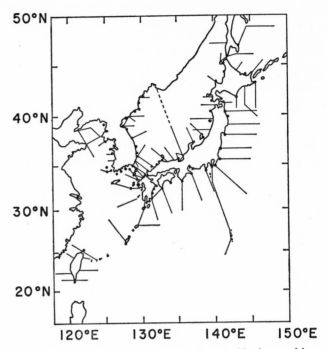

FIG. 1   Locations of coastal stations and hydrographic sections occupied by fisheries agencies since 1930.

according to this plan, and the quality of those made by the research ship *Tankai-Maru* of Hokkaido Fisheries Experimental Station and by the research ship *Misago-Maru* of Korea Fisheries Experimental Station were particularly excellent. In those days oceanographic observations by the Fisheries Experimental Station of every prefecture facing the sea were in general conducted systematically but were confined only to regions not too distant from the coast, and therefore oceanographic conditions in offshore regions were not revealed nearly as well as those near shore. In contrast, the oceanographic observations made by Hydrographic Office and the Marine Observatory were conducted mainly in offshore regions, but they were not conducted systematically with the aim of clarifying the synoptic conditions over a fairly vast area within a fairly short period of time. Multiple-ship surveys were therefore planned to fill this gap. The surveys were conducted by Uda, who was on the staff of Imperial Fisheries Experimental Station, in the Japan Sea and in the western North Pacific.

The first survey was carried out with more than 50 ships in the Japan Sea in the spring of 1932. Surveys were carried out again in autumn of 1933 and again in the spring of 1941. Some of the hydrographic sections extended across the Japan Sea. In addition, multiple-ship surveys were conducted repeatedly in the western North Pacific off the Northeastern Districts of Japan every summer for nine years from 1933 to 1941. The main hydrographic sections extended as far offshore as 1000 nautical miles. Multiple-ship surveys were also conducted extensively in the sea south of Japan from Kyushu to the Kanto Districts during the same period.

These multiple-ship surveys preceded the 1950 Operation Cabot in the Gulf Stream by 10 to 20 years. In view of this fact, the significance of such surveys can be appreciated, although substantial differences exist between the surveys by Uda and Operation Cabot with regard to purposes and to the operation of ships in relation to these purposes. After analyzing observation data from the surveys, Uda (1934, 1935, 1936a, 1936b, 1937, 1938, 1942a, 1943, 1943, 1949, 1950) published lengthy reports in which the synoptic conditions and varying processes in the Kuroshio, Oyashio and front regions were described. These descriptions of meanders and other varying processes as well as the finding of fine structures contributed greatly to the study of the Kuroshio. Although the accuracy of measurements was relatively poor compared to recent measurements or to those made by Hydrographic Office at that time, and although the sampling intervals were rough, the existence of shingle structures and multiple-current structures which were described in the Gulf Stream by Fuglister (1955) was suggested, if, as Jotaro Masuzawa pointed out, the isotherms at depth of 200 m were drawn by rearranging data from the multiple-ship surveys of the Kuroshio in 1938. In addition to clarifying variations in oceanographic conditions in the front region, the multiple-ship surveys in the western North Pacific were useful in providing information for a preliminary examination of the relation between oceanic variations and variations in the Japanese climate. Multiple-ship surveys conducted in the sea south of Japan were particularly notable because they revealed, at least in part, features of a large meander of the Kuroshio south of the Kii Peninsula and its associated cold water mass. Abnormal oceanographic conditions in the region, first described scientifically by Uda (1937), might bear some relation to continuous abnormally poor fishing since 1932. The cold water mass and associated Kuroshio meanders were considered to have existed at least during the period from 1934 to 1943. Similar phenomena have never appeared in the Gulf Stream System. In addition to Uda, Kan'ichi Koenuma also endeavored to describe the oceanographic conditions in those days. However, in contrast to Uda, who helped to reveal the varying processes and fine structures of oceanographic conditions, Koenuma made his contribution to the clarification of synoptic

features of water property distribution. The results they obtained proved to be essentially correct even after re-examination on the basis of recent, more accurate observations.

With the onset of World War II all research ships were commandeered and thus, for all practical purposes, the oceanographic observations of the fisheries agencies ceased.

Hydrographic Office

Kishindo began to carry out his grand concept of oceanographic surveys in the year following his return to the office as chief of Survey Section in 1937. The idea was to establish a survey network from which information on oceanographic conditions in the western North Pacific would be regularly available. His plan called for a newly-organized fleet of the office, under the control of civilians of the office, in addition to the existing fleet of survey ships. The new fleet would consist of six survey ships of 200 gross tons to be completed within four years from 1939 to 1942, and ten survey ships of 800 gross tons to be completed by 1945. In addition to the organization of the fleet, he planned the establishment of 25 bases for oceanographic surveys in coastal regions. The plan was adopted and was put into practice, but the onset of World War II prevented its completion.

In the summer of 1938 the office conducted a multiple-ship survey in the sea south of Japan. At that time the office did not yet possess its own fleet, and the survey was made by five chartered whaling ships. Multiple-ship surveys in this region were then conducted regularly every summer until 1942. In addition to the multiple-ship surveys, oceanographic surveys were conducted extensively in the eastern sea of Japan, the Yellow Sea, the East China Sea, the South China Sea, the Japan Sea and the western North Pacific. Many merchant ships and research ships of the fisheries stations were chartered before the war and commandeered during the war to participate in these surveys. In accordance with Kishindo's shipbuilding plan, two of the six smaller survey ships, the *Daiichi Kaiyo* and *Daini Kaiyo,* were completed in 1939, and starting in the following year, both ships made regular hydrographic sections in the sea south of Japan. The third and fourth survey ships were completed in 1942 and had some slight improvements over the first two based on experience gained at sea. Early in the following year the fifth and sixth ships were completed. These ships, though small in tonnage, were equipped with first-class navigation- and observation-instruments and were excellent research vessels for those days. They made it possible to carry out 10-day observations with 25 men, including navigation and observation staffs. Unfortunately, however, once the war had started it was impossible to use these new ships for survey purposes as they were commandeered for military purposes. The four completed survey ships were able to participate all together in regular multiple-ship surveys only once in the summer of 1942. The design for the larger survey ships was completed in 1942, but construction had to be stopped.

The plan to conduct oceanographic surveys in coastal regions made more headway and the bases for the surveys were established at Shimoda, Hachijo-jima (later transferred to Oshima), Kii-Katsuura, Tosa-Shimizu, Aburatsu, Naha and Soo (in Taiwan) in 1939, at Onagawa, Onahama, Chiba-Katsuura, Hamashima and Koniya in the period from 1940 to 1941, and at Kushiro, Urakawa, Hachinohe, Muroto, Karatsu, Fuzan (in Korea) and Shinko (in Taiwan) in the period from 1942 to 1943. However, no significant results were obtained because analyses of the surveys were not made. After 1943 oceanographic surveys declined noticeably with the routing of the Japanese Navy, and activity stopped completely in 1945 with the defeat of Japan.

More than 30000 hydrographic stations were occupied by the office during the 10 years from 1933 to 1942 and meshes of stations covered the western North Pacific. Many of the hydrographic sections were made down to deep layers. In addition, direct measurements of current by simultaneous use of two current meters, one at a reference deep layer and the other at a desired layer were extensively made. Unfortunately, these observation data were circulated only within the navy and could not be used for oceanographic research. If the data had been made available, knowledge of synoptic features and varying processes of oceanographic conditions would have made great progress. One study, possibly a rare one in those days, dealing with deep circulation in the Pacific in relation to distributions of water properties and especially of potential temperature, was made by Kishindo (1940a, 1940b). His results were supported later by Nan'niti and Akamatsu (1966). Tsuchiya (1961) revealed the existence of the Equatorial Undercurrent in the western region of the Equatorial Pacific through analysis of the data. Uda and Hasunuma (1969) also studied the Subtropical Countercurrent in the western North Pacific from the data. Thus these observations, almost all of which have been published recently by the office, are still appreciated for their usefulness in the study of oceanographic conditions in the tropical and subtropical regions of the North Pacific, for which only a few observations are available.

Meteorological agencies

The necessity for inquiries into the relationship between variations in climate and oceanographic conditions promoted the intensification of the meteorological agency's oceanographic-observation system. The first step in the intensification was the organization of a country-wide system for oceanographic observations to cover seas adjacent to Japan, and three research vessels, the *Ryofu-Maru* I (1179 gross tons), the *Asashio-Maru* (58 gross tons) and the *Kuroshio-Maru* (31 gross tons), were constructed in the period from 1936 to 1937. The former two vessels were operated by Central Meteorological Observatory and the latter was under the control of the Miyako Meteorological Observatory. The Miyako Observatory was located on the Pacific coast of the Northeastern Districts, and the inquiry was to be initiated in the adjacent region where the oceanic conditions were thought to be closely related to the rice crop. In 1941, the Central Meteorological Observatory adopted the division-section system for its activities and assumed leadership among the observatories in carrying out systematic observations. In the following year, the Oceanography Section and Chemistry Section were established at the observatory. The next step was the establishment of the Marine Observatory at Hakodate in 1941 and the construction of a new research vessel, the *Yushio-Maru* (143 gross tons), in the following year.

Just when the organization was completed, the *Ryofu-Maru* and *Yushio-Maru* were both commandeered by the navy and from 1941 could not be used for observations. Shortly before the commandeering a sampling of sea water was made from the *Ryofu-Maru* in the sea adjacent to the Ogasawara Islands by Masayoshi Ishibashi, Hidaka, Miyake and others. The purpose was to make standard sea-water for chemical analysis because imports of international standard sea water had ceased due to the war. This was probably the only outstanding contribution of that period. In 1942, the Central Meteorological Observatory published a collection of original papers to commemorate the placing of prefectural meteorological observatories under the control of the Central Meteorological Observatory, which had been realized in 1939. This publication included an interesting paper on the storm current by Kinosuke Kimura of the Imperial Fisheries Experimental Station. The storm currents are strong, warm currents which suddenly rush counterclockwise into some coastal regions east of the peninsulas north of the Kuroshio, and thus have an extreme influence upon the fisheries of the regions. Kimura analyzed surface water-temperature data from the coastal stations and studied the processes of storm currents. Close association of the currents with Kuroshio fluctuations may exist, as Kimura pointed out; however, to reveal detailed processes and structures of the currents, temperature measurements, denser both in time and space, and velocity measurements are necessary as well as simultaneous measurements of the Kuroshio variations. As mentioned before, Koenuma (1933, 1937, 1938, 1939, 1941) studied the general features of the Kuroshio by analyzing observed data. Although the data he was able to obtain were insufficient both in quality and quantity, he gave an astonishingly correct description of the essential characteristics.

3 OBSERVATIONS IN THE PERIOD AFTER WORLD WAR II

3.1 Observations at an early stage

Most oceanographic research ships were either sunk or damaged as a result of the war, and Japan, with her facilities exhausted, found it almost impossible to keep up with oceanographic observations. The lives of the Japanese people are, however, very closely associated with the ocean so that the observations were essential. Oceanographers consequently endeavored to do their best to reestablish the quality of observations. First, a conference including the fisheries agencies, Hydrographic Office and meteorological agencies was held to discuss means of cooperation in carrying out observations most effectively. In addition, the exchange of observation data, the manufacturing of standard sea water and so on, all essential in advancing oceanographic investigations, were treated at the conference. The conference was opened to any institutions and research workers who wanted to join in discussions on oceanographic observations. In spite of this endeavor, observations at this time were very poor, due on the one hand to extreme deterioration throughout the country as a result of the war and on the other to the policy of the General Headquarters of the Occupation Forces (hereafter abbreviated GHQ) to limit observations. Conditions improved gradually, and after several years observation methods and data processing improved. Information on the development of oceanography in foreign countries during and after the war began to be actively introduced and the information stimulated Japanese oceanographers to greater efforts in advancing oceanographic investigations in Japan. Among these efforts, the introduction of BT and GEK into routine observations should be noted here, since it marked an epoch in the history of

Japanese oceanographic observations. As is well known, the BT allows quick measurement of temperature profiles and the GEK quick measurement of surface water movement against the earth. In addition, these instruments can be utilized by vessels under way even in relatively rough seas. Such measurements were requisite for examining the structures and mechanisms of thermal and velocity fields, which varied extensively and quickly. In 1953 the research ship *Baird* of the Scripps Institution of Oceanography visited Japan during the TRANSPAC Expedition, and held open house at Hakodate, Tokyo and Kobe. The new instruments and facilities greatly impressed Japanese oceanographers. On leaving Japan the *Baird* presented a BT to the Hydrographic Office. Using the instrument as a model, more were manufactured with the help of the Meteorological Research Institute. The manufacture of the GEK was made with the help of the Hydrographic Office.

From about 1954, new observation methods and programs were developed in association with the introduction of BT and GEK as well as with an intensification in Japanese observation systems, and the next stage in observations was established.

Fisheries agencies

Imperial Fisheries Experimental Station of the Ministry of Agriculture and Forestry, which had lost almost all of its ships and facilities, contributed very little to observational work in the few years after the war. Some observations were carried out during this period, however, in the region east of the Northeastern Districts in July 1947 as an inquiry into the oceanographic conditions of skipjack-fishing grounds, in the Japan Sea from July to September 1947, in regions south of the Kii Peninsula and the Izu Peninsula from November to December 1947, and in the region east of the Northeastern Districts in August 1948. The Fisheries Experimental Stations of the adjacent prefectures cooperated in these observations, although the stations were also in generally poor condition.

In 1948 the national inquiry-system for fisheries was altered at the request of the GHQ. The Fisheries Bureau of the Ministry of Agriculture and Forestry was reorganized into the Fisheries Agency as an extraministerial board of the ministry. In 1949 the Imperial Fisheries Experimental Station was abolished, and eight Regional Fisheries Research Laboratories were established. Each laboratory was to devote itself to the investigation of problems in a specified region. The reorganization was not appreciated from the oceanographic point of view, because emphasis on regional studies would detract from efforts to establish participation in observations on a large scale and at a national level. Another problem was the alteration of observation policies, in which, contrary to past policies, greater emphasis was placed on biological investigation of fisheries resources. Oceanographic observations made since that time concentrated mainly on measurements of water movement and water properties directly associated with biological problems within areas of the fisheries' interests. Thus, little could be expected from the laboratories in carrying out systematic oceanographic observations to reveal physical processes not always directly connected with fisheries. In sardine-resource investigations made after 1949, the emphasis on biological investigations and neglect of observations of oceanographic conditions were marked.

This policy was later re-examined, and was altered to emphasize the importance of oceanographic observations at the start of the Tsushima-Current investigation in 1953. Once again the oceanographic observations of the Fisheries agencies entered a new stage.

Hydrographic Office

The war was over and the defeated Japanese Navy was demobilized. Because of its importance to civilian life the Hydrographic Office was retained, although greatly reduced in scale, and the superintendence of the office was transferred to the Ministry of Transportation. Toward the end of the war most of the buildings of the Hydrographic Office, except the main building, were destroyed in an air raid. Fortunately the observation instruments were saved because they had been previously moved to the country for safety. Almost all of the more than 20 survey ships had been sunk and only the *Daiyon Kaiyo* remained intact. Later, two smaller ships, the *Daiichi Tenkai* and the *Heiyo,* were returned to the Hydrographic Office. About six months later the *Daigo Kaiyo* was found abandoned in Uraga Port and was retrieved.

From 14 July to 7 August 1946, the first oceanographic cruise after the war was made by the *Daiyon Kaiyo* in the sea south of Japan. Cruises were subsequently made in the seas adjacent to Japan in September and in December in addition to a special cruise to explore the Tsushima Current. In the following year, three cruises were made in the seas adjacent to Japan. The cruises were intended to clarify the synoptic features of oceanographic conditions in the seas for use by navigators.

In May 1948, the Maritime Safety Agency was newly established as an extraministerial board of the Ministry of Transportation, and included the Hydrographic Office. For convenience the seas adjacent to Japan were divided into nine regions, each with a Regional Maritime Safety Headquarters as branch office. Each headquarters had a Hydrographic Division which was responsible for regional oceanographic observations and soundings. In the same year 10 cruises were made in the seas adjacent to Japan. Another expedition was also made in the Japan Sea from the middle of July to the latter part of September to explore the Tsushima Current. Stations occupied during the expedition were distributed throughout roughly half the area of the sea adjacent to Japan. The observation area was limited by order of the GHQ. The stations did not cover the whole area of the Japan Sea and so the synoptic features of the current system in the sea could not be fully examined. However, the observations were noteworthy in that the observed area covered the main part of the Tsushima-Current region and that the expedition presented observation data to be compared to data obtained from the expedition in 1943. At some stations hydrographic casts were carried out down to depths of more than 2000 m. Aside from oceanographic observations, the Hydrographic Office started to process a great deal of unpublished data obtained by the navy before and during the war. In September of that year the first of these data were published. In 1949, seven cruises were made, one in the sea east of the Northeastern Districts to explore uneven oceanographic conditions resulting from intensification of the north-flowing tendency of the Kuroshio. A branch of the Kuroshio flowed northward to 39°N and formed a high temperature region in conjunction with the Tsugaru Current which was an extension of a branch current of the Kuroshio and which came southward through the Straits of Tsugaru. In the front region a train of large eddies of cold and warm water appeared. In 1950, 14 cruises were made. The area observed was extended to the seas east, south and southeast of Japan, the seas southeast of Hokkaido and the Japan Sea. The three observations in the sea south of Japan indicated fluctuations of the Kuroshio path with durations of a few months or less. In 1951, 11 cruises were made. That year the Hydrographic Office was 80 years old, and an anniversary volume of the Hydrographic Bulletin was published. In this volume, Shoji (1951) discussed variations in ocean currents in the seas adjacent to Japan on the basis of analyses of observations made in the past. The main purpose of the study was to show that, contrary to the unconfirmed expectations of many oceanographers, the currents were not steady and that in a number of cases large variations in the velocity field of currents had been completed within fairly short periods. A number of examples were provided, such as current variations caused by a passing typhoon, current variations found in association with the appearance of a cold water mass south of the Kii Peninsula, short-term current variations of 10 days or less, current variations found in association with masses of cold and warm water in a front region, fluctuations in the path of the Tsushima Current and variations in the transport of the Kuroshio. Shoji emphasized the necessity of planning observations to reveal the variations and to clarify their mechanisms.

The anniversary volume also presented a description of a self-recording current-meter devised by Ono (1951). This reliable, easy-to-use meter has been utilized widely for current measurements since then. In 1952 preliminary tests connected with the manufacture of GEK began at the Hydrographic Office in cooperation with the Science Research Institute, and subsequent tests of the instrument in routine observations were made by Shoji in the seas adjacent to Japan. From about 1954, all the survey ships of the Hydrographic Office were equipped with GEK. It should be mentioned that for exploring currents in the Japan Sea drifting bottles were frequently used by the Regional Maritime Safety Headquarters in cooperation with the Hydrographic Office during the period from 1950 to 1955. Year to year variations in the geographical distribution of bottles reaching the coast were noted.

In 1952, the *Daigo Kaiyo* was damaged while exploring the Myojin-Sho eruption and, in a subsequent eruption, sank with many outstanding oceanographers aboard.

Meteorological agencies

After the war ended in 1945, it was again possible to make oceanographic observations in the region adjacent to Japan; however, no vessels were available for the observations. The *Shumpu-Maru* of the Kobe Marine Observatory had been heavily damaged. The *Ryofu-Maru* of the Central Meteorological Observatory and the *Yushio-Maru* of the Hakodate Marine Observatory had both been commandeered to transport home soldiers demobilized in foreign countries, and to ship supplies to the Nansei Islands. In spring of the next year the two commandeered vessels returned, and were used for observations in the sea east of the Northeastern Districts in order to predict the climate from oceanographic conditions. For the Japanese, who had been suffering from a food shortage, the condition of the rice crop was a subject of great interest. The observation results showed that the water temperature in the region was higher than the yearly mean temperature, leading to speculation that the ex-

pected crop would be good.

In April 1947, new marine observatories were established at Nagasaki and Maizuru, and, in addition, activities at the Kobe Marine Observatory and Hakodate Marine Observatory were intensified. This intensification was aimed at monitoring variations in synoptic features of oceanographic conditions in the sea adjacent to Japan and finding the relation between the variations and those in atmospheric conditions. Each observatory as well as the Central Meteorological Observatory was put in charge of observations in a specified regional area, which was formed by dividing the adjacent sea into five parts. GHQ was asked to remedy the shortage of research ships by transferring converted warships to the observatories. The request was refused, and instead the observatories were asked to handle maintenance of weather stations Extra and Tango. Station Extra was at 39°N, 153°E and station Tango at 29°N, 135°E. Approval was given only for observations made on the way to the stations and at the stations. The *Ryofu-Maru*, the best equipped among the few research ships possessed by the meteorological agencies, was assigned to the weather-station watch. Thus, plans for a program of regular observations had to be given up. All other observations had to be carried out temporarily in a few confined areas only. This situation continued until November 1953, when the occupation of station Extra was abandoned.

Although the temporary observations were insufficient to establish the synoptic features of oceanographic conditions in the seas adjacent to Japan, hydrographic sections made on the way to the weather stations presented detailed information on temporal variations in oceanographic conditions. On the way to station Extra a section was made along 38°16'N, from 142°E to 151°E, more or less regularly once every 20 days, and on the way to station Tango a section was made along approximately 135°E, from 33°N to 29°N, once or twice a month. As the data accumulated studies were made of the variations noted earlier by Uda and others from observations conducted by the fisheries agencies and by the Hydrographic Office before and during the War. Nan'niti (1951), Ichiye (1953), Fukuoka (1954), Marumo (1954) and Masuzawa (1954a) reported on the variations along 38°16'N during the period from 1948 to 1953. The section was in a region north of the Kuroshio front where the current flowed eastward under average conditions. The spatial interval of 1 degree in sectioning was too rough for the exploration of a complicated structure in this front region, and the sectioning in the east-west direction was not appropriate for examining fluctuations of the current flowing eastward. Moreover, the regularity of data acquisition was disrupted in winter because of exceptionally rough seas. In spite of the inadequacy of available data, the studies succeeded in revealing seasonal and short-term variations of oceanographic conditions in the region to some extent, and provided useful suggestions for future planning of observations in the region. For example, sections in later observations by the Meteorological Agency were selected to cross the front in a north-south direction. For the section along 135°E studies were made by Masuzawa (1950, 1954b), Masuzawa and Nakai (1953), Ichiye (1954) and Fukuoka and Tsuiki (1954). The section was selected to cross the Kuroshio off the Kii Peninsula, and the studies dealt with temporal variations in distribution of water properties at the section or at station Tango. The distribution of water velocity at the section was also studied.

While the Hydrographic Office devoted its efforts to introducing the GEK into observations, the meteorological agencies were more interested in introducing the BT into their observations. A replica of the BT was used during a cruise of the *Yushio-Maru* in the sea east of the Northeastern Districts of Japan in the spring of 1952. After that its use gradually became common.

In November 1953 occupation of station Extra was discontinued, and the *Ryofu-Maru* was released from weather-station watch. In spring of the next year the *Shumpu-Maru II* was completed, bringing the number of research ships to be used by the meteorological agencies to three. At that time the oceanographical activities of the agencies were thoroughly re-examined and a new yearly observation program was established.

3.2   Observations in a recent stage

Operation Cabot, a study carried out in the Gulf Stream in 1950, marked an epoch in the history of oceanographic observations. Results clearly indicated that specially designed programs were necessary in order to reveal the varying processes of ocean currents. Japanese oceanographers were greatly stimulated by such activities as Operation Cabot and subsequent explorations of the Gulf Stream which suggested new aspects of oceanographic observations, and keenly felt the necessity to re-examine observation systems and methods in Japan. To this end, the Hydrographic Office, Meteorological Agency and Fisheries Agency frequently held conferences to discuss items, times, areas and methods of observation. It was decided initially to cover the whole area of the region adjacent to Japan with observation

stations occupied by the three agencies, and secondly to occupy all the stations regularly throughout the year. Observations along these lines were carried out from 1955, and seasonal variations in the synoptic oceanographic conditions of the region, including the Kuroshio region, have been monitored ever since.

It was at this time that the need for international cooperation in carrying out oceanographic observations in the vast, unexplored regions in as short a period as possible was recognized. As a result, plans for international cooperation in such ventures as the NORPAC, EQUAPAC, IGY and CSK expeditions came into being.

The NORPAC Expedition, an international cooperative venture between Canada, Japan and the United States to study the synoptic features of the Pacific Ocean north of 20°N, was carried out in 1955. The expedition included 1002 hydrographic stations, and most of the work was done in July, August and September of 1955. This was the first time that such a large area had been covered in so short a time with such a dense array of stations. The effect of noise due to seasonal changes on observation data was minimized by reducing the observation period as much as possible and by selecting a period at or near the water-temperature maximum over most of the area. Nineteen research ships from 14 institutions joined the expedition: the Pacific Oceanographic Group of the Fisheries Research Board of Canada, the Faculty of Fisheries of Hokkaido University, Japan Meteorological Agency, Japanese Hydrographic Office, Tokyo University of Fisheries, the Faculty of Fisheries of Kagoshima University, Tokai Regional Fisheries Research Laboratory, the California Cooperative Oceanic Fisheries Investigations, the Pacific Oceanic Fishery Investigations of the United States Fish and Wildlife Service and the Department of Oceanography, University of Washington. The EQUAPAC Expedition was planned by the U.S.A., Japan and France in order to study current systems in the equatorial region of the Pacific where the equatorial undercurrent had been found to exist. In this expedition in 1956, one Japanese ship participated.

The following year, the IGY Expedition was carried out and included scientists from 64 countries. The Japan planned to conduct multiple-ship surveys of the polar front region and of the equatorial region in the western North Pacific. This plan included direct measurements of deep current. A working group was established under the Committee for International Cooperative Exploration of the Ocean at the Science Council of Japan. Kozo Yoshida was asked to participate in planning the program as a representative of young learned oceanographers and gave valuable suggestions. For the front region three multiple-ship surveys were conducted in the period from 1957 to 1958. In the past, two or three weeks had usually been required to complete the preplanned observation network for a region; this time, however, the discrepancy had been neglected. The influence of this discrepancy upon the interpretation of observation results was examined through the multiple-ship surveys. An effort was made to reveal varying processes and their structures in the front region east of Japan. The first survey included four vessels from the Japan Meteorological Agency and Japanese Hydrographic Office. In the second survey four vessels from the Japan Meteorological Agency, Japanese Hydrographic Office, Faculty of Fisheries of Hokkaido University and Tokyo University of Fisheries were included. The third survey included three vessels from the Japan Meteorological Agency, Japanese Hydrographic Office and Tokai Regional Fisheries Research Laboratory. For the equatorial region two surveys were conducted by the Hydrographic Office and the University of Kagoshima. The main purpose was to clarify synoptic features of the equatorial current system in the western North Pacific.

The Cooperative Study of the Kuroshio and adjacent region (abbreviated CSK) was organized on the basis of a proposal presented at the second Meeting of Marine Science Experts in East and Southeast Asia, Manila, February 1962. Research vessels of institutions from 11 countries—China, Indonesia, Japan, Korea, the Philippines, Thailand, Hong Kong, Vietnam, Singapore, the Soviet Union and the United States—joined in the expedition which took place from 1965 to 1969. Synoptic surveys of the regions were conducted and special surveys were made by each institution on problems of particular interest concerning the Kuroshio. The area to be covered for the synoptic surveys was the northwestern part of the Pacific west of 160°E and between 4°S and 40°N and adjacent seas such as the Japan Sea, the East China Sea and the South China Sea. To reveal seasonal variations in synoptic features of oceanographic conditions it was proposed that surveys be conducted in specified periods from the middle of June to the end of August and from the beginning of January to the end of March. Although from a statistical standpoint observations in only these two periods were inadequate for determining seasonal variations, the surveys contributed to the advancement of knowledge on the synoptic oceanographic conditions in the subtropical region of the western North Pacific. In planning the special surveys, emphasis was placed on revealing structures and

mechanisms of short-term fluctuations of the Kuroshio and clarifying varying processes of the beginning and extension of the Kuroshio.

Fisheries agencies

It is clear that fishing conditions in the sea are governed by changes in oceanographic conditions, and therefore oceanographic observations of fishing and spawning grounds are quite important for research in fisheries resources. On this basis, endeavors were made to conduct observations in connection with fisheries investigations starting in about 1951. The observations were carried out under the leadership of the Regional Fisheries Research Laboratory in each assigned region. Prefectural Fisheries Experimental Stations provided cooperation for observations made in adjacent regions.

The Tokai Regional Fisheries Research Laboratory has acted as representative for the eight laboratories of the Fisheries Agency in carrying out oceanographical business. Thus, in addition to its own business, the laboratory conducted observations which were not directly associated with its own problems but were requested by the Japanese Government. Oceanographic observations of the salmon and trout fishing grounds in the subarctic region of the North Pacific were conducted by the laboratory as well as oceanographic observations in the region off the Northeastern Districts of Japan. The latter observations were made to study the relationship between the occurrence of exceptionally cold weather in the districts and variations in oceanographic conditions in the region. Observations were made in cooperation with the Hydrographic Office and Central Meteorological Observatory during the period from 1952 to 1956. The laboratory also joined in the NORPAC Expedition with the *Soyo-Maru* in 1955, and in polar front surveys of the IGY Oceanographic Expedition in 1958. Hirano (1957a, 1957b, 1958, 1961) of the laboratory made a hydrographic analysis of the observation data obtained in the subarctic region of the western North Pacific, and was able to show some interesting characteristics of the circulation of water in the surface layer as well as the formation and movement of water in the intermediate layer.

In 1955 the International North Pacific Fisheries Commission, which was established by Canada, Japan and the United States, initiated an extensive scientific research program on the subarctic region of the North Pacific. Attention was directed mainly to research on salmon resources. A study of the oceanographic conditions of the region in which salmon spent the marine part of their existence was required. Dodimead, Favorite and Hirano (1963) joined in the study. The circulation and hydrographic structure of the water in the region were reviewed on the basis of observations made from 1955 through 1959.

The most important subject of study pursued by the Tohoku Regional Fisheries Research Laboratory was the clarification of the relation between oceanographic conditions and conditions for skipjack and Pacific saury fishing in a vast region of the Pacific east of the Northeastern Districts of Japan. Starting in about 1950, oceanographic observations were actively carried out in cooperation with Fisheries Experimental Stations of every prefecture facing the region. Kawai (1954, 1955, 1957, 1959) and Kawai and Sasaki (1962) endeavored to reveal the hydrographic structure and varying processes of water in a front zone of the region. As mentioned earlier, Uda had previously investigated the zone on the basis of data from multiple-ship surveys before the war. Uda's work was a milestone in the study of the oceanography of the region, but no subsequent analyses were carried out although comparatively extensive and frequent observations were made after the war. Kawai, encouraged by the striking progress in Gulf Stream studies, recognized the need to re-examine past observations more strictly in order to determine to what extent the hydrographic structures and varying processes could be revealed through analysis of the observations. He pointed out that cross-stream sections from past observations were scarcely available because almost all the sections had been made in the east-west direction and did not cross the front zone. He proposed zigzag crossing of the zone and conducted cruises with the *Soyo-Maru* in 1953 and 1954. Through analytical studies of data from these cruises and other cruises made by the Meteorological Agency and Hydrographic Office he was able to extend knowledge of the oceanography of the zone. In addition, studies of the Shiome in the zone and the adjacent region were also undertaken. These studies, based on extensive observations from ships and airplanes from 1955 to 1961, were conducted mainly by Kuroda (1958, 1959, 1960, 1962), who was particularly interested in temperature and density inversion.

At the Japan Sea Regional Fisheries Research Laboratory, oceanographic observations were carried out starting in 1950 in cooperation with Fisheries Experimental Stations of prefectures facing the Japan Sea. The observations were to contribute ultimately to the study of variations in fishing conditions. The observations were supported greatly by a cooperative investigation of the Tsushima Current, the branch current of the Kuroshio which flows into the Japan

Sea through the straits between Japan and Korea. The general features, varying processes and their mechanisms were not studied as extensively and frequently as the main flow of the Kuroshio; however, the investigation was designed to explore new fishing grounds in the Tsushima Current region and was conducted by the Fisheries Agency for four years, from 1953 to 1957, with the help of many oceanographers from other institutions. Hydrographic sections to be regularly made by the experimental stations of the prefectures were specified, although the sections could not be extended to reach the central part of the Japan Sea. Variations in hydrographic conditions and current patterns in the adjacent region were regularly observed.

The cooperative investigation of the Tsushima Current also advanced exploration of the western region of Kyushu, where observations had been conducted by the Seikai Regional Fisheries Research Laboratory. Prior to the start of the investigation in 1953, the purposes of the observations were examined in connection with studies on the fisheries of the region, and specifications for several sections and for the observation methods were established. In addition, as part of the T-Y Project at the Fisheries Agency, extensive oceanographic observations were carried out in the East China Sea and the Yellow Sea. For 10 years, from 1950 to 1959, 113 cruises were conducted in the region. Alteration of the water properties of the Kuroshio through mixing with coastal water in the region, the branching processes of the Tsushima Current from the Kuroshio and the variations of the Tsushima Current were studied. Tokimi Tsujita summarized the results of the investigation in a report published in 1961.

The Fisheries Agency participated in the CSK and made synoptic surveys of oceanographic conditions in the Kuroshio and adjacent regions beginning in 1965. Special surveys of the Kuroshio were also conducted under the leadership of Regional Fisheries Research Laboratories. These included a study of the diffusive properties of the Kuroshio conducted by Hirano at the Tokai Regional Fisheries Research Laboratory, an estimate of small-scale convergence in the Kuroshio conducted by Kawai at the Nansei Regional Fisheries Research Laboratory and an investigation of short-term fluctuations of the Tsushima Current conducted by Masato Kondo at the Seikai Regional Fisheries Research Laboratory. In the first study, dye diffusion in the Kuroshio was observed in connection with studies of the alteration of water properties and fish egg densities. In the second study, the divergence was measured directly in the surface mixed layer of the Kuroshio by tracking four drogues released from a quadrilateral with sides of about 1 mile or less at the beginning of the tracking. From the results, Kawai et al. (1969) and Kawai and Hisao Sakamoto (1970) concluded that when treating the distribution of drifting material, such as eggs, in the Kuroshio water, the divergence of water should be taken into account as well as the diffusion. The third study involved continuous measurement of water velocity in parallel with hydrographic casts.

Hydrographic Office

In 1954, 27 cruises were conducted by the office to make hydrographic sections in the region adjacent to Japan. In the region adjacent to the Marshall and Caroline Islands one cruise was also made to examine pollution resulting from an atomic bomb experiment by the United States. In addition, cruises for GEK measurements were conducted from that year. In parallel with these activities, Shoji (1954) examined the availability of variations in sea level as an indication of ocean-current fluctuations. He found that variations in the 25-hour running-averaged sea-level were closely related to variations in the geopotential in the vicinity of tide stations provided that the influence of atmospheric pressure variations upon the sea level was eliminated. Variations in geopotential are directly associated with Kuroshio fluctuations, so that variations in sea level are considered to be indicative of the fluctuations. From an analysis of sea level at Hachijojima, Shoji studied the period of the most prominent fluctuations and variations in strength of the Kuroshio. In the following year, 31 cruises for hydrographic sections, 24 cruises for GEK measurements and one cruise for BT section were carried out under the new program established by the three agencies for conducting regular, cooperative observations in the region adjacent to Japan. Starting that year these cruises were conducted annually on a regular basis. In 1956, a special cruise was conducted by the *Satsuma* in the southwestern region of the North Pacific as part of the NORPAC Expedition.

In March 1957 a new survey ship, the *Takuyo* (770 gross tons), was completed, the largest of the survey ships. The capabilities of the Hydrographic Office in oceanographic observations, especially in regions far from Japan, were greatly increased. From October to November the *Takuyo* participated in the Front-Zone Survey of the IGY Expedition. Hydrographic sections were made along 36°N and 38°N from 142°E to 170°E. The IGY Expedition was also conducted in the following year, in which the Hydrographic Office planned to make two expeditions including hydrographic sections as well as direct current measure-ments in northern tropical and equatorial regions of the western North Pacific between 147°E and 157°E. These expeditions were aimed at revealing the structure of the equatorial current system in the western North Pacific. The first expedition was made with the cooperation of the *Takuyo* and the *Satsuma* for about a month from the end of January to February. The ships available were too few and the period of observation too short to reveal the complex structure of comparatively vast an area, so that the observations could only present synoptic features of the oceanographic conditions. Having learned a lesson from the first expedition, the office planned the second, which was to be made in summer with the cooperation of the *Kagoshima Maru* of the University of Kagoshima, to concentrate on a more definite purpose. The plan was abandoned, however, due to an accident in which the *Takuyo* was exposed to radioactivity from fallout ash from an atomic bomb exploded by the United States (Yosida et al., 1959).

Recent observations had made clear that fluctuations of the Kuroshio and its branch currents occurred more frequently, in larger magnitudes and possibly with much more complex mechanisms than had been presumed in the past. In 1959 the office started a continuous recording of electric potential difference across the Straits of Tsugaru to reveal temporal variations in transport. From a preliminary examination made by Nitani and others (1959) a year earlier, it had become clear that the electric potential difference sufficiently reflected day-to-day and longer period variations. A special observation program including velocity measurements with towed electrodes and current meters was carried out by Terumi Kubota and others from the survey ship *Meiyo* to examine further the validity and availability of the study method used in the ocean current in the Straits. The potential recording was continued until 1961. Another special program to investigate short-term fluctuations of ocean current was conducted by Yosida (1961a) from the survey ships *Kaiyo* and *Meiyo* in the region adjacent to the Izu Islands. Measurements were made with BT and GEK at intervals of 5 to 15 miles along two longitudinal lines from 35°N to 33°N every 36 hours for one month from the end of June. Hydrographic casts were also carried out at two stations, one near Mera in Chiba prefecture and the other at Hachijojima. The measurements showed the close relation between variations in sea level and variations in geopotential. On the basis of these results, the relation among variations of thermal structures in the surface layer, of surface velocity and of sea level were examined. Sea level data from only two islands, Oshima and Hachijojima, were available at this time. To show in greater detail the processes of the Kuroshio fluctuations, sea-level data from many more islands were needed. This requirement was met later in the CSK period with the establishment of new tide stations at Miyakejima and Kozushima.

In the same year one more special expedition was conducted by Nitani (1961) on board the *Takuyo* in the western boundary region, 15°~31°N and 120°~131°E. This region, which is regarded as the place of origin of the Kuroshio, had been covered a few times by the observation network of the Japanese Navy as part of the general oceanographic surveys in the western North Pacific. However, because the surveys were not focused on revealing the oceanographic conditions of the origins of the Kuroshio, even the synoptic features have remained unknown. This expedition was the first to be conducted in the region after the war and was aimed at clarifying the transition processes of the Kuroshio from the North Equatorial Current as well as the mixing processes of the Kuroshio water with South China Sea water flowing through the Straits of Bashi. The observation plan was partly altered later, however, and the expected results could not be obtained. From an analysis of observations made in this expedition, Nitani showed changes in water characteristics of the Kuroshio as it flowed northward from Luzon to Kyushu. He also showed cyclonic and anticyclonic eddies existing east of the Kuroshio in the region.

In 1959 a large mass of cold water appeared in association with a great meandering of the Kuroshio in a region off the Kii Peninsula and Enshu Nada. This phenomenon, which has not been observed in the Gulf Stream system, greatly influenced climate and fishing conditions, to the interest of many Japanese oceanographers. As mentioned before, monitoring of variations in oceanographic conditions around Japan had been made regularly by the three agencies since 1955. Through the monitoring the process of formation of this large cold water mass was observed, although the space and time intervals of the monitor observations were not short enough to reveal the process in detail. Yosida (1961b) analyzed data from the monitoring and showed that the small disturbance generated in the Kuroshio east of Kyushu gradually developed during its eastward travel and was fixed when it reached the region off the Kii Peninsula and Enshu Nada. Similar large cold water masses and the associated meanderings of the Kuroshio, which survived in the region for a period of a few years to about 10 years, had been found twice through observations in the past starting in 1930, but the process through which they appeared was still

unclear. Hence, the monitoring was useful even if it failed to provide sufficient details. As Yosida and others pointed out, the cause of generation and the formation mechanism of the phenomenon, as well as its variations and decaying processes, remained unknown. The development of new observation methods, as well as an intensification of the monitoring process, are necessary to increase our knowledge on the subject. Unfortunately, however, these advances have not yet been realized.

In addition to large stable masses of cold water, smaller cold masses, which appeared in the same region and decayed within 6 months, were frequently observed in the past. Yosida termed the large, stable masses cold water masses of type A and the smaller, unstable ones cold water masses of type B. Because of their fairly short life, a considerable number of type-B masses may have been missed in past observations. Another type of cold water mass also appeared in the region east and west of the Izu Islands. These type-C masses, as they were termed, are quite different from the others and are not formed in association with other Kuroshio disturbances. The origin and mechanism of these unstable masses are not yet clear, but they may be very closely associated with Kuroshio variations.

In May 1962, an expedition to make three hydrographic sections across the Kurile-Kamchatka, Japan and Izu-Bonin Trench was carried out by Nitani (1963) with the *Takuyo* as part of the JEDS (abbreviation of Japanese Expedition of the Deep Sea) program. The purpose of the expedition was to obtain data on the deep current along the trench and the vertical distributions of potential temperature, chlorinity and dissolved oxygen. The estimated geostrophic current was examined in reference to that deduced by Stommel, and on the basis of the resulting geostrophic current model the observed distribution of water properties was examined.

Around this time, an experimental automatic buoy system for telemetering horizontal components of velocity, temperature, salinity and surface wave height was devised by Kinji Iwasa (1963). Experiments have continued since 1960, and steady improvements in the instrument have been made.

The CSK began in 1965. All oceanographic observations in the annual program of the Hydrographic Office were contributed to the CSK program. Special CSK programs were also conducted by the office. The synoptic features of the Kuroshio in the region adjacent to Japan have long been known, and interest has been directed to revealing varying processes and detailed hydro-thermo-dynamical mechanisms. For the region of origin and the region of extension of the Kuroshio, however, even the synoptic features are not yet well known. Observation programs to fill out this gap fall under two categories: obtaining data on varying processes and dynamical mechanisms, and surveying the regions of the origin and of the extension. A special program in the first category was planned to obtain continuous measurements of vertical distributions of water temperature, and was carried out in the vicinity of Hachijojima since 1965, in the vicinity of Miyakejima since 1966 and in the vicinity of Shionomisaki on the Kii Peninsula since 1967 (Shoji, 1966; Iwasa, 1968). These regions are in the Kuroshio, and the main purpose was to observe temperature variations occurring in association with variations of the Kuroshio. In the Hachijojima region, an array of thermistors was suspended from a buoy moored at a 40~50 m deep subsurface layer. The current in this layer is very strong and the buoy system swung with variations of the current. In order to avoid this trouble, a thermistor array was again installed, this time in an area with a steep sea bottom. This method was adopted for other regions. Thermistors were installed at depths of 5, 50 and 500 m in the Hachijojima region, 5, 100 and 200 m in the Miyakejima region and 5, 100, 200 and 300 m in the Shionomisaki region. In all regions, telemetering of temperature signals to the shore-based recorder was made through an armored cable. In the Kuroshio, continuous recording of temperature for a long period has never been done in the past. From the measurements, a spectrum of temperature variations at each depth for each region, the relation of temperature variations at different depths and the relation of temperature variations to sea-level variations are expected to be made available. The program should contribute greatly to knowledge of the variations of the temperature field associated with Kuroshio fluctuations as well as intensification of the tide-station network in the Izu Island region.

From 1964 to 1966, a special program was carried out annually for about a week in October to reveal short-term variations of the Kuroshio, using two or three vessels, in the region south of Shionomisaki (Shoji & Nitani, 1966, 1967). From statistical studies of earlier observation data, fluctuations of the Kuroshio path were expected to be small in this region and the Kuroshio was expected to flow almost eastward in the vicinity of the peninsula, so that this region seemed most appropriate for Eulerian measurement of the variations. GEK measurements and BT sections were made repeatedly from ships cruising reciprocally at about 136°E between 32°40′N and about 33°40′N, and hydrographic casts were regularly made at the ends of the cruise path. The measurements showed the predominance of semidiurnal and diurnal components among short-term variations of surface velocity in an east-west direction, although the former component prevailed during a part of the periods of measurement and the latter component prevailed during the remainder. These components and the steady component were compared to those estimated from corresponding components of steric sea-level difference. Semidiurnal and diurnal variations which were not directly due to tides had been presumed to exist in a velocity field of the Kuroshio. The results obtained gave evidence for their existence. The generation processes and detailed dynamical mechanisms have not yet been revealed, and further observations with more advanced programs are necessary. In the summer of 1965 and 1966 Nitani conducted a synoptic survey of the region of origin of the Kuroshio as a special program in the second category (Hashiguchi and Saruwatari, 1968). The aim was to make clear distributions of the thermodynamical state of sea water, such as temperature and salinity, in association with current patterns in the western part of the North Equatorial Current region and in the region of origin of the Kuroshio. Another aim was to examine the alteration of the Kuroshio water through mixing with water of the South China Sea and the East China Sea. These subjects were too complex to be studied fully during these cruises, but new knowledge was acquired. The other program in the second category, conducted by Kiyoshi Hori, was designed to explore an extension region of the Kuroshio in 1967 and 1968 (Hashiguchi and Saruwatari, 1968). Because of the *Takuyo*'s limited cruising capability, only a few zigzag crossings of the Kuroshio Extension could be made on the cruise from Japan to Hawaii, the main purpose of which was to survey water-property alteration through mixing and the decay of the Kuroshio in the extension region. Many phenomena that are still unclear, such as meandering, branching and the other varying processes of the Kuroshio, relevant to the formation of warm and cold eddies and to the mixing processes, occur in the region. In order to understand these processes, which are important in association with variations and decay of the Kuroshio Extension, more extensive and intensified programs will have to be carried out.

Meteorological agencies

Oceanographic activities at meteorological agencies were originally initiated to study large-scale atmosphere-sea interactions having profound influence upon the climate of Japan. For this reason synoptic oceanographical conditions in the region adjacent to Japan were required first. Unfortunately, oceanographic observations in the past had not been conducted systematically, so that not even synoptic conditions were available. The activities of the agencies were re-examined when their research ships were released from weather-station duty at station Extra and when new observation programs were to be designed. It was decided that their activities should be aimed primarily at predicting variations in oceanographic conditions. Efforts of this sort had already been initiated in association with long-period weather-prediction at the request of the Ministry of Agriculture and Forestry in 1948, and had been continued as a side-line. Now weather prediction became the primary official duty of the agencies.

To make reasonable predictions hydro-thermo-dynamical equations governing variations in oceanographic conditions were necessary, and to establish the equations the hydro-thermo-dynamical mechanisms of the variations had to be clarified. At the preliminary stage, observations to reveal synoptic features of conditions which were not yet well known had to be conducted. An observation program was drawn up on the basis of the number of available ships and their capabilities in the area to be covered. Observations were planned for each season of the year to allow tracking of season-to-season variations of synoptic conditions in the Kuroshio and Oyashio regions adjacent to Japan. This network covered regions of great interest to the Kuroshio study. It included the seas south and west of Kyushu where the initial conditions for the Kuroshio variations south of Japan appear and where branching of the Tsushima Current from the Kuroshio occurs, and the sea south of Shikoku and the Kii Peninsula where a large cold water mass and the associated meandering of the Kuroshio sometimes appear, as well as the sea east of the Northeastern Districts of Japan where complex varying processes of the Kuroshio and Oyashio fronts exist. With such an all-round observation-policy aiming to cover all the important regions, it was thought that the network might be meaningful even in the preliminary stage. In the second stage, where the synoptic features are already known, the policy should be altered to emphasize the mechanism of variations in oceanographic conditions. Since the oceanographic conditions in each of the regions involved are likely to vary irregularly in time and space, the observation network for the preliminary stage is too rough and the observation interval too long to reveal these variations. The observation policy was not substantially altered for a long time, however, though some improvements were made. Therefore the observations remained in the

preliminary stage and the dynamical mechanisms of the variations remained unknown.

At this time, endeavors were made to reveal detailed synoptic features of oceanographic conditions and detailed processes of seasonal variations in oceanographic conditions in the seas adjacent to Japan. As was mentioned earlier, multiple-ship surveys were conducted in the sea east of Japan by the Imperial Fisheries Experimental Station from 1933 to 1941 and synoptic features of currents and water-property distributions were obtained. Owing to an inadequate distribution of observation stations and to inadequate accuracy in measurements, however, the results showed much ambiguity in spite of the quantity of existing data. The lack of detailed information on the vertical distribution of water properties made a study of the vertical structure impossible. In the sea south of Japan, oceanographic surveys including multiple-ship surveys from 1938 to 1942 were frequently conducted by the Hydrographic Office before and during the war. Because of the lack of published data from the surveys, however, a detailed study of the oceanographic conditions was impossible. Under such circumstances observations in this period were essentially focused upon a detailed re-examination of previous works by Koenuma and Uda, and Masuzawa made important contributions in this respect.

The following describes the observation program of the agencies up to the present time in detail. In 1954, the *Shumpu-Maru II* (150 gross tons) of the Kobe Marine Observatory was completed. Thus, eight research ships, including small auxiliary ones, were available for observations. The Central Meteorological Observatory and Marine Observatories were put in charge of observations in the regions adjacent to the observatories.

Observation cruises started in the same year, and since then important cruises have been conducted annually, mainly in February, May, August and November. Observations were also made on the cruises to and from the weather station Tango from June to October every year. The BT began to be utilized fairly extensively in the following year. At that time, the first attempts were made to predict synoptic features of oceanographic conditions. In 1956, the fifth volume of *Kaiyo-Hokoku* was devoted to methods for predicting synoptic oceanographic conditions for every season, and the conditions in the seas adjacent to Japan predicted for 1956 from observations made in 1955 were reported. Very little accurate information on the varying processes of oceanographic conditions had been obtained, and reliable dynamical equations on the basis of which predictions could be made were not of course available. Hence, predictions should naturally have been made on the basis of the statistical probability with which temporal variations in oceanographic conditions would occur. But oceanographic conditions which do not vary regularly like the tides cannot be reliably predicted by the statistical method.

In the same year a GEK was installed on the *Ryofu-Maru*. The GEK together with the BT made possible the quick observation of surface velocity and temperature fields even in rough seas. The observation program was improved by replacing some of the hydrographic stations with GEK and BT stations and by making observations with smaller spatial intervals. This improvement contributed to the revealing of fine structures. For example, Masuzawa (1956) treated a cold water belt along the northern edge of the Kuroshio Extension east of Japan. The adoption of north-south hydrographic sectioning instead of east-west sectioning as in the past also contributed greatly to the possibility of studying fine structure. Preliminary experiments were made by Yoshitada Takenouti in the same year to measure subsurface currents by means of parachute drogues. This Lagrangian method made possible the accurate measurement of subsurface currents and was successfully used by John Knauss in the measurement of the equatorial undercurrent.

In 1957 IGY observations began. The observation program of that year emphasized multiple-ship surveys of short-period variations of the front and direct measurement of water movement in the cold water belt along the northern edge of the Kuroshio Extension. In addition, measurements of internal waves which could have a great influence on the accuracy of BT sections and hydrographic sections were made. The *Ryofu-Maru, Shumpu-Maru* and *Yushio-Maru* from the Meteorological Agency and the *Kaiyo* from the Hydrographic Office participated in the survey under the direction of Takenouti on board the *Ryofu-Maru*. From 8 June to 3 July five crossings of the Kuroshio were made between 142°E to 146°E at intervals of about five days. In order to reveal the shorter-period variations four crossings were made at the section of minimal shift. To examine the variations of the cold water-belt measurements of surface current in the belt were carried out from three ships by tracking parachute drogues at three layers in parallel with GEK measurements and BT sections. Measurements of isotherm fluctuations were made with BT simultaneously at two stations, one in the Oyashio region and the other in the Kuroshio region. From these measurements, which were made at half-hour intervals for 39 hours, semidiurnal and diurnal fluctuations were found to prevail over shor-

ter-period fluctuations. From analysis of the data from this multiple-ship survey, Takenouti (1958) tried to explain the formation of the cold water belt and Masuzawa (1958) treated short-term variations of the Kuroshio in the region.

From May to July 1959, a special survey to follow the entire course of the Kuroshio Extension was conducted by Masuzawa (1964). Except for observations of the current from the *Ryofu-Maru* in the NORPAC Expedition in 1955, systematic observations for such a purpose had never been conducted earlier. The emphasis of the survey at this time was put on clarifying synoptic features in the region east of 150°E, where the Kuroshio gradually weakens and finally decays. The following results were obtained. A strong stream zone, which meandered with an amplitude of about 100 miles, with peaks at 143° and 151°E, was found between 34° and 36°N in the region west of 157°E. As the Kuroshio went further eastwards beyond 157°E and its path deviated gradually southward, a thin, narrow belt of strong stream zone of the highest temperature seemed to occur. But whether or not the zone spread continuously far eastward was not clear from the observations, which were made at intervals of about 2 degrees. North of the Kuroshio, a train of warm and cold water eddies was observed. In November of the same year one more special survey was conducted in the region east of Japan. In order to estimate the geostrophic transport with reference to the 800 db, 1000 db, 1200 db and 1500 db layers and the layer of minimum oxygen, hydrographic sections were made down to the 3000 db layer from about 34° to 38°N along 144°E. Surface geostrophic currents estimated in reference to these layers were compared with those measured with the GEK. The comparison showed that the reference layer should be selected at 2000 db or deeper in the southern part and should be gradually made shallower northward in order to obtain sufficient agreement (JMA, Oeanogr. Sect. 1960). Ichiye (1960), on the other hand, determined that the layer of no motion should be selected at 1000 to 1200 db in the region under consideration. In the hydrographic sections he used, however, the spacings of the samplings were 1000 m below the 2000 db layer, so that his estimate is unreliable. Moreover, the actual idea of a no motion layer seems dubious. On the basis of his study he estimated the deep current from hydrographic sections made by the Japanese Navy from 1930 to 1939. He found that a countercurrent existed under the Kuroshio and that the transport of this countercurrent was almost equal to that of the Kuroshio. The reference level he selected was ambiguous and his results are not reliable; however, his subject was of great interest to oceanographers. Even for the region adjacent to Japan, there was little data from observations systematically conducted to explore the deep layers of the sea. The need to explore the deep sea was keenly felt, but at that time Japan had no research ship with the facilities and instruments capable of hydrographical, geological and biological studies of the deep sea. For this reason Miyake began preliminary negotiations with the Rockefeller Foundation for financial support of Japanese deep-sea researches. In April 1957 the Planning Committee for Deep Sea Research was organized by Kiyoo Wadati for the purpose of exploring the Japan Trench and the neighboring deep seas. In 1958 the Deep Sea Research Committee was established as part of the National Committee of Ocean Research (NCOR) of the Science Council of Japan to promote deep-sea research. Interested oceanographers from various agencies and institutions formed the acting staff of the Committee. Facilities and equipments for deep-sea research were obtained and mounted on the *Ryofu-Maru* in March 1959. The expedition, which was called the Japanese Expedition of the Deep Sea (JEDS), was first conducted in June 1959 and since then has been conducted repeatedly. In September of the same year the second JEDS expedition made a hydrographic section across the Kurile Trench. The geostrophic current was estimated with reference to the minimum oxygen layer as well as to the 3000 db layer. A special expedition, also aiming at making hydrographic sections down to a deep layer, was carried out in the southwestern region in the following year. Two sections, one southward from Daio-Saki to a station (23°N, 137°E) and the other from the same station to Amamioshima, were made down to the 4000 db layer.

Direct measurements of deep current by means of neutrally buoyant floats were also carried out by Takenouti et al. (1962) of the Meteorological Agency. The first measurement, which was unsuccessful because of instrumental failure, was made east of Torishima in the latter half of the third JEDS expedition in June 1960. Then, in early September of the same year, tracking of the floats was successfully carried out twice in the northern part of the Japan Trench. In the first tracking the floats at 1500 m and 2500 m depths were successfully followed for about one day and in the second tracking those at 1000 m and 3000 m depths were followed for about half a day and two and a half days, respectively. From these measurements southward currents were detected in a subsurface layer shallower than 3000 m. The periods of these measurement were not long enough to obtain mean currents by eliminating variations. This result

was in contrast with the current predicted by some oceanographers from distributions of water properties in the region. The measurements were also carried out in the region south of the Enshu Nada in November 1962 to study a current field associated with a cold water mass present there at the time. The next deep current measurement was carried out in a region southeast of Torishima near the Ramapo Deep for five days from 11 to 15 July of the following year. Nan'niti and Akamatsu (1966), in summarizing the results, proposed a new model for deep circulation in the North Pacific based on the dynamic model by Henry Stommel. It was pointed out that the proposed model seemed to support the circulation pattern deduced earlier by Kishindo from the distribution of water properties. In 1966, deep current measurements made by a similar method were carried out by Nan'niti, Akamatsu and Masashi Yasui at the 800 m layer near 39°48′N, 131°43′E for about four days.

On the basis of observations from the routine monitoring program, Shigeo Moriyasu (1956, 1958a, 1958b, 1958c, 1959, 1960, 1961a, 1961b, 1963) studied variations of oceanographic conditions in the southern sea of Japan in connection with fluctuations of the Kuroshio, and supposed that the small disturbances of the Kuroshio occurring in the southeastern region of Kyushu became the large meanders associated with the cold water mass during its eastward travel to the region south of the Kii Peninsula. He pointed out that an off-shore shift of the Kuroshio path frequently found in winter in the region southeast of Kyushu was caused by a strong monsoon and that in association with wind-stress shear in the region, an upwelling which might produce a cold water mass developed. This shift of the Kuroshio was generated almost every winter, and the mechanism by which the shift developed in the large meanders should, according to him, be studied. With respect to the meanders, he showed that a change in sea-level difference along the Kii Peninsula was closely related to the shift of the path in the region south of the Peninsula. Fluctuations of the Kuroshio south of Japan were noted along with those east of Japan.

During the CSK, the Meteorological Agency made hydrographic sections crossing the Kuroshio southeast of Kyushu regularly eight times a year to monitor the Kuroshio fluctuations. In addition, the agency conducted a multiple-ship survey with five research ships from October to November 1967, and made regular and frequent GEK and BT measurements of the Kuroshio south of the Kii Peninsula. Variations in surface velocity of diurnal and semidiurnal periods were clearly observed. A shift in the path of the strong stream, which exceeded 2 knots, was less than 10 miles. By following this shift, the temperature field also shifted (Masuzawa, 1968a). This survey was an important factor in subsequent re-examination of survey results by the Hydrographic Office. A special expedition to make a hydrographic section along 137°E from about 34°N to the equator in the Pacific was conducted by Masuzawa (1967, 1968b, 1970) every winter from 1966 to 1969 to get synoptic features and year-to-year variations of the section. The section was selected so as not to encounter islands, so that any hydrographic section obtained is expected to be typical of any other hydrographic section in the western North Pacific. Another purpose of the expedition was to obtain information on synoptic features of the western part of the North Equatorial Current. Direct current measurements were conducted along with the hydrographic section in the region northeast of the Philippines.

Other Institutions

After the war, the Tokyo University of Fisheries and the Faculty of Fisheries at the Universities of Hokkaido, Nagasaki and Kagoshima were established. These institutions, which had their own ships, conducted cruises primarily for the purpose of training students in navigation and fisheries work, which requires a knowledge of oceanography. Oceanographic observations on the cruises were conducted mainly for training; however, some of the observations were intended for a study of oceanographic conditions. For example, the *Umitaka-Maru* of the Tokyo University of Fisheries participated in the NORPAC Expedition. The *Shinyo-Maru* of the same university has conducted a cruise to study the variations in oceanographic conditions in the Kuroshio region near the Izu Islands. The *Kagoshima-Maru* of the University of Kagoshima participated in the EQUAPAC Expedition. The *Oshoro-Maru* of the University of Hokkaido has conducted a cruise to explore conditions in the subarctic region of the western North Pacific. These ships, as well as the *Nagasaki-Maru* of the University of Nagasaki, contributed to oceanographic observations by participating in the CSK. The results of their observations were included in CSK data reports. In 1962, the Ocean Research Institute was established at the University of Tokyo, and two research ships, the *Tansei-Maru* (258 gross tons) and the *Hakuho-Maru* (3226 gross tons), were completed for the cooperative use of Japanese oceanographers. Observations of short-term fluctuations and fine structures of oceanograph c conditions as well as deep current measurements have been conducted from these ships.

Concluding remarks

As has been briefly reviewed, what we have obtained from observations in the past is quite insufficient to clarify completely processes of variations in the Kuroshio and adjacent regions and to understand mechanisms of the variations. Promotion of advanced observation programs are strongly required. Advanced programs mentioned here imply programs aimed at more sharply focussed on a specified purpose; that is more sharply focussed on variations of a narrower range of temporal and spatial frequencies. Development of observation method is of course essentially important for promoting advanced observations, however more efficient use of available facilities is also an important subject as well. In this connection observation programs proposed by Stommel (1963) at the symposium of the Kuroshio can be considered to have presented an example. Maintaining routine monitoring network is naturally important, but it should bear in mind that monitoring reduces its significance greatly if it were not made at temporal intervals adequate for clarifying variations of spatial scales expected to be caught with spatial observation intervals in monitoring. And it should be taken into consideration that relation between temporal and spatial scales are finally to be confirmed through specially designed observations.

REFERENCES

Dodimead, A. J., F. Favorite and T. Hirano (1963): Salmon of the North Pacific Ocean, Part II. Review of oceanography of the Subarctic Pacific Region—*Int. North. Pac. Fish. Comm., Bull.*, **13**: 1–195.

Fukuoka, J. (1955): The variation of the polar front in the sea adjacent to Japan. *Oceanogr. Mag.*, **6**(4): 181–195.

Fukuoka, J. and I. Tsuiki (1954): On the variation of chlorinity in the Kuroshio Area (especially, on the low chlorinity in summer). *Oceanogr. Mag.*, **6**(1): 15–23.

Hashiguchi, Y. and R. Saruwatari (1968): Taking part in the cooperative study of Kuroshio (in pursuit of the origin and extension of Kuroshio). *Hydrogr. Bull.*, **85**: 13–22.

Hirano, T. (1957a): The oceanographic study on the Subarctic Region of the Northwestern Pacific Ocean I. *Bull. Tokai Reg. Fish. Res. Lab.*, **15**: 39–55.

——— (1957b): The oceanographic study on the Subarctic Region of the Northwestern Pacific Ocean II. *Bull. Tokai Reg. Fish. Res. Lab.*, **15**: 57–69.

——— (1958): The oceanographic study on the Subarctic Region of the Northwestern Pacific Ocean III. *Bull. Tokai Reg. Fish. Res. Lab.*, **20**: 23–46

——— (1961): The oceanographic study on the Subarctic

Region of the Northwestern Pacific Ocean IV. *Bull. Tokai Reg. Fish. Res. Lab.*, **29**: 11–39.

Ichiye, T. (1953): On the variation of oceanic circulation (IV). *Oceanogr. Mag.*, **5**(1): 23–44.

―――― (1954): On the variation of oceanic circulation (VII). *Oceanogr. Mag.*, **6**(1): 1–14.

―――― (1960): On the deep water in the western North Pacific. *Oceanogr. Mag.*, **11**(2): 99–110.

Iwasa, K. (1963): Apparatus for recording oceanographic conditions. *Bull. Coast. Oceanogr.*, **2**(1): 30–32.

―――― (1968): On the establishment of continual measuring instrument of water temperature with submarine cable. *Hydrogr. Bull.*, **85**: 27–28.

Japan Meteorological Agency, Oceanographical Section (1960): Report of the oceanographic observations in the sea east of Honshu in November, 1959. *The Results of Marine Meteor. Oceanogr. Obs.*, **26**: 28–36.

Kawai, H. (1955): On the polar frontal zone and its fluctuation in the waters to the northeast of Japan. (I) *Bull. Tohoku Reg. Fish. Res. Lab.*, **4**: 1–46.

―――― (1955): On the polar frontal zone and its fluctuation in the waters to the northeast of Japan. (II) *Bull. Tohoku Reg. Fish. Res. Lab.*, **5**: 1–42.

―――― (1957): On the natural coordinate system and its applications to the Kuroshio System. *Bull. Tohoku Reg. Fish. Res. Lab.*, **10**: 141–171.

―――― (1959): On the polar frontal zone and its fluctuation in the waters to the northeast of Japan. (III) *Bull. Tohoku Reg. Fish. Res. Lab.*, **13**: 13–59.

―――― (1969): A synoptic view of problems in synoptic analysis in the ocean (in Japanese). *Bull. Jap. Soc. Fish. Oceanogr.*, Special Number (Prof. Uda's Commemorative Papers), 77–79.

Kawai, H. and M. Sasaki (1962): On the hydrographic condition accelerating the skipjack's northward movement across the Kuroshio Front. *Bull. Tohoku Reg. Fish. Res. Lab.*, **20**: 1–27.

Kawai, H., H. Sakamoto and M. Momota (1969): A study on convergence and divergence in surface layer of the Kuroshio I. *Bull. Nansei Reg. Fish. Res. Lab.*, **1**: 1–14.

Kawai, H. and H. Sakamoto (1970): A study on convergence and divergence in surface layer of the Kuroshio II (in Japanese). *Bull. Nansei Reg. Fish. Res. Lab.*, **2**: 19–38.

Kishindo, S. (1940a): On the bottom temperature and the bottom current in the Pacific (in Japanese). *Hydrogr. Bull.*, **19**(8): 259–270.

―――― (1940b): On the stratification of the water masses on the deep layer currents in the Pacific Ocean (in Japanese). *Hydrogr. Bull.*, **19**(11): 351–362.

Kishi, S. (1967): Operation of the oceanographic survey by the Hydrographic Office under the Imperial Navy (Part 2). *Hydrogr. Bull.*, **82**: 39–54.

Koenuma, K. (1933): Some oceanographical conditions of the Kuroshio area (in Japanese). *Jour. Ocean-*

ogr., **5**(2): 469–476.

Koenuma, K. (1937): On the hydrography of the south-western part of the North Pacific and the Kuroshio. Part 1: General oceanographical features of the region. *Memo. Kobe Mar. Obs.*, **6**(1): 279–331.

―――― (1938): On the hydrography of the south-western part of the North Pacific and the Kuroshio. Part 2: Characteristic water masses which are related to this region, and their mixtures, especially the water of the Kuroshio. *Memo. Kobe Mar. Obs.*, **6**(4): 349–414.

―――― (1939): On the hydrography of the south-western part of the North Pacific and the Kuroshio. Part III. Oceanographical investigations of the Kuroshio area and its outer region. Development of ocean currents in the North Pacific. *Memo. Kobe Mar. Obs.*, **7**(1): 41–114.

―――― (1941): On the hydrography of the south-western part of the North Pacific and the Kuroshio. Part IV. Energy of the Kuroshio, Part V, Concluding remarks. *Memo. Kobe Mar. Obs.*, **7**(4): 399–435.

Kondo, M. (1968): Short-period fluctuations of the oceanographic condition of the Tsushima warm current (I); Fluctuations of temperature and chlorinity. *Bull. Seikai Reg. Fish. Res. Lab.*, **36**: 1–20.

Konishi, T. (1921a): Whereabouts of drift-bottles (I) (in Japanese). *Umi to Sora*, **1**(8): 86–90.

―――― (1921b): Whereabouts of drift-bottles (II) (in Japanese). *Umi to Sora*, **1**(4): 38–40.

―――― (1922a): Whereabouts of drift-bottles (IV) (in Japanese). *Hydrogr. Bull.*, **1**(3): 111–122.

―――― (1922b): Whereabouts of drift-bottles (III) (in Japanese). *Hydrogr. Bull.*, **1**(2): 59–66.

―――― (1923): Whereabouts of drift-bottles (V) (in Japanese). *Hydrogr. Bull.*, **2**(5): 31–41.

Kumata, T. (1922): Investigations of ocean current around Japan (in Japanese). *The Osaka Mainichi Press*.

Kuroda, R. (1958): Notes on the "inversion of water density" in the vertical distributions of sea water density. *Bull. Tohoku Reg. Fish. Res. Lab.*, **11**: 82–87.

―――― (1959): Notes on the phenomena of "inversion of water temperature" off the Sanriku Coast of Japan. (I) *Bull. Tohoku Reg. Fish. Res. Lab.*, **13**: 1–12.

―――― (1960): Notes on the phenomena of "inversion of water temperature" off the Sanriku Coast of Japan. (II) *Bull. Tohoku Reg. Fish. Res. Lab.*, **16**: 65–86.

―――― (1962): On the states of appearances of "Shiome" in the waters to the northeast of Japan. *Bull. Tohoku Reg. Fish. Res. Lab.*, **22**: 45–115.

Marumo, R. (1954): Relation between planktological and oceanographical conditions of a sea area east of Kinkazan in winter. *Jour. Oceanogr. Soc. Japan*, **10**(2): 77–84.

Masuzawa, J. (1950): On the intermediate water in the Southern Sea of Japan. *Oceanogr. Mag.*, **2**(4): 137–144.

―――― (1954): On the seasonal variation of the Kuroshio east of Cape Kinkazan of Japan proper (in

Japanese). *Kaiyo Hokoku*, **3**(4): 251–255.

Masuzawa, J. (1954): On the Kuroshio south off Shiono-Misaki of Japan (Currents and water masses of the Kuroshio System I). *Oceanogr. Mag.*, **6**(1): 25–33.

———— (1956): On the cold belt along the northern edge of the Kuroshio (Currents and water masses of the Kuroshio System III). *Oceanogr. Mag.*, **8**(2): 151–156.

———— (1958): A short-period fluctuation of the Kuroshio east of Cape Kinkazan. *Oceanogr. Mag.*, **10**(1): 1–8.

———— (1964): A typical hydrographic section of the Kuroshio Extension. *Oceanogr. Mag.*, **16**(1–2): 21–30.

———— (1967): An oceanographic section from Japan to New Guinea at 137°E in January 1967. *Oceanogr. Mag.*, **19**(2): 95–118.

———— (1968a): Cruise report on multi-ship study of short-term fluctuations of the Kuroshio in October to November 1967. *Oceanogr. Mag.*, **20**(1): 91–96.

———— (1968b): Second cruise for CSK, Ryofu Maru, January to March 1968. *Oceanogr. Mag.*, **20**(2): 173–185.

———— (1970): Geostrophic flux of the north equatorial current south of Japan (in Japanese). *Jour. Oceanogr. Soc. Jap.*, **26**(1): 61–64.

Masuzawa, J. and T. Nakai (1953): On the fluctuation of the Kuroshio Current off Shionomisaki from 1950 to 1952. *Rec. Oceanogr. Works Japan*, **1**(2): 25–32.

Moriyasu, S. (1956): On the fluctuation of the Kuroshio south of Honshu (1) (Statistical treatment of the displacement of the stream axis). *Oceanogr. Mag.*, **8**(2): 143–149.

———— (1958a): On the fluctuation of the Kuroshio south of Honshu (3) (A short note on the computation of the relative volume transport). *Oceanogr. Mag.*, **10**(1): 81–89.

———— (1958b): On the fluctuation of the Kuroshio south of Honshu (2); Seasonal variations of water temperature and chlorinity in the upper layer. *Memo. Kobe Mar. Obs.*, **12**(2): 1–18.

———— (1958c): On the fluctuation of the Kuroshio south of Honshu (4) (The influence of the oceanographic conditions upon the monthly mean sea level). *Jour. Oceanogr. Soc. Japan*, **14**(4): 137–144.

———— (1959): On the fluctuation of the Kuroshio south of Honshu (5): the horizontal distribution of density in circular eddy. *Memo. Kobe Mar. Obs.*, **13**(2): 12–21.

———— (1960): On the monthly sea level on the south coast of Japan. *Memo. Kobe Mar. Obs.*, **14**: 19–31.

———— (1961a): On the difference in the monthly sea level between Kushimoto and Uragami, Japan. *Jour. Oceanogr. Soc. Japan.*, **17**(4): 197–200.

———— (1961b): An example of conditions at the occurrence of cold water region. *Oceanogr. Mag.*, **12**(2): 67–76.

———— (1963): The fluctuation of hydrographic condi-

tions in the sea south of Honshu, Japan (Review). *Oceanogr. Mag.*, **15**(1): 11–29.

Nan'niti, T. (1951): On the variation of the oceanographical condition along the so-called "C" Line (38°N, from 141°E to 153°E) from August 1948 to December 1949. *Oceanogr. Mag.*, **3**(1): 27–48.

Nan'niti, T. and H. Akamatsu (1966): Deep current observations in the Pacific Ocean near the Japan Trench. *Jour. Oceanogr. Soc. Japan*, **22**(4): 154–160.

Nitani, H. (1961): On the general oceanographic conditions at the western boundary region of the North Pacific Ocean (in Japanese). *Hydrogr. Bull.*, **65**: 27–35.

———— (1963): On the analysis of deep sea in the region of the Kurile-Kamchatka, Japanese and Izu-Bonin Trench. *Jour. Oceanogr. Soc. Japan*, **19**(2): 82–92.

Nitani, H., K. Iwasa and W. Inada (1959): On the oceanic and tidal current observation in the channel by making use of induced electric potential (in Japanese). *Hydrogr. Bull.*, **61**: 14–24.

Nitani, H. and D. Shoji (1970): On the variability of the velocity of the Kuroshio II. In *The Kuroshio—A Symposium on the Japan Current* (edited by J. C. Marr), East-West Center Press, Honolulu, 107–116.

Okada, T. (1922): On the possibility of forecasting the summer temperature and the approximate yield of rice crop for northern Japan. 2nd paper. *Mem. Imp. Mar. Obs.*, **1**: 18–26.

Ono, K. (1951): New current meter (In Japanese). *Hydrogr. Bull.*, New Ser. **25**: 228–229.

Shoji, D. (1951): The variations of oceanic currents in the adjacent sea of Japan (in Japanese with English abstract). *Hydrogr. Bull.*, New Ser. **25**: 230–244.

———— (1954): On the variation of daily mean sea level and oceanographic condition (I) (in Japanese). *Hydrogr. Bull.*, New Ser. Special **14**: 17–25.

———— (1966): Moored array observations by the Hydrographic Office. *Bull. Coast. Oceanogr.*, **5**(1): 31–33.

Shoji, D. and H. Nitani (1966): On the variability of the velocity of the Kuroshio I. *Jour. Oceanogr. Soc. Japan*, **22**(5): 10–14.

Stommel, H. (1965): Some thoughts about Planning the Kuroshio survey. Proceedings of Symposium on the Kuroshio, Tokyo, 29 October 1963, Oceanographical Society of Japan and UNESCO, 22–33.

Takenouti, Y. (1958): Measurements of subsurface current in the cold-belt along the northern boundary of Kuroshio. *Oceanogr. Mag.*, **10**(1): 13–17.

Takenouti, Y., T. Nan'niti and M. Yasui (1962): The deep-current in the sea east of Japan. *Oceanogr. Mag.*, **13**(2): 89–101.

Tsuchiya, M. (1961): An oceanographic description of the equatorial current system of the western Pacific. *Oceanogr. Mag.*, **13**(1): 1–30.

Tsujita, T. (1961): The fisheries oceanography of the East China Sea and the Tsushima Strait (The oceanographic structure and the ecological character of the fishing grounds.) *Bull. Seikai Reg. Fish. Res. Lab.*, **84**: 1–47.

Uda, M. (1934): The results of simultaneous oceanographical investigations in the Japan Sea and its adjacent waters in May and June, 1932 (in Japanese). *Jour. Imp. Fish. Exper. Sta.*, **5**: 57–190.

———— (1935): The results of simultaneous oceanographical investigation in the North Pacific Ocean adjacent to Japan made in August, 1933 (in Japanese). *Jour. Imp. Fish. Exper. Sta.*, **6**: 1–130.

———— (1936a): Oceanic circulation and its variation (in Japanese). *Kagaku*, **6**(10): 449–453.

———— (1936b): Results of simultaneous oceanographic investigations in the Japan Sea and its adjacent waters during October and November, 1933 (in Japanese). *Jour. Imp. Fish. Exper. Sta.*, **7**: 91–151.

———— (1937): On the recent abnormal condition of the Kuroshio to the south of Kii Peninsula (in Japanese). *Kagaku*, **7**(9): 360–361, **7**(10): 403–404.

———— (1938): Hydrographical fluctuation in the northeastern sea-region adjacent to Japan in North Pacific Ocean. (A result of the simultaneous oceanographical investigations in 1934–1937) (in Japanese). *Jour. Imp. Fish. Exper. Sta.* **9**: 1–66.

———— (1942): On the fluctuation of oceanic current I (in Japanese). *Jour. Oceanogr. Soc. Japan*, **2**(1): 22–26.

———— (1943): On the structure of oceanic front (in Japanese). *Jour. Oceanogr. Soc. Japan*, **2**(4): 9–16.

———— (1949): On the correlated fluctuation of the Kuroshio Current and the cold-water mass. *Oceanogr. Mag.*, **1**(1): 1–12.

———— (1950): On the fluctuation of oceanic current (2nd paper) (in Japanese). *Jour. Oceanogr. Soc. Japan*, **5**(2–4): 55–69.

———— (1969): *Umi* (in Japanese). Iwanami Shinsho, 732, Iwanami Shoten: 1–242.

Uda, M. and K. Hasunuma (1969): The eastward subtropical countercurrent in the western north Pacific Ocean. *Jour. Oceanogr. Soc. Japan*, **25**(4): 201–210.

Yosida, S. (1961a): On the short period variation of the Kuroshio in the adjacent sea of Izu Islands. *Hydrogr. Bull.*, New Ser. **65**: 1–18.

———— (1961b): On the variation of Kuroshio and cold water mass off Ensyū Nada (Part 1) (in Japanese). *Hydrogr. Bull.*, New Ser. **67**: 54–57.

Yosida, S., H. Nitani and N. Suzuki (1959): Report of multiple ship survey in the equatorial region (I.G.Y.) Jan.–Feb., 1958. *Hydrogr. Bull.*, **59**: 1–30.

Notes to the chronological table and cruise track charts

(1) Around lines showing periods of main cruises, information on the observation area and observation process is given together with the ships' names. The observa-

tion area is included in the oceanic regions represented by numeral notations (see Fig. 2). In the cases in which information on observation processes and ships' names is presented but no information on observation area is pre-

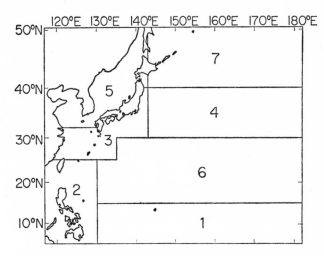

Fig. 2 Numeral notations for sub-regions of the northwestern Pacific divided for the convenience to indicate approximate location of observation area in every cruise.

sented, the areas are all in the oceanic regions represented by numeral notation 3.

Observation processes are indicated with the following notations:

○ : Hydrographic casts
● : Current measurements with towed electrodes
▲ : Temperature measurements with BT
△ : Current measurements with simultaneous use of two current meters, one at a reference layer and the other at the layer of measurement.

Ships' names are abbreviated as follows.

Odomari	OT
Matsue	ME
Koshu	KS
Yodo	YD
Manshu	MS
Komahashi	KH
Yamato	YT
Itsukushima	IS
Katsuriki	KR
Soyo-Maru	SY
Tama-Maru	TM
Daini Tama-Maru	FT
Daisan Tama-Maru	ST
Daigo Tama-Maru	GT
Dairoku Tama-Maru	RT

Dairoku Kyo-Maru	RK	*Kiji*	KG
Daishichi Kyo-Maru	NK	*Abukuma*	AB
Daihachi Kyo-Maru	HK	*Uzura*	UZ
Daiju Kyo-Maru	TK	*Miochidori*	MC
Daiichi Toshi-Maru	TS	*Kawachidori*	KD
Daini Toshi-Maru	FS	*Tenryu*	TR
Daigo Toshi-Maru	GS	*Meiyo*	MY
Daishichi Toshi-Maru	NS	*Shoyo*	YO
Daiichi Takunan-Maru	IN	*Sawachidori*	SC
Daini Takunan-Maru	FN	*Mogami*	MO
Daisan Takunan-Maru	SN	*Tsugaru*	TG
Daigo Takunan-Maru	GN	*Shinano*	NA
Daihachi Takunan-Maru	HN	*Miura*	MU
Daiju Takunan-Maru	TN	*Kuzuryu*	KU
Oshoro-Maru	OS	*Kitakami*	KT
Iwaki-Maru	II	*Oyodo*	OO
Iwate-Maru	IW	*Shunkotsu-Maru*	SU
Miyagi-Maru	MG	*Daini Koyoshi-Maru*	FO
Daiichi Kaiyo	IK	*Tone*	NE
Daini Kaiyo	FK	*Satsuma*	TU
Daisan Kaiyo	SK	*Kiso*	KO
Daiyon Kaiyo	YK	*Heiyo Eisho-Maru*	HE
Daigo Kaiyo	GK	*Okushiri*	OK
Dairoku Kaiyo	MK	*Amakusa*	MA
Tankai-Maru	TA	*Sado*	SD
Aomori-Maru	AM	*Kikuchi*	CU
Ibaraki-Maru	IG	*Asachidori*	AD
Fusa-Maru	FU	*Rishiri*	RI
Sagami-Maru	SM	*Takuyo*	AA
Siratori-Maru	SR	*Noto*	NT
Kamoi-Maru	KM	*Mikura*	MR
Kiyo-Maru	KY	*Yoshino*	YS
Kochi-Maru	KC	*Fuji*	FZ
Hyuga-Maru	HG	*Ishikari*	IR
Tonan-Maru	TO	*Nagara*	NG
Zuisho-Maru	ZH	*Kabashima*	KB
Kaiho-Maru	KI	*Murachidori*	MD
Daiichi Shonan-Maru	IO	*Miyake*	MJ
Daisan Shonan-Maru	SO	*Rebun*	RB
Dairoku Shonan-Maru	RO	*Chifuri*	CF
Daishichi Shonan-Maru	NO	*Genkai*	GE
Daihachi Shonan-Maru	HO	*Shikine*	SS
Hakuyo-Maru	HY	*Kumano*	KN
Toyama-Maru	TY	*Awaji*	AW
Yoko-Maru	YU	*Kozu*	KZ
Daiichi Hikifune	IT	*Hachijo*	HJ
Hayato-Maru	HS	*Asakaze*	AK
Shinyo-Maru	SI	*Yamazuki*	YZ
Daijusan Choyo-Maru	CY	*Ojika*	OJ
Daijugo Choyo-Maru	CZ	*Okichidori*	OD
Daihachijuyon Banshu-Maru	CB	*Shunyo-Maru*	SH
Daiichi Kaiko-Maru	KK	*Sumida*	GG
Daiichi Tenkai-Maru	DT	*Koshiki*	KK
Soei-Maru	SE	*Sendai*	LA
Sanyo-Maru	SA	*Iki*	LB
Yukari	YR	*Kuma*	LC

Daio	LD	*Hokusei-Maru*		LR
Isoshio	LE	*Yang Ming*	YM	(China)
Tenyo	LF	*Orlick*	OR	(U.S.S.R.)
Tomochidori	TD	*Argo*	AG	(U.S.A.)
Soya	LG	*Suro*	SW	(Korea)
Oki	LH	*Bering Strait*	BS	(U.S.A.)
Natori	LI	*Atlantis*	AL	(U.S.A.)
Ashiya-Maru	LJ	*U. M. Schokalsky*	SL	(U.S.S.R.)
Noshiro	LL	*Balk Du San*	BD	(Korea)
Sorachi	LM	*Nevelskoy*	NV	(U.S.S.R.)
Chikugo	LN	*Cape St. Marry*	CS	(Hong Kong)
Yamachidori	LO	*Uliana Gromova*	UG	(U.S.S.R.)
Tatsuta	TT	*Han Ra San*	HR	(Korea)
Amami	LP	*Tac Balk San*	TB	(Korea)
Heian-Maru	LQ	*Ji Ri San*	JR	(Korea)
Ryofu-Maru	RF	*Researcher I*	RC	(Philippine)
Kuroshio-Maru	BB	*Iskatel*	IL	(U.S.S.R.)
Oyashio-Maru	CC	*Buek Ak San*	BA	(Korea)
Asashio-Maru	DD			
Yushio-Maru	EE			
Ukuru-Maru	UK			
Chikubu-Maru	CK			
Shinnan-Maru	FF			
Ikuna-Maru	IA			
Shumpu-Maru	SP			
Shiga-Maru	SG			
Tsuru-Maru	TP			
Nagasaki-Maru	NP			
Atsumi	AT			
Umikaze-Maru	UM			
Chofu-Maru	TF			
Heian-Maru	HA			
Nojima	NJ			
Kofu-Maru	KF			
Seifu-Maru	SF			
Kagoshima-Maru	KA			
Nagasaki-Maru	NI			
Tansei-Maru	TI			
Keiten-Maru	KE			
Koyo-Maru	YY			
Tenyo-Maru	TE			
Hakuho-Maru	HH			
Umitaka-Maru	UT			
Tokaidaigaku-Maru	LK			

(2) Publications of oceanographic observation data are abbreviated as follows.

DRHO : Data Report of Hydrographic Observations Series of Oceanography

HB : Hydrographic Bulletin

HBS : Hydrographic Bulletin Special Number

OB : Oceanographic Bulletin

BHD : The Bulletin of the Hydrographic Department

RMMOO : The Results of Marine Meteorological and Oceanographical Observations

OI : Oceanographical Investigation

RFOO : The Results of Fisheries Oceanographical Observation

DRC : Data Report of CSK

DROOEF : Data Record of Oceanographic Observations and Exploratory Fishing

(3) For observations after 1967 by Fisheries Agency, Hydrographic Office and Meteorological Agency, data have not been published yet at these agencies at the time of compiling the Chronological table. Information on those observations described in the Chronological table and the Cruise track charts are all quoted from Data Report of CSK published by Japanese Oceanographic Data Center.

Year		1878													1880											
Month		1	2	3	4	5	6	7	8	9	10	11	12	1	2	3	4	5	6	7	8	9	10	11	12
Chronological Table	Fisheries agencies																								
	Hydrographic office																								
	Meteorological agencies																								
	Other institutions																								
Cruise track charts																									
Data publications	Fisheries agencies																								
	Hydrographic office																								
	Meteorological agencies																								
	Other institutions																								
Main events related to oceanographic observations in Japan		In Kan'no-Kyoku of the Ministry of Home Affairs (Naimusho), some officials were appointed to be in charge of fisheries businesses.												In Kan'no-Kyoku, Fisheries Section was established.											
Events in the foreign countries which had great influence on oceanographic observations in Japan																									

1885												1889												1893											
1	2	3	4	5	6	7	8	9	10	11	12	1	2	3	4	5	6	7	8	9	10	11	12	1	2	3	4	5	6	7	8	9	10	11	12

Sailing directory Vol.1 (Kasoei Suiroshi Dai-Ikkan) was published.

In Ministry of Agriculture and Commerce (Noshomu-Sho), Fisheries Bureau (Suisan-Kyoku) was established.

Report on preliminary investigation of fisheries in seas around Japan (Suisan Yosatsu Chosa Hokoku) by S. Matsubara was published.

Fisheries Inquiry Institute was established. Y. Wada conducted ocean current exploration by the use of drifting bottles.

1894												1897												1899											
1	2	3	4	5	6	7	8	9	10	11	12	1	2	3	4	5	6	7	8	9	10	11	12	1	2	3	4	5	6	7	8	9	10	11	12

M. Knudsen established silver nitrate titration method.

Fisheries Training Institute was established.

(The Sino-Japanese War was conducted in the period from 1894 to 1895)

1900												1901												1904												
1	2	3	4	5	6	7	8	9	10	11	12	1	2	3	4	5	6	7	8	9	10	11	12	1	2	3	4	5	6	7	8	9	10	11	12

Fisheries Bureau established five coastal observation stations. T. Akanuma of Fisheries Bureau began to conduct the temperature measurement at a subsurface layer in the open sea.

T. Akanuma developed the areometer.

(The Russo-Japanese War was conducted in the period from 1904 to 1905)

	1905											1909												1910											
	1	2	3	4	5	6	7	8	9	10	11	12	1	2	3	4	5	6	7	8	9	10	11	12	1	2	3	4	5	6	7	8	9	10	11	12

Ekman current meter was developed. K. Honda started the experiment on bell jar type tide gauge.

V.W. Ekman established the theory on wind driven current.

T. Kitahara and others of Fisheries Bureau commenced fisheries fundamental exploration (Gyogyo Kihon Chosa).

Wada of Central Meteorological Observatory published the chart of monthly-mean surface water temperature of the Western North Pacific for the normal year.

Dynamic computation of ocean current was developed by V. Bjerknes and by W. Sandström.

1913	1914	1918
1 2 3 4 5 6 7 8 9 10 11 12	1 2 3 4 5 6 7 8 9 10 11 12	1 2 3 4 5 6 7 8 9 10 11 12

OI. No. 1, 2, 3; Only an announcement that the *Ungo Maru* made hydrographic observations in a cruise from Hokkaido to Tokyo in a period from middle September to late October is included and no observation datum is presented.

Oceanographic Inquiry Section was established in Fisheries Training Institute.

H. Harukawa of Fisheries Training Institute introduced the chlorinity determination through silver nitrate titration.

(World War I was conducted in the period from 1914 to 1918)

1919   1920   1921

OI, No. 4,5,6,7: Temperature and density data for each observation layer averaged over observation stations in each cruise are only presented. The maximum depth of observation was 100 m.

OI, No. 8,9,10,11: Data presentation is similar to that in the preceding volumes.

OI, No. 12,13,14,15.

The first meeting of the International Council for the Exploration of the Sea was held. Kobe Marine Observatory was established.

Hydrographic Office decided to conduct systematic oceanographic survey.

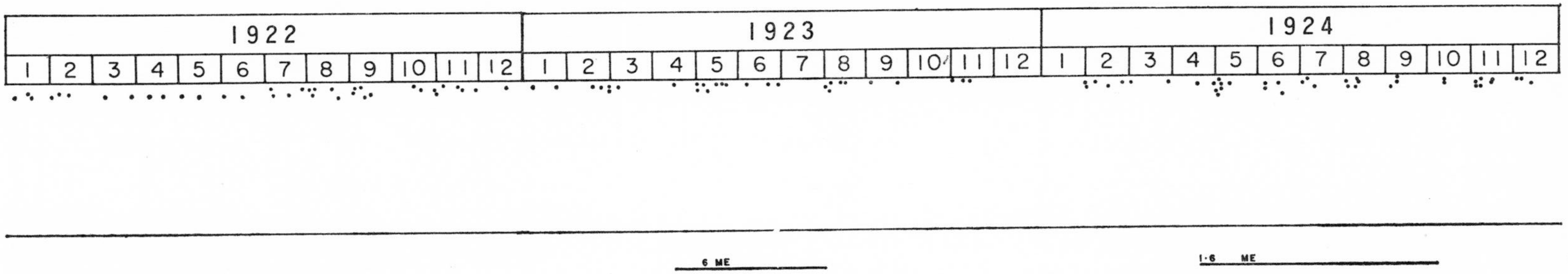

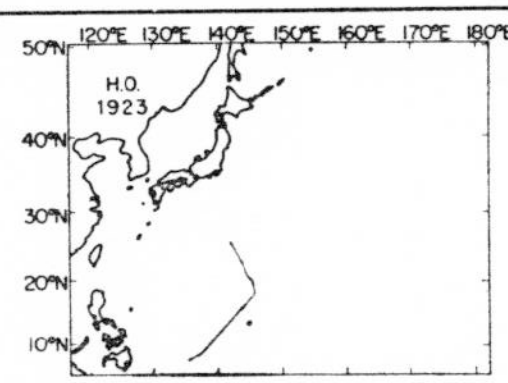

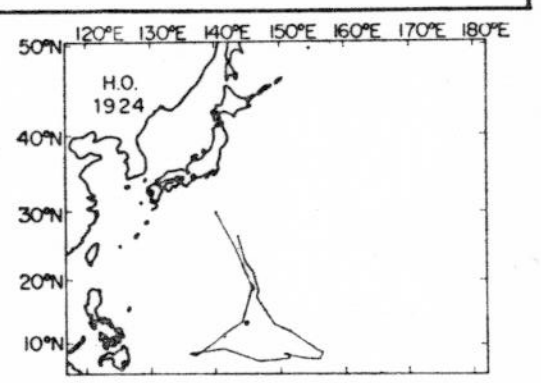

1922

| 1 | 2 | 3 | 4 | 5 | 6 | 7 | 8 | 9 | 10 | 11 | 12 |

1923

| 1 | 2 | 3 | 4 | 5 | 6 | 7 | 8 | 9 | 10 | 11 | 12 |

1924

| 1 | 2 | 3 | 4 | 5 | 6 | 7 | 8 | 9 | 10 | 11 | 12 |

6 ME

I-6   ME

H.O. 1923

H.O. 1924

OI. No. 16,17,18,19.	OI. No. 20, 21, 22, 23.	OI. No. 24, 25, 26, 27.
	OB. No. 3; Water sampling was made with Kitahara sampler, and accuracy of temperature measurements was ± 0.1°C. Data publications for observations in preceding years were reduced to ashes by a fire during the Great Earthquake of 1923.	OB. No. 3; Salinity was determined by silver nitrate titration with the accuracy of ± 0.01 %. Maximum depth of observation was 200 m and number of occupied stations was 38.
Results of ocean current investigations in the seas around Japan (Nihon Kankai Kairyu Chosa Gyoseki) edited by Y. Wada and T. Kumata was published.		

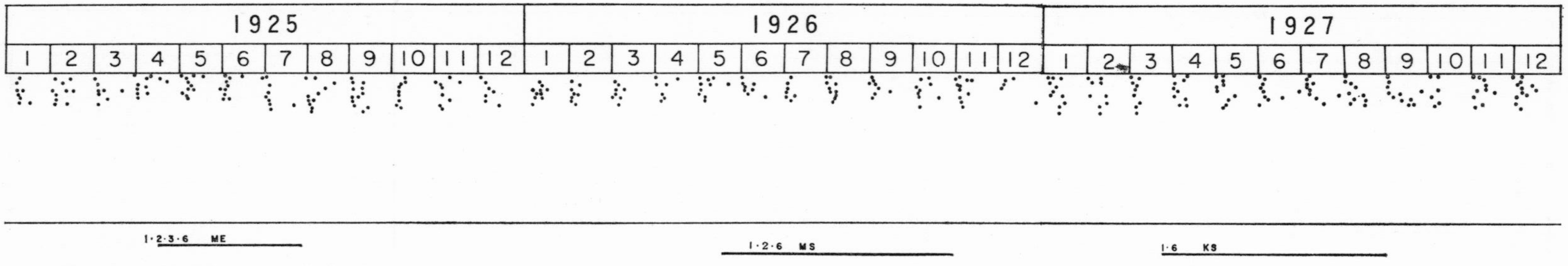

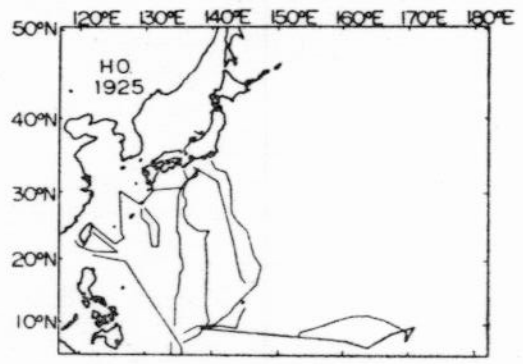

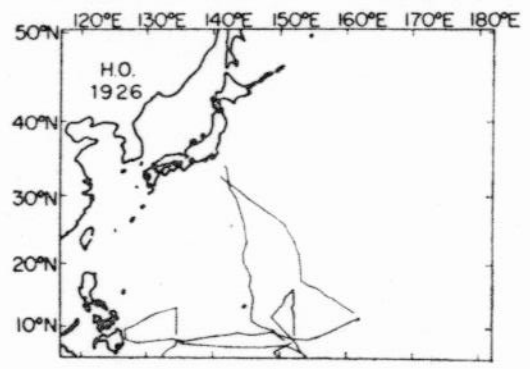

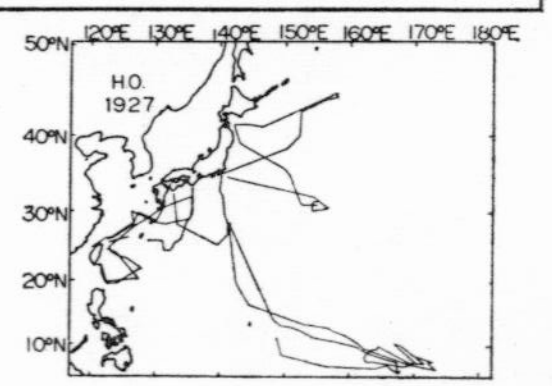

1925   1926   1927

1·2·3·6 ME

1·2·6 MS

1·6 KS

H.O. 1925

H.O. 1926

H.O. 1927

OI.·No. 28, 29, 30, 31; Density data tabulated was determined from temperature and salinity determined by silver nitrate titration, and only vertical averages of temperature and density at depths of 0, 25, 50, 100 and 200 m are presented.

Main purpose of cruises was sounding.

BHD, No.6.

OI. No. 32, 33, 34, 35.

OI. No. 36, 37, 38, 39.

OB, No. 3.

The Meteor Expedition was conducted during the period from 1925 to 1928.

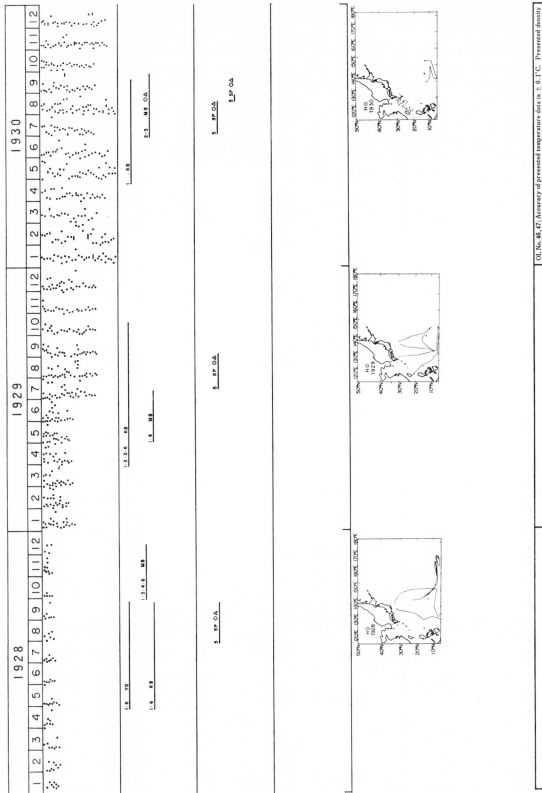

OI. No. 40, 41, 42, 43.	OI, No. 44, 45.	OI, No. 46, 47: Accuracy of presented temperature data is ± 0.1°C. Presented density data were obtained from measurement with an areometer in some cases and were estimated from chlorinity determination through silver nitrate titration in other cases. Maximum depths of measurements were 300 m as an average, and the deepest measurement was at 1200 m.
OB, No. 3: Temperature measured with accuracy of ± 0.01°C were first carried out on board the Koshu.	OB, No. 3: Accuracies of temperature and salinity measurements were 0.01°C and ± 0.01‰, respectively.	OB, No. 6: Accuracies of temperature and salinity measurements at layers deeper than 300 m were ± 0.01°C and 0.01‰, respectively. The deepest measurement was at 5700 m.
Kaiyo Jiho Vol. 2; Temperature measurements were made with accuracy of ± 0.01°C and data from current measurements with Ekman-Merz current meter are included. The first expedition of the Japan Sea was conducted. The expedition of the Japan Sea was made every year till 1932.	Kaiyo Jiho Vol. 3.	Kaiyo Jiho Vol. 4.
	Imperial Fisheries Experimental Station was established.	S. Kasuga, the director of Imperial Fisheries Experimental Station held the first conference with in Japan for promoting contact on observations by these stations.

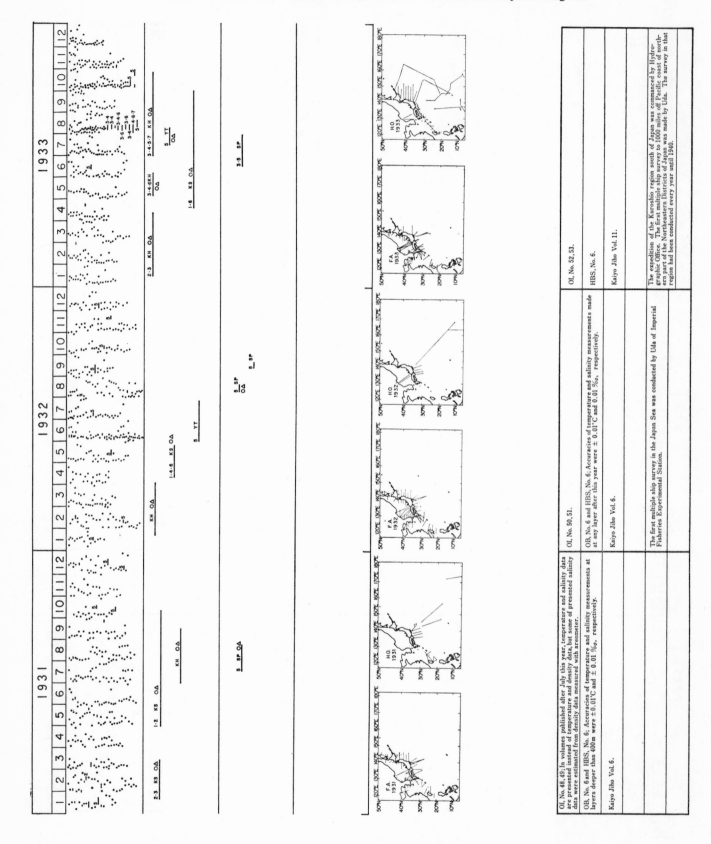

1934		1935		1936	

OI. No. 54, 55.

HBS, No. 6.

Kaiyo Jiho, Vo. 6.

OI, No. 56, 57; In cruises of period longer than 5 days, salinity data were missing for about a half of the occupied stations.

HBS, No. 8.

Kaiyo Jiho, Vol. 11, 12.

OI, No. 58, 59.

HBS, No. 8.

Kaiyo Jiho, Vol. 12.

The guide book on the method for oceanographic observations was published by Kobe Marine Observatory.

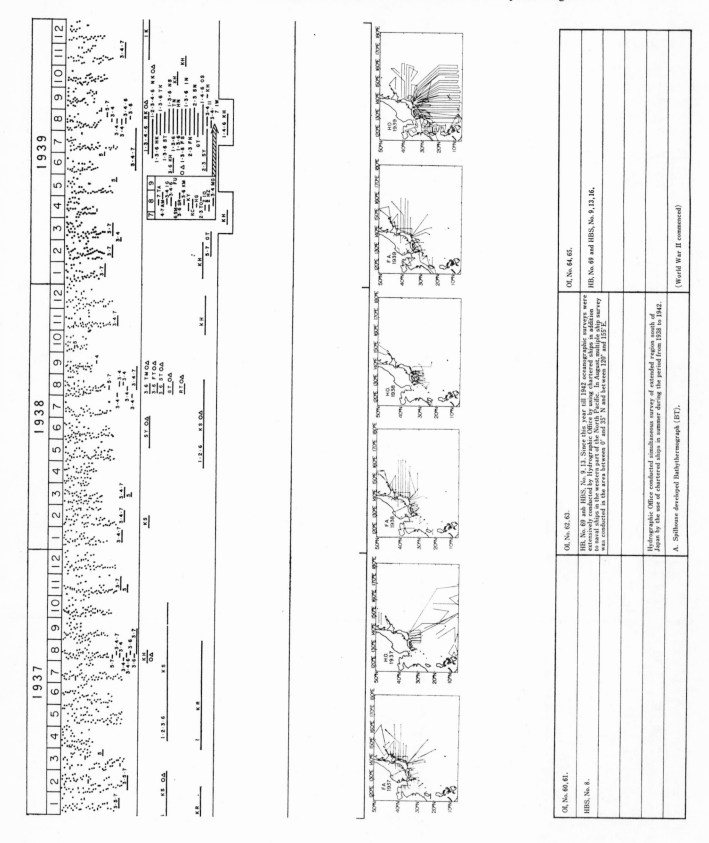

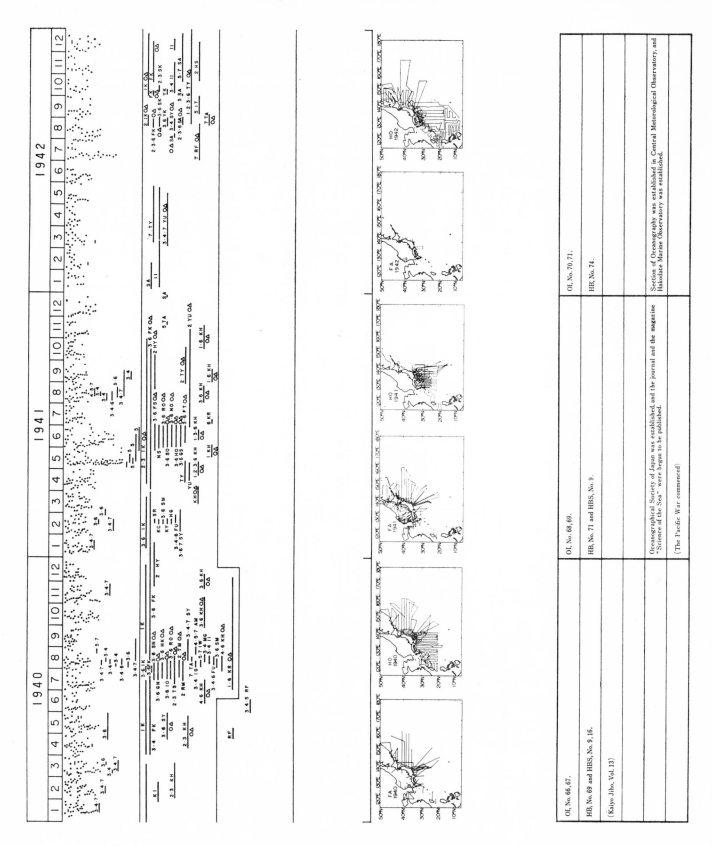

(World War II ceased)

OI, No. 72.

DRHO, No. 6.

OI, No. 72.

DRHO, No. 2.

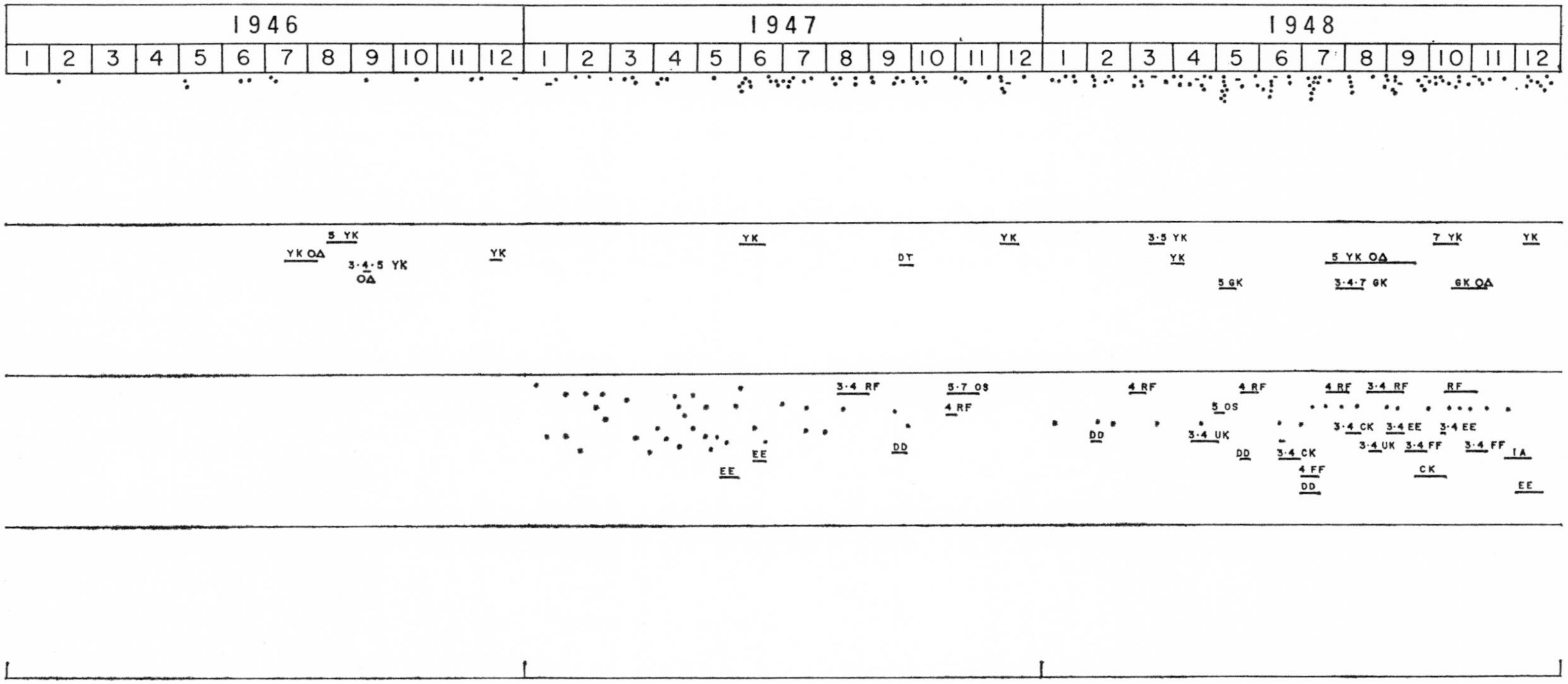

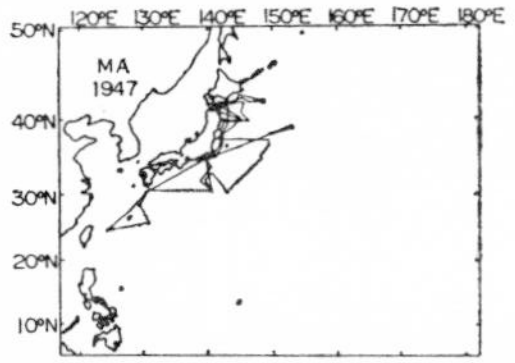

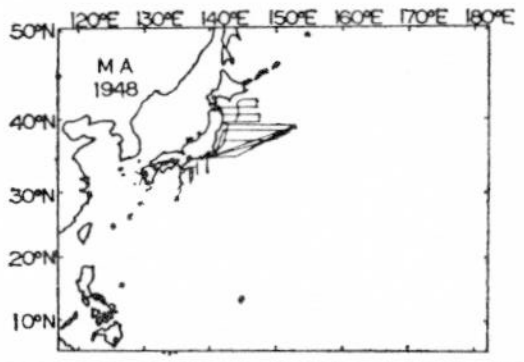

1946

| 1 | 2 | 3 | 4 | 5 | 6 | 7 | 8 | 9 | 10 | 11 | 12 |

1947

| 1 | 2 | 3 | 4 | 5 | 6 | 7 | 8 | 9 | 10 | 11 | 12 |

1948

| 1 | 2 | 3 | 4 | 5 | 6 | 7 | 8 | 9 | 10 | 11 | 12 |

M A 1947

M A 1948

	OI, No. 73: Accuracy of temperature measurements was ± 0.1°C and chlorinity data presented were estimated from density data measured with areometer with accuracy of ± 0.01 ‰.	OI, No. 73: Accuracy of temperature measurements recoverd to ± 0.01°C.	OI, No. 73.
	HBS, No. 10 and OB, No. 1, 4.	HBS, No. 10.	HBS, No. 5, 10.
	RMMOO, No. 1, 2.	RMMOO, No. 3, 4.	
		Nagasaki Marine Observatory and Maizuru Marine Observatory were established. Weather station watch at station Extra (39°N, 135°E) commenced.	Weather station watch at station Tango (29°N, 135°E) commenced. Fisheries Agency was established.

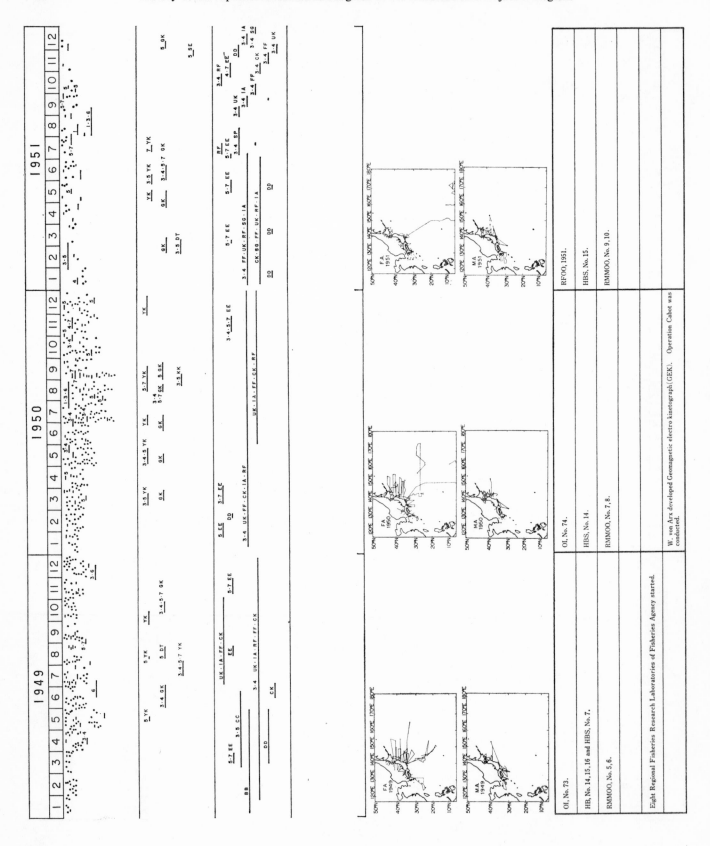

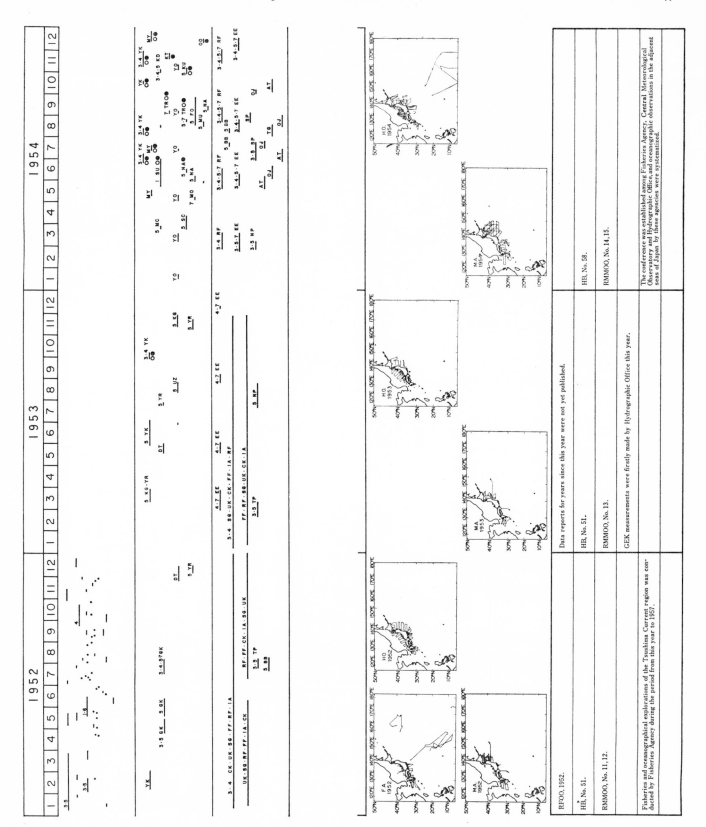

1955   1956   1957

HB, No. 58.

RMMOO, No. 16, 17, 18 (Part 1).

NORPAC Atlas.

The NORPAC EXPEDITION was conducted.

J. Swallow developed the neutrary buoyant float.

HB, No. 62.

RMMOO, No. 19, 20.

The EQUAPAC EXPEDITION was conducted. Meteorological Agency was established. The exploration of the fishing field in the North Pacific by Japan, U.S.A. and Canada was conducted during the period from this year to 1961.

HB, No. 64.

RMMOO, No. 21, 22.

DROOEF

The IGY expeditions were conducted in this year and the following year.

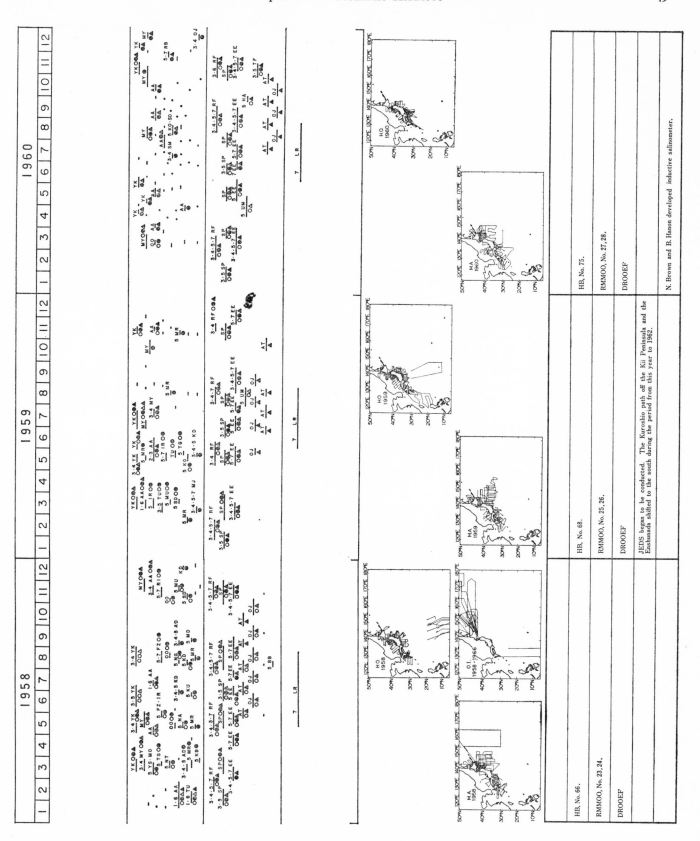

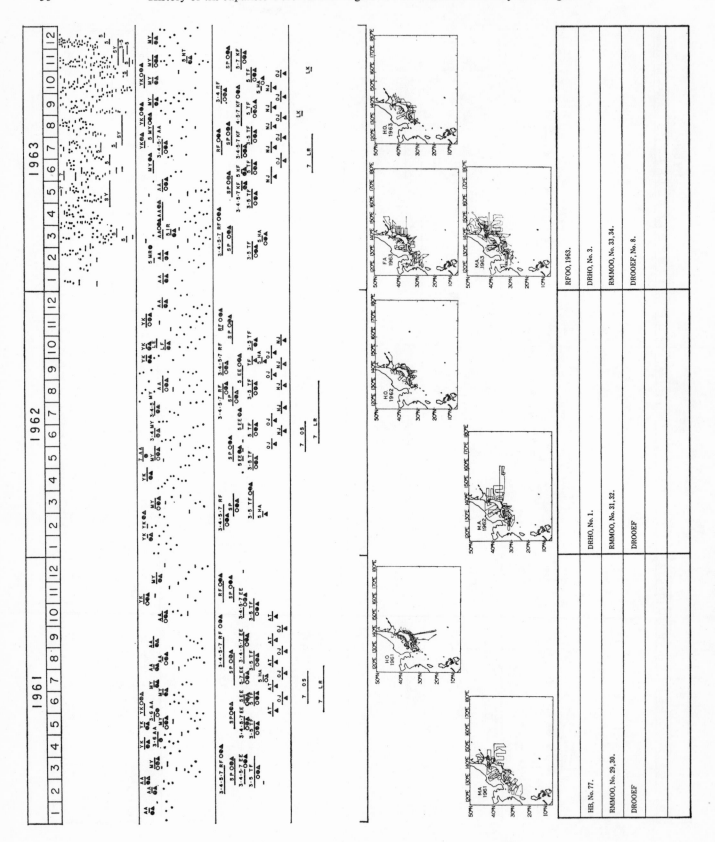

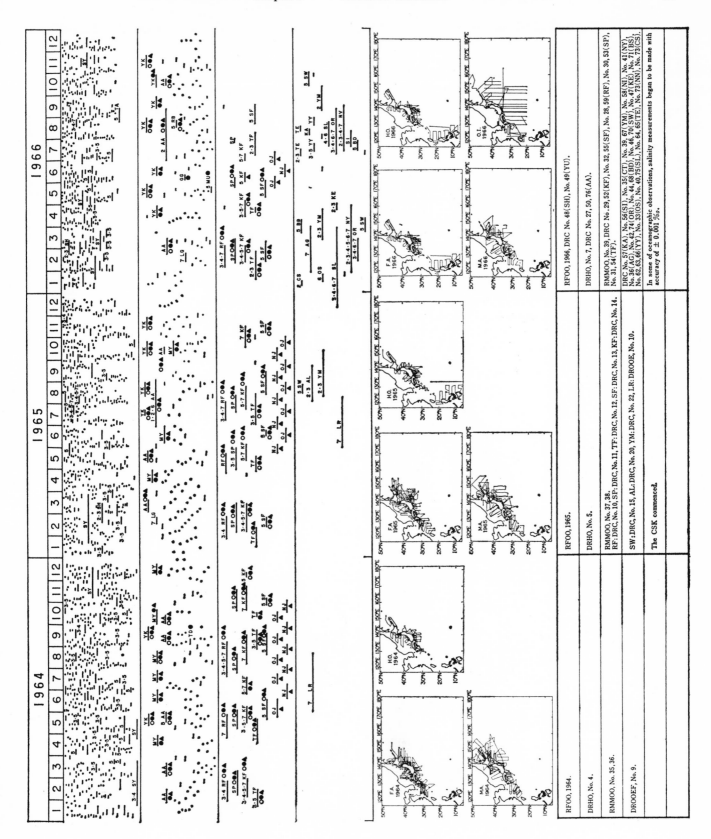

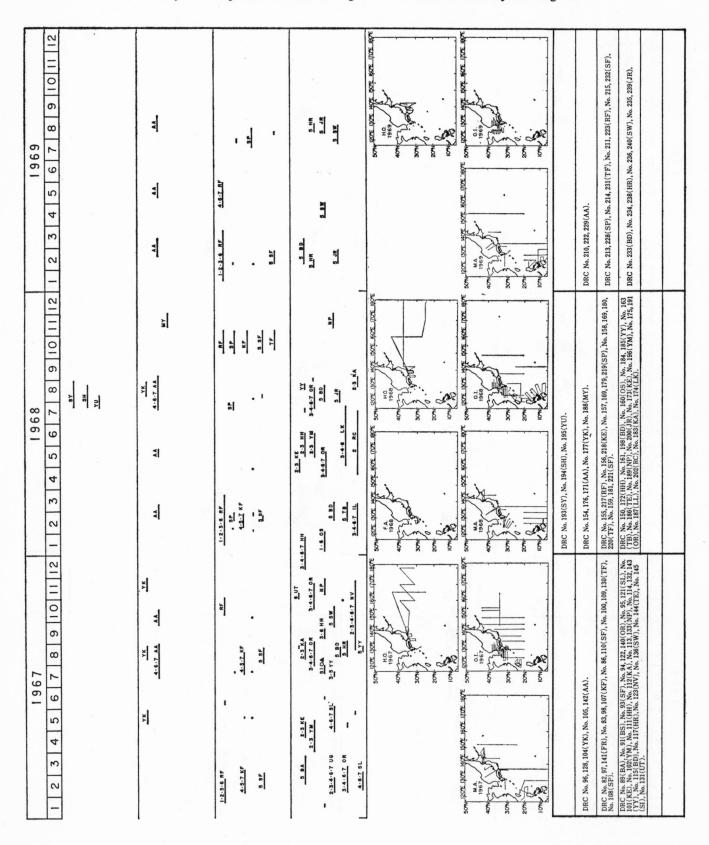

Chapter 2

BATHYMETRY OF THE KUROSHIO REGION

AKIO MOGI

The 7th Maritime Safety Head Quarters, Maritime Safety Agency, Moji, Japan.

Abstract

Trenches and their associated island arcs are the dominant features of the western Pacific Ocean. The continental arcs of the Kamchatka-Kurile Islands, Japan, the Ryukyu Islands and the Philippines embrace the marginal seas on their continental side while the oceanic arcs extending south from Japan separate the Pacific Basin proper from the Philippine Sea. The Philippine Sea is a very complicated region surrounded by the island arcs. A few ridges and seamounts not associated with trench run parallel to the Ryukyu-Philippine Arc and the Izu-Mariana Arc. These ridges close the space eastward. The Daito Ridges in the northwestern part of the Philippine Sea have an unusual trend. There are numerous seamounts, called Marcus-Wake Seamounts, in the Northwest Pacific Basin. The Kuroshio flows across these island arcs, ridges and seamounts. Therefore, the Kuroshio course may be affected by these features.

1 INTRODUCTION

A systematic bathymetric investigation of the western Pacific is being carried out by the Hydrographic Department, Maritime Safety Agency of Japan as a part of the GEBCO program. The GEBCO program is concerned with the compilation of 1:10,000,000 bathymetric charts of the world in 24 sheets. The first version was published in 1904. Since then, revisions have been made and in 1952 a resolution was made at the International Hydrographic Congress to publish the fourth version. It was agreed that hydrographic offices of 17 nations would provide 1:1,000,000 worksheets for the seas around their countries and the fourth version of GEBCO charts would be compiled on the basis of these sheets by the GEBCO Committee composed of representatives from ICSU, IAPO and IHB.

Most of the western Pacific region is covered by Chart A-III (0°–46°40′N, 90° E–180°) of GEBCO. Of the region covered by Chart A-III, the Hydrographic Department of Japan is responsible for the compilation of the area (18°N–48°N, 120° E–180°).

More than half the soundings are from Japanese sources, the majority being from the Hydrographic Department and others from the Tokyo University of Fisheries, the Japan Meteorological Agency and the Ocean Research Institute, University of Tokyo. The remainder of the soundings are from foreign sources, mainly from the United States of America and Great Britain.

Based on 1:1,000,000 work sheets for GEBCO, the Hydrographic Department published in 1965 four 1:3,000,000 Bathymetric Charts of the seas adjacent to Japan. These charts (No. 6301–6304), which replace the existing chart No. 6901, are highly reliable in comparison with existing charts of the area (Fig. 1).

Tayama (1952) and Dietz (1954) had described the submarine geomorphology of the western Pacific covered by No. 6901. The description of two sheets (No. 6301 and No. 6302) were made by the geologists of the Hydrographic Department of Japan in 1966. Also, the Hydrographic Office, U.S. Navy prepared a bathymetric chart of the area from Korea to New Guinea (H.O. 5485). An interpretation of this chart was made by Hess in 1948.

Recently, the U.S. Naval Oceanographic Office in cooperation with the Undersea Surveillance Oceanographic Center and the Scripps Institution of Oceanography published a bathymetric atlas of the northwestern Pacific Ocean to fill the need of naval planners and scientists working in the many fields of ocean science for reliable bathymetric detail in a convenient size for quick reference. This atlas contains 49 bathymetric charts and its scale is approximately 1:2,000,000.

Trenches and their associated island arcs are the dominant features of the western part of the Pacific Ocean. The arcs can be divided into two groups: the continental arcs of the Kamchatka-Kurile Islands, Hokkaido, Japan, the Ryukyu Islands and the Philippines, and the oceanic arcs extending south from Japan. The former embrace the marginal seas of Okhotsk, Japan and East China on their continental side, while the latter separates the Pacific Basin proper from the Philippine Sea.

Gutenberg and Richter (1949) state that the following sequence of structures and attendant phenomena are typical of Pacific arcuate structure: (1) a fore-deep trench; (2) shallow earthquakes and negative gravity anomalies in a narrow belt on the concave side of the trough; generally, the ocean bottom rises

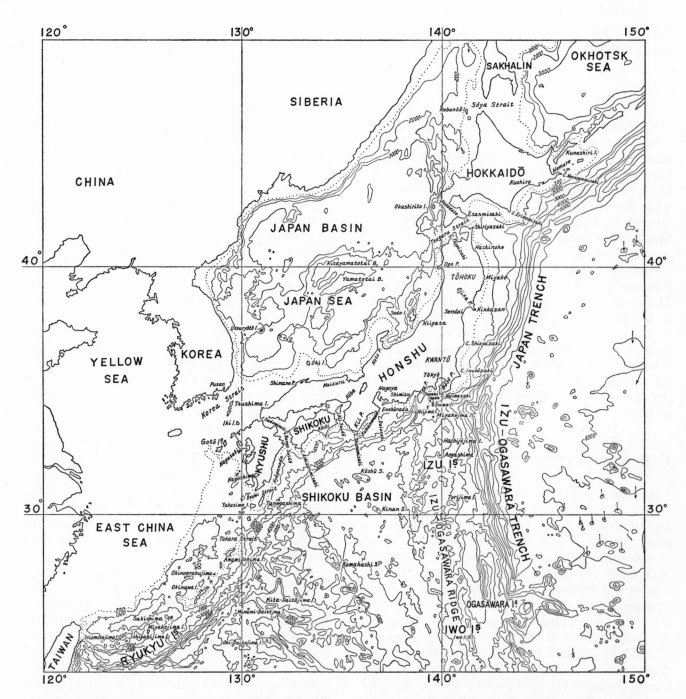

Fig. 1   Bathymetry and geographical names of the adjacent seas of the Japanese Islands. Compiled from J.H.O. chart No. 6301 and 6302.

here to a ridge with emergent small non-volcanic islands; (3) maximum positive gravity anomalies and earthquakes at a depth near 60 km; (4) the principal structural arc (Late Cretaceous or Tertiary) having active or recently extinct volcanoes with quakes at a depth of about 100 km; (5) a second structural arc with older volcanism; and (6) a belt of shocks at a depth of 300–700 km. It was widely believed that the fundamental structure of arcs is the tectogene or a downbuckle of the earth's crust into subcrustal material. Recently, Uyeda and Sugimura (1970) defined the island arc-trench system by the following characteristics: (1) volcanic activity in recent times, (2) existence of a foredeep more than 6000 meters in

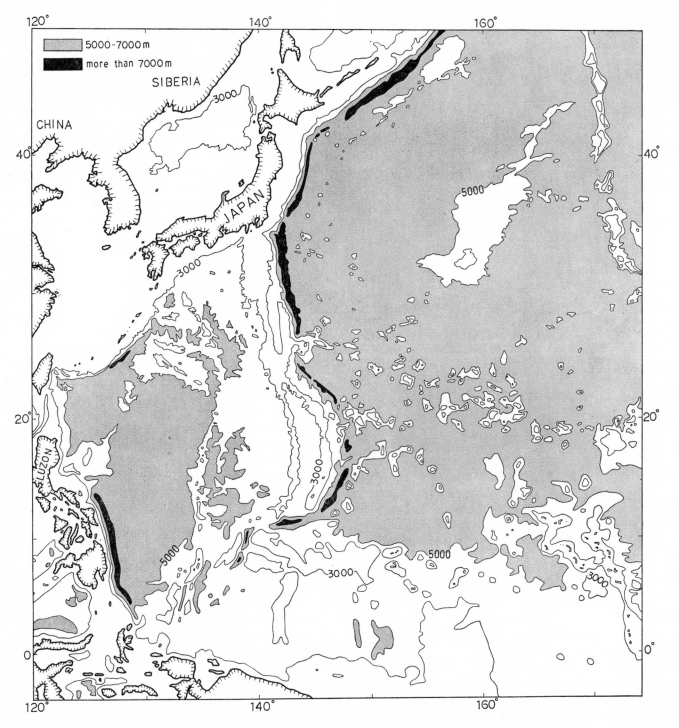

FIG. 2   Simplified depth chart of the Kuroshio region.

depth, and (3) occurrence of an earthquake at a depth more than 70 kms. They also asserted that the island arc-trench system of the western Pacific should be divided into the following two groups: (1) East Japan island arc-trench system (the Kuriles, Northeast Japan and Izu – Mariana arc), and (2) West Japan island arc-trench system (Southwest Japan, the Ryukyu Islands and Formosa – Philippine arc).

Both the Japan Sea Basin and the Kurile Sea Basin are of similar geomorphic form, having an arcuate southeast side and an almost straight northwest side and becoming narrow to the northeast. However, in the Japan Sea Basin, the deepest portion is located in the northwestern part, while in the Kurile Sea Basin it is in the southeastern part. The environs of the East China Sea are almost completely occupied by the vast continental shelf, except for the southeastern depression, the Okinawa Trough.

The Northwest Pacific contains two basins with the unusually great depth of about 6000 m. One of them is the Northwest Pacific Basin and the deepest of all ocean basins. The other lies beneath the western half of the Philippine Sea and is called the Philippine Basin (Fig. 2).

The Philippine Sea is bounded on the west by the southern half of Japan, the Ryukyu Islands, and the Philippines and on the east by the arcuate seafloor ridge extending from Japan to the Palau Islands through the Mariana Islands. This abyssally deep basin is separated into east and west basins by the Kyushu-Palau Ridge. The east basin is somewhat shallower and is divided into two parts, Shikoku Basin and West Mariana Basin, by the central narrow zone. The west basin (the Philippine Basin) is unusually deep.

The most striking aspect of the Northwest Pacific Ocean is the extensive groups of seamounts, some of which pierce the surface while others are deeply submerged. The seamounts are not randomly disposed but rather are arranged in broad belts. The linear arrangement is not as perfect as that which characterizes the volcanoes associated with the arcuate structures.

2 PHILIPPINE SEA

2.1 Morphological division of the Philippine Sea (Fig. 3)

The Philippine Sea is bounded on the west by the southern half of Japan, the Ryukyu Islands and the Philippines, and on the east by the arcuate seafloor ridge extending from Japan to the Palau Islands.

This abyssal deep basin is separated into east and west basins by the Kyushu-Palau Ridge. The east basin is somewhat shallower, and is divided by the central narrow zone at 23°N into two parts, Shikoku Basin and West Mariana Basin, which gradually rise toward the Honshu-Mariana Ridge. There are, however, several seamounts arranged NNW-SSE in the Shikoku Basin. This linear arrangement of seamounts is called the Kinan Seamount Chain. On the other hand, in the West Mariana Basin, there are several deeps arranged N-S called the West Mariana Deep.

The west basin (Philippine Basin) is unusually deep. A region of rugged topography called the Daito Ridges lie at the northernmost part of the Philippine Basin, nested between the Ryukyu Trench and the Kyushu-Palau Ridge. The mountainous area is largely submerged, but two peaks pierce the surface, one to form the north and south Daito Islands and the other to form the Oki-Daito Islands. There is a rise running in the NNE-SSW direction in the western half of the Philippine Basin. The rise is called the Daito-Luzon Rise, since it stretches from Daito Ridge to Luzon Island. There is a striking feature called the Central Basin Fault in the central part of the Philippine Basin. The Central Basin Fault, which is the low ridge associated with the trough in the NW-SE direction, separates the Philippine Basin into the north basin and the south basin.

The Honshu-Mariana Arc which forms the east boundary of the Philippine Sea extends southward from approaches to Tokyo Bay to the southern limit of the Marianas. This feature can be divided into two parts, northern and southern, at latitude 25°N. The northern or Izu-Ogasawara Ridge is straight, while the southern or Mariana Ridge is arcuate.

These features are clearly of double ridges and embrace their respective troughs. The Ogasawara Trough lies between the Ogasawara Ridge and the Iwo-To Ridge. The Mariana Trough lies between the Middle Mariana Ridge and the West Mariana Ridge.

2.2 Arcuate structure

Philippine Arc and Trench

In the Philippine Archipelago, two zones characterized by ultrabasic rocks are distributed in the NE-SW direction. One of them extends from Palawan to Mindoro and the other from the Sulu Islands to the western part of Mindanao.

The Philippine Fault Zone trends NW-SE in the Philippine Archipelago and is remarkable for its topographic features. Along the fault zone, left-lateral strike-slip faulting is active. The NW-SE fault

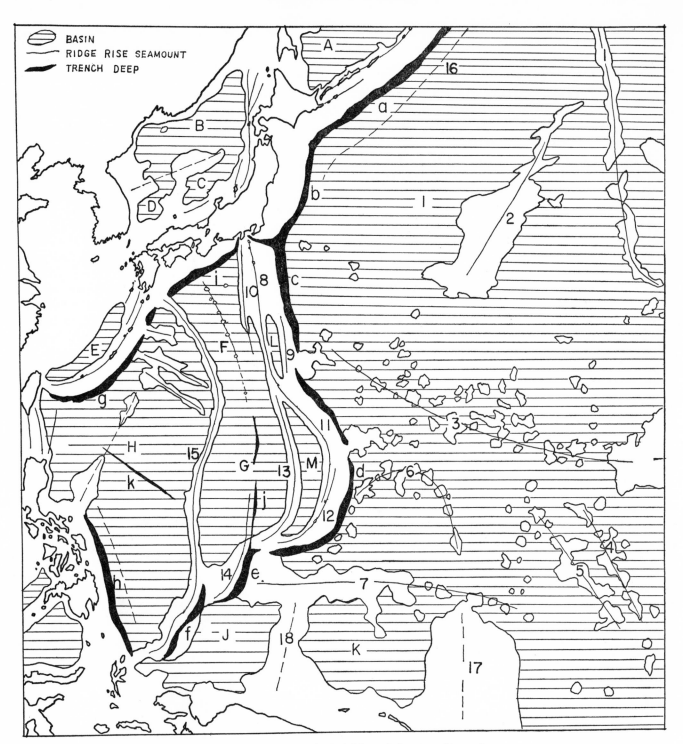

FIG. 3   Geomorphological division of the Kuroshio region.

Basin and Trough

A, Kurile Basin;   B, Japan Basin;   C, Yamato Basin;   D, Tsushima Basin;   E, Okinawa Trough;   F, Shikoku Basin;   G, West Mariana Basin;   H, Philippine Basin;   I, Northwest Pacific Basin;   J, West Caroline Basin;   K, East Caroline Basin;   L, Izu-Ogasawara Trough;   M, Mariana Trough.

Ridge, Rise and Seamounts

1, Northwest Pacific Seamounts (Emperor Seamounts);   2, Northwest Pacific Rise (Shatsky Rise);   3, Marcus-Wake Seamounts; 4, Radak Ridge;   5, Ralik Ridge;   6, Magellan Seamounts;   7, Caroline Ridge;   8, Shichito-Iwoto Ridge;   9, Ogasawara (Bonin) Ridge;   10, Nishi-Shichito Ridge;   11, Middle Mariana Ridge;   12, East Mariana Ridge;   13, West Mariana Ridge;   14, Yap Ridge; 15, Kyushu-Palau Ridge;   16, Hokkaido Rise;   17, Greenwich Ridge;   18, Eauripik Ridge.

Trench and Deep

a, Kurile-Kamchatka Trench;   b, Japan Trench;   c, Izu-Ogasawara Trench;   d, Mariana Trench;   e, Yap Trench;   f, Palau Trench; g, Nansei-Shoto(Ryukyu) Trench;   h, Philippine Trench;   i, Nankai Trough (Southwest Japan Trench);   j, West Mariana Deep; k, Central Basin Fault.

zone cuts these NE-SW trends. (Kimura et. al., 1968, Hashimoto and Sato, 1969, 1970). These features parallel the Philippine arc, the ocean trench, volcanoes and earthquake centers.

The Philippine Trench is nearly straight and its axis lies almost equally distant from the 4000 m contour at the base of the Philippine island chain. Three basins deeper than 9500 m occur: the longest one, near Siargao, is slightly more than 10,000 m deep and contains the Emden Deep, the Snellius Depth, the Cape Johnson Depth, the Planet Deep and the Galathea Deep.

The cross sections traced from echograms clearly show a V-shaped trench, having an asymmetrical profile with the slopes steeper on the west or shoreward flank than on the offshore flank. The lower walls of this trench are irregular, with small knolls or pinnacles, and particularly on the offshore side, with dam sediments or rocks slumping from further upslope. Profiles of the Philippine Trench and its vicinity taken by the *Takuyo* in 1965 and 1966 are shown in Fig. 4. According to this profile, benches exist on the wall of the shoreward side of the Philippine Trench. They are 4 to 12 km wide at depths of about 4000 to 8000 m. On the sea floor of the seaward side of the trench, a small narrow depression and step-like topography with the floor gap exist, which is interpreted as a fault origin (Iwabuchi and Saiki, 1968).

According to Kiilerich (1959), the Philippine Trench at the deepest parts has a narrow flat bottom, 1–3 km wide. Projecting the lower trench walls into an intersection and allowing for steepening with depth, a thickness of less than 300 m is indicated for the sedimentary fill in this basin. Northward the Philippine Trench ends against a broad area shallower than 5000 m which extends northeastward as the Daito-Luzon Rise. This broad swell has seamounts such as Benhom Bank and step-like topography. The trough between this swell and Luzon Island has a flat bottom buried by sediments, although it is hilly in some places. Gravity observations indicate that the tectogene structure does not partake of the southeasterly bend of the trench but continues southward to pass west of Halmahera.

The continuation of the Philippine Arc is seen between Luzon and Formosa. There are three ridges here, the east of which is the deepest ridge extending northward from Escarpada Point of Luzon. This ridge is composed of several seamounts having the minimum depth of 1710 m; the northernmost part of the ridge ends against the western part of Ryukyu Trench. The central ridge is composed of a few island chains in echelon; a ridge including Lutao Island and Lanhsu Island; the Y'Ami, Itbayat, Diogo, Batan,

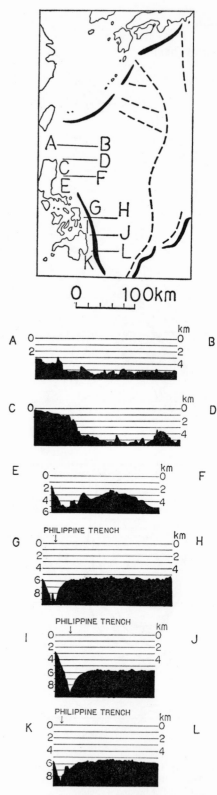

FIG. 4   Profiles of the sea floor off the east coast of the Philippine Islands.

Babutan and Sabang islands (this ridge may continue to Didicas Rock and Camiguin Island across the deep channel); and Calayan, Dolupiri and Fuga islands. The west is a shallow ridge elongated from the southernmost cape of Formosa. The troughs between these three ridges are some 4500 m and 3500 m deep, respectively.

Nansei-Shoto (Ryukyu) Arc and Trench

The Nansei-Shoto Arc is divided into a south and north arc by a depression west of the Kerama Islands. The south arc is a single arc including the Yaeyama and Miyako islands and shows a smoothly curved form. The north arc is a double arc: an outer non-volcanic ridge including the Okinawa, Amami and Osumi islands and an inner volcanic ridge of the Tokara Islands. The outer ridge is divided into two provinces by a depression north of the Amami Islands. Konishi (1965) divided the inner volcanic ridge into two zones: the Palaeo-Ryukyu Volcanic Belt of andesitic volcanic rocks of Miocene to Pliocene on the west and the Ryukyu Volcanic Belt of Quaternary volcanic rocks derived from high alumina basalt magma on the east. The former includes Iriomote, Ishigaki, Kume, Agumi, Tokara, Kusagaki and Uji islands. The latter, superposing on just on the oceanic side of the former belt, includes Torishima, Suwanose, Kuchi-no-Erabu, Iwo, Takeshima, Yokoate, Kobakura, Akuseki and Gaja islands. Many reports on submarine volcanic eruption have been made by navigators in the environs of Iwo, Suwanose Hirashima, Tokara and Iriomote islands. The Pre-Miocene basement complex mainly exposed along Sakishima, Okinawa, Amami and Osumi islands is further divided into six tectonic belts parallel to the axis of the ridge. Konishi pointed out that the zonal distribution of these tectonic belts was similar to those in the Outer Zone of Southwest Japan, and further that the axis of the island arc structure intersected somewhat obliquely the general trend of the zonal structure in the Pre-Miocene basement complex. A zone of intermediate-depth earthquakes with focal depths between 70 and 300 km follows the volcanic zone of the Ryukyu arc and can be traced to the northeast into Kyushu (Katsumata and Sykes, 1969).

Narrow shelves are seen along the Philippines to Kyushu: the shelf edge is shallower than about 100 m in the northern part of the Luzon, the Babuyan Islands and the Batan Islands; about 120 m on the east of Formosa and about 150 m along the Nansei-Shoto (Ryukyu) Islands.

The deep sea terraces are seen along the Pacific side of the Nansei-Shoto Islands at 2000 to 4000 m in depth, as already pointed out by Tayama (1952). The development of those terraces is more remarkable along the southern arc. They are composed of an outer ridge and an inner depression.

The Nansei-Shoto Trench is divided into three parts by seamounts south of Miyako Island and the elongation of the Amami Plateau. The deepest soundings are 7130 m in the west, 7881 m in the middle and 5988 m in the north trenches. The middle trench is deep and long. The north trench is enclosed by the Amami Plateau and Kyushu-Palau Ridge.

The continental margin of the Nansei-Shoto Islands is intersected by two depressions at right angles to each other: one is a depression deeper than 1000 m west of the Kerama Islands by which the Nansei-Shoto Arc is divided into the south and the north arc. The deepest area of the Nansei-Shoto Trench is situated at the southeastern extension of this depression. The other depression is an asymmetrical trough north of the Amami Islands having an approximately E-W direction.

According to reflection surveys recently carried out, thick sediment layers are present beneath the deep sea terrace off Okinawa Island. Therefore the 2000 to 2500 m terrace is a sediment trap buried by sediments some 600 to 900 m thick.

Kyushu-Palau Ridge

The Kyushu-Palau Ridge is an elongated ridge from Kyushu to the Palau Islands running midway along the Philippine Sea Basin. Its northern part beyond Lat. 18°N forms an arcuate shape convexed toward the east, while its southern part beyond Lat. 18°N is straight. Volcanism and seismicity are not reliable, but some navigators saw coloured water in this area. There are no associated trenches. The Kyushu-Palau Ridge has a single island near its midpoint, Parece Vela. Little is known of this island except that Tayama (1935a) listed it as a coral island.

Its northern part beyond Lat. 25°N is a seamount range which is a series of seamounts aligned along a swell. Between Lat. 25°N and Lat. 18°N, it is a seamount chain which is a series of seamounts in a line, rising dependently from the deep sea floor. The southern part further than Lat. 18°N shows a ridge with a N-S direction.

Near 22°N, the continuation of the ridge is interrupted as shown by the 4000 m isobath and the sill depth of the ridge is deepest there. In the north, the Kyushu-Palau Ridge terminates at the junction of the Nankai Trough and the Nansei-Shoto Trench, and continues to the Palau Islands in the south. The northernmost part of the ridge intersecting the trench at a large angle separates the shallow northern trough (Nankai) from the deep southern trench (Nansei-Shoto).

At the northern part of the Kyushu-Palau Ridge, many seamounts stand abruptly about 1500 to 2000 m high above the swell, gently rising to a height of about 1000 to 2000 m from the Philippine Sea Basin. These seamounts have depths of 200 to 2000 m. It is not clear whether they have round tops or flat tops. According to a detailed survey at Komahashi Seamount, one of the two peaks has an uneven top which has its peak at a depth of 500 m. The other peak is conical with a distinctly flat top 3.7 km in diameter and 700 m in depth. Apparently, the second seamount is a wave-truncated submerged volcano like the guyot on the Mid-Pacific Mountains (Mogi and Kato, 1964).

Daito Ridges

This includes three ridges in the WNW-ESE direction located at the north end of the Philippine Basin: the northernmost ridge has been tentatively designated as the Amami Plateau, the middle one the Minami-Daito Ridge, and the southernmost one the Oki-Daito Ridge. A famous deep drilling made at Kita-Daito Island disclosed a Miocene limestone some 400 m thick, but the basement is unknown. It is apparent that this ridge was submerged at least since pre-Miocene times (Hanzawa, 1940).

The Minami-Daito Ridge and the Oki-Daito Ridge elongate from the respective islands southeastward. The 3500 m isobath shows several parallel ridges with intervening narrow depressions. The straightness and linear reliefs of these ridges may be caused by tectonic movement. Generally, each of the ridges has a steep, linear slope and rugged features on the southern side and a step-like feature on the northern side. There are flat plains at depths of about 1500, 2000 and 3000 m on the northern side of each of the ridges (Fig. 5). These flat plains apparently originated from faulting along the tectonic line. The Amami Plateau

is connected to the continental margin of the Ryukyu Ridge at its western margin. Consequently, the Ryukyu Trench is interrupted at this site.

Southwest Japan Arc and Nankai Trough

The Japanese Islands comprise Hokkaido, Honshu, Shikoku, Kyushu and other subordinate islets, which are arranged, on the whole, in an arc shape stretching from northeast to southwest and convex towards the Pacific. A great ruptured zone called the "Fossa Magna" traverses the central part of Honshu from the Japan Sea to the Pacific, and this zone divides Japan into Southwest Japan and Northeast Japan, tectonically.

Southwest Japan consists predominantly of Palaeozoic and Mesozoic rocks, with Cenozoic rocks in lesser amounts. In Southwest Japan, a conspicuous tectonic line called the Median Line runs nearly parallel to the island-arc from central Honshu, through Shikoku to Kyushu. This line subdivides Southwest Japan into the Outer Zone (the Pacific side) and the Inner Zone (The Japan Sea side), which show striking contrasts in their geology.

There are alternations of prominent coastal salients with embayed and relatively low-lying stretches of shoreline on the south coast of Southwest Japan; some salients are Cape Ashizuri, Cape Muroto and Cape Shio.

There are four deep-sea basins beyond the continental shelves of the south coast of Southwest Japan. These basins are located off Bungo Suido, Tosa Bay, Kii Suido and Kumano-nada which form the coastal embayment. That is, the upper flank of the continental slope resembles the coastal feature. Tayama (1950) noticed the presence of deep-sea terraces in this area. He distinguished the upper at depths of 800 to 1000 m and the lower at depths of 1600 to 2000 m and called the former Tosa terrace and the latter Hyuga terrace.

Two kinds of submarine valleys can be distinguished: valleys which terminate at the deep-sea terrace surface, and valleys which undercut the deep sea terrace and terminate at the ocean floor.

Most shallow earthquake centers are located on the upper part of the continental slope. The Nankai Earthquake in 1946 originated south-east of Shikoku and affected the whole of Southwest Japan. Tayama thought that the tectonic scarp for the continental slope bounded the upper part of the deep-sea terrace. According to recent reflection surveys, thick sediment layers are known to be beneath these terraces. The flat terrace about 2000 m deep in the Kumano-Nada Sea is a sediment trap with sediments some 500 to 700 m thick, and a rocky bottom is seen on the edge

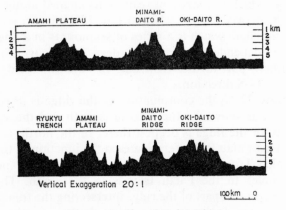

FIG. 5   Profiles of Daito Ridges.

and slope of the terrace. The flat bottom off Kii Strait, 1000 to 2000 m deep, is also buried by sediments about 750 m thick (Murauchi et al., 1968).

The Nankai Trough is roughly parallel to the general trend of Southwest Japan. It is about 600 km long and 60 km wide and its depth is in general 4000 to 5000 m. The outline of the trough is revealed by the 4200 m isobath in the northeastern half, and by the 4600 m isobath in the southwestern half. The trough bottom is deeper only by about 200 to 300 m than the floor of the Shikoku Basin. Despite this, the trough is smaller in scale and its morphology shows features just like those of other known trenches of the world (Hoshino, 1963). According to the reflection survey, the Nankai Trough south of Cape Muroto, Shikoku, is buried by sediments some 1500 m thick. The north part of Shikoku Basin is covered by sediments about 900 m thick. If these unconsolidated sediments were removed, the Nankai Trough would be a long and narrow depression with steep sides. It is considered that the Nankai Trough may be a past trench or a marginal depression of the Shimanto Geosyncline. The Nankai Trough enters Suruga Bay eastward and its continuation to the Japan Trench is interrupted by the Izu-Ogasawara Ridge.

Izu-Ogasawara Ridge and Trenches

There are three ridges rising behind the Izu-Ogasawara Trench: the Ogasawara (Bonin) Ridge including the Ogasawara Islands, the Shichito-Iwoto Ridge elongated from the Izu-Shichito Islands to the Volcano Islands, and the Nishi-Shichito Ridge elongated to the south from the Izu Peninsula.

The outline of the Ogasawara Ridge is shown by the 3000 m isobath. Although it elongates as far as Lat. 29°N, elongation further north is unknown.

The Chichijima islands are composed of agglomerate and tuff of bronzite andesite called Boninit. The Hahajima islands are composed of agglomerate and tuff of pyroxene andesite. This andesitic complex is uncomformably covered by Miocene limestone 200 m above sea level. The Ogasawara Ridge is an old volcanic chain. The west side of the ridge is limited by a straight steep slope suggesting a fault scarp, while the east side of the ridge is relatively gentle.

Many volcanoes are superposed on the Shichito-Iwoto Ridge, along which there is volcanic activity as in the Izu Islands and the Volcano Islands.

The northern part of the Izu-Ogasawara Ridge is a shallow and broad rise, while its southern part is deep and narrow as shown by the 3000 m isobath. Near Lat. 30°, the three ridges are interrupted and sill depth is the deepest in this arc. On the other hand,

the trench in front of this area is the deepest in the Izu-Ogasawara Trench, and the basin floor behind this area is the deepest in the Philippine Sea Basin. This sill of ridges is concordant with the deepest areas of the trench and basin. The axis of these depressions is parallel to the Japanese Island arc. If the Japanese Island arc is a major uplift zone as advocated by Sugimura (1966), it seems probable that these depressions are major depression zones in the west Pacific Ocean floor.

There are some remarkable features besides the volcanic belt. The first is the presence of a deep sea terrace. This terrace may be an outer arc like the Ogasawara Ridge. The second is that the minor lineations WSW are seen from volcanic cones of the Shichito-Iwoto Ridge. The western margins of these reliefs are arranged in a straight line and are called the Nishi-Shichito Ridge. Some seamounts and hills which form the relief on the western part of the volcanic belt are of volcanic origin, but no active volcanism exists.

The third is the presence of several basins. These basins are arranged closely west of the line of volcanoes in the Shichito-Iwoto Ridge (Fig. 6).

The Volcano Ridge is slightly asymmetric, with the east slope generally steeper than the west. Consequently, the Ogasawara Trough between the Ogasawara Ridge and the Volcano Ridge has steep slopes on both sides. This trough has a very flat bottom and is buried by sediments more than 400 m thick (Fig. 7).

The Izu-Ogasawara Trench continues to the Japan Trench, so that it is difficult to define the boundary. However, it can presumably be defined at the embayments of the 6500 m and 700 m isobaths toward Sagami Bay.

In the northern half of the Izu-Ogasawara Trench, the boundary between the continental slope along the Izu-Ogasawara Ridge and the trench wall is 2000 to 3000 m deep, and toward the southern half, it deepens to 4000 to 4500 m. Also, the boundary between the ocean floor and the seaward wall of the trench increases its depth in a southward direction.

The deepest portion of this trench was thought to be the Ramapo Deep (10,347 m at about Lat. 30°N). However, recent surveys revealed 9647 m as the deepest sounding. The seaward wall of the trench has narrow depressions indicating a graben-like topography, with the extreme elongation parallel to the trench axis. The marginal swell develops on the outside of the seaward wall of the trench, and abruptly terminates with the escarpment that would be expected of a fault origin at its east side. In morphological features, the Izu-Ogasawara Trench is similar to the Japan Trench, but it has deeper depths and more

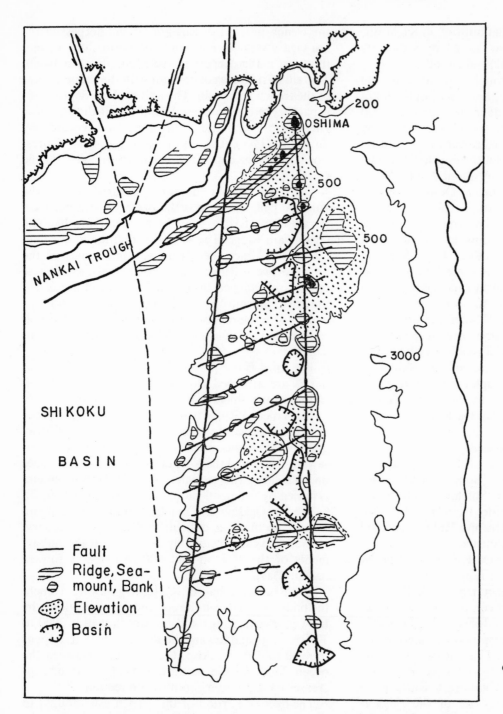

FIG. 6   Geomorphological map of Izu Ridge.

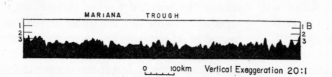

FIG. 7   Longitudial profiles through the Ogasawara and Mariana Troughs.

extensive reliefs on the whole (Iwabuchi, 1968).

Mariana, Yap, Palau arcs and Trenchs

Three ridges, termed the West Mariana, Middle Mariana and East Mariana Ridge by Tayama (1952), are recognized southward of the Izu-Ogasawara Ridge. The Middle and East Mariana Ridges extend south from the Shichito-Iwoto Ridge and the Oga-sawara Ridge, respectively. Between the Middle Mariana Ridge and West Mariana Ridge there is a large crescentic basin which may be related to the anomalous curvatures of the central part of the Mari-ana Trench and the island arc. The Middle Mariana Ridge is about 1000 to 1500 m high above the ocean floor, 3500 to 4000 m deep. The West Mariana Ridge is 1000 m high. On the West Mariana Ridge, only Aragane Reef and the 25 m reef rise above the surface. The Middle Mariana Ridge is outlined by ten vol-canic islands, including Uracas, Maug, Agrihan, Pagan, Anatahan and six reefs. All these islands and reefs are distributed along the arc. The northern ex-tremity of this ridge is linked with the volcanic chain of the Iwoto Islands. Many of these volcanoes are active. The volcanoes are built up from an alternation of andesitic or basaltic lava, agglomerate and tuff. The volcanic trend continues southward from Anata-han but is submerged. East of the volcanic axis is an outer chain of islands called the East Mariana Ridge (Tayama, 1952).

The East Mariana Ridge includes only limestone islands: Medinilla, Saipan, Tinian, Aguijian, Rota and Guam, and four reefs or banks. The foundation rocks are of igneous origin and there is evidence of volcanic activity as late as the Oligocene. According to Hess (1948), terraces are common on all the islands from Farallon de Medinilla to Guam and are tilted westward at the north end of the group and north-westward at the south end. The volcanoes west of the above island chain are submerged. Considering this and the attitude of the terraces, Hess thought that westward tilting in the southern Marianas is accom-panied by eastward shift of the geanticlinal axis.

The western slopes of the West Mariana Ridge are gentle and gradually graded to the abyssal plain of the West Mariana Basin. Its western slopes are gener-ally mountainous and comprise small mounts, some of which attain 1000 m in height, but its eastern slopes are steeper than the western slopes. Otherwise, the western slopes of the Middle Mariana Ridge are steeper than the eastern slopes. The Mariana Trough between the West and the Middle Mariana Ridge has a sea floor 3000 to 4500 m deep and is limited by the steep slopes on both sides. The trough shows a very complicated bottom of minor reliefs. The reliefs

mostly attain to 1000 m in height (Fig. 8). This trough seems like a deep sea hill basin without a sediment blanket. The deep sea hills develop on the western half of the trough and the eastern half shows a rela-tively smooth bottom.

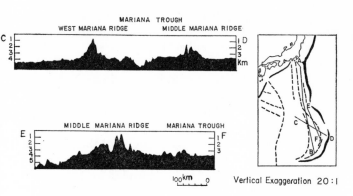

FIG. 8   Cross sections of the Middle and West Mariana Ridges.

The Mariana Trench is broken into three lengths by shallow areas at 19° and 16° N Lat.

The central section is the most shallow and the southern section the deepest, more than 10,000 m. A maximum depth of more than 10,000 m was sounded recently by the *Challenger* (1951), *Vitiaz* (1957, 1958), *Stranger* (1959) and *Trieste* (1960) at the western edge near the southern extremity. This deep is the greatest deep in the world. Although the northern and south-ern extensions of the Mariana Trench are parallel to the arc of the Mariana Islands, the central section is as far as 140 nautical miles offshore. In other words, the curvature of the Mariana Trench is greater than that of the arc of the Mariana Islands.

From the southern extremity of the Mariana arc, the Yap Islands and the Palau Islands trend south-west in echelon. An unusual feature of the Yap Islands is that the basement amphibolite and schist is exposed, and pebbles from it are found in overlying Miocene and younger conglomerates and gravels (Tayama, 1952). The Yap Trench, east of the Yap Ridge, is a continuation and its curvature is small. The maximum depth, 8527 m, was sounded in 1960.

The Palau Islands include numerous coral and volcanic islands. Tayama (1935b) found various an-desitic agglomerates and also limestones but no base-ment rocks. The Palau Trench lies in echelon on the west side of the Yap Trench. It extends north and south with a slight curvature to the east. The maxi-mum depth, 8054 m, was sounded in 1960.

2.3   Philippine Basin

The Philippine Basin is bounded on the north by the Daito Ridges, a triangular shaped area lying between the Kyushu-Palau Ridge and the Ryukyu and the Philippine trenches. The depth of this basin is 5500 to 6000 m and deeper than that of the east basins, the Shikoku and the West Mariana. Although this basin is regionally level, sounding lines show that it is mountainous in detail, suggesting the absence of a thick sedimentary blanket (Fig. 9).

The most striking feature of this basin is the elongated deep, called the Central Basin Fault by Hess (1948). This deep extends in a SE-NW direction in the center of the Philippine Basin and is flanked by much shallower water. If this trend is extended to the northwest it terminates in the junction of the Nansei-Shoto and Formosa-Philippine arc. Although Hess pointed out that if this trend is extended to the southeast, it terminates in the junction of the Marianas and

West Caroline arcs, the author prefers the site of the westward bending of the Kyushu-Palau Ridge to the junction of the Marianas and West Caroline arcs for the southeastward extremity of this deep. This may be a very important major shear zone of transcurrent faulting along which crustal blocks of the basin have adjusted themselves to the curving trends of the arcs; the northside arc of Lat. 14° and the southside arc of Lat. 14°, including the Palau arc. Another feature is the rise running from the western part of the Oki-Daito Ridge to the rise including the Benhom Bank which is located east of Luzon. This rise is tentatively called the Daito-Luzon Rise. Although the topography of this rise is not striking, isolations shallower than 4000 m deep are arranged in the NE-SW trend.

Many echograms show the abyssal hill area and the abyssal plain. Abyssal plains are developed only in limited localities, such as the west side of the northern arc of the Kyushu-Palau Ridge and the east side of Formosa.

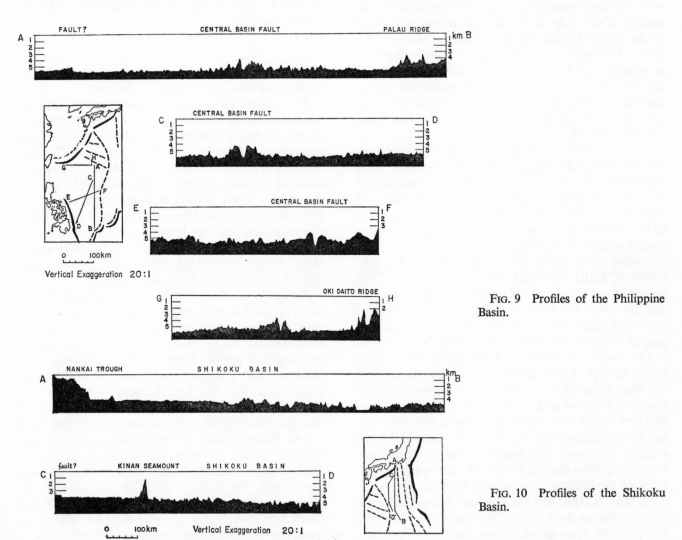

FIG. 9   Profiles of the Philippine Basin.

FIG. 10   Profiles of the Shikoku Basin.

On the Philippine Basin floor, the most extensive area is the abyssal hill region. The reliefs of the abyssal hills range from 100 to 1000 m in height. The abyssal hills have steep slopes and deeps in some places. For example, it appears that the sea scarp facing southward in the north Philippine Basin extends in a NW-SE direction parallel to the central Basin Fault. Somewhat larger reliefs are developed on the ridge along the central Basin fault and the Daito-Luzon Rise.

2.4 Shikoku and West Mariana Basins

The east basin, composed of the Shikoku and the West Mariana Basins, is bounded on the north by Southwest Japan, on the west by the Kyushu-Palau Ridge and on the east by the arcuate ridge from Japan to the Yap Islands. This basin is somewhat shallower, 4000 to 5500 m, and it rises gently toward the Izu-Ogasawara and the Mariana Ridges. The east basin is a region of rugged topography similar to the Philippine Basin. However, a smooth sea floor lies at the northern part and the eastern part of this basin.

The Shikoku Basin is composed of the smooth floor of the northeastern part and the mountainous area of the southwestern part. The depth of the former is shallower than that of the latter and the basin floor is inclined southwestward on the whole. The profile along the N-S course shows that the sea floor of the northern part is the abyssal plain or the gently undulated plain and its depth is shallower than the depth of the summit level of the abyssal hills in the southern part (Fig. 10). It is suggested that the abyssal plain or the gently undulated plain might be buried abyssal hills. Reflection surveys revealed that sediment coverage is as thick as about 900 m in the northern part, but decreases to 400–500 m southward, presumably with increasing distance from the source area. The most striking feature of this basin is the Kinan Seamount chain which runs in a NNW-SSE direction. Three northern seamounts which have deep flat tops (some 2500 m in depth at the northernmost seamount and some 1000 m at the southernmost two) were well sounded. They are guyots. From about 700 m deep on the flat tops of the latter two seamounts, bryozoan limestone of shallow sea type was dredged. From the echogram, it is surmised that there are some faults in a NW-SE or N-S direction. One extends from the depression in the east side of the northern Kyushu-Palau Ridge to the junction of the

FIG. 11 Topography of the West Mariana Deep.

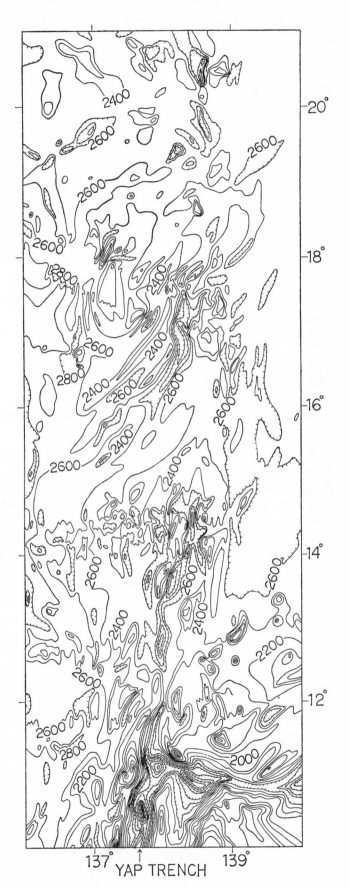

Shichito-Iwoto Ridge and the Middle Mariana Ridge. Linearities with the same trend are seen also on the 2800 m isobath of the U.S. Atlas.

The other fault is the sea scarp running in a N-S direction on the west side of the Nishi-Shichito Ridge. The abyssal plain is separated by this scarp into two levels in Fig. 10.

The West Mariana Basin is divided into two parts, a very rugged area and a smooth flat area. The eastern half of the basin is occupied by a smooth bottom and is somewhat shallower than the western half. The bottom topography of the western half is very complicated and shows abyssal hills and seamounts. The reliefs of the hills attain 1000 m in height. Some hills have a NE-SW trend. The smooth floor rises gently eastward and is graded to the west side slope of the West Mariana Ridge. There are sediments more than 450 m in thickness west of Kita-Iwo Island. The most striking feature of this basin is the West Mariana Deep. This deep extends in a N-S direction in the center of the West Mariana Basin and is interrupted at some places. Each deep is flanked by a much shallower ridge-like sea floor (Fig. 11).

It is noteworthy that the West Mariana Deep runs on the boundary between the abyssal hill region and abyssal plain. It has been suggested that this deep trapped sediment from the Mariana Ridge. Another possibility is that the Kinan Seamount Chain may be the northern extension of this deep. In my opinion the linear arrangement of seamounts and deeps suggests a large structure similar to the Mariana and the Kyushu-Palau arc. In the north seamounts may have erupted along a fault which runs from south of the Kii Peninsula to the Yap Trench in a N-S direction, and on the south, the faulting may be expressed as the trench and the deep (Fig. 12).

3 EAST CHINA SEA

The East China Sea floor is divided into two contrasting provinces: the broad continental shelf and the Okinawa Trough from Formosa to Kyushu along the inner side of the Ryukyu Island arc.

The continental shelf in the East China Sea is one of the widest shelves in the world and is called the Tunghai Shelf. The shelf edge ranges from 160 to 170 m in depth from north of Formosa to Kyushu. There are some scattered islets near its outer edge, the Senkaku Islands. Uotsurijima Island is of diorite

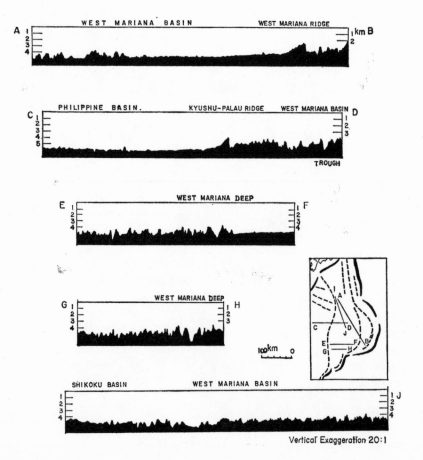

FIG. 12 Profiles of the West Mariana Basin.

covered by tertiary beds with unconformity and the Kibisho Rock is composed of basalt. Tertiary sedimentary rocks are dredged from the outer portion of the shelf and the submarine valley wall cuts the shelf edge. The reflection survey suggests that the fold zone of thick tertiary formation underlying the igneous rock extends from the Formosa Strait to the Danjo Islands (Niino, 1969).

The East China Sea shelf is mainly cloaked in terrigenous sediments brought down by the Yangtze Kiang River and the Yellow River. Despite these large sediment sources the shelf does not show an outward gradation from coarse to fine sediment. There is a 75-mile-wide belt of mud close to the shore paralleled by an even broader zone of residual sediments, mainly sand and mud. On the west side, the sediments are fine-grained consisting largely of reworked loess from the Hwang Ho and Yangtze Kiang rivers. The outer shelf sand zones consist of sands high in calcareous content, containing considerable authigenic material (glauconite and phosphorite). Distribution of sediments reflects the shore line at the glacial age.

Off the mouth of the Yangtze Kiang, there is a remarkable bank, called the Great Yangtze Bank, showing a nearly flat bottom over an extent of about a 300 km radius with a depth of about 30 m as submerged delta, covered by gray or fine, black-spotted sand. Yoshikawa (1953) thought that the development of the Great Yangtze Bank might be related to the rise in sea level in the post-Glacial Age.

There are numerous submerged ridges, running perpendicular to the west coast of Korea. A strong tidal current flows through the narrow channels which lie between these ridges.

The Okinawa Trough borders the continental shelf, and its western side is the continental slope. The deepest part near Taiwan reaches 2270 m and its floor shoals northeastward toward Japan. The floor of the trough is generally flat in cross section, but is incised by a narrow valley that probably owes its origin to turbidity currents that flowed along the axis of the trough. Sediment fill in the trough exceeds 1.2 km and contains many subbottom reflectors that may be turbidite sand layers.

According to Emery and others (1969), forming the structural framework of the region is a succession of NE-SW trending ridges that separate sediment-filled depressions. The farthest landward of these ridges is the Taihung-Great Kingan Range which was uplifted in Caledonian times. An older (Precambrian) uplift formed the Tai Shan-Laoyehling Range that is the backbone of the Shan-tung Peninsula at the head of the Yellow Sea. Next is the Fukien-Reinan Massif

which was uplifted during the Middle to Late Mesozoic Era across the mouth of the Yellow Sea. Probably this massif barred the ocean from the continent until the end of the Cretaceous Period. Evidently, the barrier was breached in Paleogene times because Paleogene and especially Neogene sediments are widespread and thick in the Yellow Sea. Deposition of these sediments was aided by the uplift of the Taiwan-Sinzi Folded Zone, which was raised probably throughout the Neogene and probably part of the Paleogene, damming sediments from the land to build the present continental shelf and to cover the floor of the Yellow Sea. Clearly, the Okinawa Trough is an area of deposition dammed by the Nansei-Shoto Ridge in much the same manner that the Fukien-Reinan Massif dams the Yellow Sea basin, and the Taiwan-Sinzi Folded Zone may dam an elongate basin beneath the continental shelf.

4   JAPAN SEA

4.1   Morphological division of the Japan Sea

The Japan Sea Basin is separated from the main Pacific floor by the Japanese Island Arc on the southeast side. However, the three straits, Tsushima, Tsugaru and Soya, which all have very shallow sill depths, connect the Japan Sea with the marginal sea and the main Pacific Ocean.

The middle of the Japan Sea is traversed by the Yamato Rise and the Korea Plateau, which also separates the Japan sea into the Japan Basin on the north and Yamato and Tsushima Basins on the south. A narrow continental shelf fronted with a continental slope fringes the Siberian coast and Korea Peninsula, while on the Japanese Island side many banks, basins and troughs developed, thus forming a complicated topography termed the Continental Borderland. These many banks are arranged in echelon, forming two rows of ridges parallel to the Japanese Island Arc. The inner ridges are termed Okushiri Ridge and Wakasa Ridge, and the outer ones, Sado Ridge and Oki Ridge (Fig. 13).

4.2   Continental Borderland off the northwest coast of the Japanese Islands

Okushiri and Sado Ridges

Many islands and banks forming Okushiri Ridge are arranged intermittently on a line extending from the offing of Niigata to the vicinity of the Musashi Bank, west of Hokkaido, via the Awa and Okushiri islands.

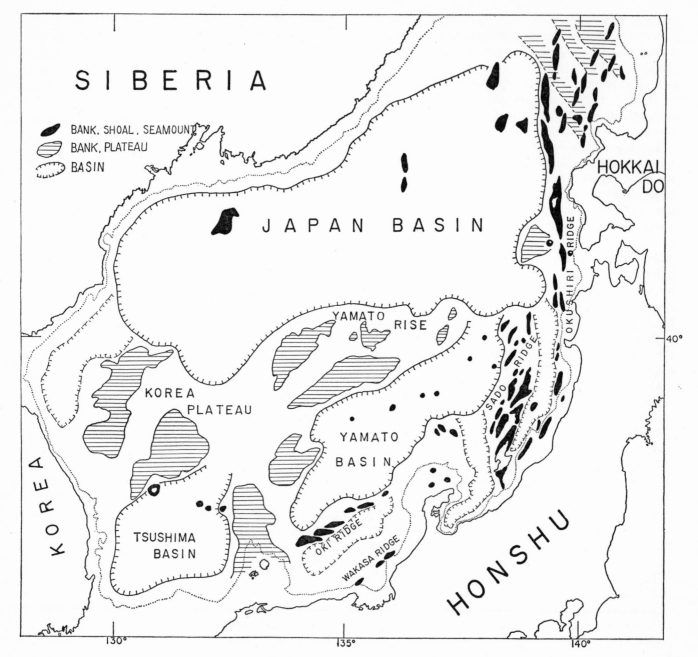

FIG. 13   Geomorphological map of the Japan Sea.

In the north of Oshima Peninsula, this ridge has two branches. An individual island or bank on the ridge generally presents a slender shape with a N-S axis and is composed of Tertiary bed rocks with volcanic rocks intruding therein. The directions of the banks coincide with those of Tertiary folds and faults developed on the coast of the Japan Sea. On the southernmost part farther than the Oga Peninsula, the bank tops in the ridge have shallow depths of about 100 m. Off the Tsugaru Strait, the ridge has a maximum depth of about 1997 m, and then tends to become shallower again to the west of Hokkaido.

A series of banks intermittently linked at the entrance to Wakasa Bay and called Wakasa Ridge, are arranged NE-SW.

Sado Ridge is a group of banks extending in a row from Sado Island to the north via Hyōtan Guri (bank), Mukai Se (bank) and Mogami Tai (bank) to the west of the Oga Peninsula. These banks are arranged in echelon, and between them ditch-shaped valleys have developed. The banks are composed of Tertiary rocks with volcanic rocks intruding therein.

The top of each bank is truncated at a depth of 100 to 170 m and is covered with numerous boulders. Bank chains form the same geanticlinal structures as the mountain ranges of the mainland, some of them being confirmed by means of acoustic reflection profiles. Bank slopes on both sides are steepened by faults and form Horst. Moreover, topographic evidence suggesting fault origin is provided by this straightness of the slope as well as the rectilineal notch along its base. In fact, some of the presumed faults continue to a known fault on the mainland. The tops of many banks in the Sado Ridge are truncated at about 100 m in depth due to wave abrasion at the time of the Pleistocene lowering in sea level. It is noteworthy that these flat tops are inclined from east to west. This means that many banks which are Horst mountains have tilted northwestward since about 100,000 years ago as land blocks in Niigata (Mogi and Sato, 1958). During the Niigata Earthquake in 1964, Awa Island on the Okushiri Ridge upheaved about 1 m and tilted westward. In addition, the sea floor adjacent to the island also upheaved and a fault scarp or flexure displacement of 2 to 3 m in height occurred on the east side of the foundation of Awa Island. This fact shows that the block movement of these ridges has continued till today, through the Tertiary Age (Nakamura et al., 1964, Mogi et al., 1964).

The Oki Ridge consists of banks in echelon in an ENE-WSW direction east of Oki Island. Their tops are truncated at depths of about 300 m and are covered with boulders.

Small Basins and Troughs

Between the banks mentioned above, there are numerous small basins and troughs. Generally, in the region north of Tsugaru Strait, many small triangular basins are developed, while in the south of the strait, Mogami, Toyama and Oki Troughs are developed. These triangular basins may have been formed by the depressions due to three fault systems in the N-S, NE-SW and NW-SE directions, respectively. They are completely separated from the Japan Basin by Okushiri Ridge.

Long, large troughs extending southward from the Tsugaru Strait are situated between Okushiri Ridge and Sado Ridge and between Oki Ridge and Wakasa Ridge, respectively, and generally have their openings into Yamato Basin at their northeastern extremities.

The formation of the depressions of basins and troughs are closely related to the building of bank-ranges, and conspicuous depressions since their birth have been filled by sediments measuring several hundred meters in thickness. The submarine fans developed along the continental slope and the floor of the basins generally tend to become deeper from the landward side to the offshore side. It seems that the main source of the detrital sediments was from the land, and an abundant supply of the sediments was made through the submarine valleys.

A number of submarine valleys are developed in the sea floor of the Japan Sea. They can be classified into the following two categories:

(1) Short valleys deeply cutting down the continental slope, opening to small basins. These valleys are terminated by various basins with various depths at their terminals. Off the western coast of Oshima Peninsula, many submarine valleys drain into several basins, cutting through the Tertiary bed and volcanic rocks.

(2) Deep sea channels reaching to a depth of 3000 m, each accompanied by a natural levee and an abyssal fan. The deep sea channels are developed in troughs. The Toyama Deep Sea Channel incises a few hundred meters downward in the floor of Toyama Trough and meanders strongly. Along the stream the development of a submarine natural levee can be found. There is also an abyssal fan at the outlet of the trough into the Yamato Basin which drains into the Japan Basin at the eastern end of the Yamato Rise through the Yamato Basin. Bottom sediments of

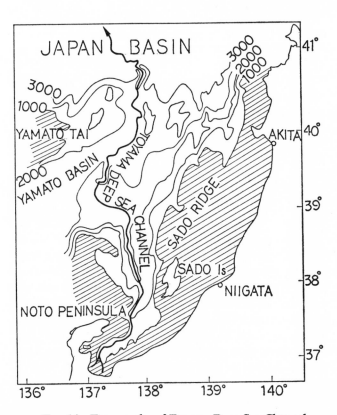

FIG. 14 Topography of Toyama Deep Sea Channel.

the natural levee and the abyssal fan accompanying it are coarser than in other areas (Iwabuchi, 1969) (Fig. 14).

4.3 Oceanic Rise and Basins

The Yamato Rise, located in the middle of the Japan Sea, is an oceanic rise with a relative height of about 2500 m, running in a ENE-WSW direction. It is divided into two blocks, Yamato Tai (bank) and Kita-Yamato Tai (bank) by the graben running in a NE-SW direction in the middle of the rise. Both banks have extensive flat tops, one at 300 m and the other at 500 m, suggesting that the banks were once islands. Dredged rocks are mostly well rounded and are of continental types resembling those found on the mainland of Japan. The oldest rocks forming the banks are probably gneissose granite and Late Mesozoic igneous rocks with welded tuff; ash flows are developed there. The direction of the major axis of each bank obliquely intersects the general direction of the whole rise, the two banks thus forming an echelon (Sato and Ono, 1964).

The Korea Plateau lies on the WSW extension of the Yamato Rise. Depths over the plateau are 700 to 1000 m, but the existence of any flat top there has not yet been made clear. This plateau is also divided into two blocks, like the Yamato Rise.

The Japan Basin, the deepest part of the Japan Sea, has a depth of 3000 to 3500 m, and the basin floor is very flat indicating a considerable fill of sediments.

The Yamato and Tsushima Basins, located on the southern side of the Korea-Yamato Rise, have depths of 2500 and 2000 m, respectively. The Basin floors are generally abyssal plains. The reflection surveys in the Yamato Basin show that it has been filled by sediments measuring several hundred meters in thickness and three or more layers deep (Murauchi, 1966). There are some features disturbing the flat floor of the basins. One is the seamount range and the others are the abyssal fan and the deep sea channel. The Utsuryo Seamount Range runs from Utsuryo Island to Shinto-sho (bank) through Take Island across the Tsushima Basin. The Meiyo Seamount Range including the Meiyo Seamount, etc., traverses the middle of the Yamato Basin.

The Toyama abyssal fan covers the basin floor at the entrance of Toyama Trough into the Yamato Basin, and the Toyama Deep Sea Channel cuts the basin floor along a NE-SW course from the abyssal fan to the Japan Basin as mentioned above.

Geophysical exploration has shown that the Yamato Rise consists of a continental layer and that the Japan Basin and the Yamato Basin on both sides of it consist of oceanic or semi-oceanic layers (Murauchi, 1966). Some geologists believe that the origin of the Japan Sea Basin is due to basaltization of a part of the continent after the continent subsided. The drifting out of the Japanese arc from the Asiatic continent is another way to account for the basin behind the arc. Accordingly, the large banks in the Japan Sea Basin may be remnant sialic blocks. Other geologists believe that the principal basins in the Japan Sea are rift basins related to the formation of large tectonic lines represented in the submarine features therein.

High values of heat flow were obtained in the whole of the Japan Sea. There are various opinions on this high heat flow: one is that the excess heat was transported by the upward intrusion of magma beneath the bottom, and another is that a tectonic deformation of the crust in the past resulted in the anomaly of heat flow at present.

5 NORTHWEST PACIFIC OCEAN

5.1 Northeast Japan Arc and Trench

Northeast Japan is separated by the "Fossa Magna" from Southwest Japan. Hokkaido Island may be considered the junction of three island arcs. Its southwestern peninsula is nothing but an extension of Northeast Japan, and belongs to the arc of Japan proper. Its central main part belongs to the arc of folded mountains stretching from Sakhalin. The eastern portion of Hokkaido has the character of the Kurile Island arc. In Northeast Japan, which is covered extensively by Cenozoic rocks, older rocks occur at many isolated places such as Kitakami, Abukuma and Kwanto mountains.

Pre-Neogene rocks were in a zonal arrangement parallel to the general trend of the Honshu arc. A new trend of the Fossa Magna appeared in Northwest Japan in the middle Tertiary. The older strata were fragmented and separated into many blocks by NNW, NNE and N-S faults.

In the western and middle parts of Northeast Japan, the younger formations have been folded in forming the Mizuho-Fossa Magna fold zone (Ōtsuka 1939). Here the folding is still going on. This folded zone is the northern extension of the Mariana arc and parallel to the distribution of the earthquake centers, volcanoes and the Japan Trench (Sugimura, 1960).

There is a continental shelf with a width of 20 to 50 km along the Pacific side of Northeast Japan. The shelf edge ranges from 100 to 200 m in depth. From Kashimanada to Sendai Bay, rock, gravel and a coarse sand bottom predominate on the shelf and several shelf channels cut down its surface (Mogi and Iwabuchi, 1961).

There is a wide, flat terrace from 20 to 100 km wide at a depth ranging from 1000 to 3000 m between the continental shelf and the Japan Trench. It was called the Japanese Pacific Sea Shelf by Nasu (1964). On the outer edge of this sea shelf, coarse sediments including over 500 pebbles and cobbles were collected by Nasu and others in 1959. The largest of these is 56 cm in diameter and has been determined from its diatom assemblage to be Miocene siltstone.

Also, at three other localities, coarse sediments including well-rounded sedimentary pebbles were obtained. Kagami and Iijima (1959) assumed that the Japanese Pacific Sea Shelf might once have been a fluvial plain or wave cut terrace overlain by coarse sediments, and then submerged to its present depths. They thought that its peneplanation might have taken place during post-Miocene time and its sub-mergence referred to the forming of the Japan Trench.

Very abundant shallow earthquake epicenters are located between the Japanese Island and the Japan Trench. These epicenters are concentrated in the two parts of the continental slope and the western trench wall. This fact shows that both side slopes of the sea shelf may be an active tectonic zone.

Some seismologists found that the aftershock areas of an earthquake are not piled on one another. The Japanese Pacific Sea Shelf is constituted of several blocks topographically and these blocks correspond to the aftershock areas.

The Japan Trench is parallel to the general trend of Northeast Japan. It is separated from the Kurile Trench by the Erimo Seamount extending from Cape Erimo in south central Hokkaido on the north. The Japan Trench is bounded on the south by the trough extending southeastward from Sagami-Nada from the Izu-Ogasawara Trench. The depth of the Japan Trench is about 7500 to 9200 m, and generally in-

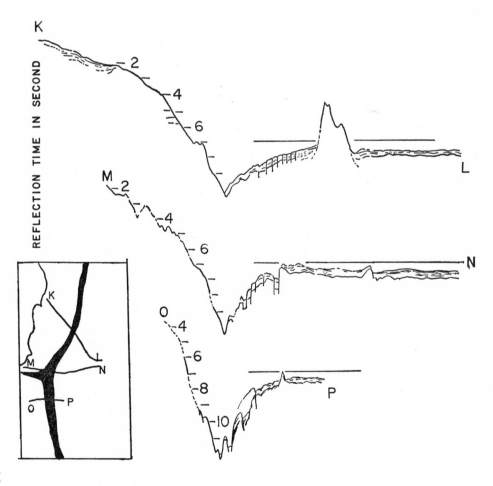

FIG. 15   Seismic reflection profiles across Japan Trench (by Ludwig, W. J. et al., 1966).

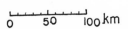

Vertical Exaggeration   25 : 1

creases southward. The cross sections traced from echograms show a V-shaped trench, having an asymmetrical profile with the west or shoreward flank having steeper slopes than the offshore flank. The flat floor of the trench bottom is narrow or absent, consequently thick sediments cannot be expected.

On the shoreward flank of the trench, locally continuous benches are found. On the outer side and in the central part of the benches, ridges having a relief of a few hundred meters are often found, and sometimes trough-shaped depressions flattened by sediments are recognized on the inner side of the ridges. On the offshore flank of the trench, several small a narrow depressions are found. These depressions are remarkably long for their depth and width and strike parallel to the trench axis. Ludwig and others (1966) found the presence of graben and step faults along the entire seaward slope of the trench by the reflection technique (Fig. 15). Conspicuous faults are recognized by the displacement of sub-bottom layers on either side of the fracture. They assumed that the faults are caused by tensional forces introduced in the convex side of the oceanic crustal plate as it is being further depressed.

5.2   Kurile Arc and Trench

The Kurile arc is a large double geanticline. The inner ridge is the Great Kurile Ridge stretching from Hokkaido to Kamchatka and is crowned by a chain of active volcanoes connected by a great deepseated fracture. The outer ridge, the Minor Kurile Ridge, extends along the outer side of the Great Kurile Ridge and is apparently somewhat older (Tertiary). The most intense volcanic and seismic activities are concentrated here (Bezrukov, 1960).

The Kurile-Kamchatka Trench is bounded on the south by the Erimo Seamount from the Japan Trench and intersects the Aleutian Trench at an acute angle on the north. On the insular slopes of the Kurile-Kamchatka Trench, benches are 20 to 30 km wide and at depths of 4000 and 7500 m. They have ridges at the outer edges, backed by smooth-bottomed basins, and continue to the deep sea terrace around the Japanese Islands on the west. According to Udintsev (1955), at least nine oblique troughs about 100 km long occur on the slope between the Kurile Islands and the Kurile-Kamchatka Trench. They are straight and trend at angles to the trench ranging between 40° to the north and 80° to the south. Structures trending parallel to the submarine troughs occurred in the Kamchatka Peninsula and were cut by the younger structure trending. Details of the Kurile-Kamchatka Trench were investigated by Udintsev

(1955) based on the soundings taken by the *Vitiaz*. The deepest sounding was reported to be 10,542 m in the area south of Urup Island, but on the cruise of the *Takuyo* in 1962, depths deeper than 9550 m could not be found (Iwabuchi, 1968).

5.3   Northwest Pacific Basin

The northwest Pacific contains two basins, the Northwest Pacific Basin and the Philippine Sea Basin, using these terms in a wide sense. In this paper the Northwest Pacific Basin is restricted to the area limited by the Emperor Seamounts to the east, the Mid-Pacific Mountains to the south and the Kurile-Japan Trench to the northwest, following the definition of the GEBCO sub-committee. The Northwest Pacific Basin has a depth of about 6000 m which is the deepest of all ocean basins. There are two rises in this basin: the Hokkaido Rise and the Northwest Pacific Rise.

The Hokkaido Rise is the marginal swell mainly along the oceanic side of the Kurile Trench. It is a broad swell about 250 km wide and 1400 km long. The depth of the swell is about 4500 to 5000 m and increases generally southwestward. It rises about 500 to 1000 m above the ocean floor of the Northwest Pacific Basin. There are several seamounts and hills on the swell. Several seamounts are located near the junction of the Kurile-Kamchatka and Japan Trenches. These seamounts appear to lack any orientation. The Ryofu Seamount on the axis of the swell has a depth of 1345 m. The Daini-Erimo Seamount on the trench side slope of the swell has a depth of 2565 m and the Erimo Seamount in the Trench has a depth of 3678 m. Recently, the Erimo Seamount, which is the site of the conjunction of the Kurile-Kamchatka and Japan Trenches, was sounded and dredged (Tsuchi, 1966). The depth of the trench bottom is about 7000 m and the top of the seamount is flat where *plesioptygmatis*, one of the late Cretaceous Nerinea fossils and some calcareous algae were dredged. Various evidence indicates that the truncation of the Erimo Guyot may be older than the formation of the trenches. On the east side of the Japan Trench, marginal swell is obscure. There are also several seamounts ranged in a line near the southern part of the Japan Trench. The line trends at about 30° to the axis of the trench here. Kashima Guyot at the southwesternmost part of the seamount chain is located in the trench and the depth of the top is 3600 m. Komukai and Nakayama (1958) found Foraminiferal fossils of the Miocene age on the flat top by dredging. There are linear low hills and troughs on the border of the Hokkaido Rise and the

Northwest Pacific Basin. These hills and troughs run parallel to the axis of the swell. This feature is similar to the feature which originated from faulting on the trench side slope of the swell. Magnetic anomalies off the Kurile-Kamchatka Trench have a fine structure parallel to the swell. This also may be a topographic expression of the block displaced by faulting. It is assumed that the marginal swell is the great structural form uplifted by the faulting on both sides of the swell.

There is a rise called the Northwest Pacific Rise or the Shatsky Rise on the central part of the Northwest Pacific Basin. The Northwest Pacific Rise is about 1000 km long and 200 km wide, and rises about 3000 to 4000 m above the adjacent sea floor. It stretches in a NE-SW direction and runs parallel to the Kurile-Kamchatka Trench and Hokkaido Rise. On the northeast, it is connected to the Emperor Seamount chain at Lat. 43°N. Cross sections show the presence of irregularities on the top and the side slopes. Apparently this rise is a great folded mountain associated with the faulting on the side slopes. A cretaceous fauna was found by Ewing et al. (1966) on this rise. This cretaceous formation, called the Opaque layer on the echogram, continues from the Northwest Pacific Rise to the Izu-Ogasawara Trench. Another striking feature of the Northwest Pacific Basin is the seamounts running ENE-WSW on the southeastern part of the basin.

Many small seamounts are arranged in a line from the southernmost of the Emperor Seamounts to the central part of the Marcus-Wake Seamounts. These seamounts may be the westward extension of the Mendocino Fracture zone. The sea floor of the Northwest Pacific Basin has a relatively smooth flat bottom associated with small irregularities. The sea floor between the Northwest Pacific Rise and the Emperor Seamounts is a particularly rugged area.

5.4 Seamounts (Fig. 16)

A great linear range of seamounts dominates the northeastern portion of the Northwest Pacific Ocean and extends from the vicinity of the intersection of the Aleutian Trench with the Kamchatka Trench to the vicinity of 30°N and 174°E. They were termed the Emperor Seamounts by Dietz (1954). The Emperor Seamounts contain many guyots. Individual seamounts are elongated in a N-S direction and are connected with each other at the base. The flat-topped depths of the guyots range from 300 m for Kanmu Guyot to 3000 m for Guyot (G52–170). Guyot 52–170 was found just at the axis of the Aleutian Trench (Mogi, 1953). Foraminiferal fossils of Neogene age

were reported by Niino (1961) from the Kanmu Guyot.

An east-west 1200-mile-long broad band of seamounts is present east of the trench system and between Lat. 17° and 21°N. The group is termed the Marcus-Wake Seamounts, which is the western part of the Mid-Pacific Mountains. The seamounts are deeply submerged, generally more than 1000 m. At least some of them display flat tops and thus are guyots. These submarine volcanoes with broad, almost level tops were first discovered in this region

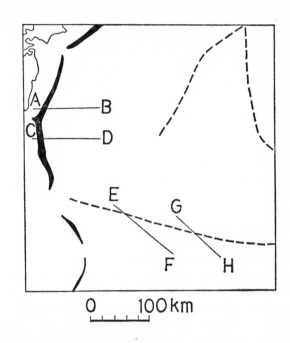

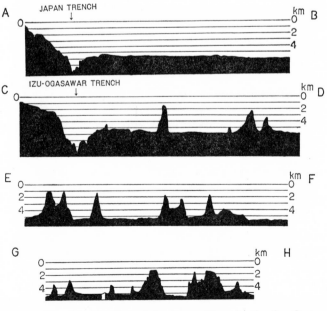

FIG. 16 Profiles of the Japan Trench, the Izu-Ogasawara Trench and the Northwest Pacific Basin.

by Hess (1946). In 1950, the flat tops were surveyed in the Mid-Pacific Mountains and rounded cobbles as well as an integrated reef fauna of Middle Cretaceous age were dredged from the guyots at a depth of about 1800 m (Hamilton, 1956). Menard (1964) proposed the hypothesis of the Darwin Rise and the evolution of oceanic rises. If the flat top of a guyot indicates the ancient sea level, the height of a guyot above the ocean floor shows the depth of the ancient ocean. He also made a paleo-submarine topography for the Mid-Pacific Basin 1 billion years ago, and discovered the paleo-oceanic rise known as the "Darwin Rise."

6 RIDGES CROSSING THE KUROSHIO COURSE

6.1 Bashi channel

There are three ridges extending from Luzon to Formosa, as mentioned earlier. The east ridge has an asymmetrical profile, the eastward slope is steeper than the westward slope and several seamounts are superposed on the ridge. The seamounts have an average depth of about 2300 m and the depth of the base is about 3400 m.

The central ridge is composed of a few island chains in echelon. It is the shallowest ridge in this strait and many islands and shoal banks are closely spaced and clustered. But there are a few depressions between each island chain having a NE-SW direction. A depression between Lutao Island and the Lanhsu Islands is about 1800 m deep. A depression between the Lanhsu Islands and Batan Islands is about 2100 m deep and between the Batan Islands and Babuyan Islands about 1400 m deep.

The west ridge is elongated southward from the southernmost cape of Formosa and its depth increases southward. The eastward slope of the ridge is steeper than the westward slope. The sill depth at Lat. 20.5°N is about 2100 m. In this region, the Kuroshio may be strongly affected by the central ridge due to the shallow depth.

6.2 Nansei Shoto (Ryukyu) Ridge

The Nansei Shoto Islands are divided into three island groups: to the north the Osumi Islands, in the center the Okinawa and Amami Islands and to the south the Sakishima Islands. These islands are surrounded by continental shelves. There are depressions between each islands. The Sakishima Islands are separated by a depression deeper than 500 m from

Formosa: the Okinawa and Amami Islands are separated by a depression deeper than 1000 m from the Sakishima Islands. The outer ridge of the Osumi Islands is also separated by a depression deeper than 1000 m from Okinawa, but the inner ridge, the Tokara Islands, is elongated southward to the west of Amami Oshima Island. Although a southernmost depression cuts the ridge in the N-S trend, other depressions cut in an E-W direction.

The Okinawa Trough along the inner side of the Nansei Shoto Ridge gradually rises toward Kyushu Island. Therefore the continental slope of the East China Sea Shelf is steeper and higher at the southern portion than at the northern portion.

6.3 Off the south coast of Southwest Japan

Alternation of prominent coastal salients with embayed and low-lying stretches of shore line is characteristic of Southwest Japan. Submerged extensions of these coastal features are seen on the continental slope at a depth of 1000 m. That is, Hyuga Basin off Bungo Suido, Tosa Basin off Tosa Bay, Kii Basin off Kii Suido, Kumano Basin in Kumano-Nada and the unnamed basin in Enshu-Nada. Low ridges between each basin stretch southward from the capes of the coast. The submarine ridge off Cape Toi is traceable at a depth of 2000 m and continues to the Kyushu-Palau Ridge beyond the Nankai Trough. The submarine ridges off Cape Ashizuri, Cape Muroto and Cape Daio are traceable at a depth of 1000 m. Around the Kii Peninsula, the continental slope steeply descends to the deep sea terrace at a depth of 1500 m beyond the narrow continental shelf.

6.4 Izu Ridge and its environs

The outline of the Izu Ridge is shown by the 3000 m isobath. The Izu Ridge is a shallow and broad rise and is pierced by volcanic islands arranged in a N-S direction. The northern volcanic islands are higher above the sea surface than the southern volcanic islands which are a part of the great submarine caldera. Hoshino (1954) pointed out that the heights of the volcanoes are approximately uniform and are determined essentially by the level of their basements. The Nishi-shichito Ridge on the western margin of the Izu Ridge is arranged in a straight line with a height of about 3000 m above the sea floor of the Shikoku Basin. Another effective feature against the Kuroshio course is the secondary ridges from volcanic cones trending southwestward in echelon. They are estimated at about 12 rows. Therefore, the Izu Ridge is cut by depressions running NE-SW between each minor ridge. Depressions between the Izu Peninsula, the Oshima-Zenisu Ridge and Miyake Island are shallower than 500 m. A depression separating Hachijo Island from Mikura Island is deeper than 1000 m. Depressions in the south further than Aogashima are all deeper than 1000 m.

There is a submarine ridge extending eastward from Cape Inubo. This ridge may be the eastward extension of the Tanzawa-Mineoka Uplifted Zone formed during the Miocene.

6.5 Korea Strait

Korea Strait is occupied by the extension of the vast continental shelf of the East China Sea. The sea floor is generally shallower than 200 m, but there is a trough-shaped depression deeper than 200 m in the north off Tsushima Island. Rocky bottoms occur on submarine hills, off rocky points and islands; at these places, strong wave and/or current action prevents deposition of fine grained sediments. Emery and Niino (1967) obtained many rock fragments identified by fossils or lithology as belonging to Neogene strata around Tsushima, the Danjo Islands and Mishima.

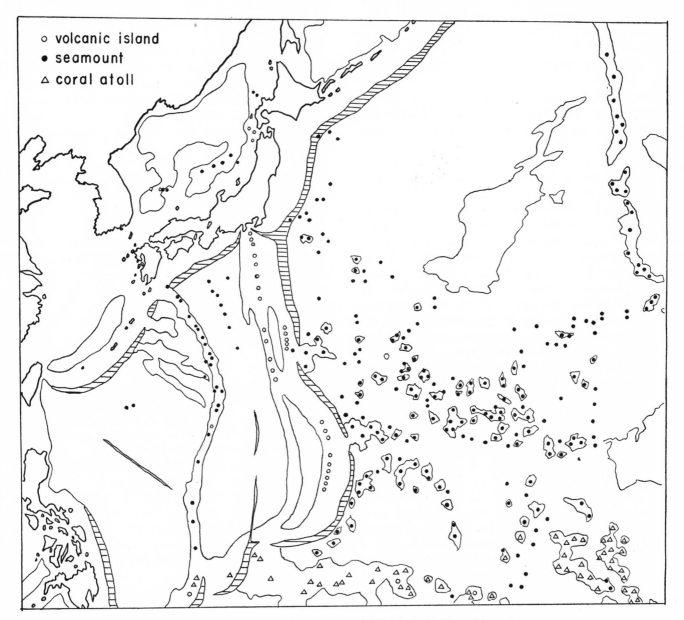

FIG. 17   Seamounts, atolls and volcanic islands in the Kuroshio region.

7. CONSIDERATION

Trenches and their associated island arcs are the dominant features of the western Pacific Ocean. The continental arcs of the Kamchatka-Kurile Islands, Japan, the Ryukyu Islands and the Philippines embrace the marginal seas on their continental side, while the oceanic arcs extending south from Japan separate the Pacific Basin proper from the Philippine Sea. The continental arcs, except for the Kurile-Kamchatka Trench, are diagonally cut in Central Japan by the new arc, which is one of the Neogene fold zones and which extends from the Kurile Islands to the Mariana Arc through Northeast Japan.

There are several old ridges paralleling the outer arcs in the marginal seas: the Shantung-Laoyeling Massif, the Fukien-Reinan Massif and the Taiwan-Sinze Folded Zone in the East China Sea, and the Korea-Yamato Rise, the Sado Ridge and the Okushiri Ridge in the Japan Sea. However, it is difficult to suppose that all the old ridges in the East China Sea continue to the Japan Sea. For the continuation is interrupted by the great tectonic line along the east coast of the Korea Peninsula. Basins between old ridges have been buried by vast sediments that are derived mostly from the large area of China that is drained by the Yellow River and the Yangtz River. Otherwise, the Korea-Yamato Rise in the Japan Sea is a remnant of the old continental block which has not been buried by small sedimentation. Continuation from the Korea Plateau to the Yamato Rise is interrupted by the 2000 m basin. On the other hand, there is a submarine plateau, called the Oki plateau, on the boundary of the Tsushima and Yamato basins which lie on the southern side of the Korea-Yamato Rise. This arrangement of ridges and basins suggests that the Tsushima and Yamato basins may be rift basins occurring along the E-W tectonic line like the Japan Basin, and that the Oki Plateau is a part of the Korea-Yamato Rise which shifted southward.

The Philippine Sea is a very complicated region surrounded by island arcs. A few ridges and seamounts not associated with trenches run parallel to the Ryukyu-Philippine Arc and the Izu-Mariana Arc: the Kyushu-Palau Ridge, Kinan Seamounts-West Mariana Trough, and Nishi-Shichito Ridge-West Mariana Ridge. These ridges close the space eastward. The Kyushu-Palau Ridge consists of two arcs, a northern arc parallel to the Ryukyu arc and a southern arc parallel to the Philippine arc. These two arcs are separated by the Central Basin Fault. Another possible fault runs from the depression in the east side of the northern Kyushu-Palau Ridge

to the junction of the Izu-Ogasawara Arc and Mariana Arc. The southern arc (Mariana) projects remarkably southeast along this fault (Fig. 19).

There are other ridges having unusual trends. The Daito Ridges have an E-W trend and the minor ridges on the Izu Ridge also have a NE-SW trend. These trends correspond to the trend of the Southwest Japan Arc.

Ridges with E-W trends are cut by ridges with N-S trends, the Kyushu-Palau and Izu-Ogasawara ridges, like the Southwest Japan Arc.

The abyssal hill provinces in the Philippine Sea usually occupy the western half of each basin. Some of the flat plains occupying the eastern half appear on the archipelagic aprons which were constructed by eruptions on the concave flanks of the arcuate ridges.

The numerous minor ridges which construct the abyssal hill region also have E-W trends in the northern part of the Philippine Sea, although trends change southward gradually to NNW-SSE in the Philippine

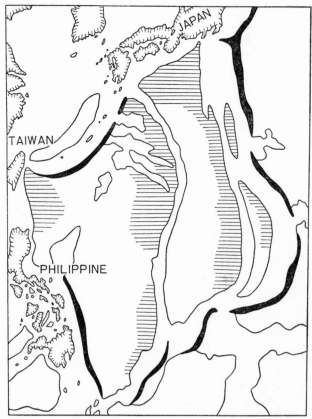

| | abyssal plain and archipelagic apron |
| abyssal hill area |

FIG. 18 Distribution of the abyssal plain (including the archipelagic apron) in the Philippine Sea Basin.

Fig. 19   Structural pattern of the Kuroshio region supposed from trenches, ridges and seamounts.

Basin and to NNE-SSW in the West Mariana Basin.

The Daito Ridges with their coral islands and wide flat plains appear to have an evolution similar to that of the guyots and atolls in the Marcus-Wake seamounts to the east. They are connected to the Marcus-Wake seamounts through abyssal hill ridges having WSW-ENE trends at the southern part of the Shikoku Basin.

The westward extension of the Marcus-Wake seamounts to the Philippine Sea is supported by geophysical data. I believe that there was a mid-oceanic rise, called the Darwin Rise by Menard, extending from the Tuamotu Ridge to the Daito Ridge during the Mesozoic Age. The Nankai Trough and its associated phenomena may have been formed by the convection current flowing northward from the Darwin Rise during the Mesozoic Age.

It is noteworthy that oceanic ridges, seamounts and great faults encounter the junction of trenches: the Central Basin Fault meets with the junction of the Ryukyu Trench and Taiwan-Luzon Arc, the Kyushu-Palau Ridge meets with that of the Nankai Trough and the Ryukyu Trench and the Erimo Seamounts meet with that of the Kurile-Kamchatka and Japan trenches. The trenches usually intersect each other at large angles. The intersection of the Japan Trench and the Nankai Trough, where the Japan Arc is crossed by the Izu-Mariana Arc stretching from the south, is obscure compared with that of other arcs. But the Japan Arc has bent structures with convex sides toward the north there. The Japan Arc, which was a simple arc originally, was destroyed by the Izu-Mariana Arc in the Middle Tertiary. Bent structures associated with strike-slip faults on both sides were formed in this age. Earth movements that formed bent structures go on to recent times (Kimura, 1967).

The flow of the Kuroshio may be affected by these island arcs and other ridges. Two southern depressions in the Ryukyu Islands, the depression between Formosa and the Sakishima Islands and the depression between the Sakishima Islands and Okinawa Islands, are important for the flow of the Kuroshio branch into the marginal sea. Submarine ridges stretching southward from the capes of Southwest Japan may influence the course of the Kuroshio.

REFERENCES

Bezrukov, P. L. (1960): Bottom sediments of the Okhotsk Sea. *Tr. Inst. Okeanol. Akad. Nauk SSSR*, **32**, 15–95.

Dietz, R. S. (1954): Marine geology of the northwestern Pacific; Description of Japanese Bathymetric Chart 6901. *Geol. Soc. Amer. Bull*, **65**, 1199–1224.

Emery, K. O. and H. Niino (1967): Stratigraphy and petroleum prospects of Korea Strait and East China Sea. *Geol. Survey of Korea. Rep. of Geophys. Explor.*, **1**, 249–263.

Emery, K. O. et al. (1969): Geological structure and some water characteristics of the East China Sea and the Yellow Sea, Econ. Comm. Asia, Far East Comm. Coord. Joint Prospect. for Mineral Resources in Asian offshore Area. *Tech. Bull.*, **2**, 3–43.

Ewing, M., Saito, T., Ewing, L. and H. Burckle (1966): Lower Cretaceous sediments from the Northwest Pacific. *Science*, **152**, 751–755.

Gutenberg, B. and C. F. Richter (1949): *Seismicity of the earth and associated phenomena*. Princeton, N. J., Princeton Univ. Press, 273.

Hamilton, E. L. (1956): Sunken islands of the Mid-Pacific Basin. *Geol. Soc. Amer. Mem.* **64**, 97.

Hanzawa, S. (1940): Micropaleontological studies of drill cores from a deep well in Kita-Daito-Zima. *Jubilee Pub. Comm. Prof. H. Yabe, M. I. A. 60th Birthday*, **2**, 643–665.

Hashimoto, W. and T. Sato (1968): A contribution to the study of geologic structure of the Philippines. *Jour. Geogr.*, **77**(2): 78–116.

Hess. H. H. (1946): Drowned ancient islands of the Pacific basin. *Am. Jour. Sci.*, **244**, 772–791.

——— (1948): Major structural features of the western North Pacific, an interpretation of H. O. 5485, Bathymetric Chart, Korea to New Guinea. *Geol. Soc. Amer. Bull.* **59**, 417–445.

Hoshino, M. (1954): Topography of the Izu Islands. *Inst. Geol. Tokyo Univ. of Education*, **3**, 243–247.

——— (1963): Southwest Japan Trench. *Marine Geology, Japan*, **1**, 10–15.

Hydrographic Office Japan (1966): *Explanatory text of J. H. O. Bathymetric Chart No. 6301*, 1–4.

——— (1966): *Explanatory text of J. H. O. Bathymetric Chart No. 6302*, 1–8.

Iwabuchi, Y. (1968): Topography of trenches east of the Japanese Islands. *Jour. Geol. Soc. Japan*, **74**, 37–46.

——— (1968b): Submarine geology of the southeastern part of the Japan Sea. *Inst. Geol. Paleo. Tohoku Univ. Contr.*, **66**, 1–76.

Iwabuchi, Y. and K. Saiki (1968): Topography of the Philippine Trench. *Rep. Hydro. Res. Japan*, **4**, 155–171.

Kagami, H. and A. Iijima (1959): On the bottom sediments off Onagawa and Kushiro, the adjacent continental slope of Japan Trench. *Oceanogr. Magazine*, **11**, 233–242.

Katumata, M and L. R. Sykes (1969): Seismicity and tectonics of the Western Pacific: Izu-Mariana-Caroline and Ryukyu-Taiwan Regions. *Jour. Geoph. Res.* **74**, 5923–5948.

Kimura, T. (1967): Structural division of Japan and the Honshu Arc. *Japanese Jour. Geol. Geogr.* **38**, 117–131.

Kimura, T. et al. (1968): Geologic Structures in the Tayabas Isthmus District, Philippines. *Geol. Paleo. S. E. Asia*, **4**, 156–178.

Komukai, R. and R. Nakayama (1958): On the bottom topography of Japan Trench. *Hydro. Bull.*, **57**, 45–52.

Konishi, K. (1965): Geotectonic framework of the Ryukyu Islands (Nansei-shoto). *Jour. Geol. Soc. Japan*, **71**, 437–547.

Kiilerich, A. (1959): Bathymetric features of the Philippine Trench. *Galathea Rep.*, **1**, 155–171.

Ludwig, W. J. et al. (1966): Sediments and structure of the Japan Trench. *Jour. Geoph. Res.*, **71**, 2121–2141.

Menard, H. W. (1964): *Marine Geology of the Pacific*. McGraw Hill, New York.

Mogi, A. (1953): On the flat-topped seamounts (guyot) in the North Pacific Ocean. *Hydro. Bull.* spec., **12**, 58–61.

Mogi, A. and T. Sato (1958): On the bottom configuration and sediments in the adjacent sea of Mogami Bank, Japan Sea. *Hydro. Bull.*, **55**, 37–53.

Mogi, A. and Y. Iwabuchi (1961): Submarine topography and sediments on the continental shelves along the coasts of Joban and Kashimanada. *Geogr. Review* Japan, **34**, 39–58.

Mogi, A. et al. (1964): Submarine crustal movement due to the Niigata Earthquake in 1964, in the environs of the Awa Sima Island, Japan Sea. *Jour. Geod. Soc. Japan*, **70**, 180–186.

Mogi, A. and T. Kato (1964): A guyot on the northern part of Kyushu-Palau Ridge. *Ann. Tohoku Geogr. Assoc.*, **16**, 203.

Murauchi, S. (1966): The UMP seismic refraction measurements in and around Japan. *Proc., 11th Pacific Sci. Cong.,* **3**.

Murauchi, S. et al. (1968): Crustal structure of the Philippine Sea. *Jour. Geoph. Res.*, **73**, 3143–3171.

Nakamura, K. et al. (1964): Tilting and uplift of an island, Awa Sima, near the epicenter of the Niigata Earthquake in 1964. *Jour. Geod. Soc. Japan*, **10**, 139–145.

Nasu, N. (1964): The provenance of the coarse sediments on the continental shelves and the trench slopes off the Japanese Pacific coast, in *Marine Geology, Shepard Commemorative Volume*, 65–101.

Niino, H. (1961): A survey of Kammu Seamount at the southern end of the Emperor Seamounts in the Central North Pacific Ocean. *The 10th Pacific Sci. Cong.*, 383.

Niino, H. and K. O. Emery (1961): Sediments off shallow portions of East China Sea. *Bull. Geol. Soc. Amer.*, **72**, 731–762.

Otsuka, Y. (1939): Tertiary crustal deformations in Japan. *Jubilees Pub. Comm., Prof. H. Yabe*, **1**, 481–519.

Sato, T. and K. Ono (1964): The submarine geology off San'in District, Southern Japan Sea. *Jour. Geol. Soc. Japan*, **70**, 434–445.

Sugimura, A. (1960): Zonal arrangement of some geophysical and petrological features in Japan and its environs. *Jour. Facul. Sci. Univ. Tokyo, sec. II*, **7**, 133–153.

———— (1966): East Japan Island arcs and Quaternary subsidence of the Japan Trench. *Proc., 11th Pacific Sci., Cong.*, **10**.

Tayama, R. (1935a): Insular disposition and submarine topography of the South Sea Islands. *Inst. Geol. Pal. Tohoku Imp. Univ., Contrib.*, **17**, 1–22.

———— (1935b): Topography, geology and coral reef of the Palau Islands. *Inst. Geol. Pal. Tohoku Imp. Univ. Contrib.*, **18**, 1–67.

————: (1950): The submarine configuration off Shikoku especially the continental slope. *Hydro Bull.*, spec. **7**, 54–82.

———— (1952): On "Depth Curve Chart of the Adjacent Sea of Japan" (chart 6901). *Hydr. Bull.*, **32**, 160–167.

———— (1952): *Coral reef in the South Seas, Bull. Hydr. Office*, **11**, 8–292.

Tsuchi, R. (1966): Discovery of Nerineid gastropoda from seamount Sysoev (Erimo), at the junction of Japan Trench and Kurile trench. *Proc., 11th Pacific Sci. Cong.*, **2**.

Udintsev, G. B. (1955): Topography of the Kurile-Kamchatka Trench. *Tr. Inst. Okeanol. Akad. Nauk SSSR*, **12**, 16–61.

Uyeda, S. and A. Sugimura (1970): *Arc Island*, Iwanami Shoten, 1–156.

Yoshikawa, T. (1953): Some consideration on the continental shelves around the Japanese Islands. *Nat. Sci. Rep. Ochanomizu Univ.*, **4**, 138–150.

Plate 1   Bathymetric chart of the Kuroshio region. Compiled from J.H.O. chart No. 6301, 6302, 6304 and No. 6901.

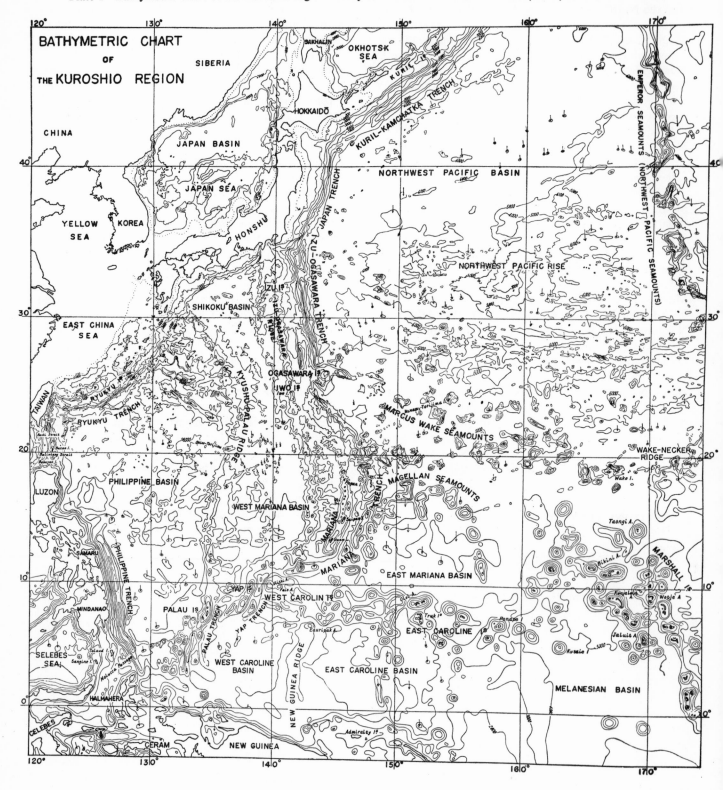

Chapter 3

DISTRIBUTION OF STATIONS, AND PROPERTIES AT STANDARD DEPTHS IN THE KUROSHIO AREA

THOMAS WINTERFELD*
HENRY STOMMEL**

* National Oceanographic Data Center, Rockville, USA.
** Massachusetts Institute of Technology, Cambridge, USA.

Our purpose in this chapter has been to make scientific use of the tabulating and sorting techniques available for the study of the hydrographic station data on file at the U.S. National Oceanographic Data Center, prior to the international program C.S.K. (the Cooperative Study of the Kuroshio and Adjacent Regions) which began in 1965. Our hope has been that this résumé of past data would be useful as a basis for present studies of C.S.K. data. The population of stations in the data files does not stay fixed. Thus if one consults the files at different times the number of stations in any area fluctuates: mostly they increase, as data from old works are gradually entered (see Fig. 5). The period during which we used the historical files began in early 1966 and ends in 1968. We have not used much C.S.K. data.

1 DISTRIBUTION OF STATIONS

The area with which we originally planned to deal was bounded by 10°N, 50°N, 120°E and 170°E. As of the date of writing this chapter, the U.S. National Oceanographic Data Center has a total of 43,372 hydrographic stations recorded for this area. Many of these are fairly shallow stations, either because they were taken in very shallow water, or because the casts in deep water were not made to the bottom. An indication of the distribution of stations where sampling extended to deeper than 950 m is given in Fig. 1. As we see from this chart, deep stations are most dense near the 950 m line nearest to the coast and the density falls off nearly exponentially away from the coast with an *e*-folding scale of 250 km. There are exceptions: for example, the high density at two weather ship positions, and on certain frequently occupied lines of stations. The density of the stations is sufficiently great to blur individual station points, so that Fig. 1 is useful only in a qualitative way.

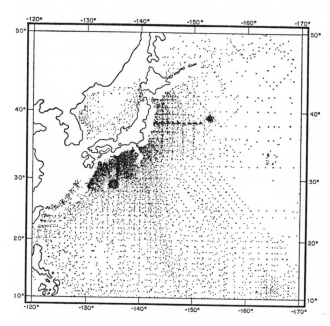

FIG. 1 Positions of all stations deeper than 950 meters whose data are on file at the U.S. National Oceanographic Data Center.

Figure 2 gives totals by five-degree squares, for all stations regardless of depth. In comparing the display offered by these two figures we recognize that Fig. 1 gives a superior spatial resolution, whereas Fig. 2 gives better information on totals of stations. Tabulations have also been made for one-degree squares, and for twenty-minute squares, but to display these fields of numbers would require larger and more numerous plates. Economy of display is another constraint to which we are subject.

In order to obtain some idea of how the density of stations depends upon time of year, we tabulated the totals of stations by one-degree squares for both temperature and salinity to at least 200 meters depth for each month. We then chose January as the month representing the period with the least density of stations per degree per month, and July as that with largest density. Instead of trying to reproduce the numerals we contoured the field with logarithmic spacing; the results are shown in Figs. 3 and 4. More than half the area has zero-density in January, and even in July a large fraction of the total area has zero

density.

In order to show other features of the distribution of data in the Kuroshio without entering into a great mass of detail, we chose "typical" one-degree areas in what appear to be regions of fairly uniform density of stations. As examples, we chose two one-degree

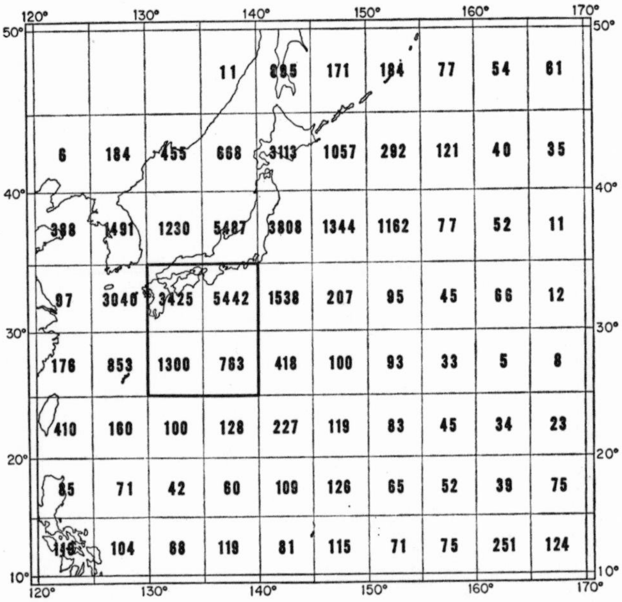

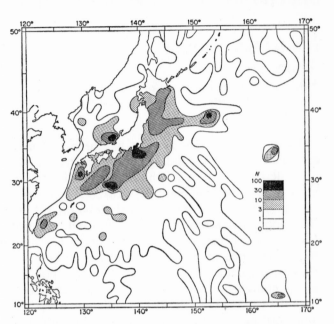

FIG. 2   Number of stations at all depths for five-degree squares for all months.

FIG. 4   Density of stations for both temperature and salinity at 200 meters, in number, per one-degree square for the month of July.

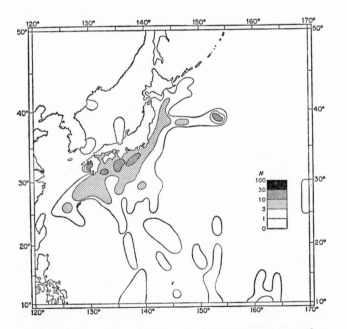

FIG. 3   Density of stations for both temperature and salinity at 200 meters, in number, per one-degree square, for the month of January.

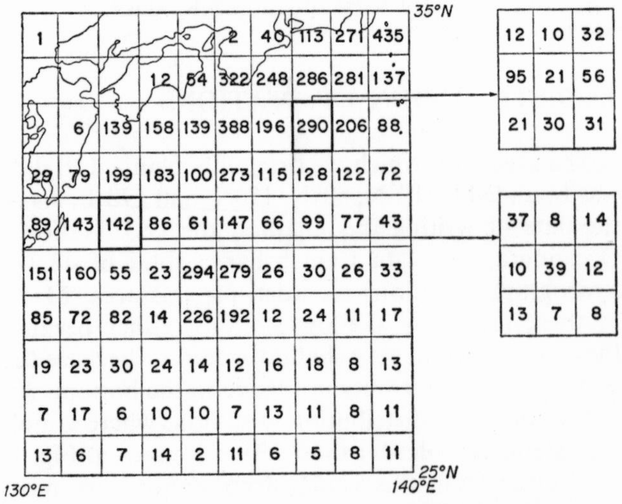

FIG. 5   Total station numbers for all months for 200 meter depth of one degree squares; and two specimen squares expanded to density by twenty-minute square.

The totals in the blocks of nine twenty-minute squares do not tally with those in corresponding one-degree squares since the tabulations were made from an up-dated version of the master data tapes. This simply illustrates a minor difficulty in working with data centers which will gradually become more familiar as time goes on.

squares in Marsden Square 131 which is heavily outlined in Fig. 2. The total number of hydrographic
stations in each one-degree square within Marsden
Square 131 with available data on both salinity and
temperature down to 200 meters for all months combined is tabulated in Fig. 5. The two "typical" heavily
outlined, one-degree squares within this Marsden
Square are squares 02 at 30°–31°N, 132°–133°E
(which has a total of 148 stations) and 27 at 32°–33°N,
137°–138°E (which has a total of 290 stations)*. Each
of these one-degree squares has neighboring squares
of similar density, and within each of them the density
is reasonably uniform as shown by the two expanded
insets, each made up of twenty-minute squares, on
the right side of the figure. The statistics on the distribution of these stations by year, month and depth
of cast can be easily displayed.

Figure 6 is a tally by year and month of the periods
when stations were taken in square 02. Observations
began in 1929 with *Kosyu Maru* station 459, and
continued with increasing frequency up until 1941.
In the years following, no observations were made
until 1946 when they were resumed at a reduced rate.
By 1954 the observation program of square 02 had
been completely resumed, and although not quite to
the degree of intensity as the 1939–40 program, it was
steadier. The apparent drop-off during the 60's is probably due to the fact that the Data Center did not yet
have all the recent data completely tabulated; the
time distribution for the year is also somewhat uneven. There are no January observations, and in
recent years no April, September, or December
observations, presumably because of weather conditions. Table 1 is a summary of these statistics. All 148
stations were taken by Japanese vessels; no other
nation has carried out sampling in this area.

One-degree square 02 is mostly deep water; the
depth ranges from 2000 to 5000 meters, and half of
the area is deeper than 4000 meters, according to the
detailed 1966 edition of the bathymetric chart 6302
of the Maritime Safety Agency. On the other hand,
the only station as deep as 4000 meters was made by
the *Manshu* in 1930; all the other stations have been
shallower than 2500 meters and all but five stations
are shallower than 1500 meters.

Although the North Atlantic Ocean as a whole has
more hydrographic stations than the North Pacific,
the density of stations within the Kuroshio is greater
than that in the Gulf Stream. Table 2 exhibits this
difference in total stations per five-degree square. As
can be seen, there are about five times as many
Kuroshio stations in the critical regions near the

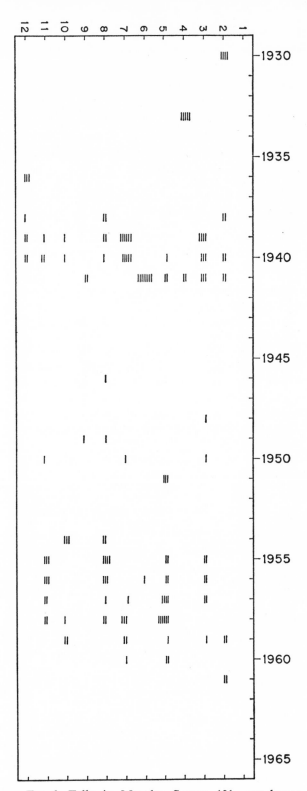

FIG. 6   Tally in Marsden Square 131, one-degree
square 02, by month and year at depths 200 meters.

---

* Statistics of square 27 are not shown here but were computed to verify that square 02 is typical and not an unusual one.

point of departure from the coast than in the case of the Gulf Stream. However, further offshore the density of stations becomes similar.

TABLE 1. Density of stations in one-degree Square 02 of Marsden Square 131.

Years	Number		Month	Number
1925–29	1		1	0
1930–34	11		2	14
1935–39	32		3	19
1940	30		4	7
1945–49	4		5	22
1950–54	11		6	11
1955–59	55		7	19
1960–65	5		8	20
			9	3
			10	10
			11	14
			12	8

Depth of Casts (m)	Number
0–500	6
500–1000	37
1000–1500	92
1500–2500	4
2500–3500	0
>3500	1

TABLE 2. Density of stations per five-degree square in the Kuroshio and Gulf Stream areas.

Gulf Stream						
40°N						
	1799	762	707	179	70	54
35°N						
	490	186	752	225	24	13
30°N						
80°W	75°W	70°W	65°W	60°W	55°W	50°W

Kuroshio						
40°N						
	5487	3808	1344	1162	77	52
35°N						
	5442	1538	207	95	45	66
30°N						
135°E	140°E	145°E	150°E	155°E	160°E	165°E

2 CHARTS SHOWING PROPERTIES AT STANDARD DEPTHS

Charts of temperature, salinity, sigma-t, and dissolved oxygen are presented in Plates 1–24. They are a selection from charts prepared for a larger number of standard depths based upon average values per one-degree square. It was decided, however, not to reproduce the deep charts because of the sparse data coverage for these depths, and because the old salinity determinations are not always very reliable.

Originally we contemplated compiling vertical sections along selected meridians across the Kuro-

shio, showing seasonal variations of properties. Preliminary compilations showed us that the amount of data available on such sections was insufficient to justify such a presentation. Sections made from individual cruises are incomparably better.

The 43,372 oceanographic stations which were utilized in the preparation of the standard depth charts comprise all available computer-processed oceanographic stations on file at the U.S. National Oceanographic Data Center (NODC) as of 1966. With few exceptions the data was processed from published or unpublished data reports containing data collected as a result of Japanese oceanographic activities. Much of the data was codified into the standard NODC format through a contractual arrangement with the Japanese Oceanographic Data Center.

A complete bibliographic index of all source documents used as data sources has been compiled by the U.S. NODC. Also compiled has been an index of the identities of all individual oceanographic stations, sorted geographically by one-degree squares with cross reference to the bibliographic index. The extreme bulk of these indexes has made it impractical to include them in the treatise; copies may, however, be obtained from NODC.

The actual data bank used for the Kuroshio study is maintained on magnetic tape. It is maintained both in the original cruise sequence and in a geographical arrangement (technically called "sort") of one-degree squares of latitude and longitude. Each station consists of various parameters at discrete observed levels, interpolated values at international standard depths, and derived values such as density (sigma-t). Almost all stations carry temperature and salinity values. Oxygen, however, is reported in general for only about 15 percent of the stations.

All data have been subjected to coarse final subjective quality control review prior to being put into the archives. Temperature and salinity values suspected of having deviated significantly from the norm of the prevailing water mass, or having an instability in the vertical density structure, have been flagged as questionable and have not been used in the preparation of the standard depth charts. A small residual error population, nevertheless, was still present in the data set but all indications are that it is within tolerable limits.

The initial plotting sheets were prepared by computer and a high-speed plotter on 20-1/2″ × 26″ work sheets. For better control in contouring, the work sheet coverage extended eastward to 160°E. All observations for each one-degree area were averaged, and the arithmetic mean (to hundredths) and the

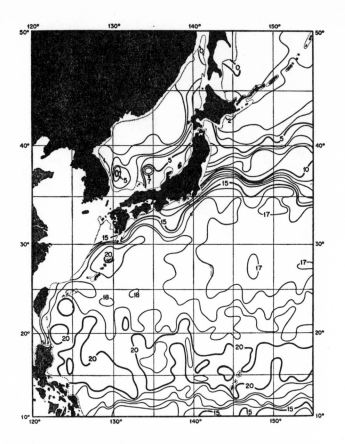

Plate 1　Temperature (°C) at 200 meters

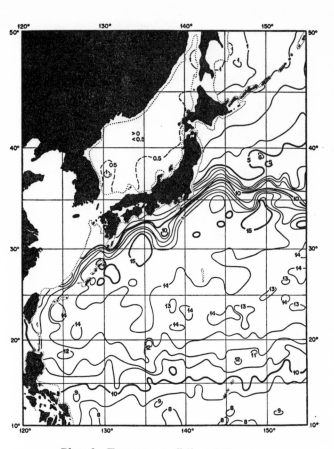

Plate 3　Temperature (°C) at 400 meters

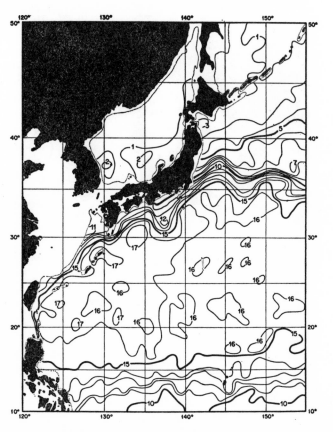

Plate 2　Temperature (°C) at 300 meters

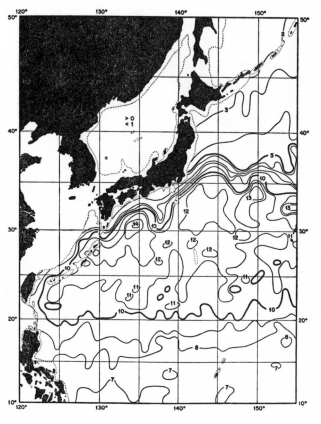

Plate 4　Temperature (°C) at 500 meters

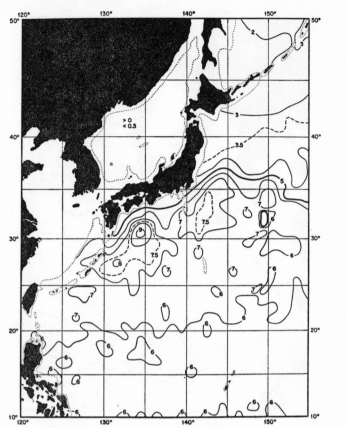

Plate 5   Temperature (°C) at 700 meters

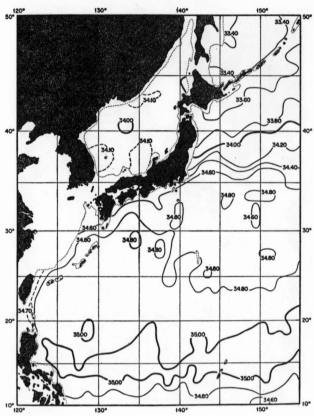

Plate 7   Salinity (‰) at 200 meters

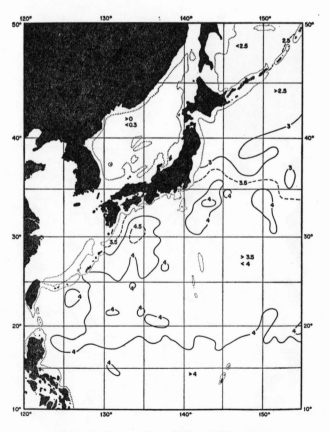

Plate 6   Temperature (°C) at 1000 meters

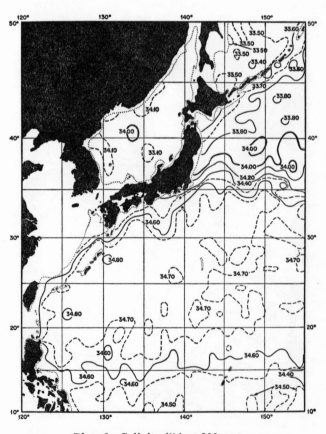

Plate 8   Salinity (‰) at 300 meters

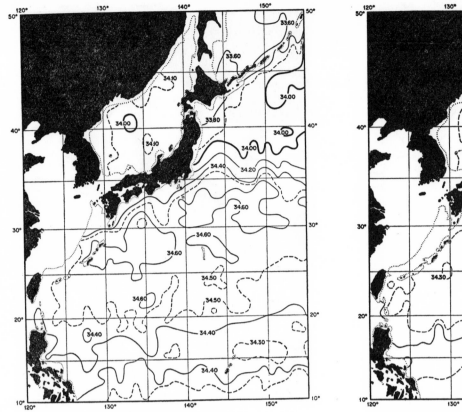

Plate 9   Salinity (‰) at 400 meters

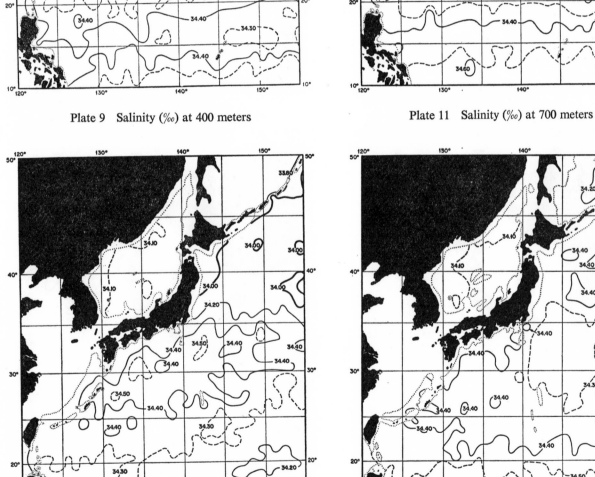

Plate 11   Salinity (‰) at 700 meters

Plate 10   Salinity (‰) at 500 meters

Plate 12   Salinity (‰) at 1000 meters

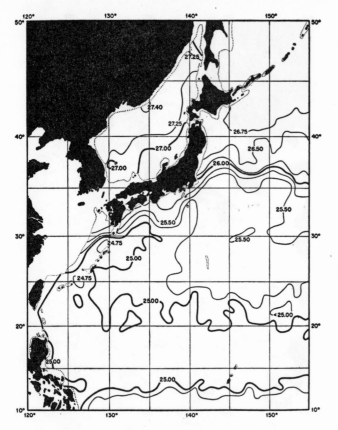

Plate 13   Density (sigma-t) at 200 meters

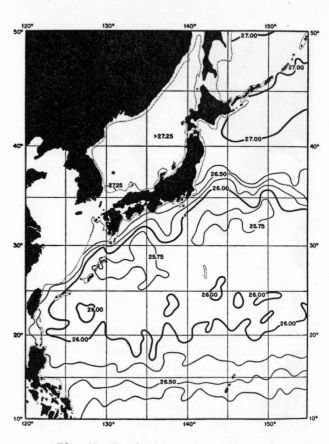

Plate 15   Density (sigma-t) at 400 meters

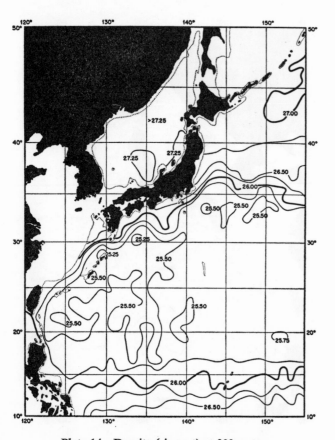

Plate 14   Density (sigma-t) at 300 meters

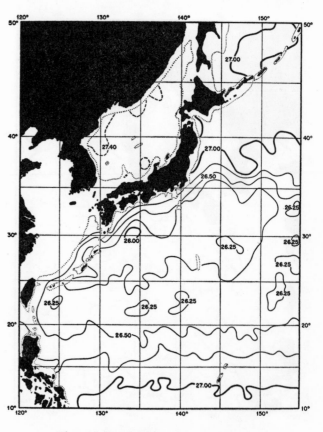

Plate 16 Density (sigma-t) at 500 meters

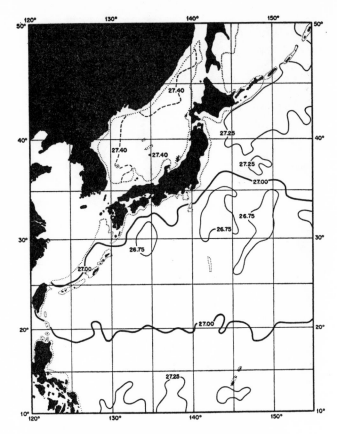

Plate 17   Density (sigma-t) at 700 meters

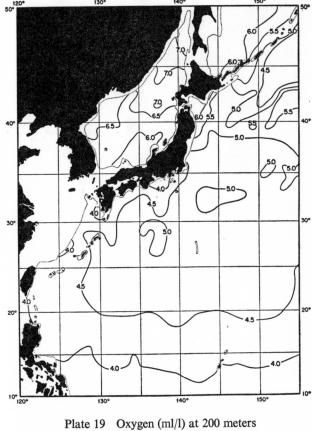

Plate 19   Oxygen (ml/l) at 200 meters

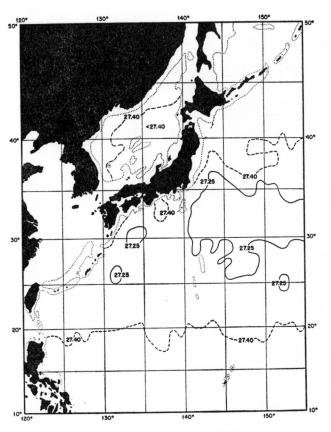

Plate 18   Density (sigma-t) at 1000 meters

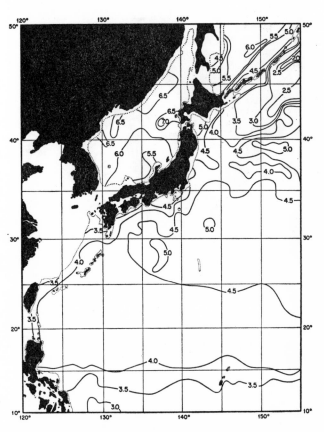

Plate 20   Oxygen (ml/l) at 300 meters

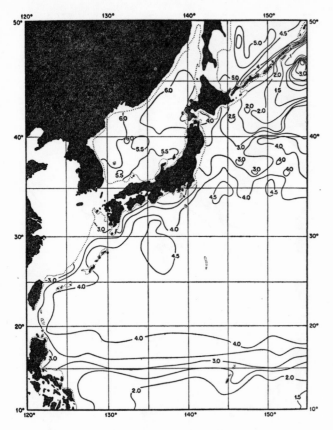

Plate 21   Oxygen (ml/l) at 400 meters

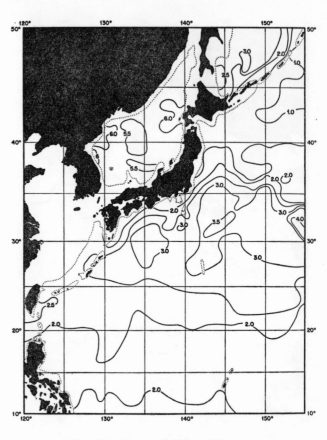

Plate 23   Oxygen (ml/l) at 700 meters

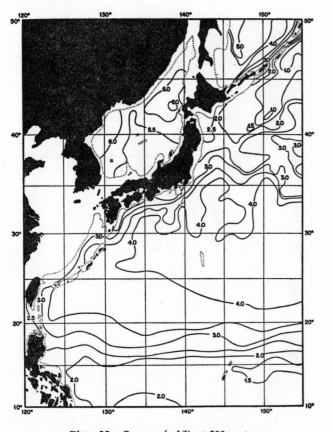

Plate 22   Oxygen (ml/l) at 500 meters

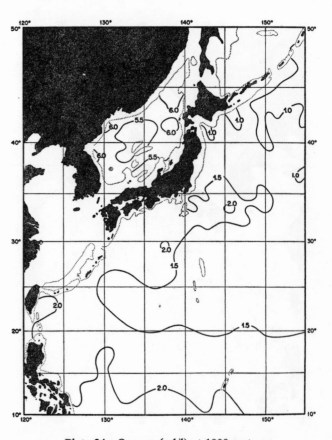

Plate 24   Oxygen (ml/l) at 1000 meters

number of observations were entered by machine. Three types of values were utilized: 1) observed values falling at standard depths; 2) values computer-interpolated to the respective standard depths from observed values using a modified 3 pt LaGrange formula; and 3) values interpolated to standard depths by the originating agency and contained in published data reports. (NOTE: Sigma-t averages were derived from the sums of the individual sigma-t values rather than computed from the average values of temperature and salinity.)

Contouring on the work sheets was done at the NODC; the isobaths ("land contours") at the respective standard depths were derived from the Maritime Safety Agency bathymetric chart 6302. The main considerations in contouring were: 1. The averages were taken to apply to the center of each one-degree square of contour. 2. Almost all values based on three or more observations were taken into account and only a small number of single observations, mainly on the salinity and oxygen charts, were considered grossly implausible and ignored. 3. With the exception of the Japan Sea where some anomalous appearing contours were verified by inspection of the source data, it was not generally possible to verify the one-degree square average values by referring to individual stations.

In the data-rich area south, southeast and east of Honshu, the contours probably depict actual long-term annual average conditions. In the data-sparse southern part of the area, however, the data is largely based on individual, generally north-south trending cruise tracks. The wavy appearance of the isopleths and some of the more pronounced north-south oriented tongues or "bubbles" may thus be more indicative of the inhomogeneity of the data in time than of a stable long-term pattern. Data coverage for oxygen is especially sparse in the easternmost and southern half of the area. Interpolation of isopleths over considerable distances was often necessary. This, coupled with the fact that oxygen values received only very cursory quality control, probably renders the oxygen contours considerably less reliable than those of the other parameters.

The greatest density of data used in the preparation of the plates is in the five-degree square 30–35N, 135–140E immediately south of Enshunada. At all depths from 200 meters to 1000 meters, there were more than 2500 stations available for temperature, salinity and density contouring. The oxygen values were approximately half as dense. Below 1000 meters, the data density falls off rapidly: thus at 2000 meters there are only 42 temperature measurements in this five-degree square. The lack of deep data is

general over the whole area of the charts, and for this reason charts are not presented for depths greater than 1000 meters. We anticipate that the results of C.S.K. will make it possible to draw better deep charts.

An example of a square with less density but still very good for contouring is the five-degree square 35–40N, 145–150E, in the region of the Kuroshio Extension near Japan. Here the temperature, salinity, and density measurements exceed 1000 down to 500 meters, and exceed 500 between 500 and 1000 meters. The oxygen values available are somewhat less than half this density.

In the Japan Sea (five-degree square 40–45N, 135–140E,) there are approximately 500 determinations of temperature, salinity and density in the upper 500 meters, and between 500 meters and 1000 meters the number is close to 150 measurements. Again the density of oxygen determinations is about one half that of temperature, etc.

The density of stations falls off rapidly as we approach the corners of the charts. Thus in the three-corner five-degree squares (the NE corner, 45–50N, 155–160E; the SE corner, 10–15N, 155–160E and the SW corner, 10–15N, 120–125E), the number of stations lies between 50 and 100 stations; and between 500 meters to 1000 meters, about half as many. The oxygen data in these corner squares are sparse between 5 and 25 stations in the NE and SE squares, at depths between 200 and 700 meters, whereas the SW corner has a slightly higher population of between 50 and 100 stations with oxygen values. At 1000 meters all corner stations have very sparse oxygen data: NE:5; SE:0; SW:10.

3   SOME FURTHER STATISTICAL FEATURES OF THE DATA

In searching for some other useful ways in which we could use compiled data we looked at various statistics of data within one-degree squares. Figure 7 shows contours of the standard deviation of temperature at 200 meters using all available data. The areas of large variability are, as might be expected, in the region of the large cold water eddy south of Enshunada and the region along the northern edge of the Kuroshio where it leaves the coast to flow eastward in the Kuroshio Extension.

We also computed histograms of temperature at selected depths for a number of squares in the region of large variability. Figure 8 shows histograms of 200-meter temperatures computed for several one-degree squares located on a line across the Kuroshio

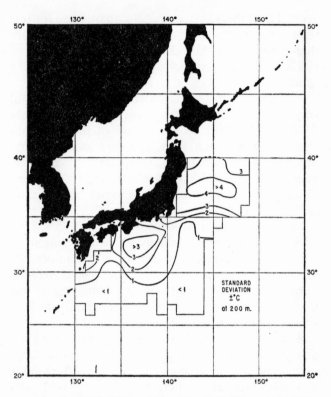

FIG. 7 Statistics on temperature at 200 meters. All available station data was used. Contours are isopleths of standard deviation in ± degrees centigrade.

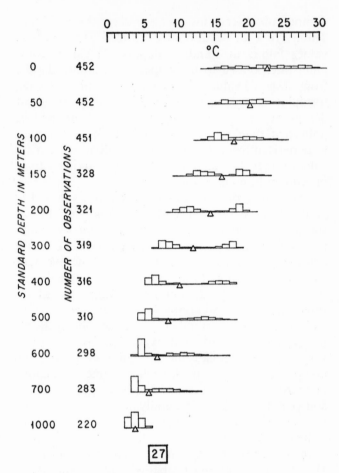

FIG. 9 Histograms of temperature in one-degree square 27, the region of the frequent cold eddy off Enshunada, at selected standard depths.

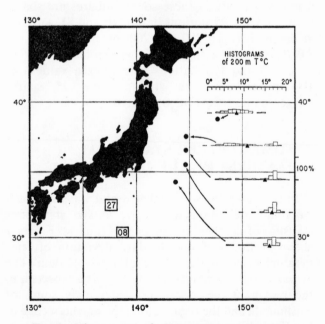

FIG. 8 Histograms of 200 meter temperature at selected one-degree squares across the Kuroshio Extension. Triangle denotes mean.

Extension. Toward the southern end of the line the distribution is centered mostly about a warm mode, but as we proceed northward cold temperatures are encountered with increasing frequency until at the northernmost degree-square the mode has shifted over to cold temperatures.

For one-degree square 27, the region of the cold eddy off Enshunada, we show in Fig. 9 a histogram of temperatures at selected depths. This histogram is quite remarkable: it shows the bimodal distribution of temperature at mid-depths corresponding to the times of presence and absence of the cold water eddy. Evidently it is a very definite on or off phenomenon, transition from one regime to the other being of short duration. For comparison, a histogram for square 08, to the south of the eddy, is also reproduced in Fig. 10. Here the histograms all have single modes. The increased spread of surface histograms is due to the seasonal effect.

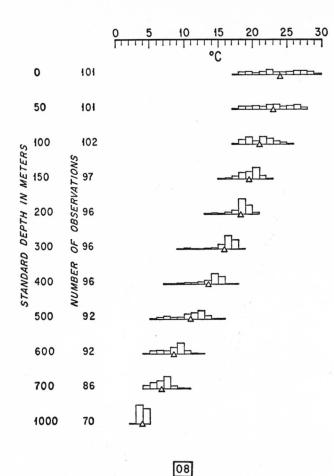

FIG. 10 Histograms of temperature in one-degree square 08, a region outside that of the cold eddy, at selected standard depths.

REFERENCES

National Oceanographic Data Center. *Reference sources for oceanographic station data.* Publication C-1, 1965. Washington, D. C.

National Oceanographic Data Center. *Inventory of archived data.* Publication C-3, 1966. Washington, D. C.

Chapter 4

WATER CHARACTERISTICS OF THE NORTH PACIFIC CENTRAL REGION

JOTARO MASUZAWA

Japan Meteorological Agency, Tokyo, Japan.

1 INTRODUCTION

The study of water characteristics of the Kuroshio and adjacent regions is not only interesting from various aspects of oceanography but is also an important source of data for meteorological and fishery activities in the North Pacific. Regional features of water characteristics of the Kuroshio, however, are fully discussed in the other chapters of this volume. In the present chapter special emphasis is placed on the gross description of water types associated with the Kuroshio in the central part of the North Pacific and of the water characteristics in the Kuroshio itself.

There are so many papers on water characteristics in the Kuroshio region, mostly by Japanese and Russians, that not all of them can be referred to in the present chapter. The inclusive bibliography in this volume is recommended for more information on the so-called water-mass analysis. The papers by Koenuma (1938, 1939) on the hydrography in the southwestern part of the North Pacific must be the first and most comprehensive study in terms of temperature and salinity. His study is based on the data collected by the *Manshu* of the Japanese Hydrographic Office from 1925 to 1927. Extensive hydrographic observations were made by the Japanese Hydrographic Office in the western North Pacific from 1933 to 1941. Unfortunately, however, the tremendous volume of data has not been studied systematically or comprehensively.

Since the early 1950's, classical observations have been repeated in the Kuroshio region near Japan. During the NORPAC Expedition (1955), the International Geophysical Year (IGY, 1957–1959), and the Cooperative Study of the Kuroshio (CSK, 1965—), extensive surveys have been made of the Kuroshio region. Descriptions in this chapter are based mostly upon data taken during these surveys. However, no overall statistical treatment has been tried in this chapter. The chapter by Stommel and Winterfeld in this volume is the first statistical representation of the existing data before the CSK. Figure 1 is an inventory chart showing the locations of important stations and sections referred to in figures in this chapter. Table 1 is a list of cruises from which data used in figures in this chapter were obtained.

Discussions cover not only the warm surface waters which are more or less directly influenced by atmospheric processes, but also the intermediate or mid-depth waters such as the salinity-minimum water and oxyty*-minimum water below the main thermocline. However, descriptions are not extended to water deeper than 1000 to 1500 m in the present chapter. The deep sphere of the North Pacific is occupied by the homogeneous Common Water (Montgomery, 1958). Details of the deep water in the western North Pacific are discussed in the chapter by Moriyasu in this volume.

First, the gross features of the North Pacific oceanography are roughly discussed in Sec. 2. Oceanic circulation (Sec. 2.1) and hydrographic conditions (Sec. 2.2) at the sea surface over the whole North Pacific are presented as background for understanding the original forms of water types in the central part of the North Pacific. The vertical distribution of water characteristics or types is demonstrated by two graphical methods: long meridional sections (Sec. 2.3) and selected station graphs (Sec. 2.4).

In Sec. 3, five important water types (Sec. 3.2 to 3.6) in the North Pacific Central Region are selected by vertical distributions of water properties and by the volume distribution by characteristics (Sec. 3.1). The water types are described chiefly by the isanosteric method.

---

* Concentration of dissolved oxygen (Montgomery, 1969).

TABLE 1. List of cruises mentioned in figures.

Vessel	Date	Expedition	Figures	Source
Atlantis	Aug. 1965	CSK	10, 12, 13, 14, 18, 19	DR-CSK, **20**, 1966
Black Douglas	Aug.–Sep. 1955	Norpac	10, 12, 14	OOP-Norpac, 1960
Bushnell	Aug. 1934	—	11	OOP-Pre1949, 1961
Charles H. Gilbert	Sep. 1954	—	10, 12, 13, 14, 18, 19	OOP-1954, 1965
Chikubu Maru	Sep. 1951	Tango	23	RMMOO, **10**, 1952
Chofu Maru	July–Sep. 1965	CSK	10, 12, 13, 14, 18	DR-CSK, **12, 26**, 1966
	July–Sep. 1966	CSK	19	DR-CSK, **54, 61**, 1967
Horizon	Aug. 1955	Norpac	10, 12, 14	OOP-Norpac, 1960
	Oct.–Dec. 1955	Eastropic	10, 12, 14	OOP-1955, 1962
	Aug. 1956	Equapac	10, 12, 13, 14	IGY-WDCA, 1961
	Oct. 1957	Downwind	10, 12, 14	OOP-1957, 1965
Hugh M. Smith	July 1950	Cr. 5	10, 11, 12, 13, 14, 19	OOP-1950, 1960
	Sep.–Oct. 1951	Cr. 11	10, 12, 14	OOP-1955, 1962
	Feb. 1952	Cr. 14	10, 12, 14	OOP–1952, 1965
	May 1952	Cr. 15	10, 12, 14	OOP-1952, 1965
	Jan. 1955	Cr. 27	11	OOP-1955, 1962
	July–Aug. 1955	Norpac	10, 11, 13, 14, 18, 19	OOP-Norpac, 1960
	Aug. 1956	Equapac	10, 12, 13, 14	IGY-WDCA, 1961
Kaiyo	Aug. 1965	CSK	10, 11, 12, 13, 14, 18, 19	DR-CSK, **3**, 1966
Komahashi	July 1939	—	10, 12, 13, 14	HB, Sup., **16**, 1955
Kyo Maru VI	Aug. 1939	—	11	HB, **69**, 1962
Meiyo	Feb. 1955	—	23	HB, **58**, 1959
Oshoro Maru	Sep. 1957	IGY	10, 12, 13, 14	DR-HU, **3**, 1959
Ryofu Maru	Aug. 1953	Tango	23	RMMOO, **13**, 1954
	Sep. 1957	IGY	10, 12, 14	RMMOO, **22**, 1958
	July–Sep. 1958	IGY	8, 9, 10, 11, 12, 13, 14, 18, 19	RMMOO, **24**, 1959
	Nov. 1959	—	17	RMMOO, **26**, 1960
	May 1965	JEDS-10	17	RMMOO, **37,**, 1968
	July 1965	CSK	10, 12, 13, 14, 18, 19	DR-CSK, **10**, 1966
	Jan.–Feb. 1967	CSK	7, 9, 17	DR-CSK, **82**, 1968
Satuma	Aug. 1955	Norpac	10, 11, 12, 13, 14, 18	OOP-Norpac, 1960
	July–Aug. 1956	Equapac	10, 11, 12, 13, 14	IGY-WDCA, 1961
	Jan.–Feb. 1958	IGY	10, 12, 13, 14	HB, **66**, 1961
Shiga Maru	Mar. 1953	Tango	23	RMMOO, **13**, 1954
Shinyo Maru	July 1965	CSK	10, 12, 13, 14	DR-CSK, **5**, 1966
Shumpu Maru	Mar. 1955	—	23	RMMOO, **16**, 1955
	Aug. 1955	Norpac	11, 23	OOP-Norpac, 1960
Soyo Maru	July–Aug. 1958	IGY	10, 11, 12, 13, 14, 19	TRFRL, 1959
Spencer F. Baird	Nov. 1953	Transpac	11	OOP-1953, 1965
	Aug.–Sep. 1955	Norpac	10, 12, 14	OOP-Norpac, 1960
	Oct. 1955	Eastropic	10, 12, 14	OOP-1955, 1962
	Aug. 1956	Chinook	11	OOP-1956, 1963
Stranger	Aug.–Sep. 1955	Norpac	10, 12, 14	OOP-Norpac, 1960
	Aug.–Sep. 1956	Equapac	10, 12, 13, 14	IGY-WDCA, 1961
Takuyo	Oct. 1957	IGY	10, 12, 13, 14	HB, **64**, 1960
	Feb.–Mar. 1958	IGY	10, 11, 12, 13, 14	HB, **66**, 1961
	July 1958	IGY	8, 10, 11, 12, 13, 14	HB, **66**, 1961
	June–Aug. 1965	CSK	10, 11, 12, 13, 14, 18	DR-CSK, **2**, 1966
	July–Aug. 1967	CSK	19	DR-CSK, **105**, 1968
Umitaka Maru	Oct. 1955	Norpac	11	OOP-Norpac, 1960
	Sep. 1957	IGY	10, 12, 13, 14	IGY-DR
	Aug. 1958	IGY	11	IGY-DR
	Aug. 1965	CSK	10, 11, 12, 13, 14	DR-CSK, **4**, 1966
Vityaz	Aug.–Sep. 1957	Cr. 25	10, 11, 12, 13, 14	IGY-DR
	Feb. 1958	Cr. 26	10, 11, 12, 13, 14	IGY-DR
	Feb. 1959	Cr. 29	10, 11, 12, 13, 14	IGY-DR
Zhymchug	Sep. 1965	CSK	10, 12, 13, 14	DR-CSK, **24**, 1966
Weather ships	June 1950–Nov. 1953	Tango	20, 21, 22	RMMOO, **8–13**, 1952–54

DR-CSK: Preliminary Data Report of CSK, Kuroshio Data Center, Maritime Safety Agency, Tokyo.
DR-HU: Data Records of Oceanographic Observations and Exploratory Fishing, Fac. Fish., Hokkaido Univ.
HB: Hydrographic Bulletin, Maritime Safety Agency, Tokyo.
IGY-DR: Unpublished Data Report for IGY.
IGY-WDCA: Oceanographic Observations in the Intertropical Region of the World Ocean During IGY and IGC, Dept. Oceanogr. Meteor., Texas A. & M., Texas, U.S.A.
OOP: Oceanic Observations of the Pacific, Univ. Calif. Press, Berkeley and Los Angeles.
RMMOO: The Results of Marine Meteorological and Oceanographical Observations, Japan Meteorological Agency, Tokyo.
TRFRL: IGY, Physical and Chemical Data by the *Soyo Maru*, Tokai Regional Fisheries Research Laboratory, Tokyo.

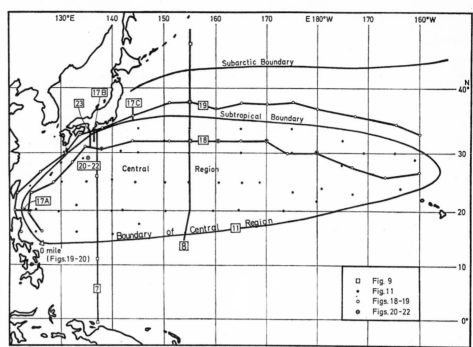

FIG. 1 Index map of stations and sections referred to in this chapter. The numbers in squares indicate figure numbers.

The last section (Sec. 4) deals with the characteristics of water carried by the Kuroshio and with their changes along the path of the Kuroshio. Examples are shown of seasonal variations of water characteristics at representative stations and sections in the Kuroshio region.

Acknowledgments

The author wishes to express his greatest appreciation to Dr. R. B. Montgomery, Johns Hopkins University, for his useful suggestions and comments. The author is also indebted to Dr. M. Tsuchiya, Scripps Institution of Oceanography, and Dr. K. Yoshida, University of Tokyo, for their comments.

2   GROSS HYDROGRAPHIC FEATURES OF THE WESTERN NORTH PACIFIC

2.1   Surface oceanic circulation

The large-scale oceanic circulation at the sea surface in the North Pacific is outlined in the map of geopotential topography by Reid (1961), or in the mpa of average geopotential topography by Burkov and Pavlova (1963). Figure 2 has been reproduced with some modifications from the latter map.

Much of the subtropical and tropical ocean is enclosed by the anticyclonic oceanic gyre, which is composed of the North Equatorial Current on the south, the Kuroshio on the west, the Kuroshio Ex-

tension and the North Pacific Current on the north and the California Current on the east. The anticyclonic gyre has been conventionally referred to as the subtropical gyre (Sverdrup, Johnson and Fleming, 1942). The subtropical gyre exhibits westward intensification and a broad eastern equatorward flow.

The Kuroshio begins east of northern Luzon as a swift and narrow segment of the western boundary current of the subtropical gyre. It flows close to the east coast of Taiwan and then into the East China Sea through the shallow strait (sill depth 800 m; 25°N) between Taiwan and the Ryukyu Islands. The portion of the Kuroshio east of Taiwan is called the Taiwan (Formosa) Current by Wyrtki (1961). In the East China Sea, the Kuroshio follows the submarine trough between the continental shelf and the Ryukyu Ridge. The current splits southwest of Kyushu into two parts; the major part passes through Tokara Strait (29°N) between Yakushima Island and Amamioshima Island into the sea south of Shikoku, and the minor branch flows north as the Tsushima Current west of Kyushu through the Korea Strait into the Japan Sea.

The Kuroshio south of Japan normally flows northeastward close to the continental slope, though it occasionally makes a detour south of Honshu. Crossing the Izu-Ogasawara Ridge approximately at 140°E, roughly 500 to 1500 m deep, the Kuroshio leaves the Japanese coast at Cape Inubozaki (36°N), and flows generally east in a remarkable meandering path. The Kuroshio east of Japan has often been

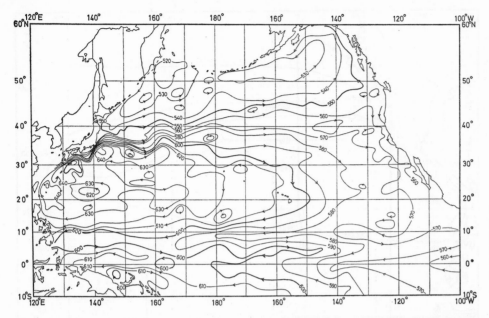

FIG. 2 Climatic geostrophic circulation at the sea surface in summer, based on 1500 db; reproduced, with modifications, from Burkov and Pavlova (1963). Contour interval is 10 dynamic centimeters.

called the Kuroshio Extension (Sverdrup, Johnson and Fleming, 1942; Kawai, 1955). The fastest sector of the Kuroshio, 3 to 5 kn, is found south and east of Honshu between 136°E and 147°E.

At longitudes of 160°E or so, the Kuroshio Extension transforms from a concentrated narrow current into the weak, broad North Pacific Current. The southern part of the North Pacific Current flows east-southeast, and much of this does not reach the coast of California but turns south just east of Hawaii. The northern part of the North Pacific Current flows east or east-northeast nearly to the coast of North America, and some turns southeast to form the California Current.

The North Equatorial Current is fed not only by the California Current but also by southward transports both east and west of Hawaii. Thus, the North Equatorial Current increases in flux downstream from east to west (Reid, 1948; Masuzawa, 1964a). East of the Philippines the North Equatorial Current splits into two branches: the northward branch is the embryo of the Kuroshio, and the southward branch, the Mindanao Current (Wyrtki, 1961; Cannon, 1969; Masuzawa, 1969a), is one of the important sources of the Equatorial Countercurrent.

There are weak currents and eddies in the tropical and subtropical region surrounded by the anticyclonic subtropical gyre. The region is referred to as the Central Region in Sec. 3 in this paper. Although the current pattern seems to be irregular in the region, careful examination reveals a semi-systematic eastward current, which was first described and named by Yoshida and Kidokoro (1967), the Subtropical

Countercurrent. The climatic current chart shown in Fig. 2, however, does not depict the Subtropical Countercurrent as a continuous current, but a train of eastward currents from 20°N, 135°E to 30°N, 165°W. Through these weak currents or eddies inside the subtropical gyre, some water is carried southward from the Kuroshio Extension and the North Pacific Current to join the North Equatorial Current.

The subarctic region north of the subtropical gyre is occupied by the cyclonic subarctic gyre, which seems to be separated into eastern and western circulations. The Oyashio, the western boundary current of the subarctic gyre, flows south along the Kurile Islands and Hokkaido and encounters the Kuroshio east of Japan. The eastward current forming the southern portion of the subarctic gyre is called the Subarctic Current (Sverdrup, Johnson and Fleming, 1942), and much of it turns north to form the Alaska Current off Canada. The Alaska Current flows westward from the Gulf of Alaska along the Aleutian Islands, the current being sometimes referred to as the Alaskan Stream (Dodimead, Favorite and Hirano, 1963; Favorite, 1967; Kitano, 1967).

In the equatorial region south of the anticyclonic gyre, on the other hand, there are long zonal currents: the eastward Equatorial Countercurrent, the westward South Equatorial Current and the eastward Equatorial Undercurrent. The equatorial currents north of New Guinea show complicated spatial and temporal variations (Tsuchiya, 1961).

There are various scales of lateral mixing between the subarctic gyre and the subtropical gyre and between the subtropical gyre and the equatorial currents, not only in the western and eastern boundary regions but also in the interior regions.

2.2   Sea-surface oceanographic conditions

The original characteristics of different water types in the subsurface or great depths are formed at the sea surface in different climatic conditions. Hence, in this section, both the temperature and salinity at the sea surface and the heat exchange across the sea surface are briefly summarized for the whole North Pacific Ocean.

Temperature (Fig. 3)

In winter, much of the Okhotsk Sea and Bering Sea is covered by sea ice, and the lowest temperature, below 0°C, is recorded in these seas. On the other hand, the low-latitude zone south of 20°N shows temperatures over 25°C west of 160°W. The meridional gradient of sea-surface temperature is greater at mid latitudes of the western North Pacific in winter than in summer. This great-gradient zone, however, is separated into two parts by a nearly isothermal belt at 30° to 35°N. The mean temperature of this isothermal zone is roughly 18°C. The part with the rather strong gradient at 20° to 30°N latitude disappears and is replaced by homogeneous water warmer than 25°C in summer.

The eastern mid-latitudes region, however, is less heated in summer than the western half. This may be due to the stronger upwelling in the California Current region in summer. Consequently, the isotherms in the central North Pacific are less zonal in summer than in winter.

The annual range of sea-surface temperature is as much as 4° to 11°C in the Kuroshio region north of 20°N (Masuzawa, Tsuchida and Inoue, 1962; Wyrtki, 1965b). This wide range is closely associated with severe winter cooling due to both high evaporation and large loss of sensible heat.

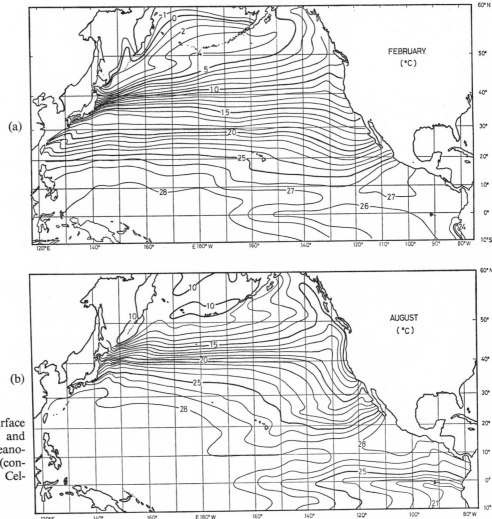

FIG. 3 Average sea-surface temperature in February and August. U. S. Naval Oceanographic Office, SP-123, 1969 (converted from Fahrenheit to Celsius).

Salinity (Fig. 4)

The salinity of the North Pacific Ocean is the lowest of any ocean. The maximum salinity in the tropical or subtropical region is slightly more than 35.5‰. The saline water of the North Pacific is separated from the South Pacific saline water by the low-salinity tongue stretching all the way across the Pacific Ocean from Central America to the Philippines. This low-salinity water is sometimes called the Equatorial Surface Water.

Heat flux through the sea surface

Wyrtki (1965a) made a computation of the annual heat flux through the sea surface in the North Pacific. His computation indicates that the average amount of heat used for evaporation from the sea surface over the whole North Pacific is as much as 198 cal cm^{-2} day^{-1}, which is equivalent to the amount of evaporation of 123 cm/yr. The annual average of evaporation heat (Fig. 5) is less than 100 cal cm^{-2} day^{-1} (evaporation 62 cm/yr) in high latitudes and in the California Current region. Evaporation exceeding 156 cm/yr (over 250 cal cm^{-2} day^{-1}) takes place in the tropical region and in the Kuroshio region. This is one of the important factors for the formation of the saline Tropical Water (Sec. 3.2) in the Central Region (Sec. 3.1). The remarkable evaporation in the Kuroshio region is chiefly due to the dryness and the high wind speed of the Northwest Monsoon in winter (Hanzawa, 1950; Masuzawa, 1952; Terada and Osawa, 1953).

Large heat gain by the ocean is found in the equatorial zone, in the tropical region and in the eastern boundary region (Fig. 6). On the other hand, the Kuroshio region shows a big heat loss from the ocean. This is due to high evaporation and large flux of sensible heat from sea to air, particularly during the winter monsoon. Thus, the North Pacific Ocean receives heat from the atmosphere in the eastern boundary region and in the equatorial and tropical region, much of the heat received is transported by

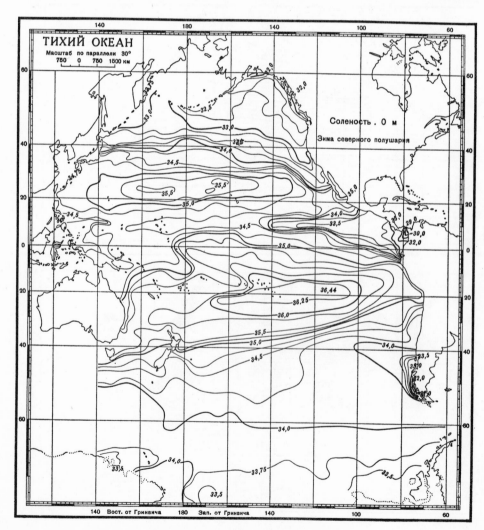

FIG. 4 Average sea-surface salinity (‰) of the norther Pacific in winter (Muromtsev, 1963).

the North Equatorial Current and the Kuroshio, and most of the heat transported is lost toward the atmosphere in the Kuroshio region. According to Wyrtki's calculation, the North Pacific Ocean as a whole gains heat through the sea surface at the rate of 3×10^{14} cal/sec (34 cal cm^{-2} day^{-1}), which is approximately 10% of the incoming radiation from sun and sky.

2.3   Meridional hydrographic sections

To demonstrate vertical distributions of water properties in the western North Pacific Ocean, two long meridional sections (Fig. 1) were chosen; a winter section at 137°E in the Philippine Sea and a summer section at 155°E in the Northwest Pacific Basin. Each section consists of (a) geopotential profiles of selected isobaric surface; (b) temperature against depth; (c) thermosteric anomaly* against depth; (d) salinity

and oxyty against thermosteric anomaly on a logarithmic scale. The representation of (d) (Masuzawa, 1967) is helpful for isanosteric* examination of water types in the vertical section. The choice of a logarithmic scale of thermosteric anomaly was done to make the distribution with respect to thermosteric anomaly resemble the distribution with respect to depth, because thermosteric anomaly roughly decreases exponentially with depth. This representation is used in this chapter for isentropic analysis as well as maps of isanosteric surfaces (Sec. 3) and station graphs (Sec. 2.2).

The 137°E section (Fig. 7)

The section at 137°E was made by the *Ryofu Maru* of the Japan Meteorological Agency in January 1967. The section extends from the coast of Japan to 1°S, and thus crosses the mature Kuroshio south of Japan

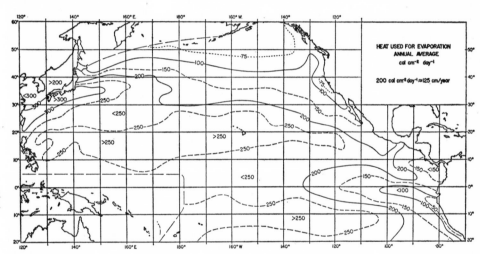

FIG. 5   Annual average latent heat of evaporation, cal cm^{-2} day^{-1} (Wyrtki, 1965).

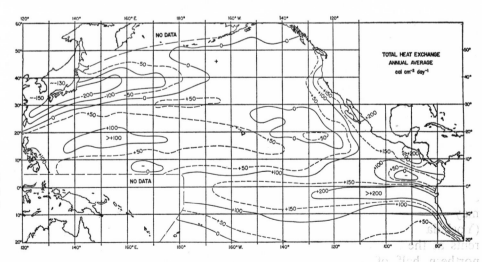

FIG. 6   Annual average heat flux, cal cm^{-2} day^{-1}, through the sea surface; positive indicates the gain by the sea (Wyrtki, 1965).

---

* Thermosteric anomaly is the specific-volume anomaly without pressure-dependent terms; the symbol used is σ_T; the unit chosen is centiliter per ton (cl/t). *Isanosteric* means uniform in specific-volume anomaly or thermosteric anomaly (Montgomery and Wooster, 1954).

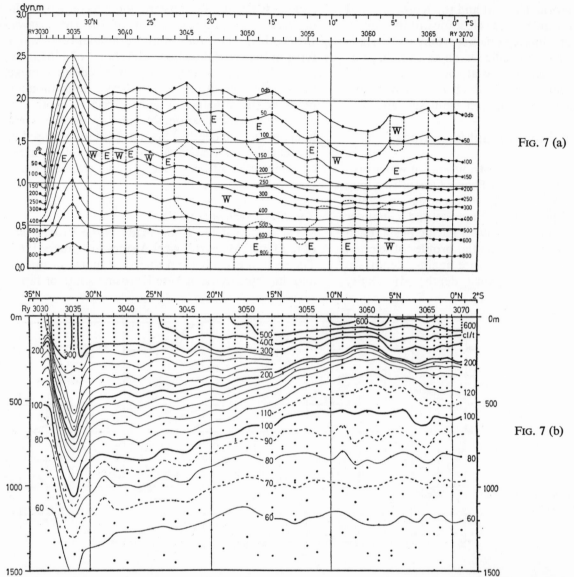

FIG. 7 (a)

FIG. 7 (b)

and the whole system of equatorial currents north of New Guinea. This section was described in detail by Masuzawa (1967).

The meridional distribution of geopotential anomaly relative to the 1000-db* surface may be used to define the limits of zonal currents through the section (Fig. 7a). The northward downslope at the northernmost portion of the section corresponds to the eastward Kuroshio. In the subtropical zone between 30°N and 20°N there appear to be weak currents alternately eastward and westward, though a rather strong eastward current is seen at 22 to 24°N, which may correspond to the Subtropical Countercurrent (Yoshida and Kidokoro, 1967). The equatorial currents at the section consist of four zones: (1) the northern half of the North Equatorial Current

between 22°N and 15°N, where the core of the westward current is seen at depths of 200 to 300 m; (2) the southern half of the North Equatorial Current between 15°N and 7°N, where the velocity is greatest at the sea surface and decreases with depth; (3) the Equatorial Countercurrent at 7 to 2°N; (4) the current zone south of 2°N where geostrophic currents are not clear, but direct current measurements suggest the existence of the Equatorial Undercurrent and of the South Equatorial Current, which was covered by the shallow eastward flow of the New Guinea Coastal Current (Masuzawa, 1967 and 1968). Wyrtki (1961) suggested that the New Guinea Coastal Current disappeared and was replaced by the South Equatorial Current in seasons other than northern winter.

*   Decibar, 10⁵ dyne cm⁻².

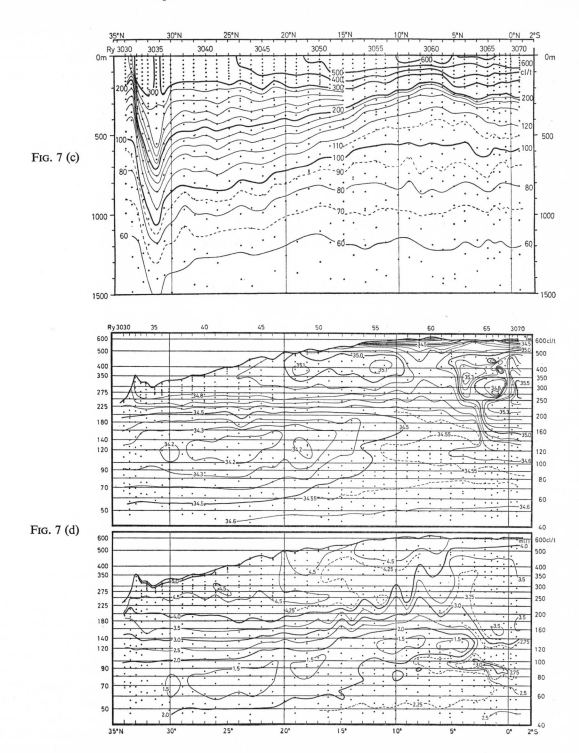

FIG. 7 (c)

FIG. 7 (d)

FIG. 7   Meridional sections at 137°E, from Japan to New Guinea (see Fig. 1), made by the *Ryofu Maru*, January 1967 (Masuzawa, 1967): (a) geopotential anomaly relative to 1000 db; (b) temperature; (c) thermosteric anomaly; (d) salinity and oxyty against thermosteric anomaly on logarithmic scale; vertical exaggeration for (b) and (c) 1480.

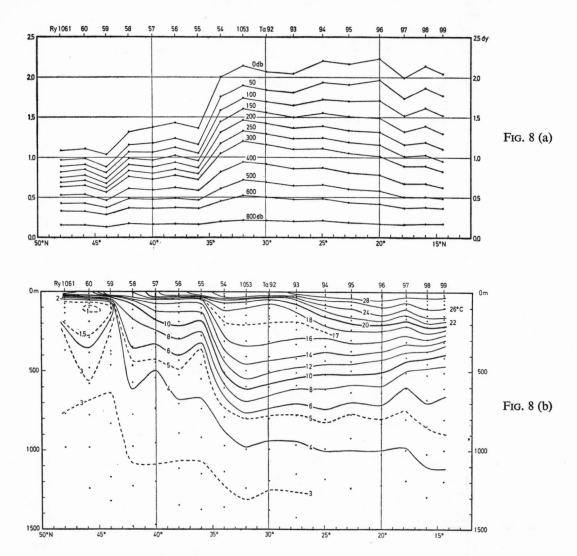

FIG. 8 (a)

FIG. 8 (b)

The main thermocline is deep at the southern edge of the Kuroshio and gradually shoals southward from 30°N to 15°N (Fig. 7b). In the vicinity of 15°N, the temperature at the core of the thermocline changes from 12°C on the north to 20°C on the south. At the boundary between the North Equatorial Current and the Equatorial Countercurrent, the thermocline is shallowest and the vertical gradient of temperature in the thermocline is largest.

The saline water with salinity exceeding 35‰ and with thermosteric anomaly 350 to 500 cl/t lies in the upper part (50 to 150 m) of the thermocline in the zone from 10°N to 20°N (Fig. 7d). The other saline water, originating in the tropical South Pacific, extends as far north as the northern edge of the Equatorial Countercurrent. These saline waters are the Tropical Waters defined by Sturges (1965) and Cannon (1966). The Tropical Water of South Pacific origin is more saline than the North Pacific Tropical Water and has lower thermosteric anomaly at the core.

The major salinity minimum can be traced from the coast of Japan beyond the southern edge of the North Equatorial Current as far south as 3°N. The minimum, however, lies on the isanosteric surface of 125 cl/t north of 15°N and on the isanosteric surface of 170 cl/t or so south of 15°N. The water forming the salinity minimum north of 15°N is commonly called the North Pacific Intermediate Water (Uda, 1935a; Sverdrup, Johnson and Fleming, 1942; Reid, 1965). The salinity-minimum water centered at 170 cl/t south of 15°N is different in origin from the North Pacific Intermediate Water (see Sec. 3). The deep salinity minimum at 80 cl/t in thermosteric anomaly is formed by the South Pacific Intermediate Water (Koenuma, 1939; Rochford, 1960; Wyrtki, 1961; Reid, 1965; Masuzawa, 1967) and can be detected as far north as 13°N, so that two salinity minima are revealed between 13°N and 3°N. The distribution of salinity in the Equatorial Countercurrent and Undercurrent is complicated and interesting (Masuzawa,

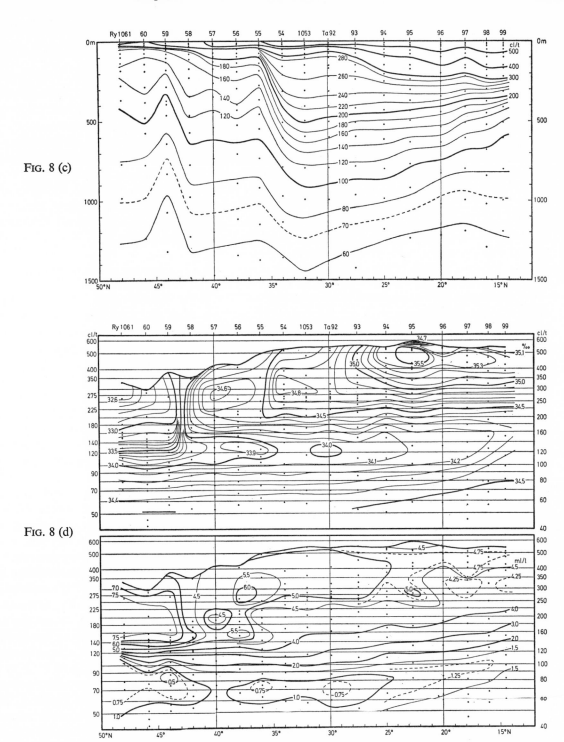

FIG. 8 (c)

FIG. 8 (d)

FIG. 8   Meridional sections at 155°E (see Fig. 1), made by the *Ryofu Maru* (Ry) and *Takuyo* (Ta), July to September 1958:
(a) geopotential anomaly relative to 1000 db; (b) temperature; (c) thermosteric anomaly; (d) salinity and oxyty against thermosteric anomaly on logarithmic scale; vertical exaggeration for (b) and (c) 1480.

1967, 1968). A detailed discussion of these currents, however, is out of the scope of the present chapter.

A pronounced minimum of oxyty is found below the thermocline from the Japanese coast to 3°N. The minimum changes in depth around 15°N where the thermosteric anomaly at the minimum jumps from 80 cl/t on the north to 124 cl/t on the south. A faint oxyty minimum is seen in the deep layer between 60 cl/t and 90 cl/t south of 13°N. Thus, there are two oxyty minima between 13°N and 3°N as well as two salinity minima.

The waters forming the major salinity and oxyty minima are transported to the west by the North Equatorial Current and then to the east by the Equatorial Countercurrent, and probably by the Undercurrent, too (Reid, 1965; Masuzawa, 1967; Tsuchiya, 1968).

The 155°E section (Fig. 8)

The section at 155°E is a composite section; the northern half was made by the *Ryofu Maru* of the Japan Meteorological Agency and the southern half by the *Takuyo* of the Japanese Hydrographic Office in the summer of 1958 (IGY). The section crosses the Oyashio region south of the Kurile Islands and the Kuroshio Extension east of Japan in the Northwest Pacific Basin. The station spacing is not so dense as in the 137°E section, so details of the 155°E section cannot be discussed.

This section is divided into two portions by the steep slope of the sea surface at 35°N (Fig. 8a). This steep slope corresponds to the Kuroshio Extension and may be referred to as the Subtropical Boundary. South of the Subtropical Boundary at 155°E, the Tropical Water is more saline, the Intermediate Water is less saline, and the Oxyty-minimum Water is less saline, and the Oxyty-minimum Water is lower in oxyty than at 137°E. It is also noted that there is a well-defined, nearly isothermal layer centered at 17°C, the Subtropical Mode Water (Sec. 3.3), and that the shallow eastward current between Ta94 (25°N) and Ta93 (27°30′N) might be the Subtropical Countercurrent.

North of the Subtropical Boundary, on the other hand, the water characteristics take on a different pattern. Another boundary is seen in the upper water at 43°N, where the Subarctic Current or the northern part of the North Pacific Current flows east. The boundary was called the Subarctic Boundary by Dodimead, Favorite and Hirano (1963). The zone between the Subarctic Boundary at 43°N and the Subtropical Boundary at 35°N is the transition region between the subarctic water and the subtropical water. That is, the upper 200- or 300-m layer in the transition region is occupied by diluted subtropical water with higher temperature and salinity, while the underlying layer is occupied by fresher Intermediate Water, which originates in the subarctic region and is centered at the depth of 400 m in the transition region (Kawai, 1955; Dodimead, Favorite and Hirano, 1963; Reid, 1965).

North of the Subarctic Boundary, the upper 200 or 300 m of water has lowest temperature with a vertical minimum (Uda, 1935b, 1963), lowest salinity and highest oxyty. Salinity increases downward all the way without a vertical minimum. Oxyty decreases downward to a pronounced minimum at 80 cl/t, less than 0.5 ml/l.

2.4   Selected station graphs

The station graph proposed by Montgomery (1954) is used for demonstrating vertical distributions of temperature, salinity and oxyty at the six stations chosen in the western North Pacific. Three of these six stations lie on the 137°E section and the other three stations on the 155°E section (Fig. 1). Each station graph was constructed of three curves having depth, salinity and oxyty drawn against temperature as common abscissa (Fig. 9).

Station Ry3070 is near the coast of New Guinea and is influenced by waters from the South Pacific. The surface mixed layer, shallower than 70 m, is relatively low in salinity and high in oxyty. The mixed water flows eastward in the New Guinea Coastal Current in the northern winter (Wyrtki, 1961; Masuzawa, 1967). In the thermocline between 100 m and 300 m there is saline water from the South Pacific (South Pacific Tropical Water), and the oxyty is fairly uniform, 3.5 ml/l. The salinity maximum (35.5 ‰) at the center of the thermocline lies near the 300-cl/t surface and is associated with a faint minimum of oxyty. Another oxyty minimum is seen near the 125-cl/l surface. This minimum seems to be connected with the pronounced oxyty minimum in the southern part of the North Equatorial Current and the Equatorial Countercurrent. The slight minimum of salinity near 80 cl/t is formed by the less saline water from the South Pacific (South Pacific Intermediate Water).

At Station Ry3056, in the southern part of the North Equatorial Current, the thermocline is indicated by the straight line between 400 cl/t and 180 cl/t on the temperature-salinity diagram. The transition from this North Pacific Thermocline Water (Sec. 3.3) to the deeper water of South Pacific origin occurs between the salinity minimum at 180 cl/t and the faint salinity maximum near 120 cl/t. This salinity maxi-

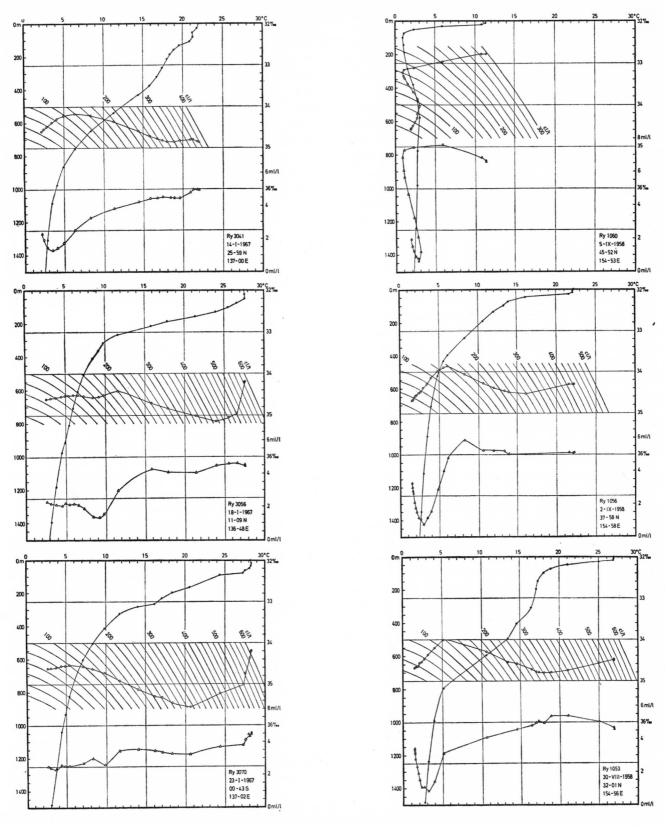

FIG. 9   Station graphs for six selected stations made by the *Ryofu Maru*, three on the 137°E section and three on the 155°E section (see Fig. 1). Depth, salinity and oxyty are plotted against temperature. Family of curves shows isanosteric lines on temperature-salinity diagram.

mum coincides with a remarkable oxyty minimum. It will be mentioned in Sec. 3 that both the salinity minimum and the oxyty minimum in the equatorial region are different both in thermosteric anomaly and in origin from the minima in the central North Pacific Ocean.

The station graph at Ry3041 is one of the representative graphs in the subtropical region of the Philippine Sea. The layer at 18°C with a small vertical gradient of temperature is separated by a small thermocline at a depth of 125 m from the surface-mixed layer. The 18°C layer is a thermostad, antonym of thermocline (Seitz, 1967), and is formed by the Subtropical Mode Water (Sec. 3.3). The salinity minimum at 125 cl/t and the oxyty minimum at 70 cl/t show the representative values: 34.2‰, 1.5 ml/l, respectively.

The station curves at Ry1053 in the subtropical region east of the Izu-Ogasawara Ridge are similar to those at Ry3041. The typical thermostad formed by the Subtropical Mode Water is separated by a pronounced seasonal thermocline from the thin surface-mixed layer, because the station was made in summer. This thermostad is characterized by 17°C, 34.8‰, 5 ml/l and 270 cl/t. The salinity minimum at 120 cl/t and the oxyty minimum at 70 cl/t are lower, both in salinity and in oxyty, at this station than at Ry3041.

Station Ry1056 is located in the transition zone between the Subtropical Boundary and the Subarctic Boundary. The salinity maximum (34.5‰) at 250 cl/t is seen at the bottom of the pronounced seasonal thermocline. This maximum is formed by the diluted Tropical Water or Subtropical Mode Water. Below this salinity maximum, the main thermocline between 14°C and 6°C is the same in temperature-salinity relation at Stations Ry1053 and 1056, but is shallower by 400 m than at Ry1053. The salinity minimum is lower in salinity, 33.8‰, and higher in thermosteric

anomaly, 140 cl/t, than the minimum at Ry1053. The oxyty minimum, less than 1 ml/l, is similar to that at Ry1053.

The station graph at Ry1060, which is situated in the subarctic water, is quite different from the other five graphs. The temperature minimum, as low as 1°C, is seen at 100 m in depth and at 140 cl/t in thermosteric anomaly below the shallow seasonal thermocline. Uda (1935b) called the minimum water the *dichothermal* water and concluded that it was formed by winter cooling of the surface-mixed layer and by subsequent summer heating of the surface part of the winter-mixed layer. Below the temperature minimum there is a temperature inversion in the mid-depths from 100 to 600 m. Thus, a slight maximum of temperature is found at the depth of 600 m or at the 100 cl/t surface. Salinity increases with depth all the way to the bottom, but oxyty shows a remarkable minimum, as low as 0.5 ml/l, at a depth of 800 m or at the surface of 70 to 80 cl/t.

3 WATER TYPES IN THE NORTH PACIFIC CENTRAL REGION

3.1 Volume distribution of upper water

The surface warm water, as can be seen in the meridional section of temperature at 155°E (Fig. 8b), spreads as a thin layer over the transition region between the Subtropical Boundary (the Kuroshio Extension) and the Subarctic Boundary. In the equatorial region, likewise, the heated surface water over the thermocline is limited to relatively shallow layers (Fig. 7b). Much of the warm-water mass lies between the Kuroshio or the Kuroshio Extension and the North Equatorial Current. The bottom of the warm water mass is formed by the main thermocline. The

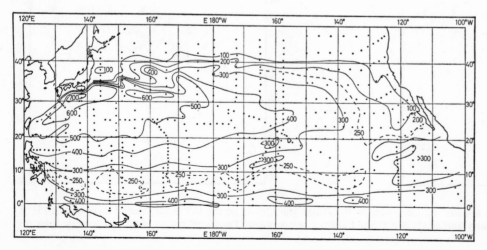

FIG. 10 Depth(m) of the 10°C surface, based on data taken mostly in summer (Table 1).

center of the thermocline can be represented, for simplicity, by the 10°C isothermal surface (Fig. 10) in the same way as in the North Atlantic (Iselin, 1936).

The lateral boundary of the warm-water mass is best defined in the Kuroshio region and worst defined in the California Current region (Fig. 10). The warm-water mass is as thick as 600 m or 700 m just south of the Kuroshio and its extension. In the Gulf Stream region the thickness is some 300 m greater (Iselin, 1936). The difference in depth between the North Pacific and the North Atlantic could be due to a difference in heat exchange through the sea surface and in the intensity of the subtropical gyre or the western boundary current. The depth of the 10°C isothermal surface is less than 250 m along the boundary between the North Equatorial Current and the Equatorial Countercurrent. At this boundary the isothermal surface lies just below the thermocline. The slight

trough about 400 m deep on and near the equator may be related to the vertical spreading of the thermocline in the Equatorial Undercurrent.

The warm-water mass floats as a lens on the bulk of cold water in the ocean. The location and shape of this warm-water lens must be associated with the anticyclonic subtropical gyre in the North Pacific. Though the lateral boundary of the warm-water lens cannot be determined entirely objectively, the current pattern (Fig. 2) or the depth of the main thermocline (Fig. 10) may roughly define the main body of the warm-water lens. Masuzawa (1969b) arbitrarily defined the region of the lens by the closed line shown in Fig. 1. The region has an area of 14.3×10^6 km² and, for simplicity, was called the Central Region.

In order to show water characteristics in the upper 1000 m of the Central Region (Masuzawa, 1969b), 50 hydrographic stations that cover the region as

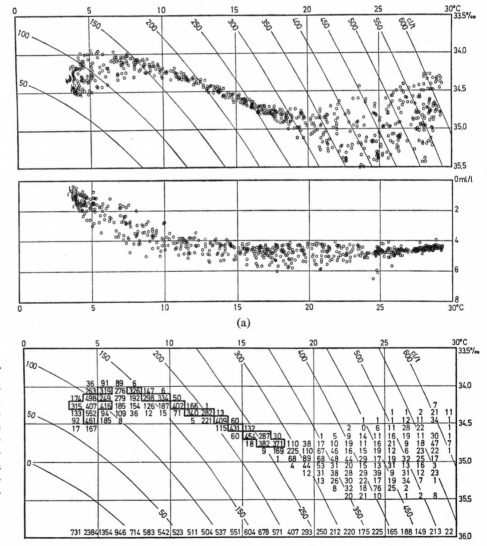

FIG. 11 The upper 1000 m of water at 50 stations in the North Pacific Central Region (Fig. 1). All except 6 stations were made in summer (Table 1). (a) correlations of temperature-salinity and temperature-oxyty; (b) the distribution of volume (10³ km³) in each class of 1°C × 0.1‰ on a temperature salinity diagram (Masuzawa, 1969b).

uniformly as possible were chosen (Fig. 1). Most of the chosen stations were made in summer after 1950. All observed values of salinity and oxty in the upper 1000 m at the 50 stations are plotted against temperature (Fig. 11a). The volume of the upper 1000 m of water in each class with a temperature range of 1°C and a salinity range of 0.1‰ was computed from the data from the stations (Fig. 11b).

The surface water warmer than 20°C or lighter than 300 cl/t exhibits a large scatter on the temperature-salinity diagram, because the surface water is much modified by different atmospheric effects at different places. For example, the saline water is large in range of thermosteric anomaly, the water more saline than 35.3‰ ranging from 350 cl/t to 550 cl/t. On the other hand, the surface water is fairly uniform in oxty.

The lower water, water colder than 8°C or with thermosteric anomaly below 150 cl/t, shows the next largest scatter. For example, the salinity of the Intermediate Water between 100 cl/t and 150 cl/t ranges from 33.9‰ to 34.4‰. The range of oxty at a given temperature in the Intermediate Water is also large, as can be seen in Fig. 11a.

An outstandingly uniform correlation between temperature and salinity is revealed in the water between the surface water and the Intermediate Water. In contrast to the uniform correlation between temperature and salinity, the range of oxty with respect to temperature is as large as 1.5 ml/l or more. This water has been commonly called the Western Central Water of the North Pacific since the definition by Sverdrup, Johnson and Fleming (1942) was established. As much of this water forms the main thermocline in the Central Region, however, the name Thermocline Water was introduced by Masuzawa (1969b).

The distribution of water volume against temperature, as can be seen at the bottom of Fig. 16b, has a remarkable mode near 16°C. This water corresponds to the 18°C Water in the North Atlantic (Worthington, 1959). The same mode can be found around 16°C in the frequency distribution by water characteristics in the whole North Pacific Ocean (Karavayeva and Radzikovskaya, 1965). For this water in the North Pacific Masuzawa (1969b) proposed the name Subtropical Mode Water instead of 18°C Water.

3.2 Tropical Water

The chart of salinity at the sea surface (Fig. 4) shows a closed area of high salinity in the central North Pacific like those in the other oceans. The maximum salinity slightly exceeds 35.5‰, which is lower than in any other ocean. It is estimated from Fig. 11b

that the water more saline than 35.0‰ forms 6.2‰ (887×10^3 km³) of the total volume (14.3×10^6 km³) of the upper 1000 m in the Central Region. Hence, the average thickness of the saline water is 62 m over the Central Region, though the maximum thickness is 200 m or more. The saline water was named Tropical Water by Sturges (1965) and Cannon (1966).

The North Pacific Tropical Water ranges in thermosteric anomaly from 300 cl/t to 550 cl/t as shown in Fig. 11a. The isanosteric surface of 400 cl/t is a proper surface to represent the geographical distribution of the Tropical Water (Cannon, 1966). The North Pacific Tropical Water is clearly separated from the more saline Tropical Water of South Pacific origin by a tongue of low salinity stretching from Central America westward in the boundary zone between the North Equatorial Current and the Equatorial Countercurrent. The low-salinity water occupying the surface-mixed layer in the zone may be called Equatorial Surface Water. The Tropical Water touches the sea surface on the north and extends toward the equator beneath the surface water of low salinity and low density, the Equatorial Surface Water (Figs. 7d and 8d), so that the bottom of the Tropical Water is deepest near 15°N. For instance, the isopycnic surface of $\sigma_t = 24.4$g/l ($\delta_T = 354$ cl/t) in Barkley's *Atlas* (1968) intersects the sea surface near 25°N in winter and sinks to depths of 200 m or more at the southern edge of the Central Region.

The Tropical Water at the sea surface is in the zone from 20°N to 30°N, which is roughly coincident with the zone of large evaporation (Wyrtki, 1965a). The high salinity of the Tropical Water may be caused both by excess of evaporation over precipitation (Wüst, Brogmus and Noodt, 1954) and by stagnation or long residence of water in the Central Region (Montgomery, 1959). The origin of the Tropical Water can be found in the vicinity of the intersection of the 400-cl/t surface with the sea surface in winter, between 20°N and 30°N (Cannon, 1966).

The Tropical Water is carried west by the North Equatorial Current, the salinity decreasing downstream, and becomes part of the Kuroshio water. The lower part of the Tropical Water forms the salinity maximum in the Kuroshio, the maximum salinity being nearly 35.0‰ in the Kuroshio near Taiwan. The Tropical Water in the Kuroshio is much diluted and replenished with oxygen in the Kuroshio Extension and farther east.

3.3 Subtropical Mode Water

A distinctive water is seen between the sea surface or the surface thermocline and the subsurface permanent thermocline in the northern half of the Central Region (Figs. 8b and 9). This water forms a nearly isothermal layer, which can suitably be called a thermostad (Seitz, 1967) in contrast to a thermocline. As mentioned in Sec. 3.1, the water shows a pronounced mode in the frequency distribution of upper water on a temperature-salinity diagram (Fig. 11b). Therefore, this distinctive water has been named Subtropical Mode Water (Masuzawa, 1969b).

The Subtropical Mode Water is the same as the so-called 18°C Water, which, for the North Atlantic, was first described by Worthington (1959) and was further studied by Schroeder, Stommel, Menzel and Sutcliffe (1959) and Istoshin (1961). The volume distribution of the whole North Pacific water (Karavayeva and Radzikovskaya, 1965) indicates a mode ranging from the class of 14.5°C, 34.55‰ to the class of 18.5°C, 34.85‰ on a temperature-salinity diagram with classes of 1°C×0.1‰. The class showing the largest volume is found at 15.5°C, 34.65‰. According to the statistics for the Central Region of the North Pacific (Fig. 11b), on the other hand, the mode seems to be characterized by 16.5°C in temperature, 34.75‰ in salinity and 250 cl/t in thermosteric anomaly.

Salinity on the 250-cl/t surface (Fig. 12) decreases

northward from the Subtropical Boundary, which corresponds to the Kuroshio Extension or the southern edge of the North Pacific Current. On the other hand, a tongue of low salinity from near California stretches south of Hawaii and then continues to the west along 10°N. In connection with this low-salinity tongue, a zone of low oxyty stretches at latitudes of 5 to 10°N as far west as 140°W. Salinity exceeding 34.8‰ appears in the Equatorial Undercurrent and in the South Equatorial Current south of 2 or 3°N (Tsuchiya, 1968). The interior of the Central Region is occupied by uniform water with salinity of 34.6 to 34.8‰, with oxyty of 4 to 5 ml/l. The uniform water is remarkably extensive not only over the whole Central Region but also over a western region reaching nearly to the equator.

Although the salinity is surprisingly uniform on the particular surface of 250 cl/t over the Central Region, the temperature at the core of the Subtropical Mode Water decreases from west to east (Fig. 18, or Masuzawa, 1969b), that is, 18.5°C south of Japan, 17.5°C just east of Japan and 16.5°C in the eastern part of the Kuroshio Extension. Therefore, the thermosteric anomaly at the core changes from west to east, too.

The depth of the 250-cl/t surface exceeds 250 m

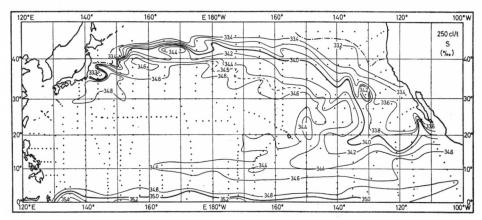

FIG. 12 Salinity on the 250-cl/t surface. The chosen stations are the same as in Fig. 10.

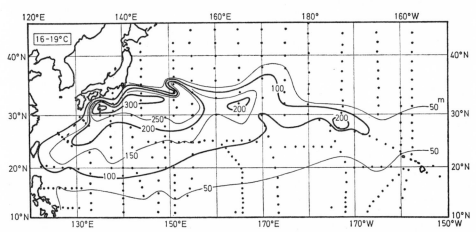

FIG. 13 Thickness(m) of the 16–19°C layer. The source of the data is the same as in Fig. 10.

over much of the Central Region, and a maximum depth as great as 400 m is seen south of the Kuroshio. Furthermore, the thickness of the Subtropical Mode Water is largest south of the Kuroshio and the Kuroshio Extension (Fig. 13).

The Subtropical Mode Water might be formed at the sea surface of the northern portion of the Central Region in winter. The chart of sea-surface temperature in winter (Fig. 3a) shows that the temperature corresponding to that of the Subtropical Mode Water is found extensively south of the Kuroshio and the Kuroshio Extension. In other words, the Subtropical

Mode Water is the coldest water which is formed at the sea surface inside the Central Region. This water is not only carried clockwise by the subtropical gyre, but also transported westward in the subsurface depths of the interior zone between 20°N and 30°N (Masuzawa, 1969b).

3.4 Thermocline Water

As described briefly in Sec. 3.1, the temperature-salinity curves are uniform and nearly linear between 18°C and 8°C, or between 300 cl/t and 150 cl/t over

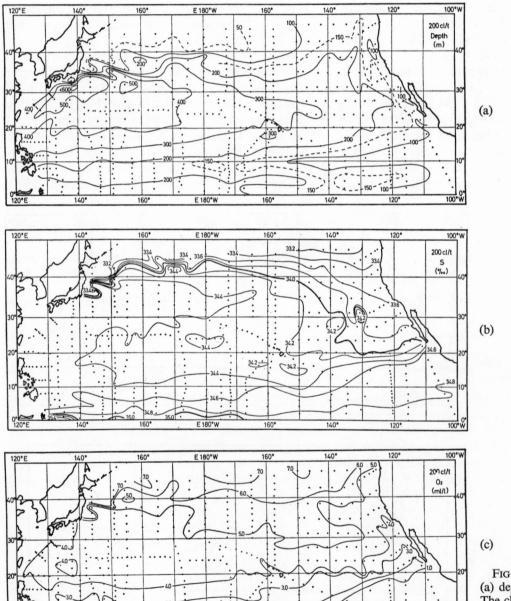

(a)

(b)

(c)

FIG. 14 The 200-cl/t surface: (a) depth; (b) salinity: (c) oxyty. The chosen stations are the same as in Fig. 10.

the Central Region (Fig. 11a). Sverdrup, Johnson and Fleming (1942) have named the *water mass* Central Water, because it is commonly seen in the central region of an ocean. But this water may better be referred to as Thermocline Water, because much of the water in the temperature range mentioned above forms the main thermocline in the Central Region (Figs. 7b and 8b).

The distribution of water volume against temperature, as shown at the bottom of Fig. 11b, indicates a minimum frequency near 12°C, so that the isothermal surface of 12°C may be assumed statistically as the center of the thermocline over the Central Region. Thus, the isanosteric surface of 200 cl/t close to the isothermal surface of 12°C was chosen to represent the distribution of water characteristics of the Thermocline Water (Fig. 14).

The depth of the 200-cl/t surface (Fig. 14) shows naturally a geographic distribution similar to the depth of the 10°C surface (Fig. 10). The boundary of the Central Region defined before (Fig. 1) coincides roughly with the isobath of 300 m on the 200-cl/t surface. Although both salinity and oxyty are fairly uniform on the 200-cl/t surface over the Central Region, there are small, but probably significant, differences in both properties over the region. The salinity is rather lower in the southeastern half of the Central Region than in the northwestern half, and the temperature must show a corresponding distribution on the 200-cl/t surface. The oxyty is highest in the northeastern corner of the Central Region and lowest in the westernmost part. This distribution may suggest that the Thermocline Water roughly circulates anticyclonically in the Central Region.

According to Sverdrup, Johnson and Fleming (1942), the so-called Central Water in the North Pacific is formed in winter in the region east of Japan, 30°N to 40°N, 150°E to 160°E. This idea must be based upon the description by Iselin (1939) that the meridional temperature-salinity correlation in winter at the sea surface near the northern boundary of the Central Region of the North Atlantic is similar to the vertical temperature-salinity correlation of the North Atlantic Central Water. However, the idea dose not seem to have been fully justified in the North Pacific Ocean. To my mind, the temperature-salinity correlation at the sea surface north of the Subtropical Boundary is not likely to be as uniform as the vertical correlation in the Central Region of the North Pacific. It seems to me that lateral mixing may play an important role for the formation of the uniform temperature-salinity correlation of Thermocline Water not only in the Kuroshio Extension and its continuation but also in the Central Region's thermocline

bounded vertically by the Subtropical Mode Water at the top and by the Intermediate Water at the bottom.

3.5   Intermediate Water

The Intermediate Water forms a pronounced vertical minimum of salinity in the central North Pacific Ocean just as the Antarctic Intermediate Water does all over the oceans of the southern hemisphere. The core of the North Pacific Intermediate Water lies on the isanosteric surface of 125 cl/t in the Central Region (Reid, 1965). In the equatorial region west of Hawaii, however, the salinity minimum is not found on the 125-cl/t surface but on the isanosteric surfaces of 170 cl/t or so (Masuzawa, 1967; Tsuchiya, 1968). A different salinity minimum in the California Current region appears at lesser depth (Reid, 1965).

According to the comprehensive description of the 125-cl/t surface by Reid (1965) (Fig. 15), the surface lies in depths of 300 to 600 m in the transition region between the Subarctic Boundary and the Subtropical Boundary and in depths of 600 m or more over most of the Central Region. The maximum depth, exceeding 800 m, is seen south of the Kuroshio near Japan. The 125-cl/t surface may not touch the sea surface near the Kurile Islands in winter (Reid, 1966) and is usually in depths of 100 to 200 m in the subarctic region.

The geostrophic circulation along the 125-cl/t surface, referred to the 1000-db surface (Fig. 15a), suggests that there is an eastward flow near the northern edge of the Central Region and that westward currents are revealed not only in the North Equatorial Current region but also in the interior zone from 20°N to 30°N. On the other hand, a counterclockwise circulation is seen in the subarctic region from northern Japan and the Kurile Islands to the Gulf of Alaska. In the equatorial region, the southern part of the North Equatorial Current west of 180° long. comes both from the California Current region and from off Central America, and the Equatorial Countercurrent on this surface is seen all the way across the ocean from west to east. Thus, the gross circulation along this surface is roughly similar to the sea-surface oceanic circulation.

The northeastern half of the Central Region is lower in salinity (Fig. 15b) and higher in oxyty (Fig. 15c) than the southwestern half of the region. The salinity at the Kuroshio region from Luzon to the Boso Peninsula (35°N) is higher than 34.2‰, which is the highest in the ocean north of 20°N except for that in the Davidson Current region near California. These features of the distribution of salinity on the

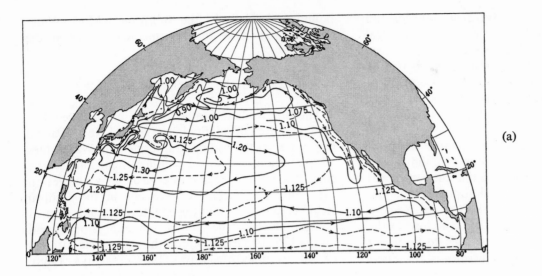

(a)

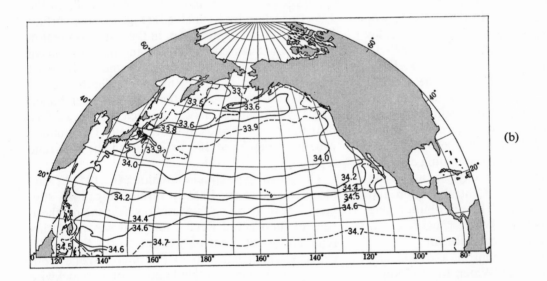

(b)

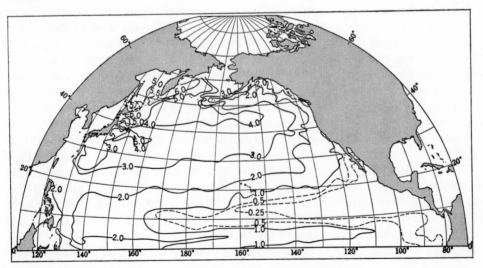

(c)

FIG. 15 The 125-cl/t surface: (a) accerelation potential relative to 1000 db, 10^5 erg/g; (b) salinity, ‰; (c) oxyty, ml/l (Reid, 1965).

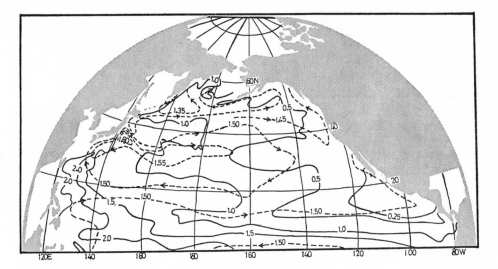

FIG. 16 Oxyty, ml/l (solid curves) and acceleration potential relative to 1500 db, 10^5 erg/g (dotted curves) on the 80-cl/t surface (Reid, 1965).

125-cl/t surface suggest that the Intermediate Water roughly circulates clockwise, its salinity increasing by mixing in the Central Region, so that the maximum salinity is seen in the Kuroshio region. The Intermediate Water originating east of Japan does not seem to flow directly underneath the Kuroshio to the south of Japan.

An outstanding low-oxyty tongue is seen to extend from Central America nearly all the way to the Philippines (Fig. 15c). This lateral oxyty-minimum water forms a vertical oxyty minimum at the 125-cl/t surface (Reid, 1965; Masuzawa, 1967). The water of the oxyty-minimum tongue flows roughly westward in the North Equatorial Current and becomes higher in oxyty and lower in salinity by inflow of the Intermediate Water from the north and by vertical mixing.

The water on the 125-cl/t surface in the subarctic region is low in salinity, less than 33.8‰, and high in oxyty. Reid (1965) concludes that this peculiar 125-cl/t water is formed hardly at all in winter at the sea surface in the Oyashio region but rather at subsurface depths in the subarctic gyre by vertical mixing.

Small tongues of low-salinity and high-oxyty water stretch from the Oyashio region to the south or southeast into the Kuroshio Extension east of Japan. This is an indication that the colder, less saline and higher oxyty Intermediate Water is transferred from the Oyashio to the Kuroshio Extension (Ichiye, 1955; Masuzawa, 1956; Kuksa, 1962, 1963; Reid, 1965). The data taken in 1957 during the IGY at the Kuroshio Extension region, 30° to 45°N, 140° to 170°E, show that the transition region between the Kuroshio Extension and the Subarctic Boundary between 150°E and 170°E is rather uniform in salinity, 33.8 to 33.9‰ on the 125-cl/t surface. This suggests strong lateral mixing in the region. The relatively uniform-salinity

water in the transition region might be an important source of the Intermediate Water in the Central Region. It is likely that the transfer of the 125-cl/t water from the subarctic gyre to the subtropical gyre is accomplished by various types of lateral mixing rather than by southward advective undercurrents which have been called the Oyashio Undercurrent (Uda, 1935a).

3.6   Oxyty-minimum Water

The oxyty of the deep water is less in the North Pacific than in the other oceans (Stommel and Arons, 1960), and the minimum below the thermocline is most pronounced in the North Pacific. The core of the low-oxyty layer lies in the range of 60 to 90 cl/t in the North Pacific excluding the equatorial-current region (Figs. 7d and 8d). Kawamoto (1955) suggests that the oxyty minimum is located at $\sigma_t = 27.3$ to 27.4 g/l ($\delta_T = 78$ to 69 cl/t) to the north of 20°N. Reid (1965) made a choice of the 80-cl/t surface for examining the Intermediate Water of the South Pacific. Fortunately, this isanosteric surface is nearly coincident with the oxyty minimum over the North Pacific Ocean except the equatorial-current region.

In the subarctic region, for instance north of 42°N at 155°E (Fig. 8d), the low-oxyty layer less than 1 ml/l is as thick as 500 m or more, the minimum value being less than 0.5 ml/l. The counterclockwise subarctic gyre is seen on the 80-cl/t surface (Fig. 16) as on the 125-cl/t surface. The depth of the 80-cl/t surface is less than 700 m inside the subarctic gyre. The 80-cl/t surface is shallower than 400 m south of the western Aleutian Islands. This thick low-oxyty water may be due not only to high productivity in the subarctic region (Miyake and Saruhashi, 1956), but also to upward motion in the deep layers (Masuzawa, 1960; Knauss, 1962; Uda, 1963).

Oxyty on the 80-cl/t surface decreases from the Japanese coast to the American coast at latitudes near 40°N (Fig. 16), where eastward geostrophic flows are found. The tongue of low oxyty stretching westward along latitudes near 30°N is roughly associated with westward geostrophic flow at 20 to 30°N (Reid, 1965). This distribution may suggest that as for a flowing water particle, at latitudes near 40°N the consumption of oxygen exceeds the replenishment by diffusive processes, and at latitudes around 30°N the oxygen supply by mixing exceeds the decrease by consumption.

Oxyty at the oxyty minimum is 1 ml/l or so over much of the Central Region, but tends to increase westward to 2 ml/l or more at the westernmost part of the region. A rather rapid change in oxyty is revealed between west and east of the Izu-Ogasawara Ridge north of 30°N (Kawamoto, 1955; Okubo, 1958): 1.5 ml/l or so to the west of the ridge in contrast to about 1 ml/l to the east.

The remarkable oxyty minimum in the southern half of the North Equatorial Current and in the Equatorial Countercurrent lies on the 125-cl/t surface, as shown in the meridional sections at 160°W (Reid, 1965) and at 137°E (Masuzawa, 1967). This was also pointed out by Kawamoto (1956). It is suggested in Fig. 15a and c that this low-oxyty water in the equatorial-current region originates off Central America and flows westward, increasing in oxyty. Along the 80-cl/t surface at low latitudes, on the

other hand, the geostrophic flow seems to be eastward (Reid, 1965) and oxyty decreases eastward presumably by vertical mixing with the oxyty-minimum water centered at 125 cl/t.

4 WATER CHARACTERISTICS OF THE KUROSHIO

4.1 Flux by water characteristics

Because the Kuroshio is located horizontally between different waters and vertically reaches great depths, the water carried by the Kuroshio has a wide range of characteristics. The correlations of temperature-salinity and -oxyty are not always uniform in the Kuroshio. In Sec. 4.1, the temperature-salinity and -oxyty relations and the distribution of geostrophic flux on a temperature-salinity diagram are demonstrated for three selected sections across the Kuroshio. From this representation (Fig. 17), we can see mutual relations among temperature, salinity, oxyty and thermosteric anomaly and the relative importance by geostrophic flux of different water types in the Kuroshio.

All three selected sections were made by the *Ryofu Maru* of the Japan Meteorological Agency: (a) south of Taiwan, 20°10'N, February 1967; (b) south of Japan, 136°30'E, May 1965; (c) east of Japan, 144°E, November 1959. The hydrographic cast did not reach below 1500 m at the 20°10'N section, so the geostrophic computation is referred to 1500 db for the

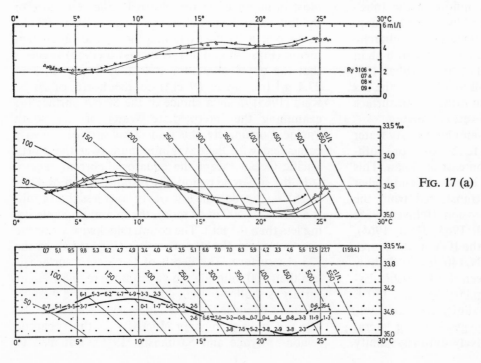

FIG. 17 (a)

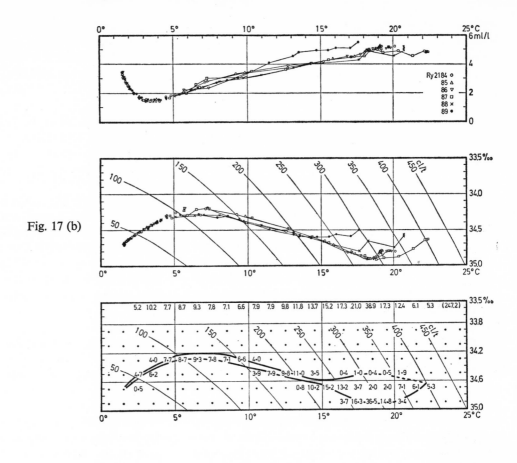

Fig. 17 (b)

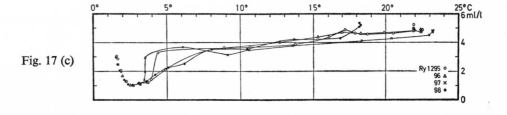

Fig. 17 (c)

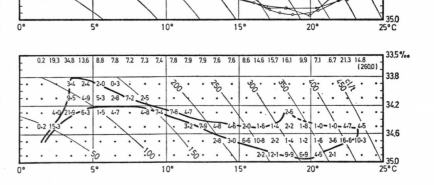

FIG. 17 Temperature-oxyty and temperature-salinity relations and geostrophic flux, km³/hr, of the upper 1500 m for (a) and 2000 m for (b), (c), of water carried by the Kuroshio, *Ryofu Maru* (Fig. 1, Table 2): (a) south of Taiwan, 20°10′N, Ry3106 (122°28′E)—Ry3109 (120°58′E), February 1967; (b) south of Japan, 136°30′E, Ry2184(32°02′N)—Ry2189(33°40′N), May 1965; (c) east of Japan, 144°E, Ry1295 (35°22′N)—Ry1298(37°01′N), November 1959. The thick curve on the figures of flux indicates the envelope of temperature-salinity correlations.

TABLE 2.   Geostrophic features of the three selected sections of the Kuroshio, made by the *Ryofu Maru*. The geostrophic computation refers to 1500 db for Section A and to 2000 db for Section B and C, respectively.

Section	Date	Range	Width km	$\Delta\phi$ dyn. cm	Flux km³/hr
A: 20°10′N	22–23 Feb. 1967	Ry 3106–3109 122°28′E–120°58′E	156	59	159
B: 136°30′E	5–6 May 1965	By 2184–2189 32°02′N–33°40′N	182	110	247
C: 144°00′E	14–15 Nov. 1959	Ry 1295–1298 35°22′N–37°01′N	200	113	260

$\Delta\phi$:   Sea-surface geopotential difference between lateral limits of the Kuroshio at each section.

20°10′N section, though to 2000 db for the sections at 136°30′E and 144°E. The cross-current features of the Kuroshio at the latter two sections were described by Masuzawa (1964b and 1965b). The lateral limits of the Kuroshio are taken at the stations where the geopotential at the sea surface shows the maximum and minimum over the section. The geostrophic features of these three sections are shown in Table 2.

The method for representing the distribution of geostrophic flux on a temperature-salinity diagram was first proposed by Montgomery and Stroup (1962) and applied to the Kuroshio by Masuzawa (1964b and 1965b). The geostrophic flux was computed for each trapezoid formed by two vertical lines at two adjacent stations and by two isobaric lines at intervals of 50 db in the upper 1 km and of 100 db in the lower 0.5 or 1 km. By use of the vertical distributions of temperature and salinity, the flux through each trapezoid was divided into temperature-salinity classes, 1°C × 0.2‰. The computed fluxes for each class were plotted on a temperature-salinity diagram and then summed over all trapezoids for the whole width of the Kuroshio.

Section 20°10′N (Fig. 17a)

The temperature-salinity and -oxyty relations are fairly uniform in this section between Luzon and Taiwan. The salinity maximum with salinity of 34.9 to 35.0‰ lies in the layers of 350 to 400 cl/t at this section, while the salinity maximum is centered near the 300-cl/t surface at the sections south and east of Japan. The primary mode in flux on a temperature-salinity diagram is formed by the surface mixed water at 24 to 26°C, the flux of which is 40 km³/hr.,[*] one-fourth of the total flux, 159 km³/hr. The faint oxyty maximum at 250 to 300 cl/t may be indicated by the rather higher-oxyty Subtropical Mode Water, which makes a faint mode in the flux distribution. The Intermediate Water is rather higher in salinity

(34.25 to 34.4‰ at the salinity minimum) at this section than at the other two sections. The oxyty at the oxyty minimum is 2 ml/l or so, which is the highest of the three sections. Another mode of flux is seen in the Oxyty-minimum Water.

Section 136°30′E (Fig. 17b)

The primary mode in the flux distribution lies near the point 18°C, 34.9‰, 280 cl/t. This mode is formed by the Subtropical Mode Water. At temperatures below 15°C, the temperature-salinity and -oxyty relations are very uniform, with the exception of the temperature-oxyty curve between 10°C and 15°C at the station nearest shore, Ry2189. Slight modes are seen in the Intermediate Water and in the Oxyty-minimum Water. The minimum salinity of the Intermediate Water is 34.2 to 34.3‰ and the minimum oxyty of the Oxyty-minimum Water is 1.5 ml/l or so.

Section 144°E (Fig. 17c)

At this section east of Japan, the temperature-salinity and -oxyty correlations are not homogeneous in the Kuroshio. Three modes in the distribution of flux against temperature can be noted near 23°C, 18°C and 3.5°C. The warmest mode corresponds to the surface-mixed layer. Similar modes cannot be seen in the section made at 136° 30′E in early May, when the surface-mixed layer was not definitely separated from the Subtropical Mode Water. The Subtropical Mode Water makes a mode at the same point on a temperature-salinity diagram as at the sections south of Japan and Taiwan. The less saline waters in the layer lighter than 200 cl/t in the northern part (Ry1297–1298) of the section come from the Oyashio region. On the other hand, the temperature-salinity relation in the surface water in the southern part is nearly the same as that in the section south of Japan.

In this section there are two different types of Intermediate Water. The Intermediate Water at the

[*]   For the unit of flux in this chapter, the cubic kilometer per hour, km³/hr (Montgomery and Stroup, 1962), is chosen instead of million cubic meter per second, 10⁶ m³/s, which has been commonly used.

two stations, Ry1296 and 1298, is nearly the same as the water at the section south of Japan, though the salinity is rather low and the oxyty is slightly high. At the two other stations, Ry1295 and 1297, the Intermediate Water is much less saline, below 34.0‰ at the minimum, and is higher in oxyty. This is apparently the fresh Intermediate Water originating in the Oyashio region. A fresh Intermediate Water like this forms not only a remarkable salinity minimum but also an oxyty maximum or occasionally a temperature minimum in the section of the Kuroshio Extension east of Japan. The importance of the less saline Intermediate Water in the Kuroshio Extension is suggested by the fact that the flux between 2°C and 5°C at this section is about 45 km³/hr larger than at the section south of Japan. The oxyty minimum in the Kuroshio Extension east of Japan is rather lower in oxyty, 1 ml/l or so, and in thermosteric anomaly, 60 cl/t, than the minimum in the Kuroshio south of Japan and Taiwan.

The warm surface water is important in flux but different in characteristics from place to place and from season to season. The Subtropical Mode Water forms sometimes a salinity maximum and a mode of flux in the Kuroshio section. That is, the Subtropical Mode Water is a mode water not only in the volume distribution of upper water in the Central Region but also in the distribution of flux by the Kuroshio. Furthermore, the characteristic water types below the thermocline, Intermediate water and Oxyty-minimum Water, are also important in the Kuroshio flux.

4.2 Longitudinal changes of the Kuroshio water

As the Kuroshio is not a river with solid sides in the ocean, the characteristics of the water flowing downstream are much changed by gain and loss of water through boundaries of the current and mixing as well as by air-sea processes that exchange heat and water through the sea surface. Unfortunately, however, few studies have been made along this line.

Comparison of the three sections described in Sec. 4.1 gives a rough idea of differences in water characteristics of the Kuroshio between the starting region, the mature stage and the extension. Another representation is now tried for showing longitudinal changes of the Kuroshio water from east of the Philippines to north of Hawaii. Hydrographic stations have been chosen at intervals of about 500 km along the Kuroshio on the right and left sides of the Kuroshio (Fig. 1). All the stations were made in summer. The right-hand edge of the Kuroshio has been se-

lected near the station with the maximum sea-surface geopotential over a section across the Kuroshio, so the station roughly coincides with the station showing the deepest main thermocline on the section. On the other hand, the choice of the left-hand edge was fairly arbitrary. Although the two longitudinal sections (Figs. 18 and 19) do not show the complete range in characteristics of water carried by the Kuroshio for a given place, the sections show the gross changes of the Kuroshio water along the path.

Right-hand section (Fig. 18)

The relatively less saline Tropical Water centered near 350 cl/t is seen west of 130°E or south of 30°N (Fig. 18c), and the Tropical Water more saline than 35.0‰ can be found east of 170°E. Between 130°E and 170°E the salinity maximum lies in or just above the Subtropical Mode Water, which is most pronounced in this region (Fig. 18b). The Mode Water, however, is different in temperature and thermosteric anomaly between the western half and the eastern half of the region from 130°E to 170°E; 18°C, 280 cl/t in the west and 17°C, 250 cl/t in the east at the core of the Mode Water (Figs. 18a and b).

The depth of the main thermocline is indicative of the strength of the Kuroshio. That is, we can see high velocities of the Kuroshio between 130°E and 150°E, where the main thermocline is deeper than anywhere else along the whole length of the Kuroshio (Fig. 18a).

The lower part (below 200 cl/t) of the Thermocline Water is lower in oxyty (Fig. 18d) west of 140°E than east of 140°E. There is a definite contrast in salinity of the Intermediate Water and in oxyty of the Oxyty-minimum Water between the regions west and east of 140°E. This contrast suggests that there is southward transfer of fresher Intermediate Water through the Kuroshio Extension east of 140°E, and that the communication of the Oxyty-minimum Water between west and east is interrupted by the Izu-Ogasawara Ridge at 140°E.

Left-hand section (Fig. 19)

The diluted Tropical Water can be seen not only at the near-shore edge of the Kuroshio west of 140°E but also at the northern edge of the Kuroshio Extension or its continuation east of 140°E. At 144°E there appears a remarkably fresh Intermediate Water which is characterized by the temperature minimum at 150 m, 130 cl/t, by the salinity minimum at 100 m, 150 cl/t, and by the oxyty maximum at 50 m, 190 cl/t. Similar fresher Intermediate Waters are sometimes found in the Kuroshio Extension (Masuzawa, 1957). East of 155°E, however, the salinity at the salinity

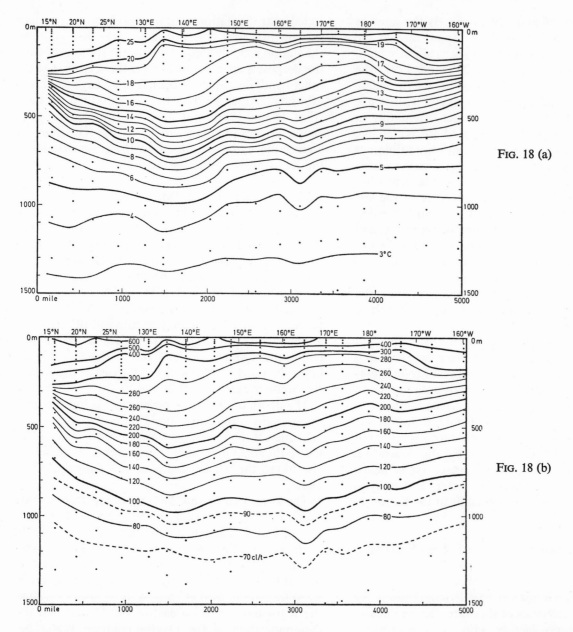

FIG. 18 (a)

FIG. 18 (b)

minimum is uniform, 33.9 to 34.0‰, and there are no fresh Intermediate Waters of lower salinity and higher oxyty.

In the eastern part east of 170°E, the weak main thermocline lies at 100 to 600 m or at 220 to 120 cl/t. This layer is formed by a mixed water having fairly uniform temperature-salinity and -oxyty correlations.

It may be suggested from the above description that the Kuroshio water changes in characteristics not only in surface layers but also in intermediate or deeper layers relatively suddenly at some critical regions, for example, at 130°E, 140°E and 170°E.

4.3 Seasonal variations of water characteristics

As outlined in Sec. 2.2, the sea-surface oceanographic conditions vary seasonally in the western North Pacific so strongly that the surface water carried by the Kuroshio shows a remarkably seasonal variation of temperature and salinity. This great variation is caused primarily by large precipitation in the rainy season (*baiu*) of the late spring and early summer and in the typhoon season of late summer and early fall and by evaporation and transfer of sensible heat from sea to air in the winter-monsoon season.

There are some indications of seasonal changes in speed and flux of the Kuroshio south of Japan (Masuzawa, 1954, 1965a), but these changes have not been well demonstrated. However, much information on

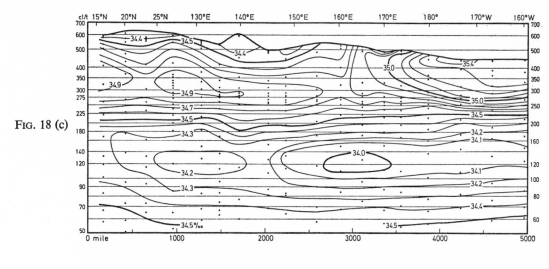

FIG. 18 (c)

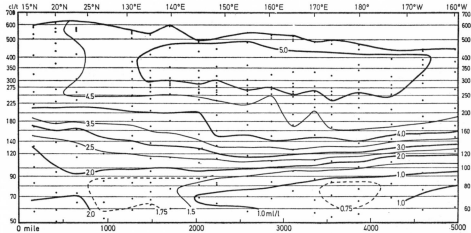

FIG. 18 (d)

FIG. 18 The longitudinal sections at the right-hand edge of the Kuroshio and the Kuroshio Extension, based on 18 stations made in summer (Fig. 1, Table 1): (a) temperature against depth; (b) thermosteric anomaly against depth; (c) salinity and oxyty against thermosteric anomaly on logarithmic scale; vertical exaggeration for (a) and (b) 3700.

the seasonal variation of water characteristics has been collected for the Kuroshio region. For instance, Suda (1938) studied the seasonal variations of temperature and salinity in the Kuroshio at the strait between Suo, Taiwan, and Yonakunijima in the Ryukyu Islands. The average temperature over the strait ranges from 28.7°C to 23.3°C at the sea surface and from 22.8°C to 20.7°C at 100 m, the maxima in September and the minima in March, and indicates no regular annual variation at 200 m. The monthly mean salinity shows a maximum in March, a major minimum in September and a secondary slight minimum in May.

Systematic observations were made at the section across the Kuroshio south of Cape Shionomisaki by weather ships for Ocean Weather Station *Tango* from June 1950 to November 1953. Koizumi (1955) showed the annual variations of temperature and salinity in the upper water at Tango.

Ocean Weather Station Tango

Ocean weather Station *Tango* (29°N, 135°E) is al-

ways located in uniform water south of the Kuroshio, so no appreciable currents are found and a seasonal variation representative of the monsoon region can be anticipated. Figure 20 to 22 showing the annual variations have been made from all the data taken by 10 to 20 standard hydrographic casts each month from June 1950 to August 1953.

The monthly mean temperature (Fig. 20) is lowest (19.5°C) in February and March and highest (28.7°C) in August at the sea surface. The annual range decreases with depth from 9.3°C at the sea surface to 1.0°C at 150 m. The difference in temperature between the maximum in July and the minimum in December is about 1°C at any mid-depth from 150 m to 600 m. A somewhat semi-annual period might be presumed at the mid-depths.

The annual variation of salinity (Fig. 21) is conspicuous in the upper 100 m; salinity is high in winter and low in summer. The annual range is nearly 0.5‰ in the upper 25 m. The rapid drop of the surface-water salinity in June is caused by the start of the rainy season, when many cyclones pass on the atmos-

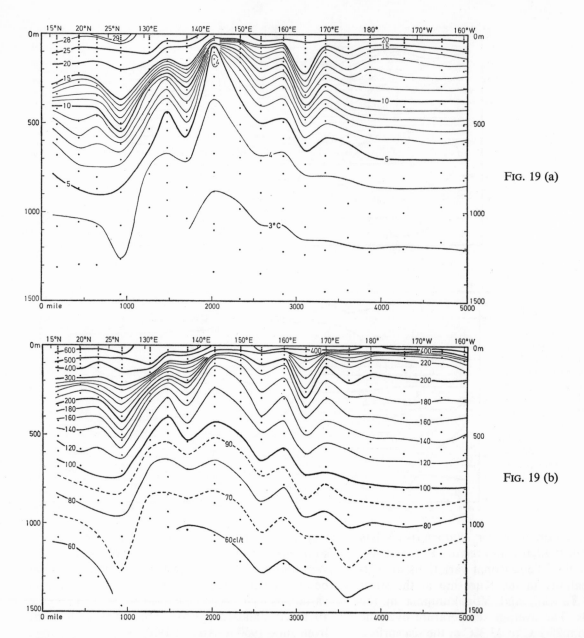

FIG. 19 (a)

FIG. 19 (b)

pheric frontal zone near Station *Tango*. The saline and thick surface-mixed layer in winter is formed by large evaporation (about one centimeter per day, Masuzawa, 1952) and by cooling in the prevailing winter monsoon. The vertical maximum from May to November, with salinity exceeding 34.8‰, seems to be formed by winter evaporation, followed by summer dilution of the surface part of the winter mixed layer.

Oxyty in the upper 150 m (Fig. 21) shows an annual variation similar to that of temperature; the highest (5.6 ml/l) in March, the lowest (4.7 ml/l) in September. An oxyty maximum is recognized at depths of 50 to 75 m from June to December. This maximum may be related to biological processes. An interrup-

tion in the regular trend of the seasonal variation is noted in May throughout the upper layer.

The seasonal variations of temperature and salinity result in a seasonal variation of steric sea level. Steric sea level is approximated in Fig. 22 by thermosteric sea level, which is computed by vertically integrating thermosteric anomaly from a standard surface. Thermosteric sea level relative to 1000 m is low in winter and high in summer to autumn, the annual range being 24 cm. Much of this annual variation is contributed by the smooth annual variation in the upper 200 m. The differences in thermosteric sea level relative to 1000 m and to 200 m are associated with the higher temperature in May to July and the lower temperature in December in the deep layers.

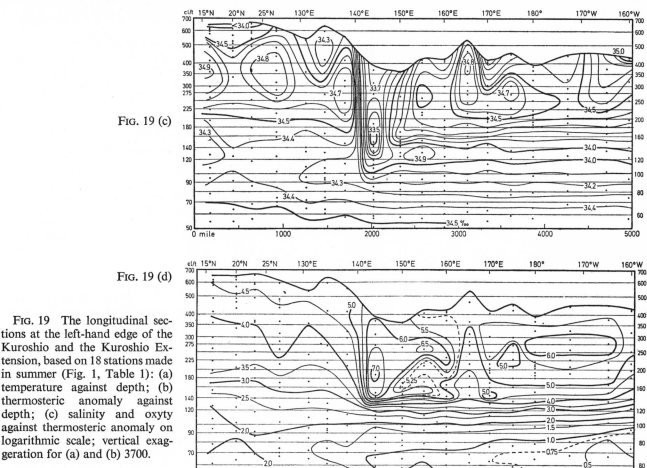

FIG. 19 (c)

FIG. 19 (d)

FIG. 19 The longitudinal sections at the left-hand edge of the Kuroshio and the Kuroshio Extension, based on 18 stations made in summer (Fig. 1, Table 1): (a) temperature against depth; (b) thermosteric anomaly against depth; (c) salinity and oxyty against thermosteric anomaly on logarithmic scale; vertical exaggeration for (a) and (b) 3700.

The annual variation of thermosteric sea level relative to 200 m agrees well with that of sea level at the coastal station Kushimoto near Cape Shionomisaki (136°E). The variation at Kushimoto represents the annual variation commonly observed at the coast from Kyushu to the Izu Peninsula (Tsumura, 1963). This suggests that mean sea level at the coast of southern Japan varies seasonally, roughly speaking, not with the change in strength of the Kuroshio but with the steric seasonal variation.

The annual variation of temperature in the upper 200 m contributes more to the annual variation of thermosteric sea level relative to 200 m than does that of salinity. Incidentally, two imaginary seasonal variations of thermosteric sea level relative to 200 m have been computed: (A) for the 200 m water column with uniform salinity 35‰ and the actual temperature: (B) for the 200 m water column with uniform temperature 0°C and the actual salinity. The annual range of thermosteric sea level is 18.2 cm for the actual water column, 16.3 cm for the imaginary (A) water column, 2.0 cm for the imaginary (B) water column.

Winter-summer differences in water characteristics of the Kuroshio

We have had no comprehensive studies of the seasonal variations of characteristics of the water carried by the Kuroshio, so in this section an example is shown of winter-summer contrast in water characteristics of the Kuroshio (Fig. 23).

Six hydrographic sections south of Cape Shionomisaki were chosen to demonstrate the gross differences in temperature-salinity correlation with geostrophic flux between winter and summer (Masuzawa, 1965b). The method for the determination of lateral edges of the Kuroshio and for the computation of geostrophic flux on a temperature-salinity diagram were described in the paper by Masuzawa (1965b). The geostrophic calculation was referred to the 1000-db surface. The flux computation has been made for three winter sections and three summer sections separately, but only the average flux for each season is demonstrated.

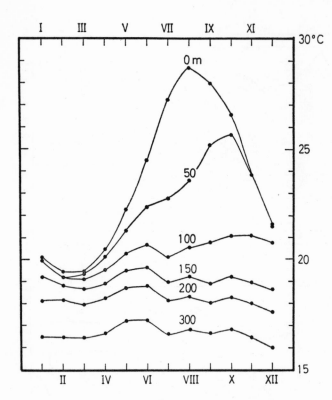

FIG. 20   Annual variation of water temperature at Ocean Weather Station *Tango* (29°N, 135°E, Fig. 1) from June 1950 to August 1953.

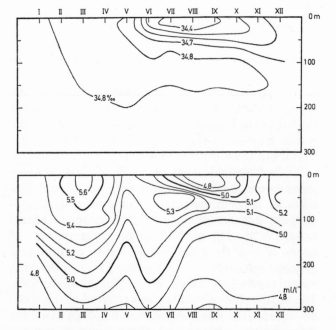

FIG. 21   Annual variation of salinity and oxyty at Ocean Weather Station *Tango* (29°N, 135°E, Fig. 1) from June 1950 to August 1953.

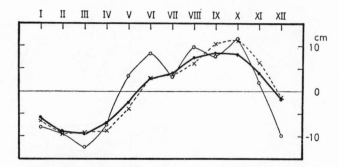

FIG. 22   Annual variation of thermosteric sea level at Ocean Weather Station *Tango* (29°N, 135°E, Fig. 1); (a) open circle, relative to 1000 db; (b) dot, relative to 200 db. Mean sea level at the coastal station Kushimoto, Cape Shionomisaki (136°E), is marked by cross; the effect by the change of atmospheric pressure is eliminated (Tsumura, 1963).

The total flux is 197 km³/hr in winter and 178 km³/hr in summer; this difference may not be significant because of many errors in the computation. The average temperature and salinity, weighted by flux, are 15.5°C, 34.63‰ in winter and 17.1°C, 34.48‰ in summer, but these values depend on the choice of reference level for geostrophic computation.

In winter, the range of salinity for a given temperature is small, 0.1 to 0.15‰, from the sea surface to 1000 m. Almost all temperature-salinity classes appearing in Fig. 23 are seen in each winter. Two modes in the flux distribution are found in the warm surface layer: one at the class of 16.5°C, 34.65‰, 260 cl/t, lying in the Subtropical Mode Water, the other at the class of 18.5°C, 34.75‰, 290 cl/t, in the surface-mixed layer. But it is doubtful that these two modes are really separate.

In summer, on the other hand, a weak mode of flux can be seen in a few classes centered at 18°C, 34.75‰. This mode spreads over the two modes in winter. The range of salinity is extremely wide in the surface layer, but all classes in the surface layer in Fig. 23 are not seen in each summer. However, it should be remembered that the surface water less saline than 34.5‰ is one of the important waters carried by the Kuroshio in summer. The flux of the surface less-saline water is 26 km³/hr, 15% of the total flux.

The distribution of flux on a temperature-salinity diagram shows a big difference between winter and summer. The saline surface-mixed water in winter is replaced by the warmer and much-less-saline surface water in summer. In layers colder than 15°C, although the flux is nearly 35% of the total flux in both seasons, for a given temperature the water is more saline in winter and less saline in summer.

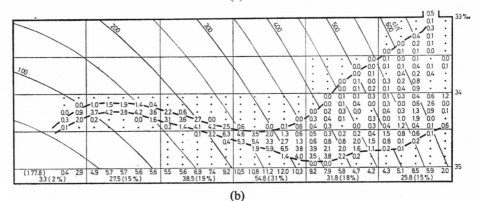

(a)

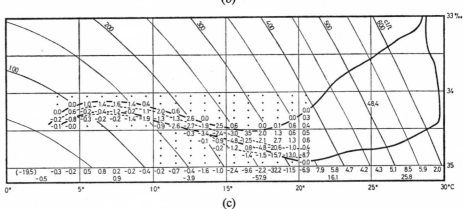

(b)

FIG. 23 Average geostrophic flux, km³/hr, of the Kuroshio south of Cape Shionomisaki (Masuzawa, 1965): (a) average of three crossings in winter; (b) average of three crossings in summer; (c) difference in flux between winter and summer. The thick curve in (a) and (b) indicates the envelope of temperature-salinity correlations. The thick curve and the dashed curve in (c) indicate the envelope of temperature-salinity correlations in summer and in winter, respectively.

(c)

REFERENCES

Barkley, R. A. (1968): *Oceanographic atlas of the Pacific Ocean.* Univ. of Hawaii Press, Honolulu.

Burkov, V. A. and Yu. V. Pavlova (1963): Geostrophic circulation on the surface of the northern part of the Pacific in summer (in Russian). *Result at y Issledovaii po Programme Mezhdunarodnogo Geofizicheskogo Goda, Okeanologicheskie Issledo vaniya, Sbornik Statey, X Razdel Programmy Magg, Izdatelstvo Akademii Nauk SSSR,* **9**: 21–31.

Cannon, G. A. (1966): Tropical waters in the western Pacific Ocean, August–September 1957. *Deep-Sea Res.,* **13**: 1139–1148.

—— (1969): Characteristics of waters east of Mindanao, Philippine Islands, August 1965, pp. 205–211. In: *The Kuroshio,* J. C. Marr, editor, East-West Center Press, Honolulu, 614pp.

Dodimead, A. J., F. Favorite and T. Hirano (1963): Review of oceanography of the subarctic Pacific Ocean. *International North Pacific Fisheries Commission (INPFC), Bull.,* **13**, Salmon of the North Pacific Ocean, Part II, 195pp.

Favorite, F. (1967): The Alaskan Stream. *International North Pacific Fisheries Commission, Bulletin,* **21**: 1–20.

Hanzawa, M. (1950): On the annual variation of evaporation from the sea-surface in the North Pacific Ocean. *Oceanogr. Mag.,* **2**: 77–82.

Ichiye, T. (1955): On the possible origin of the intermediate water in the Kuroshio. *Rec. Oceanogr. Wks. Jap.,* **2**: 82–89.

Iselin, C. O'd. (1936): A study of the circulation of the western North Atlantic. *Pap. phys. Oceanogr. Meteor.,* **4**: 1–101.

—— (1939): The influence of vertical and lateral turbulence on the characteristics of the waters at mid-depths. *Trans. Amer. Geophys. Union,* 1939, 414–417.

Istoshin, Yu. V. (1961): Formative area of 'eighteen-degree' water in the Sargasso Sea (in Russian). *Okeanologiia,* **1**: 600–607.

Karavayeva, V. I. and M. A. Radzikovskaya (1965): The volumes of the main water masses in the Northern Pacific. *Okeanologiia*, **5**: 230–234.

Kawai, H. (1955): On the polar frontal zone and its fluctuation in the waters to the northeast of Japan (I). (in Japanese). *Bull. Tohoku Reg. Fish. Lab.*, **4**: 1–46.

Kawamoto, T. (1955): On the distribution of the dissolved oxygen in the Pacific Ocean. Part 1. On the σ_t-O_2 diagram in the western North Pacific Ocean (in Japanese). *Umi to Sora*, **32**: 1–15.

—— (1956): On the distribution of the dissolved oxygen in the Pacific Ocean. Part 2. On the σ_t-O_2 diagram in the equatorial region and the eastern region of the North Pacific Ocean (in Japanese). *Umi to Sora*, **32**: 92–98.

Kitano, K. (1967): On the Alaskan Stream (in Japanese). *J. Oceanogr. Soc. Jap.*, **23**: 306–307.

Knauss, J. A. (1962): On some aspects of the deep circulation of the Pacific. *J. geophys. Res.*, **67**: 3943–3954.

Koenuma, K. (1938): On the hydrography of the southwestern part of the North Pacific and the Kuroshio. Part II. Characteristic water masses which are related to this region, and their mixtures, especially the water of the Kuroshio. *Mem., Imp. Mar. Obs., Kobe*, **6**: 349–414.

—— (1939): On the hydrography of the southwestern part of the North Pacific and the Kuroshio. Part III. Oceanographical investigations of the Kuroshio area and its outer regions: Development of ocean currents in the North Pacific. *Mem., Imp. Mar. Obs., Kobe*, **7**: 41–114.

Koizumi, M. (1955): Researches on the variations of oceanographic conditions in the region of the ocean weather station "Extra" in the North Pacific Ocean. (I) Normal values and annual variations of oceanographic elements. *Pap. Meteor. Geophys.*, **6**: 185–201.

Kuksa, V. I. (1962): On the formation of the layer of low salinity in the northern part of the Pacific Ocean. *Okeanologiia*, **2**: 769–782.

—— (1963): Basic regularity in formation and distribution of intermediate waters in the northern part of the Pacific Ocean. *Okeanologiia*, **3**: 30–43.

Masuzawa, J. (1952): On the heat exchange between sea and atmosphere in the Southern Sea of Japan. *Oceanogr. Mag.*, **4**: 49–55.

—— (1954): On the Kuroshio south of Shionomisaki of Japan. *Oceanogr. Mag.*, **6**: 25–33.

—— (1956): A note on the Kuroshio farther to the east of Japan. *Oceanogr. Mag.*, **7**: 97–104.

—— (1957): A contribution to the knowledge on the Kuroshio east of Japan. *Oceanogr. Mag.*, **9**: 21–34.

—— (1960): Western boundary currents and vertical motions in the subarctic North Pacific Ocean. *J. Oceanogr. Soc. Jap.*, **16**: 29–33.

Masuzawa, J., T. Tsuchida and T. Inoue (1962): The monthly mean sea-surface temperature in the north-

western Pacific. *Oceanogr. Mag.*, **13**: 77–87.

Masuzawa, J. (1964a): Flux and water characteristics of the Pacific North Equatorial Current: 121–128. *In: Studies on Oceanography*, K. Yoshida, editor, Tokyo. 560pp.

—— (1964b): A typical hydrographic section of the Kuroshio Extension. *Oceanogr. Mag.*, **16**: 21–30.

—— (1965a): A note on the seasonal variation of the Kuroshio velocity (in Japanese). *J. Oceanogr. Soc. Jap.*, **21**: 117–118.

—— (1965b): Water characteristics of the Kuroshio. *Oceanogr. Mag.*, **17**: 37–47.

—— (1967): An oceanographic section from Japan to New Guinea at 137°E in January 1967. *Oceanogr. Mag.*, **19**: 95–118.

—— (1968): Second cruise for CSK, *Ryofu Maru*, January to March 1968. *Oceanogr. Mag.*, **20**: 173–185.

—— (1969a): The Mindanao Current (in japanese). *Bull. Jap. Soc. Fish. Oceanogr. Dr. Uda's Vol.*, 38–43.

—— (1969b): Subtropical Mode Water. *Deep-Sea Res.*, **16**: 463–472.

Miyake, Y. and K. Saruhashi (1956): On the distribution of the dissolved oxygen in the ocean. *Deep-Sea Res.*, **3**: 242–247.

Montgomery, R. B. (1954): Analysis of a *Hugh M. Smith* oceanographic section from Honolulu southward across the equator. *J. mar. Res.*, **13**: 67–75.

Montgomery, R. B. and W. S. Wooster (1954): Thermosteric anomaly and the analysis of serial oceanographic data. *Deep-Sea Res.*, **2**: 63–70.

Montgomery, R. B. (1958): Water characteristics of Atlantic Ocean and of World Ocean. *Deep-Sea Res.*, **5**: 134–148.

—— (1959): Salinity and residence time of subtropica, oceanic surface water. *Rossby Memorial Volume*, 143–146.

Montgomery, R. B. and E. D. Stroup (1962): Equatorial waters and currents at 150°W in July–August 1952. *Johns Hopkins Oceanogr. Studies*, **1**: 68pp.

—— (1969): The words naviface and oxyty. *J. mar. Res.*, **27**: 161–162.

Muromtsev, A. M. (1963): Atlas of temperature, salinity and density of water in the Pacific Ocean. *Mezhduvedomstvennuy Geofizicheskii Komitet pri Prezidiume Akademii Nauk*, SSSR, Moscow, 120pp.

Okubo, A. (1958): The distribution of dissolved oxygen in the northwestern part of the North Pacific Ocean in the aspect of physical oceanography. Part 1. General features of the oxygen distribution (I). *Oceanogr. Mag.*, **10**: 137–156.

Reid, J. L. Jr. (1961): On the geostrophic flow at the surface of the Pacific Ocean with respect to the 1000-decibar surface. *Tellus*, **13**: 489–502.

—— (1965): Intermediate waters of the Pacific Ocean. *Johns Hopkins Oceanogr. Studies*, **2**: 85pp.

—— (1966): Zetes Expedition. *Trans., Amer. Geophys.*

Union, **47**: 555–561.

Reid, R. O. (1948): The equatorial currents of the eastern Pacific as maintained by the stresses of the wind. *J. mar. Res.*, **7**: 74–99.

Rochford, D. F. (1960): The intermediate depth waters of the Tasman and Coral Seas. I. The 27.20 σ_t surface, II. The 26.80 σ_t surface. *Aust. J. mar. Fish. Res.*, **11**: 127–165.

Schroeder, E., H. Stommel, D. Menzel and W. Sutcliffe (1959): Climate stability of eighteen degree water at Bermuda. *J. geophys. Res.*, **64**: 363–366.

Seitz, R. C. (1967): Thermostad, the antonym of thermocline. *J. mar. Res.*, **25**: 203.

Stommel, H. and A. B. Arons (1960): On the abyssal circulation of the world ocean—II. An idealized model of the circulation pattern and amplitude in oceanic basins. *Deep-Sea Res.*, **6**: 217–233.

Sturges, W. (1965): Water characteristics of the Caribbean Sea. *J. mar. Res.*, **23**: 147–162.

Suda, K. (1938): On the variations of the oceanographical state of the Kuroshio in the original region (Part 1). *Geophys. Mag.*, **11**: 373–410.

Sverdrup, H. U., M. W. Johnson and R. H. Fleming (1942): *The Oceans, their physics, chemistry and general biology.* Prentice-Hall, New York, 1087pp.

Terada, K. and K. Osawa (1953): On the energy exchange between sea and atmosphere in the adjacent seas of Japan. *Geophys. Mag.*, **24**: 155–170.

Tsuchiya, M. (1961): An oceanographical description of the equatorial current system of the western Pacific. *Oceanogr. Mag.*, **13**: 1–30.

——— (1968): Upper waters of the intertropical Pacific Ocean. *Johns Hopkins Oceanogr. Studies*, **4**: 50pp.

Tsumura, K. (1963): Investigation of the mean sea level and its variation along the coast of Japan (Part I)—Regional distribution of sea level variation (in Japanese). *J. Geodetic Soc. Jap.*, **9**: 49–90.

Uda, M. (1935a): The results of simultaneous oceanographical investigation in the North Pacific Ocean adjacent to Japan made in August, 1933 (in Japanese). *J. Imp. Fish. St.*, **6**: 1–130.

——— (1935b): On the distribution, formation and movement of the dichothermal water off the northeast of Japan (in Japanese), *Umi to Sora*, **15**: 445–452.

——— (1963): Oceanography of the Subarctic Pacific Ocean. *J. Fish. Res. Bd., Canada*, **20**: 119–179.

Uda, M. and K. Hasunuma (1969): The eastward Subtropical Countercurrent in the western North Pacific Ocean. *J. Oceanogr. Soc. Jap.*, **25**: 201–210.

Worthington, L. V. (1959): The 18° water in the Sargasso Sea. *Deep-Sea Res.*, **5**: 297–305.

Wüst, G., W. Brogmus and E. Noodt (1954): Die zonale Verteilung von Salzgehalt, Niederschlag, Verdunstung, Temperatur, und Dichte an der Oberfläche der Ozeane. *Kieler Meeresforschungen*, **10**.

Wyrtki, K. (1961): Physical oceanography of the Southeast Asian Waters. Scientific results of marine investigations of the South China Sea and the Gulf of Thailand, 1959–1961. *Naga Report*, **2**: 195pp.

——— (1965a): The average annual heat balance of the North Pacific Ocean and its relations to ocean circulation. *J. geophys. Res.*, **70**: 4547–4559.

——— (1965b): The annual and semiannual variation of sea surface temperature in the North Pacific Ocean. *Limnology and Oceanography*, **10**: 307–313.

Yoshida, K. and T. Kidokoro (1967): A Subtropical Countercurrent in the North Pacific—An eastward flow near the Subtropical Convergence. *J. Oceanogr. Soc. Jap.*, **23**: 88–91.

Chapter 5

BEGINNING OF THE KUROSHIO

HIDEO NITANI

Hydrographic Department, Maritime Safety Agency, Tokyo, Japan.

1 INTRODUCTION

The term "beginning of the Kuroshio" often means the current east of Luzon Island and Taiwan which is a continuation of the North Equatorial Current. In this chapter, however, the current in the somewhat broad region from 130°E in the Pacific to west of the Luzon Strait, and from the south boundary of the North Equatorial Current to the north boundary of the Kuroshio in the East China Sea are treated as the area of the beginning of the Kuroshio.

The Kuroshio just south and east of Japan has been studied since 1893; in that year Wada was the first in Japan to study the Kuroshio with use of drift bottles. Since that time, many observations and studies of the Kuroshio have been carried out mainly by Japanese oceanographers. However, in the regions of the North Equatorial Current west of about 130°E, east of the Philippines, the Luzon Strait and east of Taiwan, oceanographic observation data have been rather scarce, especially after World War II.

Kishindo (1931,1932), Shigematsu (1932), Wüst (1936), Koenuma (1936–41), Sverdrup et al. (1942) and some other oceanographers studied the beginning of the Kuroshio. Koenuma's papers were very extensive studies of water characteristics. After World War II, Uda (1955, 1964), Wyrtki (1961), Nitani (1961), Yamanaka et al. (1965), Yoshida and Kidokoro (1967) and Takahashi and Chaen (1967) studied this area, but not enough to know the beginning of the Kuroshio in detail.

In this area, occasional oceanographic observations have been made in the past years, but broad and systematic observations of the beginning of the Kuroshio on a synoptic basis, including the sea east of the Philippines, the Luzon Strait, the South China Sea, the sea off Taiwan and the East China Sea, have never been carried out. In the western subtropical and equatorial regions, fragmentary observations on

board the *Snellius, Dana* and *Albatross* were made. Before World War II, fairly systematic, but not complete, observational programs operated by the Japanese Hydrographic Office using the R/V *Manshu* (1925–1928) and many other vessels (1933–1942) were carried out. The R/V *Komahashi, Katsuriki, Kosyu, No. 1,2,3 Kaiyo, Soyo Maru* and *Toyama Maru* and many catcher boats were mainly engaged in these observations. The most systematic synoptic observation was carried out in the summer of 1939, but the vessels used were mainly chartered catcher boats. The level of accuracy of observation by catcher boats is lower than that by research vessels, and the data obtained by them seem to be inappropriate for detailed discussions.

After the War, only EQUAPAC (1956) and IGY (1958) cruises were systematic, but the main areas covered by these were to the east of long. 130°E. A few scattered observations extending to the Philippine coast were carried out on board the *Spencer F. Baird* (1948–1949) and the *Takuyo* (1959). Of course from many data cited above, the fact that the Kuroshio forms one part of the North Pacific Circulation continuing from the North Equatorial Current is well known, but detailed knowledge about its origin, speed, volume transport and many other characteristics are still obscure.

Observations of the Kuroshio in the East China Sea have been carried out more frequently than south of the Ryukyu Islands in the periods before and after World War II, and many papers have been published (by the authors mentioned above, and Koizumi, 1962, 1964; Fujii and Kimura, 1960; Fujii, 1961; Asaoka and Moriyasu, 1966; etc.). Regular seasonal oceanographic observations in the East China Sea have been carried out since 1955 by the Nagasaki Marine Observatory.

The Cooperative Study of the Kuroshio and Adjacent Regions (CSK) started in the summer of 1965 under the recommendation of UNESCO, and the first program of synoptic observation in these areas was carried out. Since that time, synoptic observations have been made twice a year, in summer and in winter, and they will continue until 1970 or later. These observations are probably the most systematic oceanographical observations of these areas since World War II. The data obtained from these new synoptic observations will probably be more excellent in accuracy and method than those obtained before the War, for example, in the summer of 1939. Therefore, in this chapter, much use is made of data

from the cruises of the CSK. For example, the oceanographic stations of the CSK synoptic observations in the summer of 1965 and in the winter of 1966 are shown in Fig.1(a), (b). In general, the stations are more dense in summer than in winter, and the data from the area just east of Luzon and the Luzon Strait are lacking in winter.

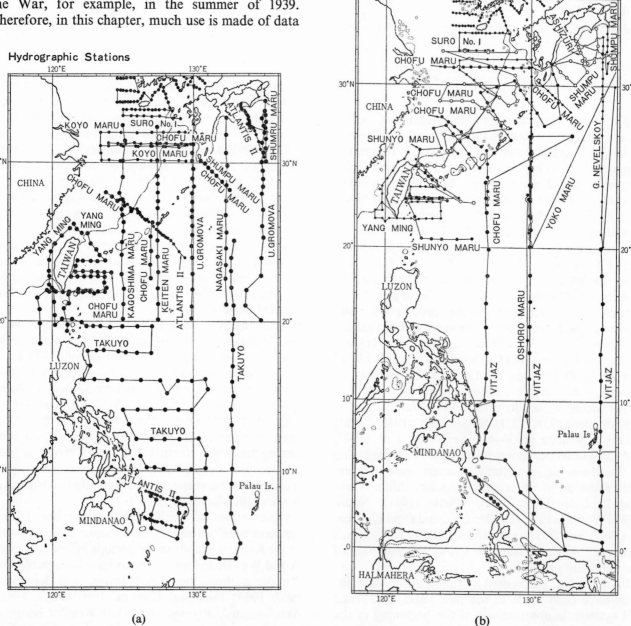

(a)   (b)

FIG. 1   Oceanographic stations of CSK synoptic observations in the northwestern Pacific, (a) in the summer of 1965 and (b) in the winter of 1966. Both are reprinted from the CSK ATLAS for summer 1965 and the CSK ATLAS vol. 2 for winter 1965–1966, respectively, issued by the Japanese Oceanographic Data Center.

2 SURFACE CURRENTS IN THE SOUTH-WESTERN REGION OF THE NORTH PACIFIC AND IN THE EAST CHINA SEA

2.1 Current charts published in 1935

Figure 2(a), (b) shows the current charts for summer and winter published as a classified publication in 1935 by the Japanese Hydrographic Office of the Imperial Navy (before 1945, the Hydrographic Office in Japan belonged to the Navy). These charts show the mean current derived from data obtained between 1924 and 1934 by naval research vessels and reports of Japanese warships and merchant ships. In these figures, the northward current appears to increase its velocity in the sea east of southern Taiwan. Off the eastern coast of Luzon, the current is not so distinct owing to the lack of data, though the northward current along the coast exists. The existence of the southward Mindanao Current is clearly shown in these figures. The clockwise eddy is seen on the right-hand side of the Kuroshio between 15°N and 25°N, and the anti-clockwise eddy, having its center near 8°N, 130°E, exists to the east of Mindanao. The

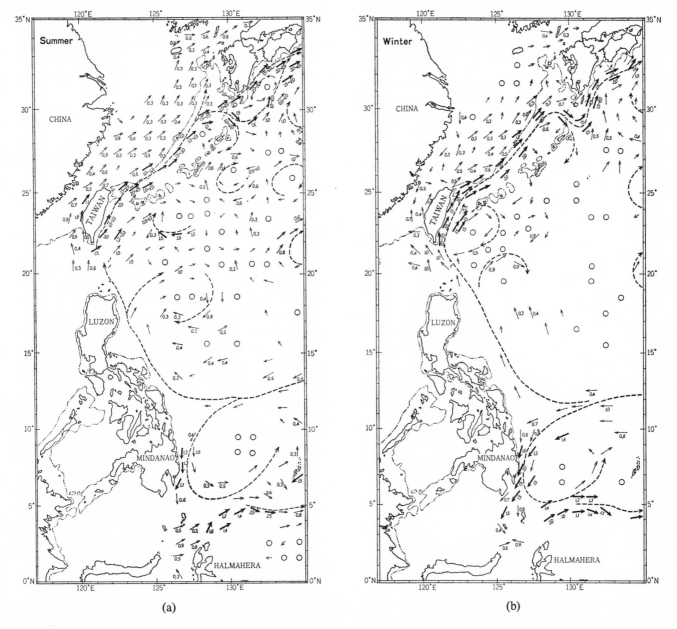

(a)   (b)

FIG. 2   Current charts in knots, (a) in the summer and (b) in the winter, published in 1935 by Japanese Hydrographic Office of the Imperial Navy.

distinct differences of the currents in two different seasons are seen. In summer, the velocity of the Mindanao Current is lower than in winter. The Osumi branch of the Kuroshio, flowing between Yakushima Island and Osumi Peninsula, is stronger in summer than in winter. In winter, the surface water of the Pacific enters the South China Sea through the Luzon Strait, but this could not be ascertained for summer.

2.2 Separation of the North Equatorial Current off the Philippines

The results of GEK observations in the summer of 1965 are shown in Fig. 3. The dynamic height anomalies at the sea surface referred to 1200 db in the summers of 1965 and 1966 are shown in Fig. 4(a), (b). From Figs. 2, 3 and 4, the figures in Wyrtki's paper (1961) and those published by the U.S. Navy in 1945 and 1964 as Pub. No. 236 and 237, the outline of the currents in these regions will be explained.

The North Equatorial Current flows westward and is divided into two currents, moving southward and northward, off the coast of southern Luzon or Samar. The variation of this region of divergence at the sea

TABLE 1. Region of divergence of the North Equatorial Current east of the Philippines.

Year	Month	Region of divergence at the sea surface	Vessel
1934	Dec.	11°N	*Komahashi*
1935	Dec.	12.5°N	*Katsuriki*
1939	Aug.	12°N	*No. 3 Takunan Maru*
1942	Nov.	13°N	*Toyama Maru*
1949	June	12°N	*Spencer F. Baird*
1965	July	14.5°N	*Takuyo*
1966	Aug.	12.5°N	*Takuyo*
1967	Jan.	12°N	*Ryofu Maru*
1968	Mar.	12–13°N	*Ryofu Maru*

surface ranges from about 11°N to 14.5°N as shown in Table 1. Seven out of nine cases are between 12°N and 13°N. This place, as a rule, shifts to the north with increasing depth. For example, the observation of the *Takuyo* in Aug. 1966 shows that it is at lat. 12.5°N for the sea surface and 15°N for below 300 meters.

The southward current is the Mindanao Current. It flows southward along the coast of Mindanao Island and into the Equatorial Countercurrent. The South Equatorial Current which joins the Countercurrent comes from north of New Guinea. According to Figs. 2(a) and (b), and geostrophic calculation, the Mindanao Current has a velocity of about 1–2 knots.

2.3 Currents in the beginning of the Kuroshio

According to GEK observations, the northward current flows along the coast of Luzon Island with a maximum velocity of 1.5–2.0 knots and enters in the Luzon Strait. Then it turns to the northwest and its main axis reaches as far west as 121°E, with a maximum velocity of about 3.0 knots. One branch often goes westward at 20°N and enters the South China Sea, and most parts of it go round the warm eddy and then return to the main axis of the Kuroshio. The Kuroshio in the Luzon Strait comes again into the Pacific Ocean and flows north-northeast along the eastern coast of Taiwan with a maximum velocity of about 3.0 knots or more.

According to the above-mentioned average current charts, the velocities of the northward current off Luzon and in the Luzon Strait are 0.5–1.0 and about 1.0 knot, respectively. The axis of the current in the Luzon Strait is comparatively simpler than those obtained by GEK observations and geostrophic flows. The decrease in velocity and the simplicity of the current pattern on the average charts may be due to the character of the averaging process.

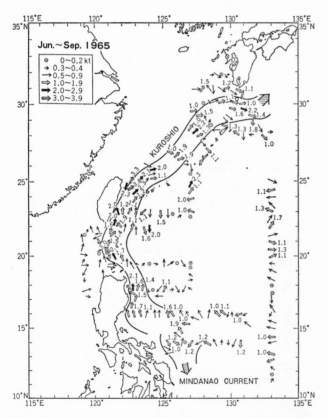

FIG. 3   Results of GEK observations carried out by the R/V *Takuyo* and *Chofu Maru* in the summer of 1965.

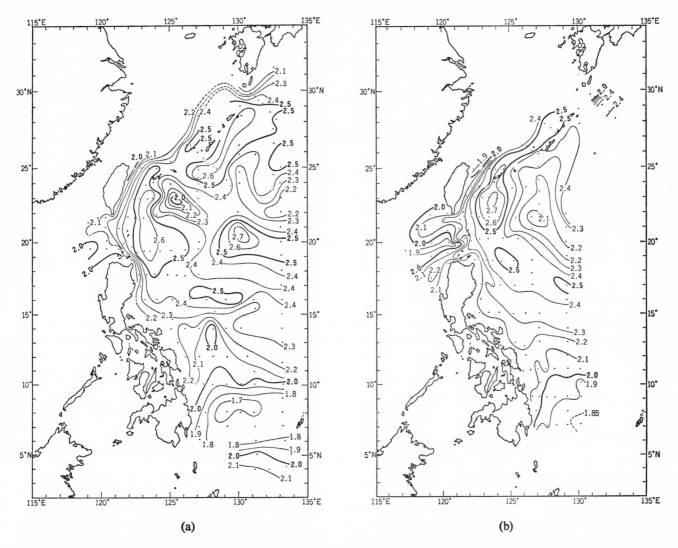

(a)   (b)

FIG. 4   Geopotential topography at the sea surface (in dynamic meter) referred to 1200 db surface in the summer (a) 1965 and (b) 1966.

2.4   Countercurrent associated with warm eddies at the right-hand side of the Kuroshio in the Luzon Strait

There is a noticeable countercurrent having a velocity of 1.0–2.0 knots, associated with the warm eddy located to the east of the Kuroshio in the Luzon Strait. The existence of such a warm eddy was pointed out by Shigematsu (1932) and Koenuma (1936) from the results of the R/V *Manshu* operations in 1925–1928, and by Nitani using the data of the R/V *Takuyo* operation in May 1959. Observations on board the *No. 1 Kaiyo* (Aug.–Sept., 1942) and the *Ryofu Maru* (Feb., 1967) also show the existence of the warm eddy (Fig. 5). When the warm eddy is strong, it extends as a belt from about 130°E or farther east to east of Taiwan along the right-hand side of the

Kuroshio. This warm eddy seems to exist permanently, though its scale and position vary from time to time. It is difficult to say whether the countercurrent associated with this warm eddy corresponds to Munk's shown in his wind-driven current (1950), because there are several clockwise eddies along the right-hand side of the Kuroshio.

2.5   Kuroshio in the East China Sea

The Kuroshio east of Taiwan enters the East China Sea through the passage between Taiwan and Yonakunijima Island, the western tip of the Ryukyu Islands. The Kuroshio in the East China Sea flows northeasterly along the continental slope and the southern end of the continental shelf with a maximum velocity of about 1.5–3.0 knots (Fig. 6). From

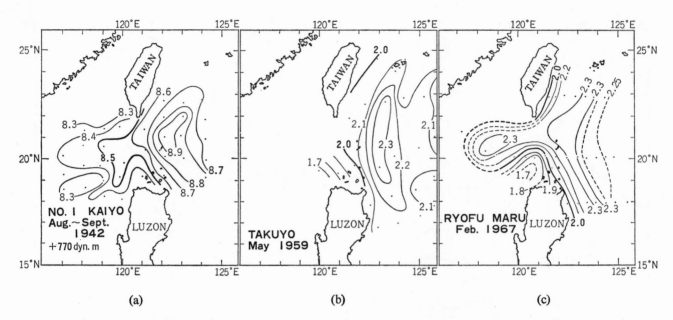

FIG. 5   Geopotential topography in and near the Luzon Strait: (a) in Aug.–Sept. 1942 based on 800 db surface (No. 1 *Kaiyo*); (b) in May 1959 based on 1000 db surface (*Takuyo*); (c) in Feb. 1967 based on 1200 db surface (*Ryofu Maru*).

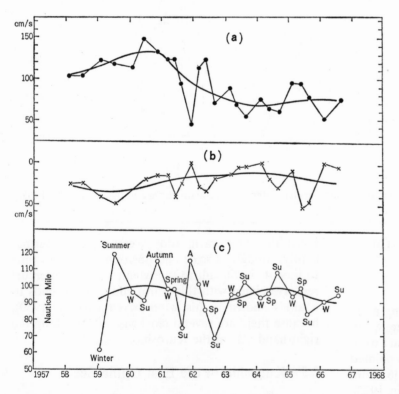

FIG. 6   (a) Maximum geostrophic velocity of the Kuroshio northwest of Okinoerabujima Island; (b) Maximum geostrophic velocity of the countercurrent between the Kuroshio in the East China Sea and the Ryukyu Island; (c) Distance of the axis of the Kuroshio from Okinoerabujima Island.

the results of half-yearly observations which have been carried out by the Nagasaki Marine Observatory since 1955, some characteristics of the Kuroshio in this region can be described. The mean position of the Kuroshio axis is located about 90–100 nautical miles northwesterly from Okinoerabujima Island which is located at 27°25′N lat. and 128°30′E long.

The position of the axis of the Kuroshio fluctuates by about 20 nautical miles. The countercurrent associated with the warm eddy between the Kuroshio and Ryukyu Islands is seen frequently, and its maximum velocity amounts to about 1.0 knot and 0.5 knots at a maximum and on the average, respectively (Fig. 6).

The Tsushima Current separates from the Kuro-

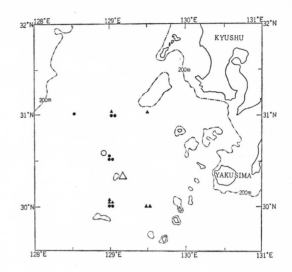

FIG. 7   Region of divergence of the Tsushima Current from the Kuroshio in summer (▲) and in winter (●), determined by GEK observation and temperature distributions at depths of 100 and 200 meters. The mean summer and winter positions are shown by a large triangle and a large circle.

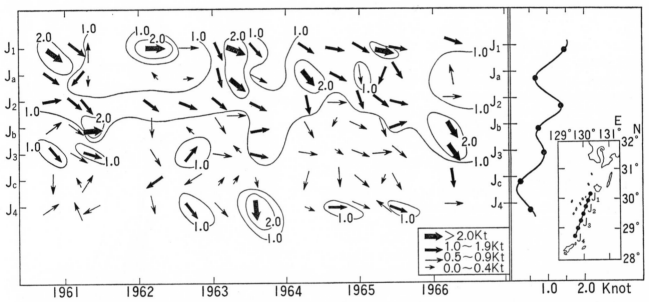

FIG. 8   Left; Surface current in the Tokara Strait obtained by GEK observation from 1960 to 1966. The stations are shown in the right portion of this figure.
Right; Velocity component perpendicular to the Tokata Strait.

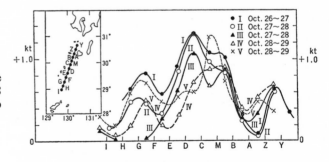

FIG. 9 (a)   Velocity component perpendicular to the Tokara Strait obtained by GEK observations, Oct. 26–29, 1968 on board the *Satsuma*. Each value is averaged between two succesive stations shown in the inserted chart.

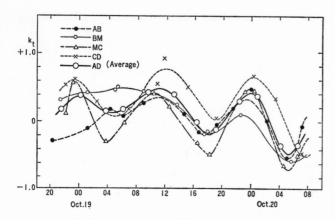

FIG. 9(b) Mean velocities perpendicular to the Toka-ra Strait obtained by GEK between two succesive sta-tions and between station A and station D in the figure. This observation was made on board the *Takuyo* and the *Satsuma*, Oct. 18–20, 1968.

shio at a mean location of about 30°30′N lat. and 129°E long., though this region of divergence is not so distinct. The variation is shown in Fig. 7, indicat-ing no significant seasonal variation. After separation of the Tsushima Current, the Kuroshio flows east or east-southeast and enters into the Pacific through the Tokara Strait. The axis of the Kuroshio at the Tokara Strait is in its northern half and is frequently very close to Yakushima Island. In Fig. 8, the results of GEK observations at the Tokara Strait in 1960–1966 are shown. The average velocity component per-pendicular to the Tokara Islands is shown at the right side of this figure. The current has some maxima and minima in velocity, and therefore it has a striped character. This character is confirmed by the observation of the short periodic variation of the Kuroshio at the Tokara Strait made in Oct. 26–29, 1968 on board the *Satsuma* (Nitani, unpublished), (Fig. 9a). The striped character is probably due to the effects of some islands and bottom topography in the Tokara Strait, because this type of character is hardly seen in the Kuroshio at the section south-east of Yakushima Island, downstream 30–40 nauti-cal miles from the Tokara Strait. Another result obtained from the above-mentioned observation in the Tokara Strait is shown in Fig. 9(b). The GEK observations were made on board the *Takuyo* and the *Satsuma* at the northern half of the Tokara Strait on Oct. 18–20, 1968. The mean surface velocity com-ponent between Yakushima Island and Nakanoshima Island perpendicular to the Tokara Islands, which was obtained every three hours, shows the semidiurnal variation. Diurnal or semidiurnal variations are often seen in the Kuroshio south of the Kii Peninsula (Shoji and Nitani, 1966; Masuzawa, 1968; and Nitani

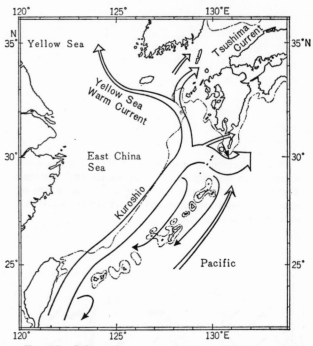

FIG. 10 Schematic representation of the current system in the East China Sea.

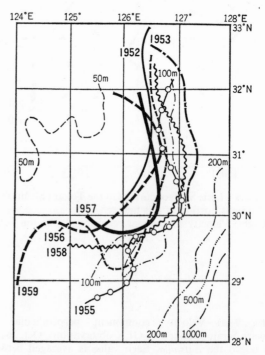

FIG. 11 Variation of the isoline of 19.00‰ of mean chlorinity in the upper 75 m, February 1952–1959 (after KATO, 1959; Fig. 7).

and Shoji, 1970), and this semidiurnal variation in the Tokara Strait may not be the pure tidal current as in the open sea. Some parts of the Kuroshio pass through the Osumi Strait frequently, and enter into the Pacific. According to Fujii and Kimura (1960), this Osumi Branch disappears accompanied by a decrease of temperature at the north coast of Yakushima Island when a northerly or northwesterly wind prevails. Most parts of the Tsushima Current including the surrounding cold and less saline water in the East China Sea and the Yellow Sea enter into the Japan Sea through the Tsushima Strait, but some parts of the current branch off east of the Goto Islands before reaching the Tsushima Strait, and turn northwest and enter into the Yellow Sea.

A narrow countercurrent flows just southeast of the Ryukyu Islands, and at the offshore side of this countercurrent a northeasterly current flows.

A schematic representation of the current system in the East China Sea is shown in Fig. 10. The main feature of the current system varies with the abundance and decay of the Yellow Sea Cold Water. Kato (1959) showed fluctuations of 19.00‰ isoline of mean chlorinity in the upper 75 m in February, which

is a good indicator of the boundary between the Kuroshio, the Tsushima Current and the Yellow Sea Cold Water in the East China Sea (Fig. 11). Koizumi (1964) studied the standard deviation of the surface temperature (σ) in the East China Sea and the Yellow Sea, and showed that this quantity in the region of the Yellow Sea Cold Water reaches an average of about 1.0°C and is about two times that in the Kuroshio region, indicating the large effect of the outflowing continental waters. Further, σ in the region of the Tsushima Current in the East China Sea falls between the values of σ in the region of the Yellow Sea Cold Water and that in the Kuroshio region, and σ is largest in the region of the Yellow Sea Warm Current corresponding to its instability.

2.6   Definition of the Kuroshio

Kishindo (1931) defined the Kuroshio as follows:

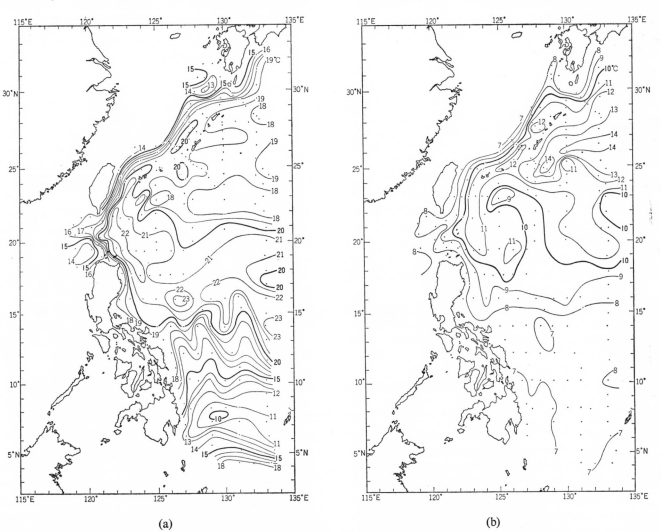

(a)   (b)

FIG. 12   Distributions of temperature at the depths of (a) 200 m and (b) 500 m in the summer of 1965.

Takuyo (Jun.~Jul. 1965)

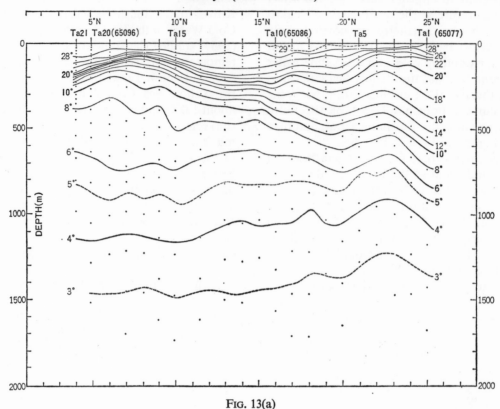

FIG. 13(a)

Takuyo (Aug. 1965)

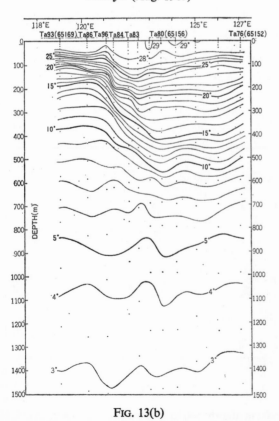

FIG. 13(b)

"The northward or eastward current, of which the characteristic water is pelagic, existing from offshore east of Taiwan to offshore of the Tohoku district of Japan, and which is to be called the Kuroshio." Sverdrup et al. (1942) treated the Kuroshio as a part of the Kuroshio system, and defined it as follows: "The current running northeast from Taiwan to Ryukyu and then close to the coast of Japan as far as lat. 35°N." On the other hand, Wyrtki (1961) called the flow east of Taiwan the Formosa Current. Wüst (1936) compared the Kuroshio in the East China Sea and in the Tokara Strait with the flow in the Caribbean Sea and the Florida Strait, respectively.

The separation into two parts off the eastern coast of the Philippines is clear and the northward-moving branch of the current grows strong enough to be recognized as a distinct current off the eastern coast of Luzon and in the Luzon Strait (Figs. 3, 4). We come to the conclusion that the current from just north of the place where the North Equatorial Current is separated into two branches offshore east of the Philippines to the east of Japan where the current veers away from land can be called the Kuroshio in a broad sense, or that this current can be divided into two or three in nomenclature: for example, the Luzon Current, the Taiwan Current and the Kuroshio. I prefer the former definition. Strictly speaking, the

Atlantis II (Aug. 1965)

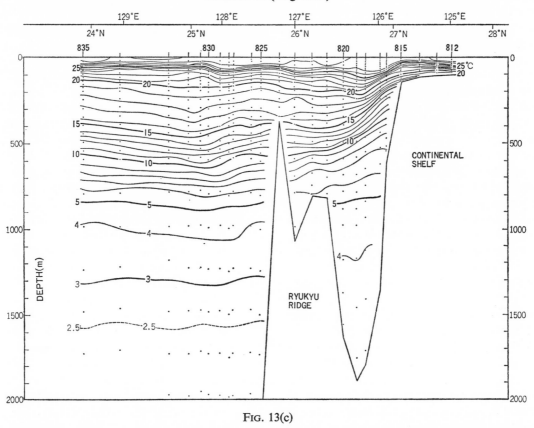

FIG. 13(c)

FIG. 13   Temperature sections in the summer of 1965: (a) across the North Equatorial Current (*Takuyo*) on the 133°E line; (b) across the Kuroshio in the Luzon Strait (*Takuyo*) on the 19°30′N line; (c) across the Kuroshio in the East China Sea (*Atlantis* II).

beginning of the Kuroshio is in the region off the east coast of Luzon, but it may be appropriate to define the Kuroshio between east of Luzon and Taiwan and, sometimes, in the East China Sea as the beginning of the Kuroshio in a general sense.

3   DISTRIBUTION OF TEMPERATURE AND SALINITY

3.1   Temperature

The horizontal temperature distributions at depths of 200 m and 500 m in the summer of 1965 are shown in Fig. 12(a), (b). The distribution of the temperature at 200 m is similar to the dynamic height anomaly at the sea surface, but at 500 m it is similar to that in the region north of 15°N. In the region south of 15°N, the temperature at 500 m is nearly uniform corresponding to the shallowness of the current in this region.

Of course, the horizontal temperature gradient at 200 m is very large in the region of the Kuroshio. The approximate 18°C isotherm at the depth of 200 m

is a good indicator of the main axis of the Kuroshio at the sea surface in the regions from the Luzon Strait to the Tokara Strait. The 15–17°C isotherms are its counterpart south of Japan.

As mentioned in the preceding section, a long, warm core, like a belt, whose temperature is higher than 22°C in the region south of Taiwan and higher than 20°C in the East China Sea, exists having its center line on the right side of the Kuroshio. It continues to the east along the north part of the North Equatorial Current and lies between 15–20°N. This feature is seen in the statistical treatment of temperature in this region carried out by Winterfeld and Stommel (Chapter 3, Plate 1). The phenomenon is well known in the Kuroshio region south of Japan. The Kuroshio Countercurrent flows along the right side of this warm core. There is a tendency for this warm core to become an isolated warm eddy east of the Luzon Strait and off Okinawa Island accompanying a clockwise countercurrent. Then, on the right side of the Kuroshio, there are some isolated warm eddies, the temperatures of which are higher than those of the surrounding warm core along the Kuroshio and which can be likened to peaks in a mountain chain. Below the depth of 300 m in this region, the warm core or eddy is very weak or disappears, and its depth is shallower than that near Japan.

In the Luzon Strait, the 16°C isotherm is an indicator of the left-hand boundary of the Kuroshio, and intrusions of a cold eddy from the South China Sea and a warm eddy from the Pacific are seen at the south and north parts of the Luzon Strait, respectively; the Kuroshio meanders corresponding to these eddies. The 11°C isotherm at the depth of 200 m is a good indicator of the boundary between the North Equatorial Current and the Equatorial Countercurrent. There is a cold eddy associated with the upwelling of deep water. This cold eddy is surrounded by the North Equatorial Current, the Mindanao Current and the Equatorial Countercurrent. Shigematsu (1932), Takahashi (1959), Wyrtki (1961) and Reid (1961) showed the existence of this eddy. According to Takahashi, it corresponds to Munk's western boundary vortex between the North Equatorial Current and the Equatorial Countercurrent.

The vertical sections of temperature at 133°E across the North Equatorial Current (*Takuyo*), at 19°30′ N across the Luzon Strait (*Takuyo*) and across the Ryukyu Islands (*Atlantis* II) observed in the summer of 1965 are shown in Fig. 13(a), (b), (c), respectively. In section 133°E, the thermocline is shallowest at 7.5°N, and the vertical gradient of temperature is largest here reaching about 0.1°C/m. In section 19°30′N, the largest slope of the isotherms is found between 121°E and 123°E corresponding to the Kuroshio, and the largest vertical gradient is found west of 120°E having a value of about 0.06°C/m. In the section of the Ryukyu Islands, the largest slope of the isotherms is found on the continental slope in the East China Sea, and the largest vertical gradient is about 0.08°C/m. In the East China Sea, the temperature decreases very slowly with depth below about 1200 m, and the bottom temperature is 3.7°C which is lower by about 1.5°C than at the same depth in the Pacific in the same section due to the Ryukyu Ridge.

The so-called "18°C water" seen in the western subtropical region of the North Pacific (Masuzawa, 1965, 1967; Takahashi and Chaen, 1967) and of the North Atlantic (Worthington, 1959) appears in the region north of 20°N on the 133°E line, east of 123°E on the 19°30′N line and southeast of 26°30′N and 126°30′E in the section of the *Atlantis* II. Masuzawa (1967) called this water "Subtropical Mode Water."

3.2 Salinity

The pattern of salinity distribution at the 200 m layer in the summer of 1965 (Fig. 14a) is similar to the dynamic height anomaly of the sea surface, as is the temperature distribution. At the depth of 200 m, the isolines of 34.8–34.9‰ east of Luzon, 34.7–34.8‰ north of the Luzon Strait and 34.6–34.8‰ in the East China Sea correspond to the main part of the Kuroshio at the sea surface.

A low salinity eddy is found at 7.5°N east of Mindanao. The high-salinity water comes from the South Pacific to about 6°N, near the northern limit of the Equatorial Countercurrent. The salinity of the South China Sea near the Luzon Strait is about 0.3‰ lower than that of the Pacific. A salinity maximum which is higher than 34.9‰ extends between 15°N and 20°N corresponding to that of temperature, but north of 20°N the existence of the salinity maximum is not so clear as that of temperature.

At the depth of 500 m (Fig. 14b), the main salinity minimum which is lower than 34.3‰ extends between 15°N and 20°N. In the region south of 10°N, the salinity distribution is comparatively uniform, and this layer nearly corresponds to the apparent second maximum layer of salinity in this region. In the South China Sea, the salinity is about 0.1‰ higher than that just east of the Luzon Strait, because most of the North Intermediate Waters are prevented from passing the Luzon Strait across the Kuroshio.

The vertical sections of salinity at 133°E across the North Equatorial Current, at 19°30′N across the Luzon Strait and the line across the Ryukyu Islands observed in the summer of 1965 are shown in Fig. 15(a), (b), (c). On the section of 133°E, a salinity maximum of about 35.00‰ is found at a depth between 100 m and 200 m, and this saline water reaches to 6–7°N, the boundary between the North Equatorial Current and the Equatorial Countercurrent, decreasing with distance southward. Another salinity maximum of about 35.3‰ extending from the South Pacific is seen clearly and reaches to 5–7°N. The minimum layer slopes upward to the south, and it reaches to about 4°N or farther south. A weak salinity minimum extending from the south is seen at a depth of 600–700 m, and the minimum value of about 34.53‰ reaches to 10°N. Because of the presence of two minimum layers extending from the North and South Pacific, an apparent maximum is seen between two minimum layers, though it is not significant from the viewpoint of water-mass analysis as Wyrtki (1961) points out.

In the vertical section at 19°30′N, the salinity maximum and minimum layers slope up suddenly to the west from 123°E associated with the Kuroshio, and reach to the South China Sea, but their values are extensively weakened because the bottom topography in the Luzon Strait and the Kuroshio play the role of a barrier against invasion of the Pacific waters. In the South China Sea, there is less saline continental surface water whose salinity is less than 34.0‰.

In the East China Sea, the above-mentioned two waters of high and low salinity come from the Pacific across the passages between the Ryukyu Islands. There are probably two main passages, one between Yonakunijima Island and Taiwan (Koenuma, 1938) and the other between Miyakojima Island and Okinawa Island. Saline water having the value of 34.70‰ reaches the continental shelf and creeps up along the bottom. The Yellow Sea Water whose salinity is less than 34.0‰ comes from northwest at the surface. In the deep layer of the East China Sea, there is uniform bottom water whose salinity is 34.44‰ at 1200 m and 34.45‰ near the bottom.

4 WATER CHARACTERISTICS

4.1 Water masses in the western North Pacific

Koenuma (1939) subdivided the southwestern part of the North Pacific Ocean north of 20°N into thirteen regions, seven of which were in the Kuroshio region, considering the individual water characteristic and showing typical T-S curves in each subdivided region. On the other hand, from a broader viewpoint, Sverdrup, Johnson and Fleming (1942) showed three principal water masses in upper layers, the Western North Pacific Water, the Western South Pacific Water and the Pacific Equatorial Water, in the western Pacific Ocean. The main waters which form these water masses are the North and South Pacific high-salinity waters and the North and South Pacific intermediate waters.

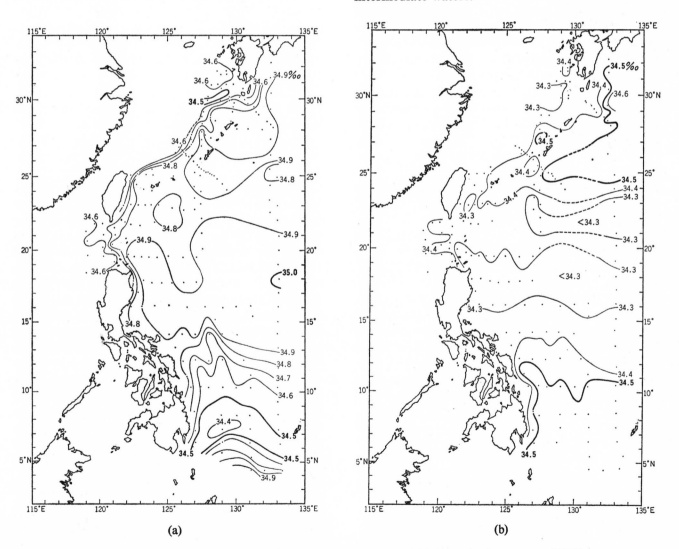

(a)   (b)

FIG. 14   Distribution of salinity at the depths of (a) 200 m and (b) 500 m in the summer of 1965.

Takuyo (Jun.~Jul. 1965)

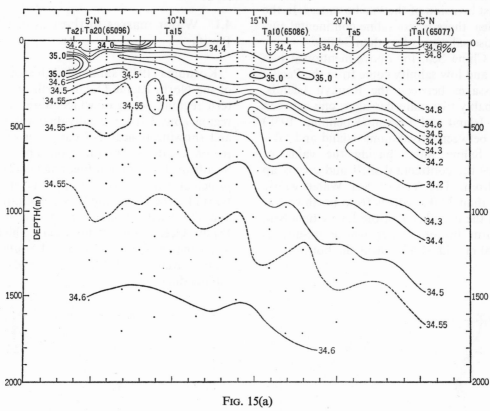

FIG. 15(a)

Takuyo (Aug. 1965)

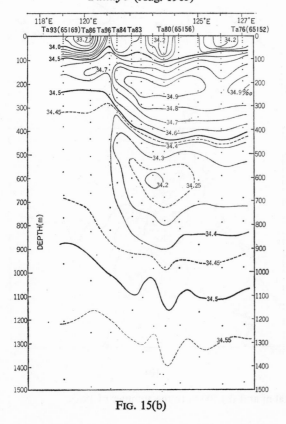

FIG. 15(b)

4.2   Tropical Water

According to Koenuma (1938), Wyrtki (1961), Matsudaira (1964) and Masuzawa (1964), the North Pacific high-salinity water is formed at the sea surface in winter by strong evaporation in the regions whose centers are at 20–30°N and 160–180°E, 25°N and 165°W, 25°N and 180°E and 22–26°N and 172°E, respectively, nearly adjoining each other and having a salinity of 35.5‰. This water is called the North Subtropical Lower Water (Wyrtki, 1961), the North Tropical Water (Cannon, 1966) and the North Tropical Saline Water (Masuzawa, 1967). The author uses the name "North Tropical Water" in this book for convenience. This water is carried west by the North Equatorial Current chiefly along the isanosteric surfaces of 300–400 cl/t, decreasing its value with distance from the origin, and becomes one of the main waters of the Kuroshio. The salinity of this North Tropical Water in the Kuroshio east of Luzon is about 34.9‰.

From Fig. 16, it can be seen that a large area of the western North Pacific is occupied by water of 34.9–35.0‰ salinity at the layer of salinity maximum, and that the maximum value decreases in the southern part of the North Equatorial Current, in the South

Atlantis II (Aug. 1965)

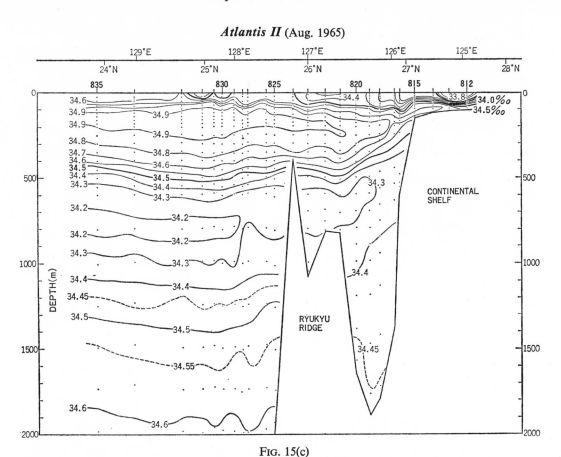

FIG. 15(c)

FIG. 15   Salinity sections in the summer of 1965: (a) across the North Equatorial Current (*Takuyo*) on the 133°E line; (b) across the Kuroshio in the Luzon Strait (*Takuyo*) on the 19°30′N line; (c) across the Kuroshio in the East China Sea (*Atlantis II*).

China Sea and in the East China Sea where the intrusion of this saline water is limited. The lowest salinity maximum having the value of 34.75‰ in the southern part of the North Equatorial Current is in the 6–8°N zone off Mindanao. The lowest salinity maximum came from the Pacific and is about 34.60‰ in the South China Sea and in the East China Sea. In general, the value of the salinity maximum and its depth decrease in the left side of the main currents (the Kuroshio, the North Equatorial Current and the Mindanao Current).

According to Wyrtki (1961) and Masuzawa (1964), the high-salinity water of South Pacific origin, the South Tropical Water, is formed in regions the center of which are at 18°S and 150–120°W and at 20°S and 120°W, respectively. The intrusion of this water across the equator reaches to 5–7°N having a value higher than that of the North Tropical Water. According to Wyrtki (1961), a difference between the North and South Tropical Waters is evident based on the values of their dissolved oxygen. The North Tropical Water has a higher value by about 0.5–

1.0 ml/l than the South Tropical Water.

4.3   Intermediate water

The North Pacific Intermediate Water formed in the subarctic region flows into the western North Pacific Ocean forming a Subtropical Gyre (Reid, 1965). Figure 17 shows the values of the salinity minimum, the isoline of which is comparatively similar to the pattern of stream lines at the sea surface in the western North Pacific. The lowest minimum exists between 20°N and 25°N and its value is less than 34.15‰. The highest minimum exists between 5°N and 10°N and its value is higher than 34.50‰. The value of the salinity minimum is 34.20–34.30‰ in the northern half of the North Equatorial Current and in the Kuroshio east of Luzon, and is 34.40‰ in the South China Sea. Its value is about 34.25–34.40‰ in the left half of the Kuroshio just east of Taiwan and in the East China Sea. The exchange of waters between the South China Sea and the Pacific by the mixing process or advection is not negligible.

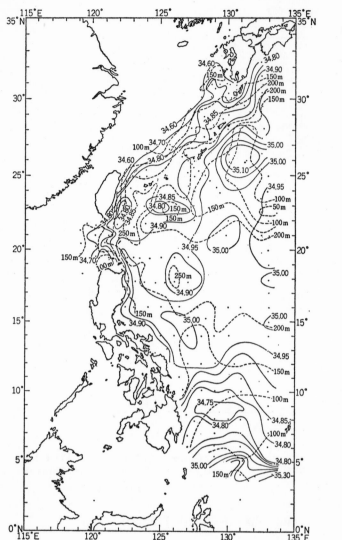

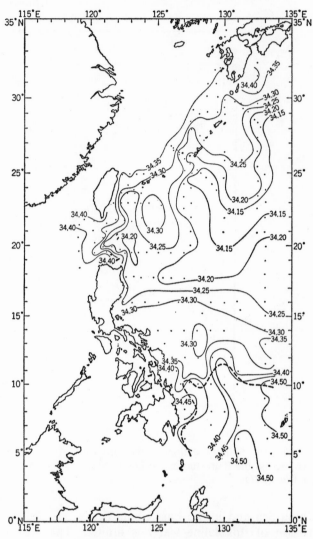

FIG. 16 Distribution of maximum salinity in the core of the Tropical Water and its depth (dotted line) in the summer of 1965.

FIG. 17 Distribution of minimum salinity in the core of the North Pacific Intermediate Water in the summer of 1965. Northern limit of the South Pacific Intermediate Water lying below that water, shown by the thick dashed line.

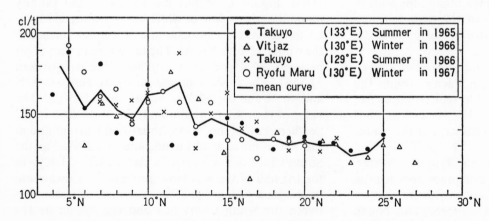

FIG. 18 Thermosteric anomaly at the salinity minimum near 130°E meridian in 1965–1967.

The salinity-minimum water in the North Pacific reaches to 4°N or beyond, but its value increases to 34.45–34.50‰. On the other hand, the South Pacific Intermediate Water reaches to about 10°N or beyond, lying below that water, and its northern limit is shown in Fig. 17 by a thick dotted line. According to Reid (1965) and Masuzawa (1967), the core layer of the North Intermediate Water having the thermosteric anomaly of about 125 cl/t coincides with the salinity-minimum layer in the region north of about 15°N. To the south of this latitude, the thermosteric anomaly of the salinity-minimum water jumps to the value of 175 cl/t. This salinity-minimum water extends from 15°N to 3°N and comes from the California Current region, and it is to be distinguished from the major North Pacific Intermediate Water north of about 15°N. The thermosteric anomalies of the salinity-minimum water on four longitudinal sections near 130°E observed in 1965–1967 are shown in Fig. 18. Though the fluctuations are not so small and the average values in the northern and southern parts of the North Equatorial Current are a little larger and smaller than 125 cl/t and 175 cl/t, respectively, owing to the mixing with surrounding waters, the discontinuity is not so sharp. Figure 18 appears to partly support the descriptions of Reid and Masuzawa mentioned above.

Figure 19 shows the depth of the salinity minimum and the velocity of water in this layer assuming the geostrophic motion referred to 1200 db. The calculation was carried out at every hundred meters rounding up the depth fractions of salinity minimum smaller than 50 meters. The greatest depth of salinity minimum is about 800 m in the region south of Japan and it slopes upward towards south and west. In the Luzon Strait, in the South China Sea and in the East China Sea, its depth decreases to 500 m, 400 m and 500 m, respectively. At the boundary zone between the North Equatorial Current and the Equatorial Countercurrent, the depth becomes shallower than 200 m relating to upwelling of the lower water.

The directions of the flows at the salinity-minimum layer agree fairly well with the surface currents. In general, the shallower depth of the salinity-minimum layer associates with the higher value of salinity minimum in contrast to the Tropical Water, and the intermediate current flows on its left flank, looking in the direction of the current, as the Tropical Water does.

The maximum velocity reaches 30–40 cm/s and the strong intermediate current associates with the strong surface current. The North Intermediate Water forms the lower part of the northern half of the North Equatorial Current and the Kuroshio. The Kuroshio flows

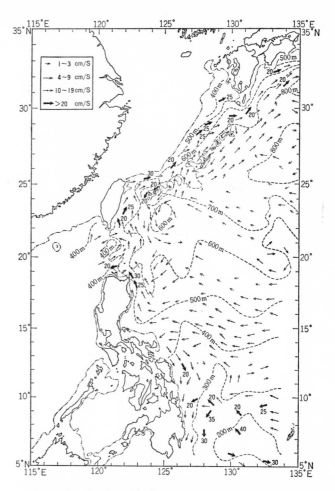

FIG. 19   Geostrophic flow based on 1200 db surface at the salinity minimum and its depth shown by the dotted line in the summer of 1965. The calculation of velocity was carried out at every hundred meters rounding up the depth fluctuations of salinity minimum smaller than 50 meters.

along the coast in the western North Pacific and, subsequently, the intermediate current is strong near the coast.

A fairly strong intermediate current enters into the South China Sea along the northern coast of Luzon, and the Intermediate Water in the South China Sea flows out to the Pacific along the southern coast of Taiwan after complicated meandering in the Luzon Strait. This may explain the comparatively higher salinity minimum in the left-hand part of the Kuroshio just east of Taiwan.

The North Intermediate Water in the Pacific enters into the East China Sea through the passage between Taiwan and Yonakunijima Island as the lower part of the Kuroshio and through the passage between Miyakojima Island and Okinawa Island with the velocity of about 10–20 cm/s as mentioned in the preceding section (3.2). The North Pacific Intermedi-

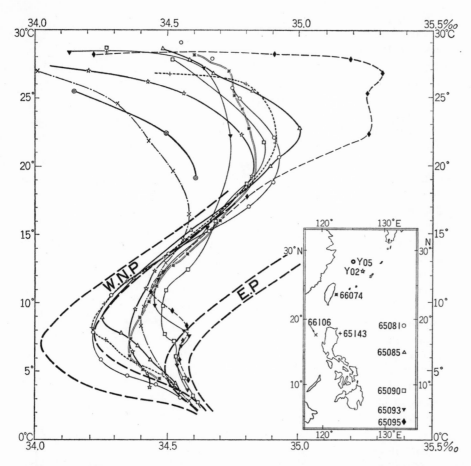

FIG. 20 T-S diagrams in the northwestern Pacific in the summer of 1965 and 1966. The inserted chart shows the location of the stations used. The thick dotted bands show the Western North Pacific Water and the Equatorial Pacific Water defined by Sverdrup et al. (1942).

ate Water in the East China Sea flows out to the Pacific through the Tokara Strait. There may be two small circulations of intermediate water surrounding the Ryukyu Islands. A small portion of the water flows out through the Tokara Strait, returns back along the southern coast of the Ryukyu Islands, and then enters into the East China Sea through the passage between Miyakojima Island and Okinawa Island together with the North Intermediate Water which comes from the Pacific as mentioned above, forming a small circulation. A small portion of the intermediate water which comes into the East China Sea through the passage between Taiwan and Yona-kunijima Island flows out to the Pacific through the passage between Ishigakijima Island and Miyakojima Island forming a small circulation.

4.4   T-S diagram

Figure 20 shows the T-S diagram of several stations in the western North Pacific. The representative T-S curves of the Western North Pacific Water and the Equatorial Pacific Water (Sverdrup, Johnson and Fleming, 1942) are indicated by dotted bands. This figure shows the T-S curves of three stations (St.

65081, 65085 and 65090) in the northern, middle and southern parts of the North Equatorial Current, of three stations (St.65143, 66074 and Y02) in the Kuroshio east of Luzon, Taiwan and in the East China Sea, of two stations (St.65093 and 65095) in the Equatorial Countercurrent, of one station (St. 66106) in the South China Sea and of one station (St.Y05) on the continental shelf in the East China Sea together with an insert chart showing their positions.

Station 65081 is located to the north of the warm belt continued from the right-hand side of the Kuroshio east of the Luzon Strait. It has a representative T-S curve at the western North Pacific which is composed of the surface water, the North Pacific Tropical Water, the North Pacific Intermediate Water and the Pacific Deep Water. The water passing through St.65085 in the middle part of the North Equatorial Current joins the Kuroshio off Luzon, and thus the T-S curve of St.65085 is similar to that of St.65143 in the Kuroshio just east of Luzon. The T-S curves at these two stations are nearly similar to the representative T-S curve in the western North Pacific. The water near St.65085 is the original water of the beginning section of the Kuroshio. The water

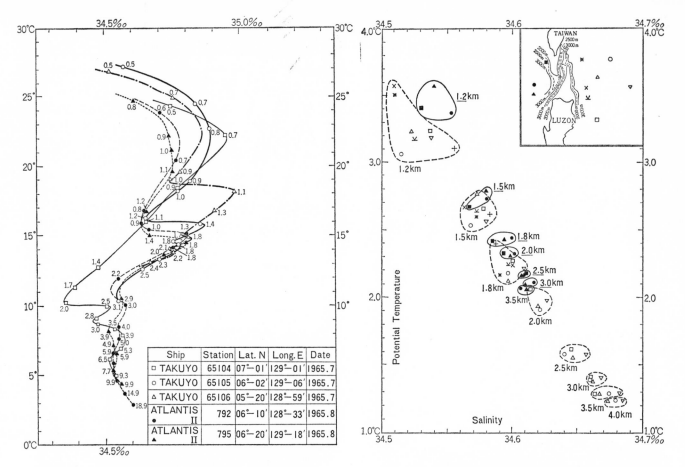

FIG. 21 T-S relation near 6°N and 129°E north of Halmahera Island obtained on board the R/V *Takuyo* and *Atlantis* II in the summer of 1965.

FIG. 22 Relations between potential temperature and salinity of the waters at both sides of the Luzon Strait, the Pacific and the northern South China Sea. Location of the stations used and the bottom topography in the Luzon Strait are shown in the inserted chart.

at St.65090 in the southern part of the North Equatorial Current is the origin of the Mindanao Current. Below the depth of 500 m, the T-S curve approaches the Equatorial Pacific Water.

A remarkable difference between St.65093 and St.65095 in the northern and southern parts of the Equatorial Countercurrent is evident in the maximum of salinity. At St.65093, the layer of the salinity maximum is filled with the North Pacific Tropical Water having the lower maximum of salinity. At St.65095, however, the South Pacific Tropical Water occupies the upper part of this station with the higher salinity maximum of 35.30‰. At these two stations and St.65090, there are two salinity minima corresponding to the North and South Pacific Intermediate Waters.

As a result of the overlapping and mixing of two water masses, the Western North Pacific Central Water and the Equatorial Pacific Water, complicated

T-S diagrams are seen in the region near 5–6°N and 128–129°E north of Halmahera Island as pointed out by Wyrtki (1961). According to observations carried out on board the *Takuyo* and the *Atlantis II* in the summer of 1965 (Fig. 21), these T-S curves have pronounced bends at depths of 100–200 m accompanied by new salinity minima and maxima. In conclusion, the T-S curves here have three salinity maxima and minima, as shown in Fig. 21. Of three maxima, the first maximum is in the North Tropical Water, the second associates with the bend mentioned above and the third is between the North and South Intermediate Water. The first minimum associates with this bend, the second and the third are in the North and the South Intermediate Water, respectively. The depth of these pronounced bends seems to increase and the bends become weaker with increasing latitude.

At St.66106 in the northern South China Sea, there

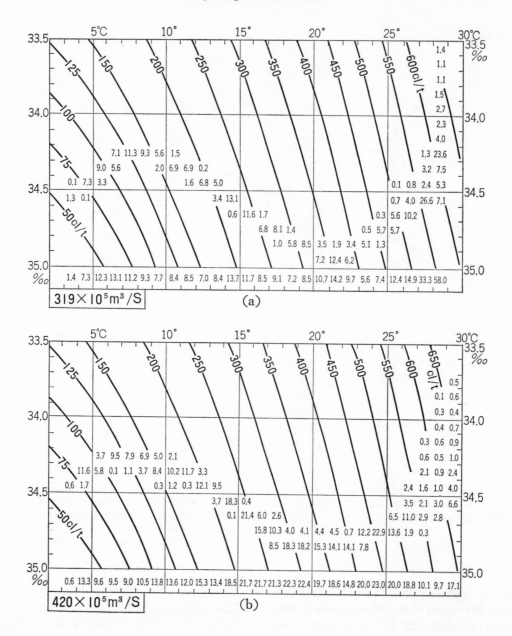

is less saline surface water mixed with the continental water and the low-salinity-maximum and high-salinity-minimum waters owing to the presence of the Kuroshio in the Luzon Strait. Below the depth of 1000 m, the water is nearly similar to that in the Pacific. This fact means that the Kuroshio is very weak in the deep layer and has little influence on the invasion of the Pacific water into the South China Sea.

Station 66074 in the Kuroshio east of Taiwan and St. Y02 in the East China Sea have a higher salinity minimum and a little lower salinity maximum than St. 65143 and St.65085. This indicates mixing with waters from the South China Sea and the East China Sea. The T-S curve of St.Y05 on the conti-

nental shelf in the East China Sea indicates the existence of the continental water or the Yellow Sea Water whose temperature and salinity are lower than those of St.Y02 for the surface layer. The major difference between St.66074 and St.Y02 is that the salinity in the surface layers is lower at St.Y02 than at St.66074 owing to mixing with the continental water in the East China Sea.

All T-S curves in Fig. 20 except that of St.Y05, on the continental shelf in the East China Sea, converge to a range of 14–16°C in temperature and 34.55–34.65‰ in salinity, which corresponds to about 230 cl/t in thermosteric anomaly. This means that the water of 230 cl/t in thermosteric anomaly is made up of uniform water of about 15°C in temperature and

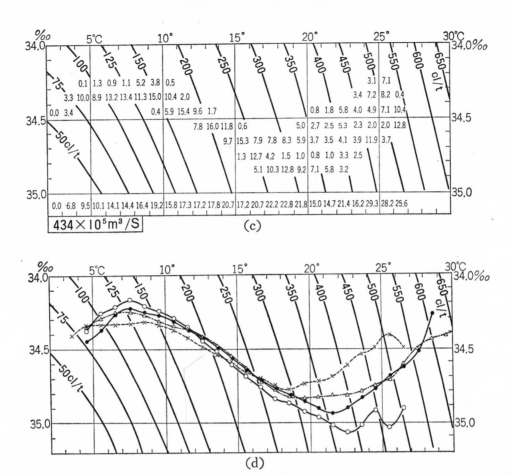

FIG. 23   Geostrophic fluxes in 10⁵m³/s based on 1200 db surface in classes of 1°C×0.1‰ on T-S diagram at three stations across the Kuroshio: (a) east of Luzon; (b) east of Taiwan; (c) southeast of Yakushima Island. Under each figure, the geostrophic fluxes with respect to every one degree of temperature are shown. (d) shows average temperature-salinity relations of the fluxes at three stations and of the northern half of the North Equatorial Current, the last of which was computed by Masuzawa's figure (1967, Fig. 10b).

● east of Luzon; △ East of Taiwan; × southeast of Yakushima Island; ○ northern half of the North Equatorical Current (137°E).

34.6‰ in salinity, and this characteristic water spreads over extensive areas in the western North Pacific including some portions of the South China Sea and the East China Sea. This uniform water may correspond approximately to the central portion of Masuzawa's "Thermocline Water" (Chap. 4) or the Western North Pacific Central Water (Sverdrup et al., 1942).

4.5   Deep water in the South China Sea

Nitani (1970) showed the relationship between the deep water in the South China Sea and in the Pacific. In Fig. 22, the relation between potential temperature and salinity at depths greater than 1200 m in the Pacific and the South China Sea are shown. A small difference in salinity is seen at the depth of 1200 m. The two groups of water, the Pacific and the South

China Sea, lie on the same T-S curve at depths greater than 1500 m, but a given T-S point appears at different depths.

The T-S relation at the depth of 1500 m in the South China Sea and in the Pacific are nearly the same. Those at 1800, 2000, 2500, 3000 and 3500 m in the South China Sea, however, correspond to those at about 1700, 1750, 1850, 1900 and 1920 m in the Pacific, respectively. The bottom water at the depth of 4000 m, which is nearly the maximum depth in the South China Sea, may be the same as the water at 1900–1950 m in the Pacific by extrapolating the T-S curve in the South China Sea. This phenomenon is explained by the bottom topography in the Luzon Strait (Fig. 22). The submarine valley in the Luzon Strait runs northeast to southwest (from the Pacific to the South China Sea), and the maximum depth is about 3800 m at its center. The greatest sill depth at

TABLE 2.   Characteristic elements in three sections of the Kuroshio.

Section	Period	Vessel	Area of section (m²)	Total flux (m³/s)	Average velocity (cm/s)	Average temperatre of total flux (°C)	Average salinity of total flux (‰)
East of Luzon	Aug. 1965	*Takuyo*	178 × 10⁶	31.9 × 10⁶	18.0	19.2	34.52
East of Taiwan	Aug. 1966	*Chofu Maru*	217	42.0	19.2	17.8	34.61
Southeast of Yakushima Island	July 1966	*Chofu Maru*	268	43.4	16.2	16.4	34.53

TABLE 3.   Percentages of fluxes of the warm, middle and cold water, and characteristic elements of the Tropical Water and the Intermediate Water in the Kuroshio.

Section	Flux of warm water (>20°C)	Flux of middle water (10–20°C)	Flux of cold water (>10°C)	Average temperature, salinity and thermosteric anomaly in the core of the Tropical Water			Average temperature, salinity and thermosteric anomaly in the core of the Intermediate Water		
East of Luzon	52%	29%	19%	21.5°C	34.94‰	361 cl/t	7.5°C	34.23‰	129 cl/t
East of Taiwan	41%	43%	16%	19.5°C	34.84‰	317 cl/t	7.0°C	34.25‰	121 cl/t
Southeast of Yakushima Island	35%	44%	21%	18.5°C	34.78‰	297 cl/t	8.0°C	34.33‰	129 cl/t
Northern half of the North Equatorial Current (after Masuzawa)				22.5°C	35.08‰	378 cl/t	7.5°C	34.20‰	132 cl/t

the Pacific may be, though this is not clear, in the range of 2000–2500 m. The deepest water entering into the northern end of the valley from the Pacific over the sill may have the same characteristics as the water at 2000–2500 m in the Pacific. But when it reaches the southern end of this submarine valley, the deepest water will become equivalent to the water at a depth of 1900–1950 m in the Pacific owing to the mixing process during its long trip along the valley.

4.6   Geostrophic fluxes on T-S diagrams of some sections at the beginning of the Kuroshio

The geostrophic fluxes corresponding to 1200 db on T-S diagrams of three sections of the Kuroshio, east of Luzon (17°45′N), Taiwan (23°00′N) and southeast of Yakushima Island in the summers of 1965 and 1966 were calculated (Fig. 23a, b, c). The calculations by class intervals of 1°C temperature and 0.1‰ salinity were carried out graphically by use of overlapped charts of isothermal, isohaline and velocity distributions in three sections in the Kuroshio whose boundaries were determined geostrophically such that the velocity perpendicular to the section was zero.

Table 2 shows some characteristic elements in the three sections mentioned above. The area of the section and the flux of the Kuroshio increase from south to north, indicating the admixture of the surrounding waters during the passage of the Kuroshio water. The average velocity of the upper 1200 m layer of the Kuroshio is about 15–20 cm/s, and the differences between three sections are not so significant because the absolute velocities at the reference level of each section may not be uniform. The average temperature of the Kuroshio decreases from south to north as we would presuppose. As shown in Table 3, the percentage of the flux of warm water whose temperature is higher than 20°C in the Kuroshio east of Luzon exceeds 50% of the total flux, and its value decreases as the Kuroshio flows northwards in accordance with the average temperature of the total flux. Consequently, the percentage of the flux of middle water whose temperature is in the range of 10–20°C increases gradually to the north.

According to Masuzawa's investigation based on the cruise of the *Ryofu Maru* in the winter of 1967 (Masuzawa, 1967), in the section of the northern half of the North Equatorial Current at 137°E, the Tropical Water has salinity exceeding 35.00‰ and the Intermediate Water has salinity below 34.20‰. In these three sections of the Kuroshio, however, there is no saline water higher than 35.00‰ and no less saline water lower than 34.20‰ except for the surface water. As a rule, the salinity of the flux above the Thermocline Water decreases with increasing latitude, and below that the opposite occurs. This is clear from Fig. 23 (d) in which the average temperature-salinity relations of the flux at three sections of the Kuroshio and of the northern half of the North Equatorial Current, the last of which was computed from Masuzawa's figure (1967, Fig. 10), are shown. The characteristic classes of the flux with respect to temperature and salinity are scattered in order of the section southeast of Yakushima Island, east of Taiwan and east of Luzon. This is especially remarkable in the layer above the core of the Tropical Water. These changes of water characteristics among the three sections of the Kuroshio mentioned above may be attributed to the inflow of surrounding waters into the Kuroshio and the mixing between the waters of the Kuroshio and the surrounding waters at both sides. The meteorological effect, however, is not negligible in some respects, especially for the change in temperature.

In the sections east of Luzon and Taiwan, there are less saline surface waters with salinity less than 34.00‰ caused by land water. In the section southeast of Yakushima Island, however, there is no water of salinity less than 34.00‰ in spite of the existence of very low-salinity waters in the East China Sea. This less saline surface water of the Kuroshio in the East China Sea may not pass through the Tokara Strait because it flows into the Japan Sea with the Tsushima Current and into the Osumi Strait with the Osumi Branch.

The flux of the saline Tropical Water is most saline in the section east of Luzon, and the temperature at its core changes from 21.5°C east of Luzon to 18.5°C southeast of Yakushima Island while the salinity changes from 34.94‰ to 34.78‰ and the thermosteric anomaly from 361 cl/t to 297 cl/t as shown in Table 3. The changes from the section east of Luzon to the section east of Taiwan are about twofold of that from the section east of Taiwan to southeast of Yakushima Island being inversely proportional to the distances between the three sections. This suggests two causes, the first is a large exchange and inflow of waters in the Luzon Strait and the second is the branching off the Tsushima Current and the Osumi Branch from the Kuroshio in the East China Sea. The flux of the core of the Intermediate Water is nearly the same in temperature and thermosterict anomaly with a range of 7–8°C and 120–130 cl/t in the three sections, but its salinity increases northwards slowly from 34.23‰ to 34.33‰.

In the three sections, there are some modes of flux with respect to temperature. In the section east of Luzon, the modes are centered in the classes of 28–29°C, 21–22°C, 14–15°C and 6–7°C corresponding to the core of the surface water, the Tropical Water, the Thermocline Water and the Intermediate Water, respectively. In the section east of Taiwan, the modes are in the classes of 29–30°C, 24–25°C and 15–20°C. The third mode may correspond to the combination of the Tropical Water and the Subtropical Mode Water (18°C Water) or the Thermocline Water. In the last section, the modes are in the classes of 24–26°C, 22–23°C and 17–20°C. The modes correspond-

ing to the Tropical Water and the surface water are common in the three sections of the Kuroshio, though the temperatures at the center are different.

5 VELOCITY OF THE CURRENTS IN THE WESTERN NORTH PACIFIC

5.1 Velocity section across the North Equatorial Current

Many oceanographers have used the 1000–2000 db surfaces; for example, Wyrtki (1961) used 1000–1200 db, and Masuzawa (1967) 1000 db as the level of no motion for calculations of the geostrophic flow in the region of the Kuroshio and the North Equatorial Current. The author uses the 1200 db surface as the reference level considering the data available.

The velocity section along the longitude 133°E across the North Equatorial Current, which was obtained from the observation carried out in the summer of 1965 by the R/V *Takuyo*, shows some characteristic features of this current, though the details of the current vary with time and space (Fig. 24). The southern boundary of the North Equatorial Current is at latitude 7°N corresponding to the temperature section. On the other hand, its northern boundary is difficult to determine because the east and west components of the velocity appear alternately in the northern part, having nearly the same water characteristics and velocities. The boundary between westward and eastward currents, the latter of which is located at latitude 20°25′N and, recently, has been called the Subtropical Countercurrent (Uda and Hasunuma, 1969; Yoshida and Kidokoro, 1967), is regarded as the northern limit of the North Equatorial Current for convenience (i.e., 20°N in the case of Fig. 24), though it would be more appropriate from the physical viewpoint to regard the boundary between the second eastward and westward bands (see Section 5.2), which corresponds to the central line of the warm core continuing to the right side of the Kuroshio east of the Luzon Strait, as the northern limit of the North Equatorial Current.

The maximum velocity of the North Equatorial Current is found at latitude 10°N, having the value of 100 cm/s which belongs to the highest of all classes observed in this current. In general, the velocities of westward currents in the northern half of the North Equatorial Current are smaller than in the southern half, and on the other hand, the velocities of eastward currents in the northern half are larger than in the southern half. At the depth of 500 m, the maximum westward speed is about 10 cm/s, and the eastward currents are shallower than the westward currents. The rate of decrease of velocity with increasing depth is larger in the southern half than in the northern half of the North Equatorial Current. The North Equatorial Current is deeper than the Equatorial Countercurrent and is shallower than the Kuroshio which becomes deeper with increasing latitude relating to the conservation of the potential vorticity in the western boundary current.

The North Equatorial Current is divided into two portions, the northern and southern halves, and each portion is frequently subdivided into two or three portions. From the figures of the vertical sections across the North Equatorial Current obtained by Yamanaka, Anraku and Morita (1965) and Masuzawa (1967), it is shown that the North Equatorial Current is subdivided into three or four portions. There are two possible causes for this phenomenon. The first is the existence of continuous currents parallel to the latitude and the second is the existence of isolated cyclonic or anticyclonic eddies. It is difficult to determine which is true not only from one section, but also from a few sections obtained in one month or so because of the variation of sea conditions during the cruise considering the comparatively small velocities of eastward current in the southern portion of the North Equatorial Current.

As to the northernmost eastward current in the North Equatorial Current region, Yamanaka et al. (1965) concluded from their observations on the line near 135°E from 1958 to 1964 on board the R/V *Shunyo Maru*, that this eastward current near 18°N

might continue to the easterly current near the same latitude in the Central Pacific, the existence of which was pointed out by Cromwell (1950) and Mao and Yoshida (1955), as well as the fact that it developed during the period of the Southwest Monsoon. The existence and continuity of this eastward current were pointed out already in the preceding section as the easterly current, which is associated with the warm core like a belt on the right-hand side of the North Equatorial Current and the Kuroshio, very often forming isolated warm eddies off the Luzon Strait, Taiwan and the Ryukyu Islands.

Recently, Yamanaka et al. (1965) and Uda and Hasunuma (1969) pointed out the existence of the eastward current associated with the thermal front near the Subtropical Convergence at latitudes between 20°N and 25°N in the western North Pacific. Moreover, Charnell and Seckel (1966) found a new eastward current just to the north of 20°N near the Hawaiian Islands. Yoshida and Kidokoro (1967a, b) deduced theoretically that a trough in the anticyclonic wind-stress vorticity located near the boundary between the trades and the westerlies was responsible for this eastward current. The eastward current near 20–22°N in Fig. 24 may correspond to the so-called Subtropical Countercurrent. Very little is known about the other several westward and eastward components in the region of the North Equatorial Current.

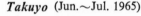

Takuyo (Jun.~Jul. 1965)

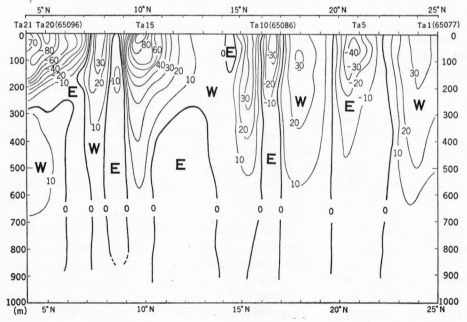

FIG. 24 Velocity section at 133°E in July 1965 calculated geostrophically based on 1200 db surface.

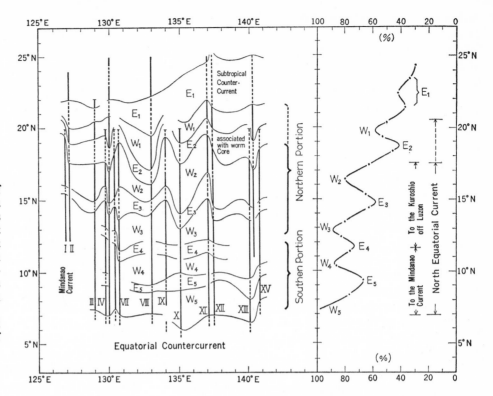

FIG. 25 Left; Appearance of west and east currents at the sea surface in fifteen sections across the North Equatorial Current between 127°E and 140°E. West and east components are shown by thick and dotted lines, respectively. The detailed elements for each section are shown in Table 4.

Right; Frequency of westward current in every one degree of latitude.

5.2 Statistical treatment of the North Equatorial Current

A statistical investigation regarding the velocity in the region of the North Equatorial Current was made by Nitani (unpublished) to determine whether these westward and eastward components in the North Equatorial Current were the continuous zonal currents or not. When many eddies, whether cyclonic or anticyclonic, exist, the east and west components appear alternately from north to south in the meridional sections, and there may be no statistical modes of east and west components with respect to the latitude. Figure 25 shows the distribution of east and west components in the fifteen sections, some of which are limited to the south of 20°N because of the lack of data, across the North Equatorial Current in the region between 127°E and 140°E. The sections chosen here, based on considerations of accuracy (Table 4), are those obtained after World War II.

According to Fig. 25, the region of the North Equatorial Current and just to the north of it can be divided into five west and east component bands in general, though the separations in the southern half are not so clear as in the northern half especially to the west of 130°E (near the region of the Mindanao Current). In the case of the same season in the same year, the separation into several bands in Fig. 25 was carried out such that the difference of average dynamic height anomalies referred to the 1200 db surface in the same east component band at different meridians is smaller than 0.05 dynamic meters except for the winter of 1968. If there is a zonal current band, the dynamic height anomalies in it must be nearly the same at different meridians considering the time difference of each observation. Some uncertainty connected with making each band still exists because we have no definite reasons and appropriate means for connecting the successive eastward or westward components in different seasons or years. The maxima and minima in the frequency of westward components at every degree of latitude suggests the existence of continuous westward and eastward current bands though the maxima and minima are smoothed owing to the variation of their locations with time (right portion of Fig. 25). The appearance of the observed east component in each assumed continuous eastward band occurs at a frequency of 100% for the first and second (counted from north to south and designated E_1, E_2, E_3, etc.) eastward band currents, and it decreases in the southern band currents, but does not drop below 50% (Table 5). If it is limited to east of 130°E in the southern half of the North Equatorial Current, its frequency will increase more.

From these facts, the North Equatorial Current appears to have a striped rather than mosaic character, and this is more certain in the northern portion than in the southern portion. If the eastward com-

TABLE 4. Elements for each section across the North Equatorial Current.

No.	Long.	Period	Research Vessel	Cruise
I	127°E	Jan. 1968	*Ryofu Maru*	CSK
II	127°E	Feb. 1966	*Vitjaz*	CSK
III	129°E	Aug.–Sept. 1966	*Takuyo*	CSK
IV	130°E	Aug. 1956	*Kagoshima Maru*	EQUAPAC
V	130°E	Jan. 1967	*Ryofu Maru*	CSK
VI	130°E	Feb.–Mar. 1968	*Ryofu Maru*	CSK
VII	130°E	Feb. 1966	*Vitjaz*	CSK
VIII	133°E	July 1965	*Takuyo*	CSK
IX	134°E	Feb. 1968	*Ryofu Maru*	CSK
X	135°E	Jan. 1966	*Vitjaz*	CSK
XI	137°E	Jan. 1967	*Ryofu Maru*	CSK
XII	137°E	Feb. 1968	*Ryofu Maru*	CSK
XIII	140°E	Jan. 1966	*Vitjaz*	CSK
XIV	140°E	Aug. 1956	*Satsuma*	EQUAPAC
XV	140°E	Feb. 1959	*Takuyo*	IGY

TABLE 5. Some results relating to the banded structure of the North Equatorial Current.

Band current	Width (mile)	Average of max. vel. (cm/s)	Average of mean vel. (cm/s)	Surface transport (mile-Knot)	Freq. of appearance of east comp. in each assumed band (%)
W_1*	100	18**	11	2.2	—
W_2	150	24	15	4.5	—
W_3	110	37	16	5.7	—
W_4	120	46	30	7.2	—
W_5	110	56	39	8.6	—
Total	590	36	22	28.2	—
E_1*	145	21	17	4.9	100
E_2	80	17	12	1.9	100
E_3	65	8	6	0.8	87
E_4	60	13	8	0.8	50
E_5	60	16	11	1.3	57
Total except E_1	265	14	9	4.8	—

* W_n: nth westward band current countered from the north
    E_n: nth eastward band current countered from the north
** Velocities were obtained from the geostrophic calculation

ponent is nearly zero or a weak westward component is present instead of the eastward band, we can still say that the North Equatorial Current has essentially a banded character. Some results relating to the banded structure of the North Equatorial Current are shown in Table 5. The mean width of eastward band currents is about one half of that of westward band currents. The average maximum and mean velocities of westward band currents increase southwards and those of eastward band currents are small in the middle and southern portions relating to lower confidence for the existence of a banded structure in the southern portion than for the northern portion. The eastward and westward surface transports, the product of mean velocity at the surface and width, are calculated by use of a unit of Mile Knot, MK. The eastward surface transports are very small except for the first eastward band current which is the Subtropical Countercurrent. The first (from the north) westward band current of the North Equatorial Current, W_1, does not join the Kuroshio just east of

Luzon. The second and third westward band currents join the Kuroshio east of Luzon, and these two separated bands appear off Luzon frequently. The sum of these two surface transports is about two thirds of that of the fourth and fifth bands which join the Mindanao Current. This ratio in the surface transport is a little smaller than that in the total volume transport which will be mentioned later.

In conclusion, as mentioned already, the first and second eastward band currents in this classification correspond to the Subtropical Countercurrent and the easterly current near 18°N associated with the warm core to the north of the North Equatrial Current, respectively. The possibility of existence of the third eastward band current is fairly large. However, the existence of the fourth and the fifth eastward band currents and the causes of these three (E_3, E_4, and E_5) are still obscure at present. The seasonal variations of position, width, thickness and velocity are not clear because of lack of data. Rapid, accurate, dense and widely distributed simultaneous oceanographical

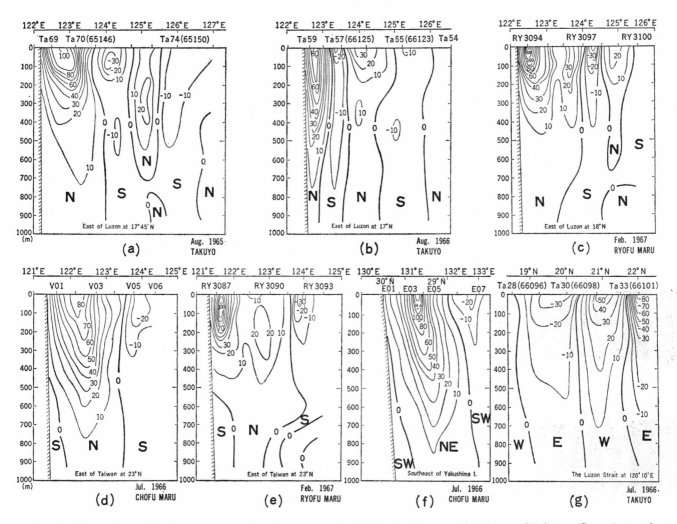

FIG. 26   Velocity sections in cm/s across the Kuroshio: (a), (b), (c) east of Luzon; (d), (e) east of Taiwan; (f) southeast of Yakushima Island; and (g) at 120°10′E in the Luzon Strait.

and meteorological observations in every season are desirable to obtain accurate structures, variations and their causes.

5.3   Velocity sections across the Kuroshio at its beginning

The geostrophic velocity sections across the Kuroshio off Luzon, Taiwan and Yakushima Island and in the Luzon Strait, assuming that 1200 db is the reference level, show the some features of the Kuroshio at its beginning (Fig. 26). In the area just east of Luzon, the geostrophic surface current in Aug. 1965 agrees fairly well with the current obtained by GEK (cf. Fig. 3). The three sections in this area (Fig. 26a, b, c), though made at different times, indicate the similar pattern of the Kuroshio having two northward bands and having the strongest part at the westernmost part close to the coast. The Kuroshio here still may have a banded structure like the North Equatorial Current. A narrow southward current in Aug. 1966 and a weak northward current in Feb. 1967, located just east of the strongest northward current, forming a boundary zone between two northward currents, may correspond to the third eastward band current in the North Equatorial Current mentioned above (5.2), since the average dynamic height anomalies at the boundary zones are nearly the same as those of the third eastward bands. In the case of Feb. 1967, moreover, the pattern of the Kuroshio off Luzon is very similar to that off Taiwan (Fig. 26c, e), having the same average dynamic height anomaly in each boundary zone. Subsequently, the strongest current portion in the Kuroshio just east of Luzon may be connected to the third westward band current in the North Equatorial Current. For Aug. 1965, the Kuroshio shows the band character too, but the connection to the North Equatorial Current is not so clear.

The locations of the southward countercurrent associated with the warm core or eddy east of the Luzon Strait may be variable with time in a range of less than 100 nautical miles.

In the section east of Taiwan (Fig. 26d,e), the structure of the Kuroshio becomes simple in general, but there are some exceptions like the case in Feb. 1967. The Kuroshio flows to the north or north by northeast close to the coast of Taiwan. The center of the countercurrent is located near 124°E at the section of 23°N, having its left hand boundary near 123.5°E. The weak southward current along the coast is seen frequently below the depth of several hundred meters, though its existence is not so consistent because of its weakness and many assumptions for geostrophic calculations. As a rule, the flow here becomes simpler, deeper and more invariable in location than that off Luzon. This might be the reason that Kishindo (1931), Sverdrup et al. (1942) and other oceanographers restricted the Kuroshio to the north of Taiwan.

In the section southeast of Yakushima Island, there have been many observations, especially eight times a year since 1965 as part of the CSK program, mainly carried out by the R/V *Chofu Maru*. The typical velocity section is shown in Fig. 26(f). The Kuroshio reaches this section through the Tokara Strait, of which the sill depth is less than 650 meters in the northern half and is nearly 800 meters in the southern half, after mixing with the continental water and branching off the Tsushima Current and the Osumi Branch in the East China Sea. The velocity here is a little larger than that at the Tokara Strait, but it is not so strong as south of the Japanese mainland. The shift of the axis of the Kuroshio to the right facing downstream with increasing depth becomes more remarkable than those in the section off Luzon and Taiwan, though this is a common character of the Kuroshio and the Gulf Stream. The shift reaches about 25–30 nautical miles at the depth of 500 m. At the left side of the Kuroshio, a weak countercurrent along the coast exists below the depth of about 200 m reaching a more shallow depth than that off Taiwan. This countercurrent is seen at the sea surface very often off the Japanese mainland. The Kuroshio here has a character intermediate to the Kuroshio off Taiwan and off the Japanese mainland not only in velocity distribution but also in water characteristics, and this section can be regarded as the end of the beginning of the Kuroshio.

The vertical section of velocity on the line of 120°10′E, the entrance of the South China Sea, observed in July 1966 shows the alternate distribution of the velocity components, west (weak), east, west and east from the south to the north (Fig. 26g). These alternate components, especially the northern three, associate with the cold and warm eddies stretching to the Luzon Strait from the South China Sea and from the Pacific, respectively, as mentioned in 3.1. These patterns of velocity are variable with time, for example, each velocity component in the summer of 1965 (Nitani, 1970) is about one half of that in the summer of 1966. These current patterns associated with the cold and warm eddies appear frequently or permanently.

6 VOLUME TRANSPORTS OF THE NORTH EQUATORIAL CURRENT AND THE KUROSHIO AT ITS BEGINNING

6.1 Assumptions for the geostrophic calculation

Geostrophic volume transport depends upon the selection of a reference level in calculation. The average transports with respect to 800 db, 1000 db, 1200 db and 1500 db for several sections across the North Equatorial Current, the Kuroshio east of northern Luzon, Taiwan and southeast of Yakushima Island were calculated in percentages assuming that the volume transports referred to 1200 db are 100% in each section (Fig. 27). The data for these computations were adopted from the CSK cruises by the *Takuyo*, *Ryofu Maru*, *Chofu Maru*, *Shumpu Maru* and *Vitjaz* in 1965, 1966 and 1967. In each section, the data of three to five observations were averaged though fourteen for southeast of Yakushima Island. The individual transport of each section fluctuates comparatively and some of them become nearly constant or decrease with the increasing depth of the reference level suggesting the existence of layers below which the vertical gradient of mean horizontal velocity perpendicular to the section vanishes or where it changes sign. In general, an average volume transport in each section, however, increases monotonously with the increasing depth of the reference level. But the increasing rate decreases with increasing depth corresponding to the decrease of deep current. The increasing rate of the volume transport with increasing depth of reference level is smallest in the North Equatorial Current and is largest in the Kuroshio southeast of Yakushima Island. This suggests that the reference level must be adopted at the deeper layer in the north rather than in the lower latitudes owing to the deep penetration of the current in the northern latitudes. So far as there is no direct measurement of absolute velocity, however, we cannot say so much as to the best choice for the reference

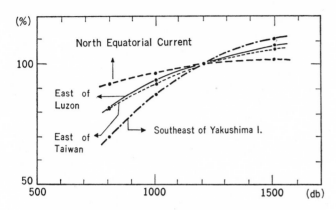

FIG. 27 Average geostrophic transports in percentages based on 800 db, 1000 db, 1200 db and 1500 db surfaces assuming that the volume transports referred to 1200 db surface are 100% in each section.

level. Considering the facts mentioned above and the data available for convenience, a surface of 1200 db is used as a reference level for the calculations of geostrophic volume transport, but we have to bear in mind such variations of transport as shown in Fig. 27.

When there is a countercurrent associated with an isolated eddy in contact with the current, a calculated total volume transport contains the contribution by the eddy. The net transport is to be reduced by this volume, though the estimation of the volume transport to be subtracted is difficult. When the scale of a countercurrent associated with this eddy is large and long like a belt, however, no volume transport has to be reduced, because the main current and the countercurrent should be regarded as a current system.

When the current flows along the coast like the Kuroshio east of Luzon and Taiwan, an extrapolated correction must be made for the transport between the coast and the station nearest to the coast.

In the calculation of the volume transport on the section across the current, the left- and right-hand boundaries are to be determined such that the boundaries are on the places where dD/dy is zero, where D is the dynamic height anomaly of each depth referred to 1200 db surface and y is the distance across the current. In general, the boundary of the current may vary with depth. In the North Equatorial Current, however, there are several banded eastward components as mentioned before (5.2), which do not always reach the deep layer, and then the net volume transport is to be reduced by the amount of such eastward components. When the eastward component outside the current does not reach the deep layer, there are some difficulties in determining the limit of the current for the geostrophic calculation.

Considering the above-mentioned conditions, it is difficult and rather meaningless to expect to obtain

the precise volume transport of the current. Only a rough estimate of the volume transport is possible and meaningful.

6.2 Volume transport of the North Equatorial Current

Sverdrup et al. (1942) showed a schematic transport chart based on 1500 meters in the North Pacific using the data available. The transport of the North Equatorial Current in the central and western parts of the Pacific amounts to 45 million cubic meters per second. 10×10^6 m³/s of this transport is separated as the Mindanao Current, and 20×10^6 m³/s and 15×10^6 m³/s of it join the Kuroshio through the section just east of Luzon and the section far east of Taiwan, respectively.

On the other hand, Wyrtki (1961) estimated the transports of the North Equatorial Current, the Mindanao Current and the Formosa Current—the latter corresponds to the Kuroshio off Taiwan—for every two months by means of the observed velocity, the width and depth of the currents. Though the data and the method used for the estimation are not described in detail, according to Wyrtki, the mean transports of the North Equatorial Current, the Mindanao Current and the Formosa Current are 38, 10 and 29 million m³/s.

Masuzawa (1964) calculated the geostrophic fluxes of the North Equatorial Current based on the 1000 db surface at six sections across the current between 158°W and 130°E observed in summers between 1939 and 1956. The main results selected from his tables are shown in Table 6. The mean value is 48 million m³/s and the transport increases from east to west. Masuzawa showed that this increase could be explained using Ekman's drift-current theory and Sverdrup's wind-driven current theory applied to the region bounded by 130°E, 160°W, 10°N and 20°N, and that these two theories were valid for the western tropical Pacific.

He pointed out that the exceptionally large transport in the section 130°E was to be attributed to the contributions from the eddies in the southernmost and northernmost parts of this section.

The geostrophic volume transport of the North Equatorial Current was calculated by use of reliable data obtained in the CSK (Table 7). A schematic transport chart in summer was shown by Nitani (1970). Figure 28 is an improved transport chart made by using the data listed in Table 7, 8 and data from the East China Sea and the sea southeast of Yakushima Island.

In general, the seasonal variation of transport

TABLE 6. The Transport of the North Equatorial Current (Masuzawa, 1964).

Section	Vessel	Year	Range	Net westward flux in the whole range	Km³/hr, (10⁶m³/s) 10°N–20°N
158°W	*Hugh M. Smith*	1950	9–21°N	120 (33)	77 (21)*
172°W	*Hugh M. Smith*	1950	10–21°N	101 (28)	95 (26)
165°E	*Komahashi*	1939	9–19°N	93 (26)	88 (24)
151°E	*Satsuma*	1956	9.5–20°N	208 (58)	200 (56)
140°E	*Satsuma*	1956	9.5–18.5°N	143 (40)	123 (34)
130°E	*Kagoshima Maru*	1956	7–20°N	381 (106)	238 (66)
Average				172 (48)	137 (38)

* The numerical values in brackets show the transport in 10^6 m³/s.

TABLE 7. Volume transport in 10^6m³/s of the North Equatorial Current.

Year	Month	Lat.	Volume trans.	Volume trans. except W₁ band	To the Mindanao Current	To the Kuroshio	Vessel
1965	VI–VII	133°E	92	68	40	52(28)*	*Takuyo*
1966	II	130°E	78	73	44	34(29)	*Vitjaz*
1966	VIII–IX	129°E	80	67	37	43(30)	*Takuyo*
1967	I	130°E		74	37	(37)	*Ryofu Maru*
1968	II–III	130°E		66	27	(39)	*Ryofu Maru*
1967	I	137°E	68	52			*Ryofu Maru*
1968	I	137°E	69	54			*Ryofu Maru*
Average		129°E–133°E	83	70	37	(32)	
		137°E	69	53			

* The numerical values in brackets show the transport excluding the W₁ band.

TABLE 8. Volume transport in 10^6m³/s of the beginning of the Kuroshio.

Section	Period	Lat.	Volume transport of the Kuroshio	Volume transport of the Kuroshio Countercurrent	Vessel
	1959—V	18°20′N	27	12	*Takuyo*
East of Luzon	1965—VIII	17°45′N	36	12	*Takuyo*
	1966—VIII	17°00′N	20	6	*Takuyo*
	1967—II	18°00′N	28	6	*Ryofu Maru*
	Average		28	9	
	1959—V	20°00′N	55	17	*Takuyo*
In the Luzon Strait	1965—VIII	19°30′N	48	23	*Takuyo*
	1966—VII	20°15′N	42	11	*Takuyo*
	1967—II	20°10′N	45	8	*Ryofu Maru*
	Average		47	15	
	1965—VII	23°00′N	(47)*		*Chofu Maru*
	1966—I	23°00′N	45	(18)*	*Chofu Maru*
East of Taiwan	1966—VII	23°45′N	34	15	*Takuyo*
	1966—VII	23°00′N	46	(10)*	*Chofu Maru*
	1967—II	23°00′N	29	5	*Ryofu Maru*
	Average		40	12	
	1942—X	25°00′N	29		*No. 3 Kaiyo*
Entrance of the	1961—X	25°00′N	33		*Chofu Maru*
East China Sea	1965—VIII	25°00′N	36	9	*Chofu Maru*
	1966—I	25°00′N	33	2	*Chofu Maru*
	1966—VII	25°00′N	32		*Rhofu Maru*
	Average		33		

* The numerical values in brackets are not certain.

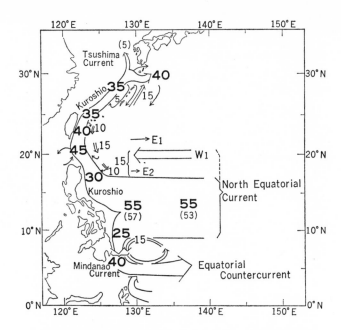

FIG. 28   Schematic chart of volume transport based on 1200 db surface in million m³/s in the western North Pacific and the Kuroshio region by geostrophic calculation. The values are rounded out to five or ten except for the values in the bracket in the North Equatorial Current which show the increase of transport towards west. The numerical value in the bracket for the sum of the Tsushima Current and the Osumi Branch is estimated from the other transports of the currents near the East China (see Section 6.4).

cannot be discussed because of the lack of data and the uncertainty for the estimation of volume transport. In the North Equatorial Current, two values are calculated: one is the total transport, and the other the transport obtained by reducing the first westward band, W_1, and second eastward band, E_2, considering the sign mentioned in 5.2. This transport to be subtracted is 13 (130°E) and 16 (137°E) million m³/s, and this transport does not pass the sections just east of Luzon and Taiwan. An average total transport is 83×10^6 m³/s near 130°E and 69×10^6 m³/s at 137°E, and the transport which minus the value, W_1 minus E_2, is 70×10^6 m³/s and 53×10^6 m³/s, respectively. An increase of average transport from 137°E to 130°E is 14×10^6 m³/s for the total transport and 17×10^6 m³/s for the transport minus the value, W_1 minus E_2, and this increase may be principally attributed to the contribution from the cold eddy just east of Mindanao Island having its center near 130°E. Some of this increase may also be due to the increase in Sverdrup's transport between 137°E and 130°E. According to the least square method applied to Masuzawa's six sections in Table 6, this increase between 137°E and 130°E can be

estimated as 4×10^6 m³/s.

In conclusion, the transport of 13 million m³/s at 130°E, which can be obtained by subtracting 4×10^6 m³/s from the increasing transport of 17×10^6 m³/s, is associated with the eddy mentioned above, and then the net transport of the North Equatorial Current at 130°E is about 70 million m³/s for the total transport or 57 million m³/s for the transport minus the value, W_1 minus E_2. Then the volume transport of about 25×10^6, 30×10^6 and 15×10^6 m³/s flow as the net Mindanao Current, the Kuroshio just east of Luzon and as the algebraic sum of the first westward band W_1 and the second eastward band E_2, respectively. The net transport of the North Equatorial Current is about one and a half or two times those estimated by Sverdrup et al. (1942) and Wyrtki (1961) and nearly equal to the averaged transport between 140°E and 130°E calculated by Masuzawa (1964), though the period of each observation is not the same. The transport of the Mindanao Current, both in net and apparent transport, is larger than those estimated by Sverdrup et al., Wyrtki using the old data (before World War II) and Munk (1950) using his wind-driven theory.

6.3   Volume transports of the Kuroshio off Luzon and Taiwan

In Table 8, the transports of the Kuroshio from east of Luzon to the entrance of the East China Sea are shown with the transports of its countercurrent. An average volume transport of the Kuroshio just east of the northern part of Luzon may be of the order of about 30 million m³/s because the average transport obtained by geostrophic calculation is 28 million m³/s (Table 8) and that estimated from the continuation of the North Equatorial Current is 32 million m³/s (Table 7). The southward countercurrent at the right side of this section has a transport of 10^6 million m³/s, and some of this transport is to be reduced from 30 million m³/s when the net transport of the Kuroshio just east of Luzon is required, because some of 10 million m³/s is associated with the warm eddy centered at the right side of the Kuroshio in the Luzon Strait.

The Kuroshio in the Luzon Strait has a transport of about 45 million m³/s. The main cause of this large value may be attributed to the contributions from the warm eddy mentioned above and from the cold eddy existing in the southern part of the Luzon Strait in general.

In the section east of central Taiwan, the transports of the Kuroshio and its countercurrent are 40 and 12 million m³/s on the average, respectively. Considering these values and the contributions from two eddies in the Luzon Strait, the inflow of the Kuroshio to the South China Sea or the outflow of the water in the South China Sea to the Pacific is not so large on the average, though the seasonal variation of the current is seen in the upper layer at the Luzon Strait associating with the change of monsoon (Wyrtki, 1961; Shigematsu, 1932). At the entrance of the East China Sea, the average transport of the Kuroshio is about 33 million m³/s and is less than east of central Taiwan owing to the disappearance of the distinct eddy to the right of it. From Table 8 it is seen that the transport of the Kuroshio Countercurrent is largest at the section east of the Luzon Strait, meaning that the center of the warm eddy is here. In this region, the transports of the Kuroshio in the summer of 1965 are larger than those in the summer of 1966 and in the winter of 1967 by about 10 million m³/s on the average, and this tendency is similar in the transports of the Kuroshio Countercurrent. This tendency is seen, moreover, not only on the average but also in each section of the Kuroshio and its countercurrent. It seems that when the transport of the Kuroshio increases, the warm core is strong and subsequently the transport of the countercurrent increases.

6.4 Volume transports of the Kuroshio in the East China Sea and southeast of Yakushima Island

The transports of the Kuroshio northwest of Okinoerabujima Island, as the representative in the East China Sea, and southeast of Yakushima Island and of the section east or southeast of Amamioshima Island are shown in Fig. 29. The values of the former two sections were calculated by the Nagasaki Marine Observatory, and of the latter section by the author using some approximations when the observations were limited to the depth of 500 m. The average transport is about 33, 38 and 16×10^6 m³/s, respectively. The transports in these sections indicate the long periodic variations. The phase of this periodic variation of the Kuroshio southeast of Yakushima Island is nearly the reverse of that northwest of Okinoerabujima Island, and is nearly similar to that of the northeastward current southeast of the Ryukyu Islands which may be the continuation of the first westward band of the North Equatorial Current W_1 mentioned before. The amplitude at the section southeast of Yakushima Island is the largest of the three having about 10×10^6 m³/s, and it is about one-half to one-third of the transports at the other two sections. It is difficult to illustrate the above phase relations from these data only.

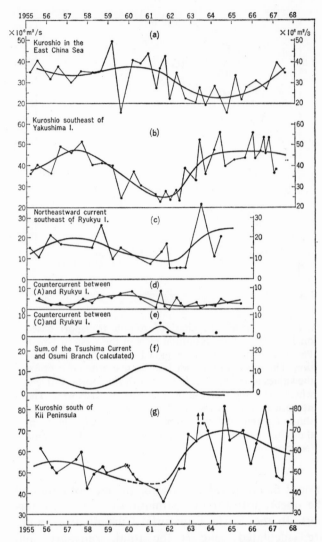

FIG. 29 Volume transport of the Kuroshio and the adjacent regions (a) Kuroshio northwest of Okinoerabujima Island in the East China Sea; (b) Kuroshio southeast of Yakushima Island; (c) northeastward current southeast of the Ryukyu Islands; (d) Countercurrent between (a) and the Ryukyu Islands; (e) Countercurrent between (c) and the Ryukyu Islands; (f) sum of the Tsushima Current and Osumi Branch; (g) Kuroshio south of Kii Peninsula.

The author calculated the transports of the Kuroshio Countercurrent between the Kuroshio and Ryukyu Islands referring to the 800 db surface, of the countercurrent between the Ryukyu Islands and the northeastward current mentioned above, of the Kuroshio south of Kii Peninsula referring to the 1000 db surface and of the sum of the Tsushima Current and the Osumi Branch shown in Fig. 29, but the last of which was deduced from the other transports.

To illustrate these relations, an assumption is made. The transport of the Kuroshio southeast of Yakushima Island can be obtained by subtracting the following four transports, those of countercurrents between the Kuroshio in the East China Sea and the Ryukyu Islands, between the Ryukyu Islands and the northeastward current mentioned above, of the Tsushima Current and the Osumi Branch from the sum of the transports of the Kuroshio in the East China Sea and the northeastward current southeast of the Ryukyu Islands. The transports of the Osumi Branch and of the Tsushima Current cannot be obtained by geostrophic calculation with sufficient accuracy owing to the shallow bottoms. The sum of transports of the Tsushima Current and the Osumi Branch, which can be computed from the above assumption if it is true, is about 5 million m³/s on the average. This volume transport is not so inappropriate for the Tsushima Current as shown in Chap. 9. The transport of the Osumi Branch may be on the order of 1–2 million m³/s considering the GEK values and the topography of Osumi Strait. The phase of variation is almost the same with the observed mean surface velocity or surface transport obtained by GEK in these two currents. From these, though the amplitude of the computed transport of the Tsushima Current seems to be large, the above assumption appears to be valid in the accuracy of geostrophic calculation, and the phase relations between currents in these regions may be explained by this assumption. The variations of transport can be classified approximately in two reverse groups, having a period of about seven-nine years, with regard to their phases. The variation of the Kuroshio in the East China Sea, the countercurrent southeast of it and the Tsushima Current belong to the first group, and those of the northeastward current southeast of the Ryukyu Islands and of the Kuroshio southeast of Yakushima Island belong to the second group. The variation of the volume transport of the Kuroshio southeast of Yakushima Island appears to be accompanied by that of the northeastward current southeast of the Ryukyu Islands, not by that of the Kuroshio in the East China Sea. The reason for these relations, however, is not clear at present. In general, the amplitudes of the variation are nearly proportional to the volume transports of the currents. The transport of the Kuroshio southeast of Yakushima Island is smaller by about 20 million m³/s than that south of Kii Peninsula. One reason for this is the joining of surrounding water between these two sections to the Kuroshio, and the other may be the overestimation in the geostrophic calculation of the Kuroshio south of Kii Peninsula owing to the frequent existence of a large warm eddy south of it. The long periodic variation in the section south of Kii Peninsula is nearly

the same as in the section southeast of Yakushima Island.

6.5 Vertical distribution of volume transport in the Kuroshio at its beginning

The vertical distributions of the transport per 100 m thickness in the Kuroshio east of Luzon, Taiwan and southeast of Yakushima Island are shown in Fig. 30 with the average percentage of transport above the selected depths. In the upper 500 m, the increase of transport between sections east of Luzon and Taiwan is comparatively large, and in the upper 300 m, the transport southeast of Yakushima Island is about 3 million m³/s smaller than east of Taiwan, probably owing to the branching of the Tsushima Current and the Osumi Branch. Below the depth of 500 m, the transport increases gradually with increasing latitude. This tendency may be seen from the current chart of the North Intermediate Water (Fig. 19), in which it is seen that the deep current joins the Kuroshio in a more straightforward manner than the surface current with increasing latitude as one part of the Subtropical Gyre.

Stommel (1953) and Charney (1955) used the depth of 10°C isotherm as the indicator of the thickness of the Gulf Stream in their studies of inertial aspects, but here the depths above which 50% and 75% of total transport are included will be used as indicators of the thickness of the Kuroshio for the sake of convenience. If the Kuroshio is inertially controlled, the potential vorticity, $(f+\zeta)/D$ has to be conserved in the Kuroshio, where f is the Coriolis parameter, ζ the relative vorticity and D the thickness of the Kuroshio. As ζ is not so large in comparison to f,

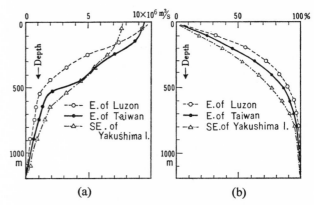

FIG. 30 (a) Vertical distribution of volume transport per 100-m thickness in the Kuroshio east of Luzon, east of Taiwan and southeast of Yakushima Island.

(b) Percentages of volume transport above the depth to which the volume transport is integrated from the sea surface.

the potential vorticity can be approximated by f/D. According to Fig. 30 (b), the values of D corresponding to 50% and 75% of total geostrophic transport based on 1200 db surface are 175,225 and 290 m for 50% and 330, 395 and 490 m for 75% in three sections, respectively. The values of f/D are 2.44, 2.59, 2.48×10^{-9} for 50% and 1.30, 1.46, 1.47×10^{-9} for 75% in three sections, respectively. This means that the Kuroshio at its beginning may have an inertial character like the Gulf Stream (Charney, 1955; Morgan, 1956; and Stommel, 1958).

REFERENCES

Asaoka, O. and S. Moriyasu (1966): On the circulation in the East China Sea and the Yellow Sea in winter. (Preliminary Report). *Oceanogr. Mag.*, **18**: 73–81.

Cannon, G. A. (1966): Tropical waters in the western Pacific Ocean, August–September 1957. *Deep-Sea Res.*, **13**: 1139–1148.

Charnell, R. L. and G. R. Seckel (1966): The circulation on the Trade Wind zone of the central North Pacific. *Proc. XIth Pacific Science Congress, Tokyo, 1966,* Vol. **2.** (Abstract).

Charney, J. G. (1955): The Gulf Stream as an inertial boundary layer. *Proc. Nat. Acad. Sci. Wash.*, **41**: 731–740.

Cromwell, T. (1951): Mid-Pacific oceanography, January through March, 1950. *Spec. Sci. Rep. U.S. Fish. Wildl. Serv., Fish.*, No. 54, 9pp.

Ekman, V. W. (1905): On the influence of the earth's rotation on ocean currents. *Ark. f. Mat. Astr. Fys.*, **131**: 1–13.

Fujii, M. and M. Kimura (1960): Concerning the relation between the variation of Kuroshio in Osumi Gunto Area and that of the water temperature in Yakushima coastal area (in Japanese). *Jour. Oceanogr. Soc. Japan*, **16**(2): 55–58.

——— (1961): Drifting of 20,000 current bottles, released in the sea southwest of Kyushu, July, 1960 (in Japanese). *Hydr. Bull.*, **67**: 58–62.

Kato, T. (1958): Analysis of distribution and correlation of water masses in the East China Sea in summers 1954–1956 (in Japanese). *Jour. Meteor. Res.*, **10**: 693–697.

——— (1959): Oceanographic conditions of the East China Sea in winters (in Japanese). *Jour. Meteor. Res.*, **10**: 743–751.

Kikuchi, S. (1958): Researches on the sea surface temperature of the East China Sea (in Japanese). *Jour. Meteor. Res.*, **11**: 169–183.

——— (1964): On the abnormal low sea temperature in the East China Sea due to the cold wave (in Japanese). *Jour. Meteor. Res.*, **16**: 84–95.

Kishindo, S. (1931): On the method of ocean current observation now used by the Hydrographic Department and some results obtained. Part 3 (in Japanese). *Hydr. Bull.*, **109**: 483–503.

——— (1932): On the method of ocean current observation now used by the Hydrographic Department and some results obtained. Part 4 (in Japanese). *Hydr. Bull.*, **111**: 60–63.

Koenuma, K. (1938): On the hydrography of the southwestern part of the North Pacific and the Kuroshio. Part 2: Characteristic water masses which are related to this region, and their mixtures, especially the water of the Kuroshio. *Imper. Marine Observ., Memoirs*, **6**(4): 349–414.

——— (1939): On the hydrography of the southwestern part of the North Pacific and the Kuroshio. Part 3: Oceanographical investigations of the Kuroshio area and its outer region. Development of ocean currents in the North Pacific. *Imper. Marine Observ., Memoirs*, **7**(1): 41–114.

Koizumi, M. (1962): Seasonal variation of surface temperature of the East China Sea. *Jour. Oceanogr. Soc. Japan, 20th Anniv. Vol.*, 321–329.

——— (1964): On the standard deviation of the surface temperature of the East China Sea (in Japanese). *Studies on Oceanography* (dedicated to Prof. Hidaka in commemoration of his sixtieth birthday), Tokyo, 140–144.

Mao, H. L. and K. Yoshida (1955): Physical oceanography in the Marshall Island area. *Geological Survey Professional Paper*, **260-R**: 645–684.

Masuzawa, J. (1964): Flux and water characteristics of the Pacific North Equatorial Current. *Studies on Oceanography* (dedicated to Prof. Hidaka in commemoration of his sixtieth birthday), Tokyo, 121–128.

——— (1965): Water characteristics of the Kuroshio. *Oceanogr. Mag.*, **17**: 37–47.

——— (1967): An oceanographic section from Japan to New Guinea at 137°E in January 1967. *Oceanogr. Mag.*, **19**: 95–118.

——— (1968): Cruise report on multi-ship study of short-term fluctuations of the Kuroshio in October to November 1967. *Oceanogr. Mag.*, **20**: 91–96.

Matsudaira, Y. (1964): The meaning of the high salinity region in the ocean (in Japanese). *Studies on Oceanography* (dedicated to Prof. Hidaka in commemoration of his sixtieth birthday), Tokyo, 85–88.

Morgan, G. W. (1956): On the wind-driven ocean circulation. *Tellus*, **8**: 301–320.

Munk, W. H. (1950): On the wind-driven ocean circulation. *Jour. Meteor.*, **7**: 79–93.

Nitani, H. (1961): On the general oceanographic conditions at the western boundary region of the North Pacific Ocean (in Japanese). *Hydr. Bull.*, **65**: 27–35.

——— (1970): Oceanographic conditions in the sea east of the Philippines and Luzon Strait in summers of 1965 and 1966. *In The Kuroshio—A Symposium on the Japan Current* (edited by J. C. Marr), East-West Center Press, Honolulu, 213–232.

Nitani, H. and D. Shoji (1970): On the variability of the velocity of the Kuroshio—II. *In The Kuroshio—A Symposium on the Japan Current* (edited by J. C. Marr), East-West Center Press, Honolulu, 107–116.

Reid, J. L. Jr. (1961): On the geostrophic flow at the surface of the Pacific Ocean with respect to the 1000 decibar surface. *Tellus*, **13**: 489–502.

—— (1965): Intermediate waters of the Pacific Ocean. *Johns Hopkins Oceanographic Studies*, **2**: 85pp.

Shigematsu, R. (1932): Some oceanographical investigation of the results of oceanographic survey carried out by H.I.J.M.S. Mansyu from April 1925 to March 1928. *Rec. oceanogr. Wks. Japan*, **4**(1): 151–170.

Shoji, D. and H. Nitani (1966): On the variability of the velocity of the Kuroshio—I. *Jour. Oceanogr. Soc. Japan*, **22**(5): 192–196.

Stommel, H. (1953): Examples of the possible role of inertia and stratification in the dynamics of the Gulf Stream System. *Jour. mar. Res.*, **12**: 184–195.

—— (1958): *The Gulf Stream—A physical and dynamical description*—Univ. of California Press, Berkeley and Los Angeles, 202pp.

Sverdrup, H. U., M. W. Johnson and R. H. Fleming (1942): *The oceans, their physics, chemistry and general biology*. Prentice-Hall, New York, 1087pp.

Sverdrup, H. U. (1947): Wind-driven currents in a baroclinic ocean with application to the equatorial currents of the eastern Pacific. *Proc. Natl. Acad. Sci.*, 318–326.

Takahashi, T. (1959): Hydrographical researches in the western equatorial Pacific. *Mem. Fac. Fish., Kagoshima Univ.*, **7**: 141–147.

Takahashi, T. and M. Chaen (1967): Oceanic condition near the Ryukyu Island in summer of 1965. *Mem. Fac. Fish., Kagoshima Univ.*, **16**: 63–75.

Uda, M. (1955): On the Subtropical Convergence and the currents in the northwestern Pacific. *Rec. Oceanogr. Wks. Japan*, **2**: 141–150.

—— (1964): On the nature of the Kuroshio, its origin and meanders. *Studies on Oceanography* (dedicated to Prof. Hidaka in commemoration of his sixtieth birthday). 89–107.

Uda, M. and K. Hasunuma (1969): The eastward Subtropical Countercurrent in the western North Pacific Ocean. *Jour. Oceanogr. Soc. Japan*, **25**(4): 201–210.

Worthington, L. V. (1959): The 18° Water in the Sargasso Sea. *Deep-Sea Res.*, **5**: 297–305.

Wüst, G. (1936): Kuroshio and Golfstrom. Eine vergleichende hydro-dynamische Untersuchung. *Veröff. des Inst. f. Meereskn.* Neufolge A., **29**: 69pp.

Wyrtki, K. (1961): Physical oceanography of the Southeast Asian Waters. Scientific results of marine investigations of the South China Sea and Gulf of Thailand, 1959–1961. *Naga Report*, **2**: 195pp.

Yamanaka, H., N. Anraku and J. Morita (1965): Seasonal and long-term variations in oceanographic conditions in the western North Pacific Ocean (in Japanese). *Rep. Nankai Regional Fish. Res. Lab.*, **22**: 35–70.

Yoshida, K. and T. Kidokoro (1967): A Subtropical Countercurrent in the North Pacific.—An eastward flow near the Subtropical Convergence. *Jour Oceanogr. Soc. Japan*, **23**(2): 88–91.

—— (1967): A Subtropical Countercurrent (II).—A prediction of eastward flows at lower subtropical latitudes. *Jour. Oceanogr. Soc. Japan*, **23**(5): 231–246

DATA SOURCES

Data Report of Hydrographic Observations, Maritime Safety Agency, Tokyo. 4.

Hydrographic Bulletin, Maritime Safety Agency, Tokyo. 62, 68, 74, 77 and Spec. No. 6, 8, 13.

Oceanic Observation of the Pacific, Calif. Press, Berkeley & Los Angels. Pre 1949.

Oceanographic Observation made during the EQUA-PAC, 1956. Kagoshima Univ., Japan. 1957.

Preliminary Data Report of CSK, Kuroshio Data Center, Maritime Safety Agency, Tokyo. 1, 2, 6, 7, 8, 12, 20, 25, 26, 31, 43, 50, 54, 57, 60, 61, 82, 155.

The results of Marine Meteorological and Oceanographical Observations, Japan Meteorological Agency, Tokyo. All from 17 to 39.

Chapter 6

CHARACTERISTICS OF THE FLOW OF THE KUROSHIO SOUTH OF JAPAN

BRUCE TAFT

Scripps Institution of Oceanography, La Jolla, USA.

1 INTRODUCTION

Modern scientific investigations of the Kuroshio began about 50 years ago when the Japanese Hydrographic Office initiated a program of offshore hydrographic observations (Uda, 1964). Scientific papers based on these observations described the water characteristics and the geostrophic flow of the Kuroshio. Although these studies provided a good basis for a general description of the properties of the Kuroshio, a need was felt for more systematic coverage of the Kuroshio so that long-period fluctuations in the transport of the current and changes in the path of the current could be described. Monitoring of the Kuroshio on a regular basis began in 1955. Every year following 1955, at least four surveys of the waters south and east of Japan have been carried out. During these surveys surface currents were measured with the GEK (geomagnetic electro-kinetograph) and hydrographic sections across the current were made. As often as possible, sections on the quarterly surveys were repeated at the same location to provide for intercomparison between surveys. These data are unique in that they are the only long-term series of systematic surveys of a western boundary current. For example, the Gulf Stream has never been studied in this way (Stommel, 1965).

Quarterly charts of the results of these surveys have been published for the period 1955 to 1964 by the Japan Maritime Safety Agency (no date). The charts show the individual GEK current vectors and the contours of temperature at three levels: sea surface, 100 m and 200 m. The maps depict many features of the Kuroshio. An analysis of these data will produce a statistical description of the characteristics of the current. Data from the surveys of the period 1956 to 1964 in the region south of Japan, i.e., 130° to 141°E, have been selected for analysis. Because of the complex nature of the flow of the Kuroshio east of Japan, data east of Japan were not included in this study. Data from this region would need to be analyzed differently from the data south of Japan.

Primary data for this study are the GEK surface current measurements. They are used to locate the position of the Kuroshio and to measure its strength. Relative volume transports are computed from the hydrographic data and are used to estimate the change in transport along the current. Time variations in the maximum surface current speed and relative volume transport of the Kuroshio over the period 1956 to 1964 are treated in the final section. In the first section there is a brief discussion of the bottom topography off Japan.

An earlier version of some results of the analysis of the data used in this study may be found in Taft (1970).

2 ACKNOWLEDGMENTS

Work on this study was initiated while I was in residence at the Geophysical Institute of the University of Tokyo during 1966 and 1967 as a participant in the United States-Japan Cooperative Science Program. Financial support during my stay in Japan came from the National Science Foundation Grant GF-231. The paper was completed at the Scripps Institution of Oceanography of the University of California. Financial support for this research also came from the Office of Naval Research and the Marine Life Research Program of the Scripps Institution of Oceanography.

I would particularly like to express my gratitude to Professor Kozo Yoshida of the University of Tokyo for his assistance to me throughout the course of this study. Summaries of surface current data from the data file of the National Oceanographic Data Center in Washington, D.C. were kindly provided by Mr. William L. Molo. Dr. Harold Solomon translated several papers written in Japanese.

3   BOTTOM TOPOGRAPHY SOUTH OF JAPAN

A map of the bottom topography of the western Pacific may be found in the paper by Mogi elsewhere in this treatise (p. 80). Geographical names used in this paper conform to those employed by Mogi.

A large-scale map of the bottom topography north of 28°N and between longitudes 130° and 143°E is reproduced in Fig. 1. The Shikoku Basin is bounded on the west and north by the continental slope off Japan and on the east by the Izu-Ogasawara Ridge. The 4 km isobath, which marks the base of the continental slope, will be used as the boundary of the Shikoku Basin. North of 30°N there are three sea-

mounts in the basin that may influence the path of the Kuroshio. The largest of these seamounts is located at about 31°30′N and 136°E.

The continental slope offshore of Kyushu and Shikoku roughly parallels the coastlines of the islands. The trend of the continental slope changes from northeast off Kyushu to northeast by east off Shikoku. South of the Kii Peninsula the trend of the continental slope is east by north. The southernmost extension of the Kii Peninsula is Shionomisaki; this cape will be referred to often in the discussion of the measurements. If one takes the 1 and 4 km isobaths as the upper and lower boundaries of the continental slope, then the width of the slope decreases from about 60 miles off Kyushu to about 35 miles off Shionomisaki. There are two places where the width of the slope changes abruptly: near 133°E where the orientation of the slope changes from northeast to northeast by east; and near 134°30′E where the orientation changes from northeast by east to east by north. The narrowest section of the continental slope

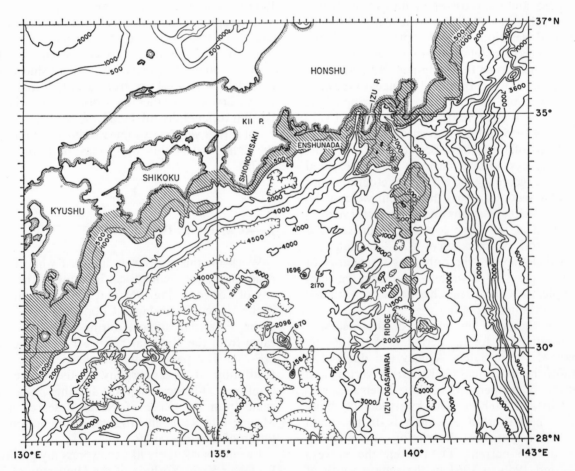

FIG. 1   Bathymetric chart of region south of Japan. Contours are in meters. The contours were taken from Sheets 1 and 2 of Bathymetric Chart of the Adjacent Seas of Nippon published in 1966 by the Maritime Safety of Japan. Depths less than 1,000 m are indicated by slanting lines. The scale is 1 : 92×10^5 at 35°N.

is between 134°30' and 135°E; the depth increases 3 km in only 20 miles. It is important to note that the deeper part of the continental slope does not follow the abrupt turn to the north of the coastline east of Shionomisaki. There is a northward turn of the 1 km isobath east of Shionomisaki, but the trend of the 2, 3 and 4 km isobaths does not change, so that there is a plateau on the upper part of the slope between 136° and 137°E.

At about 34°30'N the continental slope meets the western edge of the Izu-Ogasawara Ridge. Except for a narrow channel located at 34°N, which is deeper than 1 km, the minimum depth of the ridge is less than 1 km everywhere north of 32°30'N. If the Kuroshio west of the ridge extends to a depth greater than 1 km, then the northern section of the Izu-Ogasawara Ridge presents a barrier to the deep flow of the current. The ridge deepens gradually to the south so that at 30°N the minimum depth of the ridge is greater than 2 km. There are isolated elevations on the ridge south of 33°30'N that extend above 1 km. These isolated shallow areas are distributed so that there are two unobstructed eastwest passages on the ridge with depths greater than 1 km: one centered at 32°N; and the other centered at 30°50'N. East of the ridge the depth increases rapidly in the Izu-Ogasawara Trench. North of 30°N the axis of the trench lies approximately along 142°E.

4   PATH OF THE KUROSHIO

4.1   Description of Current Charts

A large number of direct measurements of surface current velocity were made with the GEK in the Kuroshio during the nine-year period 1956 to 1964. During each year at least four surveys were conducted; in the years 1960 to 1964 the frequency of cruises increased to five or six surveys in one year. Current velocity measurements from the Kuroshio have been represented in a series of charts which are found in Appendix 1.

The charts of the surface current velocity distribution have been simplified so as to emphasize the position of the axis of the Kuroshio, i.e., the position of maximum current speed. Only the positions of the velocity vectors with speeds greater than 0.9 knots (kn) are specifically represented. A solid isoceles triangle gives the position and current direction of the maximum current speed on each GEK transect; the adjacent number gives the maximum speed in knots. The solid line is the portion of the ship's track over which the current speed away from the Kuroshio

axis was greater than 0.9 kn. Current vectors away from the maximum are not shown but the positions of current measurements with speeds greater than 0.9 kn are indicated by dots. The length of the solid line represents the width of the region of strong flow of the Kuroshio. Because the transects were not oriented normal to the current, current widths determined in this way tend to be larger than the true widths. Where the current speed falls below 1.0 kn, the ship's track is represented by a dashed line and the positions of the GEK measurements are not indicated. In some instances there are regions outside of the main flow of the Kuroshio where the currents are relatively strong. In order to indicate the presence of these strong currents, individual current vectors have been plotted on the dashed lines where two adjacent stations show currents in excess of 0.9 kn. These currents may or may not be flowing in the direction of the Kuroshio. On some transects measurements appeared to have been made somewhere within the Kuroshio, but the transects were not oriented so that a good transverse distribution of velocity across the Kuroshio was obtained, e.g., Chart 22. On these transects individual current vectors were plotted, but no designation of the Kuroshio axis was made.

The axis of the current for a survey has been represented by a heavy line drawn through the locations of the velocity maxima on the GEK transects that cross the Kuroshio. This line is referred to as the *path of the Kuroshio*. It should be emphasized that this current path is different from the continuous paths that have been drawn for the Gulf Stream (Fuglister and Voorhis, 1965). Fuglister and Voorhis have described a method of tracking the position of the 15°C isotherm at 200 m along the current and have considered the resulting line to be a representation of the path of the Gulf Stream. The continuous paths derived in this way will show more detail than the estimated paths shown on the charts in Appendix 1. The necessity of interpolating the path over relatively large distances gives a heavily smoothed version of the path. In neither representation does the estimated path correspond to the instantaneous path of the current. Kawai (1969) states that the expression *stream axis of the Kuroshio* is preferred by Japanese oceanographers to *path of the Kuroshio*. A satisfactory term does not exist for the representation of the axis adopted in this paper. The expression *path of the Kuroshio* will be used in this paper because of its convenience, not because of its superiority as a descriptive term.

If there were transects on a survey separated by less than 20 miles, the path has been drawn through the average position of the maxima at that location rather than through the individual maxima. The path has not been drawn parallel to the direction of the velocity vector at the current axis; there can be a large component of the current normal to the drawn path. On those surveys where the GEK transects are not sufficiently close together to justify interpolation of a path or the transects are oblique to the current so the position of the maximum current is not clearly established, only the velocity measurements are represented and no path is drawn, e.g., Charts 18 and 22. Fuglister (1963) has pointed out the difficulty in correctly inferring the path of the current from data collected on north-south transects across the Gulf Stream. The path drawn on the chart does not correctly represent the curvature of the path of the Kuroshio, but it does show the approximate location of the region of high current speed in the Kuroshio during the period of each survey.

It has not been shown in the open ocean that the GEK accurately measures the surface current distribution. In order to check on the validity of the GEK measurement as an indication of the maximum surface speed, a comparison was made between the position of maximum current speed measured by GEK and the position of maximum speed determined from geostrophic current calculations using 800 db as a reference pressure. On the 32 sections which were used, the largest displacement between the maximum GEK current and the maximum geostrophic current was 2 miles and the average displacement was 1 mile. The close agreement in location of the axis of the current by the two independent methods is striking. Since the sampling interval for the GEK measurement was often smaller than the usual 20-mile spacing of the hydrographic stations, the degree of agreement must be somewhat fortuitous. However, the important point is that the maximum GEK current was always located between the two hydrographic stations with the largest horizontal gradient in geopotential computed relative to 800 db. These data strongly suggest that there would not be a significant difference between the location of the current axis which was determined by the GEK and the location which was determined by geostrophic calculation.

Masuzawa and Nakai (1955), in an analysis of vertical geostrophic current profiles from the Kuroshio, have shown that the current axis at depth is displaced offshore from the axis at the surface. The displacement increases with depth and Masuzawa and Nakai indicate that the maximum current at a depth of 500 m may be displaced offshore 20 to 40 miles from the maximum current at the surface. For this reason the current paths determined in this paper

probably differ systematically from those that would be derived from subsurface current observations.

Vertical sections of temperature across the Kuroshio were used to supplement the GEK measurements in the choice of the axis of the current. The relationship between the velocity and temperature fields in the Kuroshio south of Japan has been discussed by many authors (Masuzawa and Nakai, 1955; Shoji, Watanabe, Suzuki and Hasuike, 1958; Endo, 1961; and others). Uda (1964) summarized the results of these studies and concluded that the position of the current axis at the surface is highly correlated with the position of the strongest horizontal temperature gradient at 200 m. Uda also states that the 15 C isotherm consistently lies in the center of the strongest horizontal temperature gradient at 200 m. In a statistical study of temperature data and surface current measurements Kawai (1969) essentially confirms Uda's conclusion. Kawai finds the average 200 m temperature underneath the maximum GEK surface current is 16.3 C between 133.5° and 134.5°E and 14.9 C between 137.5° and 138.0°E. A representative bathythermograph temperature section across the Kuroshio is shown in Fig. 2. This section shows the typical relationship between the position of the maxi-

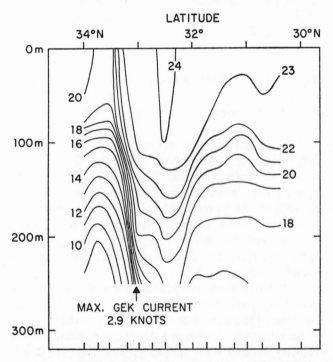

FIG. 2 Bathythermograph temperature (C) section across Kuroshio made by *Meiyo* on November 10–11, 1963. Location of the maximum GEK surface current speed is indicated by the arrow at the bottom of figure. Positions of bathythermograph observations are given by marks at bottom of figure. Vertical exaggeration is 1850.

mum GEK surface current and the position of the 15 C isotherm at 200 m. From the period 1956 to 1964 there are available 577 GEK transects across the Kuroshio. On approximately 10 percent of these transects a velocity maximum was not defined clearly for one of the following reasons: the GEK transect did not extend all the way across the current; the transect appeared to be oriented oblique to the current; or there was not a single significant current speed maximum. When there was some ambiguity in the interpretation of the velocity measurements, bathythermograph temperature sections were drawn and the position of the 15 C isotherm at 200 m was determined. The velocity measurement nearest to the ocation of the 15 C isotherm at 200 m was chosen to represent the current axis.

4.2 Statistical Characteristics of the GEK Velocity Measurements

The interval between GEK observations on most of the transects was between 10 and 20 miles. Closely spaced measurements of the surface current velocity across the Kuroshio show secondary maxima in current speed over distances less than 10 miles (Shoji and Nitani, 1966; Shoji et al., 1958). Features of this scale will be smoothed out because of the sampling interval employed on the transects used in this study. In general, the GEK transects show a single well-defined maximum so that the path usually could be drawn unambiguously.

The maps do not show the cross-stream velocity structure of the Kuroshio. The data were examined to determine the number of sections where the profile departed from a simple smooth distribution, i.e., a single maximum speed decreasing monotonically to speeds less than 1.0 kn on either side of the current. A difference in speed of 0.5 kn was used as the criterion for defining a second velocity maximum within the Kuroshio. On 9 percent of the transects the

velocity profile showed a second maximum. Table 1 gives the distribution with longitude of the relative frequency of transects with a secondary maximum. Between 131° and 139°E the relative frequency varies from 0 to 10 percent; between 139° and 141°E it is 2 to 3 times higher than between 131° and 139°E. The increased frequency of secondary maxima in the cross-stream profile occurs where the current is crossing the Izu-Ogasawara Ridge. The complexity of the current profile on the ridge probably can be attributed to the effect of the complicated topography of the ridge on the flow of the Kuroshio.

A measure of the variability of the position of the current axis within the period of a survey can be obtained by computing the difference in latitude of the location of the maximum GEK current speed on individual north-south transects which were repeated along approximately the same meridian within a 30-day period. Table 2 shows the average differences in latitude of the current axis for different time intervals between the measurements. Transects east of 139°E were excluded from the calculations because there was significantly more variability of the axis of the current on the Izu-Ogasawara Ridge than elsewhere. About 60 percent of the transects used in Table 2 were located south of Shionomisaki at longitudes between 135°40′ and 135°50′E. Table 2 shows that the average displacement of the current is not a function of the difference in time. The overall average displacement is 12 miles. The average uncertainty in the position of the current axis during a survey period is no more than the typical distance between GEK observations along the transect. Masuzawa (1968) in an analysis of 15 days of GEK observations at 5-mile intervals along 136°30′E concluded that the location of the maximum surface current speed of the Kuroshio also varied about 10 miles.

Figures 3 and 4 show almost all of the Kuroshio paths during the period 1956 to 1964. The paths have been transcribed from the charts in Appendix 1 and

TABLE 1. Relative frequency (percent) of secondary velocity maxima at different longitudes on GEK velocity transects across the Kuroshio. A secondary velocity maximum is defined so that its speed differs from adjacent speeds by at least 0.5 kn.

	131°–132°	132°–133°	133°–134°	134°–135°	135°–136°	136°–137	°137°–138°	138°–139°	139°–140°	140°–141°
Number of transects with secondary maxima	1	2	3	2	11	0	3	4	19	8
Total number of transects	27	34	53	48	115	62	80	59	76	23
Relative frequency ×100	4	6	6	4	10	0	4	7	25	38

have been grouped into two time periods: Fig. 3 - March 1956 to March 1959 and May 1963 to November 1964; and Fig. 4 - July 1959 to November 1962. West of 135°E there is overlap in the paths from the two periods. Between 137° and 138°E the paths in Fig. 4 are displaced to the south about 2 degrees of latitude with respect to those in Fig. 3. The paths are

not distributed randomly in time between the two periods. There were clearly two different mean paths for the Kuroshio during this nine-year period and each of them was stable for a period longer than three years. Paths from surveys in May-June 1959 (Chart 14) and February-March 1963 (Chart 34) are not included in Figs. 3, 4 and 5 because they appeared to

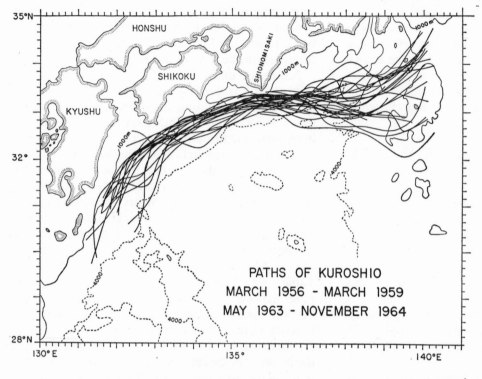

PATHS OF KUROSHIO
MARCH 1956 – MARCH 1959
MAY 1963 – NOVEMBER 1964

FIG. 3 Composite of Kuroshio paths taken from charts 1 to 13 and charts 33 to 43 in Appendix I. Charts 1 to 13 cover the period March 1956 to March 1959; Charts 33 to 43 cover the perod May 1963 to November 1964. The 1,000 and 4,000 m isobaths are shown. The scale is $1 : 92 \times 10^5$ at 35°N.

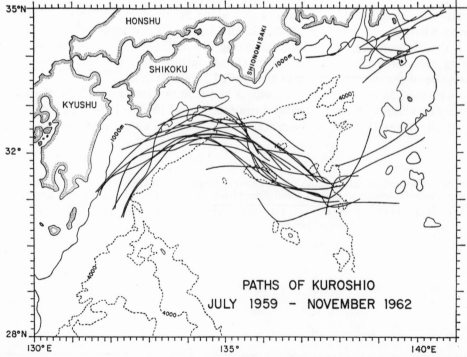

PATHS OF KUROSHIO
JULY 1959 – NOVEMBER 1962

FIG. 4 Composite of Kuroshio paths taken from Charts 15 to 31 in Appendix I. These charts cover the period July 1959 to November 1962. The 1,000 and 4,000 m isobaths are shown. The scale is $1 : 92 \times 10^5$ at 35°N.

TABLE 2. Average and maximum displacement of axis of Kuroshio at a fixed longitude for different time differences between GEK transects. *n* is the number of observations in each time difference class. Transects are located between 131° and 139°E.

Difference in Time Between GEK Transects

Displacement of axis (miles)	0–4	5–9	10–14	15–19	20–24	25–29
			(days)			
Maximum displacement	31	29	36	34	38	27
Average displacement	10	11	14	12	12	12
n	20	20	22	11	14	10

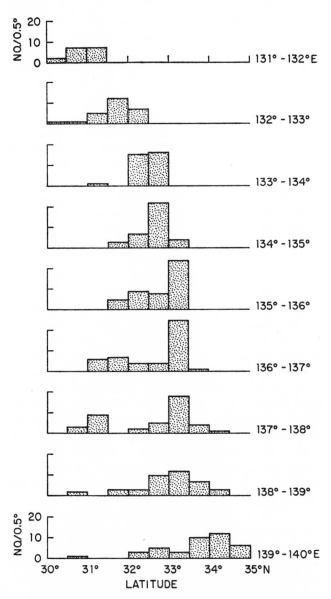

FIG. 5 Frequency distributions (number per 0.5 latitude) of latitude of maximum GEK current speed of Kuroshio for the period March 1956 to November 1964. The data are grouped by 1° of longitude.

be transitions between the two characteristic mean paths. Similar figures which show the envelopes of the paths for the Kuroshio for different periods of time have been published by Uda (1951), Masuzawa (1960), Uda (1964) and Masuzawa (1965a).

The frequency distributions of the latitude of the Kuroshio axis at different longitudes for the period 1956 to 1964 are shown in Fig. 5. West of 135°E the distributions are strongly peaked with a small range of latitudes represented. The relatively narrow range of latitudes is in part due to the orientation of the transects; the transects west of 134°30′E are northwest-southeast in orientation so that a movement of the current will not be solely expressed by a change in latitude whereas further east the transects are generally north-south. At longitudes 136°–137°E and 137°–138°E the distribution of the axis is bimodal. The southward displacement of the Kuroshio is particularly marked at longitudes 137°–138°E where the distribution is discontinuous as well as bimodal.

The southward displacement of the current is well known and has been extensively discussed. The first scientific description of the meander and the associated Cold Water Mass of Shionomisaki was given by Uda (1937, 1940). More recent discussions of the meander are found in Yosida (1961a, b), Moriyasu (1963), Uda (1964), Shoji (1965) and Masuzawa (1965a). When the Kuroshio path is displaced to the south, there is a roughly circular region north of the Kuroshio with low temperatures at its center (Fukuoka, 1958; Moriyasu, 1961). The circulation around this feature is counterclockwise (cyclonic) with the Kuroshio forming the southern part of the circulation. Because the southward movement of the Kuroshio is accompanied by the development of this large cold water region north of the Kuroshio, the phenomenon of the displacement of the Kuroshio is often referred to as the Cold Water Mass (or Cold Cyclonic Eddy) off Enshunada (Moriyasu, 1963). Uda (1964) and other writers have used a different terminology. They have referred to the southward displacement of the Kuroshio and the accompanying development of the cold water region to the north of the current as an anomalous pattern. Following the terminology of Uda, Taft (1970) distinguished between the normal and abnormal paths of the Kuroshio. When the Kuroshio was in its northern position the path was referred to as normal and when it was displaced to the south the path was referred to as abnormal.

Neither of the above conventions seem to be desirable. As will be pointed out later in this paper, there is a general tendency for the isotherms west of Shionomisaki to deepen toward the north so that there is almost always a region of relatively cold water north of the Kuroshio. The size of the cold water region changes markedly when the Kuroshio is displaced to the south, but the distribution of temperature is similar whether the Kuroshio is displaced or not. In the choice of the name for the phenomenon primary importance should be given to the change in the position of the current and not to the temperature distribution north of the Kuroshio. The use of the words *abnormal* (Taft, 1970) or *anomalous* (Uda, 1964) implies that the displacement is infrequent. As Shoji (1965) and Masuzawa (1965a) point out, the southward displacement is a characteristic feature of the Kuroshio. Over the period of about 40 years that data have been systematically collected south of Japan, the displacement has occurred three times: 1934 to 1944, 1953 to 1955 and 1959 to 1963. Thus, over a period of 40 years the displacement occurred during 17 years so that to refer to the phenomenon as either abnormal or anomalous does not seem correct. The southward displacement should be considered as a large-scale meander of the current that is stable for a long period of time. Masuzawa (1965a) and Shoji (1965) refer to the displacement as a large-scale stable meander of the Kuroshio and I will follow their terminology in this paper.

Uda (1949) gives a discussion of the distribution of properties associated with the meander in the period 1934 to 1944. There were no surveys in the offshore waters during 1953 to 1955 but Masuzawa (1954) and Ichiye (1954a) discuss measurements made along a single section south of Honshu at three-week intervals during the period of the meander. Although there is very little offshore data available before 1925, Uda (1964) mentions that there was evidence of a southward displacement of the Kuroshio and a growth of the Cold Water Mass off Enshunada during the years 1906 to 1907 and 1917 to 1919.

Conditions during the time of occurrence of the southward displacement of the current in 1959 were reasonably well described by the offshore surveys. A change in the flow of the Kuroshio was observed in May 1959 (Moriyasu, 1961). In May the current moved into deep water off Kyushu instead of remaining on the continental slope (Chart 14). Subsequently, a meander in the current developed and was observed to move eastward at an approximate speed of 10 miles per day (Shoji, 1965). In August 1959 (Chart 15) the southward displacement of the current east of Shionomisaki was first observed and the meander was found to be fully developed in November 1959 (Chart 16). This pattern then remained almost stationary until May 1963 when the current moved back on to the continental slope south of Shionomisaki

(Chart 34); the Kuroshio remained there through the end of 1964.

The discussion of the path of the Kuroshio is divided into two time periods: first, the period when the meander was absent and the current remained on the continental slope until it passed Shionomisaki; and second, the period when the meander was present and the current was in the deep water south of Shionomisaki.

4.3 Path of Kuroshio with no Meander (March 1956–March 1959; May 1963–November 1964)

West of Shionomisaki the Kuroshio consistently was located on the upper portion of the continental slope. Between the west end of Shikoku (about 133°E) and Shionomisaki (135°43′E) the axis of the current was always on the continental slope at depths between 0.8 and 2.0 km. The width of the envelope of all paths over this distance does not exceed 32 miles (Fig. 3). The small range of depths where the maximum current was found demonstrates clearly that the current was closely bound to the upper portion of the shelf.

Near Kyushu the current appeared to be located in deep water at the base of the continental slope on two surveys. The February–March 1957 (Chart 5) and February–March 1959 (Chart 13) charts show the maximum current speed to be about 70 miles offshore at a depth of 4 km. On both these surveys the axis of the current further to the east was on the upper portion of the slope; the current apparently moved up the slope into shallower water as it moved eastward. East of Shikoku the current paths for these two months do not differ from all other paths with no meander. From these two sets of data the conclusion might be drawn that if the current is displaced away from the upper portion of the continental slope off Kyushu, then the tendency will be for it to return to the upper continental slope further to the east and remain there until it passes Shionomisaki. However, a different sequence of events occurred in May 1959 (Chart 14). In this instance the displacement of the current into deep water off Kyushu was followed by the establishment of the Kuroshio meander. These measurements are discussed in the next section.

After the Kuroshio passes Shionomisaki it begins to move into deeper water and the characteristics of its path change. The 1 km isobath turns northward at about 136°E, but the current does not make a similar turn. Between 136° and 137°E the path becomes more zonal, and it tends to parallel the 2 km isobath. At about 137°E, where the continental slope turns northward, the current flows down the slope into deep water. In all but three instances (Charts 10, 11 and 43) the current is in water of depth greater than 4 km near 138°E. As the current goes into deeper water, the angle between the contours of the continental slope and the current path is usually less than 40°. After the Kuroshio moves off of the slope into deep water, the path is more variable and the envelope of the path widens by more than a factor of two (Fig. 3). The following extreme excursions of the current were observed: in May 1958 (Chart 10) the current moved northward along the 2 km isobath to 34°N at 138°E; and in November 1956 (Chart 4) and May 1964 (Chart 39) the current moved southward off of the slope to about 32°45′N at 138°E.

On the Izu-Ogasawara Ridge at 139°E all of the current paths are south of 34°N. The northernmost portion of the ridge, which has depths less than 1 km, apparently acts as a barrier to the current; the current axis was never found on this shallow portion of the ridge when the meander was absent. Even when the current west of the channel was located north of 34°N (Chart 10), the current crossed the ridge in the channel south of 34°N (Fig. 1).

There seem to be two routes for the Kuroshio to cross the Izu-Ogasawara Ridge when the meander is absent. The current most often crossed the ridge in the channel which is located on the northern section of the ridge. This channel provides a continuous passage across the ridge with depths greater than 1 km (Fig. 1). However, in 5 of the 22 surveys the Kuroshio did not cross the ridge in the channel but went over the ridge on the shallower area south of the channel (Charts 4, 34, 35, 36, 40). The most southerly crossings were in November 1956 (Chart 4) and May 1964 (Chart 40). These two southerly crossings took place when the Kuroshio, west of the ridge, was displaced farther to the south than usual.

When the Kuroshio flows across the ridge in the channel, it turns toward the north as it moves to the east. The average northward displacement of the current axis between 138° and 140°E was 42 miles. The amount of northward displacement of the current is dependent upon the latitude at which the current enters the channel. The more southerly the latitude of entry, the larger is the displacement. Once the current enters the channel its path seems to be determined by the boundaries of the channel. The width of the channel decreases to the east, and accordingly the position of the current after it has left the channel is less variable than the position of the current at the entrance to the channel (Fig. 3). There are only two examples when the current entered the channel but did not remain there. In November-Decem-

ber 1964 the current at 139°E was in the center of the channel but at 140°E the current was south of the channel at a depth less than 0.5 km (Chart 12). There is only a time difference of two days between the velocity transects at 139° and 140°E, so it seems clear that the current did leave the deeper water of the channel and went over the ridge in relatively shallow water. The other example is in February-March 1964 when the current was in the southern portion of the channel at 139°E but was found slightly south of the channel at 140°E (Chart 39).

Whether the Kuroshio enters the channel or not appears to be related to the position of the current west of the ridge. In every instance when the current was north of 33°N at 138°E the current turned northward into the channel. When the Kuroshio was south of 32°50′N on the approach to the ridge (Charts 4, 34, 40), it crossed the ridge south of the channel. In the three instances when the current west of the ridge was between 32°50′ and 33°N at 138°E, two times the current did not enter the channel (Charts 35, 36). The critical latitude for determining how the current will cross the ridge may lie between 32°50′ and 33°N.

It should be noted that almost always there is a tendency for the Kuroshio to turn cyclonically as it flows over the ridge. The northward turning of the current in the channel has already been discussed. When the current is not in the channel but flows over the ridge south of the channel, it shows the same tendency to turn cyclonically toward the north (Fig. 3). The only exception to the cyclonic turning of the current south of the channel is September 1963 (Chart 36). The measurements show the current path dipping to the south in the middle of the ridge.

4.4 Path of the Kuroshio with Meander (May 1959–March 1963)

In May 1959 the Kuroshio shifted offshore from Kyushu into the deep channel at the base of the continental slope (Chart 14). Instead of flowing along the continental slope offshore from Kyushu, Shikoku and Honshu, the Kuroshio was in deep water at 132°40′E and was flowing roughly toward the east. After the May survey a special series of weekly cruises was undertaken to describe subsequent changes in the position of the Kuroshio (Yoshida, 1961a; Moriyasu, 1961; Shoji, 1965). On each of the weekly cruises the current appeared to flow toward the east off of the continental slope and then to turn abruptly toward the north at a longitude well to the east of the slope. Between 5 May and 4 June 1959, the longitude of northward turning of the current moved eastward from 134°30′ to 137°15′E suggesting that a meander in the current was moving eastward. On the May 1959 survey, GEK transects east of 137°E show that the current was located on the continental shelf off Enshunada, and it crossed the Izu-Ogasawara Ridge on the shallow area north of the channel (Chart 14). The path of the current in May 1959, from Kyushu east to the Izu-Ogasawara Ridge, was different from the paths for the previous three years.

On the next survey, in August 1959, ship coverage of the region between Shionomisaki and the Izu-Ogasawara Ridge was not adequate to establish the position of the Kuroshio (Chart 15). The November 1959 GEK measurements show that the axis of the Kuroshio west of the Izu-Ogasawara Ridge was located at about 31°N (Chart 16). Flow on the western flank of the ridge in November was northward and the current crossed the ridge north of 34°N on the shallow area south of the Izu Peninsula (Fig. 1).

This general pattern of flow persisted through March 1963. The position of the Kuroshio between 136° and 138°E was stable during 1960 and 1961 and the first three quarters of 1962. For example, during this period the axis of the current at 137°40′E was bounded by latitudes 31°25′ and 30°45′N. The displacement of the Kuroshio had begun to decrease by October 1962 (Chart 32). In May 1963 (Chart 34) the Kuroshio had returned roughly to where it had been during the period March 1956 to March 1959. The Kuroshio in March 1963 (Chart 33) was clearly in an intermediate position with respect to the two envelopes of paths (Figs. 3 and 4), and has not been plotted on either figure. As has been pointed out by Masuzawa (1965a), the set-up time for the meander is probably considerably shorter than the dissipation time for the meander. The northward return of the Kuroshio to the continental shelf may take as long as nine months but it appears likely that the meander is established in about three months.

The position of the Kuroshio west of 134°E did not differ greatly from its position when there was no meander. Except for May 1959 (Chart 14) and August 1959 (Chart 15), the axis of the current west of 134°E was always on the continental slope. The longitude at which the Kuroshio left the continental slope and flowed into deep water did not vary greatly during the period November 1959 through March 1963 (Fig. 4). Due to the spacing of the transects, it is not possible to determine the position of the southward turning of the current on all of the surveys. However, on those surveys where a sufficient number of transects exist between 133° and 136°E to permit an estimate of where the current turned southward, the maximum latitude of the current appears to have occurred always between 134° and 135°E. There is more spread

in the abnormal paths than in the normal paths at 134°E. The normal paths are concentrated on the upper portion of the slope at depths less than 1.5 km whereas roughly one half of the abnormal paths at 134°E occur at depths between 2 and 3 km (Figs. 3 and 4). The location of the southward turning of the current is associated with a change in the bottom topography. It was pointed out previously that both the orientation and the width of the continental slope change abruptly at about 134°30′E. The slope narrows to its minimum width and the orientation becomes more east-west.

In general, the position of the Kuroshio west of the Izu-Ogasawara Ridge was clearly marked by the GEK transects and the current appeared to be coherent from transect to transect as a well-defined region of high velocity. The velocity measurements in May 1960 (Chart 19) show a more complex pattern and it is unlikely that a simple path existed during this month. During May 6 to 8, GEK measurements were made on two transects oriented north-northwest and spaced about 40 miles apart. These transects probably crossed the Kuroshio at its southernmost position slightly west of the ridge. Velocity measurements on the two transects show strong southeastward currents on the western transect and strong northwestward currents on the eastern transect (Fig. 6). The data shown in Fig. 6 are not represented on Chart 19.

The pattern of current vectors is consistent with the interpretation that a cyclonic eddy with a relatively small radius was present. The apparent dimensions of the eddy are roughly comparable to those observed in cyclonic eddies south of the Gulf Stream east of Cape Hatteras (Fuglister and Worthington, 1951;

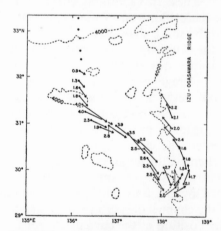

FIG. 6   GEK surface current vectors on two sections made by *Kaiyo* on May 6–8, 1960. Numbers at the base of each vector give current speeds in knots. Dots indicate current speeds less than 0.9 kn. The 4,000 m isobath is shown. The scale is 1 : 92 × 10⁵ at 35°N.

Fuglister, 1963). West of Cape Hatteras cyclonic eddies of this scale have not been reported. Data from the region east of Inubosaki often have shown cold cyclonic eddies south of the Kuroshio (Masuzawa, 1957) which appear to be similar in scale to those described by Fuglister. South of Cape Hatteras cyclonic eddies in the Gulf Stream similar to that indicated in Fig. 6 have not been reported.

The observations do not define whether or not there was a detached eddy; they do suggest that an eddy might have been in some stage of detaching itself from the Kuroshio. Unfortunately, there are not enough observations in this area during the following 30 days to determine how this feature evolved in time. Observations made on May 12, 22, and 31 at about 138°E, i.e., near the position of the feature shown in Fig. 6, show the Kuroshio axis at 31°20′, 30°40′ and 31°00′N, respectively (Chart 19). The direction of the current was approximately eastward on all these sections, and the velocity distributions on these transects do not resemble the distribution of current vectors in Fig. 6. It is possible that an eddy had become detached from the Kuroshio and had moved out of the area before the later measurements were made.

It would be interesting to know how frequently eddies of this scale occur in the Kuroshio. Fuglister (1963) has pointed out the inadequacy of meridional sections for detection of meanders in the Gulf Stream. The usual spacing of north-south sections across the Kuroshio on these surveys was too broad to describe a feature of the scale shown in Fig. 6 so the frequency of occurrence of these eddies is unknown.

The position of the Kuroshio as it flows southward in the Shikoku Basin may be influenced by an elliptically shaped seamount whose center is located at about 31°30′N, 136°E (Fig. 1). There are two isolated regions of minimum depth on the seamount—one with a depth of 2210 m and the other with a depth of 2180 m. The current axis was consistently very close to this seamount in 1960 (Fig. 7). Of the 32 velocity transects in the vicinity of the seamount in 1960, 23 transects show the axis within 15 miles of the longitudinal axis of the seamount. The extent of the southward deflection of the current may be determined by the dynamical effect of the seamount on the path of the current. Only on three transects, which were made in May 1960, was the Kuroshio clearly located south of the seamount. Transects made on May 27 and 29 show the current axis as much as 50 miles south of the seamount (Chart 19). The anomalous position of the current south of the seamount in late May 1960 may be related to the eddy that appeared to be present east of the seamount in early May (Fig. 6). It is possi-

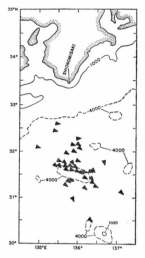

FIG. 7   Positions of Kuroshio axis in 1960 near seamounts located approximately at 31°30′N, 136°E. Triangles show the location of the maximum current and point in the direction of the GEK current vector with maximum speed. The 1,000 and 4,000 m isobaths are shown. The scale is $1: 92 \times 10^5$ at 35°N.

ble that the consistent location of the current north of the seamount is fortuitous and not related to topographic steering of the current by the seamount. The data are merely suggestive that when the meander is present there might be a connection between the seamount and the location of the Kuroshio.

In 1961 and 1962 the Kuroshio was north of the seamount on all of the velocity sections. When the southward displacement of the Kuroshio off Shionomisaki decreased in late 1962, there was no longer any possibility of interaction between the current and the seamount. For example, the axis of the Kuroshio was about 80 miles north of the seamount in August and October 1962 (Charts 31 and 32).

The pattern of flow in the vicinity of the ridge when the meander is present is not well-defined by the surveys. Many of the north-south sections show high velocities but the location of the maximum current is not clear because the currents usually have a strong component parallel to the transects. For this reason path lines on the charts are often not extended across the ridge. Another source of uncertainty in determining the current path on the ridge is the limited southern extent of the transects. Only 7 of the 20 transects on the ridge extend as far south as 32°N. It is possible that some southern crossings of the ridge were missed because of the short length of the transects.

On most of the surveys the Kuroshio on the west flank of the ridge appeared to flow northward, roughly parallel to the axis of the ridge. The current usually crossed the ridge on the shallow area south of the Izu Peninsula, e.g., see Chart 20. There is no tendency for the current to turn eastward at the opening to the channel and then flow across the ridge in the channel. When the meander is present, the current on the west side of the ridge crosses the bottom contours at a very small angle. On the other hand, when the meander is absent, the current approaches the ridge directly from the west and the current direction is roughly normal to the bottom contours. The different approach of the current to the entrance of the channel when the meander is present may be the reason why the current does not turn eastward and enter the channel. Measurements on one survey (Chart 23) suggest that the current did not flow along the west side of the ridge but might have turned northwest and reentered deep water before crossing the ridge. This proposed path is consistent with the presence of the Kuroshio off Enshunada at a latitude north of 34°N. Since the coverage on this survey was incomplete, the path of the current between the transect south of Shionomisaki and the transects off Enshunada is unknown. Except for May 1959 (Chart 14), this survey is the only one which showed the Kuroshio in such a northern position west of the ridge.

There are three instances when the Kuroshio clearly did not turn northward west of the ridge but instead flowed across the ridge at a more southern latitude with a strong eastward current component (Charts 19, 31, and 32). This more southern flow across the ridge appears to be exceptional.

5 AXIAL CURRENT SPEED OF THE KUROSHIO

The change in the speed of the Kuroshio along its path south of Japan may be determined by analyzing GEK axial surface current speeds on the transects across the current. Longuet-Higgins, Stern and Stommel (1954) give an extensive discussion of the theory of the measurement of current velocity by the method of towed electrodes (GEK). They conclude that the most serious potential source of bias in the GEK measurement of the surface current in the deep ocean is the effect of finite ocean depth. The condition that must be satisfied if the GEK is to give an unbiased measure of the surface current is that the vertically average velocity is much smaller than the surface current. If the velocity in the deep water is small relative to the velocity in the surface layer and the ocean is also deep, then the vertically averaged speed will be much less than the surface speed and the GEK output will be a good measure of the surface current.

In a western boundary current, like the Kuroshio and the Gulf Stream, the current may extend to the bottom and the above condition may not be satisfied. South of Japan, the Kuroshio flows over bottom topography where the depth of water under the current changes by 3 km so that it is possible some of the observed GEK surface current speed changes are due to changes in the average velocity between the surface and the bottom. For example, if the Kuroshio flowed without change of surface current speed or transport from the upper continental slope, where the depth is 1 km, down in the Shikoku Basin, where the depth is over 4 km, the predicted effect on the GEK measurement would be that the GEK would record a higher current speed in the basin than on the upper slope.

In order to adjust raw GEK measurements so that they give an unbiased estimate of the true current speed, an empirical k-factor is used. The k-factor is the ratio of the actual current speed to the GEK current speed (von Arx, 1962). The current speed is often determined by measurement of ship drift during the period of time that a GEK transect is being run. In general, the k-factor is not a constant. According to von Arx, k probably varies across the Gulf Stream from 1.05 in the Sargasso Sea to 1.40 at the current maximum. The published Kuroshio GEK data are based on a single k-factor for the region (Dr. Daitaro Shoji, personal communication). If the k-factor is not a constant south of Japan, then some of the apparent changes in GEK current speed discussed here may not be real. For example, it is quite possible that a different k-factor should be used when measurements are made on the continental slope than when measurements are made in the deep water of the Shikoku Basin.

The data have been sorted into two sets so that a comparison can be made between the downstream change of axial current speed in the Kuroshio when the meander was present and when it was absent. Longitudinal distributions of average axial current speed are plotted in Fig. 8. Both curves show an eastward increase in average speed between 132° and 137°E. The speed of the current offshore from Kyushu at 132°15′E and offshore from Shikoku at 134° 35′E are significantly higher when the meander was present. However, the speed was not uniformly higher west of 135°E when the meander was present; there was not a significant difference in average speed at 133°25′E. There is a particularly large increase in maximum current speed between 134°35′ and 136° 55′E when the meander was absent. The average current speed increased 1.0 kn in 100 miles. The curve for the period when the meander was present does

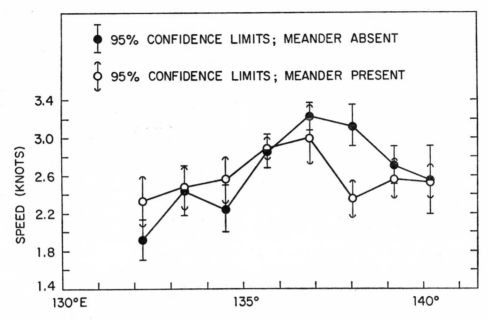

FIG. 8   Longitudinal distribution of axial GEK surface current speed of Kuroshio. Data have been divided into two sets: March 1956 to March 1959 and May 1963 to November 1964, when the meander was absent—designated by a dot; and July 1959 to November 1962, when the meander was present—designated by a circle. Ninety-five percent confidence limits are marked on the vertical lines by horizontal lines for the first set of data and by semi-circular arcs for the second set.

not show such an abrupt increase in speed; the rate of increase of speed with longitude appears to have been constant between 132°15′ and 136°55′E.

The average longitude at which the Kuroshio enters water deeper than 3 km can be estimated from the envelopes of the paths (Fig. 3 and 4). When the meander is present the Kuroshio leaves the slope between 134°30′ and 135°E and when the meander is absent the current enters deep water at about 137°E. Average current speed would be expected to decrease if the current extended to the bottom and the current did not change its transport. Surface current speed increases at the longitude where the current leaves the continental slope and enters deep water to form the meander. When there is no meander there is no change in surface current speed where the current leaves the continental slope. If the axial surface current speed can be used as an indicator of the vertically averaged speed then the GEK data suggest that either the current plotted in Fig. 8 does not extend to the bottom everywhere or that the transport of the current is not constant. Of course this inference presumes that the GEK gives a valid measure of changes in surface current when the depth of water is changing. As has been already pointed out, this assumption is questionable.

The increase in speed of 1 kn between 134°35′ and 136°55′E occurred while the Kuroshio was on the continental slope. There was not a significant change in depth of water under the axis of the current be-

tween these longitudes but the width of the continental slope over which the current is flowing decreases by about 50 percent (Fig. 1). The position of the axis of the current as shown on the current maps does not indicate that the current axis was shifted offshore where the continental slope narrowed; it remained on the upper part of the slope (Fig. 3). In order to determine whether or not there was a concomitant change in current width, geopotential anomaly of the sea surface relative to 500 db was plotted and the distance between selected isopleths of geopotential along the path of the current was measured. Isopleths were chosen so as to approximately span the region of high current speed measured by GEK. Geopotential difference between the selected isopleths is either 0.3 or 0.4 dynamic meters. Hydrographic sections across the current east of 134°E are in general oriented in a north-south direction and do not necessarily cross the current at a right angle, so that the width of the current can be seriously overestimated if the angle between the section and the streamlines of the current departs very much from 90°. When the current stays on the slope as it flows past Shiono-misaki, the path is approximately east-west (Fig. 3) so the change in the average width should be rather accurately determined.

Convergence of the isopleths of geopotential indicates a decrease in width of the current and should correspond to an increase in the relative geostrophic surface current. Figure 9 is a plot against longitude

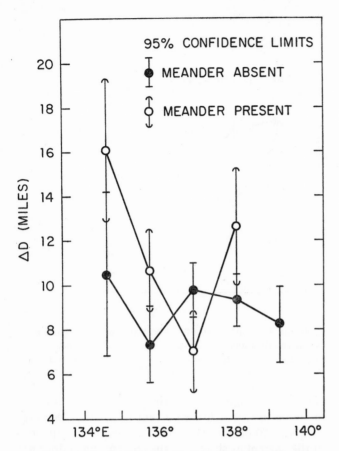

FIG. 9 Longitudinal distribution of average distance (ΔD) between selected values of geopotential anomaly of the sea surface relative to 500 db in Kuroshio. Data were divided into two sets: March 1956 to March 1959 and May 1963 to November 1964, when the meander was absent—designated by a dot; and July 1959 to November 1962, when the meander was present—designated by a circle. Ninety-five percent confidence limits are marked on the vertical lines by horizontal lines for the first set of data and by semi-circular arcs for the second set.

of the average distance between selected values of the geopotential, where the distance is weighted inversely to the absolute difference in geopotential. When there was no meander, the average current width decreases 30 percent between 134°35′ and 135°45′E. It should be noted that the width decreased on all nine surveys for which measurements were available in this range of longitude. The 1 kn increase of GEK current speed is associated with the constriction of the current south of Shionomisaki and an increase in relative geostrophic current speed. However, the increase in width of the current east of Shionomisaki is not associated with a corresponding decrease in GEK speed; the speed continues to increase to the east to a maximum value at 136°55′E. The lack of agreement between the change in GEK current speed and the change in the geostrophic current speed may be ex-

plained in several ways. After the Kuroshio passes Shionomisaki there may be a change in the vertical distribution of current. Even though there is an increase of surface current speed as measured by GEK, there might be a decrease in the relative speed between the surface and 800 db. A second possibility is that there is a change in the horizontal distribution of current speed across the current east of Shionomisaki. The maximum speed, as measured by the GEK, may increase but the average geostrophic surface current, as measured by the gradient in geopotential anomaly across the current, may decrease. A final possibility is that there is a change in the GEK speed owing to the fact that the current is moving into deeper water after it passes Shionomisaki. As was mentioned before, the change in water depth might tend to increase the GEK output and indicate a higher current speed. The inconsistency cannot be explained; it should be emphasized that the GEK measurements can only be interpreted with caution.

West of the Izu-Ogasawara Ridge both curves show a decrease in axial speed of the Kuroshio (Fig. 8). The decrease in speed occurs more abruptly and further to the west when the meander is present. The largest decrease in speed takes place between 136°55′ and 138°05′E, where the maximum southward displacement of the current occurs. There is little change in the depth of water under the current where the surface speed decreases. The current is located on the relatively flat Shikoku Basin; bottom slopes are small in the basin, except in the immediate vicinity of the sea mounts (Fig. 1). Volume transport of the current relative to 800 db does not change significantly between 136°55′ and 138°05′E (Fig. 10). If the volume transport of the current and the depth are constant and the average speed of the current decreases, then the width of the current must increase. In Fig. 9 the plot of the width of the region of high current speeds for the period when the meander was present shows that between 134°35′ and 136°55′E, where the axial current speed is increasing, the width of the current decreases but between 136°55′ and 138°05′E, where the maximum current decreases sharply, the width increases by about 80 percent. Since the path of the current became more east-west in orientation between 136°55′ and 138°05′E, the increase in width is probably not due to a distorted measurement of the width owing to the curvature of the current path. The sharp decrease in GEK current speed does seem to have been associated with a pronounced broadening of the current and therefore a decrease in relative geostrophic current speed.

Between 138° and 140°30′E the Kuroshio goes from deep water, where the depth is in excess of 4 km, over the rather shallow crest of the Izu-Ogasawara Ridge. The current always crossed the ridge where the depth was less than 1.5 km and in most instances the current axis was located in water whose depth was less than 1 km (Figs. 1, 3 and 4).

When the Kuroshio west of the ridge was in its northern position and the flow was directed roughly perpendicular to the trend of the isobaths along the ridge, the average current speed at the axis decreased about 0.7 kn as the current flowed up the western flank and crossed the crest of the ridge. Most of the change in speed occurred as the current flowed over the steep slope on the western side of the ridge. It is perhaps surprising that the speed of the current decreases on the ridge. If the current west of the ridge extends to depths greater than the crest of the ridge and the volume transport of the current does not change on the ridge, then the current would either have to increase its average speed and keep its width constant or increase its width and decrease its speed. Since the maximum surface current speed decreases on the ridge, the implication is that the current would become broader and one would expect the measurements of the width would show a change. However, Fig. 9 does not show a significant change in current width on the western flank of the ridge where there is a decrease in current speed. In an analysis of the behavior of a constant-transport jet that conserves potential vorticity, Robinson and Niiler (1967) concluded that such a jet would tend to broaden and decrease its strength as it flowed into shallow water. The decrease in current speed is consistent with the theory of Robinson and Niiler but the measurements do not also show a broadening of the current.

The effect of the ridge on the speed of the Kuroshio when the meander is present appears to be different from that described for the other path with no meander. There is no decrease of speed as the current flows over the west flank of the ridge. There may even be a slight increase of speed between 138°05′ and 139°15′ E. The GEK velocity transects suggest that the flow roughly parallels the bottom contours until the current reaches about 33°30′N. North of 33°30′N the current begins to turn eastward and cross the bottom contours on the ridge. On the west flank of the ridge the change in water depth along the path of the current is not large. The difference in the response of the current to the ridge when the meander was present and when it was absent may be due to the difference in rapidity of change in depth that the current undergoes as it approaches the crest of the ridge. It should be pointed out again that the GEK measurements may be influenced by the change in the depth of water as the current goes over the ridge and therefore the GEK

may not give a reliable measure of the surface current speed.

6 RELATIVE TRANSPORT OF KUROSHIO

Hydrographic sections south of Japan can be used to compute the geostrophic relative transport of the Kuroshio. There are two problems that arise in the selection of the data to be used in the transport computations. The first problem is the choice of a reference pressure. Even though the computation of the geostrophic velocity in the surface layer may not be very sensitive to the choice of the reference pressure, the estimate of the transport of the current is strongly dependent on which reference pressure is chosen. If the velocity at some pressure is known, then the absolute transport can be computed. Unfortunately, very few direct subsurface current measurements have been made off Japan; the velocity in the deep water beneath the Kuroshio is unknown (Nan'niti and Akamatsu, 1966; Istoshin and Sauskan, 1968). At the present time a rational basis for choosing the reference pressure does not exist. On most of the stations the maximum sampling depth of the hydrographic casts was below 1000 m, but on some casts the deepest bottle lay between 800 and 1000 m. A reference pressure of 800 db was chosen in order to obtain the largest number of transport values. Transport calculations referred to a greater pressure, e.g., 1500 or 2000 db, show a larger transport for the Kuroshio than calculations which use 800 db as a reference (Sverdrup et al., 1942, p. 727; Masuzawa, 1965b). Transport calculations in this paper do not present the total transport of the Kuroshio and will be treated solely as an indicator of variations of the transport. The distribution of transport with depth may change along the current (or with time) so that variations in transport cannot be adequately followed by using a 800 db reference surface. There are not enough deep sections across the Kuroshio to decide whether or not the proportion of the total transport that takes place above 800 db changes in a systematic way.

The second problem concerns the definition of the width of the Kuroshio. On seven sections which were located between 137° and 139°E there was a change in the sign of the slope of the geopotential of the sea surface both north and south of the Kuroshio axis. On these sections the width of the current for the calculation of transport could be determined objectively. In many instances the northern boundary of the Kuroshio was indicated by a minimum in the geopotential profile but a corresponding maximum at

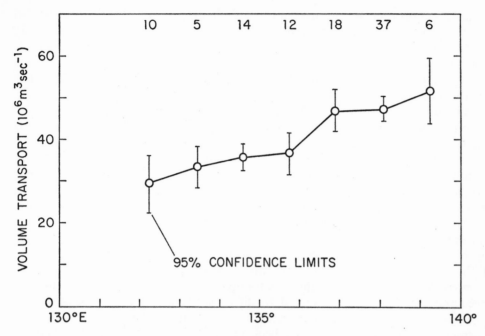

FIG. 10 Longitudinal distribution of relative volume transport of Kuroshio during period March 1956 to November 1964. Dynamic computations were done with 800 db as reference. Number of values averaged in each longitude range is given along the top of the figure. Ninety-five percent confidence limits are given by vertical lines.

the southern end of the section was seldom present. Since the velocity gradient on the southern side of the current is weak (Masuzawa and Nakai, 1955; Masuzawa, 1969), the absence of a well-defined maximum in geopotential there is to be expected. Calculations for most of the sections required making an assumption about the width of the current. When one or more of the boundaries of the current could not be determined from the geopotential profile, hydrographic stations for the transport calculations were chosen to span the region where the surface current, as measured by GEK, was greater than 1 kn. This procedure tends to systematically underestimate the transport of the Kuroshio. For example, average transport for the seven sections on which there was a minimum in geopotential north of the current and a maximum south of the current is 12 percent higher than the average transport for all sections in the same longitude range.

One hundred and two sections in the period 1956 to 1964 were used to calculate the volume transport of the Kuroshio relative to 800 db. Locations and dates of the section are listed in Appendix 2.

The relationship between average relative volume transport and longitude is shown in Fig. 10. The mean transport for the periods when the meander was present and when it was absent do not differ significantly, so it will be assumed that the downstream change in relative transport is the same during both periods. It should be noted that there are no values of transport at 135°45′E when the Kuroshio was on the upper portion of the slope south of Shionomisaki. The current axis was located so close to the boundary

at Shionomisaki that, with the usual station spacing of 20 miles, the difference in geopotential across the current could not be accurately estimated.

There was an eastward increase in relative transport of the Kuroshio. Inspection of the confidence limits in Fig. 10 indicates that there were two levels of approximately constant transport with an abrupt increase in transport occurring between 135°45′ and 136°55′E. A linear increase in transport with longitude is not consistent with the data because there is no overlap in the limits computed for estimates of transport east and west of 136°E. Mean transport between 136°55′ and 139°15′E was about 30 percent higher than it was between 132°15′ and 135°45′E. Knauss (1969) has concluded that there is a smooth exponential downstream increase (7% per 62 miles) in total transport of the Gulf Stream north of the Florida Straits. The Kuroshio relative transport data do now show either a smooth linear or exponential increase of transport.

The amount of the increase in relative transport east of Shionomisaki is about 30 percent. Does it represent an increase in the volume of water transported by the Kuroshio or is it an artifact arising from the way in which the calculations were done? If the proportion of water transported by the current outside the arbitrary speed limits of 1 kn differs east and west of Shionomisaki, then the apparent increase in transport might result from an underestimation of the transport west of Shionomisaki. On the *Atlantis II* section normal to the continental slope off Shikoku, there is a large northeastward geostrophic transport of water offshore from the region of strong flow of the Kuroshio (see Worthington and Kawai, p. 377). Forty percent of the total transport, calculated with respect to the bottom, occurs offshore from the region where sea-surface geostrophic current speeds are more than 1 kn. If a portion of this water entered the high-speed region of the current downstream, then the transport between the 1 kn limits would appear to increase even though the total transport need not change. Since the GEK surface current speeds do not show a comparable increase at the longitude of the transport increase (Fig. 8), it does not seem likely that the change in transport can be accounted for by a sudden change at Shionomisaki in the distribution of velocity across the current.

The apparent increase in volume transport east of 136°E may be due to vertical redistribution of the velocity of the current. The proportion of the total transport which is found in the layers above 800 db may increase east of 136°E. This possibility was mentioned earlier in the discussion of the GEK surface current measurements. The Kuroshio east of Shionomisaki is either flowing off of the slope into deep water (no meander) or has already left the slope and is in the deep water of the Shikoku Basin (with meander). It does not seem reasonable that the current would become more concentrated in the shallow layers as it flows into deeper water. It seems more likely that the current would deepen as it entered deeper water so that a smaller proportion of its transport would occur in the upper layers.

The possibility that water is being added to the Kuroshio from the sides at about 136°E is discussed in the next section.

7 SURFACE CURRENT SPEED AWAY FROM KUROSHIO AXIS

The GEK measurements may be used to determine the characteristics of the surface flow north and south of the axis of the current. The data have been sorted according to distance from the axis of the current, and the mean resultant currents computed (Tables 3 and 5).

7.1 North of Kuroshio

In the region between Shionomisaki and the Izu-Ogasawara Ridge, isotherms in the thermocline north of the Kuroshio usually deepen to the north. As a consequence of the deepening of the thermocline to the north, on level surfaces above 400 m there is an area of minimum temperature north of the current axis. This region of minimum temperature is very obvious when the Kuroshio is displaced southward and is then often referred to as the Cold Water Mass of Enshunada (see Table 1). Moriyasu (1963) estimated the average radius of the Cold Water Mass to be 55 miles with a standard deviation of 11 miles. When the Kuroshio stays on the continental slope until it passes Shionomisaki, there is also a region of minimum temperature north of the Kuroshio, although its size is much smaller than when the current is displaced to the south (Fig. 2). This temperature distribution results in the geopotential of the sea surface relative to 300 db being high close to Enshunada. Westward geostrophic flow relative to 800 db would be expected adjacent to Enshunada.

Analysis of the GEK measurements north of the Kuroshio, when the meander was not present, shows a significant mean resultant westward component of current of 0.15 kn directly south of Honshu between longitudes 135°40′ and 138°40′E (Table 3).* The GEK measurements are consistent with the geostrophic current that would be expected from the temperature distribution. Since the charts in Appendix 1 generally do not show currents with speeds less than 1 kn, the westward flow would rarely be indicated on these charts. It does appear to be present on Charts 5, 7, 8, 9, 34, 37 and 42. The same analysis of the GEK data, when the Kuroshio was displaced to the south, did not show significant average zonal components north of the Kuroshio. The absence of a significant average flow at any range of latitude may only mean that the current is variable in space as well as time. Inspection of the charts when the meander was present shows that there is often a region of westward flow north of the Kuroshio but that its position shifts, e.g., see Charts 19, 25, 21 and 32.

Relative volume transport of the westward flow along the boundary is considerably less than the transport of the Kuroshio. On 16 hydrographic sections across the Kuroshio the stations extend north to the coast of Honshu so that the transport of the westward flow can be calculated. Table 4 gives the values of transport relative to 800 db for the Kuroshio and the westward transport along the boundary. The average eastward transport of the Kuroshio was 48×10^6 m^3 sec^{-1} and north of the Kuroshio the average westward transport was 9×10^6 m^3 sec^{-1}. It is

---

* The statistical test of significance was carried out by computing the standard vector error of the mean current and the corresponding 95% confidence circle (Brooks and Carruthers, 1953). This computation assumes that the current vector has a circular normal distribution. If the error circle did not overlap zero the current component was taken to be significant.

interesting that the average ratio of transports (V_N/V_K) is the same when the eddy was present and when it was absent. If the ratio of the transports does not change when the current path changes, then the net transport of the cyclonic circulation south of Honshu does not vary with the horizontal scale of the region of minimum temperature. However, the small sample size does not make it possible to detect a significant difference in transport between the two periods, unless the difference were quite large. A true difference of 20 percent between two estimates of the geostrophic relative transport probably would not be detectable (Warren and Volkmann, 1968).

The plot of relative volume transport of the Kuro-

TABLE 3. Distribution of mean resultant GEK surface current north of Kuroshio axis between 135°40′ and 138°40′E. Components with asterisks are significantly different from zero at the 95 percent level. $|V|_R$ is the mean resultant current speed, θ_R is the mean resultant current direction, and u and v, respectively, are the mean east, (positive) and mean north (positive) components of velocity, and n is the number of observations averaged.

Latitude	Meander present					Meander not present				
	$|V|_R$	θ_R	$\bar{u}$	$\bar{v}$	n	$|V|_R$	θ_R	$\bar{u}$	$\bar{v}$	n
	(kn)	(°)	(kn)	(kn)		(kn)	(°)	(kn)	(kn)	
34°29′–34°00′N	0.15	070	0.14	0.05	33	0.15	251	−0.15*	−0.05*	135
33°59′–33°30′N	0.06	227	−0.04	−0.04*	33	0.13	145	0.07	−0.10	138
33°29′–33°00′N	0.24	185	−0.02	−0.24	138	0.98	091	0.98*	−0.02	83
32°59′–32°30′N	0.38	163	0.11	−0.36*	112					
32°29′–32°00′N	0.71	135	0.51*	−0.50*	93					
31°59′–31°30′N	0.71	106	0.68*	−0.20*	32					

TABLE 4. Volume transport of Kuroshio (V_K) and volume transport north of Kuroshio (V_N) relative to 800 db. V_N is directed westward, i.e., opposite to the transport of the Kuroshio (V_K).

Year	Date	Longitude	$10^6 m^3 sec^{-1}$		V_N/V_K	Ship
			V_K	V_N		
1956	11/12–13	138°30′°E	39	13	0.33	*Shumpu Maru*
1957	5/13	137°00′	39	4	0.10	*Meiyo*
	8/6	137°40′	57	11	0.19	*Shumpu Maru*
	8/10	138°00′	58	6	0.10	*Takuyo*
	12/9	137°40′	51	14	0.27	*Shumpu Maru*
1958	8/9–10	137°40′	51	1	0.02	*Shumpu Maru*
1959	5/18	135°40′	34	3	0.09	*Shumpu Maru*
	11/15–16	137°40′	54	15	0.28	*Shumpu Maru*
1960	3/20	135°40′	46	9	0.20	*Shumpu Maru*
	8/5–6	137°40′	47	15	0.32	*Shumpu Maru*
	10/21–22	135°40′	37	4	0.11	*Shumpu Maru*
1961	11/5	137°40′	54	11	0.20	*Shumpu Maru*
1962	5/17–18	136°40′	49	9	0.18	*Shumpu Maru*
	5/27	138°30′	48	5	0.10	*Shumpu Maru*
	8/22	137°00′	59	7	0.12	*Takuyo*
1963	5/13–14	137°40′	44	9	0.20	*Shumpu Maru*
	8/3	137°40′°E	53	12	0.23	*Shumpu Maru*

shio against longitude indicates a 30 percent increase in transport between 135°45′ and 136°55′E (Fig. 10). Since the mean paths of the Kuroshio (Figs. 3 and 4) show that the current is adjacent to the boundary west of 135°E, the water transported westward along the boundary between 138° and 135°40′E must be incorporated into the flow of the Kuroshio somewhere east of 135°E. It seems likely that the increase in transport shown in Fig. 10 is largely made up of water which is flowing westward south of Enshunada; this water, because of the orientation of the coast, would flow southward toward the Kuroshio east of the Kii Peninsula. When the Kuroshio is displaced to the south, the confluence between this water, which is moving southward, and the Kuroshio would probably occur farther to the west than when the Kuroshio remains on the continental slope until it passes Shionomisaki.

If the increase in relative transport of the Kuroshio at 136°E were to be accounted for by water being added to the current from the north, then the geopotential of the sea surface along the northern boundary of the Kuroshio would be expected to decrease to the east at the location where the inflow was taking place . . . on the assumption that the southward flow was in approximate geostrophic balance. In order to estimate the longitudinal slope of the sea surface relative to 800 db along the northern and southern edges of the Kuroshio, geopotential anomalies of the sea surface, which were used in the computation of the relative volume transport of the Kuroshio (see section 6), were averaged between 134° and 136°20′E and between 136°21′ and 138°40′E. These ranges of longitudes are centered east and west of Shionomisaki. Mean values of the geopotential anomaly of the sea surface relative to 800 db at the northern and southern edges of the Kuroshio are shown in Fig. 11. At the southern boundary of the Kuroshio there is not a significant variation with longitude of the geopotential anomaly. However, there is a significant eastward decrease along the northern boundary, both when the meander was present and when it was absent. The magnitude of the decrease in geopotential is possibly greater when the meander is absent because of somewhat higher values of the geopotential anomaly west of Shionomisaki. This eastward decrease of the geopotential anomaly is consistent with a relative geostrophic current direction toward the Kuroshio. The average vertical distribution of geopotential anomaly can be computed east and west of Shionomisaki and then used to calculate the average volume transport relative to 800 db toward the Kuroshio. The transport toward the Kuroshio is 12×10^6 m³ sec⁻¹ which agrees very well with the increase of $11 \times$

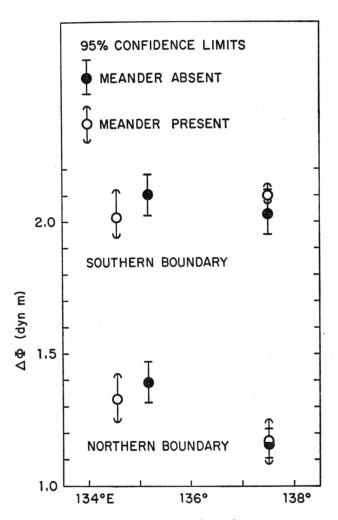

FIG. 11 Longitudinal distribution of average geopotential anomaly ($\Delta\Phi$) of the sea surface relative to 800 db at northern and southern boundaries of Kuroshio. Data have been divided into two sets: March 1956 to March 1959 and May 1963 to November 1964, when the meander was absent—designated by a dot; and July 1959 to November 1962, when the meander was present—designated by a circle. Ninety-five percent confidence limits are marked on the vertical lines by horizontal lines for the first set of data and by semi-circular arcs for the second.

10^6 sec⁻¹ in relative transport of the Kuroshio east of 136°E (Fig. 10). The increase in the transport of the Kuroshio appears to be accounted for by considering that it is transport added to the Kuroshio by flow from the north. There was an average westward transport of water along the coast of Honshu of 9×10^6 m³ sec⁻¹ (Table 4). This westward-moving water apparently follows the coastline of the Kii Peninsula and turns southward to join the Kuroshio east of Shionomisaki. Where this water joins the Kuroshio, the Kuroshio relative transport increases by about 30 percent. This inflow might be substantiated by a study of the changes in flux of temperature, salinity and oxygen characteristics of the Kuroshio similar to the Gulf Stream study of Sturges (1968).

Computations of the distribution of geopotential along the seaward edge of the Gulf Stream between Florida and Cape Hatteras by Sturges (1968) have indicated a geostrophically balanced inflow into the Gulf Stream, which is consistent with the downstream increase in transport reported by Knauss (1969). The computations for the Kuroshio shown in Fig. 11 do not indicate there is a similar inflow in the region south of Shionomisaki.

Additional evidence for a mean flow to the south in the area east of Shionomisaki is found in the GEK current measurements summarized in Table 3. Directly north of the Kuroshio meridional current components are consistently southward. However, the meridional components are significantly different from zero only when the meander is present. Between 33°30′ and 32°30′N the average southward component is 0.30 kn. The GEK measurements suggest, as does the dynamic topography along the northern boundary, that when the meander is present, water is added to the Kuroshio east of Shionomisaki by advection across its northern boundary. When the meander is not present the currents are weaker and the GEK measurements do not show a significant meridional component. It was pointed out that the longitudinal slope of the sea surface along the northern boundary of the Kuroshio appeared to be less when the meander was not present (Fig. 11). The smaller slope would indicate that the current normal to the Kuroshio is weaker when the meander is absent.

The source of water for the westward flow along the coast of Honshu is not clear. There does not appear to be a decrease of transport of the Kuroshio relative to 800 db as the current crosses over the Izu-Ogasawara Ridge (Fig. 10). The relative transport computations do not show that the westward flow represents the northern limb of a closed cyclonic circulation with water entering the Kuroshio east of Cape Shionomisaki and then being recirculated back to the west as the current approaches the ridge. If good estimates of the absolute transport could be made, then it is likely that the transports could be balanced and the source of the water flowing along the northern boundary could be determined. One possibility is that there may be a change in the vertical distribution of transport as the Kuroshio approaches the ridge and the water depth decreases. If this occurred and the transport above 800 db were increased, then some of this water might be recirculated to the north of the Kuroshio with no change in the

transport relative to 800 db of the Kuroshio. It is not possible to do more than offer this as a speculation because of the lack of data on the true transports.

7.2 South of the Kuroshio

In the Gulf Stream it has long been known that there is a surface countercurrent offshore from the main surface flow of the Gulf Stream (Stommel, 1965; von Arx, 1962). Von Arx states that the countercurrent occurs some 40 to 60 nautical miles from the axis of the Gulf Stream and can have speeds up to 1.5 kn. A similar feature has been reported to occur in the Kuroshio (Wüst, 1936; Sverdrup et al., 1942; Masuzawa, 1954; Defant, 1961; Uda, 1964). Sverdrup et al. (1942) note that the Kuroshio Countercurrent can be seen on both geostrophic current profiles and on charts of currents derived from ship's drift and that a typical speed is 0.4 kn. Masuzawa (1954), in a study of 32 hydrographic sections south of Shionomisaki, found that a countercurrent was present about 110 miles south of the Kuroshio axis. Since the Kuroshio flows eastward, the presence of a countercurrent would be indicated by a region of westward current south of the Kuroshio. Inspection of the surface current maps in volumes 1 and 2 of the State of the Adjacent Seas of Japan (Japan Maritime Safety Agency, no date) shows variable and weak currents south of the Kuroshio; it is difficult to decide whether or not there is a countercurrent on any individual GEK traverse. Only a few of the charts in Appendix 1 show evidence of a westward current south of the Kuroshio. There are no charts that show a westward countercurrent that can be traced from one transect to the other. Since only currents greater than 0.9 kn are represented, the charts would not show the countercurrent if it were weak. Uda and Hasunuma (1969) have published seasonal charts of the zonal component of current in one-degree squares which are based on ship-log data. These charts show a band of westward components south of the Kuroshio in all seasons which the authors refer to as the Kuroshio Countercurrent.

In order to remove some of the variability in the data all of the GEK current measurements south of the Kuroshio between 134° and 138°E have been grouped according to distance from the current axis and the mean resultant current computed (Table 5). The eastward component decreases regularly with distance away from the Kuroshio axis. In the class interval 120 to 159 miles, the current vectors have a statistically significant mean resultant west component of 0.22 kn. This region of westward flow would appear to be the average position of the Kuroshio

TABLE 5. Distribution of mean resultant GEK surface current south of Kuroshio axis between 134° and 138°E. Components with asterisks are significantly different from zero at the 95 percent level. $|V|_R$ is the mean resultant current speed, θ_R is the mean resultant current direction, and $\bar{u}$ and $\bar{v}$, respectively, are the mean east (positive) and mean north (positive) components of velocity, and n is the number of observations.

Distance from Axis (miles)	$|V|_R$ (kn)	θ_R (°T)	$\bar{u}$ (kn)	$\bar{v}$ (kn)	n
1–29	1.63	098	1.61*	−0.23*	338
30–59	0.77	104	0.75*	−0.19*	294
60–89	0.37	113	0.34*	−0.15*	241
90–119	0.16	133	0.11*	−0.11	102
120–159	0.22	264	−0.22*	−0.02	74

Countercurrent. Seventy percent of the current measurements at a distance of 120 to 159 miles from the Kuroshio axis had westward components. Average speed of the currents with westward components was 0.6 kn and the maximum observed speed was 1.2 kn. The northern boundary of the region of westward components in the charts of Uda and Hasunuma (1969) appears to lie about 2 to 3 degrees south of the probable axis of the Kuroshio during the period of time (1924 to 1934) in which the ship-log data were collected. There is rough agreement between the position of the countercurrent in the GEK data and its position on the average charts of Uda and Hasunuma.

The Kuroshio Countercurrent appears to lie approximately 60 miles farther seaward of the axis of the main current than does the Gulf Stream Countercurrent. Wüst (1936) also concluded in his study of selected sections across the Kuroshio and the Gulf Stream that the countercurrent off Honshu occurred at a greater distance from the axis of the main current than did the countercurrent off Chesapeake Bay.

Stommel (1965, p. 123) considered that the surface countercurrent to the right of the Gulf Stream is dynamically linked to the warm core, or region of maximum temperature, which is located seaward of the Gulf Stream. Between the center of the warm core and its offshore limit, the horizontal pressure force relative to 200 db will be directed to the right of the Gulf Stream and if the relative surface current is in geostrophic balance, the current direction will be opposite to that of the Gulf Stream. The same type of temperature distribution is found in the Kuroshio; an example is found in Fig. 2. On this section isotherms between 32°30′ and 31°20′N slope up toward the south so that south of the Kuroshio there is a region of maximum temperature with its axis at 32°30′N. The warm core is almost always present, but a description of its statistical characteristics has not been made.

In order to determine whether the Kuroshio Countercurrent is located in the warm core, as has been proposed for the Gulf Stream by Stommel, 122 bathythermograph sections across the Kuroshio between 134° and 138°W were examined and the position of the axis of the warm core and its offshore limit of the warm core has been taken to be the place where the isotherms in the thermocline begin to deepen offshore. The mean position of the center of the warm core on these sections was 28 miles offshore from the maximum GEK current speed of the Kuroshio and the average distance between the axis of the warm core and its outer limit was 71 miles. If the average offshore surface current was in geostrophic balance, then the surface countercurrent would be found between 28 and 99 miles seaward of the Kuroshio axis. The range in distances between the Kuroshio axis and the outer limit of the warm core was 20 to 120 miles. Since the westward flow was found 120 to 159 miles to the right of the axis of the Kuroshio, it is clear that the average westward flow south of the Kuroshio lies outside the region of the warm core. It would be interesting to know whether a similar statistical analysis of the GEK data from the Gulf Stream would show the association between the

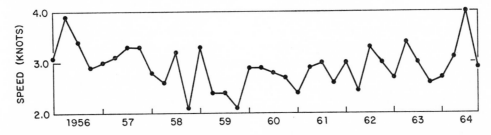

FIG. 12   Quarterly average axial GEK surface current speed of Kuroshio between 135°10′ and 138°40′E for the period March 1956 to November 1964. The first, second, third and fourth quarters correspond to the following times: first quarter—February 1 to March 15; second quarter—May 1 to June 15; third quarter—August 1 to September 15; and fourth quarter—November 1 to December 15.

warm core and the countercurrent which was proposed by Stommel (1965).

8 TIME CHANGES OF THE FLOW OF THE KUROSHIO

The Kuroshio surveys provide a relatively long-time series of measurements with an average sampling interval between cruises of three months. Cruises usually were not distributed at equal time intervals during the year. Typically cruises were made in March, May, August and November. On the average maximum time interval between cruises was 4 months and the minimum interval was 2 months. Since the data were taken roughly at quarterly intervals, the data have been grouped into four equally spaced 90-day periods and the quarterly means calculated.

8.1 Year-to-Year Variation in Speed

In Fig. 12 are plotted the quarterly average GEK axial current speeds of the Kuroshio between 135°-10′E and 138°40′E for the period 1956 to 1964. Current measurements in this range of longitude were selected for analysis because data is available from this region for each quarter of every year. Outside of this region quarterly sampling was not continuous

for the nine-year period. The general impression of Fig. 12 is that the average speed was lower from the second quarter of 1959 to the third quarter of 1962. The meander of the Kuroshio was set up in May 1959 and the current did not return to its northern position until May 1963. During most of the period of the meander, the current speed was lower than average. The differences in speed between years are more easily examined in Fig. 13. In Fig. 13(a) anomalies from the quarterly means are plotted by year. Anomalies from the quarterly means were summed and plotted by year in Fig. 13(b). The annual anomalies in Fig. 13b change from large positive values in 1956 and 1957 to a large negative value in 1959. The speed of the current generally increases with time after 1959, although the rate of increase does not appear to be constant. The main features of interest are the lower values of speed which occurred in 1959 when the meander was set up, and the persistence of low values during the following two years. By 1962 the speed had returned to the nine-year average. The relatively low current speeds when the meander was present also may be seen in Fig. 8.

Examination of Fig. 13(a) shows that the large decrease in speed for 1959 is accounted for by changes in the last three quarters of the year; during the first quarter, before the Kuroshio moved off the continental slope into deep water, the speed appeared to

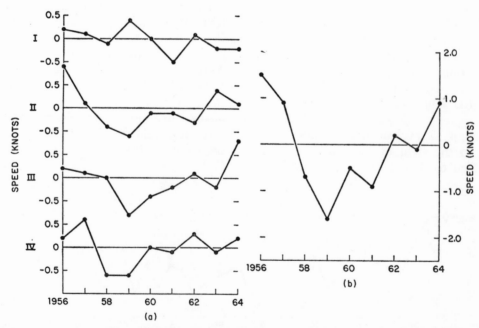

Fig. 13   Axial GEK current speed of Kuroshio between 135°10′ and 138°40′E for the period March 1956 to November 1964. (a) Anomaly from nine-year quarterly average axial GEK surface current speed. (b) Summation for each year of four quarterly anomalies of axial GEK current speed plotted in (a). The first, second, third and fourth quarters correspond to the following times: first quarter (I)—February 1 to March 15; second quarter (II)—May 1 to June 15; third quarter (III)—August 1 to September 15; and fourth quarter (IV)—November 1 to December 15.

be above normal. The association of the early stages of the meander with the decrease in speed of the Kuroshio suggests that there may be a causal relationship between the occurrence of the meander and the decrease in speed. However, another possibility is that the change in speed is merely a reflection of the change in position of the current. The average speed of the current east of Shionomisaki may decrease when the meander is present, relative to its average speed when the meander is absent, because the current is in deep water rather than on the continental slope. The average speed would have to decrease if the Kuroshio extended to the bottom everywhere and the transport were to remain a constant. Unfortunately, the transport of the current cannot be estimated from the available data; all that we have are estimates of relative transport. In the analysis of the data on volume transport relative to 800 db, a significant difference in transport between the period when the meander was present and when it was absent is not found. If the current were to increase its depth and its transport were to remain constant, the transport of the current, relative to 800 db, would have to decrease. In the discussion of the longitudinal distribution of axial GEK surface current (Fig. 8), it was pointed out that the average surface current speed did not decrease as the current entered deeper water. It seems unlikely that the explanation of the year-to-year differences in average maximum speed lies solely in the change in position of the current.

The other possibility is that the formation of the meander was linked to the decrease in current speed. Models of the steering effect produced by the bottom topography over which the Gulf Stream flows have been published by Warren (1963) and Niiler and Robinson (1967). In these models a change in the distribution of velocity will affect the way in which the current responds to the topography. Only GEK surface currents and vertical shear of the geostrophic velocity component above 800 db are available so that one cannot describe the changes in the velocity near the bottom. Unfortunately knowledge of the bottom velocity is critical for the theoretical prediction of the path of the current. Even though we do not have data on the bottom currents, it is worthwhile to document changes in the flow of the Kuroshio at the surface since it may be that there are similar changes in the bottom currents. In order to determine whether or not this is true, direct measurements of bottom currents under the Kuroshio will have to be made.*

Namias (1970) finds evidence that the year-to-year change in current speed of the Kuroshio (Fig. 13(b)) is related inversely to the year-to-year change in the average surface temperature in the subtropical western Pacific (20°–35°N). Namias reasons that the year-to-year variation in surface temperature is larger in the subtropics than in the tropics so that the difference in surface temperature between the tropics and subtropics is largely dictated by temperature changes in the subtropics. Geostrophic volume transport of the North Equatorial Current is related to the temperature difference across the current. To a rough approximation a temperature increase in the subtropics would be correlated with a decrease in transport of the North Equatorial Current, which then might be reflected in a decrease in speed and volume transport of the Kuroshio. Namias' temperature data suggest that the decrease in speed of the Kuroshio in 1959 (Fig. 13(a)) may have been correlated with a decrease in the transport of the North Equatorial Current. If the meander is a response of the Kuroshio to a change in its speed, then these data suggest that the creation of the meander could be traced back to a change in the intensity of the circulation in the North Pacific gyre. Uda (1949, 1964) has suggested that there is a correlation between the occurrence of an extremely cold year off northern Japan, with an associated stronger southward flow in the Oyashio Current, and the development of the Cold Water Mass off Enshunada five years later. It is quite possible that the weaker flow in the tropics and the stronger flow of the Oyashio are not independent events, but are expressions of a change in the whole circulation of the western Pacific. A thorough analysis of the relationship between the meander and the oceanic and atmospheric circulation in the western Pacific is needed. At present, the mechanism of formation of the meander is one of the most intriguing problems in the dynamics of the Kuroshio.

8.2   Seasonal Changes in Speed and Transport

Since the average time interval between Kuroshio surveys is quarterly, it is natural to look for evidence of a seasonal variation of the speed and transport of the Kuroshio. Inspection of Fig. 12 shows that there is no obvious periodicity in the GEK maximum speeds. There is no quarter in which there is a consistent maximum or minimum. The only consistent relation between the average speeds within a year is

---

* Work is now underway with Dr. A. R. Robinson to investigate the possible dynamical relationship between a change in the speed of the Kuroshio and the set up of the meander.

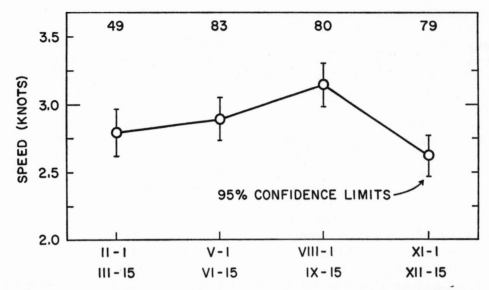

FIG. 14   Quarterly averages of axial GEK current speed of Kuroshio between 135°10′ and 138°40′E for the period March 1956 to November 1964. Ninety-five percent confidence limits are given by the vertical lines. The number of values averaged in each quarter is given along the top of the figure. The first, second, third and fourth quarters correspond to the following times: first quarter—February 1 to March 15; second quarter—May 1 to June 15; third quarter—August 1 to September 15; and fourth quarter—November 1 to December 15.

that the speed in the fourth quarter of the year appears to be less than the speed in the preceding third quarter. In eight of the nine years the maximum surface current speed decreased between the third and fourth quarters; in the other year there was no change in speed between the third and fourth quarters.

Quarterly averages of maximum speed for the nine-year period are plotted in Fig. 14. Highest average maximum speed is in the third quarter. The differences in speed between the second and third and third and fourth quarters are significant. On the other hand there is not a significant difference in speed between the first and the second quarters or between the first and fourth quarters. There does not seem to be evidence of a pure seasonal cycle (12 months) since the maximum and the minimum speeds are separated by only three months. This could mean that there also is a semiannual (six months) component of variation in the speed of the Kuroshio which, when combined with an annual component, produced the observed seasonal curve. Shoji and Nitani (1966) and Masuzawa (1969) have reported daily fluctuations in the axial speed of the Kuroshio near Shionomisaki of about 25 percent of the mean axial speed. Because there probably has been aliasing of high frequencies as low frequencies, harmonic terms will not be fit to the data to estimate the semiannual and annual components. The data will be analyzed to determine whether or not the speed is greater in one quarter than another and no attempt will be made to

analyze the data for periodicities. The sampling interval is too broad and the record is too short to yield good estimates of the spectrum of the speed fluctuations.

Masuzawa (1965c) analyzed the GEK data between Shionomisaki (135°43′E) and Irozaki (139°00′E) for the ten-year period 1955 to 1964. Masuzawa chose only data from GEK cross-sections with 10-mile spacing between measurements so that his average axial speeds are higher than those shown in Fig. 14, but the shape of the seasonal curve is the same. The seasonal curve for the axial speed of the Kuroshio between 137° and 139°E for the years 1955 to 1962 of Uda (1964) differs from Fig. 14. Uda's curve shows maxima in both the first and third quarters. It is not clear why Uda's seasonal curve differs from the ten-year curve of Masuzawa (1965c) or the nine-year curve shown in Fig. 14. The data used in the computations were essentially the same in all three studies, the only differences being that Uda used eight years of observations, instead of nine or ten years, and the longitudes were slightly different.

Seasonal curves of current speed in the Kuroshio south of Japan were computed from the ship-log surface current data on file at the National Oceanographic Data Center in Washington, D.C. The data were averaged by 1° square and month. Data from longitudes 130° to 135° and 135° to 140°E were selected for analysis. Monthly averages in the five 1° squares with the highest speeds in each 1° band of longitude between 135° and 140°E were averaged and

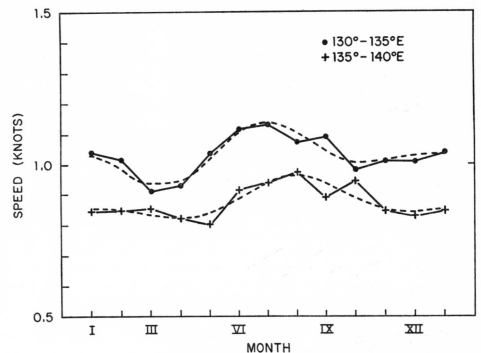

FIG. 15 Seasonal curves of speed of Kuroshio from ship displacement observations for two ranges of longitude: 130° to 135°E (●) and 135° to 140°E(+). Dashed lines are curves which have been fitted to the data; fitted curves are the summation of harmonic components with semiannual and annual periods.

the results are plotted in Fig. 15. Each of the average current speeds plotted (lower curve) in Fig. 15 is based on at least 135 observations. Between 135° and 140°E the maximum speed is in August and the minimum speed is in May. Average speed is essentially constant between November and March. This curve resembles Fig. 14 in that there is a maximum speed in the summer but the minimum speed is in the early part of the second quarter of the year rather than in the fourth quarter, as was indicated by the quarterly averages. The significance of the lack of agreement in the time of the minimum speed is not clear. Since the two sets of data are not strictly comparable, there is no reason to expect complete agreement between the two curves. Data used in the ship displacement calculations come from a great many years and the calculated speed is an average speed across the current, whereas the GEK observations are maximum speeds from a nine-year period. Ship displacement measurements of current are influenced in an unknown way by the wind so they are not a direct measure of current. The significant feature of Fig. 14 is that there is only one maximum in current speed and it occurs in the summer.

Maximum and minimum speeds on both curves are not six months apart but appear to be closer to four months apart. Dashed curves in Fig. 15 represent the sum of harmonic functions with semiannual and annual periods which have been fitted to the monthly averages. The fitted curves describe the

observed values fairly well and the agreement between observed and calculated curves might be considered to be evidence for a semiannual component of variation in current speed. Fuglister (1951) fit annual and semiannual harmonic components to ship displacement data from the Gulf Stream and found a degree of agreement between his fitted curves and the monthly averages which is comparable to that shown in Fig. 15. All that can be said from studies of this type is that the data are not inconsistent with the hypothesis that there is a semiannual component in the time variation of the Kuroshio current speed.

Pavlova (1964) has analyzed geostrophic current speeds computed at two sections across the Kuroshio south of Japan; a northwest-southeast section located off the southern tip off Kyushu and a north-south section off the eastern end of Shikoku at 134°30′E. Pavlova used all the available temperature and salinity observations from the period 1906 to 1960 to compute the average vertical distribution of specific volume in each 1° square for every month and then she computed geostrophic current from the average profiles in adjacent squares. In discussing the seasonal curve of geostrophic current speed, Pavlova refers to a double maximum in velocity of the Kuroshio but the existence of a double maximum is doubtful. There is a great deal of variability in the monthly averages and it is not possible to pick out a significant maximum in any month.

A comparison can be made with the results of Pavlova by examining the average currents computed from ship displacements between 130° and 135°E, upper curve (Fig. 15). The upper curve resembles the curve for longitudes 135° to 140°E, except that the maximum and minimum speeds are shifted one to two months earlier in the year; the minimum occurs in March and the maximum occurs four months later in July. Ship displacement data from the Kuroshio both east and west of Shionomisaki show a single rather than a double maximum of speed.

The curve for longitudes 130° to 135°E (Fig. 15) is similar in shape to the seasonal curve of geostrophic current speed published by Ichiye (1954b). Ichiye used monthly averages of geostrophic current speed computed at five sections across the current between 132° and 136°E. There were eleven sets of five sections and all the observations were made in 1940. The maximum speed was in July and the minimum speed in April. Ichiye also concluded that there was a second maximum in January and a second minimum in November. However, the secondary maximum and minimum do not appear to be statistically significant.

The relative volume transport calculations, using a reference of 800 db, can be sorted into quarterly periods and the seasonal curve for transport calculated for the nine-year period. Data does not exist in each quarter of each of the nine years so only the average seasonal curve for longitudes 135°10′ to 138°40′E is presented (Fig. 16). The relative transport was a maximum in the third quarter. There is no significant difference in transport between the first, second and fourth quarters, but the third quarter average transport is 25 percent higher than the transport in any other quarter. Comparison with the seasonal curve for axial current speed (Fig. 14) shows that the third quarter maximum is much more pronounced in transport than it is in current speed. There is no evidence for a fourth-quarter minimum in transport as was indicated for maximum GEK speed. Confidence limits for the fourth-quarter average transport are wider than for the other quarters; there is more uncertainty in the mean value of transport for the fourth quarter than for the other quarters.

There are no other calculations that can be compared directly with Fig. 16. Pavlova (1964) computed the average transport relative to 1500 db for each month across two sections: one located off Kyushu and the other at 134°30′E. These sections are west of the region from where the data were taken to compute the seasonal curve of transport shown in Fig. 16. The calculations of Pavlova show month-to-month fluctuations which are of the same magnitude as the range in means in Fig. 16; there is not an obvious maximum in transport in either August or September. Since there is no statistical analysis applied to the transports in Pavlova's paper and there is large month-to-month variability, it is not possible to decide whether or not significant differences between months exist. Masuzawa (1954) has published the results of volume transport calculations relative to 1000 db which were performed on 32 sections made across the Kuroshio during 1952 and 1953 south of Shionomisaki (135°43′E). Masuzawa states that there is marginal evidence that there are two maxima per year: one in the spring and another in the fall. The fall maximum may correspond to the August–September maximum of Fig. 16. Visual inspection of the data of Masuzawa indicates that the pattern of two maxima per year does not really repeat itself. It does not seem possible to draw any firm conclusions about the mean seasonal curve from such a short record of transport at a single section.

Wyrtki (1961) has published seasonal curves of current speed derived from surface current charts and of geostrophic volume transport relative to 800

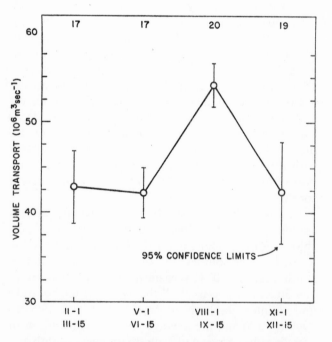

FIG. 16 Quarterly averages of volume transport of Kuroshio relative to 800 db between 135°10′ and 138°40′E for the period March 1956 to November 1964. Ninety-five percent confidence limits are given by the vertical lines. The number of values averaged in each quarter is given along the top of the figure. The first, second, third and fourth quarters correspond to the following times: first quarter—February 1 to March 15; second quarter—May 1 to June 15; third quarter—August 1 to September 15; and fourth quarter—November 1 to December 15.

db in the northwest Pacific south of 25°N. Since there is a continuous flow of water northward along the western boundary at all seasons of the year, the Kuroshio off Japan must receive a significant proportion of its water from the south. It is of interest to see what is the time relation between maxima or minima in the speed and transport of the Kuroshio south of 25°N. Wyrtki's seasonal curve of current speed of the Kuroshio east of Formosa shows a maximum in speed in May and a minimum in speed in November, December and January. The seasonal curve for relative transport, which is based on data from every second month, shows the highest transport in April and lowest transport in December. Apparently the highest relative transport off Formosa occurs earlier in the year than the highest transport off Japan. The time difference cannot be estimated very accurately because only quarterly values of the speed and transport are available in the Kuroshio off Japan. If one takes 1.5 kn for the average speed of water in the Kuroshio, then the time required for the water to travel from Formosa to the Izu-Ogasawara Ridge is only about one month. It would appear that the third quarter maximum off Japan probably is not in phase with the seasonal maximum in transport off Formosa, i.e., the larger transport off Formosa is not expressed off Japan at a later time which can be predicted from the average current speed. The third quarter maximum off Japan probably is at least partially due to water that is added to the current during its transit from Formosa to Japan. These conclusions are speculative since the seasonal curve off Japan is not based on even monthly averages so the timing of the maximum is not determined very well. Also the reliability of Wyrtki's curve of transport off Formosa is not known. Wyrtki does not give the number of observations used in the computation of average transport or indicate the variability of the transport values. Because of the small number of sections that probably were available to him, the seasonal curve off Formosa may not be very precisely determined. It would be interesting to compare variations in transport of the North Equatorial Current with the Kuroshio transport variations in order to evaluate the speculations of Namias (1970). Unfortunately, sufficient subsurface data do not exist to do this (Wyrtki, 1961).

9 CONCLUSIONS

The surveys of the waters south of Japan during the nine years 1956 to 1964 encompass a period when the meander of the Kuroshio was present. Charts of the surface current velocity (Appendix 1) represent the position of the current at the surface both when the meander was present and when it was absent. The axis of the current has been identified with the position of maximum surface current velocity measured by GEK and an approximate path for the Kuroshio has been drawn on the charts, wherever there was an adequate distribution of transects across the current.

1) Surveys during the periods March 1956 to March 1959 and May 1963 to November 1964 show that the Kuroshio did not meander south of Shionomisaki. During these two periods the Kuroshio remained on the upper portion of the continental slope until it was 60 miles east of Shionomisaki. West of Shionomisaki the width of the envelope of all paths was no more than 32 miles. The path became more variable as the current left the continental slope east of Shionomisaki and entered deep water. Usually the Kuroshio moved into the deep water at the northern end of the Shikoku Basin at about 137°E but remained in deep water only a short distance before it began to cross the Izu-Ogasawara Ridge. On most surveys when the Kuroshio did not meander, the Kuroshio crossed the Izu-Ogasawara Ridge in the channel located at about 34°N. The critical latitude determining whether or not the Kuroshio entered the channel was approximately 32°55′N. If the Kuroshio west of the ridge was south of this latitude, the current did not turn northward into the channel but crossed the ridge south of the channel.

2) The cyclonic meander of the Kuroshio south of Shionomisaki was probably established by August 1959. In May 1959 the Kuroshio east of Kyushu was displaced off the continental slope; there also appeared to be an eastward moving cyclonic meander in the current. By November 1959 the stable meander of the Kuroshio was clearly present; the current axis west of the Izu-Ogasawara Ridge was located at 31°N. Current paths west of 134°E did not differ significantly between periods when the meander was present and when it was absent. The current turned southeast off the continental slope between 134° and 135°E and entered deep water. At the point where the current turned off of the slope, the continental slope changes orientation and its width is a minimum. In contrast to the period when the current approached the ridge directly from the west, the current did not turn into the channel on the ridge. The current crossed the ridge most often on the shallow region directly south of the Izu Peninsula.

3) On only one survey (May 1960) is there evidence of a small-scale cyclonic eddy developing in the current. This eddy had approximately the same size as those that have been found south of the Gulf

Stream east of Cape Hatteras (Fuglister and Worthington, 1951; Fuglister, 1963). Because of the broad spacing of the transects across the Kuroshio, the frequency of occurrence of these eddies in the Kuroshio cannot be ascertained from the surveys.

4) Axial surface current speed of the Kuroshio between 132°W (Kyushu) and 137°E (east of Shionomisaki), as measured by the GEK, increased to the east. The increase in axial speed is most marked when the meander was not present; axial speed at about 135°E increased 1.0 kn over a distance of 100 miles. Associated with the acceleration of the current is a narrowing of the continental slope over which the current is flowing and a concomitant decrease in the width of the high-velocity region of the current. When the meander was present, the axial current speed between 137° and 138°E decreased 0.9 kn and the width of the current increased. This decrease in speed occurred at the point of maximum southern displacement of the current.

When the Kuroshio was not displaced to the south and the current approached the Izu-Ogasawara Ridge directly from the west, the axial surface current speed decreased about 0.4 kn as the current flowed over the steep western margin of the ridge. When the meander was present, there was no change in speed of the current as it flowed over the ridge.

5) Volume transport of the Kuroshio relative to 800 db was computed on 102 sections. Average relative transport was 42×10^6 m^3 sec^{-1} and the range was 17 to 64×10^6 m^3 sec^{-1}. Between 135°45′ and 136°55′E there was an abrupt increase of 30 percent in relative transport. There was not a significant difference at any longitude in relative transport between the periods when the eddy was present and when it was absent.

6) Surface currents south of Enshunada had a mean westward component when there was no meander. Average westward volume transport relative to 800 db south of Enshunada was 9×10^6 m^3 sec^{-1}, or about 20 percent of the relative transport of the Kuroshio at the same longitudes. The geopotential of isobaric surfaces relative to 800 db along the northern boundary of the Kuroshio decreased east of Cape Shionomisaki, approximately at the longitude where the relative transport of the Kuroshio increased by 30 percent. The southward transport of water into the Kuroshio associated with the eastward decrease in geopotential was 11×10^6 m^3 sec^{-1}. It appears that the downstream increase of relative transport of the Kuroshio could be accounted for by considering that water moving westward off Enshunada turns southward and joins the Kuroshio east of Cape Shionomisaki.

7) South of the axis of the Kuroshio GEK measurements show a region of average westward flow, which appears to be an offshore Kuroshio Countercurrent similar to the one found seaward of the Gulf Stream. The average distance between the countercurrent and the Kuroshio axis was 135 miles. In the western Atlantic the countercurrent lies only 40 to 60 miles seaward of the axis of the Gulf Stream and it appears to be associated with the warm core to the right of the Gulf Stream. The Kuroshio Countercurrent was roughly twice as far offshore as the Gulf Stream Countercurrent. The Kuroshio Countercurrent does not appear to coincide with the warm core; it probably lies about 60 miles seaward of the warm core.

8) During the early part of the period when the meander was present, the maximum speed of the Kuroshio between 135°40′ and 138°40′E was markedly lower than the nine-year average. After the meander was established, the speed increased with time so that the average maximum speed in 1962 was equal to the nine-year average. The decrease in speed of the Kuroshio may be dynamically linked to the movement offshore of the Kuroshio and the subsequent establishment of the meander. Namias (1970) has shown that there is a strong negative correlation between sea-surface temperature in the subtropics in the western Pacific and the maximum speed of the Kuroshio south of Honshu. If variations in the temperature gradient across the North Equatorial Current are primarily determined by variations in the temperatures in the subtropics, it may be argued that the low current speeds of the Kuroshio south of Japan were associated with a low transport of the North Equatorial Current.

9) Quarterly means of the axial current speed and the relative transport of the Kuroshio show maxima in the summer quarter (August–September). The summer maximum in relative volume transport is more pronounced than the maximum in current speed. The maximum speed and maximum relative transport in the Kuroshio south of Japan occurs about three months later in the year than the maximum speed and transport of the Kuroshio east of Formosa. There does not seem to be evidence for a significant double maximum in the seasonal curve for current speed or transport.

APPENDEX   1
Charts of surface current velocity in Kuroshio south of Japan

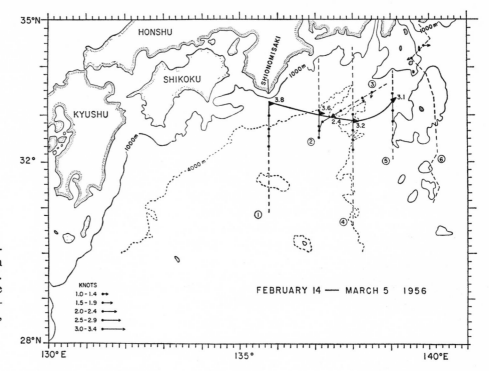

Chart 1. Distribution of surface current velocity (knots) south of Japan *II/14 thru III/5/1956*. Dates of individual transects are as follows: ① II/18–19, ② II/14–15, ③ II/14, ④ II/20–21, ⑤ III/2, ⑥ III/3–5.

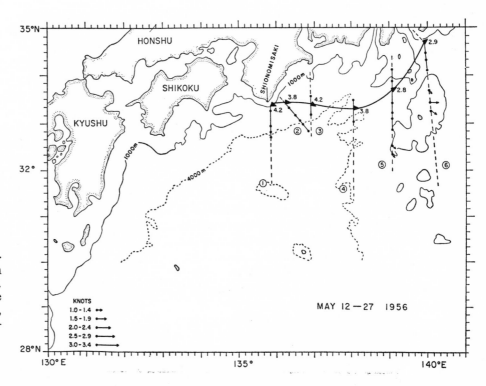

Chart 2. Distribution of surface current velocity (knots) south of Japan *V/12 thru V/27/1956*. Dates of individual transects are as follows: ① V/14, ② V/17–18, ③ V/17–18, ④ V/12–13, ⑤ V/23–25, ⑥ V/25–27.

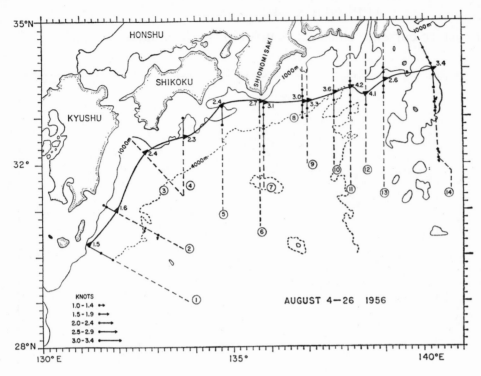

Chart 3. Distribution of surface current velocity (knots) south of Japan *VIII/4 thru VIII/26/1956.* Dates of individual transects are as follows: ① VII/7–8, ② VIII/9–10, ③ VIII/4, ④ VIII/4–5, ⑤ VIII/13, ⑥ VIII/14–15, ⑦ VIII/5–6, ⑧ VIII/4–5, ⑨ VIII/19–20, ⑩ VIII/20–21, ⑪ VIII/7–8, ⑫ VIII/25–26, ⑬ VIII/12–13, ⑭ VIII/13–14.

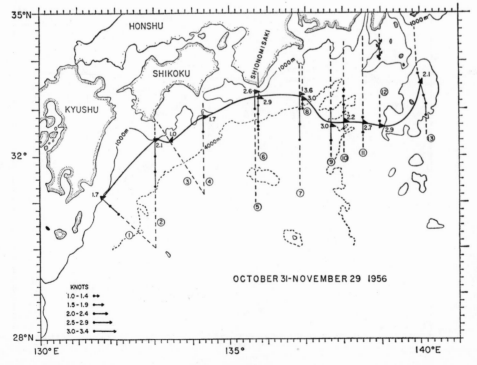

Chart 4. Distribution of surface current velocity (knots) south of Japan *X/31 thru XI/29/1956.* Dates of individual transects are as follows: ① XI/28–29 ② XI/27–28, ③ XI/26, ④ XI/25, ⑤ XI/19–20, ⑥ XI/2–3, ⑦ XI/18–19, ⑧ X/31, ⑨ XI/13–14, ⑩ XI/3–4, ⑪ XI/12–13, ⑫ X/10–12, ⑬ XI/12–13.

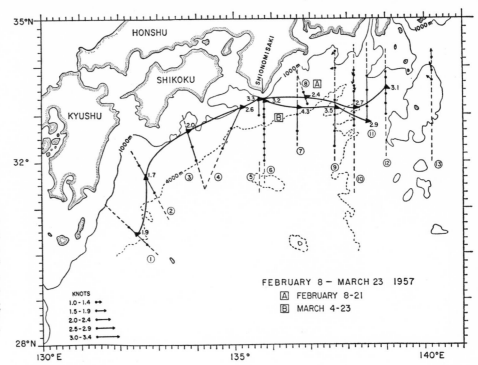

Chart 5. Distribution of surface current velocity (knots) south of Japan *II/8 thru III/23/1957*. Dates of individual transects are as follows: ① III/4–5, ② III/5–6, ③ III/9, ④ III/10, ⑤ III/13–14, ⑥ II/12–13, ⑦ III/15–16, ⑧ II/8, ⑨ III/20–21, ⑩ II/13–15, ⑪ III/22–23, ⑫ II/20–21, ⑬ III/21–22.

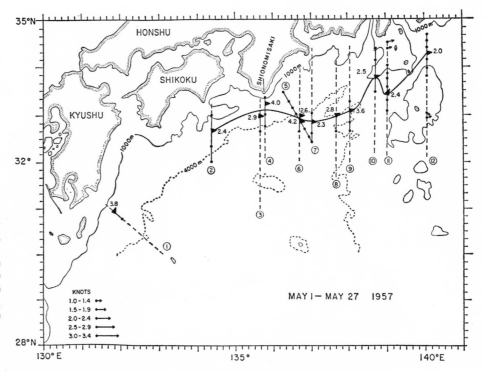

Chart 6. Distribution of surface current velocity (knots) south of Japan *V/1 thru V/27/1957*. Dates of individual transects are as follows: ① V/1–2, ② V/8–9, ③ V/9–10, ④ V/16, ⑤ V/14, ⑥ V/13–14, ⑦ V/13–14, ⑧ V/14–15, ⑨ V/17–18, ⑩ V/15–16, ⑪ V/23–24, ⑫ V/25–27.

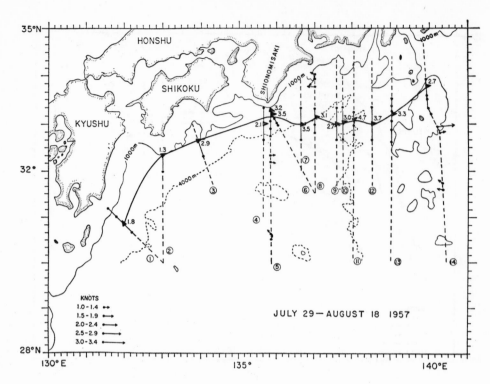

JULY 29 — AUGUST 18 1957

Chart 7. Distribution of surface current velocity (knots) south of Japan *VII/29 thru VIII/18/ 1957*. Dates of individual transects are as follows: ① VII/29–30, ② VII/30, ③ VII/31, ④ VIII/1, ⑤ VIII/8–9, ⑥ VIII/3, ⑦ VIII/4–5, ⑧ VIII/2–3, ⑨ VIII/10–11, ⑩ VIII/5–6, ⑫ VIII/9–10, ⑬ VIII/ 15–18, ⑭ VIII/17–18.

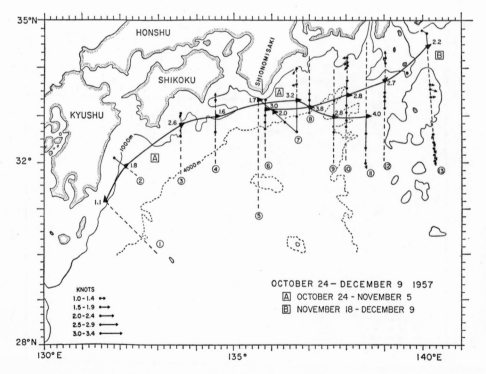

OCTOBER 24 — DECEMBER 9 1957
Ⓐ OCTOBER 24 - NOVEMBER 5
Ⓑ NOVEMBER 18 - DECEMBER 9

Chart 8. Distribution of surface current velocity (knots) south of Japan *X/24 thru XII/9/1957*. Dates of individual transects are as follows: ① XI/18–19, ② XI/22, ③ XI/29–30, ④ XI/23, ⑤ XI/30–XII/1, ⑥ X/27, ⑦ XII/1–2, ⑧ X/24, ⑨ XII/8–9, ⑩ X/28–29, ⑪ XII/7–8, ⑫ XI/3–4, ⑬ XI/4–5.

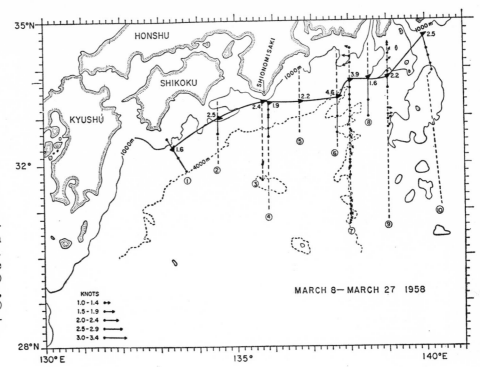

Chart 9. Distribution of surface current velocity (knots) south of Japan *III/8 thru III/27/1958*. Dates of individual transects are as follows: ① III/8, ② III/9, ③ III/14–15, ④ III/16–17, ⑤ III/16, ⑥ III/20–21, ⑦ III/18–20, ⑧ III/21–22, ⑨ III/24–25, ⑩ III/26–27.

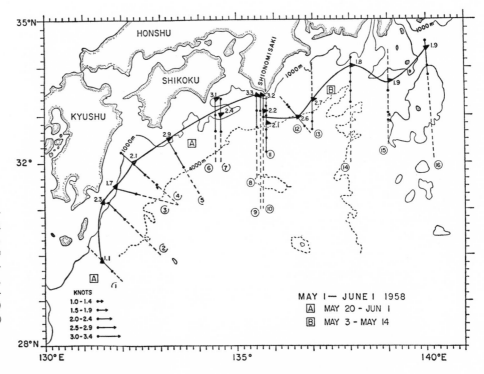

Chart 10. Distribution of surface current velocity (knots) south of Japan *V/1 thru VI/1/1958*. Dates of individual transects are as follows: ① V/20–21, ② V/20–21, ③ V/21–22, ④ V/24, ⑤ V/25, ⑥ V/30–31, ⑦ V/27–28, ⑧ V/31–VI/1, ⑨ V/28–29, ⑩ V/13–14, ⑪ V/8–9, ⑫ V/11, ⑬ V/11–12, ⑭ V/7–8, ⑮ V/3, ⑯ V/1–2.

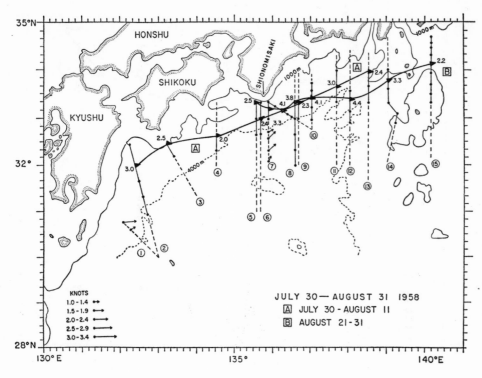

Chart 11. Distribution of surface current velocity (knots) south of Japan *VII/30 thru VIII/31/1958*. Dates of individual transects are as follows: ① VII–31–VIII/1, ② VII/30–31, ③ VIII/3, ④ VIII/3–4, ⑤ VIII/30–31, ⑥ VIII/4–5, ⑦ VIII/28–29, ⑧ VIII/29–30, ⑨ VIII/5–6, ⑩ VIII/31, ⑪ VIII/9–10, ⑫ VIII/27–28, ⑬ VIII/11, ⑭ VIII/22–23, ⑮ VIII/21–22.

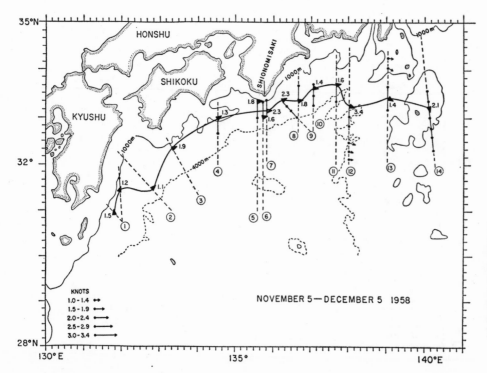

Chart 12. Distribution of surface current velocity (knots) south of Japan *XI/5 thru XII/5/1958*. Dates of individual transects are as follows: ① XI/5–6, ② XI/6–7, ③ XI/7–8, ④ XI/20–21, ⑤ XI/29–30, ⑥ XI/21–22, ⑦ XI/21–22, ⑧ XI/22–23, ⑨ XI/22, ⑩ XI/22–23, ⑪ XI/27–28, ⑫ XI/20–21, ⑬ XII/3–4, ⑭ XII/4–5.

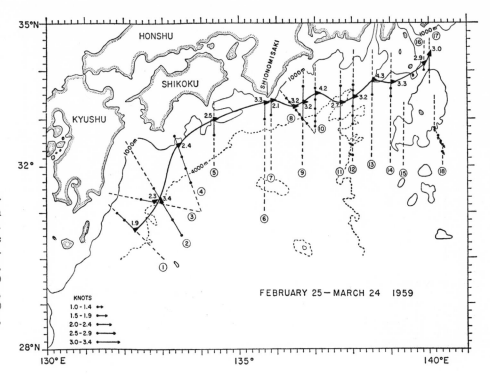

Chart 13. Distribution of surface current velocity (knots) south of Japan *II/25 thru III/24/1959.* Dates of individual transects are as follows: ① II/26–27, ② II/25–26, ③ III/2–3, ④ III/3, ⑤ III/4, ⑥ III/5, ⑦ III/14, ⑧ III/8, ⑨ III/9–10, ⑩ III/7–8, ⑪ III/10–11, ⑫ III/15–16, ⑬ III/19–20, ⑭ III/19–20, ⑮ III/20–21, ⑯ III/24, ⑰ III/21, ⑱ III/21–22.

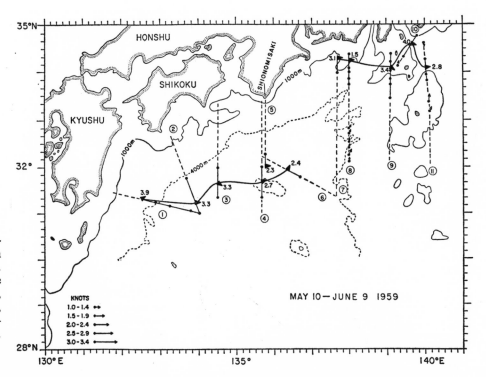

Chart 14. Distribution of surface current velocity (knots) south of Japan *V/10 thru VI/9/1959.* Dates of individual transects are as follows: ① V/11, ② V/10–11, ③ V/16–17, ④ V/17–18, ⑤ VI/5, ⑥ V/25–26, ⑦ V/24–25, ⑧ VI/3–4, ⑨ V/30, ⑩ VI/9, ⑪ V/28–29.

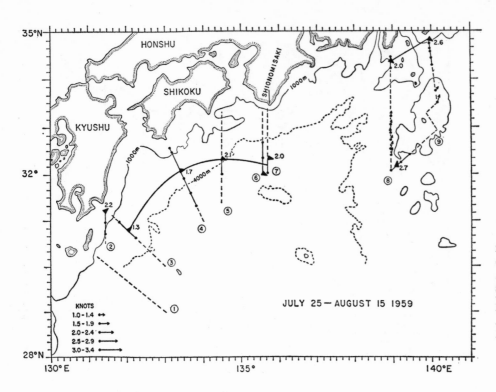

Chart 15. Distribution of surface current velocity (knots) south of Japan *VII/25 thru VIII/15/1959.* Dates of individual transects are as follows: ① VII/30–31, ② VII/31–VIII/1, ③ VIII/10–11, ④ VIII/11–12, ⑤ VIII/14–15, ⑥ VIII/12–13, ⑦ VII/27–28, ⑧ VII/25–27.

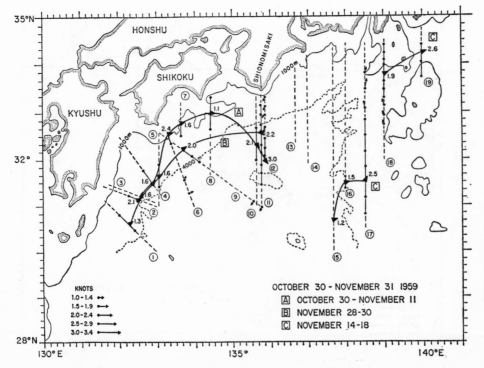

Chart 16. Distribution of surface current velocity (knots) south of Japan *X/30 thru XI/30/1959.* Dates of individual transects are as follows: ① X/31–XI/1, ② XI/28, ③ XI/4, ④ XI/4–5, ⑤ XI/28–29, ⑥ X/30–31, ⑦ XI/5–6, ⑧ XI/9–10, ⑨ XI/29, ⑩ XI/10–11, ⑪ XI/29–30, ⑫ XI/7–8, ⑬ XI/11–12, ⑭ XI/8–9, ⑮ XI/15–17, ⑯ XI/14–16, ⑰ XI/17–18, ⑱ XI/16–17, ⑲ XI/18.

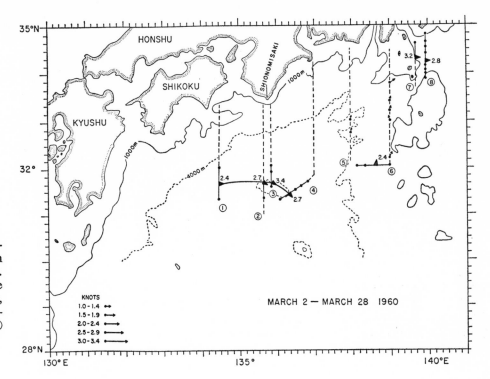

Chart 17. Distribution of surface current velocity (knots) south of Japan *III/2 thru III/28/1960.* Dates of individual transects are as follows: ① III/18, ② III/19–20, ③ III/2–3, ④ III/3–4, ⑤ III/10–11, ⑥ III/11–12, ⑦ III/28, ⑧ III/23.

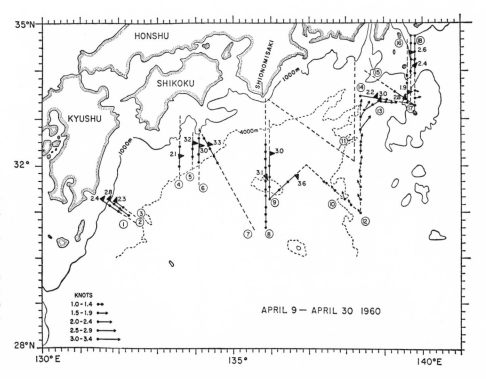

Chart 18. Distribution of surface current velocity (knots) south of Japan *IV/9 thru IV/30/1960.* Dates of individual transects are as follows: ① IV/26, ② IV/15–16, ③ VI/23–24, ④ IV/14, ⑤ IV/23, ⑥ IV/26–27, ⑦ IV/22–23, ⑧ IV/21, ⑨ IV/27–28, ⑩ IV/28, ⑪ IV/10, ⑫ IV/28–29, ⑬ IV/19, ⑭ IV/29, ⑮ IV/10, ⑯ IV/18–19, ⑰ IV/29–30, ⑱ IV/9–10.

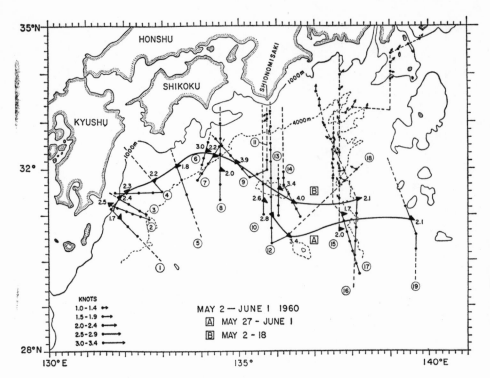

Chart 19. Distribution of surface current velocity (knots) south of Japan *V/2 thru VI/1/1960*. Dates of individual transects are as follows: ① V/11–12, ② V/2, ③ V/15, ④ V/7–8, ⑤ V/12–13, ⑥ V/3, ⑦ V/14–15, ⑧ V/16–17, ⑨ V/3, ⑩ V/17–18, ⑪ V/3–4, ⑫ V/27–29, ⑬ V/14, ⑭ V/6–8, ⑮ V/22–23, ⑯ V/11–13, ⑰ V/31–VI/1, ⑱ V/29, ⑲ V/30.

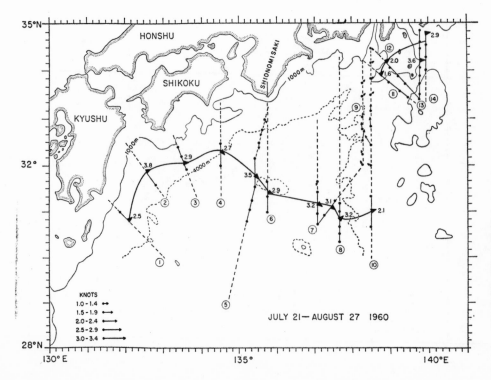

Chart 20. Distribution of surface current velocity (knots) south of Japan *VII/21 thru VIII/27/1960*. Dates of individual transects are as follows: ① VII/23, ② VII/21–22, ③ VII/24–25, ④ VII/29–30, ⑤ VIII/1–2, ⑥ VII/30–31. ⑦ VII/30, ⑧ VIII/5–6, ⑨ VII/29–30, ⑩ VIII/7–8, ⑪ VII/29, ⑫ VIII/27, ⑬ VIII/26–27, ⑭ VII/28.

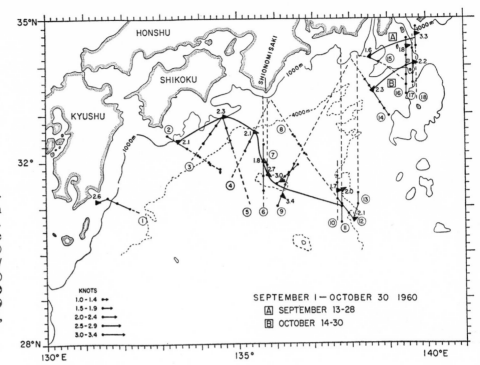

Chart 21. Distribution of surface current velocity (knots) south of Japan *IX/1 thru X/30/1960.* Dates of individual transects are as follows: ① IX/5, ② IX/9, ③ IX/3, ④ IX/9, ⑤ IX/2–3, ⑥ X/21–22, ⑦ X/11–12, ⑧ X/29, ⑨ IX/1–2, ⑩ X/30, ⑪ X/23–24, ⑫ IX/11-12, ⑬ IX/12, ⑭ X/30, ⑮ IX/13, ⑯ IX/27–28, ⑰ X/14, ⑱ IX/13.

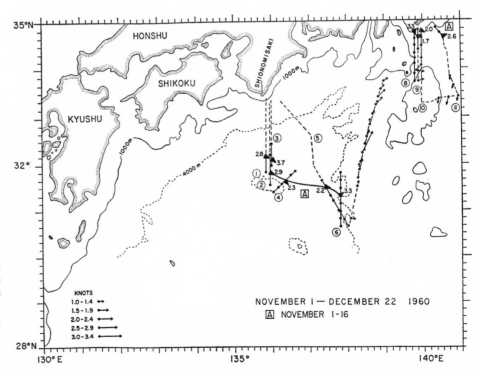

Chart 22. Distribution of surface current velocity (knots) south of Japan *XI/1 thru XII/22/1960.* Dates of individual transects are as follows: ① XII/21–22. ② XI/1–2, ③ XII/8, ④ XI/6, ⑤ XI/9–10, ⑥ XI/6–8, ⑦ XI/10–11 ⑧ XII/19–20, ⑨ XII/10, ⑩ XI/15, ⑪ XI/15–16.

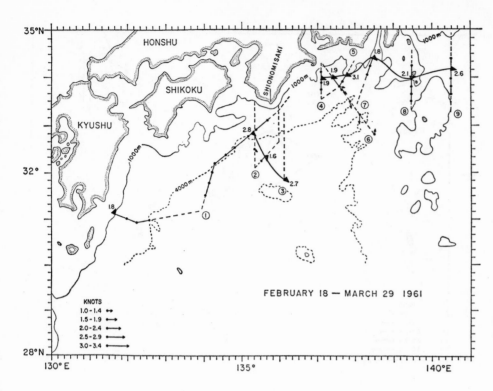

Chart 23. Distribution of surface current velocity (knots) south of Japan *II/18 thru III/29/1961.* Dates of individual transects are as follows: ① II/18–19, ② III/14–15, ③ II/28, ④ III/18, ⑤ III/13, ⑥ III/22–23, ⑦ II/28–III/1, ⑧ III/25–26, ⑨ III/28–29.

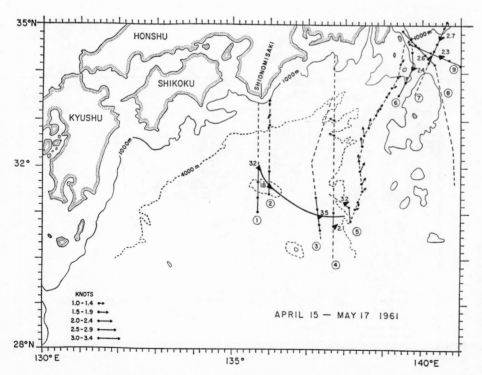

Chart 24. Distribution of surface current velocity (knots) south of Japan *IV/15 thru V/17/1961.* Dates of individual transects are as follows: ① V/12–13, ② IV/24–25, ③ IV/18–19, ④ V/14–15, ⑤ V/16–17, ⑥ IV/26–27, ⑦ IV/15, ⑧ V/2–4, ⑨ IV/20–21.

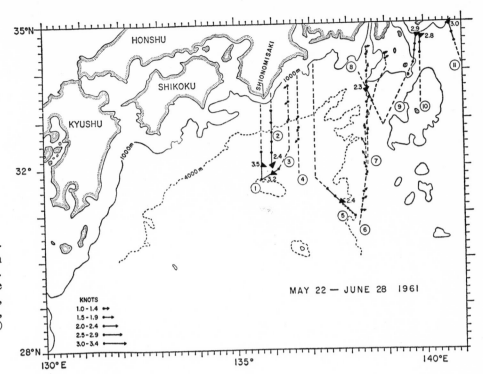

Chart 25.   Distribution of surface current velocity (knots) south of Japan *V/22 thru VI/28/1961*. Dates of individual transects are as follows: ① V/20–31, ② VI/28, ③ V/31, ④ V/22–23, ⑤ VI/5–6, ⑥ VI/6–7, ⑦ V/25, ⑧ V/27, ⑨ V/26–27, ⑩ VI/11, ⑪ VI/12.

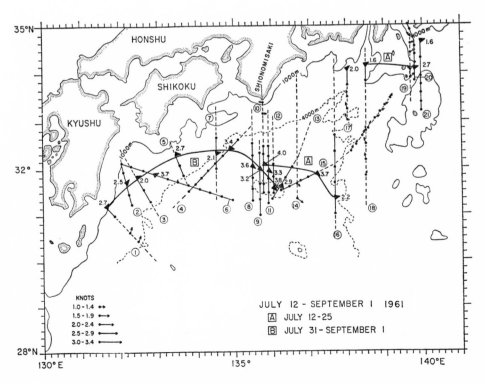

Chart 26.   Distribution of surface current velocity (knots) south of Japan *VII/12 thru IX/1/1961*. Dates of individual transects are as follows: ① VIII/10–11, ② VIII/4–5, ③ VIII/11–12, ④ VIII/11, ⑤ VIII/12–13, ⑥ VIII/4, ⑦ VIII/9, ⑧ VIII/3–4, ⑨ VIII/18–19, ⑩ VII/12–13, ⑪ VIII/12–13, ⑫ VIII/31–IX/1, ⑬ VIII/13, ⑭ VIII/5, ⑮ VII/13–14, ⑯ VII/24–25, ⑰ VIII/15, ⑱ VII/22–24, ⑲ VIII/30–31, ⑳ VII/14, ㉑ VII/31–VIII/1.

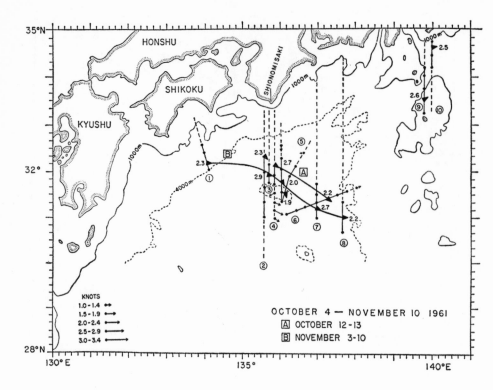

Chart 27. Distribution of surface current velocity (knots) south of Japan *X/4 thru XI/10/1961*. Dates of individual transects are as follows: ① XI/3, ② XI/5–10, ③ XI/2, ④ X/12–13, ⑤ XI/7–8, ⑥ XI/3–5, ⑦ X/13, ⑧ XI/5–6, ⑨ X/4–5, ⑩ X/29–30.

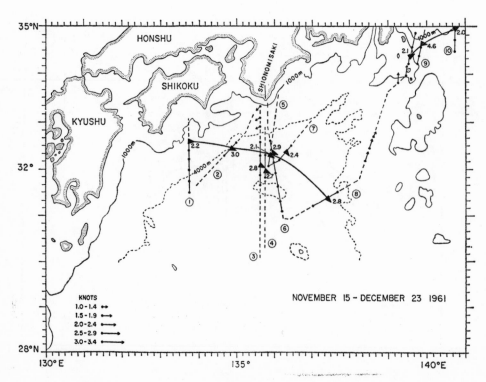

Chart 28. Distribution of surface current velocity (knots) south of Japan *XI/15 thru XII/24/1961*. Dates of individual transects are as follows: ① XI/17–18, ② XI/19, ③ XI/15, ④ XI/16, ⑤ XI/28, ⑥ XI/27, ⑦ XII/22–23, ⑧ XII/23–24, ⑨ XI/26, ⑩ XII/10–11.

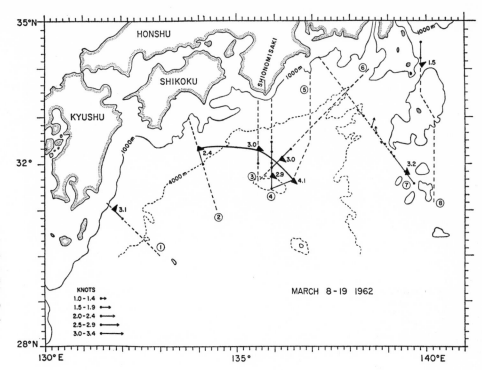

Chart 29.   Distribution of surface current velocity (knots) south of Japan *III/8 thru III/19/1962.* Dates of individual transects are as follows: ① III/10–11, ② III/11–12, ③ III/17–18, ④ III/8–9. ⑤ III/9–10, ⑥ III/18–19, ⑦ III/12–13, ⑧ III/18–19.

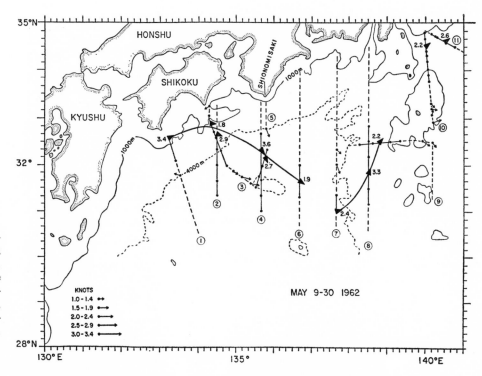

Chart 30.   Distribution of surface current velocity (knots) south of Japan *V/9 thru V/30/1962.* Dates of individual transects are as follows: ① V/10–11, ② V/9–10, ③ V/16–17, ④ V/16, ⑤ V/18, ⑥ V/17–18, ⑦ V/24–26, ⑧ V/26–27, ⑨ V/22–25, ⑩ V/26, ⑪ V/30.

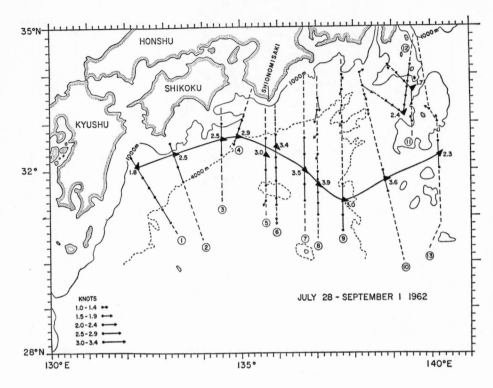

Chart 31. Distribution of surface current velocity (knots) south of Japan *VII/28 thru IX/1/1962.* Dates of individual transects are as follows: ① VII/29–30, ② VII/28–29, ③ VIII/4–5, ④ VIII/30, ⑤ VIII/6–7, ⑥ VIII/18–19, ⑦ VIII/10–11, ⑧ VIII/21–23, ⑨ VIII/12–13, ⑩ VIII/28–30, ⑪ VIII/16–17, ⑫ VIII/13–14, ⑬ VIII/30–IX/1.

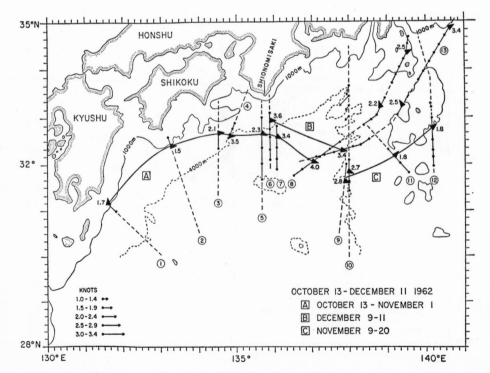

Chart 32. Distribution of surface current velocity (knots) south of Japan *X/13 thru XII/11/1962.* Dates of individual transects are as follows: ① X/31–XI/1, ② X/26–27, ③ X/21–22, ④ X/13, ⑤ X/20, ⑥ XII/9, ⑦ X/15, ⑧ X/13–15, ⑨ XI/14, ⑩ XI/9–11, ⑪ XI/19, ⑫ XI/19–20, ⑬ XII/9–11.

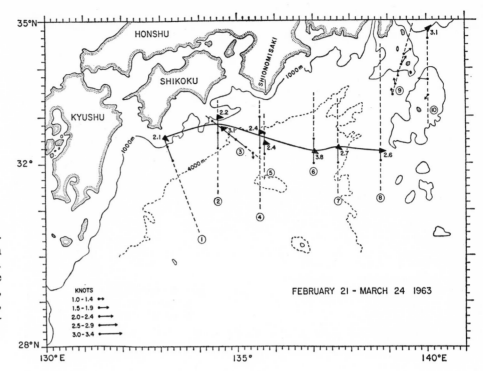

Chart 33. Distribution of surface current velocity (knots) south of Japan *II/21 thru III/24/1963*. Dates of individual transects are as follows: ① II/12–22, ② III/4–5, ③ II/25, ④ III/5–6, ⑤ III/14–15, ⑥ III/15, ⑦ III/18–19, ⑧ III/21–22, ⑨ III/11–12, ⑩ III/23–24.

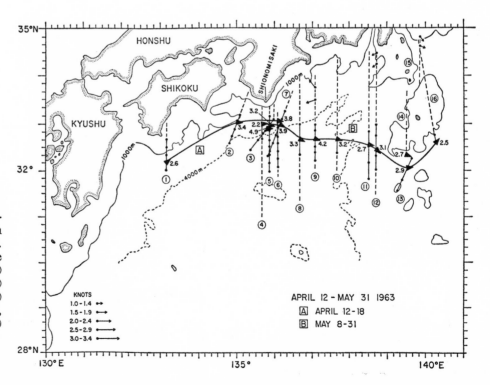

Chart 34. Distribution of surface current velocity (knots) south of Japan *IV/12 thru V/31/1963*. Dates of individual transects are as follows: ① IV/14, ② IV/12, ③ V/29–30, ④ V/8, ⑤ V/30–31, ⑥ IV/18, ⑦ V/28–29, ⑧ V/9–10, ⑨ V/29–30, ⑩ IV/17, ⑪ V/14–15, ⑫ V/24–26, ⑬ V/24, ⑭ V/19, ⑮ V/1, ⑯ V/23–24.

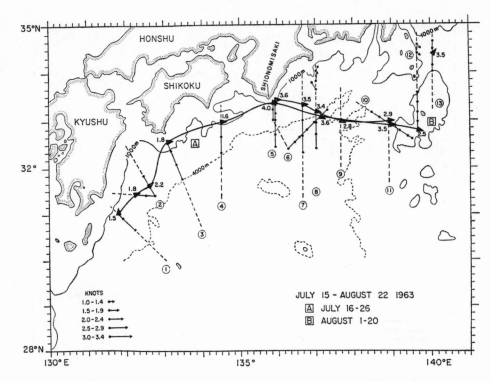

Chart 35. Distribution of surface current velocity (knots) south of Japan *VII/15 thru VIII/22/1963*. Dates of individual transects are as follows: ① VII/20–21, ② VII/21–22, ③ VII/19–20, ④ VII/25–26, ⑤ VIII/14–15, ⑥ VII/16–17, ⑦ VIII/1–2, ⑧ VIII/16, ⑨ VIII/3, ⑩ VII/16, ⑪ VIII/20, ⑫ VII/15–16, ⑬ VIII/21–22.

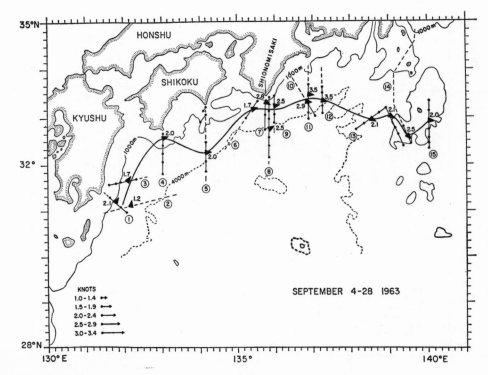

Chart 36. Distribution of surface current velocity (knots) south of Japan *IX/4 thru IX/28/1963*. Dates of individual transects are as follows: ① IX/20, ② IX/18, ③ IX/23, ④ IX/23–24, ⑤ IX/17–18, ⑥ IX/8, ⑦ IX/13, ⑧ IX/24–25, ⑨ IX/12, ⑩ IX/4, ⑪ IX/18, ⑫ IX/4, ⑬ IZ/27–28, ⑭ IX/10–11, ⑮ IX/28.

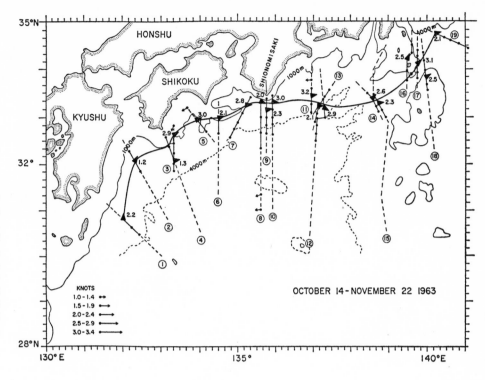

Chart 37.   Distribution of sur-face current velocity (knots) south of Japan *X/14 thru XI/22/1963*. Dates of individual transects are as follows: ① XI/9–10, ② XI/6, ③ XI/12–13, ④ XI/4–5, ⑤ X/20, ⑥ XI/1–2, ⑦ XI/7–8, ⑧ X/30–XI/1, ⑨ X/16, ⑩ XI/8–9, ⑪ XI/13, ⑫ XI/10–11, ⑬ X/15, ⑭ X/15, ⑮ XI/16–17, ⑯ XI/5–6, ⑰ X/14–15, ⑱ XI/19–20, ⑲ XI/22.

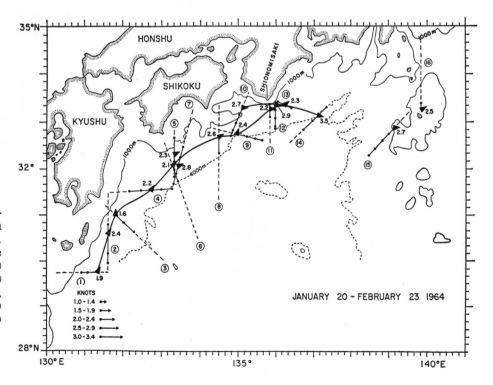

Chart 38.   Distribution of sur-face current velocity (knots) south of Japan *I/20 thru II/23/1964*. Dates of individual transects are as follows: ① II/7, ② II/7–8, ③ II/13–14, ④ II/8, ⑤ II/23, ⑥ II/14–15, ⑦ II/8–9, ⑧ II/18–19, ⑨ II/9–10, ⑩ I/20, ⑪ II/10, ⑫ I/27–28, ⑬ II/13, ⑭ I/27, ⑮ II/13–14, ⑯ I/26.

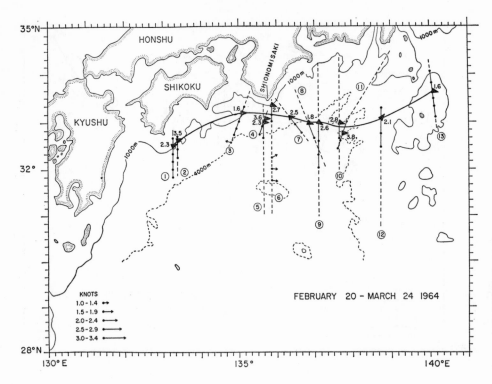

Chart 39. Distribution of surface current velocity (knots) south of Japan *II/20 thru III/24/1964*. Dates of individual transects are as follows: ① III/24, ② III/11, ③ III/3–4, ④ III/3, ⑤ II/20–21, ⑥ III/5–7, ⑦ III/3–4, ⑧ II/26–27, ⑨ III/11–12, ⑩ II/27–28, ⑪ III/2–3, ⑫ III/12–13, ⑬ III/18–19.

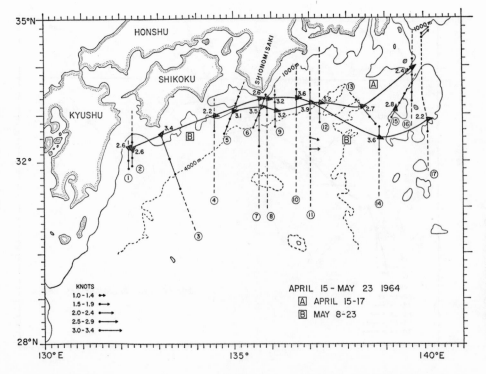

Chart 40. Distribution of surface current velocity (knots) south of Japan *IV/15 thru V/23/1964*. Dates of individual transects are as follows: ① V/19, ② V/8–9, ③ V/9, ④ V/8, ⑤ IV/15, ⑥ IV/15, ⑦ V/12–13, ⑧ V/15–16, ⑨ IV/16, ⑩ V/14–15, ⑪ V/16–17, ⑫ IV/17, ⑬ IV/15–16, ⑭ V/21, ⑮ V/12–13, ⑯ IV/15, ⑰ V/22–23.

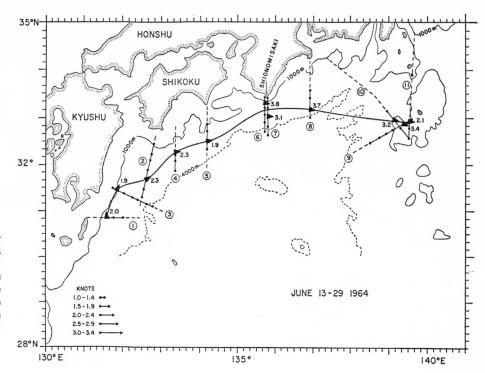

Chart 41. Distribution of surface current velocity (knots) south of Japan *VI/13 thru VI/29/1964*. Dates of individual transects are as follows: ① VI/24, ② VI/25, ③ VI/18–19, ④ VI/20, ⑤ VI/18, ⑥ VI/16, ⑦ VI/28–29, ⑧ VI/15, ⑨ VI/29, ⑩ VI/14–15, ⑪ VI/13–14.

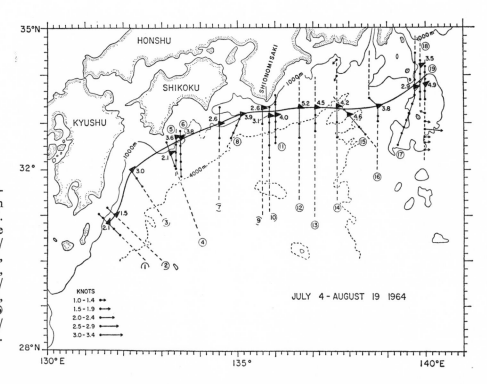

Chart 42. Distribution of surface current velocity (knots) south of Japan *VII/4 thru VIII/19/1964*. Dates of individual transects are as follows: ① VIII/7–8, ② VII/25–26, ③ VIII/8–9, ④ VII/24–25, ⑤ VII/4, ⑥ VIII/9, ⑦ VIII/3–4, ⑧ VIII/11, ⑨ VIII/5–6, ⑩ VIII/9–10, ⑪ VII/12–13, ⑫ VIII/9–10, ⑬ VIII/10–11, ⑭ VIII/10–12, ⑮ VIII/7–8, ⑯ VIII/15–16, ⑰ VIII/6–7, ⑱ VIII/18–19, ⑲ VII/26–27.

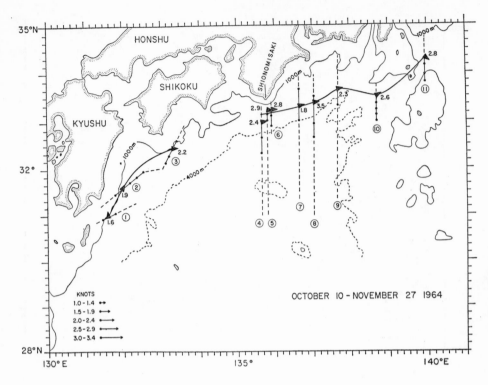

Chart 43. Distribution of surface current velocity (knots) south of Japan *X/10 thru XI/27/1964.* Dates of individual transects are as follows: ① X/10, ② X/13–14, ③ X/14, ④ XI/6–7, ⑤ XI/15–16, ⑥ X/18, ⑦ X/29–30, ⑧ XI/16–17, ⑨ X/31–XI/1, ⑩ XI/25–26, ⑪ XI/26–27.

APPENDIX 2

Volume transport of the Kuroshio (V_K) relative to 800 db. Ships are designated as follows: S=Shumpu Maru; T=Takuyo; M=Meiyo; K=Kaiyo; R=Ryofu Maru. Sections that are not oriented north and south are designated by the longitudes of the first and last stations on the section.

Year	Date	Longitude	V_K 10^6m^{3}sec^{-1}	Ship
1956	II/20–21	138°00′E	52	M
	III/2	139°00′	46	M
	V/17	136°55′	42	M
	V/12–13	138°00′	47	M
	VIII/19–20	136°50′	50	S
	VIII/20–21	137°40′	55	S
	VIII/7–8	138°00′	57	K
	VIII/25–26	138°30′	59	S
	XI/13–14	137°40′	41	S
	XI/3–4	138°00′	45	M
	XI/12–13	138°30′	39	S
1957	III/4–5	132°02′–133°00′	29	S
	II/14	138°00′	39	K
	V/1–2	131°37′–133°00′	26	S
	V/13–14	137°00′	39	M
	V/17–18	138°00′	37	M
	VIII/5–6	137°40′	49	S
	VIII/10	138°00′	59	T
	VIII/9–10	138°30′	53	S
	XII/8–9	137°40′	51	S
	X/28–29	138°00′	64	K

Year	Date	Longitude	V_K 10^6m^{3}sec^{-1}	Ship
1958	III/9	134°30′	31	S
	III/19–20	138°00′	52	K
	VIII/3–4	134°30′	41	S
	VIII/5–6	136°40′	53	S
	VIII/10	137°40′	51	S
	VIII/27–28	138°00′	53	K
	XI/6–7	132°19′–133°20′	21	S
	XI/20–21	134°30′	25	S
	XI/22–23	136°40′	28	S
	XI/23–24	137°00′	33	M
	XI/27–28	137°40′	38	S
	XI/20–21	138°00′	32	M
1959	II/26–27	131°36′–133°00′	36	S
	III/7–8	137°00′	48	K
	III/11	137°40′	53	S
	III/15–16	138°00′	50	K
	III/19–20	138°30′	43	S
	III/19–20	139°00′	49	K
	V/17–18	135°40′	34	S
	VII/31–VIII/1	131°36′–133°00′	21	S
	X/31–XI/1	131°36′–133°00′	17	S
	XI/10–11	135°40′	31	S
	XI/16–17	137°40′	54	S
1960	III/19	135°40′	46	S
	V/12–13	133°13′–134°00′	32	S
	VII/23	131°39′–133°00′	49	S
	VII/29–30	134°30′	37	S
	VIII/6	137°40′	47	S
	X/21–22	135°40′	37	S

Year	Date	Longitude	V_K $10^6 m^3 sec^{-1}$	Ship
1961	V/14–15	137°40′	41	S
	VIII/11–12	132°12′–133°00′	36	S
	VIII/12	133°18′–133°43′	34	S
	VIII/9	134°30′	33	S
	XI/17–19	133°40′	37	R
	XI/6–10	135°40′	29	R
	XI/5–6	137°40′	54	R
1962	III/10–11	131°37′–132°30′	30	S
	III/18	135°40′	32	S
	V/9–10	134°30′	38	S
	V/16	135°40′	43	S
	V/17–18	136°40′	49	S
	V/25–26	137°40′	49	S
	V/26–27	138°30′	48	S
	VIII/4–5	134°30′	44	S
	VIII/6–7	135°40′	47	S
	VIII/10–11	136°40′	45	S
	VIII/21–22	137°00′	59	T
	VIII/29–30	138°31′–139°02′	54	T
	X/21–22	134°30′	30	S
	X/20	135°40′	21	S
1963	III/1–5	134°30′	40	S
	III/5–6	135°40′	43	S
	III/14–15	135°48′	35	T
	III/18	137°40′	41	S
	III/22	138°45′	40	T
	V/8	135°40′	39	S
	V/9–10	136°40′	44	S
	V/14	137°40′	44	S
	V/24–25	138°30′	39	T
	VII/25–26	134°30′	37	S
	VIII/2–3	137°40′	53	S
	XI/9–10	131°35′–133°00′	37	S
	XI/1–2	134°30′	28	S
	XI/11	137°10′	56	M
	XI/16–17	138°50′	57	M
1964	II/13–14	132°00′	27	S
	II/15	133°26′	27	S
	II/18–19	134°30′	38	S
	III/11–12	137°05′	29	T
	II/27	137°40′	29	S
	V/8	134°30′	34	S
	V/16–17	137°00′	48	K
	V/21	138°50′	44	K
	V/22–23	139°57′	30	K
	VIII/9–10	136°40′	60	S
	VIII/10–11	137°00′	65	M
	VIII/10–11	137°40′	58	S
	VIII/15–16	138°39′	57	M
	X/29–30	136°40′	46	S
	X/31–XI/1	137°37′	47	S

REFERENCES

Brooks, C.E.P. and N. Carruthers (1953): *Handbook of Statistical Methods in Meteorology.* London: Her Majesty's Stationery Office, 412pp.

Defant, Albert (1961): *Physical Oceanography.* New York: Pergamon Press, Vol. **1**: 707pp.

Endo, Hiroshi (1961): On the correlation between the surface water temperature and current axis in the Kuro-shio region (in Japanese with English abstract and legends). *Hydrogr. Bull., Tokyo,* **65**: 42–47.

Fuglister, F. C. (1951): Annual variations in current speeds in the Gulf Stream system. *J. Mar. Res.,* **10**: 119–127.

——— (1963): Gulf Stream '60. *Progr. Oceanogr.,* M. Sears (ed.), **1**: 265–373.

Fuglister, F. C. and A. D. Voorhis (1965): A new method of tracking the Gulf Stream. *Limnol. Oceanogr.,* **10** (supplement): R115–R124.

Fuglister, F. C. and L. V. Worthington (1951): Some results of a multiple ship survey of the Gulf Stream. *Tellus,* **3**: 1–14.

Fukuoka, Jiro (1958): The variations of the Kuroshio Current in the sea south and east of Honshu. *Oceanogr. Mag.,* **10**: 201–213.

Ichiye, Takashi (1954a): On the variation of oceanic circulation (VII). *Oceanogr. Mag.,* **6**: 1–14.

——— (1954b): On the variation of oceanic circulation (VI). *Geophys. Mag.,* **25**: 185–217.

Istoshin, Yu. V. and E. M. Sauskan (1968): On the counter-currents of the Kuroshio (in Russian). *Okeanologiya,* **8**: 949–959.

Japan Maritime Safety Agency (no date): *State of the adjacent seas of Japan.* Vol. **1** (1955–1959). Vol. **2** (1960–1964). Tokyo.

Kawai, Hideo (1969): Statistical estimation of isotherms indicative of the Kuroshio axis. *Deep-Sea Res.,* **16** (supplement): 109–115.

Knauss, J. A. (1969): A note on the transport of the Gulf Stream. *Deep-Sea Res.,* **16** (supplement): 117–123.

Longuet-Higgins, M. S., M. E. Stern and Henry Stommel (1954): The electrical field induced by ocean currents and waves, with applications to the method of towed electrodes. *Pap. Phys. Oceanogr. Meteor.,* **8**: 1–37.

Masuzawa, Jotaro (1954): On the Kuroshio south off Shiono-Misaki of Japan (Currents and water masses of the Kuroshio System I). *Oceanogr. Mag.,* **6**: 25–33.

——— (1957): An example of cold eddies south of the Kuroshio (Currents and water masses of the Kuroshio System VII). *Rec. Oceanogr. Wks. Japan,* **3**: 1–7.

——— (1960): Statistical characteristics of the Kuroshio current. *Oceanogr. Mag.,* **12**: 7–15.

——— (1965a): Meandering of the Kuroshio (The cold water mass south of Honshu) (in Japanese). *Kagaku,* **35**: 588–593.

——— (1965b): Water characteristics of the Kuroshio. *Oceanogr. Mag.,* **17**: 37–47.

——— (1965c): A note on the seasonal variation of the Kuroshio velocity (in Japanese). *J. Oceanogr. Soc. Jap.,* **21**: 117–118.

——— (1968): Cruise report on multi-ship study of short-term fluctuations of the Kuroshio in October to November 1967. *Oceanogr. Mag.,* **20**: 91–96.

——— (1969): A short note on the Kuroshio stream axis

(in Japanese). *J. Oceanogr. Soc. Japan*, **25**: 259–260.

Masuzawa, Jotaro and Toshisuke Nakai (1955): Notes on the cross-current structure of the Kuroshio (Currents and water masses of the Kuroshio System V). *Rec. Oceanogr. Wks. Jap.*, **2**: 96–101.

Mogi, Akio (1972): Bathymetry of Kuroshio Region. *KUROSHIO—Its Physical Aspects*. Univ. Tokyo Press: 53–80.

Moriyasu, Shigeo (1961): An example of the conditions at the occurrence of the cold water region. *Oceanogr. Mag.*, **12**: 67–76.

——— (1963): The fluctuation of hydrographic conditions in the sea south of Honshu, Japan (Review). *Oceanogr. Mag.*, **15**: 11–29.

Namias, Jerome (1970): Macroscale variations in sea-surface temperatures in the north Pacific. *J. Geophys. Res.*, **75**: 565–582.

Nan'niti, Toshio and H. Akamatsu (1966): Deep current observations in the Pacific Ocean near the Japan Trench. *J. Oceanogr. Soc. Jap.*, **22**: 154–160.

Niiler, P. P. and A. R. Robinson (1967): Theory of free inertial jets II. A numerical experiment for the path of the Gulf Stream. *Tellus*, **14**: 601–619.

Pavlova, Yu. V. (1964): Seasonal variations of the Kuroshio current (in Russian). *Okeanologiya*, **4**: 625–640.

Robinson, A. R. and P. P. Niiler (1967): Theory of free inertial currents I. Path and structure. *Tellus*, **19**: 269–291.

Shoji, Daitaro (1965): Description of the Kuroshio (Physical Aspect) in *Proc. Symp. on the Kuroshio*. Oceanogr. Soc. Jap. and UNESCO: 1–11.

Shoji, Daitaro, Ryuzo Watanabe, Naruji Suzuki and Katsumi Hasuike (1958): On the "Shiome" at the boundary zone of the Kuroshio and the coastal waters off Shionomisaki. *Rec. Oceanogr. Wks. Jap.*, Spec. No. 2: 78–84.

Shoji, Daitaro and Hideo Nitani (1966): On the variability of the velocity of the Kuroshio-I, *J. Oceanogr. Soc. Jap.*, **22**: 192–196.

Stommel, Henry (1965): *The Gulf Stream—A physical and dynamical description*. 2nd Edition. Univ. of Calif. Press, Berkeley and Los Angeles, Cambridge Univ. Press, London, 248 pp.

Sturges, Wilton (1968): Flux of water types in the Gulf Stream (abstract). *Trans. Am. Geophys. Un.*, **49**: 198.

Sverdrup, H. U., M. W. Johnson and R. H. Fleming (1942): *The oceans; their physics, chemistry and general biology*. New York: Prentice Hall. 1087pp., 7 charts.

Taft, B. A. (1970): Path and transport of the Kuroshio south of Japan, *The Kuroshio—A Symposium on the Japan Current:* East-West Center Press, 185–196.

Uda, Michitaka (1937): On the recent abnormal condition of the Kuroshio to the south of the Kii Peninsula (in Japanese). *Kagaku*, **7**: 360–361, 403–404.

——— (1940): On the recent anomalous hydrographical conditions of the Kuroshio in the south waters off Japan proper in relation to fisheries (in Japanese with English abstract). *J. Fish. Exp. Sta. Tokyo*, **10**: 231–278.

——— (1949): On the correlated fluctuation of the Kuroshio Current and the cold water mass. *Oceanogr. Mag.*, **1**: 1–12.

——— (1951): On the fluctuation of the main stream axis and its boundary line of Kuroshio. *J. Oceanogr. Soc. Jap.*, **6**: 181–189.

——— (1964): On the nature of the Kuroshio, its origin and meanders, in *Stud. Oceanogr.*, Kozo Yoshida (ed.), Univ. Tokyo Press: 89–107.

Uda, Michitaka and K. Hasunuma (1969): The eastward Subtropical Countercurrent in the western North Pacific Ocean. *J. Oceanogr. Soc. Japan*, **25**: 201–210.

von Arx, W. S. (1962): *An introduction to physical oceanography*. Massachusetts: Addison-Wesley. 402pp.

Warren, B. A. (1963): Topographic influences on the path of the Gulf Stream. *Tellus*, **15**: 167–183.

Warren, B. A. and G. H. Volkmann (1968): Measurement of volume transport of the Gulf Stream south of New England, *J. Mar. Res.*, **26**: 110–126.

Worthington, L. V. and Hideo Kawai (1972): Comparison between deep sections across the Kuroshio and the Florida Current and Gulf Stream. *KUROSHIO—Its Physical Aspects*. Univ. Tokyo Press: 371–385.

Wüst, Georg (1936): Kuroshio und Golfstrom. *Veröff. d. Inst. f. Meeresk. an d. Univ. Berlin*, N. F. Reihe A; Heft 29. 69pp. +1 map.

Wyrtki, Klaus (1961): Physical oceanography of the southeast Asian waters. *NAGA Rept.*, Vol. **2**, Scripps Instn. Oceanogr., Univ. Calif., 195pp. +44 pls.

Yosida, Shozo (1961a): On the short period variation of the Kuroshio in the adjacent sea of the Izu Islands (in Japanese with English abstract). *Hydrogr. Bull.*, **65**: 1–18.

——— (1961b): On the variation of Kuroshio and Cold Water Mass off Ensyu-Nada (Part 1) (in Japanese with English abstract). *Hydrogr. Bull.*, **67**: 54–57.

Chapter 7

TIME VARIATION OF THE KUROSHIO SOUTH OF JAPAN

DAITARO SHOJI

Hydrographic Department, Maritime Safety Agency, Tokyo, Japan.

1 INTRODUCTION

In the preceeding chapter, the statistical characteristics of the Kuroshio south of Japan are described by Dr. Taft based mainly on the data obtained between 1955 and 1964. In this chapter, time dependent phenomena of the Kuroshio south of Japan will be treated. Off the south coast of Japan, the Kuroshio has been observed and studied by many Japanese oceanographers since scientific and more or less systematic observation of the oceans began about 50 years ago in Japan. However, it must be admitted that knowledge about the nature and extent of the time dependent phenomenon of the Kuroshio still remains very fragmentary and incomplete and that theories which explain the phenomenon are only speculations in most cases. Therefore, the description in this chapter will be in the nature of a case study.

The properties of sea water in this region are described in a previous chapter by Dr. Masuzawa. As shown there they are rather simple and it can be said that the Kuroshio generally flows between coastal upwelling and the North Pacific Central Water, except near the Izu Islands where cold water of Oyashio origin is found from time to time.

The bottom topography of this region was described in Chapter 2 and is shown in Fig. 1 of Chapter 6. No doubt the bottom topography has a very large influence on the variation of the Kuroshio, especially the ridge south of Kyushu, the Kii Peninsula and the Izu-Ogasawara Ridge and Izu Islands (Fukuoka, 1958, Nan'niti, 1960), but their roles are not yet well understood.

The spectrum of time variation of the Kuroshio extends from more than several years to a few hours. It has been pointed out by many authors that there are periods of variation of the Kuroshio of 15 years, 11 years, 9 years, about 5 years, annual, semi-annual, a few months, a few weeks, diurnal and semi-diurnal.

Of course there may be periods on the order of 40–50 years or more or several minutes or less. We have no data for such phenomena. An important thing in this respect is that the amplitudes of these variations are all nearly of the same order and that there is no predominant one like the annual cycle in surface temperature. It seems rather strange that in spite of the fact that the temperature of the surface layer and sea level along the coast of Japan show distinct annual changes, the corresponding seasonal variation of the Kuroshio is rather difficult to detect from observations.

2 LARGE MEANDER OF THE KUROSHIO OFF THE SOUTH COAST OF JAPAN

2.1 General Description

The most conspicuous feature of the Kuroshio in this area is its large and long-lived meander. This meander of the Kuroshio is closely related with a cold upwelled water region surrounded by the meander. Therefore, the term "Cold Water Mass off the Kii Peninsula or Enshunada" has been used by many oceanographers.

In 1935 a naval operation was conducted by the Japanese Navy in this area and a scheduled encounter of opposing fleets did not take place due to the bad weather and unexpected direction of the strong current of the Kuroshio. This incident caused an intensive study of the Kuroshio by the Japanese Navy before the War. Fisheries oceanographers have also shown much interest in it. Uda (1937) first described the anomalous pattern of the Kuroshio. According to Yosida (1961), the meander or unexpected westward flow off Shionomisaki was first reported in May and June, 1934, by merchant ships. This meander continued for about 10 years until 1943 or 1944. Because observations were very scarce at that time, the last stage of this meander is not clear. The dynamic topography of the sea surface of this period was published as a publication titled "Current Chart in the Adjacent Seas of Japan" by the Hydrographic Department of Japan in 1951. Similar meanders of the Kuroshio developed between 1953 and 1955, and between 1959 and 1963. It must be pointed out that in view of the fact that the meander existed for more than 15 out of 40 years since 1930, it is not an anomalous phenomenon, but rather one of two relatively stable patterns of the Kuroshio in this area (Masuzawa, 1965).

Figures 1 and 2 show the position of the Kuroshio by distances measured from several capes of Japan. These figures were prepared by Yosida from various sources including navigation reports of naval and merchant vessels as well as data obtained by observation vessels of various institutions.

The starting times of the meander are marked by

A1, A2 and A3 respectively. From these figures several major characteristics are readily seen:

1)   The north-south translation of the Kuroshio is as much as 200 miles off Daiozaki and Omaezaki.

2)   The meander develops in a rather short time, i.e., in a few months.

3)   Prior to the development of the meander off Shionomisaki and Daiozaki, a small departure of the Kuroshio occurs off Toimisaki, Kyushu. However, this departure of the current off the east coast of Kyushu does not last long.

4)   The meander decays slowly, i.e., in several years, shifting eastwards.

5)   The life of a meander is 3 to 10 years.

As seen from Fig. 2, the density of the observation in this area greatly increased after the spring of 1960. The third large meander mentioned above existed from 1959 to 1963. Figure 3, no. 1–no. 15 were prepared to show the variation of the Kuroshio from 1959 to 1965 in more detail. Figure 2 and Figure 3,

⟨1924–1945⟩

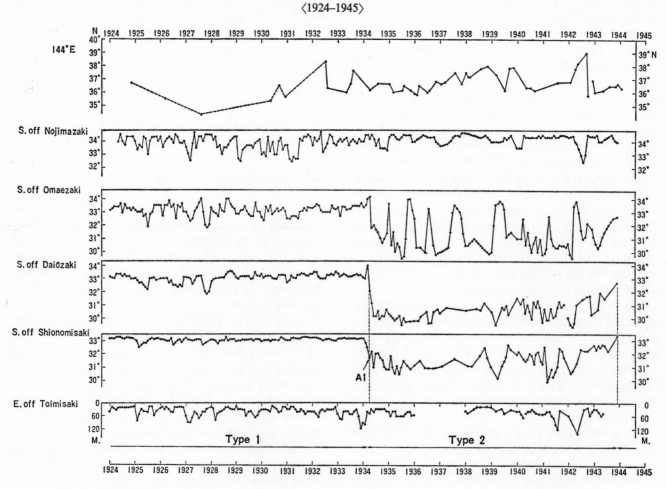

FIG. 1   The position of the Kuroshio as measured by the distance from several capes of Japan and at 144°E.

no. 1 indicate that the meander developed in the spring and summer of 1959. Details concerning this development are given below.

In the first half of 1960, the Kuroshio extended as far south as 30°N, at about 137°E off the Kii Peninsula. From June 1960 to December 1961, the Kuroshio in this region was rather stable with a meander off Enshunada, and the southern limit of the Kuroshio was at about 31°N and between 137°E and 138°E. In 1962, the northward flow of the Kuroshio west of the Izu Islands became unstable and the Kuroshio often crossed the Izu-Ogasawara Ridge south of Hachijojima. The Kuroshio rarely reached 31°N in the latter half of 1962. In 1963 the Kuroshio was very unstable near the Izu Islands and its position off Enshunada was at about 33°N. From October, 1963, it can be said that the Kuroshio returned to its so-called normal state, which means the Kuroshio flows almost parallel to the coast of Japan.

From March to October 1964 and from May to December 1965 the Kuroshio was again very unstable near the Izu Islands, but it never reached 30°N.

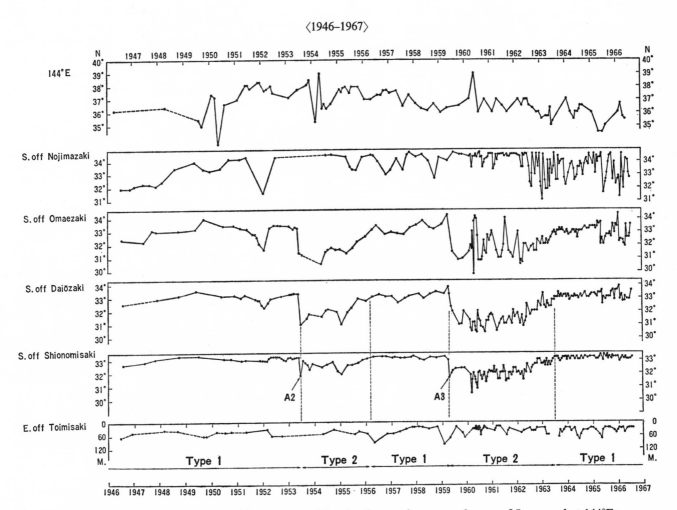

FIG. 2 The position of the Kuroshio as measured by the distance from several capes of Japan and at 144°E.

FIG. 3   no. 1~15
Patterns of the Kuroshio from January 1959 to December 1965.

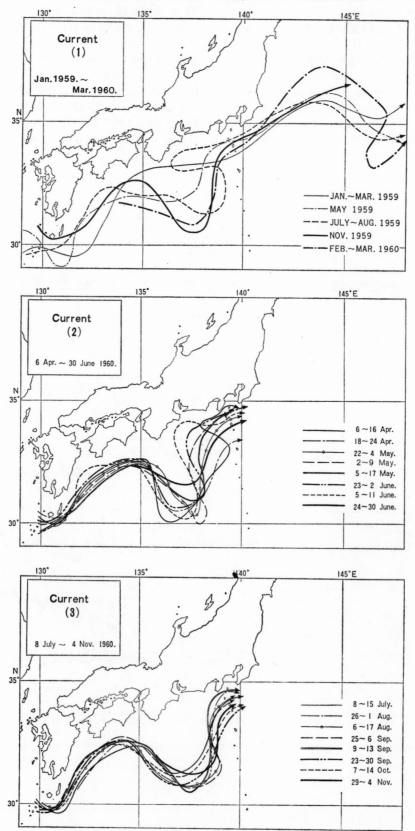

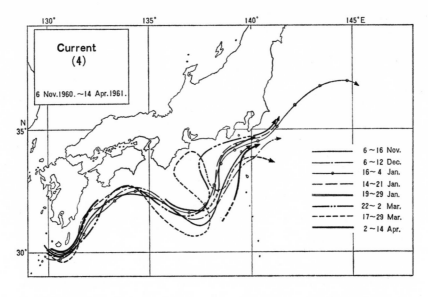

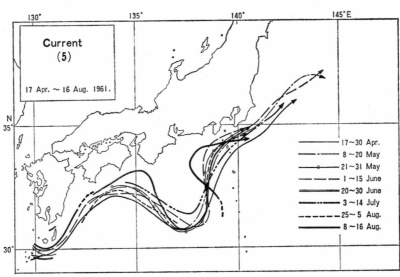

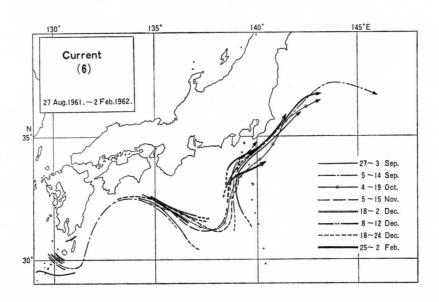

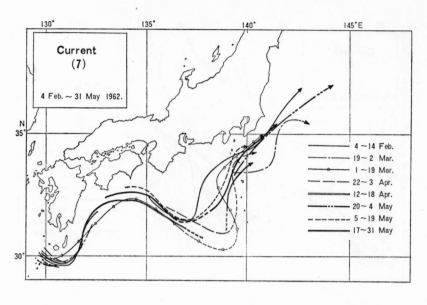

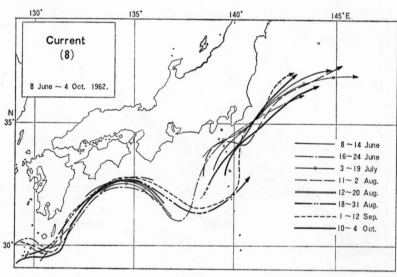

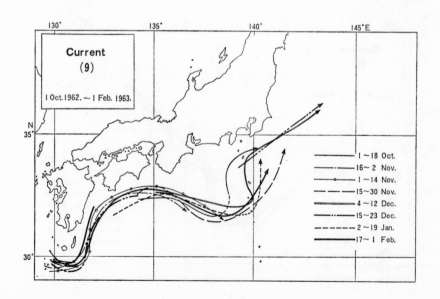

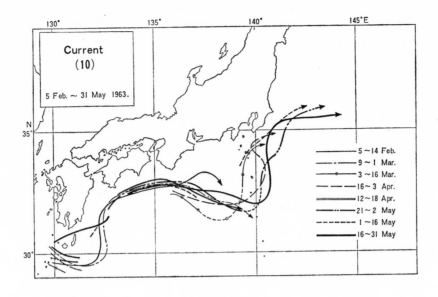

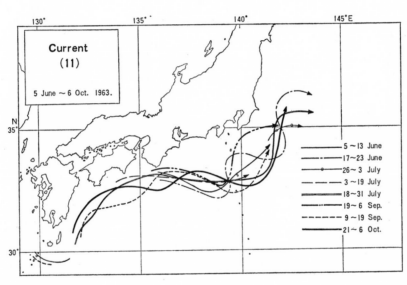

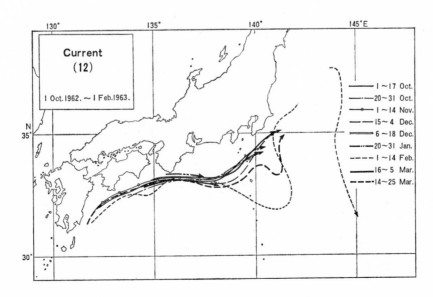

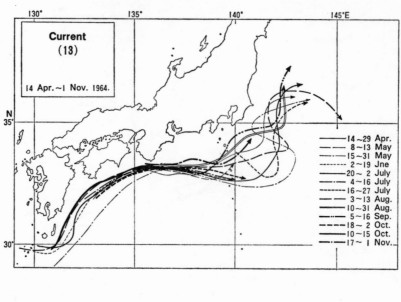

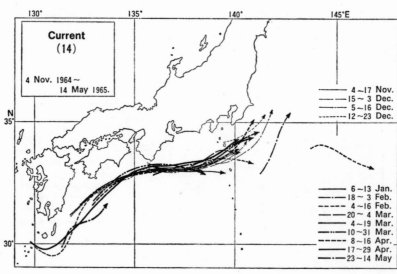

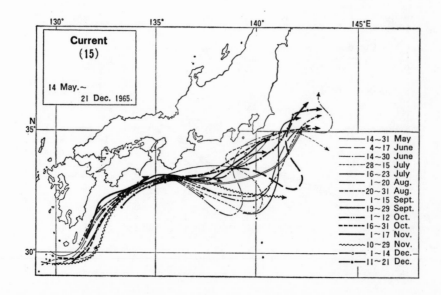

2.2 Generation and decay of the meander.

Yosida (1961) and Moriyasu (1961) investigated the generation of the meander in 1959; Fig. 4 was prepared by Yosida. It was observed in March, 1959, that the Kuroshio shifted off-shore off the east coast of Kyushu and the northward portion of the Kuroshio proceeded with a speed of about 3 miles per day up to off Omaezaki in June, 1959. In 1934 and 1953 observations were insufficient to show this kind of phenomenon. However, as pointed out in the preceding section, both in 1934 and 1953 departure of the Kuroshio from the coast of Kyushu prior to the development of a meander off the Kii Peninsula was observed. According to Ichiye (1954)

and Masuzawa (1954), the Kuroshio off Shiono-misaki divided into three branches in June 1953. In 1969, the progression of the meander of the Kuroshio off the south coast of Japan was observed again (Nakabayashi, 1970). The change in the pattern of flow is shown in Fig. 5. The speed of the progression of the pattern was the same as in the case of 1959, and a meander of the Kuroshio off Enshunada developed after that.

As pointed out by Masuzawa (1965), a departure of the Kuroshio from the east coast of Kyushu is often observed, especially in winter. Usually it does not grow to a large meander such as the one described here. There is no satisfactory explanation yet for the conditions which determine the development of a

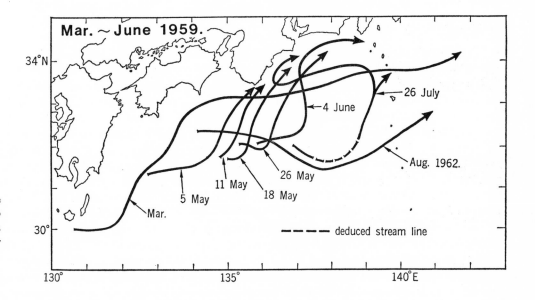

FIG. 4 Change of the path of the Kuroshio prior to the generation of the meander, March–June 1959.

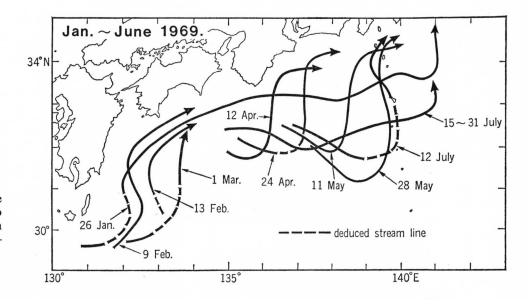

FIG. 5 Change of the path of the kuroshio prior to the generation of the meander, January–June 1969.

large scale meander off the south coast of Japan.

Uda (1949) investigated the meander of 1934–43. He explained that the primary cause of the phenomenon would be the increased southerly invasion of the cold Oyashio Water (subarctic water) which prevailed in 1934 and 1935 in the northeastern sea of Japan. The Oyashio undercurrent was assumed to reach this area through the deep portion of the Izu-Ogasawara Ridge near Torishima with a speed of 3 centimeters per second.

Suda (1943) suggested that the extremely destructive Muroto Typhoon which struck Japan at Muroto-misaki and Osaka on September 21, 1934 might be the cause of the meander. However, this does not agree with Fig. 1, though there are several cases known where meteorological disturbances caused a major change in oceanic conditions (Shoji, 1951; Moriyasu, 1963).

Nan'niti (1958, 1960) presented a theory concerning the mechanism of the generation of the cold water region in this area. He concluded that when the maximum velocity of the Kuroshio is less than 2.5 knots the upwelling and overflow of cold intermediate water takes place off Enshunada. He suggested that the strengthening and weakening of the Kuroshio might be caused by the growth and decay of the western North Pacific Central Water. Shoji (1964) pointed out that in fact in the spring of 1963 when the meander of 1959–1963 was disappearing the velocity of the Kuroshio was stronger than in 1960 and 1961 based upon actual and sea level data. But he attributed the stronger current of the Kuroshio to the stronger upwelling along the Japanese coast.

Several attempts were made to link the meander of the Kuroshio with world-wide meteorological conditions such as the westerly index and the variation of vorticity of wind field over the Pacific Ocean (Nan'niti, 1959). Solar activity (number of sun spots) was also often quoted as a cause of the variation of the Kuroshio (Uda, 1949; Outi, 1966).

Uda (1962) pointed out upon examination of the data from the period between 1911 to 1960 that the rise and fall of surface temperature along the coast of Japan occurs almost conversely to that along the North American coast.

He interpreted this phenomenon according to the changes in distribution of low and high pressure over the Northern Hemisphere. He also suggested that the occurrence of the meander of the Kuroshio might be related to the occurrence of the El Nino off the coast of South America.

It must be noted here that all these investigations suggest that the generation of the large-scale meander of the Kuroshio off the south coast of Japan must be treated from the viewpoint of circulation in the whole Pacific. However, it is also true that we have not yet enough data for a complete understanding of this phenomenon.

As seen from Figs. 1, 2 and 3, the meander decays slowly, though there are many fluctuations which overlap the general trends of decay. Ichiye (1954) and Moriyasu (1958, 1959) computed the radius. angular momentum, vorticity and kinetic energy of the eddy surrounded by the meander and discussed its change with time. Moriyasu suggested the conservation of vorticity and a critical value for the radius of the eddy of 90 kilometers which may be significant from a dynamic view point. However, it seems that due to the complexity of the phenomena no definite conclusion was obtained.

It is specially noted that in the first half of 1963, extremely low water temperature and mass mortality of fish due to it were reported along the coast of southern Japan. The temperature was 4° to 5°C lower than mean values and at some places more than 10° lower (Fujimori, 1964). Relations between the coastal low temperature and the disappearance of the meander and violent fluctuation of the Kuroshio near the Izu Islands, as seen from Fig. 3, no. 9 to no. 11, were discussed. It was pointed out (Shoji, 1964) that in 1944 and in 1956 when the previous meanders were disappearing, the temperature was very low along the coast of Japan, too. However, it should be pointed out that the meteorological conditions in 1963 were also extraordinary (Hanzawa, 1964). Hanzawa emphasized the effects of the abnormal atmospheric circulation upon oceanic conditions and general warming of sea water temperature in the northern part of the northwestern Pacific and along the American west coast in contrast to general cooling of sea water temperatures in the seas adjacent to Japan.

3 SHORT-TERM FLUCTUATIONS

In this section fluctuation of the Kuroshio over several months or several weeks to several days will be treated. Because of the difficulty in observing this kind of phenomenon, here again a few special cases are described.

3.1 Small scale meander

From 1963 to 1968 there was no large-scale meander of the Kuroshio as described in the previous section. However, there were small but rapid fluctuations around the Izu Islands. Nitani (1969) classified the pattern of the Kuroshio into four types as shown

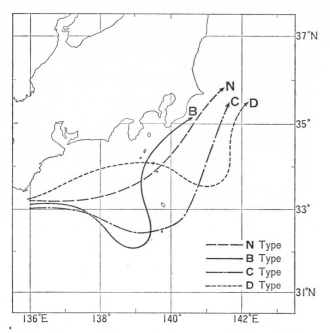

FIG. 6   4 types of the patterns of the Kuroshio near the
Izu Islands.

dynamic height of the nearby sea surface (Shoji, 1954;
Yosida, 1961). Therefore, the variation in sea level
indicates the change of distribution of density in the
sea, which is closely connected with the Kuroshio in
this area.

In Fig. 7 there are 3 major particular features,
marked by A, B and C, repectively. A in the middle of
May is the high peaks at Kozu (18) and Miyake (19)
of about 55 centimeters. (Numbers in brackets are
the station numbers in Fig. 9). Peaks of lesser height
are also present at Hachijo (20) and Oshima (17). B
indicates a sharp rise in sea level at all stations at the
end of May. C indicates a sharp rise in sea level at
Hachijo (20), Miyake (19) and Kozu (18) islands. It
should be noted here that the difference in sea level
between the right and left side of the Kuroshio is
estimated to be about 80 centimeters.

Case A

Very high peaks of more than 30 centimeters are
also recognized at Shionomisaki (7) and Kainan (6)
prior to the peaks at Kozu (18) and Miyake (19).
Smaller peaks appear at Omaezaki (11), Onisaki (10),
Toba (9) and Uragami (8) after Kozu (18).

On May 3 and 4, a severe storm hit Japan through
the Strait of Kii. Winds of more than 40 meters per
second were measured at Shionomisaki and Muroto-
misaki. It is possible that these strong winds caused
a strong intrusion of the Kuroshio in the Strait of
Kii. Actually it was reported that very warm and
saline water of the Kuroshio was observed off the
west coast of the Kii Peninsula after the storm.

Figure 10 shows the current chart in early May. It
should be noted that on May 12, when the sea level
was very high at Shionomisaki, there was a very
strong current of 4.3 knots very near the cape. There
was no strong current near Kozu on May 9, but on
May 13 a very strong current was observed between
Kozu and the Izu Peninsula.

It might be assumed that the abrupt change of the
Kuroshio near Shionomisaki propagated to the east
and caused the high peaks of A at Kozu (18) and
Miyake (19). As the highest peaks at Shionomisaki
(7) and the two islands occurred about 3 days apart,
the speed of propagation is estimated to have been
about 3 miles per hour. This speed is comparable to
that of the Kuroshio. The fact that sea level at Kozu
(18) rose one day earlier than Miyake seems to support this assumption.

It can not be said that the above-mentioned propagation of the disturbance caused by weather is
proved very well. Nevertheless, it presents a very
interesting case.

in Fig. 6 and analyzed statistically the occurrence of
these patterns. He found that type N has the longest
mean life of about 2 months and the other 3 types
have a mean life of about 1 month. In the majority of
cases types change in the order of N-B-C-D. It was
noted that type N frequently occurs in April and
October–November and is seldom seen in January
and June–July. This might be an expression of seasonal change of the Kuroshio. As this study was
based upon twice monthly observations in this area,
the conclusion may be modified in future.

3.2   Variation of daily mean sea level and the Kuroshio

Variations of ocean current over a few days or
several weeks are very difficult to observe with observation vessels. In order to supplement the observation vessels, some kind of continuous measurement
is very useful. For this purpose, sea level observations at Hachijo, Miyake and Kozu islands have been
introduced. Fig. 7 shows the sea level at the three
islands and Oshima for the period of April to September in 1965. The sea level at the mainland coast of
Japan for the same period is shown in the next figure.
The location of tide stations is given in Fig. 9.

Adjusted daily mean sea level means that the
atmospheric effect is corrected by assuming that one
millibar of air pressure corresponds to one centimeter
of sea level. The direct effect of wind can be neglected.
Adjusted sea level is found to change parallel to the

It should be noted here that the peak at Shiono-misaki seems to have propagated westward with decreasing height to Nishino-Omote (1) and that a peak also seems to have propagated westward along the coast from Omaezaki (11) up to Uragami (8) after the peak at Kozu (18). There have been several discus-sions (Shoji, 1961; Isozaki, 1968) concerning this phenomenon of peak propagation along the coast. This case is the first example of a variation of the Kuroshio causing a shelf wave. The interaction between oceanic current and shelf wave has not been discussed very fully yet.

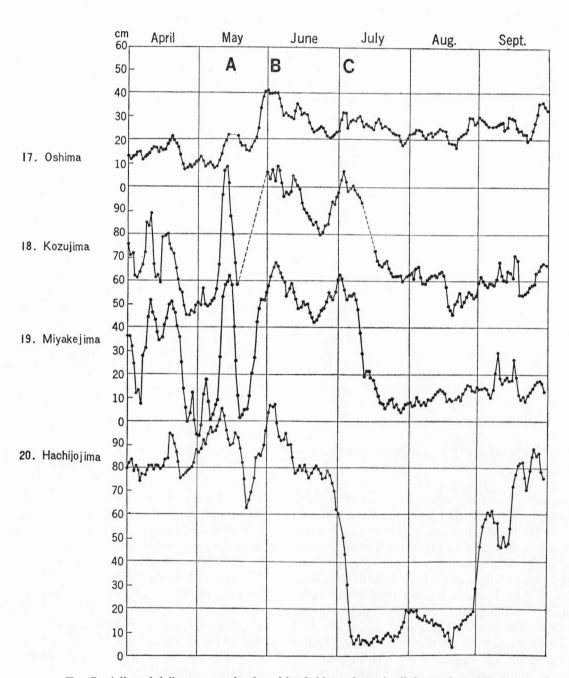

FIG. 7   Adjusted daily mean sea levels at island tide stations, April–September 1965.

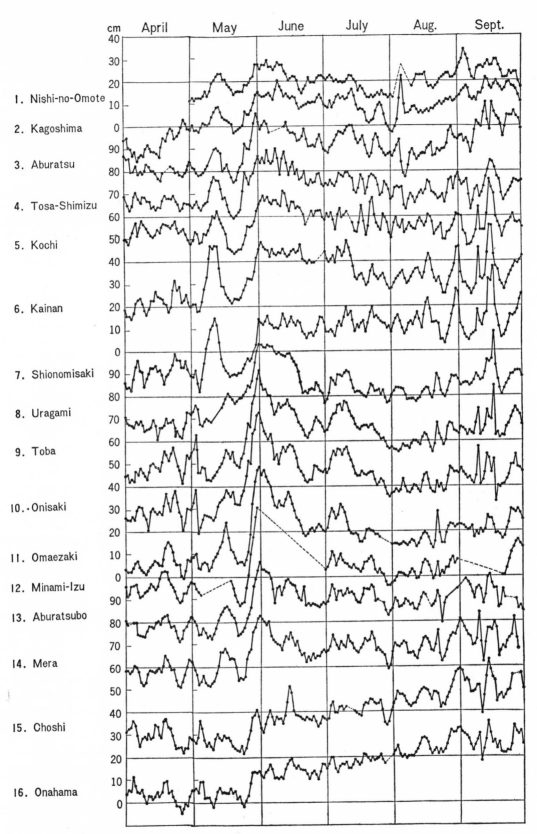

FIG. 8   Adjusted daily mean sea levels at coastal tide stations, April–September 1965.

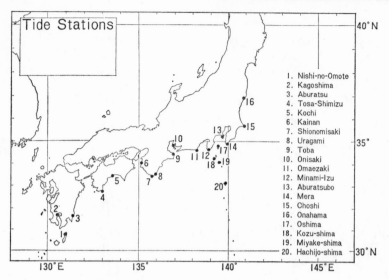

FIG. 9 Location of tide stations.

Current velocity measured by GEK.

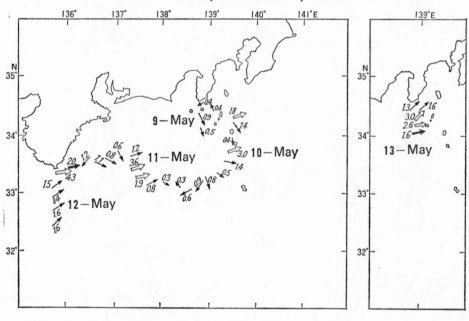

FIG. 10 Current chart in early May, 1965.

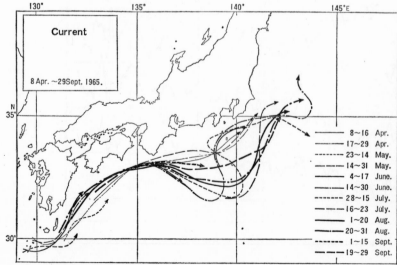

FIG. 11 Patterns of the Kuroshio south of Japan, April–September, 1965.

Case B

A sharp rise of sea level at almost all stations at the end of May is very noticeable. Along the mainland coast the rise is largest between Minami-Izu (12) and Toba (9). Figure 11 shows the current patterns for this period, and it can be seen from the figure that a meander developed off the coast between Minami-Izu and Toba. It was reported that a strong westerly current was observed along the east coast of the Kii Peninsula at the end of May until about June 9. From the fact that it took about 10 days to reach the peak, it can be assumed that the meander developed in about 10 days. After the sea level reached its peak in the beginning of June, it decreased generally through June and July, especially at Toba (9), Onisaki (10) and Omaezaki (11). This can be interpreted as an indication of the decay of the meander. It is emphasized that although the rise in sea level was largest along the coast between Minami-Izu and Toba affected directly by the meander, the rise was observed at every station under study. This indicates that the development of the meander affects the whole area and is not limited to the area where the meander is actually observed.

Case C

At the beginning of July, the sea level of island stations dropped sharply. It is known from Fig. 11 that the meander moved to the east and the Kuroshio was flowing south of Hachijojima in July and August. The sea level at Hachijojima (20) again rose in early September, but the sea level at Miyake (19) and Kozu (18) remained low. This indicates that the Kuroshio was located between Hachijojima and Miyakejima.

There were several typhoons in June and August which affected this area. However, the influence of these typhoons was not clear. High peaks in the middle of September were due to the typhoons.

4　VERY SHORT-TERM FLUCTUATION

There have been a number of observations which indicate that the Kuroshio changes its velocity in a very short time, i.e., in several hours or in a day (e.g., Shoji et al., 1958). To investigate this kind of phenomenon, a carefully planned multiple ship survey is necessary.

The first multiple ship survey in Japan in recent years was conducted off the east coast of northern Japan in 1957 as a project of the International Geophysical Year, 1957–58. Masuzawa (1958) analyzed the data and suggested a change in the velocity of the Kuroshio of more than 1 knot in less than 24 hours.

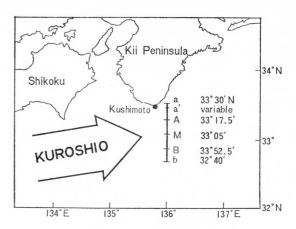

Fig. 12　Observation line, October 1964.

The Japan Hydrographic Department conducted multiple ship surveys off the Kii Peninsula in 1964 and 1965. The data were analyzed by Shoji and Nitani (1966, 1970). Figure 12 shows the observation line and stations in October 1964. Measurement of velocity was carried out by means of G.E.K., and BT and serial Nansen bottle observations were made at the stations. Two observation vessels moving in opposite directions were employed in order to obtain the average velocity across the 50 mile width of the Kuroshio as rapidly as possible. It was possible to get 1 cross-section in about 3 hours. The results of the measurements are shown in Fig. 13. Unfortunately, the observation was interrupted on October 6 and 7 owing to a storm.

The maximum east component of the velocity of the Kuroshio observed on the cross-section ranged between 2.5 knots and 4.5 knots. The other five curves show the mean values of the east component of the velocity between a′ and A, A and M, B and b, and the overall mean between a′ and b. The symbol a′ designates the northern limit of the Kuroshio. The location of a′ moved a little during the period of the measurement, but the shift was less than 5 miles.

According to this figure, diurnal variation of the east component of velocity was predominant in the first period of measurement, and the amplitude of fluctuation of mean velocity was about 0.5 knots, which is roughly 25 percent of the overall mean of 2 knots. During the second observation period no well-defined period was apparent, though a semidiurnal fluctuation may have been present.

In October 1965, the same type of observation was carried out with 3 vessels only a few miles east of the line shown in Fig. 12. In this study the main purpose of observation was to investigate the relation between the change of velocity and the surface gradient. The results obtained were:

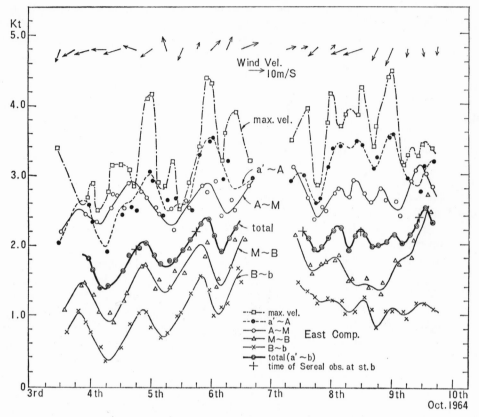

FIG. 13 Very short-term variation of the velocity of the Kuroshio, October 1964.

1) Semi-diurnal variation of the velocity was predominant.

2) Variation of parallel and perpendicular components to the mean direction was nearly of the same order and the velocity vector of the fluctuation rotated clockwise with a semi-diurnal period.

3) The observed velocity fluctuation was apparently nearly geostrophic. The change of sea level or dynamic height was observed at both sides of the Kuroshio.

The Japan Meteorological Agency made a multiple ship survey in October and November 1967 with its fleet of 5 observation vessels (Masuzawa, 1968). Masuzawa noted that diurnal fluctuation of the velocity was predominant with an amplitude of 1 to 1.5 knots. He suggested that this diurnal variation might be caused by inertial motion, the period of which is about 22 hours in this area. He also observed that larger values of the average speed of the Kuroshio appeared at intervals of 4 or 5 days.

There are several fundamental problems to be answered in future concerning these short period fluctuation of the Kuroshio. They are: What is the mechanism of generation of these fluctuations? Are these fluctuations progressive? If progressive, in what direction, upstream or downstream? Yasui (1961)

suggested that an internal wave in the frontal zone could be confined in the stream. In order to study these problems, arrays of observation buoys may be most useful, but for us they are not yet available.

5 CONCLUDING REMARKS

In this very short paper, the very vast problem of time variation of the Kuroshio was discussed. There are many important subjects which have not been described here.

One is the seasonal variation of the Kuroshio which was described in the chapter on Statistical Characteristics. However, it is noted here that the large-scale meander may suffer from seasonal variation in its shape as well as in strength, as suggested by Ichiye (1954). Also, the meander may arise in a particular season, probably in spring.

Another subject is the very strong coastal current which occurs unexpectedly in consequence of the meander of the Kuroshio. Such an incident in 1966 was described by Ishino (1967) and Uda (1953). Interaction of coastal water and the Kuroshio is very important from the practical point of view.

A clockwise vortex appears far off Shikoku, southwest of the large scale meander discussed in Section 2. This warm water vortex may be a twin of the cold water vortex off Enshunada and it may be necessary to treat it as such (Uda, 1949).

It was noted by Yamashita and Yamauchi (1967) that temperature deeper than 600 meters changed prior to the change in position of the Kuroshio. If this is true, it could be a method to predict the path of the Kuroshio.

It must be borne in mind that every fluctuation may play an important role in the variation of a longer period. Case A of Section 3 might be a forerunner of the development of the meander described thereafter as Case B.

It is quite true that the variation of the Kuroshio cannot be understood without understanding the whole Pacific, including the motion of deep water and the atmosphere over it. This is very difficult indeed. But it is also true that oceanographers in Japan and in the world can enjoy themselves in discussing and solving the many problems it presents.

REFERENCES

Fujimori, T. (1964): Preliminary report on the oceanographic anomaly in the early half of 1963 in the water adjacent to Kanto district. *Bull. Tokai Reg. Fish. Res. Lab.*, **38**: 77–98.

Fukuoka, J. (1958): On the Kuroshio near the Izu Islands. *J. Oceanogr. Soc. Jap.*, **14**: 11–14.

Hanzawa, M. (1964): Preliminary report on the abnormal oceanic conditions in the sea adjacent to Japan in the winter of 1963. *Studies on Oceanography*, 59–67.

Ichiye, T. (1954a): On the variation of oceanic circulation (VI). *Geophys. Mag.*, **25**: 185–217.

—— (1954b): On the variation of oceanic circulation (VII). *Oceanogr. Mag.*, **6**: 1–14.

Ishino, M., K. Otsuka, A. Setoguchi and K. Motohashi (1967): Courants aux environs du Japon étudiés au point de vue de l'océanographie de pêche I (in Japanese with French abstract). *La mer.*, **5**: 244–250.

Isozaki, I. (1969): An investigation on the variations of sea level due to meteorological disturbances on the coast of Japanese Islands (III). *J. Oceanogr. Soc. Jap.*, **25**: 91–102.

Masuzawa, J. (1954): On the Kuroshio south off Shiono-misaki of Japan. *Oceanogr. Mag.*, **6**: 25–33.

—— (1958): A short-period fluctuation of the Kuroshio east of Cape Kinkazan. *Oceanogr. Mag.*, **10**: 1–8.

—— (1965): Meanders of the Kuroshio (in Japanese). *Kagaku*, **35**: 588–593.

—— (1968): Cruise report on multiple-ship study of short-term fluctuations of the Kuroshio in October to November 1967. *Oceanogr. Mag.*, **20**: 91–96.

Moriyasu, S. (1958): On the fluctuation of the Kuroshio south of Honshu (2). *Mem. Kobe Marine Obs.*, **12**: 1–18.

—— (1959): Supplementary note on the dynamical property of the cold water region. *Oceanogr. Mag.* **11**: 13–19.

—— (1961): An example of the conditions at the occurrence of the cold water region. *Oceanogr. Mag.*, **12**: 67–76.

—— (1963): The fluctuation of hydrographic conditions in the sea south of Honshu, Japan (Review). *Oceanogr. Mag.*, **15**: 11–29.

Nakabayashi, S. (1970): (personal communication).

Nan'niti, T. (1958): A theory of the mechanism of the generation of the cold water region in the offing of Enshunada. *Pap. Met. Geophys.*, **8**: 317–331.

—— (1959): A supplementary note to the previous paper "A theory of the mechanism of the generation of the cold water region in the offing of Enshunada." *Pap. Met. Geophys.*, **10**: 51–53.

—— (1960): Long-period fluctuations in the Kuroshio. *Pap. Met. Geophys.*, **11**: 339–347.

Nitani, H. (1969): On the variation of the Kuroshio in recent several years (in Japanese). *Bull. Jap. Soc. Fish. Oceanogr.*, **14**: 13–18.

Nitani, H. and D. Shoji (1970): On the variability of the velocity of the Kuroshio—II. In *The Kuroshio—A Symposium on the Japan Current* (edited by J. C. Marr), East-West Center Press, Honolulu, 107–116.

Outi, M. (1966): Long-term variation of the atmospheric and oceanic conditions. *Sp. Contributions, Geophys. Inst. Kyoto. Univ.*, **6**: 69–78.

Shoji, D. (1951): The variation of oceanic currents in the adjacent seas of Japan (in Japanese). *Hydrogr. Bull.*, **25**: 230–244.

—— (1954): On the variation of the daily mean sea level and the oceanographic condition (in Japanese with English abstract). *Hydrogr. Bull.*, Sp. No. **14**: 17–25.

—— (1961): On the variation of the daily mean sea levels along the Japanese Islands. *J. Oceanogr. Soc. Jap.*, **17**: 21–32.

—— (1964): On abnormal low water temperature and the Kuroshio (in Japanese) *Bull. Jap. Soc. Fish. Oceanogr.*, **4**: 31–40.

Shoji, D., R. Watanabe, N. Suzuki and K. Hasuike (1958): On the "Shiome" at the boundary zone of the Kuroshio and the coastal water off Shionomisaki. *Rec. Oceanogr. Works, Jap.*, Sp. No. **2**: 78–84.

Shoji, D. and H. Nitani (1966): On the variability of the velocity of the Kuroshio—I. *J. Oceanogr. Soc. Jap.*, **22**: 10–14.

Suda, K. (1943): *Kaiyo-kagaku* (in Japanese). 720pp.

Uda, M. (1937): On the anomalous condition of the Kuroshio off Kii Peninsula in recent years (in Japanese). *Kagaku* 7(9): 360–361.

Uda, M. (1949): On the correlated fluctuation of the Kuroshio Current and the cold water mass. *Oceanogr. Mag.* **1**: 1–12.

——— (1953): On the stormy current ("Kyutyo") and its prediction in the Sagami Bay. *J. Oceanogr. Soc. Jap.*, **9**: 15–22.

——— (1962): Cyclic, correlated occurrence of world-wide anomalous oceanographic phenomena and fisheries conditions. *J. Oceanogr. Soc. Jap., 20th Year Anniv. Vol.*, 368–376.

Yamashita, Y. and S. Yamauchi (1967): On the Kuro-shio in recent years (in Japanese). *Hydrgr. Bull.*, **82**: 9–22.

Yasui, M. (1961): Internal waves in the open ocean, an example of internal waves progressing along the oceanic frontal zone. *Oceanogr. Mag.*, **12**: 157–183.

Yosida, S. (1961a): On the short period variation of the Kuroshio in the adjacent sea of Izu Islands (in Japanese with English abstract). *Hydrgr. Bull.*, **65**: 1–18.

——— (1961b): On the variation of Kuroshio and cold water mass off Enshu Nada (in Japanese with English abstract). *Hydrgr. Bull.*, **67**: 54–57.

Chapter 8

HYDROGRAPHY OF THE KUROSHIO EXTENSION

HIDEO KAWAI

Department of Fisheries, Kyoto University., Kyoto, Japan.

1 INTRODUCTION

The Kuroshio Extension, the Pacific counterpart of the Gulf Stream, is characterized not only by being free of the continental shelf and encountering Oyashio cold water or its degenerate, but also by very complicated hydrographic conditions: distorted meanders of its path and numerous eddies of various size and duration, especially on its north side. These features have captured the interest of oceanographers as well as fishermen because of the abundance of fishery resources around the Kuroshio Extension.

The description in this chapter is not restricted to the hydrography of the Kuroshio Extension, but extends to that of waters a few hundred nautical miles or more north and south of the Kuroshio Extension. To bring features of the Kuroshio Extension into relief, moreover, an attempt has been made to compare the Kuroshio Extension with the Kuroshio south of Japan and with the Gulf Stream.

In this section, a brief description of the main current system, geography and bottom topography around the Kuroshio Extension is given.

1.1 Main current system around the Kuroshio Extension

By introducing some substantial modifications into Munk's (1950) classification of ocean currents, it is possible to consider the Kuroshio and Oyashio in a wide sense as the western boundary currents and their eastward extensions composing parts of the subtropical and subarctic gyres, respectively, in the North Pacific. This general nomenclature is comparatively suitable when dealing with the Kuroshio and Oyashio in the traditional sense as currents, but it will probably be useful for the following description to determine the implications of some common terms at the beginning of this chapter. Reference to the

significance of the words *kuro* (黒, black) and *oya* (親, parent) in Kuroshio (黒潮) and Oyashio (親潮), respectively (Uda, 1941, 1955, 1964), has lost its novelty, but the duality of the Chinese character *shio* (潮) included in both Kuroshio and Oyashio has rarely been discussed. In ancient China, this character originally meant tide, but later acquired two additional meanings: sea water and current. The terms Kuroshio and Oyashio as commonly used by maritime people also imply waters: northwestern subtropical and southwestern subarctic waters inside the two gyres, respectively. These terms should, strictly speaking, be limited to the currents, but the meaning of water is at times convenient for practical oceanography in the same way as is the usage of Kuroshio and Oyashio Areas and first and second Oyashio Intrusions (Sec. 2). From the scientific standpoint, the terms Kuroshio and Oyashio waters or Kuroshio and Oyashio Areas should be used instead of the Kuroshio and Oyashio unless they indicate the currents.

A similar confusion between current and water has arisen in the term Gulf Stream (Fuglister and Worthington, 1951). The confusion may be ascribed to the fact that people generally prefer to relate terms to easily and directly perceptible properties, such as temperature and color of sea water, rather than to currents which are hardly perceptible in the ocean. The definition of waters, however, is scientifically difficult because of a sharp horizontal gradient in the properties of waters across the Gulf Stream and the Kuroshio Current as well as the Oyashio Current and because of fluctuations in these properties.

The term Japan Current (日本海流) noted by Berghaus (1837) on his geographical map is synonymous with the Kuroshio Current, and the term Kuril Current (千島海流) used by Schrenck (1873) is synonymous with the Oyashio Current (Wüst, 1936). In the past these terms were occasionally used because of geographical familiarity, but they seem to have lost general currency in Japan except in some elementary school textbooks. In the USSR, the term Kuroshio Current seems to be widely used, but not Oyashio Current. For instance, the term Kuril Current was used by Muromtsev (1958), but both the terms Oyashio Current and Kuril-Kamchatka Current are found in Bogorov (1958) and both the terms Oyashio Current and Kuril Current in the "Atlas Sahalinskoi Oblasti" (1967).

The name Kuroshio in the narrow sense has been commonly applied to a continuous band of strong currents flowing toward the northeast, close along the continental slope between Taiwan (Formosa) and Inubozaki, Honshu located at 35°42′N, 140°52′E. The band consists of a sheaf of parallel flows, a few thousand meters deep. The Kuroshio Extension (黒潮続流, Hidaka, 1955) was first defined by Sverdrup, Johnson and Fleming (1942) as the eastward continuation of the Kuroshio beyond Inubozaki, and the North Pacific Current (北太平洋海流) as the further continuation of the Kuroshio Extension which flows toward the east, sending branches to the south and reaching probably as far as 150°W. Ac-

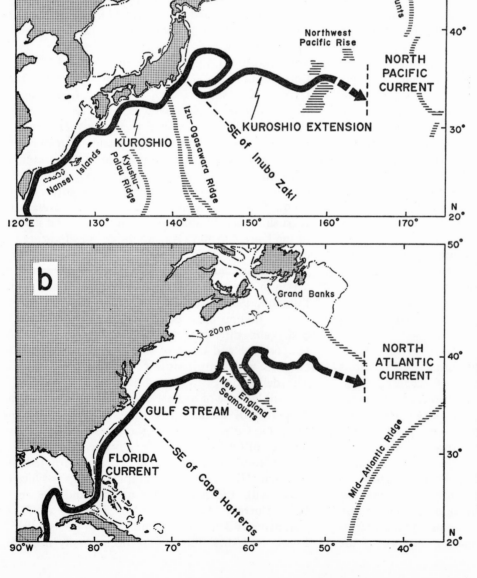

FIG. 1 Subdivisions of the main path of the Kuroshio System (a) and the Gulf Stream System (b). Principally based on results from Norpac in the North Pacific (Norpac Committee, 1960), Operation Cabot in the North Atlantic (Fuglister and Worthington, 1951) and a detailed survey in the Gulf of Mexico (Nowlin, Hubertz and Reid, 1968), the paths are greatly oversimplified, showing neither branch, confluence, countercurrent, nor eddy.

cording to the nomenclature of the Gulf Stream System established by Iselin (1936), the Kuroshio in the narrow sense seems equivalent to the Florida Current, because of the similarity in flowing close along the continental slope toward the northeast; the Kuroshio Extension to the Gulf Stream northeast of Cape Hatteras, because of the similarity in leaving the shelf and flowing nearly due east; the North Pacific Current to the North Atlantic Current, because of the similarity in fanning out before reaching the great seamounts (Figs. 1 and 41e). These similarities, however, are geographical and need to be checked in other ways, such as by comparison of the *T-S* relation of waters on opposite sides of the two streams (Sec. 4.5). It is also important to compare how far east the meanders are affected by the bottom topography (Sec. 7.6).

According to Morgan's (1956) idea about the ocean circulation, the Kuroshio Extension is included in the northern region of decay of the narrow boundary current, where nonlinear, viscous and nonsteady effects may be important. Observations have been made for many years by Japanese ships and since the 1950s by the USSR's (Muromtsev, 1958). Nevertheless, the actual features of the Kuroshio Extension are not sufficiently well known to be interpreted dynamically, owing to defects in the sampling programs and lack of accuracy and indefiniteness in the measurements, which is inevitable at the present level of oceanographic technology.

The Oyashio Current, which comes near the Kuroshio Extension in the inshore Tohoku Area (Sec. 1.2), is less readily described than the Kuroshio because of its smaller scale, more marked fluctuation and proximity to foreign territory. According to Dodimead, Favorite and Hirano (1963, Fig. 109), the name Oyashio Current was given to a southwesterly current along the southeast coast of Hokkaido and the southern Kuril Islands south of the Bussol Strait (Fig. 3), and the name East Kamchatka Current was given to a southwesterly current along the southeast coast of the Kamchatka Peninsula and along the northern Kuril Islands north of the Krusenstern Strait (Sec. 1.3). Reid (1966) used the term "Kamchatka Current" instead of East Kamchatka Current, but this usage is not adopted in this chapter to avoid confusion between currents with similar names, such as the "Kamchatka Current" used to indicate the northward protrusion of warm water at about 160°E south of the Kamchatka Peninsula (Japan Imperial Hydrographic Office, 1918, p. 64) and the West Kamchatka Current (Kurashina, Nishida and Nakabayashi, 1967) used by Japanese oceanographers to indicate the northward current entering from the

Pacific into the Okhotsk Sea through the Onnekotan Strait (also called the Fourth Strait in the USSR) and flowing along the west coast of the Kamchatka Peninsula. According to Dodimead et al. (1963, Fig. 109), the East Kamchatka Current and the Oyashio Current are connected by a detour current in the Okhotsk Sea, and no name was given to the shortcut current on the Pacific side of the central Kuril Islands between the two straits mentioned above.

The current close along the east coast of the Kamchatka Peninsula, therefore, may be called the East Kamchatka Current. The current close along the southern Kuril Islands and along Hokkaido should be called the Oyashio Current in the narrow sense, and the weak shortcut current on the Pacific side of the central Kuril Islands may in a broader sense also be included in the Oyashio Current. On the basis of the terminology given above, the East Kamchatka and Oyashio Currents seem to be equivalent to the Labrador Current in the North Atlantic because of the similarity in flowing south as the western boundary current in the subarctic gyre.

1.2 Geographical subdivision of the Tohoku Area

The sea which will be described in this chapter has often been referred to as the Tohoku Area; it is familiar to fishermen and people concerned with maritime affairs, but is sometimes misunderstood as

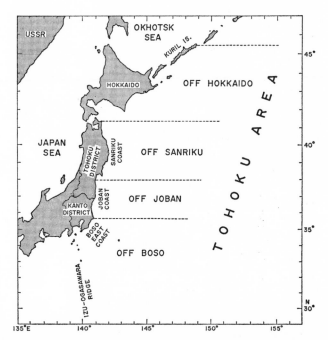

FIG. 2 Geographical subdivision of the Tohoku Area.

the sea along the Pacific coast of the Tohoku District (東北地方), the northeastern district of Honshu. It is difficult to discover where the name Tohoku Area was first used, but Uda and Okamoto (1930) used this name to mean the sea east of Honshu from Inubozaki up to Shiriyazaki (41°26′N, 141°28′E). Kimura (1949) used the name in a wider sense than Uda and Okamoto (1930). Following Kimura's (1949) usage with some modifications, it may be reasonable to interpret Tohoku Area (東北海区) as the northeastern portion of the sea along the Pacific coast of Japan, bounded on the southwest by the Izu-Ogasawara Ridge, on the west by the east coast of Honshu and on the northwest by Hokkaido and the southern Kuril Islands, but with the offshore boundary indefinite (Fig. 2). The "sea east of Honshu" and the "sea east of Japan" are alternate terms, and the latter seems to cover most of the Tohoku Area.

The term Sanriku Coast (三陸海岸) is also familiar to most Japanese, but it is sometimes mistakenly used to mean the Pacific coast of the entire Tohoku District. Strictly speaking, its use should be limited to the Pacific coast of the northern Tohoku District north of about 38°N, because Sanriku is the abbreviation for three provinces—Rikuzen (陸前), Rikuchu (陸中) and Mutsu (陸奥)—established on the east and north coasts of the northern Tohoku District in 1868. Similarly, the term Joban Coast (常磐海岸) is useful for the geography of the Kuroshio Extension. Since Joban is the abbreviation for two provinces—Hitachi (常陸) and Iwaki (磐城)—established on the east coast of the northern Kanto District and of the southern Tohoku District in 646 and in 1868, respectively, the Joban Coast is defined as the coast from Inubozaki up to about 38°N. The Boso East Coast (房総東岸) is defined as the coast from Inubozaki down to Nojimazaki (34°54′N, 139°53′E), because Boso is the abbreviation for three provinces—Shimousa (下総), Kazusa (上総) and Awa (安房)—established in the southeastern Kanto District in the seventh or eighth century. By using the names of the

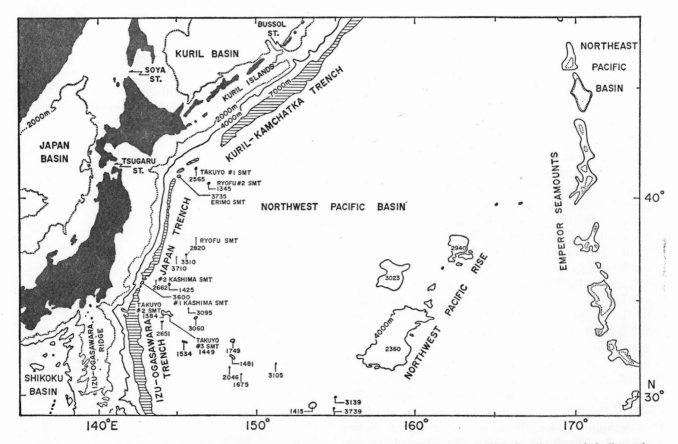

Fig. 3   Topographic features of the Tohoku Area, based on JHO charts 6301–6304 and 1006. Small numerals indicate the depths in meters at the tops of the seamounts. Dotted and solid lines indicate 2000 and 4000 m depth contours, respectively. Hatched belts indicate trenches deeper than 7000 m.

TABLE 1.   Codes used for main surveys and expeditions cited in this chapter.

Code	Period	Agency or Institution	Remarks
Fishpac '33s	summer 1933	JIFES, PFES's, JIHO	12 boats in main area 24 boats in adjacent seas
Ricepac '35w–'38f	winter, summer and fall 1935–1938	JIFES, PFES's	About 10 boats
Navypac '38s–'43s	summer 1938–1943	JIHO, JIFES, PFES's	'38s, south of Japan '39s–'43s, east of Japan
C-Sec	Oct. 1948–Nov. 1953	CMO	Twice a month, 38°N (Fig. 49)
Extra	Oct. 1948–Nov. 1953	CMO	39°N, 153°E (Fig. 49) Weather ship station Stopped in 1953
G-Sec	Oct. 1948–Nov. 1953	CMO	Twice a month, 136°E (Fig. 49)
Tango	Oct. 1948–Nov. 1953	CMO	29°N, 135°E (Fig. 49) Weather ship station Active now only in s & f
Norpac	July–Sept. 1955	International	
IGY Kuront '57	June–July 1957	JMAM, JHO, HMO, KMO	*Ryofu Maru, Kaiyo,* *Yushio Maru, Shumpu Maru*
IGY Poront '57	Aug.–Dec. 1957	HUF, JMAM, JHO, TUF	*Oshoro Maru, Ryofu Maru,* *Takuyo, Umitaka Maru*
IGY Poront '58	June–Sept. 1958	JMAM, TKRFL, HUF	*Ryofu Maru, Soyo Maru,* *Oshoro Maru*
Fish Kuront '58–'60	Apr.–July 1958–1960	THRFL, PFES's	
Jeds 1–	1959–	JMAM, JHO	*Ryofu Maru, Takuyo*
CSK '65s–	summer and winter 1965–	International	
Boreas	Jan.–Mar. 1966	SIO (USA)	*Argo*

s, w and f stand for summer, winter and fall, respectively.

CMO: Central Meteorological Observatory (reorganized into JMA in 1956)
HMO: Hakodate Marine Observatory
KMO: Kobe Marine Observatory
JMA: Japan Meteorological Agency
JMAM: Marine Division, Japan Meteorological Agency

JIHO: Japan Imperial Hydrographic Office (reorganized into JHO in 1948)
JHO: Japan Hydrographic Office
JMSA: Japan Maritime Safety Agency
JODC: Japan Oceanographic Data Center

HUF: Faculty of Fisheries, Hokkaido University
TUF: Tokyo University of Fisheries

JIFES: Japan Imperial Fisheries Experimental Station (reorganized into 8 Regional Fisheries Research Laboratories in 1949)
PFES: Prefectural Fisheries Experimental Station
HKRFL: Hokkaido Regional Fisheries Research Laboratory
NSRFL: Nansei Regional Fisheries Research Laboratory
THRFL: Tohoku Regional Fisheries Research Laboratory
TKRFL: Tokai Regional Fisheries Research Laboratory

SIO: Scripps Institution of Oceanography, U.S.A.

coasts mentioned above, the Tohoku Area is geographically subdivided into four seas off Hokkaido, Sanriku, Joban and Boso (Fig. 2). It is also sometimes convenient for fisheries statistics as well as for hydrography to separate the sea along the Izu-Ogasawara Ridge from the Tohoku Area as the Izu-Ogasawara Area (伊豆・小笠原海区).

In response to the need for research in littoral fisheries, Kawai (1965) worked out a more detailed subdivision of the east coast of Honshu, mainly on the basis of landform, but such a detailed subdivision seems unnecessary for the offshore hydrography in this chapter.

1.3 Topographic features of the Tohoku Area

To give a brief picture of the bottom topography around the Tohoku Area, only the depth contours of 2000, 4000 and 7000 m are shown in Fig. 3 (see Chap. 2 for further details).

After crossing the Izu-Ogasawara Ridge (伊豆・小笠原海嶺), a threshold into the Tohoku Area, the Kuroshio Extension leaves the continental shelf, only five to thirty nautical miles wide out to the 200 m depth contour (Fig. 1). The continental shelf is bounded on the east by the west slope of three of the greatest trenches in the world: the southern Kuril-Kamchatka Trench (千島・カムチャツカ海溝) off Hokkaido, the Japan Trench (日本海溝) off Sanriku and Joban and the northern Izu-Ogasawara Trench (伊豆・小笠原海溝) off Boso. All three trenches exceed a depth of 6000 m everywhere except where they pass through the vicinity of the Erimo Seamount (襟裳海山) and the No. 1 Kashima Seamount (第一鹿島海山), which form sills between the trenches. To the east of the trenches, there is the Northwest Pacific Basin (北西太平洋海盆), 5000–6000 m deep except at a number of seamounts, the tops of which are at various depths between 1300 and 3800 m. The Northwest Pacific Basin is bounded on the east by the Emperor Seamounts (天皇海山列), which form the western boundary of the Northeast Pacific Basin (北東太平洋海盆). The Northwest Pacific Rise (北西太平洋海膨) lends variety to the bottom topography of the Northwest Pacific Basin. Only two passages, the Bussol Strait (ブソーリ海峡) or North Urup Channel (北得撫水道) and the Krusenstern Strait (クルーゼンシュテルン海峡) or Mushiru Strait (牟知海峡), exceed a depth of 1000 m through the Kuril Islands chain, which blocks deep water exchange between the Kuril Basin (千島海盆) in the southern Okhotsk Sea and the Northwest Pacific Basin.

There is a big difference in the bottom topography between the Kuroshio Extension and Gulf Stream areas. No counterparts of the Izu-Ogasawara Ridge and the three big trenches are found in the Gulf Stream area, but the New England Seamounts, the Grand Banks and its long tail, and the Mid-Atlantic Ridge have important effects upon the Gulf Stream meanders (Fig. 1; Sec. 7.6).

1.4 Codes used for main surveys and expeditions

Listed in Table 1 are the codes used for the main surveys and expeditions cited in this chapter. Some of them have been widely used; the remainder are newly devised for this chapter. A list of abbreviations for the names of organizations follows Table 1.

2 NOMENCLATURE

There is a Japanese proverb that says, "The name should show the reality"; the advocacy of new nomenclature in the hydrography of the Kuroshio Extension would be of little use unless there were substantial progress in its hydrographic description, but there should be no hesitation in adopting new terms if the old are not realistic.

Some special words exist for the hydrography of the Kuroshio Extension: some were established before World War II (Uda, 1935a), most of them after the War (Kawai, 1955a, 1959; Dodimead et al., 1963), and the remainder were newly proposed here. A three-dimensional structure of upper water in the Tohoku Area is schematically shown in Fig. 4 as an illustration of the following description. In naming a hydrographic structure, one takes various things into consideration: its configuration, the water characteristics it retains, the geographical or hydrographic location at the time under consideration or at the time when the characteristic features of the structure were built up, traditional usage and so on. Because only structures with some substantial or sustained features, which depend upon the present network of sampling, can be named, the nomenclature given here will improve with the increase in understanding of the hydrography. Some terms that cannot be given without special comment are introduced in the respective sections of this chapter.

2.1 Hydrographic subdivision of the Tohoku Area

There are two remarkable thermal fronts at subsurface depths in the Tohoku Area; the southern is called the Kuroshio Front (黒潮前線), which is nothing but the stream axis of the Kuroshio Extension at the surface; and the northern the Oyashio Front (親潮前線), along which weak easterly flows are once in a while found in fragments. At first the Oyashio Front was defined as the stream axis of a strong easterly current by Kawai (1955a), but it should be redefined as a thermal front or thermal fronts around the western Pacific subarctic water with salinities below 33.60 ‰ at depths of 50–200 m rather than as the axis (Kawai, 1959; Sec. 4.3), because a faster easterly current is occasionally found at the northern edge of the anticyclonic eddies, the edge being sometimes located a little south of the Oyashio Front. The water in the Tohoku Area is hydrographically divided by the two fronts into three areas: the Kuro-

shio Area (黒潮水域), the Perturbed Area (混乱水域) and the Oyashio Area (親潮水域). The southern limits of the Kuroshio Area and the northern limits of the Oyashio Area are indefinite, but it is convenient for practical oceanography to include in them a few hundred nautical miles from the two fronts.

The term "polar front" or "polar front zone" seems to be inadequate for the hydrography of the Tohoku Area because the water north of the front is called neither the polar water nor the arctic water but the subarctic water. I am not in favor of the term "subarctic convergence" because there has been no direct measurement of horizontal convergence along the boundary.

Uda (1935a) called the area almost equivalent to the Perturbed Area *Kongo-suiiki* (混合水域), which

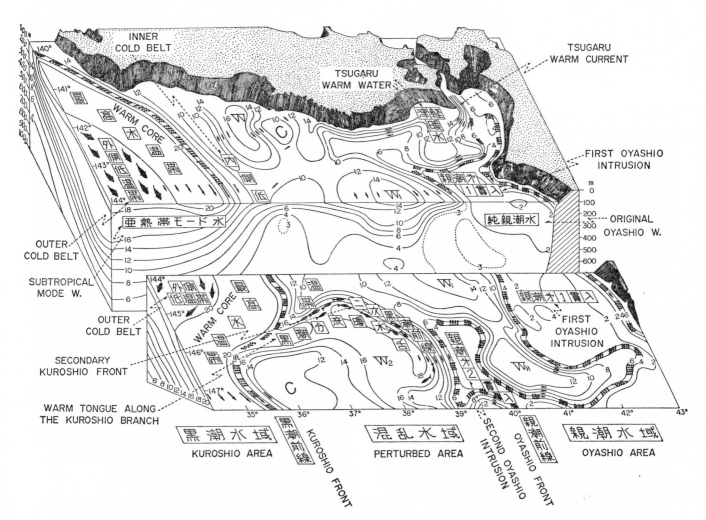

FIG. 4   Block diagram showing the schematic structure of upper water in the Tohoku Area. After the land above sea level and the water in the upper 100 m layer are removed, the block is separated by the meridional section at 144°E into two pieces. Since no deep section was made at the eastern edge of the offshore piece, the piece is intentionally cut down diagonally. Based on observations by JHO, CMO, HMO, THRFL, HKRFL and a couple of PFES's in the summer of 1954.

means a mixing or mixed area; rather than literally translating *Kongo-suiiki*, he used the term "Transition Area" (Uda, 1938a). His usage of these terms was applied to the hydrographic subdivision of the Tohoku Area by Kawai (1955a). However, the new term *Perturbed Area* is proposed here because of the fundamental difference in meaning between *Kongo-suiiki* and "Transition Area," because of poor information on the mixing process in the area, and because the term "transition" lacks specificity.

2.2   Secondary Kuroshio Front

At the southern margin of the Perturbed Area, the Kuroshio Front meanders as the axis of a narrow band of strong current of about three knots at its maximum (Table 7). From some portions of the Kuroshio Front, meandering convex to the north, other fronts branch off occasionally. To distinguish them, the former is sometimes called the primary Kuroshio Front (主黒潮前線) and the latter the secondary Kuroshio Front (二次黒潮前線). The secondary Kuroshio Front has a speed of about two knots at the axis, in the shape of a loop-line surrounding a large body of warm water which often breaks off as large anticyclonic eddies. The horizontal temperature gradient across the primary Kuroshio Front is large even at a depth of 400 m (1.4°C per ten nautical miles on the average, Table 7, but that across the secondary Front is very small at that depth, say about half of the former. In other words, the primary Kuroshio Front has a deeper root than the secondary Front. Along the secondary Kuroshio Front a surface warm tongue is likely to protrude northward. It is called the warm tongue along the Kuroshio branch (黒潮分流暖水舌). The secondary Kuroshio Front as well as the warm tongue along the Kuroshio branch is important for the northward migration of skipjack, *Katsuwonus pelamis*, into the Perturbed Area (Kawai and Sasaki, 1962).

According to Dodimead et al. (1963), the subarctic boundary, which seems to be regarded as a boundary between the subarctic and subtropic regions (their application of the name subtropic region is inadequate), ". . . is defined by the almost vertical isohaline of 34.0‰ which extends from the surface to a depth of about 200 to 400 meters." It may not be an extension of the Oyashio Front but a branch of the secondary Kuroshio Front in the Tohoku Area. The latter seems to be equivalent to the axis of the Slope Water Current (Fuglister, 1963; Mann, 1967) in the North Atlantic. The northern part of eastward currents between the Aleutian Islands and the subarctic boundary has been called the "Subarctic Current"

(Sverdrup et al., 1942; Dodimead et al., 1963). The connection between the secondary Kuroshio Front and the subarctic boundary mentioned above or between the Oyashio Front and the "Subarctic Current" is not clear, and must be investigated by extensive multi-ship surveys in the future.

2.3   Warm core

In most cases remarkable streaky structures are maintained on either side of the Kuroshio Front. The warm core is the most distinct structure. Fuglister and Worthington (1951) defined the warm core in the Gulf Stream as follows:

"The 'warm core' is defined here as that part of the Gulf Stream containing water warmer than the water at the same depth to the right, facing down stream, of the current. This 'warm core' is generally 300 to 400 meters deep with the maximum temperature anomalies at a depth of about 100 meters."

The "characteristic band of high water temperature," a thermal structure equivalent to the warm core, was discovered by Uda (1940, 1943, 1949) in the Kuroshio south of Japan. He regarded it as "a secondary phenomenon due to the invasion of warm water from the southern seas transported by the strong current of the Kuroshio." Fuglister and Worthington (1951) also confirmed that "temperatures of the warm core indicate that the Gulf Stream transports water from south of Cape Hatteras to the longitude of the Grand Banks." The depth of the warm core in the Kuroshio and the Kuroshio Extension is about 200 m, shallower than in the Gulf Stream.

In the same way as in the blocking action of the upper westerlies in the atmosphere (Rex, 1950a, 1950b), the warm core seemed to be broken down by much distorted meanders of the Kuroshio Extension in the summer of 1942, in the spring of 1954 and in the summer of 1960 (Sec. 7.3), when it was difficult to delineate the warm core. Because of its transient character, the investigation of this possibility must be the subject of well-planned synoptic surveys in the future.

The mean temperature in the upper 200 m layer at the axis of the warm core does not decrease monotonically downstream, but shows submaxima at intervals of a few hundred nautical miles along the axis. Fuglister and Worthington (1951) called a fragmental procession of the oval bodies of water warmer than the water either upstream or downstream in the warm core the "gobs." Inspection of Fuglister and Worthington's figure (1951, Fig. 2) shows that there are not necessarily anticyclonically circulating flows around the gobs, but rather there are flows parallel to the Gulf Stream. This may suggest a downstream translation of gobs. The sampling network has been too rough with respect to both time and space, preventing a detailed description of the gobs in the Kuroshio Extension.

2.4   Inner and outer cold belts and cold core

The inner cold belt (内側低温帯) is a belt of cold and low-salinity water, from a few to thirty or forty nautical miles wide, found at a depth of about 100 m on the left-hand side, facing downstream, of the Kuroshio Extension. On the basis of surveys in 1952 and 1953, Kawai (1955b) suggested that "the inner cold zone [belt] has the pattern as if the Second

Branch of the Oyashio [second Oyashio Intrusion, Sec. 2.5] is stretched towards the east by the jet stream along the Kuroshio Front," and mistakenly interpreted this fact "as an illustration of Sverdrup's opinion that 'the Intermediate Water of lower salinity does not flow directly south but flows toward the east' (Sverdrup et al., 1942)." Because of the difference in thermosteric anomaly between the inner cold belt as a whole and the North Pacific Intermediate Water (Sec. 4.4), it is, strictly speaking, not the inner cold belt but the cold core in the Kuroshio Extension (defined below) that gives plain indication of eastward flows of the Intermediate Water proposed by Sverdrup et al. (1942). A core of cold, low-salinity and high-oxyty water at depths between 200 and 500 m, appearing in Masuzawa's (1955a, Fig. 4; 1955b, Fig. 7; 1956a, Fig. 2; 1957a, Fig. 5) and Masuzawa

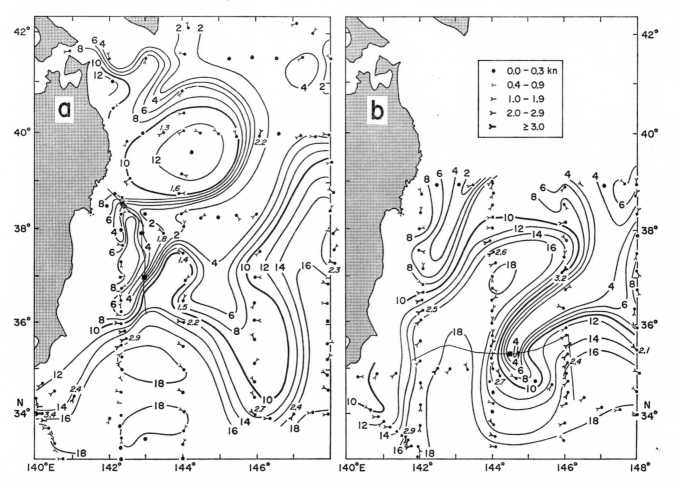

FIG. 5   Surface current from GEK and 200 m temperature from BT in August 1967 (a) and in September 1966 (b). The thin lines indicate locations of Nagata's (1970) detailed temperature section with BT (a) and of LaFond's (1968) temperature section with thermister chain (b). Solid squares indicate locations of the shallow cold cores with temperatures below 2°C (a), 3°C (west one in b) or 4°C (east one in b). The neighbor synchronous index $\Delta d/\delta t$ (Sec. 7.5) is greater than 0.1 kn for all stations except for Nagata's section (a), around which the isotherms may be deformed owing to small values of $\Delta d/\delta t$, and LaFond's section (b).

and Nakai's (1955, Fig. 7) detailed sections, may be called the cold core in the Kuroshio Extension (Sec. 5.4) to distinguish it from the inner cold belt at a depth of about 100 m. Both of these may be associated with each other, but a dislocation, not only in depth but also in relative position to the Kuroshio Extension, is frequently found between them. As pointed out by Masuzawa and Nakai, the cold core in the Kuroshio Extension flows with speeds of twenty to fifty centimeters per second in the same direction as the warm water in the Kuroshio Extension; the water at the southern edge of the inner cold belt flows in the same direction as the Kuroshio Extension, but the water at the northern edge flows as a countercurrent to the Kuroshio Extension (Sec. 5; Masuzawa, 1956b, Fig. 8). Accordingly, the inner cold belt seems to be an important constituent of the multiple current (Sec. 3.6).

With records from BT lowerings at intervals of about one nautical mile at 143°E in the inshore Tohoku Area, Nagata (1970) presented an extremely detailed temperature profile. He reported that "Especially, very cold water is located in both margins of the cold-water belt and is accompanied by pronounced temperature inversion layers, with magnitudes exceeding 2°C," and that the very cold water at the southern margin ". . . is located along, and just below, the sharp thermocline consisting of the Kuroshio front." He also assumed that "the very cold water and its surrounding water in the southern margin travels eastward to the eastern end of the cold-water belt and then travels westward along the northern margin of this belt," becoming more diffuse and splitting into several cores. He did not show any isotherms at subsurface depth but only cited surface isotherms from a prompt report of the JHO to relate his detailed thermal structure to the hydrography around it. Comparison of his temperature profile with the isotherms at a depth of 200 m in August 1967 (Fig. 5a) leads to a different interpretation of the hydrography. What he interpreted as the cold-water belt along the northern edge of a broad Kuroshio Extension may be an extension of the second Oyashio Intrusion (Sec. 2.5) north of a winter-beaten anticyclonic eddy (Sec. 7.3) in the Perturbed Area north of the Kuroshio Extension.

Inspection of continuous temperature profiles from 35°20′N, 144°02′E to 35°26′N, 145°13′E in September 1966, prepared by LaFond (1968, p. 19) with data from the thermister chain shows that very cold water below 3 or 4°C, the cold core, is located at depths of 130–200 m at both margins of the second Oyashio Intrusion or its extension appearing in the isotherms at a depth of 200 m in September 1966 (Fig. 5b). Thus,

the very cold waters appearing in both Nagata's and LaFond's profiles are interpreted as the cold cores at both margins of the Oyashio Intrusions or their extensions. Since these cold cores are centered at shallower depths than the cold core in the Kuroshio Extension is, the former may be called the shallow cold core and the latter the deep cold core. The deep cold core is a result of advection of the Original Oyashio Water (Sec. 5.5) or North Pacific Intermediate Water along the 125-cl/ton surface, but whether the thermosteric anomaly of the shallow cold core is larger than 125 cl/ton is a question that must await further surveys with close-spaced STD lowerings.

Although the "cold low salinity water along the edge of the Gulf Stream" (Ford, Longard and Banks, 1952) is located at depths shallower than 200 m, it may correspond to the deep cold core in the Kuroshio Extension, because it is entrained in, and in fact a part of, the Gulf Stream. The temperature of about 6°C observed in the cold core in the Gulf Stream seems to be higher than that of the cold core in the Kuroshio Extension, even though its fluctuation is expected. The difference in temperature as well as in depth between the cold cores in both the streams is ascribed to different hydrographic conditions.

The outer cold belt (外側低温帯) is a broad belt of relatively cold water on the right-hand side, facing downstream, of the warm core, appearing at a depth of about 100 m. The warm core and the outer cold belt are the obverse and reverse of the same phenomenon, but the latter is more superficial and less definite. In the inner cold belt the vertical gradient of temperature in the upper 200 m layer is very large, while in the outer cold belt it is relatively small, because the outer cold belt is situated in the northern portion of the North Pacific Subtropical Mode Water (Sec. 4.1). In late winter or early spring, a thick mixed layer is found in the outer cold belt, especially to the south of Japan (Sec. 4.1). The outcrop of the outer cold belt in the cold season is veiled with surface warm water above the summer thermocline in the warm season. The outer cold belt is also important for the formation of very productive fishing grounds of albacore tuna, *Thunnus alalunga*, caught with pole-and-line (Kawai, 1955b; Kawasaki, 1957).

2.5 First and second Oyashio Intrusions

At the northern margin of the Perturbed Area, very much distorted meanders of the Oyashio Front are found. So far, two southward intrusions of Oyashio water, one approaching within a distance of 100 nautical miles from the Sanriku Coast, and the other going southward offshore of warm water bodies surrounded by the meandering Oyashio Front convex to the north, have been conventionally called the "First Branch of the Oyashio" (親潮第1分枝) or "Inshore Branch of the Oyashio" (親潮接岸分枝) and the "Second Branch of the Oyashio" (親潮第2分枝) or "Offshore Branch of the Oyashio" (親潮沖合分枝), respectively. The names, however, are not suitable for hydrographic description, because they have often been mistakenly interpreted as if two branches of the Oyashio Current flowed along the thermal ridges of the two intrusions, though no significant currents exist along the ridges if the state is steady. Even in a nonsteady state, the flow speed along the ridges is estimated at roughly 0.1–0.2 kn (Sec. 7.4). For this reason, I would like to propose calling them the first Oyashio Intrusion (親潮第1貫入) and the second Oyashio Intrusion (親潮第2貫入), respectively.

The Oyashio Front does not seem to be a single front, but seems to be divided into twig-like detour paths and shortcuts, especially around the southern ends of the Oyashio Intrusions, though the details cannot be observed with the present network (Fig. 6d). Observations along a couple of watch sections with stations spaced at intervals of ten nautical miles off the Sanriku Coast maintained by the Iwate PFES for many years show that the width of the first Oyashio Intrusion parallel to the Sanriku Coast is at times very narrow, accompanied by a cold core with temperatures below 2°C, as suggested by Nagata (1970). For instance, an apparently isolated cyclonic eddy off the southern Sanriku Coast often communicates by a bottleneck-like first Oyashio Intrusion with the Oyashio Area on the north. This is of importance to the southward autumn migration of Pacific saury, Cololabis saira. As mentioned in Sec. 2.4, there may be a cold core along the boundary of the Oyashio Intrusion, especially along the western boundary, and this must be investigated by close-spaced synoptic surveys in the future.

2.6 Tsugaru Warm Current

A warm current flowing from the Japan Sea into the Pacific through the Tsugaru Strait has been called the Tsugaru Warm Current (津軽暖流). According to Hata, Hosoda and Yamamoto's (1964) estimation based on direct measurement with current drogues at the surface and at 150 m and some assumptions, the volume transport through the western inlet of the strait amounted to 1.4×10^6 m³/sec in the summer of 1962. The Current keeps its direction between Shiriya-zaki and Erimomisaki (41°55′N, 143°15′E) for some distance, then turns clockwise, and finally flows

southward, close along the Sanriku Coast. The warm water surrounded by the clockwise bending of the Current path is called the Tsugaru Warm Water, and is about 200 m thick owing to the shallow sill depth of the strait of about 130 m (Chap. 2). The temperature of the Warm Water changes annually, but its boundary at a depth of 100 m seems to coincide with the 33.80-‰ isohaline all the year round (Sugiura, 1958; Kawai, 1958a). The speed at the current maximum in the vicinity of the boundary is about one knot in winter and spring, and about two knots in summer and fall (Sugiura, 1958). The Tsugaru Warm Water occasionally comes in contact with the first Oyashio Intrusion in the seas south of Erimomisaki and east of the Sanriku Coast, and it may produce a faster current between the two waters.

2.7 Perturbed Area and eddies

The most complicated hydrographic conditions are encountered in the Perturbed Area, where numerous eddies and thermohaline fronts are irregularly distributed. To account for a wide variety of observed features in the Perturbed Area, Barkley (1968) suggested the concept of the compound vortex street, which consists of two Von Kármán vortex streets arranged side by side. It is difficult to comment upon his hypothesis because of the incompleteness of observation and theory.

The cold eddy off Enshunada shows little displacement except in the phases of generation and expiration (Chap. 7). Eddies in the Tohoku Area, however, wander about so much that naming eddies on the basis of their geographical or hydrographic location makes little sense because of the difficulty in identification. Nevertheless, some of the names, for instance the warm eddy off Hokkaido (北海道沖暖水渦), which is also known as the warm eddy off Kushiro (釧路沖暖水渦) and the cold eddy off Boso (房総沖冷水渦), are useful for practical hydrography.

It is a characteristic feature of the Perturbed Area that there are at times detached large-scale anticyclonic eddies with a diameter greater than 100 nautical miles (Secs. 7.3 and 9.1), never found in other regions of the Kuroshio System. This is due to a wide space from the Kuroshio Extension to the northern continental shelf. A similar eddy may be found in the sea southwest of Kyushu, but its diameter and thickness are far smaller. In the Gulf Stream System, large-scale anticyclonic eddies of similar diameter are detached in the Gulf of Mexico (Nowlin, Hubertz and Reid, 1968), in the Slope Water (Iselin and Fuglister, 1948) and in the sea east of the Grand Banks and its long tail (Worthington, 1962;

Mann, 1967), where a condition similar to that in the Perturbed Area in the North Pacific is satisfied (Fig. 1).

Although we can hardly observe the process in detail with the present techniques, four ways of supplying warm water into the Perturbed Area are to be expected:

—(i) Anticyclonic eddies detached from the upper water in the Kuroshio Area.

—(ii) Countercurrents separated from the main path of the Kuroshio Extension (Sec. 3.6).

—(iii) Horizontal spreading of warm water through the upper mixed layer above the summer thermocline.

—(iv) Inflow of warm water by the Tsugaru Warm Current.

3 CHANGE IN SCHEMATA OF THE KUROSHIO EXTENSION

There have been various ideas about schemata for the Kuroshio System as well as the Gulf Stream

System, including Fuglister's (1955) study entitled "Alternative analyses of current surveys." To review the historical change in ideas, seven typical schemata of the upper structure of currents and waters of the Kuroshio Extension are presented in Figs. 6a–g. It is outside the scope of the present chapter to deal in any detail with the broad and vague current schemata given in early days by Schrenck (1873) and Wada (1916), the originals being unavailable to me.

Generally speaking, hydrography schemata may evolve with improvement of the sampling network of synoptic surveys (Sec. 10). But no matter how close in both space and time the network might become, it would leave gaps to filter through the mesh; and no matter how gridlike and synchronized the network might become, the mesh would remain more or less deformed with respect to both space and time. Differences in interpolation and interpretation of the data for the gaps and for the deformations of meshes would produce a variety of schemata. In the construction of any schema, furthermore, one should omit minute or transient features and exaggerate some substantial or sustained features hidden by turbu-

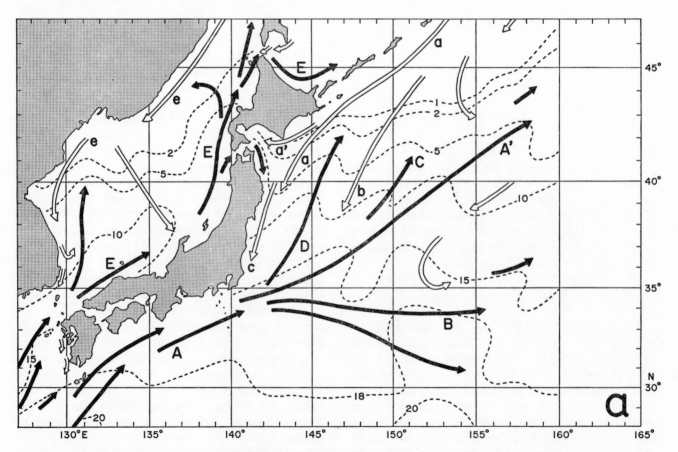

FIG. 6(a)   Alternately meshed fingers (Kajiyama, 1920, Fig. 35), a schema of the upper structure of currents and waters around the Kuroshio Extension. Dashed lines indicate surface temperature (°C) in February.

lence; differences in this adoption and rejection lead to further variation. One of the causes of misinterpreted schema may be a paucity of information on the relation between temperature or salinity field and current field.

3.1 Alternately meshed fingers
(Kajiyama, 1920)

According to Uda (1955), several textbooks of oceanography were published in Japanese from the end of the nineteenth century to the early twentieth century. The appearance of Kajiyama's "Japanese Oceanography" in 1920 was significant, because of its efforts to give the author's own interpretation of the hydrography adjacent to Japan. On the basis of JIHO's compilation of surface currents reported by Japanese ships from 1882 to 1910 and of observations in the sea adjacent to Hokkaido on board the *Tankai Maru* of the Hokkaido PFES, Kajiyama (1920, Figs. 34 and 35) prepared somewhat schematic current charts in August and February. Figure 6a is an adaptation from his February chart.

This schema appeared with the following background: Since the days prior to World War I (1914–1918), the Japanese government had adopted a shipping policy designed to maintain or extend regular service lines in the high seas and to support shipbuilding industries. Of course, the British Empire held the first position in the shipping world, but Japan's shipping circles had taken a turn for the better. Also, since the first fishing boat equipped with an oil engine was built in the Shizuoka PFES in 1906, the skipjack fishing grounds had been extending offshore in the Tohoku Area until the outbreak of World War II (1939). Measurements of surface temperature along the great circle route between Japan and North America and along nearly east–west tracks between fishing ports and grounds are apt to result in overlapped meandering isotherms in the Tohoku Area. The schema seems to have tacitly assumed that a series of tonguelike isotherms are caused by a single current along a ridge or trough line of the meandering isotherms (indicated by dashed lines in Fig. 6a). The axes of convex and concave isotherms correspond to the Kuroshio and Oyashio branches in

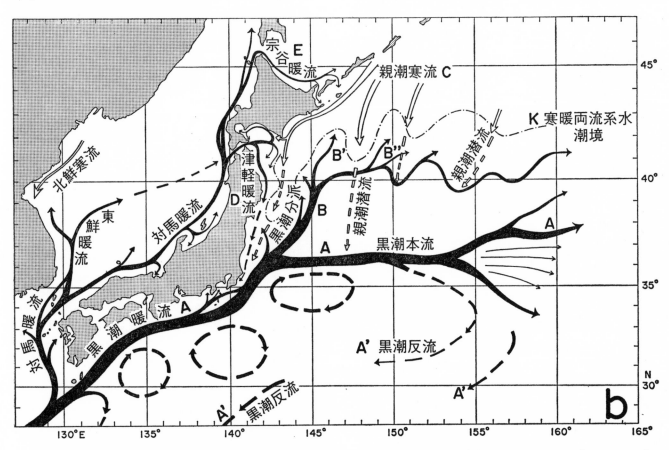

FIG. 6(b)   Forklike currents without confluence (Uda, 1935a, Fig. 49b), a schema of the upper structure of currents and waters around the Kuroshio Extension.

Fig. 6a, respectively. This is an example of the misinterpretation caused by a paucity of information on the relation between surface temperature and current distributions.

Schemata of the current system given by Uda and Okamoto (1930, Fig. 4), not extending beyond the inshore Tohoku Area, and current charts given by three Japanese textbooks of oceanography published early in the 1930s (Nomitsu, 1931, Fig. 118; Marukawa, 1932, Fig. 55; Suda, 1933, Fig. 101) showed patterns similar to Kajiyama's chart.

3.2 Forklike currents without confluence (Uda, 1935)

Around the 1930s there were some attempts to prepare temperature, salinity and oxyty (Montgomery, 1969; Lyman, 1969) profiles approximately along meridional sections in the western North Pacific. On the basis of data from the *Challenger*, the *Planet* and so on, Wüst (1929, 1930) prepared long and deep temperature, salinity and oxyty profiles along two meridional sections in the western and central Pacific. Using data averaged by season, by station and by depth, Uda (1929, 1933) prepared temperature and σ_{15} profiles approximately along the Kuroshio path from Taiwan to Nojimazaki and along the east coast of Japan from Nojimazaki to Etorofu Island, and touched upon the difference in stratification between waters along the south and east coasts of Japan. Kishindo (1940) prepared temperature, salinity and oxyty profiles approximately along a meridional deep and long section across the Tohoku Area at about 150°E. Because they were based on data from various seasons or years, and because the stations were spaced widely, they were not suitable for the purpose of discussing the structure of the Kuroshio Extension. Since few other cross-sections were made at that time, it was vaguely believed that a broad thermal front crossed the Tohoku Area.

During the summer of 1933 a large-scale survey of the Kuroshio System in the Tohoku Area and its neighboring seas was made with the cooperation of many research ships of the JIFES, several PFES's and the JIHO in order to study hydrographic conditions

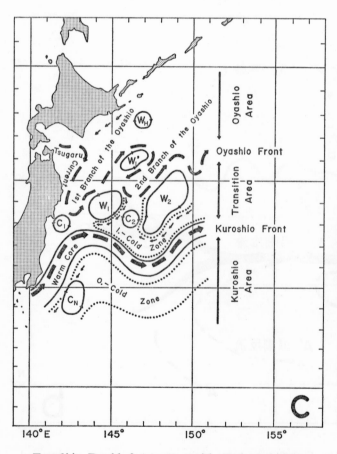

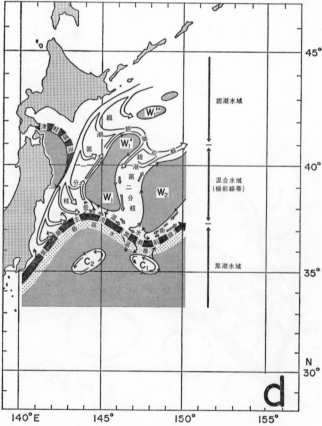

FIG. 6(c)   Double front system with streaks and eddies (Kawai, 1955a, Fig. 2), a schema of the upper structure of currents and waters around the Kuroshio Extension.

FIG. 6(d)   Modified double front system with streaks and eddies (Masuzawa, 1957b, Mimeogr., Fig. 1), a schema of the upper structure of currents and waters around the Kuroshio Extension.

in the fishing grounds which extended as far east as a thousand nautical miles or more from the coast (Fishpac '33s in Table 1). This survey marked an epoch in hydrographic researches on the Kuroshio and its Extension. On the basis of the survey results Uda (1935a, Fig. 49b) proposed a schematic picture of the Kuroshio System, shown for the most part in Fig. 6b, which until recently was regarded as one of the most authoritative expressions of the system.

Uda noticed from the temperature and salinity at several standard depths that:

"There is a clear discontinuous boundary zone of an undulating form between the Oyashiwo and the Kuroshiwo water-masses in the layers above a depth of 500 m. to the east, off the Sanriku coast. The boundary lies on a zone at about 41°–42°N on the surface of the sea [indicated by K in Fig. 6b], the positions changing in a southerly direction as we descend deeper below the surface, and on a zone between about 36°–37°N at a depth of 400 m."

He called the eastward current off Joban, indicated by A in Fig. 6b, the main Kuroshio Current and the current which flows toward the northeast along the Joban Coast and bends toward the east off the Sanriku Coast, indicated by B in Fig. 6b, the Kuroshio branch. The southern current is comparatively steady, but the northern seems to fluctuate much in transport and in position, year by year and month by month. The Kuroshio Countercurrent, indicated by A' in Fig. 6b, stems from the main Kuroshio Current at about 155°E, at first turning to the south and afterwards going to the west. The Kuroshio branch forms a complex pattern of offshoots, as indicated by B' and B'' in Fig. 6b. The Oyashio, indicated by C in Fig. 6b, flows southward in three branches off the Sanriku Coast, the first at about 50 nautical miles' distance from the coast, the second at about 147°E and the third at about 150°E. Uda believed that these Oyashio branches dive at the boundary line K and turn into the "Oyashiwo Undercurrent," indicated by open dashed arrows in Fig. 6b, two of them reaching the main Kuroshio Current. The Tsugaru Warm Current, indicated by D in Fig. 6b, is distinguishable in the sea west of a straight line through Erimomisaki and Samekado (40°32'N, 141°35'E).

In a comparative study of the Kuroshio and the

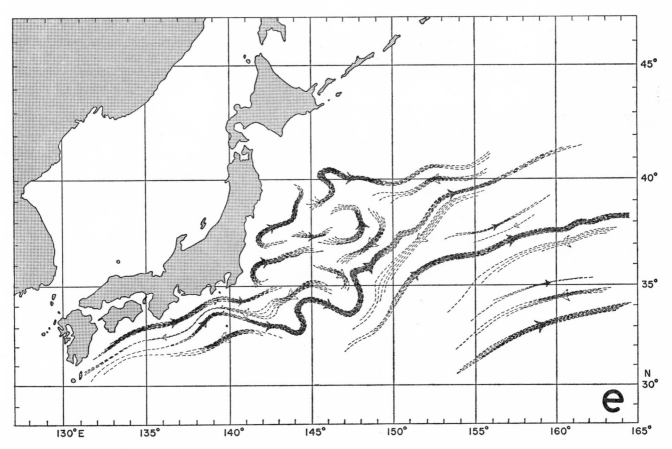

FIG. 6(e)   Turbulent shingle filaments (Fuglister, 1955, Chart 4d), a schema of the upper structure of currents and waters around the Kuroshio Extension.

Gulf Stream, Wüst (1936, Tafel zu S. 17) showed a schema with the Kuroshio branch after Uda's. Incidentally, Wüst's schema of the East China Sea, Yellow Sea and Japan Sea was based on another of Uda's schemata (1934, Fig. 48b).

Sverdrup et al. (1942) followed Uda's interpretation fairly closely in their description of the Kuroshio Extension, which they regarded as a major branch of the Kuroshio:

"In lat. 35°N, where the Kuroshio leaves the coast of Japan, it divides into two branches; one major branch turns due east and retains its character as a well-defined flow as far as approximately long. 160°E, and one continues toward the northeast as far as lat. 40°N where it bends toward the east."

Uda's interpretation, however, is characterized by forklike currents excluding any significant confluence on the north side of the Kuroshio and its Extension, and only giving confluence produced by great anticyclonic eddies on the south side. This interpretation appears to be based on the assumption that the Kuroshio Extension is a current decaying monotonically downstream. Making a comparison of various interpretations possible in contouring synoptic charts with alternative points of view, Fuglister (1955) added comments upon Uda's schema:

"A third type of analysis, usually used only to cover very large areas, tends to show that, although numerous currents exist in an area, they all stem from a single current."

In 1934, farmers in the Tohoku District had an extremely poor rice crop owing to a very cool summer. As it had been proposed by many authors that the cool summer was associated with cool water in the sea, a series of large-scale surveys were made in the Tohoku Area from 1935 to 1938 (Ricepac '35w–'38f). Following Ricepac a further series of large-scale surveys were made in the Tohoku Area every summer from 1939 to 1943 under the leadership of the JIHO (Navypac '39s–'43s), the data being kept confidential, but no other schema of the current system was given than Uda's (1938a) simplified schema, cited in the following, on the basis of data from Ricepac.

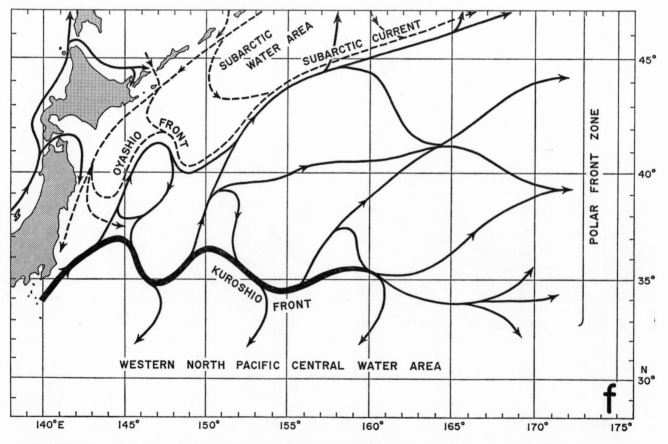

FIG. 6(f)　Entangled arabesque design (Masuzawa, 1960a, Fig. 206), a schema of the upper structure of currents and waters around the Kuroshio Extension.

3.3 Double front system with streaks and eddies (Kawai, 1955; Masuzawa, 1957)

Prior to the appearance of the profiles mentioned in the beginning of Sec. 3.2, during the summer of 1927, the *Manshu* of the JIHO occupied two rather meridional sections: one from Nojimazaki to a station 500 nautical miles east of Etorofu Island, the other from a station off Kinkazan Island (38°16′N, 141°35′E) to about 31°N, 153°E (Shigematsu, 1932). Temperature profiles along the sections show two subsurface thermal fronts corresponding to the Oyashio Front in the northern part of the first profile and to the Kuroshio Front in the second. The reason why the two fronts did not appear in the first profile is that the section was almost parallel to the Kuroshio Front.

On the basis of the isotherms at the surface and at a depth of 100 m contoured by Uda (Oceanographical Charts, No. 168, 1933, JIFES) and the temperature profiles prepared on the basis of data from the *Manshu*, the same data that Shigematsu (1932) used, Suda (1936) suggested that in summer there were two

fronts in the Tohoku Area, the "polar front" in the northern area and the "secondary polar front" in the southern area, but that in winter or even in summer below a depth of 150 m the polar front retreated southward in parallel with the secondary polar front, which was supplanted by the polar front.

In researching current-rips in the waters adjacent to Japan, Uda (1938a, Fig. 1) showed two separate lines of boundary between different waters in the Tohoku Area, which were extremely simplified, indicating neither streaky structure nor eddies. He named the northern boundary "subarctic convergence," but to the southern gave no name. He called the sea between the two boundaries the "Transition Area," which he considered to be filled with a mixture of Kuroshio warm water and Oyashio cold water. In another paper (Uda, 1938b), which followed almost the same schema as that of the forklike current without confluence (Uda, 1935a), he confirmed, on the basis of data from Ricepac '35w–'37w, that there were two eastward currents through the year in the velocity profiles along the meridional section at about 144°E, and that the southern current did not fluctuate

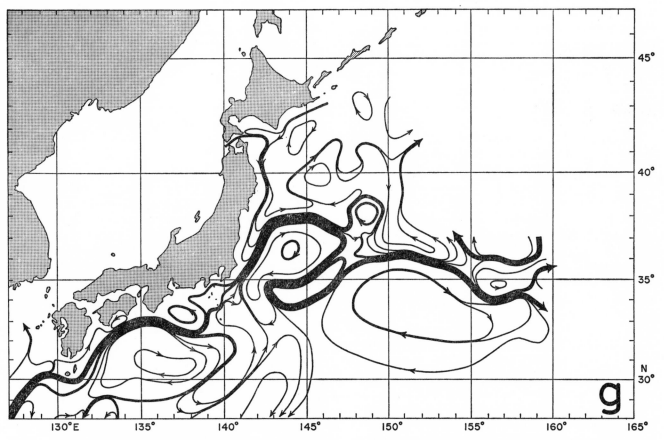

FIG. 6(g)   Modified multiple currents, a schema of the upper structure of currents and waters around the Kuroshio Extension. The width of streamline is proportional to the vorticity transport at the surface.

very much, but the northern did year by year and season by season, in general shifting northward in summer and southward in winter. On the basis of many temperature profiles along two sections south of Kushiro (42°58′N, 144°23′E) and between Nojima-zaki and Kushiro, Uda (1943) again showed a two-boundary structure and proposed two new terms: the "Pure Kuroshio Area" and "Pure Oyashio Area."

The studies mentioned above and the information cited below gave rise to the schema of the double front system with streaks and eddies shown in Fig. 6c (Kawai, 1955a, Fig. 2).

In June 1952, a very productive fishing ground of albacore tuna appeared in the sea east of Inubozaki. Plenty of surface temperature records were collected from the logbooks of a number of fishing boats which made close-spaced measurements of surface temperature on their crisscrossing tracks between the ports and grounds. Synoptic analysis of the surface temperature for every five-day period led to the conclusion that the warm core existed off Joban, and to the prediction that a jetlike stream might exist along the warm core by some analogy with the Gulf Stream (Fuglister and Worthington, 1951). On the basis of the above analyses and of the data from the zigzag survey of upper thermal structure tracking the warm core in the Kuroshio Extension in 1953 and 1954 on board the *Soyo Maru* (TKRFL), this schema was proposed (Kawai, 1955a, 1955b). Strenuous and thorough zigzag-type surveys on board the *Ryofu Maru* (CMO), starting in November 1954, also confirmed it (Masuzawa, 1955a, 1955b, 1956a, 1956b). The schema in Fig. 6d suggested by Masuzawa (1957 b, Fig. 1) is similar to that in Fig. 6c, but it differs in showing offshoots of the Oyashio Front and another cyclonic eddy south of the warm core.

The schema in Fig. 6c is characteristic of streaks of the warm core and the inner and outer cold belts along the Kuroshio Front (Sec. 2) and of several eddies. Until the appearance of the schema, no one had suggested the existence of the warm core in the Kuroshio Extension, because the Kuroshio Extension had been interpreted, on the basis of hydrographic data from sections almost parallel to it except for a crosscurrent section south of Kushiro made on board the *Soyo Maru* before World War II, as a spreading current system with velocities much weaker than that of the Kuroshio south of Japan. The Kuroshio branch off Sanriku in Uda's schema (Fig. 6b) is substituted by the northward currents composed of a flow along the Oyashio Front at the eastern edge of the first Oyashio Intrusion and segments of flows at the northwestern edges of two anticyclonic eddies W_1 and W_1' in Fig. 6c. A Kuroshio branch on a scale

as large as shown in Uda's schema occurs rarely though one on a smaller scale may occur more often. The schema for a branching phase such as occurred in May 1954 was omitted in the papers (Kawai, 1955a, 1955b) because of its transient character.

There is another difference between the two schemata. In Uda's schema (Fig. 6b), no southward flow at the surface is shown between two eastward currents, the main Kuroshio Current and the Kuroshio branch flowing east at about 40°N, though a southward current below the surface, the "Oyashiwo Undercurrent" at about 147.5°E, is shown. In the schema of Fig. 6c, two systems of southward flows at the surface are shown in the Perturbed Area east of the first Oyashio Intrusion. One is composed of segments of flows between the eastern edges of two anticyclonic eddies W_1' and W_1 and the western edge of the second Oyashio Intrusion and its extension, the other is the flow at the eastern edge of a big anticyclonic eddy W_2.

This schema did not show explicitly any way of replenishment of warm water from the south into the Perturbed Area except by detaching the anticyclonic eddies and any return flow from the Perturbed Area to the Kuroshio Front. In other words, the schema laid undue emphasis upon zonal structure along the Kuroshio Front as a reaction against Uda's interpretation. Furthermore, the definition of Oyashio Front in this schema is not correct, as discussed in Sec. 2.1.

3.4   Turbulent shingle filaments
(Fuglister, 1955)

Figure 6e is Fuglister's schema of the Kuroshio System based on an interpretation of the 200 m temperature field obtained from the same data that Uda used in his analysis. The schema essentially dismisses the idea that there is necessarily a continuous single current through the zone which separates the cool and warm water areas (Fuglister, 1955, Chart 4d), introducing a turbulent distribution of disconnected and overlapped filaments associated with intervening filaments of relatively weak countercurrents. The possibility of a filament structure for the Gulf Stream System was first suggested by Fuglister (1951a) in his multiple current hypothesis derived from considerations of temperature data alone. Von Arx, Bumps and Richardson (1955) described current filaments of smaller scale with a length of 500 to 100 km or less on the basis of data from the Stommel-Parson airborne radiation thermometer (Stommel, von Arx and Parson, 1953) together with photographic and visual observations to determine the position of the frontal outcrop on the sea surface and from serial sections on board the *Caryn*. They called a series of current filaments overlapping each other "shingle" structure, although they did not show any evidence of counter-currents.

Stommel's (1965) opinion was that "the interpretation [in Fig. 6e] is a bit forced, from an attempt to spread isotherms apart wherever possible," but the isotherms, which gave rise to the schema in Fig. 6e, seem to be closed up into each current filament. From the results of "Gulf Stream '60," however, Fuglister (1963) admitted that an extremely complicated schema such as shown in Fig. 6e is certainly not a correct interpretation of the data.

The interpretation is based on the 200 m temperature field in the summer of 1933 alone, and no comparison could be made between the 200 m temperature field, distribution of surface current measured with GEK and the dynamic topography of the surface, owing to the limitations of the data at that time. This, as well as an inadequate orientation of hydrographic sections, may have resulted in an unrealistic pattern for the Kuroshio Extension which seemed somewhat curious to Japanese oceanographers, although current filaments with a length of 500 nautical miles or more are certainly probable, as will be shown in Sec. 3.6.

No matter how curious the schema seems, Fuglister's study should be appreciated for the fact that it suggests concretely that the schema for ocean currents varies with the time and space scale employed in the analysis and that the possibility of turbulent shingle filaments should receive due consideration in any analysis.

3.5   Entangled arabesque design
(Masuzawa, 1960)

Figure 6f shows most of Masuzawa's schema (1960a, Fig. 206) appearing in "Encyclopedia of Oceanography" (*Kaiyo no Jiten*) published in Japanese principally with the cooperation of JMA oceanographers. The schema serves as an illustration of current around the so-called oceanic polar front east of Japan, extending as far east as 170°E and as far north as 50°N. He described the current distribution as follows: The north–south distance between the Kuroshio and Oyashio Fronts is 200–300 nautical miles wide in the inshore Tohoku Area, but it becomes increasingly wider with increasing distance from the east coast of Japan, and trebles at about 180° long. To the west of 155°E warm and cold waters are involved with each other, but with increasing dis-

tance beyond 155°E the waters become rapidly mixed, producing comparatively homogeneous water spread over the Perturbed Area. The North Pacific Intermediate Water is regarded as Oyashio water moving into the Perturbed Area and spreading extensively at intermediate depths. It does not go straight southward as the "Oyashiwo Undercurrent," but seems to be transported with the upper water from the inshore Tohoku Area to the offshore, subject to mixing, probably spreading from the far offing through various paths to the south as suggested by Sverdrup et al. (1942). Masuzawa also considered that the Subarctic Current described by them is an extension of the Oyashio Front.

The above is an extract from Masuzawa's description under the catchword "polar front," but he said nothing about the peculiar pattern in the schema. The current pattern west of 150°E is similar to the schema in Fig. 6d, but the confluence and branching at about 35°N, 160°E and at about 41°N, 164°E seem somewhat curious to me. Masuzawa gave no explanation to justify his schema, but perhaps he derived these singular streamline points in order to produce a repeated north–south exchange of warm and cold waters in the course of flowing offshore. One of the reasons why his schema shows a unique pattern is that it outlines a large-scale feature of the current pattern (cf. Fuglister's comment on Uda's schema, Sec. 3.2). In Fig. 6f, Masuzawa put the name Western North Pacific Central Water area to the region south of the Kuroshio Front, but it is inadequate (Sec. 4.2).

3.6   Modified multiple currents

In the summer of 1955, the *Ryofu Maru* of the CMO made an extensive and detailed survey in the Tohoku Area while participating in the Norpac Expedition, pursuing the Kuroshio Extension as far as 160°E with GEK and BT at close intervals. As reported by Fujii, Ichiye, Masuzawa, Okubo and Marumo (1956) and Masuzawa (1956a), three important features were found (Fig. 37):

—(i)   Maintenance of a narrow strong current as far as 160°E.

—(ii)   Branching of the main current between 146°E and 148°E as a transient stage to detach a cyclonic eddy into the Kuroshio Area.

—(iii)   Countercurrent to the north of the main current between 154°E and 160°E.

Feature (i) had been expected from the existence of a sharp temperature gradient comparable with that in the Gulf Stream (Sec. 3.3). It should be noted, however, that in this survey it was for the first time measured directly.

There are hardly any surveys that have simultaneously had both Nansen-bottle casts reaching a depth
of 1000 m and current measurements with GEK in a
close network of stations. Even during the *Ryofu
Maru*'s cruise for the Norpac Expedition, the Nansen-
bottle stations were more than 50 nautical miles
apart. Preparation of surface streamlines (Kawai,
1957, Fig. 3), which follow constant values of the sum
of the dynamic height and kinetic energy around the
Kuroshio Extension with the aid of values of dynamic
height at BT lowerings estimated from the mean
temperature in the upper 200 m layer and the temperature at 200 m, showed not only the above three
features but also the existence of two or three current
filaments. The surface streamlines derived in the
above also showed a more marked crowding in the
Kuroshio Extension than the dynamic isobaths of
the sea surface. Figure 6g shows a schema of the
Kuroshio System in an extended form of the interpretation given above, based on the surface streamlines, the thermal structure of the upper layer, the
surface current measured with GEK and the dynamic
height of the surface from the *Ryofu Maru* and other
ships participating in the Norpac Expedition.

This schema is characteristic of separation of
streamlines (流線剝離) from the Kuroshio Extension,
countercurrents and confluence. Part of the streamlines, having followed the northern edge of the main
path of the Kuroshio Extension for some distance,
become separated from it at some points, singly or in
a bunch. The streamlines separated from different
positions tend to make the countercurrent along the
northern edge of the inner cold belt. The countercurrent flows almost westward at first, and then turns
northward. Some of the streamlines flow back into
the main path of the Kuroshio Extension, surrounding warm water in the Perturbed Area. Marked
separations of streamlines from the northern edge of
the main path of the Kuroshio Extension were found
in the summer of 1955, at least at three positions: at
147°E, 154°E and 158°E.

The streamlines separated from the southern edge
of the main path of the Kuroshio Extension turn
clockwise loosely, making a big anticyclonic eddy on
the south of the Kuroshio Extension. The westward
current along the southern edge of the anticyclonic
eddy has been called the Kuroshio Countercurrent
(Uda, 1935a).

A similar interpretation was given by Fuglister
(1951a, Fig. 1) in his schematic chart of temperature
at a depth of 200 m in the Gulf Stream System. He
called a series of overlapping currents separated by
relatively weak countercurrents the "multiple cur-

rents" (多重海流).

4 CHARACTERISTICS OF WATERS APPEARING IN AND AROUND THE TOHOKU AREA

According to Sverdrup et al. (1942), "a *water mass*
is defined by a *T-S* curve [of a water column], but in
exceptional cases it may be defined by a single point
in a *T-S* diagram; that is, by means of a single temperature and a single salinity value." This does not
seem to give an explicit definition of the term "water
mass," but rather seems to give a definition of a
characteristic of a water mass. With suitable modification of the definition of an air mass in meteorology,
a water mass is better defined as a large body of sea
water having virtually uniform conditions of (potential) temperature, salinity, and so on along each uniform (potential) density surface in the body.

"A *water type*, on the other hand, is defined by
means of single temperature and salinity values,"
again according to Sverdrup et al. (1942). There is
another concept of water condition, related to extreme values in stratification of various water types,
such as salinity minimum and oxyty maximum traced
in vertical profiles. This is usually called "intermediate water." Because of trouble in discriminating the
usages of these terms, mostly the world *water* is used
in this chapter.

In and around the Tohoku Area, there seem to be
present five typical waters: the North Pacific Sub-
tropical Mode Water, the Western North Pacific
Central Water, the Original Oyashio Water, the
North Pacific Intermadiate Water and the Western
North Pacific Deep Water (Chap. 11), as shown in
Fig. 7.

4.1 North Pacific Subtropical Mode Water

In winter a thick upper mixed layer outcrops
at the center of a huge anticyclonic eddy in the
Shikoku Basin. In general, the waters in the thermostad (Seitz, 1967) enclosed in huge anticyclonic eddies
south of the Kuroshio Front, including their relics in
the warm season, may be regarded as the North
Pacific counterpart of "18° Water" in the North
Atlantic (Worthington, 1959). According to Masuzawa's (1969a) analysis using summer data alone, the
Subtropical Mode Water, which is an alternate name
for the 18° Water proposed by him, is characterized
by 16.5°C in temperature, 34.75‰ in salinity and 250
cl/ton in thermosteric anomaly (Chap. 4).

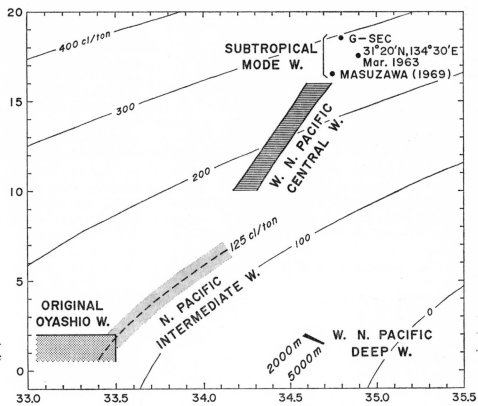

Fig. 7 Characteristics of typical waters in and around the Tohoku Area.

Table 2. Example of a very thick upper mixed layer at stations along G-Sec, where differences between temperatures read to the nearest 0.1°C at each observed depth from 10 to 300 m and at a depth of 10 m are less than or equal to 0.1°C.

Shiga Maru, 62nd cruise
Sta. G6, 19 Feb. 1953
31°20'N, 135°50'E
(RMMOO, 13, 1954, p. 43–44)

Depth (m)	Temp. (°C)	Sal. (‰)	δ_T (cl/t)	Oxyty (ml/L)
0	18.3	34.81	290	5.58
11	18.37	—	—	5.67
27	18.38	78	294	5.78
53	18.40	79	294	5.64
80	18.38	79	293	5.50
96	18.45	78	296	5.67
150	18.46	78	296	5.54
204	18.39	78	294	5.62
312	18.44	78	296	5.55
419	17.53	76	276	5.20
529	15.62	67	239	5.22
700	10.61	34	168	4.30
941	5.56	25	103	2.69
1183	3.80	45	69	1.77
1425	2.95	51	57	1.98

Shinnan Maru, 64th cruise
Sta. G6, 28 Mar. 1953
31°31'N, 135°36'E
(RMMOO, 13, 1954, p. 47)

Depth (m)	Temp. (°C)	Sal. (‰)	δ_T (cl/t)	Oxyty (ml/L)
0	18.7	34.79	301	5.79
10	18.64	83	297	5.81
24	18.62	81	297	6.22
49	18.61	85	294	5.70
73	18.58	83	295	5.66
98	18.58	85	294	5.75
146	18.60	83	296	5.67
189	18.60	83	296	5.71
287	18.50	81	295	5.64
384	18.04	76	287	5.20
486	16.18	76	245	4.91
588	14.18	56	218	4.57

RMMOO stands for "Results of Marine Meteorological and Oceanographical Observations" published by the CMO.

Time sections of monthly mean temperature, salinity and thermosteric anomaly at Station M, a fictional station with the maximum thickness of dynamic height between 0- and 1000-db surfaces along G-Sec (Fig. 49) south of Shionomisaki (33°26′N, 135°45′E) crossing the eddy in the Shikoku Basin, show that a thick upper mixed layer with temperatures of about 18.8°C, salinities of about 34.80‰ and thermosteric anomalies of 300–305 cl/ton reaches a depth of 300 m in February (Figs. 51, 52 and 53). Because of the small sample size in calculating the monthly mean values at Station M and because of the difference in position between the station with the thickest dynamic height and the station with the thickest mixed layer, the time sections are apt to miss the thickest state of the upper mixed layer. Table 2, listing values of water characteristics at observed depths at stations

from G-Sec with a very thick upper mixed layer (the difference in temperature between each observed depth from 10 to 300 m and a depth of 10 m is less than or equal to 0.1°C), shows that the water characteristics of 18.5°C in temperature, 34.80‰ in salinity and 295 cl/ton in thermosteric anomaly may be appropriate for an identification of the Subtropical Mode Water south of Japan.

As an extreme example, a thick mixed layer with temperatures of about 17.5°C, salinities of about 34.90‰ and thermosteric anomalies of about 265 cl/ton reached a depth of 400 m at 31°20′N, 134°30′E on the 5th of March, 1963 (observed on board the *Shumpu Maru* of the KMO), when Japan had a severe winter (Figs. 8 to 11). Even this, however, falls short of the depth of 500 m of the mixed layer at Station 197 of the *Chain* made on the 26th of April,

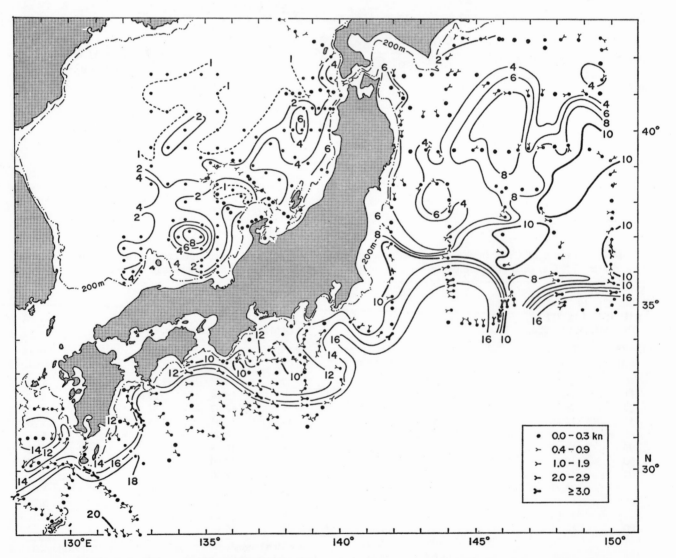

FIG. 8   Surface current from GEK and 200 m temperature (°C) in February–March 1963 around Japan. The neighbor synchronous index $\Delta d/\delta t$ (Sec. 7.5) is greater than 0.1 kn.

1960 at 37°28′N, 52°25′W (Fuglister, 1963). The formation of such a thick upper mixed layer seems to be related to the accumulation of warm water from its surroundings because the thickest upper mixed layer is usually found in a thermal trough penetrating below a depth of 1000 m.

4.2 Western North Pacific Central Water

According to Sverdrup et al. (1942), the Western North Pacific Central Water is characterized by the *T-S* relation in Table 3. Because it is stratified along the permanent thermocline, Masuzawa (1969a) proposed to call most of it the "Thermocline Water" (Chap. 4). Figure 12 shows the *T-S* relation and the salinity deviation of the water at a depth of 10 m in February and March 1963, when Japan had a very

cold winter, from the average salinity of the Central Water at the same temperature. Since the average salinity of the Central Water is given only at four selected temperatures (Table 3), a linear interpolation of the average salinity against observed temperatures has been made in the computation of the salinity deviation in Fig. 12. Because of a lack of data offshore of 150°E, the result of this analysis does not directly support the supposition of Sverdrup et al. (1942, p. 714) that the Central Water is formed in latitudes 30°N to 40°N and longitudes 150°E to 160°E in February, but they do prove its fair possibility.

Time sections of salinity deviation of the waters with temperature and salinity averaged by depth and by month at a fictional Station M in the North Pacific Subtropical Mode Water (Sec. 4.1) and at the weather ship Stations Tango (29°N, 135°E) and Extra (39°N,

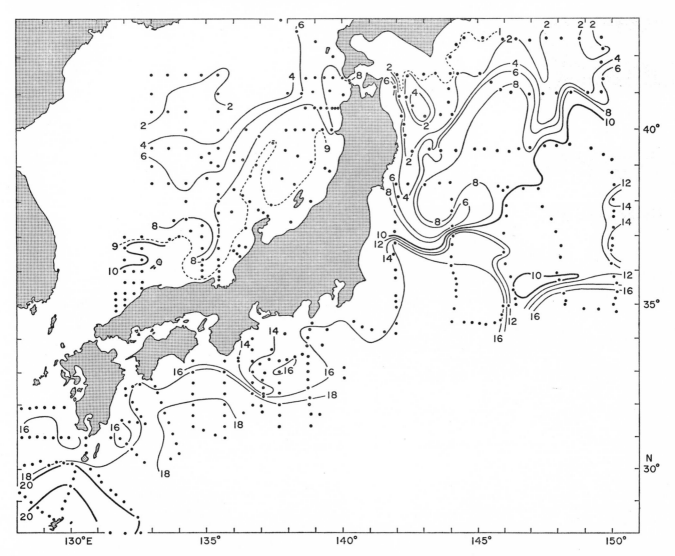

FIG. 9 Temperature (°C) at a depth of 10 m in February–March 1963 around Japan. The neighbor synchronous index $\Delta d/\delta t$ (Sec. 7.5) is greater than 0.1 kn.

TABLE 3. Average salinities of the Western North Pacific Central Water at selected temperatures and the maximum deviations from the averages (based on Table 89, Sverdrup et al., 1942) supplemented with values of thermosteric anomaly.

Temperature (°C)	10	12	14	16
Salinity (‰)	34.24±0.07	34.38±0.06	34.52±0.06	34.67±0.07
Thermosteric anomaly (cl/ton)	166±6	189±5	217±5	248±5

153°E), listed in Tables 13a, 13b and 13c, from the average salinity of the Central Water at the same temperature are presented in Figs. 13a, 13b and 13c. They show that the surface water from February to May and the subsurface water at depths of about 100–200 m all year round at Station Extra and the subsurface water at depths of about 350–500 m at Station Tango and about 450–600 m at Station M

all year round have the *T-S* relation within the envelope of the Western North Pacific Central Water.

From this consideration we can state the following: Station Extra is located within the limits of the surface formative area of the Central Water, but Stations Tango and M are beyond the limits though the subsurface water at the stations is actually the same as the Central Water. Descriptions of the Kuroshio

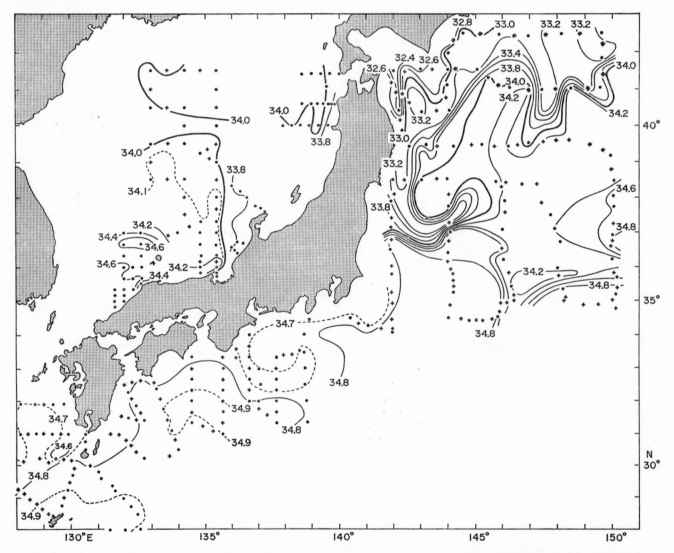

FIG. 10 Salinity (‰) at a depth of 10 m in February–March 1963 around Japan. The positions of salinity values at the surface, used supplementarily in contouring the 10 m isohalines, are indicated by crosses. The neighbor synchronous index $\Delta d/\delta t$ (Sec. 7.5) is greater than 0.1 kn.

Front or Kuroshio Extension axis as the north limit of the Western North Pacific Central Water are inadequate (Sec. 3.5). Similar care must be taken in describing the relation between the distribution of the North Atlantic Central Water and the location of the Gulf Stream axis (Sec. 4.5).

As pointed out by Sverdrup et al. (1942), the Central Waters of the Western North Pacific and the North Atlantic are quite different, the former having lower salinities and the latter higher, with a difference of more than 1‰ in salinity at the same temperature. These researchers made a list of average salinity values for the North Atlantic Central Water at temperatures between 8 and 16°C, and for the Western North Pacific Central Water at temperatures between

10 and 16°C. The lower boundary of the Western North Pacific Central Water reaches a depth of 700 m south of Japan, while that of the North Atlantic with the same temperature limit of 10°C reaches a depth of 1000 m in the Sargasso Sea.

4.3 Original Oyashio Water

Sverdrup et al. (1942) contributed greatly to the representation of a gross feature in distribution of the North Pacific Intermediate Water (Sec. 4.4). Their description of the Pacific subarctic water itself, however, does not seem to be applicable in detail to the western Pacific subarctic water, because they considered their Pacific subarctic water to be distributed

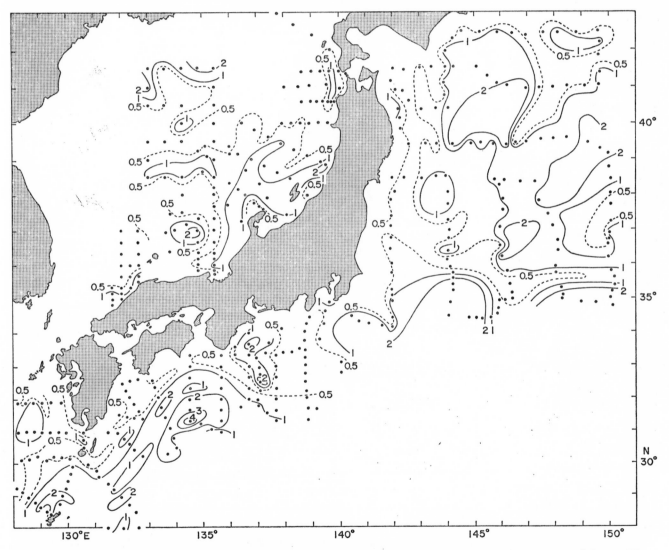

FIG. 11   Depth in hectometers to the bottom of the upper mixed layer in February–March 1963 around Japan. The difference between temperatures read to the nearest 0.1°C at 10 m and at each observed depth from 10 m to the bottom of the mixed layer is less than or equal to 0.1°C. When two values are obtained at the same position from a Nansen-bottle cast and a BT lowering, the greater depth is used for the contouring. The neighbor synchronous index $\Delta d/\delta t$ (Sec. 7.5) is greater than 0.1 kn.

mainly in the northern and eastern North Pacific and because they did not show any *T-S* relation of water column in the vicinity of the Oyashio Current but only showed the one at the *Carnegie* station 120 (47°02′N, 166°20′E, 9 July 1929) located southeast of the Kamchatka Peninsula (cf. the insertion of Fig. 196 in Sverdrup et al., 1942) as an example more of central Pacific subarctic water than of western.

According to Dodimead et al. (1963), the permanent features of the salinity structure in the subarctic Pacific region are "(1) an *upper zone* from 0 to about 100 m depth; (2) a *halocline* in which the salinity increases by about 1‰ in the depth interval from about 100 to 200 m; and (3) a *lower zone* in which the

salinity increases gradually but uniformly to about 34.4‰ at 1000 m depth." "The bottom of the halocline (or top of the lower zone) can be defined with considerable accuracy (±0.1‰) by the isohaline surface of salinity=33.8‰." "The bottom of the upper zone (top of the halocline) is the limit of winter convection or turnover." As pointed out by them, however, "extreme modified structures occur in the Okhotsk Sea and the island passages (Kuril and Aleutian)."

They subdivided the upper zone of the Pacific subarctic region into five domains: (1) Western Subarctic, (2) Transitional, (3) Alaskan Stream, (4) Central Subarctic, and (5) Coastal. Those relating

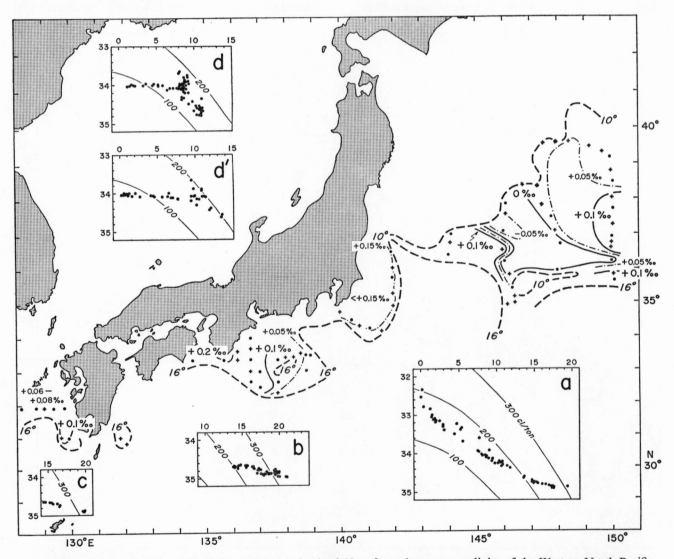

FIG. 12   Salinity deviation (‰) of the waters at a depth of 10 m from the average salinity of the Western North Pacific Central Water (Sverdrup et al., 1942) at the same temperature in February–March 1963. The crosses are the same as in Fig. 10. The dashed line indicates 10 or 16°C isotherm at a depth of 10 m. The neighbor synchronous index $\Delta d/\delta t$ (Sec. 7.5) is greater than 0.1 kn. The *T-S* diagrams are for the water at a depth of 10 m in the Tohoku Area (a), in the sea south of Japan (b), in the East China Sea (c) and in the Japan Sea (d) in February–March 1963. The *T-S* diagram (d′) in the Japan Sea, based on data from the *Seifu Maru* cruise in February–March 1967, is for control of some erratic values in the diagram (d).

directly to the hydrography of the Tohoku Area are the Western Subarctic and Transitional Domains. Roughly speaking, the former corresponds to the Oyashio Area, and the latter to the northernmost Perturbed Area. According to them, near the bottom of the upper zone in the Western Subarctic Domain, salinities are between 33.0 and 33.4‰. The extent of the Western Subarctic Domain can be adequately

defined by temperatures less than 3.5°C at the bottom of the upper zone, while the northern edge of the Transitional Domain can be identified by the 7°C isotherm at the bottom of the upper zone.

Because the indexes of 3.5°C in temperature and 33.4‰ in salinity given above seem inadequate to specify a homogeneous water limit in the far western subarctic Pacific, the Original Oyashio Water (純親

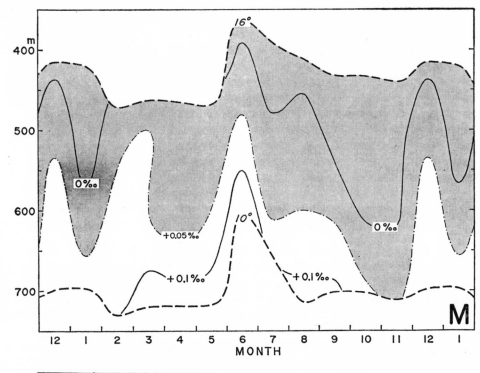

FIG. 13(a) Time section of salinity deviation (‰) of the water at Station M from the average salinity of the Western North Pacific Central Water (Sverdrup et al., 1942) at the same temperature.

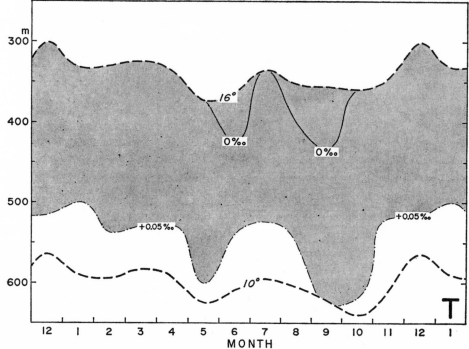

FIG. 13(b) Time section of salinity deviation (‰) of the water at Station T (Tango) from the average salinity of the Western North Pacific Central Water (Sverdrup et al., 1942) at the same temperature.

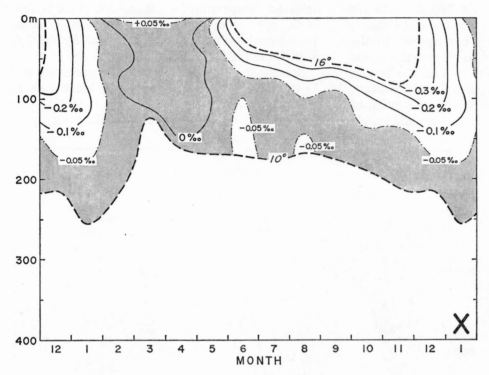

FIG. 13(c) Time section of salinity deviation (‰) of the water at Station X (Extra) from the average salinity of the Western North Pacific Central Water (Sverdrup et al., 1942) at the same temperature.

潮水) is defined here as cold and low-salinity water with temperatures below 2.0°C and with salinities below 33.50‰ in the far western subarctic Pacific, based on experiences in preparing many charts of isotherms and isohalines at a depth of 100 m in the northern Tohoku Area and in analyzing Figs. 14 to 16. The first two of these figures show depth contours of the bottoms of upper cold water with temperatures below 2.0°C and upper low-salinity water with salinities below 33.50‰, prepared by using winter data from the Boreas Expedition (SIO, 1966), an extensive winter survey in the western subarctic Pacific which has never been very successful up to the present, and from a CSK '66w survey. Figure 16 shows depth contours of the outcropping Original Oyashio Water with temperatures and salinities given above. In the Okhotsk Sea and in the adjacent sea to the Kuril Islands, this depth is equal to the depth to the bottom of the upper low-salinity water in Fig. 15, but in other seas it is exactly or nearly equal to the depth to the bottom of the upper cold water in Fig. 14.

The Original Oyashio Water defined above seems to be the subarctic counterpart of the Subtropical Mode Water (Sec. 4.1) in the western North Pacific.

4.4 North Pacific Intermediate Water and inter-cool water

As pointed out by Sverdrup et al. (1942), within the subarctic water no salinity minimum is present, but throughout the middle and lower latitudes of the North Pacific there is present the Intermediate Water, characterized by a salinity minimum or minima. Reid (1965) reaffirmed the definition of the term "Intermediate Water" to distinguish water relatively cool, low in salinity, and high in oxyty from other waters at intermediate depths without these characteristics, and "to identify it where it lies near or in the upper mixed layer." Using a new terminology in Sec. 2, the description of the distribution of Intermediate Water as given by Sverdrup et al. (1942, p. 717) can be rewritten as follows: Both to the north and to the south of the Kuroshio Extension an Intermediate Water is present, characterized by a salinity minimum. In the Kuroshio Area the salinity minimum is found at a depth of about 800 m, and the lowest salinity values are between 34.0 and 34.1‰, whereas in the Perturbed Area to the north of the Kuroshio Extension the salinity minimum is found at a depth of 300 m, the lowest salinities being less than 33.8‰. The oxyty of the Intermediate Water in the Perturbed Area is considerably higher than that in the Kuroshio Area.

As pointed out by many authors (Reid, 1965), the North Pacific Intermediate Water is centered very closely at a surface on which the δ_T (thermosteric anomaly) is 125 cl/ton (σ_t is 26.807 g/L). On the basis of an extensive study of the Intermediate Waters of the Pacific, Reid (1965) proposed the following:

"The characteristics of the North Pacific Inter-

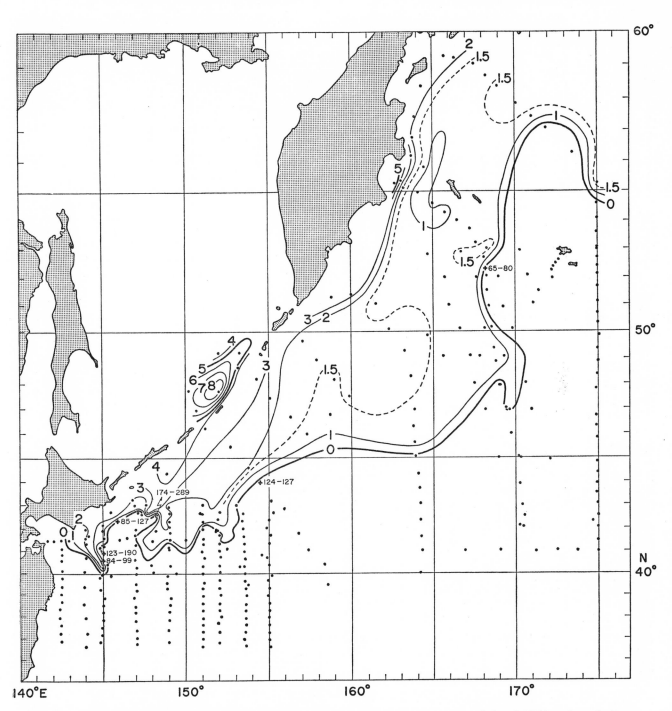

Fig. 14 Depth in hectometers to the bottom of the upper cold water with temperatures below 2.0°C based on the Boreas Expedition and CSK '66w in February–March 1966. The small numerals beside the cross indicate the depth (m) of the upper and lower boundaries of the inter-cool water with temperatures below 2.0°C.

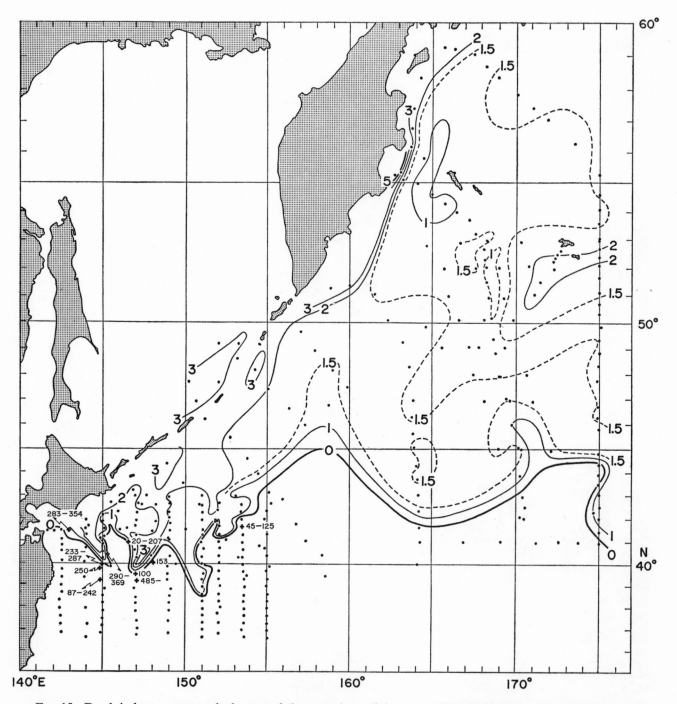

FIG. 15 Depth in hectometers to the bottom of the upper low-salinity water with salinities below 33.50‰ based on the Boreas Expedition and CSK '66w in February–March 1966. The small numerals beside the cross indicate the depth (m) of the upper and lower boundaries of the inter-low-salinity water with salinities below 33.50‰. A single numeral without a dash indicates the depth (m) of the subsurface salinity minimum with a salinity of 33.50‰.

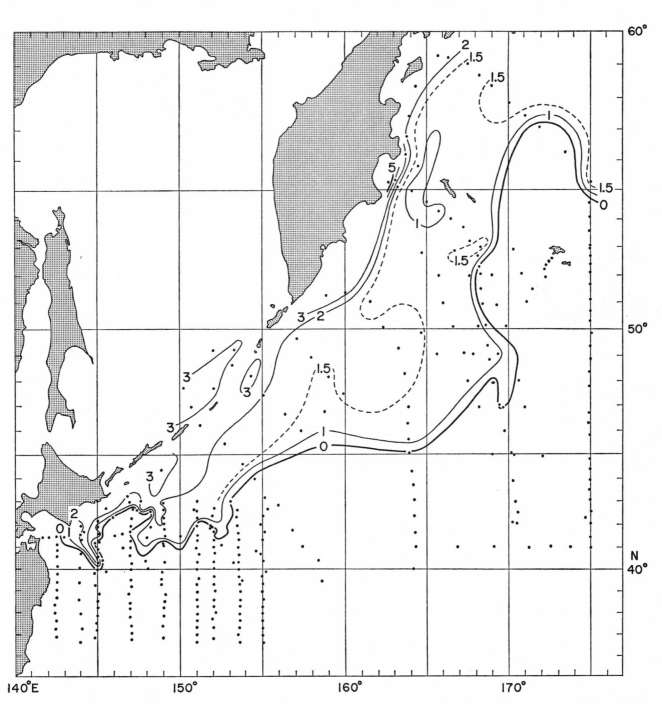

FIG. 16 Depth in hectometers to the bottom of the outcropping Original Oyashio Water with temperatures below 2.0°C and salinities below 33.50‰ in February–March 1966, based on Figs. 14 and 15.

mediate Water are formed in high latitudes by vertical mixing through the pycnocline. This mixing makes the waters in the pycnocline cold, low in salinity, and rich in oxygen in the subarctic region, and through lateral mixing (along density surfaces) and circulation these characteristics are transmitted equatorward from gyre to gyre. As a result, within the northern subtropical anticyclonic gyre the waters with density corresponding to that of the subarctic pycnocline are characterized by salinity less than that of the water above and below."

In the Tohoku Area, subsurface temperature minima and maxima are frequently found. Waters with these extreme values of temperature are called "dichotherm water" or "inter-cool water" (中冷水) and "mesotherm water" or "inter-warm water" (中暖水), respectively (Uda, 1938a). Studies of the temperature minimum in the western North Pacific started about a century ago, and the conclusion was reached that there is a deeper temperature minimum (at δ_T values from 80 to 130) besides the shallow temperature minimum near a depth of 100 m, as pointed out by Reid (1965). Uda (1935b, 1938a) noted that the inter-cool water with temperatures of about 1°C or below, salinities of about 33.3‰ and oxyties of about 7 ml/L is found in summer at depths of 50–100 m in the Oyashio Area off Hokkaido and off the Kuril Islands and in the Okhotsk Sea, and that the temperature of the inter-cool water rises by as much as 3°C or more with the increasing depth occupied by the water from north to south. The former corresponds to the shallow temperature minimum, and the latter to the deeper one.

As suggested by many authors, the occurrence of a shallow temperature minimum can be explained by summer heating of surface water in the winter mixed layer. Since the cold core is nothing but a narrow belt of inter-cool water, the occurrence of a deep cold core may be explained to some extent (Secs. 2.4 and 5.5), but the occurrence of a deep inter-cool water in a broader sense cannot be clearly explained (Reid, 1965).

The inter-warm water is located beneath the inter-cool water or beneath the surface cold and low-salinity water. As a remarkable example of inter-warm water, Uda (1935b, 1938a) cited the Tsugaru Warm Water beneath Oyashio cold and low-salinity water observed in winter off Erimomisaki, Hokkaido. This special inter-warm water had temperatures of 3 to 11°C at depths of 50–200 m, according to him. For the inter-cool and -warm waters in the vicinity of the boundary between the Tsugaru Warm Water and Oyashio water, see Sugiura (1958).

The inter-cool and -warm waters are associated with temperature inversion. Studies of the shallow inversion were made by Kawai (1956) and Kuroda (1959, 1960), and developed by Nagata (1967a, 1967b, 1968, 1970).

4.5 Comparison of the *T-S* relation between the Tohoku Area and Slope Water—Sargasso Sea Area

In the introduction, it was suggested that the similarity between the Kuroshio Extension and the Gulf Stream should be checked by a comparison of the *T-S* relation of waters on opposite sides of both stream axes. To compare the change in the *T-S* relation from inshore to offshore, the Tohoku Area and its offing between 140°E and 168°E and the Slope Water—Sargasso Sea Area between 74.5°W and 46.5°W, both of them being 28° wide in longitude, were divided into five regions, *a–e* and *a′–e′*, as shown in Fig. 17. From the data listed in Tables 4 and 5 each region was divided into three sub-regions at temperatures of 10 and 14°C at a depth of 200 m in the Pacific and at temperatures of 10 and 15°C in the Atlantic, by using different marks in Figs. 17, 18 and 19. The reason for this is as follows. The 14°C isotherm at 200 m is a good indicator of the stream axis of the Kuroshio Extension (Sec. 6), while the 15°C isotherm at 200 m marks the left-hand edge of the Gulf Stream at that depth (Fuglister and Voorhis, 1965). The 10°C isotherm at 200 m runs almost along the front of warm eddies detached from the Kuroshio Area into the Perturbed Area and from the Sargasso Sea into the Slope Water Area. In other words, the isotherm is approximately indicative of the secondary Kuroshio Front and, as pointed out by Fuglister (1963), the Slope Water Current.

To exclude the upper water under the influence of atmospheric conditions and deep water with small variations in temperature and salinity, the *T-S* diagrams in Figs. 18 and 19 are prepared for the water below a depth of 190 m, warmer than 2°C for deep water in the Tohoku Area and its offing, and warmer than 4°C in the Slope Water—Sargasso Sea Area, the colder deep waters in the two areas being located below a depth of about 2000 m.

As shown in Figs. 18 and 19, the difference in the *T-S* diagrams between the water in the Kuroshio Area (solid circles) and the warm water in the Perturbed Area (crosses) is slight, as it is between the water in the Sargasso Sea (solid circles) and the warm water in the Slope Water Area (crosses). The differences in the waters of the upper 200 m layer, however, are large.

In region *a′* of the Slope Water—Sargasso Sea

Area, the *T-S* diagrams in sub-regions on opposite sides of the 10°C isotherm at 200 m (crosses and open circles) are quite similar. This is consistent with the results obtained by Iselin (1936) in the western North Atlantic. In region *a* of the inshore Tohoku Area, however, they are quite different. In the Slope Water Area, a situation similar to that in region *a* of the Tohoku Area is found in regions *d'* and *e'* south of the Grand Banks, far east of the U.S. coast. Apparently the Perturbed Area in the North Pacific is much affected by subarctic water even quite close to the east coast of Japan, while in the Slope Water Area, the effect of the subarctic water is partly blocked by the Grand Banks and its tail.

The scattered dots on the left-hand side of the Central Waters (Sec. 4.2) in the two areas indicate that the heated or diluted water in the upper layer reaches below a depth of 190 m. This seems to be remarkable in the inshore region in the two areas. Incidentally, the Central Water (Sverdrup et al., 1942) seems to be properly defined for the western North Pacific, but not for the North Atlantic (Figs. 18 and 19).

The change in temperature and salinity of the Original Oyashio Water (Sec. 4.3) in the *T-S* diagrams (Fig. 18) is consistent with the thickness distribution of the outcropping Original Oyashio Water shown in Fig. 16. There is a marked eastward change of the *T-S* relation in the Kuroshio Area indicated by black solid dots. The salinity of the North Pacific Intermediate Water in the Kuroshio Area, represented by the salinity at the intersection of an envelope of the black dots with the 125-cl/ton line, decreases gradually from region *a* to region *e*. This supports Sverdrup and others' (1942) interpretation of the spreading of the Intermediate Water in the Kuroshio Area (Sec. 4.4).

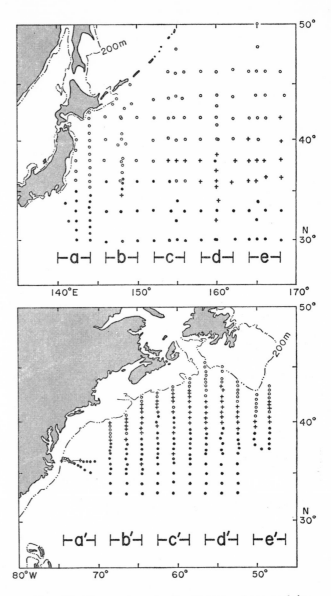

FIG. 17 Five regions in the Tohoku Area and its offing of the North Pacific and the Slope Water—Sargasso Sea Area of the North Atlantic for comparison of the *T-S* diagrams in Figs. 18 and 19. Solid circles indicate the stations with 200 m temperatures higher than 14°C (upper) and 15°C (lower). Crosses indicate the stations with 200 m temperatures between 10 and 14°C (upper) and 10 and 15°C (lower). Open circles indicate the stations with 200 m temperatures lower than 10°C.

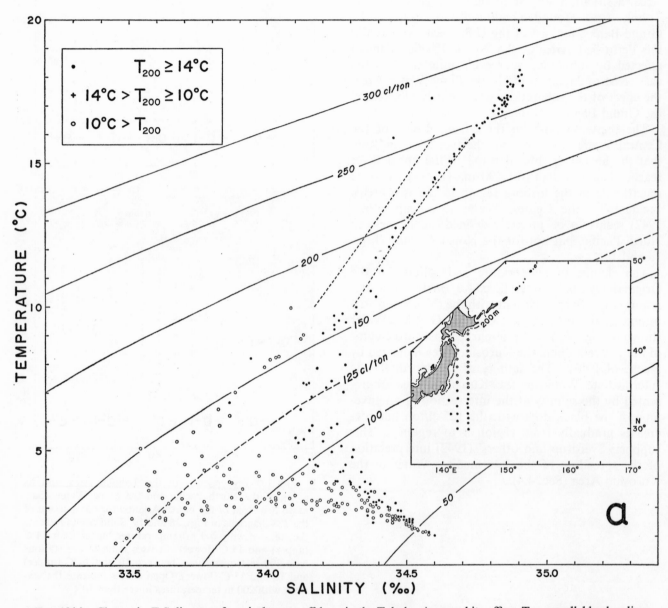

FIG. 18(a)   Change in *T-S* diagrams from inshore to offshore in the Tohoku Area and its offing. Two parallel broken lines indicate the maximum deviations from the average salinities of the Western North Pacific Central Water (Sverdrup et al., 1942).

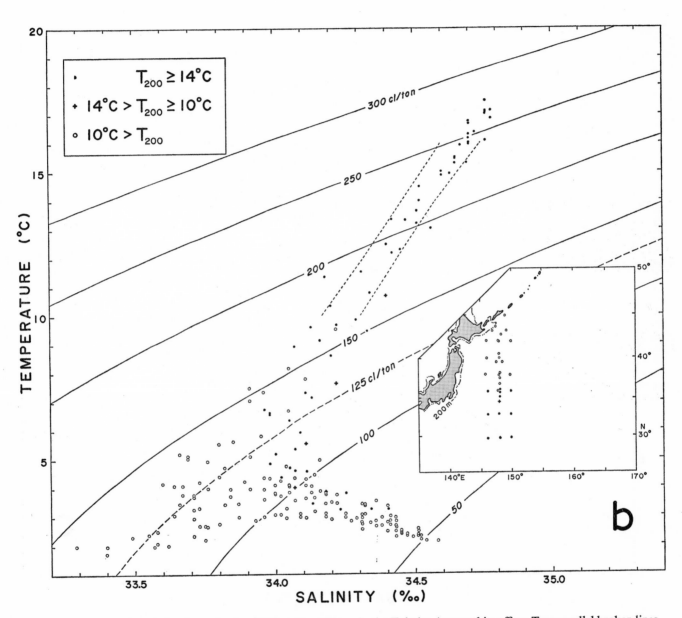

FIG. 18(b)   Change in *T-S* diagrams from inshore to offshore in the Tohoku Area and its offing. Two parallel broken lines indicate the maximum deviations from the average salinities of the Western North Pacific Central Water (Sverdrup et al., 1942).

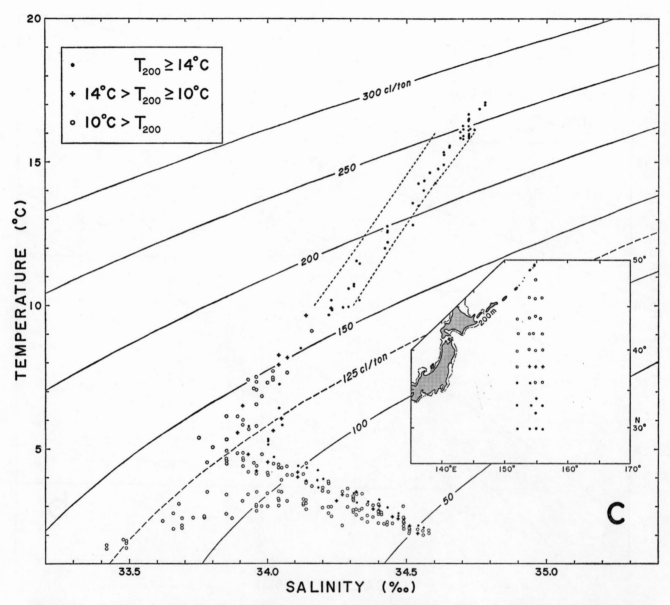

FIG. 18(c)   Change in *T-S* diagrams from inshore to offshore in the Tohoku Area and its offing. Two parallel broken lines indicate the maximum deviations from the average salinities of the Western North Pacific Central Water (Sverdrup et al., 1942).

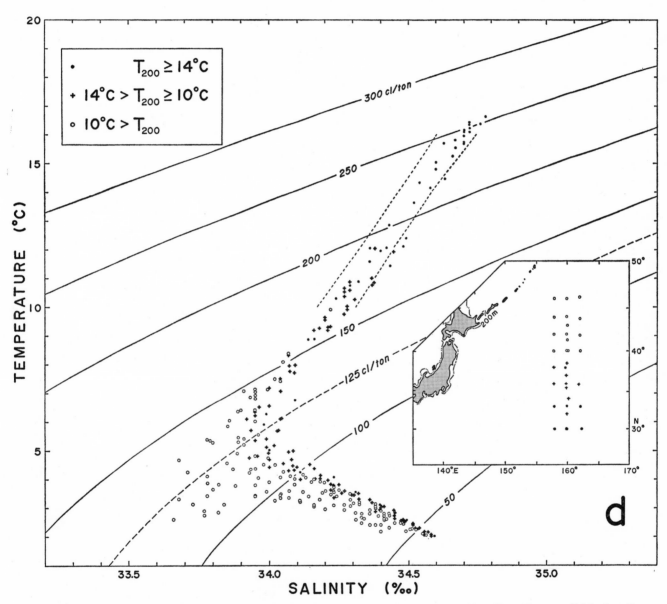

FIG. 18(d)   Change in *T-S* diagrams from inshore to offshore in the Tohoku Area and its offing. Two parallel broken lines indicate the maximum deviations from the average salinities of the Western North Pacific Central Water (Sverdrup et al., 1942).

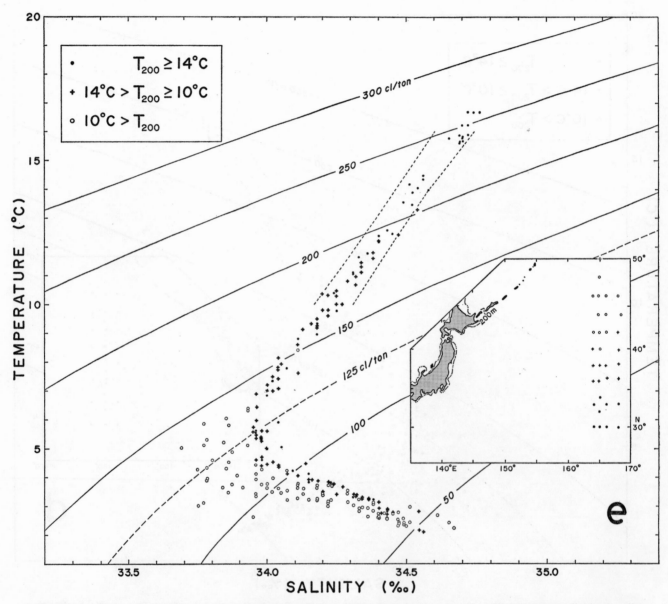

Fig. 18(e)   Change in *T-S* diagrams from inshore to offshore in the Tohoku Area and its offing. Two parallel broken lines indicate the maximum deviations from the average salinities of the Western North Pacific Central Water (Sverdrup et al., 1942).

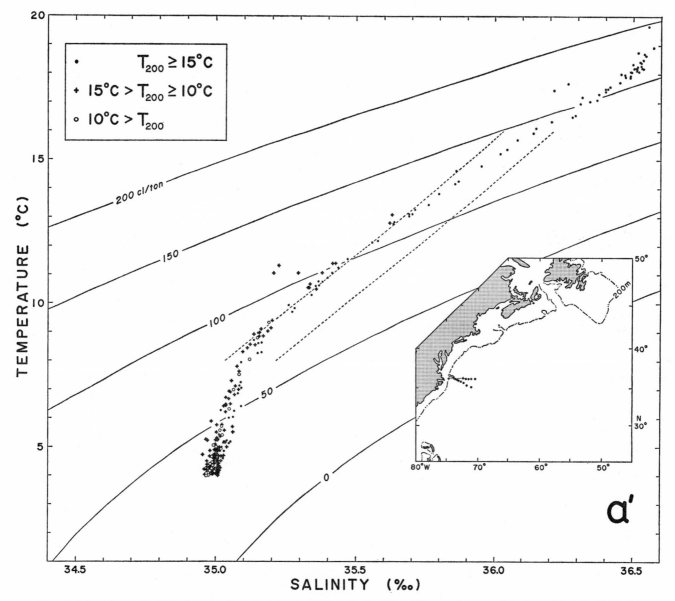

FIG. 19(a′) Change in *T-S* diagrams from inshore to offshore in the Slope Water—Sargasso Sea Area. Two parallel broken lines indicate the maximum deviations from the average salinities of the North Atlantic Central Water (Sverdrup et al., 1942).

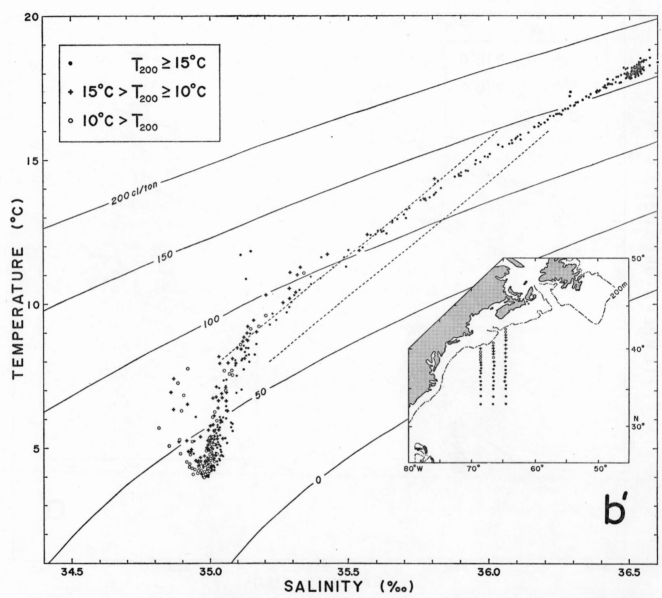

FIG. 19(b′)   Change in *T-S* diagrams from inshore to offshore in the Slope Water—Sargasso Sea Area. Two parallel broken lines indicate the maximum deviations from the average salinities of the North Atlantic Central Water (Sverdrup et al., 1942).

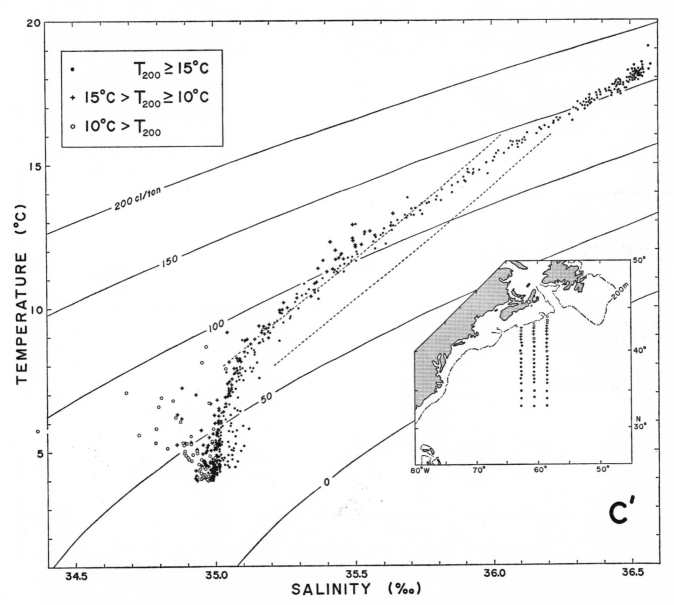

FIG. 19(c′)   Change in *T-S* diagrams from inshore to offshore in the Slope Water—Sargasso Sea Area. Two parallel broken lines indicate the maximum deviations from the average salinities of the North Atlantic Central Water (Sverdrup et al., 1942).

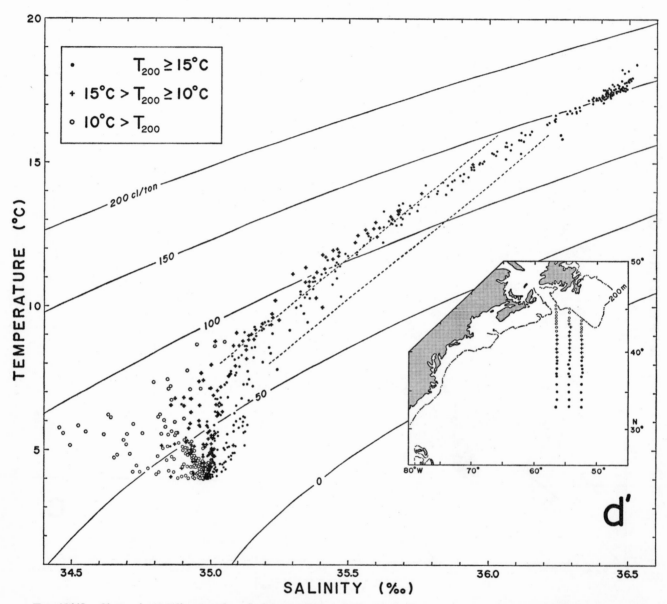

FIG. 19(d′)   Change in *T-S* diagrams from inshore to offshore in the Slope Water—Sargasso Sea Area. Two parallel broken lines indicate the maximum deviations from the average salinities of the North Atlantic Central Water (Sverdrup et al., 1942).

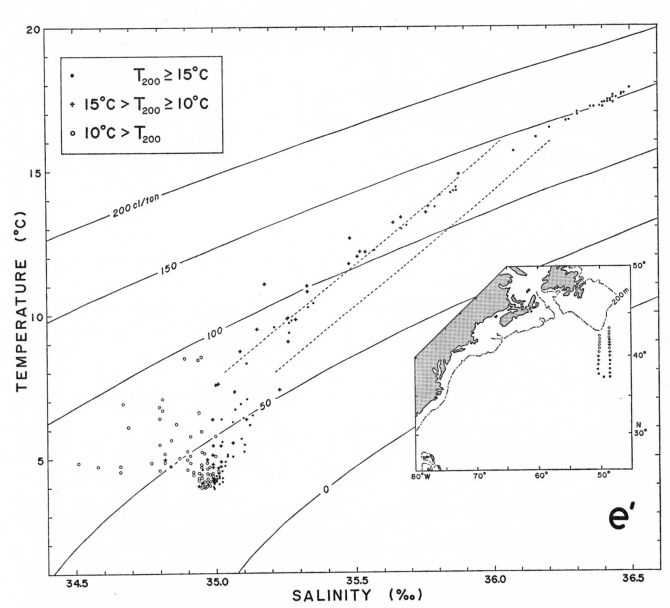

Fig. 19(e′)   Change in *T-S* diagrams from inshore to offshore in the Slope Water—Sargasso Sea Area. Two parallel broken lines indicate the maximum deviations from the average salinities of the North Atlantic Central Water (Sverdrup et al., 1942).

TABLE 4. The stations used in the T-S diagrams of Fig. 18 in the Tohoku Area and its offing.

Region	Ship	No. of Stas. used	Stations	Dates	Location of the Stas.	Expedition	Data source
a 140-144°E	Kaiyo	13	18-19, 21, 23, 25, 27, 29-35	24-29 Aug. 1965	141, 142°E	CSK '65s	DRHOSO 5 (1967)
	Ryofu Maru	20	1-20	7-13 July 1965	144°E	CSK '65s	RMMOO 38 (1968)
		Total 33					
b 146-150°E	Kofu Maru	5	281-282, 285-286, 294	3, 8, 11 Sept. 1964	43-45°N	—	RMMOO 36 (1967)
	Ryofu Maru	11	2032-38, 2048-49, 2067, 2071	15-17, 20, 29-30 Aug. 1964	148°E	—	RMMOO 36 (1967)
	Ryofu Maru	5	845-846, 868-870	17, 29-30 Sept. 1957	42, 40°N	IGY Poront '57	RMMOO 22 (1958)
	Takuyo	6	2-4, 25-27	7, 15 Oct., 5-6 Nov. 1957	38, 36°N	IGY Poront '57	HB 64 (1960)
	Umitaka Maru	6	3-5, 26-28	30 Oct., 2-3 Dec. 1957	33, 30°N	IGY Poront '57	available from WDCA
		Total 33					
c 152-156°E	Ryofu Maru	10	1052-61	29 Aug.-7 Sept. 1958	155°E	IGY Poront '58	RMMOO 24 (1959)
	Oshoro Maru	5	1-2, 22-24	26, Aug., 9-10 Sept. 1957	46, 44°N	IGY Poront '57	DROOEF 3 (1959)
	Ryofu Maru	6	848-850, 865-867	19-20, 28 Sept. 1957	42, 40°N	IGY Poront '57	RMMOO 22 (1958)
	Takuyo	6	5-7, 22-24	16-17, 24-25 Oct. 1957	38, 36°N	IGY Poront '57	HB 64 (1960)
	Umitaka Maru	6	6-8, 23-25	31 Oct.-1 Nov., 1-2 Dec. 1957	33, 30°N	IGY Poront '57	available from WDCA
		Total 33					
d 158-162°E	Takuyo	9	3-11	23-28 July 1968	160°E	CSK '68s	DRCSK 176 (1969)
	Oshoro Maru	6	3-5, 19-21	27-28 Aug., 6-7 Sept. 1957	46, 44°N	IGY Poront '57	DROOEF 3 (1959)
	Ryofu Maru	6	851-853, 862-864	21, 26-27 Sept. 1957	42, 40°N	IGY Poront '57	RMMOO 22 (1958)
	Takuyo	6	8-10, 19-21	17-19, 23-24 Oct. 1957	38, 36°N	IGY Poront '57	HB 64 (1960)
	Umitaka Maru	6	9-11, 20-22	1-2, 29-30 Nov. 1957	33, 30°N	IGY Poront '57	available from WDCA
		Total 33					
e 164-168°E	Ryofu Maru	11	1041-51	19-27 Aug. 1958	165°E	IGY Poront '58	RMMOO 24 (1959)
	Oshoro Maru	5	6-8, 16-17	28-29 Aug., 3-4 Sept. 1957	46, 44°N	IGY Poront '57	DROOEF 3 (1959)
	Ryofu Maru	5	854, 856, 859-861	22-23, 25-26 Sept. 1957	42, 40°N	IGY Poront '57	RMMOO 22 (1958)
	Takuyo	6	11-13, 16-18	19-22 Oct. 1957	38, 36°N	IGY Poront '57	HB 64 (1960)
	Umitaka Maru	6	12-14, 17-19	2-3, 28-29 Nov. 1957	33, 30°N	IGY Poront '57	available from WDCA
		Total 33					

Abbreviation for data source

DRCSK Data Report of CSK, JODC.
DRHOSO Data Report of Hydrographic Observations, Series of Oceanography, JMSA.
DROOEF Data Record of Oceanographic Observations and Exploratory Fishing, HUF.
HB Hydrographic Bulletin, JMSA.
RMMOO The Results of Marine Meteorological and Oceanographical Observations, JMA.
WDCA World Data Center, A (Washington, D.C.).

TABLE 5. The stations used in the T-S diagrams of Fig. 19 in the Slope Water—Sargasso Sea Area.

Region	Ship, cruise	No. of Stas. used	Stations	Dates	Location of the Stas.	Expedition	Data source
a' 74.5-70.5°W	Atlantis 215	12	5294-5305	9-11 June. 1955	ESE of Chesapeake Bay	—	Fuglister (1960)
	Chain 7	15	19-33	19-22 Apr. 1959	36°N	—	Fuglister (1960)
		Total 27					
b' 68.5-64.5°W	Atlantis 255	51	5874-5905, 5908-5926	9-26 Apr. 1960	68.5, 66.5, 64.5°W	Gulf Stream '60	Fuglister (1963)
c' 62.5-58.5°W	Crawford 40	61	810-870	9-26 Apr. 1960	62.5, 60.5, 58.5°W	Gulf Stream '60	Fuglister (1963)
d' 56.5-52.5°W	Chain 12	66	136-179, 181-202	8-24 Apr. 1960	56.5, 54.5, 52.5°W	Gulf Stream '60	Fuglister (1963)
e' 50.5-46.5°W	Evergreen	27	7286-7312	3-10 Apr. 1960	50.5, 48.5°W	Gulf Stream '60	Soule, Morrill & Franceschetti (1961)

5 PROFILES

One of the most important sections in the Tohoku Area seems to be that at about 144°E, because it crosses the Kuroshio and Oyashio Fronts nearly at a right angle and because it encounters two of the most essential waters in the Tohoku Area, the Subtropical Mode Water and the Original Oyashio Water. Since the *Soyo Maru* of the JIFES first occupied this section in February 1935, numerous observations have been made, initially probably without any consciousness of the importance of the section.

Unfortunately, Japanese have never made such detailed and deep Nansen-bottle profiles across the Kuroshio Extension as Worthington (1954) made across the Gulf Stream. One rather detailed profile across the Kuroshio Extension is that made by the CMO at 144°E on board the *Ryofu Maru* in November 1954 (Masuzawa, 1955a), though the bottom Nansen-bottle reached a depth of 1000 m at half the stations. Since the profile neither crossed the Oyashio Front nor contained nutrient salts, the data taken by the HMO at 144°E on board the *Kofu Maru* in February 1967 was analyzed together with the *Ryofu Maru*'s. Hereafter, these will be referred to as the November and February profiles, respectively.

Since the February profile had rather wide-spaced stations, observed properties in the upper 250 m layer were supplemented at between-station BT lowerings after Montgomery and Stroup's (1962) method. The first step was to reproduce the bathythermogram for each between-station BT lowering. The second was to draw salinity/temperature, oxyty/temperature, phosphate-P/temperature and silicate-Si/temperature curves for each between-station BT lowering with data from adjacent Nansen-bottle stations or stations with a similar bathythermogram, though the plot against temperature in subarctic water is not useful because of the small variation in temperature with depth as pointed out by Reid (1965). The third was to read the depths of the chosen values of salinity, oxyty, phosphate-P and silicate-Si by a combination of the two kinds of curves prepared in the first and second steps. The depths of the chosen values of thermosteric anomaly were read from a thermosteric-anomaly/depth curve prepared by using values of salinity supplemented in the above and of temperature from the BT. The velocity profile was prepared without supplemented values because of the far shallowness of a BT lowering compared with a pressure of 1200 db as the reference surface of geostrophic computation.

In the profiles, two different ratios of vertical exaggeration were used. A length of 1000 m on the depth scale corresponds to 400 km for the November profile and to 400 nautical miles for the February profile on the horizontal scale, a vertical exaggeration of 400 to 1 and 740 to 1, respectively. The locations of all Nansen-bottle samples are shown by dots, while those of between-station BT lowerings in the February profile are shown by short vertical lines. To indicate the position of subsurface maxima and minima in temperature, salinity, and so on, observed mid-depth values at and below a depth of 100 m and near the bottom are given in the respective profiles.

5.1 Surface current on a traverse at 144°E

Before describing the profiles, the distribution of surface current at 144°E is discussed on the basis of GEK data from cruises different from those that furnished the profile data.

Figure 20 was prepared on the basis of measurements on board the *No. 4 Kaiyo Maru* of the JHO in August 1954 (adapted from Kawai, 1955b) and the *Ryofu Maru* of the JMAM in November 1959 (adapted from Masuzawa, 1964), both on a traverse at 144°E across the Kuroshio Extension. The direction of the current maximum in both measurements is 90°T, which is exactly perpendicular to the base course. They show that the surface velocity on opposite sides of the current maximum is distributed asymmetrically (Fig. 20, upper panel). The left-hand side of the current maximum, facing downstream, with an abrupt change in velocity with lateral distance, has been called the *shear-zone* because it has

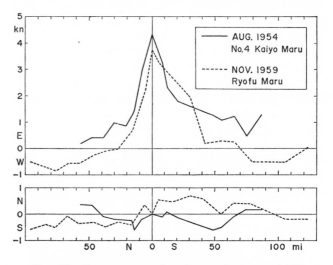

FIG. 20 Components of surface velocity at 144°E across the Kuroshio Extension, based on GEK measurements on board the *No. 4 Kaiyo Maru* of the JHO in August 1954 (solid) adapted from Kawai (1955b, Fig. 15) and the *Ryofu Maru* of the JMAM in November 1959 (dashed) adapted from Masuzawa (1964, Fig. 1).

a large lateral velocity shear amounting to some 10^{-4} sec^{-1}, while the right-hand side with a gradual change is called the *transition zone* (Haurwitz and Panofsky, 1950; von Arx, 1952). The distribution of cross-current components (Fig. 20, lower panel), though the details may be insignificant, shows that a maximum of horizontal convergence may occur in the shear zone if the change in velocity with distance along the stream axis contributes only to a trifling extent to the horizontal convergence or divergence. The location of the maximum horizontal convergence at the surface may correspond to the region where lines of floating matter, slicks and current-rips are found in most cases (Kawai, Sakamoto and Momota, 1969; Kawai and Sakamoto, 1970).

According to von Arx (1962), the countercurrent, when present, is usually found in a region of the Sargasso Sea, 40 to 60 nautical miles seaward of the Gulf Stream front. The measurement in 1959 shows two weak surface countercurrents on both sides of the Kuroshio Extension, one in a region 70 to 120 nautical miles toward the right from the maximum, the other 30 to 100 nautical miles toward the left. The measurement in 1954, however, shows no countercurrent in such a region near the current maximum (Fig. 20, upper panel). To describe the countercurrent in detail, a careful analysis of the data is necessary.

5.2   Velocity profile

It is widely known that there is a fairly good coincidence between the directly-measured velocity and the velocity computed from the geostrophic equation for such large-scale and strong oceanic currents as the Kuroshio and Gulf Stream. The Kuroshio Extension also exhibits a geostrophic nature, but the depth of the reference surface in the geostrophic computation of current velocity is still unknown because of the lack of information from direct current measurements with the neutrally buoyant float. Though a series of such float measurements in the sea adjacent to Japan were made by Jeds cruises, they skipped the Kuroshio and the Kuroshio Extension in order to avoid technical difficulties. Though the measurements were made in the summer of 1965 on board the *Atlantis II* as a part of the CSK survey, they were limited to the Kuroshio south and southeast of Japan (Chap. 10).

In the past, reference surfaces of about 2000 db were in favor with Japanese oceanographers, and the deepest was the 3000-db surface which Takenouchi, Nan'niti and Yasui (1962) adopted in their analysis of measurements from Jeds 4 in the southeastern portion of a big anticyclonic eddy in the Perturbed Area in the summer of 1960 (Sec. 7.3). The

results from this survey and from "Gulf Stream '60" (Fuglister, 1963) lead us to suppose that the current of the Kuroshio Extension reaches the bottom or a depth greater than 3000 m at the minimum.

The reference surfaces of the velocity profiles in Figs. 21b and 22b were taken at pressures of 1000 and 1200 db, respectively, because of the depth limit of Nansen-bottle casts in their cruises. Since the geostrophic current computed by the Nansen-bottle casts indicates a mean velocity between two adjacent stations, an expression different from smoothed isotachs was adopted in this chapter to avoid uncertainty in contouring and difficulty in matching the total transport by contouring to that by the difference in the vertically-integrated dynamic height between stations. The style of expression is essentially similar to that adopted by Wüst (1936).

The current maximum in the November profile (Fig. 21b) is about 120 cm/sec (2.3 kn), while that measured with a GEK is 3.4 kn (175 cm/sec) on the average at this longitude (Table 7). The difference is attributed primarily to rather wide spacing of the stations and secondarily to the shallow reference surface of 1000 db for the geostrophic computation. Countercurrents are seen on either side of the current maximum.

The current maximum in the February profile (Fig. 22b) does not reach a speed of 60 cm/sec even in the vicinity of the Kuroshio Front, less than half of the maximum in the November profile. This is attributed primarily to a wide spacing of the stations, nearly twice that of the November profile, and secondarily to the split of the Kuroshio Extension (cf. the strong current developed along the secondary Kuroshio Front between stations 20 and 21). The easterly current between stations 4 and 18 is accompanied by the Oyashio Front. The westerly current between stations 15 and 16 corresponds to the Oyashio Current flowing southwest close along the southeast coast of Hokkaido.

5.3   Thermosteric anomaly profile

Figures 21c and 22c show the thermosteric anomaly profiles in November 1954 and in February 1967. There are two kinds of *stericline*: permanent and seasonal. In the November profile (Fig. 21c), the permanent stericline is centered at depths of 500–600 m on the right-hand side of the Kuroshio Extension and goes up to a depth of 100 m on the left-hand side in accord with the seasonal stericline. In the February profile (Fig. 22c), no seasonal stericline is found, and the permanent stericline is depressed to a depth of about 500 m in the half-detached anticyclonic eddy (stations 21 and 22) surrounded by the secondary Kuroshio Front and in the isolated anticyclonic eddy (station 18).

In the upper 100 m layer between the Kuroshio Front and a spot thirty to sixty nautical miles south of it there is a core of light water in both the profiles (Figs. 21c and 22c), corresponding to the warm core in the temperature profile (Figs. 21d and 22d), but no counterpart of the cold core in the Kuroshio Extension is found in the thermosteric anomaly profile.

According to Masuzawa (1964), the horizontal gradient of thermosteric anomaly is consistently detectable to a depth of 2000 m or more under the Kuroshio Extension. Though it may reach the bottom much as in the Gulf Stream (Fuglister, 1963), this remains unconfirmed because Nansen-bottle casts reaching the bottom have never been made across the Kuroshio Extension.

The 125-cl/ton surface (Sec. 4.4), indicated by a chain line in both the thermosteric anomaly profiles, is located beneath the bottom of the permanent thermocline. To show the isentropic spreading of the North Pacific Intermediate Water, the depth of the 125-cl/ton isopleth is reproduced by a dashed line in the following profiles.

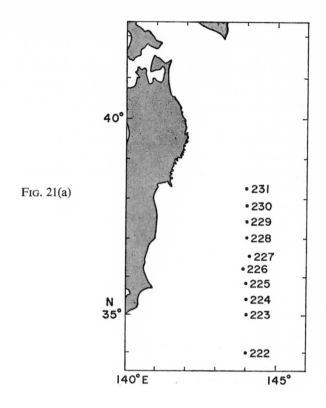

FIG. 21(a)

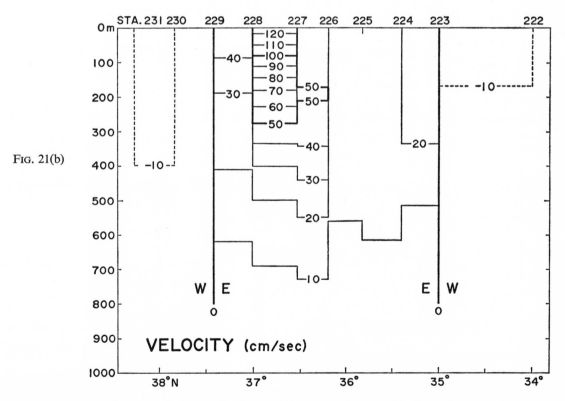

FIG. 21(b)

FIG. 21(a, b) Location of stations (a) and velocity profile (b) at 144°E across the Kuroshio Extension, based on data from the *Ryofu Maru* of the CMO in November 1954. The vertical exaggeration of the profile is 400 to 1.

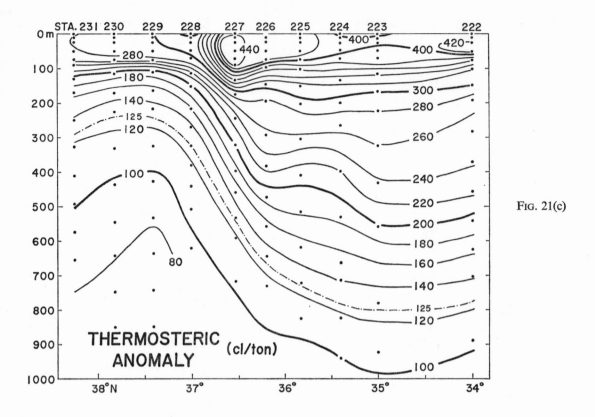

FIG. 21(c)

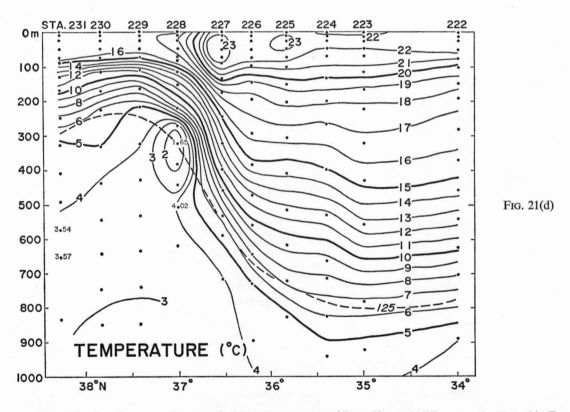

FIG. 21(d)

FIG. 21(c, d) Thermosteric anomaly (c) and temperature (d) profiles at 144°E across the Kuroshio Extension, based on data from the *Ryofu Maru* of the CMO in November 1954. The vertical exaggeration of the profiles is 400 to 1. The dashed line in (d) indicates the depth of the 125–cl/ton isopleth reproduced from (c).

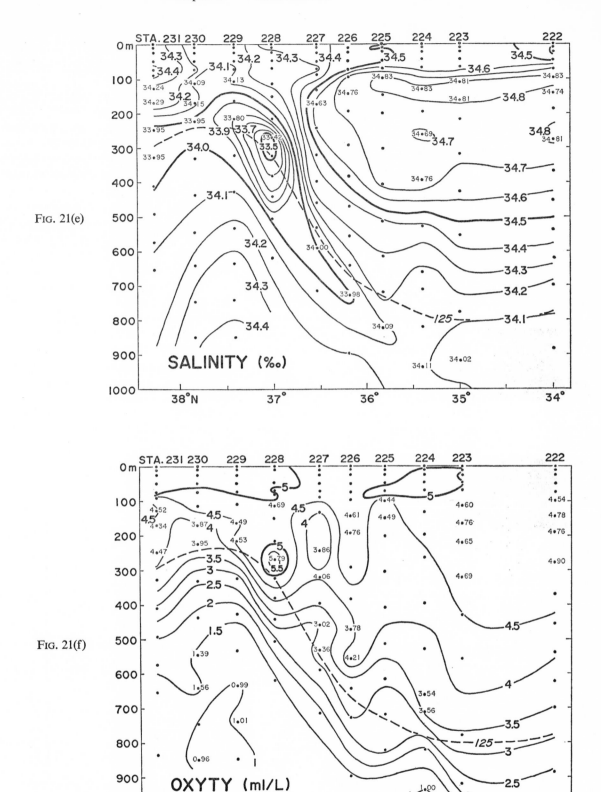

FIG. 21(e, f)   Salinity (e) and oxyty (f) profiles at 144°E across the Kuroshio Extension, based on data from the *Ryofu Maru* of the CMO in November 1954. The vertical exaggeration of the profiles is 400 to 1. The dashed line in (e) and (f) indicates the depth of the 125-cl/ton isopleth reproduced from (c).

5.4   Temperature profile

The permanent and seasonal thermoclines and the warm core in the temperature profiles (Figs. 21d and 22d) are found in the same position as their counterparts in the thermosteric anomaly profiles (Figs. 21c and 22c). At the southernmost station, the permanent thermocline is centered at depths of 500–600 m, while that in the Sargasso Sea exceeds a depth of 900 m. The warm core is outlined with the 23 and 18°C isotherms in the November and February profiles, respectively. The seasonal variation in 100 m temperature of the warm core will be shown in Sec. 8.3. It is difficult to detect a definite relation in location between the warm core and stream axis from comparisons between Figs. 21b and 21d and between Figs. 22b and 22d, but many data from GEK and BT measurements show that the stream axis of the Kuroshio Extension is located in the vicinity of the northern edge of the warm core appearing at a depth of 100 m (Fig. 4). The Subtropical Mode Water (Sec. 4.1) is beneath the seasonal thermocline on the right-hand side of the Kuroshio Extension in the November profile, but outcrops in the February profile.

Although the horizontal gradient of both temperature (Fig. 22d) and salinity (Fig. 22e) is great across

the Oyashio Front located between stations 4 and 15, between stations 17 and BT55 and between stations 19 and BT55, the gradient of thermosteric anomaly is very small in the upper 200 m layer. This is ascribed to the small slope of the observed T-S curve for the upper 200 m water on either side of the Oyashio Front, relative to the δ_T curve on the T-S diagram (Fig. 23). The Oyashio Front is a boundary between waters with different temperatures and salinities, but is not a definite boundary with respect to density of the upper water (Kawai, 1959).

A similar situation is encountered in the temperature, salinity and thermosteric anomaly profiles along a section at 175°E prepared on the basis of data from the Boreas Expedition (SIO, 1966, Figs. 3a–c). The subarctic boundary and a northern front between the Transitional Domain and the Western Subarctic Domain (Dodimead et al., 1963) correspond to the two themohaline fronts between stations 124 and 123 and between stations 117 and 116, respectively, in the profiles at 175°E, but neither distinct horizontal gradient of thermosteric anomaly is found in the upper mixed layer across the two fronts. Below the upper mixed layer, the horizontal gradient of thermosteric anomaly across each of the fronts is not so minute as that in the upper mixed layer, and a weak

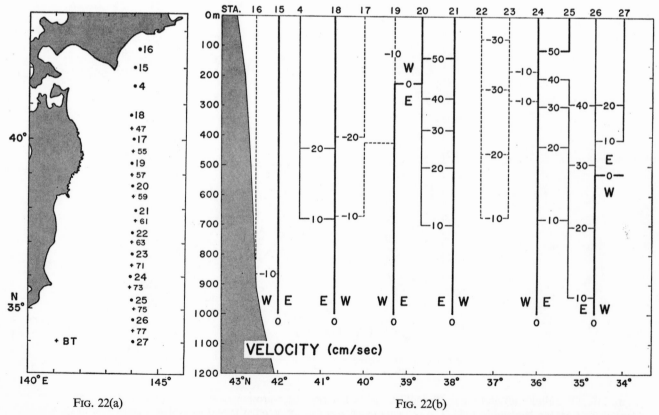

FIG. 22(a)   FIG. 22(b)

FIG. 22(a, b)   Location of stations (a) and velocity profile (b) at 144°E across the Kuroshio Extension and the Oyashio Front, based on data from the *Kofu Maru* of the HMO in February 1967. The vertical exaggeration of the profile is 740 to 1.

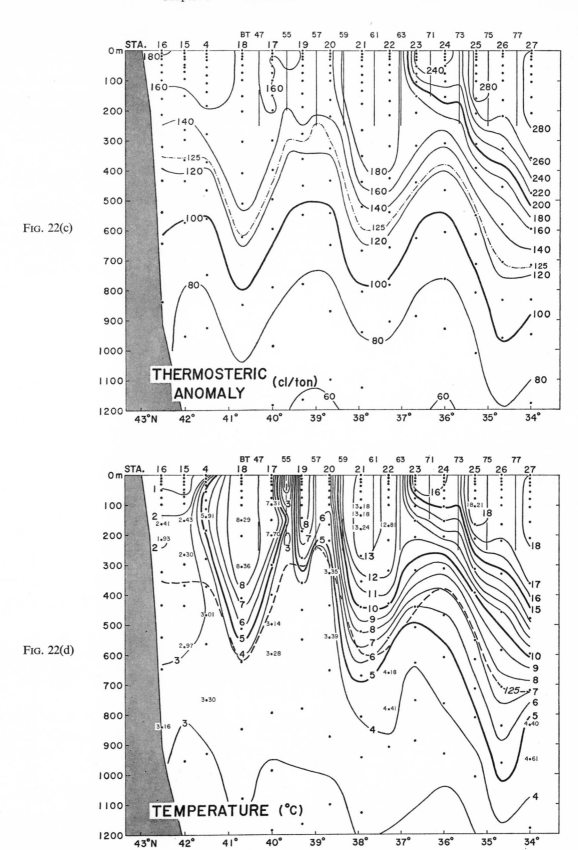

Fig. 22(c)

Fig. 22(d)

FIG. 22(c, d)   Thermosteric anomaly (c) and temperature (d) profiles at 144°E across the Kuroshio Extension and the Oyashio Front, based on data from the *Kofu Maru* of the HMO in February 1967. The vertical exaggeration of the profiles is 740 to 1. The dashed line in (d) indicates the depth of the 125-cl/ton isopleth reproduced from (c).

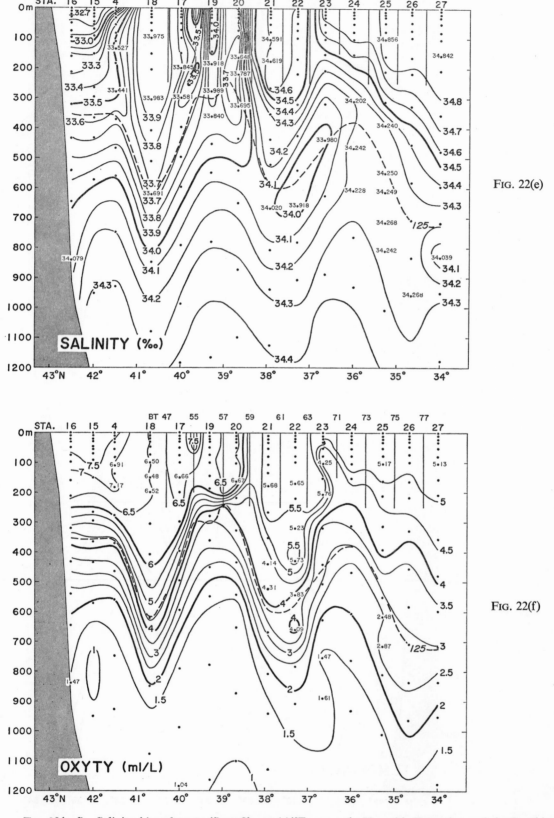

FIG. 22(e)

FIG. 22(f)

FIG. 22(e, f)   Salinity (e) and oxyty (f) profiles at 144°E across the Kuroshio Extension and the Oyashio Front, based on data from the *Kofu Maru* of the HMO in February 1967. The vertical exaggeration of the profiles is 740 to 1. The dashed line in (e) and (f) indicates the depth of the 125-cl/ton isopleth reproduced from (c).

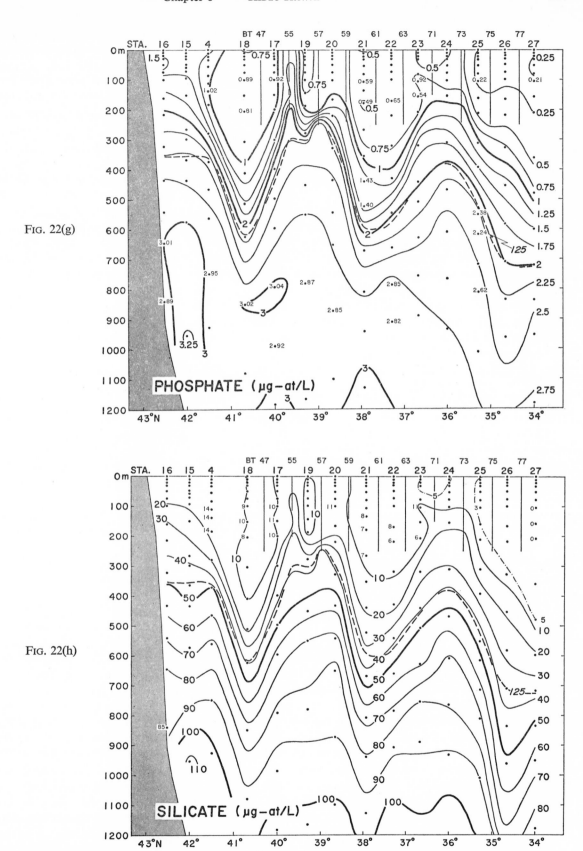

FIG. 22(g)

FIG. 22(h)

FIG. 22(g, h)   Phosphate-P (g) and silicate-Si (h) profiles at 144°E across the Kuroshio Extension and the Oyashio Front, based on data from the *Kofu Maru* of the HMO in February 1967. The vertical exaggeration of the profiles is 740 to 1. The dashed line in (g) and (h) indicates the depth of the 125-cl/ton isopleth reproduced from (c).

but nearly constant gradient seems to penetrate downward through a depth of 1500 m.

As pointed out by Reid (1965), there are two kinds of temperature minima, shallow and deep (Sec. 4.4). Owing to the winter season a distinct shallow temperature minimum is not yet formed, but the water from which it is produced is found as a very cold and low-salinity water, with temperatures less than 2°C and salinities less than 33.0‰, in the upper water surrounding the isolated anticyclonic eddy located at stations 4 to 17 (Figs. 22d and 22e).

In the November profile, the cold core in the Kuroshio Extension, which is well outlined with the 2°C isotherm, is centered at depths of 300–400 m under the tilting thermocline, flowing in the same direction as the upper and right-hand side warm water (Sec. 2.4). The thermosteric anomaly at the center of the cold core has a value of 125 cl/ton, exactly equal to the value of thermosteric anomaly along which the North Pacific Intermediate Water is found. In the February profile, no definite cold core in the Kuroshio Extension is found. The inner cold belt is not distinct in the November profile, and this may be attributed either to an underdeveloped condition of the Oyashio Intrusion, to a warm water intrusion from the Kuroshio Area across the inner cold belt or to escape through a wide mesh along the sampling profile.

There is another thermostad called the 3.5°C (in potential temperature) water, high in oxyty and accompanied by a slight minimum of salinity, located at depths of 1000–2000 m in the Slope Water Area in the North Atlantic. According to Worthington and Metcalf (1961) this water originates from the Labrador Basin, but no counterpart is found in the Perturbed Area in the North Pacific.

5.5   Salinity profile

The halocline is located at the same depth as the permanent thermocline (Figs. 21e and 22e), but the counterpart of the seasonal thermocline is not distinct (Fig. 21e). The salinity in the warm core of the November profile is lower than that of the water at the same depth on its right-hand side, but higher than that of the water at the same depth on the left; the situation is similar to that in the Gulf Stream in April 1960 (Fuglister, 1963). The salinity in the warm core in the February profile does not differ from that at the same depth on its right-hand side, and this is attributed to the difference in season.

There is a *halostad* (Seitz, 1967) with salinities of about 34.8‰ corresponding to the Subtropical Mode Water, on top of which a subsurface salinity maxi-

mum more than 34.8‰ is found in the November profile, but no salinity maximum is found owing to the absence of low-salinity water in the surface layer in the February profile (Sec. 8.2).

As pointed out by Masuzawa (1955a) there is a core of low-salinity water, which is outlined with the 33.7-‰ contour (Fig. 21e), almost at the same position as the cold core in the Kuroshio Extension of the November profile (Fig. 21d). In the November profile, the salinity minimum below 34.1‰ extends both ways from the low-salinity core, spreading northward at nearly the same depth of 300 m as that of the core and southward under the permanent thermocline to a depth of 900 m at the southernmost station (Fig. 21e). In the February profile, the salinity minimum below 33.7‰ is centered at depths of 600–650 m under the isolated anticyclonic eddy (station 18), and the salinity minimum below 34.1‰ is centered at depths of 450–850 m under the half-detached anticyclonic eddy and the halocline in the Kuroshio Area, distributed discretely between stations 23 and 27 (Fig. 22e). The salinity minima in both the profiles nearly coincide with the 125-cl/ton surface (Reid, 1965). The T-S relation of the water in the deep salinity minimum is characteristic of the North Pacific Intermediate Water, while that in the cold core in the Kuroshio Extension is characteristic of the Original Oyashio Water (Sec. 4.3). Accordingly, the cold core in the Kuroshio Extension must be an indication of sluggish advection of the Original Oyashio Water from the northern Oyashio Area, while the salinity minimum in general is not necessarily brought about by the advection.

5.6   Oxyty profile

The oxyty profile in November (Fig. 21f) shows that the water in the warm core has less oxyty than the water at the same depth to either side, as in the Gulf Stream (Fuglister, 1963), but the water in the February profile (Fig. 22f) does not show such low oxyty. Since neither two profiles are based on summer data, the oxyty maximum at the depth corresponding to the lower part of the winter mixed layer is not so distinct as that in Reid's (1962) time sections.

As pointed out by Masuzawa (1955a), there is a high-oxyty core, which is outlined with the 5-ml/L contour in the November profile (Fig. 21f), nearly at the same location as the cold core in the Kuroshio Extension; but no such core is found in the February profile (Fig. 22f). According to Masuzawa's (1964) deep profile across the Kuroshio Extension with rather wide-spaced stations, the lower portion of the salinity minimum is bounded by the oxyty minimum,

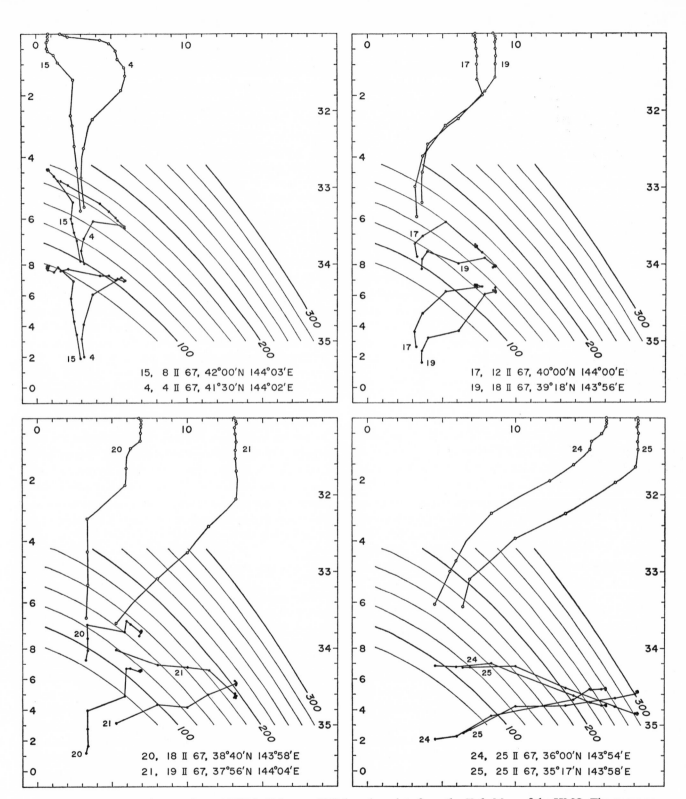

FIG. 23   Typical station graphs at 144°E in February 1967 based on data from the *Kofu Maru* of the HMO. The common abscissa is temperature in °C. In each graph, the upper curve with open circles shows depth in hectometers (scale at left); the middle curve with solid circles shows salinity in ‰ (scale at right); and the lower curve with solid circles shows oxyty in ml/L (scale at left). The group of oblique curves represents thermosteric anomaly in cl/ton. The location of stations is shown in Fig. 22a.

but only the top—outlined with the 1.0-ml/L contour —appears in Figs. 21f and 22f because most of it is located below a depth of 1000 m. In and beneath the permanent thermocline in the Kuroshio Area and in the half-detached anticyclonic eddy, there are some irregularities of oxyty different from those associated with the oxyty minimum.

In sections across the Gulf Stream, the oxyty minimum—outlined with the 3.5-ml/L contour—is centered at about the same depth as the center of the permanent thermocline, and the halocline is centered approximately 100 m shallower than the permanent thermocline (Fuglister, 1963); in other words, the lower portion of the halocline is bounded by the oxyty minimum. In addition to the big difference in the oxyty minimum between both cross-stream profiles as mentioned above, the minimum in the profile across the Gulf Stream is thinner and more distinct than across the Kuroshio Extension. The thickness between two *oxyclines* (Montgomery, 1969; Lyman, 1969) bracketing the oxyty minimum is 200–600 m in the former and greater than 1000 m in the latter.

5.7 Phosphate-P and silicate-Si profiles

Figures 22(g) and 22(h) show inorganic phosphate-P and silicate-Si profiles in February. They were determined by the Denigès-Atkins method and Diènert Wanderbulcke method (RMMOO, no. 41, 1970). Lowest values of phosphate-P (lower than 0.5 μg-at/L) and silicate-Si (lower than 5 μg-at/L) are found in the subtropical Mode Water. Above the permanent thermocline, in general, both nutrient elements have relatively low values (lower than 1.5 μg-at/L for phosphate-P and than 20 μg-at/L for silicate-Si). Below the upper mixed layer in this section, phosphate-P increases downward to the maximum value, at depths of 650–1200 m in the Oyashio and Perturbed Areas and at depths below 1200 m in the Kuroshio Area. Silicate-Si increases monotonically downward to the maximum value at depths of 2000–2500 m. The difference in depth of the maximum values for both elements is related to the fact that a relatively large vertical gradient of silicate-Si is found even below the permanent thermocline (below a depth of 700 m). The general distribution of both the elements closely parallels that of temperature or thermosteric anomaly.

6 ISOTHERMS INDICATIVE OF THE KUROSHIO AXIS

It has long been a matter of interest to Japanese oceanographers to find out how and why the Kuroshio axis shifts in position year by year. Temperature measurements are far more numerous in the Kuroshio than velocity measurements, but before they are put into full use for a detailed study of the shifting of the axis, it will be necessary to determine how to fix the position of the stream axis precisely using a quantity of existing temperature data.

Kawai (1955a) had tried to determine representative isotherms of the Kuroshio Front, on the basis of prewar data taken along 144°E by the *Soyo Maru* of the JIFES. At that time, existing data appropriate for such an analysis were very few in the Tohoku Area, and a representative isotherm appeared to be something with a value that was the mean of the temperatures on either side of the subsurface thermal front near the stream axis. Uehara (1962) reported that, in Enshunada and its offing, a narrow band with a sharp front around the 15°C isotherm at a depth of 200 m coincided well with the stream axis of the Kuroshio. Uda (1964) also stated that, "Thus established fact that '15°C-line at 200 m depth is a good indicator of Kuroshio path' is a general law applicable to every season and every year." On the basis of observations made in November 1962 and 1966, Konaga, Shuto, Kusano and Hori (1967) suggested that off Shikoku the axis coincided with the 16 or 17°C isotherm at a depth of 200 m.

In tracking the Gulf Stream with the V-Fin, Fuglister and Voorhis (1965) assumed that "the 15°C-line at 200 m marks the left-hand edge of the current at that depth."

There are various ways of dealing with this kind of problem, but basically one should compare the location of the axis with that of an isotherm at a particular depth, either by eye or by some statistical technique. So far, scarcely any statistical method has been applied to the problem. In this section a result of statistical estimation of isotherms indicative of the Kuroshio axis is introduced; the result comes from a comparison between the positions of isotherms, as determined by BT lowerings or Nansen-bottle casts and the position of the stream axis, as indicated by GEK measurements (Kawai, 1969, 1970).

6.1 Statistical method

Statistical computations were made independently in five separate areas (Fig. 24) using data from the "Results of Marine Meteorological and Oceanographical Observations" published by the JMA and the "Hydrographic Bulletin" and the "Data Report of Hydrographic Observations" published by the JMSA.

Since oceanographic data are not sampled uniformly with respect to time and space, some criteria must be established for selecting data suitable for statistical treatment. The following conditions were required of all data used in the present analysis.

—(i)   The temperature and GEK measurements were made successively along a cross-section with a single ship. This condition reduces relative errors in positions between temperature features and velocity features, and reduces differences in time of measurement between adjacent stations.

—(ii)   The distance between adjacent stations was less than 25 nautical miles for the GEK and 200 m temperature; and for the 400 m temperature, it was less than 50 nautical miles in Areas I–IV, and less than 60 nautical miles in Area V. This condition reduces errors in the position of the axis and error arising from the linear interpolation in temperature.

—(iii)   The angle between the section line and the velocity vector at the current maximum was not less than about 45°. This condition helps assure a cross-section of the current.

—(iv)   The section included substantial warm and cold water on the opposite sides of the stream axis. This condition eliminates cases of branching of the Kuroshio, cases of eddies cut off from the Kuroshio, and cases of omitting the inner cold belt (Sec. 2.4).

—(v)   Temperature decreased monotonically with latitude in the vicinity of the stream axis.

—(vi)   When a BT lowering and a Nansen-bottle cast were made at the same position, the 200 m temperature obtained with the BT was used for the analysis. Even with a small difference in position (up to 5 nautical miles) between the BT lowering and the Nansen-bottle cast, the BT temperature was adopted.

Temperatures at depths of 200 m and 400 m were plotted against distance along the section for each crossing. The position of each whole-degree isotherm was then read off by linear interpolation, as exemplified in Fig. 25. The location of the Kuroshio axis was marked by an ordinate at the latitude of the maximum eastward current as measured with the GEK. When the maximum current speed actually appeared at two successive stations, the middle point between them was adopted as the location of the axis. If the positions of temperature and GEK measurements did not coincide, the 200 m and 400 m temperatures at the axis were easily read by linear interpolation. The distance (in minutes of latitude) of each whole-degree

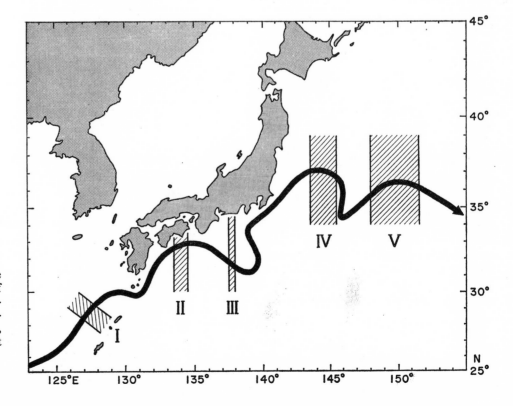

FIG. 24 Five areas for the statistics and the main path of the Kuroshio and its Extension. The path is oversimplified, showing neither branch, confluence, nor countercurrent (Kawai, 1969, Fig. 1).

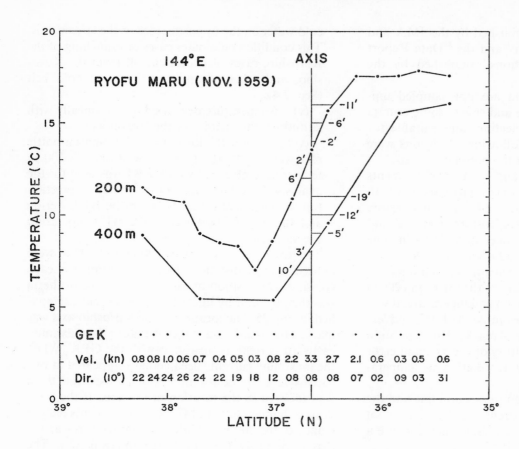

FIG. 25 An example of the data treatment for a crossing along 144°E in November 1959 on board the *Ryofu Maru* of the JMAM (Kawai, 1969, Fig. 2).

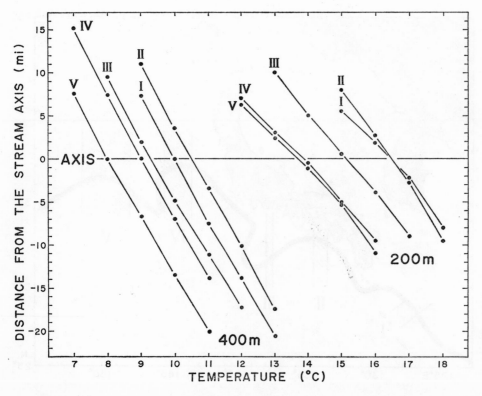

FIG. 26 Mean values, by areas and by depths, of the distance between each degree isotherm and the stream axis (Kawai, 1969, Fig. 3).

TABLE 6.   An example of statistics at 200 m in Area IV.

Sample No.	Month and Year	East Long. (deg.)	Ship	200 m Temp. at the axis	0 m Vel. at the axis	12°C	13°C	14°C	15°C	16°C
							Distance from the axis (mi)			
1	Aug. 1955	144.0	*Ryofu Maru*	16.2	2.9	11	9	6	4	1
2	Feb. 1956	144.0	*Ryofu Maru*	4.8	3.1	−16	−18	−20	−25	−32
3	Nov. 1956	144.0	*Ryofu Maru*	10.2	4.9	− 3	− 5	− 6	− 9	−10
4	May 1957	144.0	*Ryofu Maru*	2.8	3.9	−12	−14	−19	−24	−28
5	June 1957	144.0	*Shumpu Maru*	14.6	3.1	4	2	1	− 2	− 7
6	June 1957	144.5	*Ryofu Maru*	7.1	4.1	− 3	− 4	− 6	− 9	−10
7	June 1957	143.5	*Shumpu Maru*	7.1	4.5	− 3	− 4	− 4	− 6	− 7
8	June 1957	145.0	*Shumpu Maru*	15.2	2.9	9	6	3	0	− 4
.										
.										
.										
.										
.										
29	Feb. 1967	144.0	*Kofu Maru*	16.6	3.2	45	20	14	9	3
30	May 1967	144.0	*Kofu Maru*	13.9	4.3	11	5	− 1	− 5	−10
31	June 1967	144.0	*Ryofu Maru*	10.6	2.6	− 5	− 9	−12	−15	−18
	Mean			12.6	3.4	7.0	3.0	−0.5	−5.0	−9.5
	S.D.			3.7	0.8	13.7	11.1	10.0	9.0	9.1

TABLE 7.   Summary of complete temperature and velocity statistics for the five areas of the Kuroshio and its Extension.

Area		I Sections G and I	II 133.5°E −134.5°E	III 137.5°E −138.0°E	IV 143.5°E −145.5°E	V 148.0°E −151.5°E
200 m temperature at the axis	Sample size	34	40	37	31	24
	m and r (°C)	16.4±0.7	16.3±0.6	14.9±0.8	12.6±1.4	13.7±0.8
	s (°C)	2.1	1.8	2.4	3.7	2.0
400 m temperature at the axis	Sample size	15	23	41	12	10
	m and r (°C)	10.0±0.6	10.5±0.7	9.3±0.5	8.7±1.1	8.2±0.9
	s (°C)	1.1	1.6	1.5	1.8	1.2
	Temperature (°C)	16.5	16.5	15.1	13.9	13.7
Indicative isotherm at 200 m	Sample size	34	40	37	31	24
	s of distance between the isotherm and the axis (mi)	9	9	8	10	8
	Temperature (°C)	10.0	10.5	9.3	9.0	8.0
Indicative isotherm at 400 m	Sample size	15	23	41	12	10
	s of distance between the isotherm and the axis (mi)	8	11	9	10	10
Velocity at the axis	Sample size	34	40	37	31	24
	m and r (kn)	1.9±0.2	2.3±0.1	3.0±0.3	3.4±0.2	2.8±0.3
	s (kn)	0.4	0.5	0.9	0.8	0.7
Mean horizontal temperature gradient	At 200 m (°C/10 mi)	2.4	1.9	2.2	2.9	2.9
	At 400 m (°C/10 mi)	1.3	1.4	1.5	1.4	1.4
	Average (°C/10 mi)	1.9	1.7	1.8	2.1	2.1

m: mean s: standard deviation r: range of the 95% confidence interval of the mean
Sections G and I point in the direction of 305° (T) from the Amami Islands (Fig. 24).

TABLE 8.   Practical value of the isotherm at a depth of 200 m indicative of the Kuroshio axis in main watch-sections.

Area	Watch-section	Value of the isotherm (°C)
I	WNW of the Amami Islands	16.5
	S and SE of Toimisaki	16.5
	S and SSE of Ashizurimisaki	16.5
II	S of Murotomisaki	16.5
	S of Shionomisaki	16.0
	S of Daiozaki	15.5
III	S of Omaezaki	15.0
	S of Irozaki	15.0
IV	142–146°E	14.0
V	148–152°E	13.5

isotherm from the axis was read as positive on the north side of the axis, and negative on the south (Fig. 25).

Table 6 gives an example of the statistics in Area IV. The 200 m temperature and the surface velocity at the axis are shown in the fifth and sixth columns, and the distance of each whole-degree isotherm from the axis is tabulated in the seventh through eleventh columns. The mean and standard deviations are shown at the bottom of Table 6 for each column.

6.2   Results of statistics

The mean values, by areas and by depths, of the distance between each whole-degree isotherm and the axis are plotted in Fig. 26. Table 7 summarizes the complete results of the statistics. The values of indicative isotherms listed in Table 7 were read from the intersecting points of each slanting line with the stream axis in Fig. 26, and the mean horizontal temperature gradients (Table 7) were calculated from the slope of the slanting lines.

The value of the isotherm indicative of the Kuroshio axis, both at 200 m and at 400 m, reaches a maximum off Murotomisaki, Shikoku (Area II, Fig. 24), and decreases gradually downstream (Table 7). There is a small difference between temperatures at the axis and values of indicative isotherms, but both change along the Kuroshio path in a similar manner. The horizontal temperature gradient is so sharp that the standard deviation of distance between the indicative isotherm and the axis is only about ten nautical miles at both depths. There may be annual variation in the value of the indicative isotherm at 200 m, but much less would be expected at 400 m. Because of the depth limitation of the common BT, however, the indicative isotherm at 200 m is more convenient for practical use at the present time.

To watch fluctuations in the position of the Kuro-

shio axis, some sections have been established since prewar days though they have been dislocated occasionally. Practical values of the indicative isotherm at 200 m in the watch-sections are listed in Table 8. They are given to the nearest half-degree, and were obtained by linear interpolation against longitude from the indicative isotherm values listed in Table 7.

The value of surface velocity at the axis (Table 7) is basically underestimated, because GEK measurements made at intervals of some distance cannot catch the maximum current. Accordingly, the discussion to follow is limited to its relative variation.

The surface velocity at the axis reaches a maximum after the Kuroshio passes Inubozaki (Area IV, Fig. 24), and afterwards decreases downstream. The velocity increase seems to correlate very roughly with the change in the average of the horizontal temperature gradients at 200 m and 400 m (Table 7). The Kuroshio in Area I is confined within a trough of the Okinawa Basin, the depth of which is shallower than 1500 m in Area I; exclusion of Area I, because of this different situation, makes the correspondence a little better. The velocity variation also suggests that the threshold to the decay region suggested by Morgan (1956) lies, strictly speaking, in Area IV, where a meander in the Kuroshio path, convex to the north, is found in most cases.

6.3   Bias in estimates of the values owing to discrete sampling

Asymmetric distribution of surface currents on opposite sides of the current maximum (Sec. 5.1), however, gives rise to a bias in the estimates of position of the Kuroshio axis and of values of isotherms indicative of the axis based on measurements of current and temperature made at intervals of some finite distance. The bias owing to this discrete sampling will be exemplified in the following sampling experiment from population curves of surface current and 200 m temperature (Kawai, 1970).

Since no continuous measurements of current and temperature have been made across the Kuroshio, Masuzawa's (1969b, Figs. 1 and 2) averaged cross-stream distributions of the east–west component of surface current and 200 m temperature across the Kuroshio south of Japan at 136.5°E are used here as the population curves for the sampling experiment. His averaged distributions, reproduced by two curves in the upper and lower panels of Fig. 27, were based on a multi-ship survey in October–November 1967 (Masuzawa, 1968). Five research ships made 15 crossings of the Kuroshio high-velocity belt with BT lowerings and GEK fixes at 15 points spaced five nautical miles apart. The average was taken by adjustment of the cross-stream scale in such a manner that the current maximum in each of the crossings coincides with the zero-mile point of the scale, though the position of the maximum fluctuated over ten nautical miles during the 15 crossings.

A sampling experiment was made with three different intervals: 10, 20 and 30 nautical miles. The histogram of position of the sampled current maximum shows a square distribution as shown in the middle panel of Fig. 27. The mean positions of the sampled current maxima for the three sampling intervals, that is, the medians of the histograms indicated by black solid diamonds in the middle panel, shift 2.5, 5.5 and 7.5 nautical miles southward from the position of the current maximum of the population, respectively, owing to asymmetry in the population curve of current velocity. The maximum deviations of position of the sampled current maxima from the position of the current maximum of the population, given by the distances between the vertical line through the zero-mile point and the three verticals on the right hand in the upper panel of Fig. 27, amount to 7.5, 15.5 and 22.5 nautical miles for the three sampling intervals, respectively. The means of the sampled current maxima are 2.9, 2.5 and 2.2 kn, being 90, 78 and 69% of the current maximum of the population 3.2 kn, respectively.

In this experiment, the 200 m temperature at the axis itself is treated for simplicity, though a somewhat troublesome method was used to derive the values of indicative isotherms in Table 7. The range of 200 m temperatures at the sampled current maxima for each sampling interval is indicated by a column in the lower panel of Fig. 27. The mean temperatures at the sampled current maxima for the three sampling intervals are 13.8, 14.7 and 15.1°C, respectively, while the 200 m temperature at the current maximum of the population curve is 12.9°C.

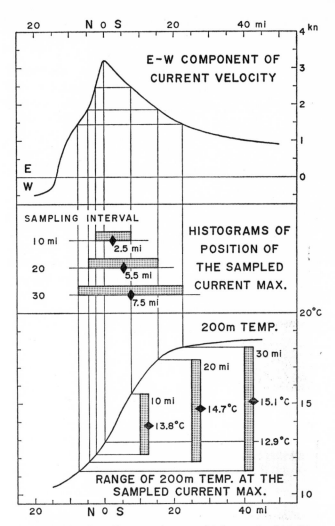

FIG. 27 Sampling experiments with intervals of 10, 20 and 30 nautical miles from the population curves of surface current and 200 m temperature across the Kuroshio.

(Upper) Population curve of east–west component of surface current across the Kuroshio reproduced from Masuzawa (1969b, Fig. 1). (Middle) Histograms of position of the sampled current maximum. Black solid diamonds indicate the medians of the histograms. Numerals beside them indicate their positions measured from the population current maximum. (Lower) Population curve of 200 m temperature across the Kuroshio reproduced from Masuzawa (1969b, Fig. 2). Columns indicate ranges of 200 m temperatures at the sampled current maxima for the three sampling intervals. Numerals beside black solid diamonds indicate the mean temperatures at the sampled current maxima.

The values of the above-mentioned mean temperatures 14.7 and 15.1°C for the sampling intervals of 20 and 30 nautical miles are very close to the value of 14.9°C listed in Table 7 as the 200 m temperature at the axis in Area III close to 136.5°E at which the population curves were obtained. Since it was required that the distance between adjacent stations be less than 25 nautical miles for the GEK and 200 m temperature for all data used in the statistics (Sec. 6.1), the closeness of the values might afford evidence to prove the reality of the above sampling model. This tempts us to adopt values about 2°C lower than those shown in Table 8 as the values of isotherms indicative of the Kuroshio axis; but this correction is not practical for a mixed use of temperature and GEK data for the determination of the Kuroshio axis, there being no way to make adequate corrections for the latter data from a discrete sampling.

The current distribution as well as 200 m temperature distribution across the Kuroshio or its Extension is not always as simple as given in Fig. 27. It occasionally shows a streaky structure with two or three zones of high velocity separated by zones of lower velocity as presented in Worthington's (1954) detailed cross-sections of the Gulf Stream. No sampling experiment from such a streaky distribution was made, but the bias in the statistical estimates would be more complicated.

The mean of the maximum and minimum temperatures of the indicative isotherm at a depth of 200 m in Table 8 is equal to 15°C, which coincides with Uda's (1964) value of the indicative isotherm cited in the beginning of this section. Because the mean temperature gradient at a depth of 200 m across the Kuroshio and its Extension is 0.2–0.3°C/mi as shown in Table 7, a small change in the value of the indicative isotherm shifts the estimated Kuroshio axis a short distance in most cases. It may be trifling for the purpose of most data analyses, but the adoption of 15°C as the value of the indicative isotherm at a depth of 200 m throughout the Kuroshio and its Extension from upstream to downstream may produce some inconvenience. The temperature gradient is not always as great as the mean; the gradient of about 0.1°C/mi occurs once in every ten times approximately, and that of 0.05°C/mi does rarely.

7 MEANDERS AND EDDIES

There are so many meanders and eddies with various scales in space and time in the Tohoku Area, that it is necessary to restrict any description to some meanders and eddies with specified scales of space and time. As a lower limit in the scales one may take some 100 km in wavelength and some a month in duration (Stommel, 1963a, 1963b). Meanders and eddies with scales smaller than the above criterion are generally beyond the scope of detailed description. To observe these meanders and eddies, careful consideration should be given to the sampling program.

7.1 Translational speeds

Let C_n denote the speed of horizontal translation of an isopleth such as an isotherm or isohaline on a certain level. Let the two lines A and B in Fig. 28 indicate the locations of an isopleth at two successive instances t_o and $t_o + \delta t$, respectively. The magnitude of C_n is easily calculated by the ratio of displacement of the isopleth, measured along the normal common to the two lines, Δn to δt, where the positive C_n is directed from A to B. The definition of C_n is given below together with three other kinds of translational speeds, where f indicates a certain property of sea water like temperature or salinity. As illustrated in Fig. 29, the *speed of spatial translation of an equiscalar surface* is given by

$$C_q = \left(\frac{\partial q}{\partial t}\right)_f = \frac{-\dfrac{\partial f}{\partial t}}{\dfrac{\partial f}{\partial q}}, \qquad (1)$$

q being the length measured along the normal to the surface; the *speed of vertical translation of an equiscalar surface* is given by

$$C_z = \left(\frac{\partial z}{\partial t}\right)_f = \frac{-\dfrac{\partial f}{\partial t}}{\dfrac{\partial f}{\partial z}}, \qquad (2)$$

z being the vertical coordinate; the *speed of horizontal translation of an isopleth on a certain level* is given by

$$C_n = \left(\frac{\partial n}{\partial t}\right)_f = \frac{-\dfrac{\partial f}{\partial t}}{\dfrac{\partial f}{\partial n}}, \qquad (3)$$

n being the length measured along the horizontal normal to the isopleth and directed to the vertical projection of q on a level surface; the *speed of linear translation of an isopleth*, that is, the speed of intersection of an isopleth on a certain level with a straight line on the level is given by

$$C_x = \left(\frac{\partial x}{\partial t}\right)_f = \frac{-\dfrac{\partial f}{\partial t}}{\dfrac{\partial f}{\partial x}}, \qquad (4)$$

x being the coordinate along the straight line. The velocities of spatial and horizontal translations are given by the vector notations

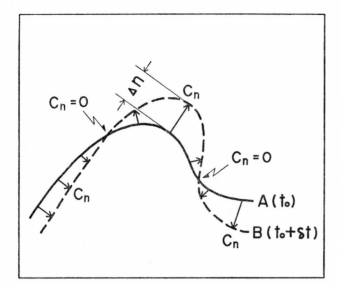

FIG. 28   Illustration of the speed (C_n) of horizontal translation of an isopleth on a certain level. Lines A and B indicate two locations of an isopleth at two successive instances t_o and $t_o + \delta t$. The symbol Δn indicates the displacement of the isopleth, measured along the normal common to the two lines A and B.

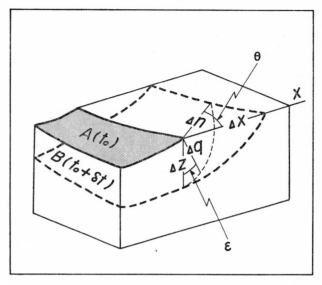

FIG. 29   Relation between the four kinds of translational velocity. Surfaces A and B indicate two locations of an equiscalar surface at two successive instances t_o and $t_o + \delta t$. The symbol Δq indicates the displacement of the equiscalar surface measured along the normal common to the two surfaces A and B. For other symbols, see the text.

$$C_q = \frac{-\dfrac{\partial f}{\partial t} \nabla f}{(\nabla f)^2} \tag{5}$$

and

$$C_n = \frac{-\dfrac{\partial f}{\partial t} \nabla_h f}{(\nabla_h f)^2}, \tag{6}$$

where ∇_h indicates the horizontal component of gradient operator.

There are some relationships between the four kinds of translational speeds (Fig. 29):

$$\frac{C_n}{C_x} = \left(\frac{\partial n}{\partial x}\right)_f = \cos \theta, \tag{7}$$

$$\frac{C_q}{C_n} = \left(\frac{\partial q}{\partial n}\right)_f = \sin \varepsilon, \tag{8}$$

$$\frac{C_q}{C_z} = \left(\frac{\partial q}{\partial z}\right)_f = \cos \varepsilon \tag{9}$$

and

$$\frac{1}{C_x{}^2} + \frac{1}{C_y{}^2} + \frac{1}{C_z{}^2}$$
$$= \frac{1}{C_n{}^2} + \frac{1}{C_z{}^2} = \frac{1}{C_q{}^2}. \tag{10}$$

Equation (7) means that any supposition of speed of translation of an isopleth on a certain level from sequences of observations along a single watch-section becomes inevitably overestimated unless the section is cross-isopleth. It is noteworthy, as shown in the above relations, that the subscript of the translational speed does not indicate the vector component.

For the meanders and eddies, qualified in the beginning of this section, one may write empirically (Sec. 7.4)

$$C_n \sim 0.1\text{–}0.2 \text{ kn (5–10 cm/sec)}$$

and

$$\sin \varepsilon \sim 10^{-4}\text{–}10^{-2}.$$

With the aid of eqs. (7), (8) and (9) we get

$$C_q \sim C_z \sim 10^{-3}\text{–}10^{-1} \text{ cm/sec}$$

and

$$C_q \leq C_z < C_n \leq C_x.$$

If f is a conservative property and effects of diffusion and air–sea interactions are negligibly small, we have

$$C_q = V_q$$

and

$$C_n = V_n + w \cot \varepsilon,$$

where V_q, V_n and w indicate the current components in the directions of q, n and z, respectively. This means that, under the above conditions, C_q is equal to V_q, but that C_n is different from V_n unless w vanishes.

As shown by eq. (10), C_q and C_n are functions of C_x, C_y and C_z. On the basis of eqs. as given by (2) and (4), values of C_x, C_y and C_z can be calculated by data from hydrographic soundings at three stations located at apices of a triangle at two successive instances. The evaluations of C_q and C_n above are similar to the Eulerian method in fluid dynamics, while those based on Fig. 28 are similar to the Lagrangian method. In Secs. 7.2 and 7.3, evaluations of the translational speed of an isotherm indicative of a stream axis or front will be made on the basis of the latter method.

The *speed of horizontal translation of an eddy* on a certain level is defined as the horizontal speed of the center of gravity of domain inside a closed isotherm or isohaline indicative of the current maximum at the boundary of the eddy. If the horizontal cross-section under consideration is taken at a depth with small seasonal variation of temperature and salinity, for example at a depth of 200 m, tracking of the eddy by the closed isotherm or isohaline is possible for a long time.

Although the *speed of horizontal translation of meanders* cannot be defined as easily as the above, one may regard the speed of crest or trough of the meanders as the speed, if the meander pattern does not change rapidly. Since such meanders are propagated as a wave, there is a relation between the propagation speed C_p and the speed of horizontal translation of an isopleth indicative of the meanders C_n. Successive positions of an isopleth travelling with a constant shape in the direction of x-positive are expressed by

$$y = Y (x - C_p t),$$

where Y denotes any function. The relation between C_p and C_n is given by

$$C_n{}^2 = \frac{\left(\dfrac{\partial y}{\partial x}\right)^2}{1 + \left(\dfrac{\partial y}{\partial x}\right)^2} C_p{}^2. \tag{11}$$

This equation leads to an inequality $C_n{}^2 < C_p{}^2$. When $(\partial y/\partial x)^2 \gg 1$, we have $C_n{}^2 \doteqdot C_p{}^2$. When $(\partial y/\partial x)^2 \ll 1$, we have $C_n{}^2 \ll C_p{}^2$. The latter case cannot easily be observed, because it needs a detailed sampling network. The former case, in which C_n is smaller than C_p but is of the same order of magnitude, can be observed.

7.2 Evaluation of the speed of horizontal translation: IGY Kuront '57

A multi-ship survey, carried out around the Kuroshio Front with the cooperation of four research ships of the JMAM, JHO, HMO and KMO during the period from the 8th of June to the 3rd of July, 1957 as a part of the IGY survey, is referred to by the code name IGY Kuront '57 in this chapter (Table 1). The survey consists of several phases: synoptic measurements with BT, GEK and Loran; tracking of four parachute drogues between depths of 50 and 600 m and successive lowerings of BT at intervals of half an hour at a fixed station. The synoptic measurements were reiterated five times intermittently, each of them lasting a few days; these are called CM, GS$_1$, GS$_2$, GS$_3$ and GS$_4$ by Takenouchi, Masuzawa and Yasui (1957), Masuzawa (1958) and Takenouchi (1958).

Figure 30 shows the analysis used to obtain the position of the 10°C isotherm at a depth of 200 m at each of the synoptic measurements, the neighbor synchronous index $\Delta d/\delta t$ (Sec. 7.5) being greater than 0.5 kn. The reasons why the 10°C isotherm was chosen for the analysis instead of the 14°C isotherm, which was statistically estimated in Sec. 6 as indicative of the Kuroshio Extension axis in this area, are that most ships' tracks in the survey did not cross the Kuroshio Front far enough to reach the 14°C isotherm and that the value of the indicative isotherm during the survey was lower than 14°C (Sec. 6.3). The 10°C isotherm was contoured by interpolating linearly along the ships' tracks and by orienting almost parallel to the current direction. The nearest hour at each crossing of a ship's track with the 10°C isotherm was estimated by linear interpolation along the ship's track.

Figure 31 shows the analysis used to obtain the speed of horizontal translation of the 10°C isotherm contoured in Fig. 30. Passing through the intersection of the 10°C isotherm in GS$_2$, which is the middle of five sequences of synoptic measurements, with each meridian of half degree, a normal common to the 10°C isotherms in GS$_2$ and GS$_1$ (or GS$_2$ and GS$_3$) was drawn. Next, passing through the intersection of the 10°C isotherm in GS$_1$ (or GS$_3$) with the normal drawn above, a normal common to the 10°C isotherms in GS$_1$ and CM (or GS$_3$ and GS$_4$) was drawn. The horizontal displacement Δn was measured by the above procedure in the vicinity of each of the half degree meridians. On the basis of time at both ends of each normal (their positions are shown by dots in Fig. 30) estimated by interpolating linearly along each of the 10°C isotherms, the time difference

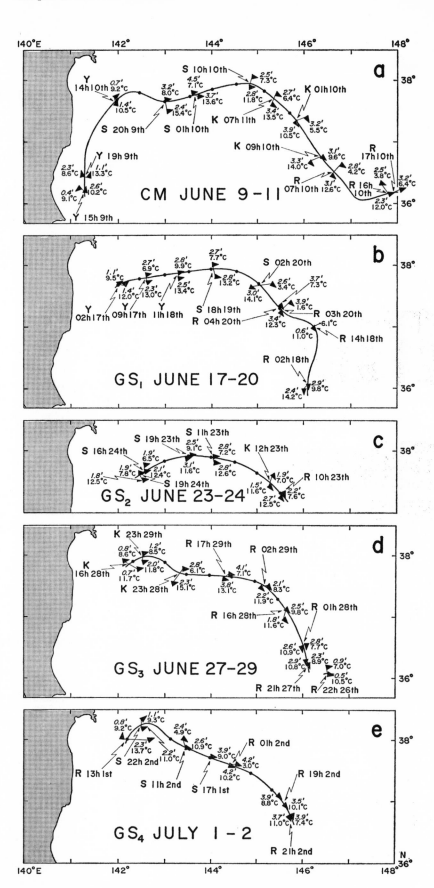

FIG. 30 Analysis obtaining position of the 10°C isotherm at 200 m, based on data from IGY Kuront '57. Triangles indicate directions of surface current measured with GEK. Small numerals in italic and roman types beside the triangles indicate speeds of the surface current and 200 m temperature measured with BT, respectively. *R, Ryofu Maru* of JMAM; *K, Kaiyo* of JHO; *Y, Yushio Maru* of HMO; *S, Shumpu Maru* of KMO. Numerals beside the ship's mark indicate the nearest hour, estimated by linear interpolation, at each crossing of ship's track with the 10°C isotherm. The neighbor synchronous index $\Delta d/\delta t$ (Sec. 7.5) is greater than 0.5 kn.

between both ends of the normal, δt, was computed. The direction of the normal and the values of Δn, δt and C_n, which was computed as a ratio $\Delta n/\delta t$, are listed in Table 9. Conclusions from the above analysis are that the speed of horizontal translation of the isotherm reaches about 0.2 kn at most, but that histograms of this speed are shaped like an inverted letter J with a mode at the left-end class of an interval between 0 and 0.025 kn (Fig. 31, upper right-hand corner).

TABLE 9.   Results of an analysis to obtain the horizontal translational velocity of the 10°C isotherm at 200 m during IGY Kuront '57 based on Figs. 30 and 31.

Longitude (E)		142°00′	142°30′	143°00′	143°30′	144°00′	144°30′	145°00′	145°30′
CM	Time	13h 10th	05h 10th	20h 9th	00h 10th	04h 10th	08h 10th	23h 10th	10h 10th
	Direction of n	140°	180°	000°	350°	350°	180°	220°	230°
	Δn (mi)	2′	2′	11′	8′	5′	2′	9′	14′
	δt (hour)	157	170	196	212	228	230	219	234
	C_n (kn)	0.01	0.01	0.06	0.04	0.02	0.01	0.04	0.06
GS₁	Time	02h 17th	07h 17th	00h 18th	20h 18th	16h 19th	22h 19th	02h 20th	04h 20th
	Direction of n	—	160°	210°	000°	180°	200°	220°	050°
	Δn (mi)	—	8′	1′	2′	3′	5′	3′	1′
	δt (hour)	—	179	150	119	92	85	82	78
	C_n (kn)	—	0.04	0.01	0.02	0.03	0.06	0.04	0.01
GS₂	Time	—	18h 24th	06h 24th	19h 23rd	12h 23rd	11h 23rd	12h 23rd	10h 23rd
	Direction of n	—	330°	020°	190°	180°	190°	200°	220°
	Δn (mi)	—	17′	3′	13′	13′	10′	4′	2′
	δt (hour)	—	103	122	126	143	148	138	130
	C_n (kn)	—	0.17	0.02	0.10	0.09	0.07	0.03	0.02
GS₃	Time	—	01h 29th	08h 29th	01h 29th	11h 29th	15h 29th	06h 29th	20h 28th
	Direction of n	—	320°	030°	020°	010°	190°	200°	220°
	Δn (mi)	—	8′	9′	9′	2′	4′	9′	7′
	δt (hour)	—	60	69	67	54	59	75	92
	C_n (kn)	—	0.13	0.13	0.13	0.04	0.07	0.12	0.08
GS₄	Time	—	13h 1st	05h 2nd	20h 1st	17h 1st	02h 2nd	09h 2nd	16h 2nd

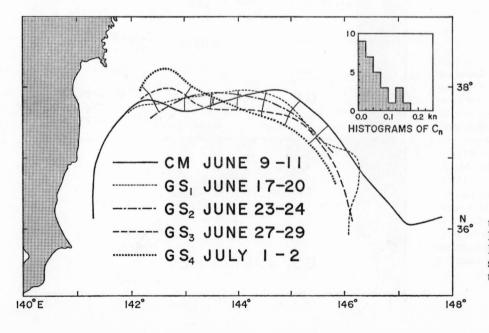

FIG. 31   Change in position of the 10°C isotherm at 200 m during IGY Kuront '57 reproduced from Fig. 30 and histograms of the speed of horizontal translation of the 10°C isotherm (Table 9).

7.3 Evaluation of the speed of horizontal translation: Detaching a big anticyclonic eddy in 1960

As mentioned in Sec. 2.7, it is difficult to show the process of detachment of an eddy in detail, because the shifting of isotherms for detachment of an eddy is very rapid. A rough representation of the process of detachment of a big anticyclonic eddy into the inshore Perturbed Area, observed once in a while (Sec. 9), however, is not as difficult as for smaller eddies, because the accompanying meanders of the Kuroshio Extension are larger in scale and because their evolutions are perhaps slower. To show the gross features of the process of detachment of a big anticyclonic eddy in the summer of 1960, the change in pattern of isotherms at a depth of 200 m or at the surface and that of currents at the surface measured with GEK during the period from February 1960 to August 1961 is presented in Fig. 32. Fish Kuront '60 covered the detachment area, but, unfortunately, some of the measurements did not reach a depth of 200 m. To use the shallow temperature data from Fish Kuront '60 together with other data, Hata (1969b) showed the gross features of change during a similar period in isotherms at a depth of 100 m, though these are subject to a marked seasonal variation at this depth.

In Fig. 32, some crowded stations are omitted where they do not satisfy the neighbor synchronous condition (Sec. 7.5)

$$S_n = \frac{\Delta d}{\delta t} \geq C_n.$$

Here C_n indicates the speed of horizontal translation of isotherms assumed to be 0.1 or 0.2 kn. The stations within dotted envelopes in Figs. 32h, 32l and 32n do not satisfy the condition, but are taken into consideration in contouring the isotherms because of their important location for the eddy.

The axis of the Kuroshio Extension, well indicated with the 14°C isotherm at a depth of 200 m (Sec. 6), showed an increasing amplitude of meanders from February to July 1960, and had detached a big anticyclonic eddy by August 1960. The 200 m temperature at the central portion of the eddy was higher than 16°C till about November 1960, but then decreased to a temperature of about 11°C by the winter turnover. The surface temperature after the winter again exceeded a temperature of 11°C, but the 200 m temperature remained at about 11°C, showing a small and gradual decrease probably owing to horizontal mixing. The 200 m temperature inside the anticyclonic eddy in the Perturbed Area,

therefore, may offer a useful way for discriminating whether the eddy had passed a winter or not.

Figure 33 shows a change in position of the 14°C isotherm contoured in Fig. 32. The north–south shifting of the crest of meanders of the Kuroshio Extension axis, as inferred from the change of latitude at the north limit of the 14°C isotherm at 200 m in Fig. 33, usually has a speed less then 0.1 kn (Table 10). The speed for the phase immediately after the detachment of the eddy, however, cannot be evaluated, because the shifting of the isotherm is geometrically discontinuous. For this reason, the value in parentheses for the phase in Table 10 is kinematically meaningless.

7.4 Evaluation of the speed of horizontal translation: Summary

Table 11 summarizes results of evaluation of the horizontal translational speed of a stream axis or eddy based on several surveys in the Kuroshio as well as in the Gulf Stream, including results from the two surveys mentioned in Secs. 7.2 and 7.3.

With data from IGY Kuront '57 (Sec. 7.2), Masuzawa (1958) estimated that the coldest parts of the "cold belt" (inner cold belt) appearing in the mean temperature of the upper 200 m layer moved along the northern edge of the Kuroshio Extension at speeds of about one knot, in accord with the drift speed of parachute drogues at depths of 100–200 m in the cold belt (Table 11). With data from Fish Kuront '59, Kawai and Sasaki (1961) estimated that a cold patch in the inner cold belt appearing at a depth of 50 m moved at speeds of 0.1–0.2 kn (Table 11). They ascribed the difference in speed between Masuzawa's and their estimates to different hydrographic conditions around the inner cold belt, but I would like to ascribe it to speculative ways of identifying the patch with extreme water characteristics among sequences of quasi-synoptic surveys for both the estimates.

Excluding the two estimates mentioned above, the meaningless estimate of 0.3–0.4 kn from Table 10 relating to the abrupt southward shifting of the Kuroshio Extension axis immediately after detaching the big anticyclonic eddy and the value of about 0.5 kn obtained by Fuglister and Worthington (1951), perhaps relating to a transient stage detaching the cold eddy "Edgar," one may assume, from Table 11, that the speed of horizontal translation is usually less than 0.2 kn, for distances as great as some 100 km and periods as long as some a month. If the time interval between sequences of synoptic measurements is shortened, however, a greater speed may be ob-

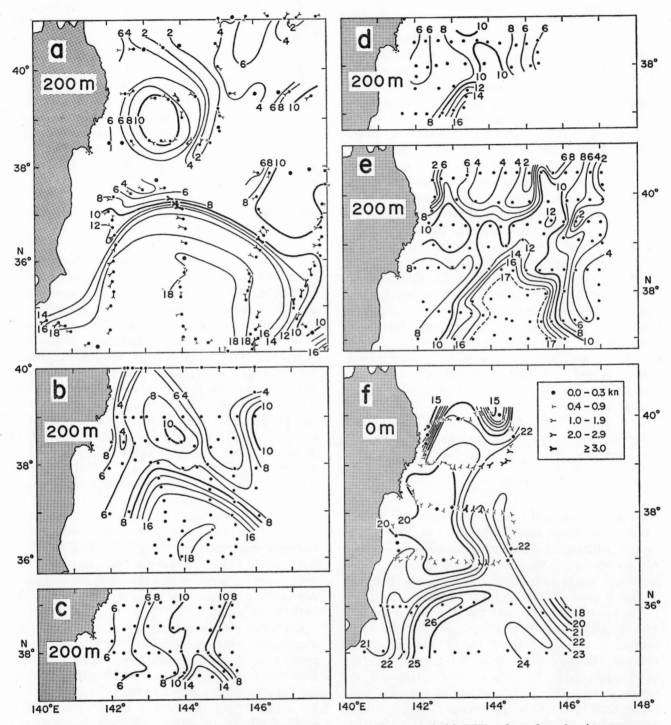

FIG. 32 Temperature at a depth of 200 m and temperature and current measured with GEK at the surface, showing processes of detaching and decaying of a big anticyclonic eddy from February 1960 to August 1961. The neighbor synchronous indexes $\Delta d/\delta t$ (Sec. 7.5) are greater than 0.2 kn for Figs. (b), (c), (d), (f), (i), (j), (k) and (m); and than 0.1 kn for Figs. (a), (e), (g), (h), (l) and (n).

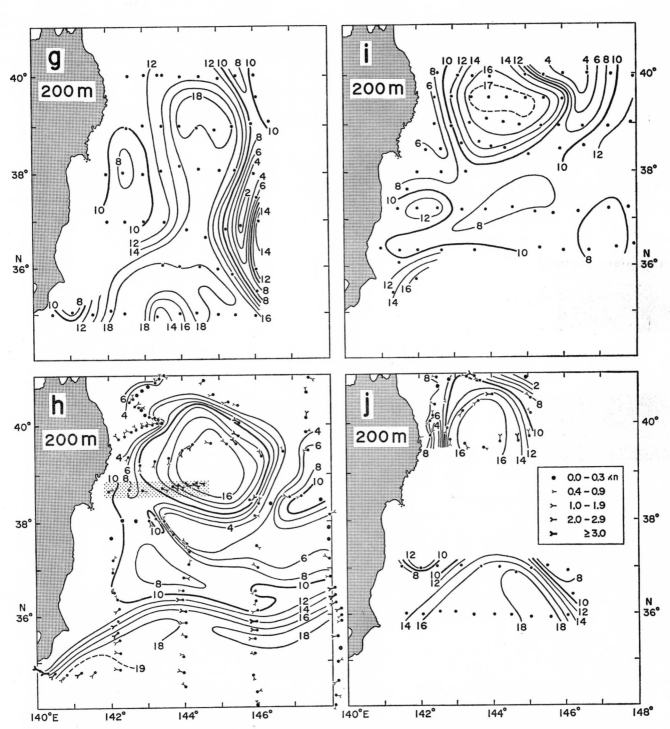

The stations within the dotted envelope do not satisfy the condition. (a), 17 Feb.–9 Mar. 1960; (b), 17–28 Apr. 1960; (c), 7–12 May 1960; (d), 16–22 May 1960; (e), 10–21 June 1960; (f), 5–10 July 1960; (g), 1–21 July 1960; (h), 12–21 Aug. (for dotted area, 3–4 Sept.) 1960; (i) 5–10 Oct. 1960; (j), 6–17 Nov. 1960; (k), 16–24 Dec. 1960; (l), 15 Feb.–14 Mar. (for dotted area, 8–9 Mar.) 1961; (m), 22 July–1 Aug. 1961; (n), 3–27 Aug. (for dotted area, 25–26 Aug.) 1961.

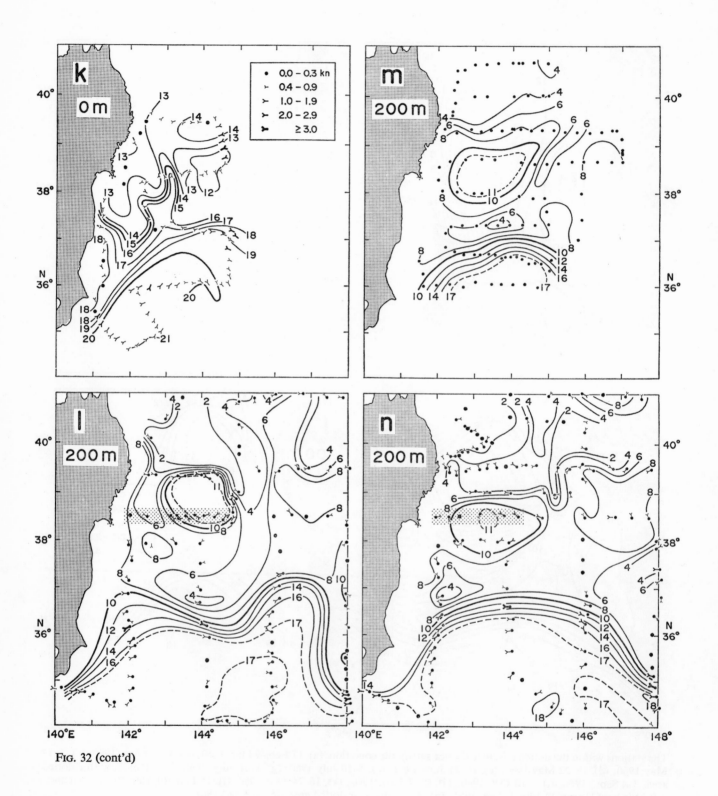

FIG. 32 (cont'd)

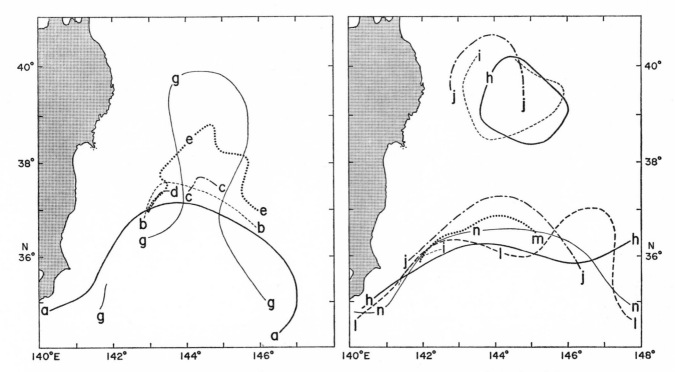

FIG. 33   Change in position of the 14°C isotherm at 200 m, reproduced from Fig. 32, as indicative of the Kuroshio Extension axis or a boundary of warm water inside the detached eddy from February 1960 to August 1961. The letters *a–n* beside the lines correspond to those in Fig. 32. (Left) Before detaching the big eddy in Fig. 32. (Right) After detaching the big eddy in Fig. 32.

TABLE 10.   Results of an analysis to obtain the north–south speed of the crest of meanders of the Kuroshio Extension axis before and after detaching a big anticyclonic eddy, as inferred from the change of latitude at the north limit of the 14°C isotherm at 200 m from February 1960 to August 1961, on the basis of Figs. 32 and 33.

Date	Lat. of the north limit	Displacement* (mi)	Days interval	Speed (kn)	Reference Figs.
19 Feb. 1960	37°10′N				32 a
		25	68	0.02	
27 Apr. 1960	37°35′N				32 b
		5	14	0.01	
11 May 1960	37°40′N				32 c
		70	35	0.08	
15 June 1960	38°50′N				32 e
		65	34	0.08	
19 July 1960	39°55′N				32 g
		(−220)	27	(0.34)	
15 Aug. 1960	36°15′N				32 h
		60	85	0.03	
8 Nov. 1960	37°15′N				32 j
		−55	100	0.02	
17 Feb. 1961	36°20′N				32 l
		30	158	0.01	
25 July 1961	36°50′N				32 m
		−15	12	0.05	
6 Aug. 1961	36°35′N				32 n

*   A negative sign indicates a southward displacement.

TABLE 11. Values of velocity of horizontal translation.

Survey (Reference)	Period	Characteristic pattern	Displacement (mi)	Time interval (hour)	Translational velocity	
					Direction	Speed (km)
IGY Kuront '57 (Masuzawa, 1958)	June–July 1957	Coldest parts of the inner cold belt in the mean temp. of the upper 200 m layer moving along the northern edge of the Kuroshio Extension	150–180	190–220	ESE	0.8
IGY Kuront '57 (Table 9)	June–July 1957	Shifting of the 10°C isotherm at 200 m	<17	60–230	—	<0.2
Fish Kuront '59 (Kawai & Sasaki, 1961)	May–July 1959	Cold patch at 50 m moving along the inner cold belt	20–40	200	mostly ENE–SE	0.1–0.2
(Yosida, 1961; Moriyasu, 1961; Shoji, 1965)	May–June 1959	Phase propagation of the Kuroshio meanders south of Japan	20–40	150–200	E	0.1–0.2
Fish Kuront '60 and others (Table 10)	Feb. 1960–Aug. 1961	Shifting of large-scale meanders of the Kuroshio Extension before and after detaching a big anticyclonic eddy — Usually: / Immediately after the detaching:	5–70 / (220)	300–2400 / 320	N–S / N–S	0.1 / (0.3–0.4)
Operation Cabot (Fuglister & Worthington, 1951)	June 1950	Shifting of the Gulf Stream off Cape Hatteras	4.5	24	SE	0.2
		All changes in the position of the Gulf Stream	<11	24	—	<0.5
Gulf Stream '60 (Fuglister, 1963)	Apr.–June 1960	Every observed change in the position of the Gulf Stream	?	?	—	<0.1
V-Fin '64 (Fuglister & Voorhis, 1965)	June 1964	Shifting of meanders of the Gulf Stream	20	200	E	0.1
V-Fin '65 & '66 (Hansen, 1970)	Sept. 1965–May 1966	Phase propagation of the Gulf Stream meanders	50–200	600–840	E	0.1–0.2
Deep current measurements in the Bay of Biscay (Gould, 1971)	Oct.–Dec. 1967	Propagation of a disturbance in current pattern	—	—	W	0.01–0.05

tained, especially for displacements of near-surface fronts and convolutions (Bratnick, 1970).

In Table 11, the propagation speeds of meanders C_p of the Kuroshio and Gulf Stream axes are evaluated. The speeds of horizontal translation of isotherms indicative of the axes C_n associated with the meanders propagation are inevitably smaller than these values of C_p as discussed in the end of Sec. 7.1.

7.5   Criterions in sampling program

When the difference in observation time between two neighboring stations a and b, with a distance of Δd, is δt (Fig. 34), the smaller Δd is the smaller δt should be. The reasonalbe relation between Δd and δt depends on the speed of horizontal translation of an isopleth C_n. Let an isopleth change location from A to B in a very short interval of time δt, with a speed of horizontal translation C_n in a direction of θ to the line ab. If a condition

$$C_x = \frac{C_n}{\cos\theta} \leq \frac{\Delta d}{\delta t} \qquad (12)$$

is satisfied, the apparent shifting of the isopleth along the line ab due to the time difference does not violate the stations beyond a range of Δd, resulting in a smaller deformation of isopleths. When θ approaches a right angle, it becomes difficult to satisfy the condition (12).

If one relaxes the condition (12), it becomes necessary to adopt the following inequality as a general criterion in the sampling program

$$S_n \geq C_n, \qquad (13)$$

where S_n is defined by

$$S_n \equiv \frac{\overline{\Delta d}}{\delta t}. \qquad (14)$$

Since δt and $\overline{\Delta d}$ are the difference in observation time and the distance between two neighboring stations, respectively, let us call S_n the *neighbor synchronous index* (隣接同時性示数). Accordingly, inequality (13) may be called the *neighbor synchronous condition* (隣接同時性条件). The condition (13) means that the apparent shifting of the isopleth due to the time difference may violate some stations on a section parallel to the isopleth, but the shifting normal to the isopleth remains within the distance of Δd.

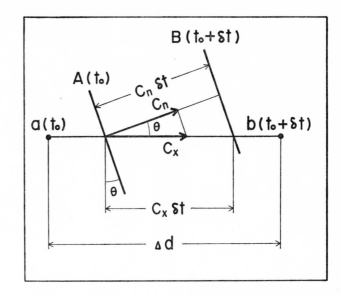

FIG. 34   Illustration of a criterion in the hydrographic sampling program to be satisfied with respect to three quantities: difference in measurement or sampling time δt, station interval Δd and speed of horizontal translation of a certain isopleth C_n. Lines A and B indicate two locations of a segment of an isopleth on a certain level at two successive instances t_0 and $t_0 + \delta t$ when the stations a and b, respectively, are occupied.

The neighbor synchronous condition (13) is only a local condition, which is apt to produce isopleths extensively deformed by the overlap of apparent displacement of isopleths due to the difference in observation time. For the strictly synoptic analysis, let us introduce the *area synchronous index* (地域同時性示数) defined by

$$S_a \equiv \frac{\overline{\Delta d}}{\delta t_a}, \qquad (15)$$

where $\overline{\Delta d}$ is the mean mesh size of a homogeneous sampling network in an area, and δt_a is the time required for occupying all stations in the area. If the mesh size changes greatly, one must divide the area into parts where the mesh size is almost constant.

By using S_a, the *area synchronous condition* (地域同時性条件) is expressed by an inequality

$$S_a \geq C_n. \qquad (16)$$

Since S_a is far smaller than S_n in most cases, it becomes difficult to satisfy the condition (16) in a conventional sampling program. For instance, if the stations are arranged in a rectangular net of 20-nautical-mile meshes, the permissible time δt_a should

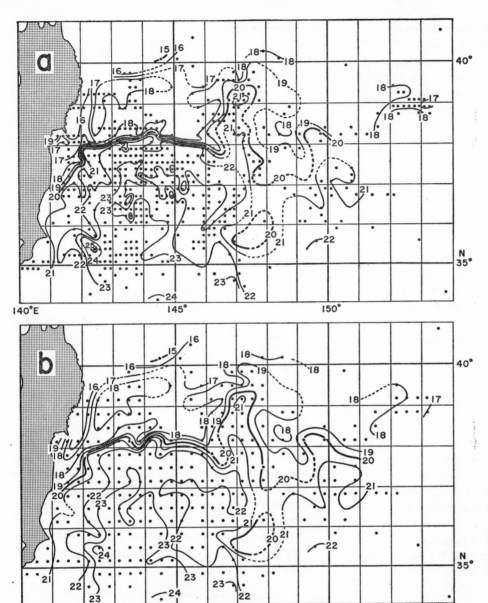

FIG. 35 Difference in the pattern of surface isotherms during 11–15 July 1953 due to the mesh size (Kawai, 1958b, Figs. 3–6). (a), 10-minute-square; (b), 20-minute-square; (c), 30-minute-square; (d), one-degree-square. Dots indicate centers of each square with averaged values.

be less than 100 hours, which is calculated by the condition (16) with the use of 0.2 kn for C_n (Sec. 7.4), regardless of the size of the area covered by stations.

To check the effect of sampling intervals in space upon contouring meanders and eddies, Fig. 35 was prepared on the basis of surface temperatures during July 11–15, 1953, taken by five research ships and cooperating skipjack fishing-boats equipped with pole-and-lines (Kawai, 1958b). Since the surface temperature was measured hourly under way on most ships, voluminous data were obtained after excluding errors. Four different contourings were made by taking averages by squares with different dimensions of $10' \times 10'$ (Fig. 35a), $20' \times 20'$ (Fig. 35b), $30' \times 30'$ (Fig. 35c) and $1° \times 1°$ (Fig. 35d) in latitude and longitude. In Fig. 35a there are insignificant

irregularities owing to errors in measurements of temperature and position, to diurnal and daily fluctuations in temperature resulting from the difference in measurement time and to the difference in degree of disturbance of the surface film of hot water by ships and apparatus for the measurements. In Fig. 35d the isotherms are so simplified that an isolated warm eddy northeast of Kinkazan Island appearing in Figs. 35a, 35b and 35c is connected with warm water at the northern edge of the Kuroshio Extension. The warm eddy is also seen distinctly in the thermal structure at a depth of 100 m (Fig. 36). To delineate eddies with a diameter of 100 km, using average values of temperature obtained by squares with dimensions of about $20' \times 20'$ in latitude and longitude may be suitable.

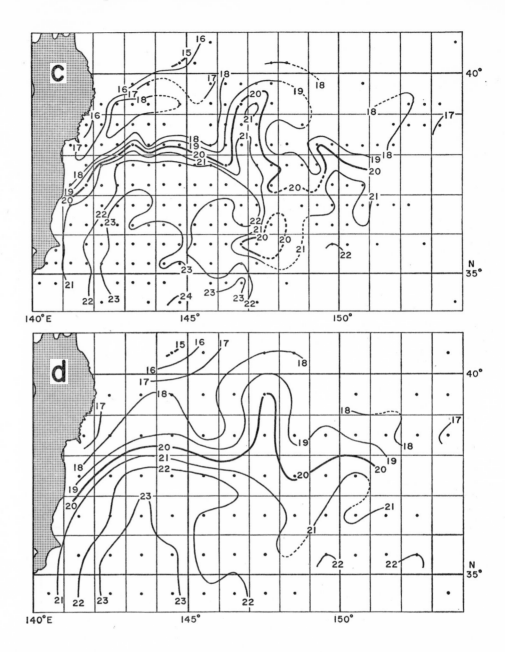

FIG. 35 (cont'd)

7.6 Stationary meanders

Figures 37 and 38 show two typical cases of the meanders of the Kuroshio Extension path. The 14°C isotherm at 200 m, which is indicative of the axis of the Kuroshio Extension (Sec. 6), runs along the sharp horizontal gradient of 200 m temperature. The path in August 1955 (Fig. 37) shows much distorted meanders, while the path in May 1959 (Fig. 38) shows a regular sine curve with a wavelength of about 350 nautical miles accompanied by two big anticyclonic eddies at two northward crests of meanders.

Compiling a table of quantities relative to meanders of the Kuroshio path on the basis of the quarterly oceanographic charts entitled "State of the Adjacent Seas of Japan" from 1955 to 1963 published by the JHO, Uda (1964) stated that the wavelengths of the meanders south and east of Japan are between approximately 300 and 500 nautical miles.

Figure 39, based on working charts of 200 m temperature and of surface currents measured with GEK, shows fluctuations in longitudes at the north limits of two northward crests of meanders of the Kuroshio Extension, assigning values for each season from 1955 to 1969. Fixing the longitude is less accurate than the latitude, which will be given later (Sec. 9), because the path takes a flat shape in the vicinity of the northward crest. Accordingly, the

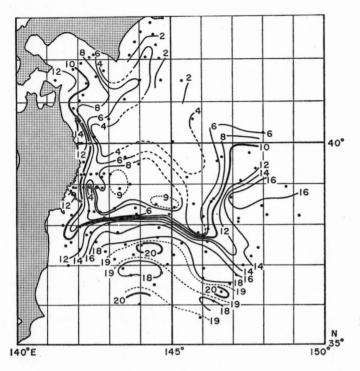

FIG. 36 Temperature at 100 m in the middle ten-day period of July 1953 (Kawai, 1958b, Fig. 7).

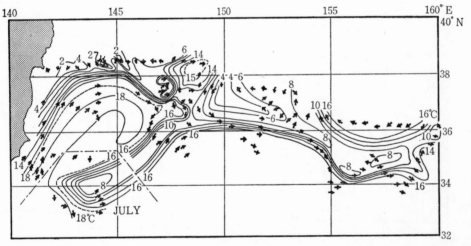

FIG. 37 Surface current measured with GEK and 200 m temperature in August 1955 based on data from the *Ryofu Maru* of the CMO (Masuzawa, unpublished).

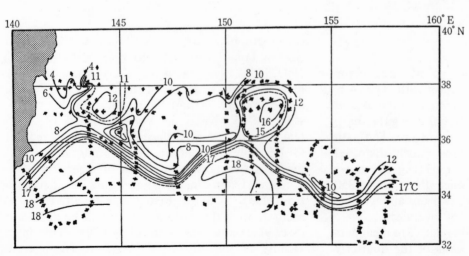

FIG. 38 Surface current measured with GEK and 200 m temperature in May 1959 based on data from the *Ryofu Maru* of the JMAM (Masuzawa, unpublished).

longitude is read to the nearest half degree. The longitude of trough of the meanders is not shown in Fig. 39 because of its ambiguity. Histograms of the longitudes in the lower panel of Fig. 39 show a mode for the inshore crest at 144°E and for the off-shore at 149.5°E. The wavelength measured as a difference between the two longitudes amounts to 5.5°, which is about 270 nautical miles. Histograms of the wavelength fixed for each individual axis of the Kuroshio Extension (Fig. 40) also show a mode at 5.5° in longitude. Thus, we may conclude that the first and second crests are located at about 144°E and 149.5°E, respectively and that the wavelength is about 270 nautical miles on the average. Several attempts have been made to explain the wavelength as a result of the stationary planetary wave, but variability in the mean speed or the mean total transport, which is subject to differences in measuring the depth and width of the Kuroshio Extension, gives rise to a wide possible range of wavelengths.

Recent fluctuations in the longitudes of the crests (Fig. 39), however, show no uncommon occurrence of wavelengths far shorter than 270 nautical miles, which might be related to the recent trend of southward shifting of the north limit of the Kuroshio Extension axis in the inshore Tohoku Area (Sec. 9.1). Since meanders with such short wavelengths seem to move rapidly (Hansen, 1970), Fig. 39 must have missed some of them.

There are two seamounts (*Takuyo* #2 SMT 1384 m, *Takuyo* #3 SMT 1449 m; Fig. 3) in the vicinity of the southern edge of the "cold sock" at about 34°N, 144°E in Fig. 37, but more examples are necessary to confirm the relation.

According to Fukuoka's (1958a, 1958b) study, the

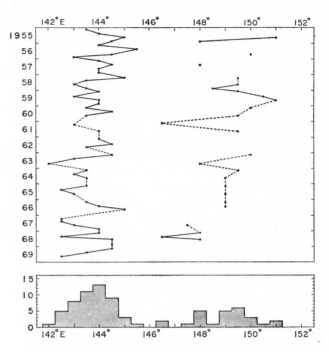

FIG. 39 Fluctuations in longitudes at the north limits of two northward crests of meanders of the Kuroshio Extension and histograms of the longitudes.

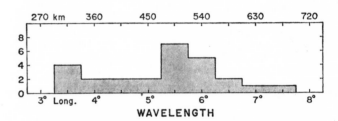

FIG. 40 Histograms of the wavelength measured as the difference in longitude between two northward crests of meanders of the Kuroshio Extension.

TABLE 12.   Main expeditions around the Northwest Pacific Rise.

Code	Period*	Agency or Institutions*	Ship*	Reference Figs.
Navypac '39s	Aug. 1939	JIHO Aomori PFES Miyagi PFES Fukushima PFES	*Aomori Maru* *Miyagi Maru* *Iwaki Maru*	41a
Navypac '40s	Aug. 1940	JIHO Aomori PFES Miyagi PFES Fukushima PFES	*Aomori Maru* *Miyagi Maru* *Iwaki Maru*	41b
Norpac	Aug. 1955	CMO	*Ryofu Maru*	41c
IGY Poront '57	Sept.–Nov. 1957	JMAM JHO TUF	*Ryofu Maru* *Takuyo* *Umitaka Maru*	41e, 42, 43, 44
Jeds-5	May 1962	JMAM	*Ryofu Maru*	41d
CSK '67s	July 1967	JHO	*Takuyo*	45
CSK '68s	July 1968	JHO	*Takuyo*	———

* Only those around the Northwest Pacific Rise are included.

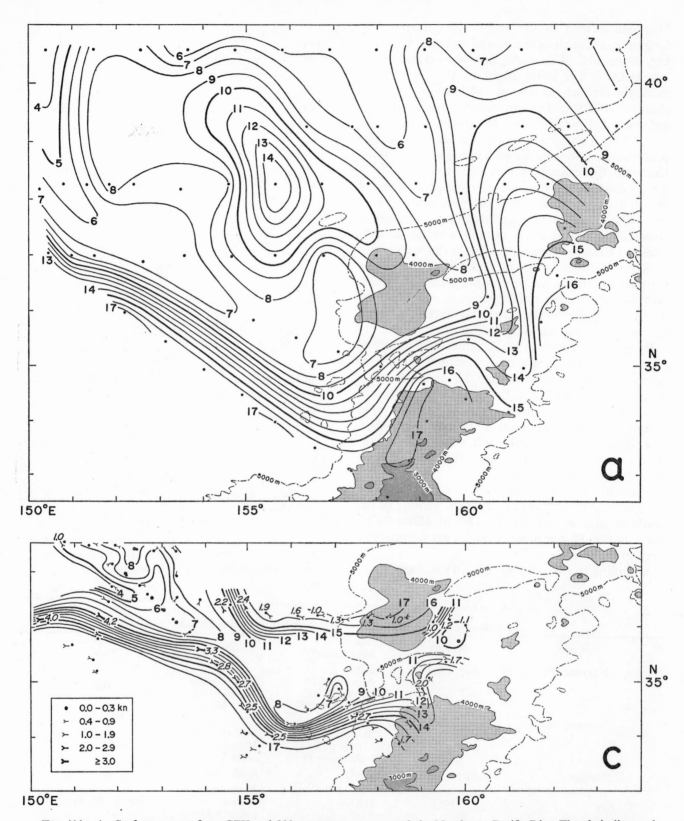

FIG. 41(a, c)   Surface current from GEK and 200 m temperature around the Northwest Pacific Rise. The chain line and thin line indicate 5000 m and 4000 m depth contours of the bottom, respectively. Hatched area indicates where the bottom is shallower than a depth of 3000 m. (a), Based on Navypac '39s in 11–27 August 1939 (Table 12). The neighbor synchronous index $\Delta d/\delta t$ (Sec. 7.5) is greater than 0.2 kn. (c), Based on Norpac in 21–28 August 1955 (Table 12). The neighbor synchronous index $\Delta d/\delta t$ (Sec. 7.5) is greater than 0.1 kn.

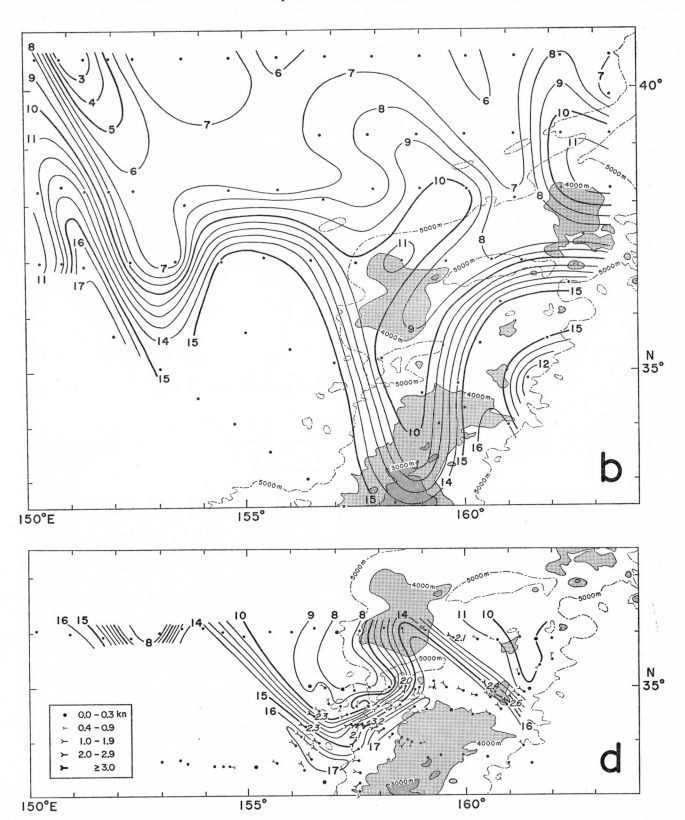

FIG. 41(b, d)   Surface current from GEK and 200 m temperature around the Northwest Pacific Rise. The chain line and thin line indicate 5000 m and 4000 m depth contours of the bottom, respectively. Hatched area indicates where the bottom is shallower than a depth of 3000 m. (b), Based on Navypac '40s in 10–29 August 1940 (Table 12). The neighbor synchronous index $\Delta d/\delta t$ (Sec. 7.5) is greater than 0.2 kn. (d), Based on Jeds-5 in 11–28 May 1962 (Table 12). The neighbor synchronous index $\Delta d/\delta t$ (Sec. 7.5) is greater than 0.1 kn.

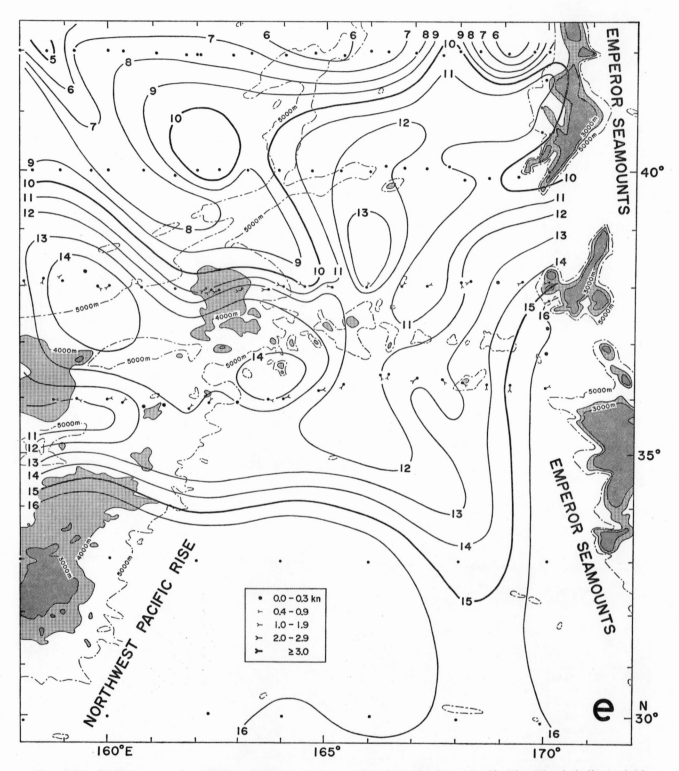

FIG. 41(e)   Surface current from GEK and 200 m temperature around the Northwest Pacific Rise. The chain line and thin line indicate 5000 m and 4000 m depth contours of the bottom, respectively. Hatched area indicates where the bottom is shallower than a depth of 3000 m. Based on IGY Poront '57 in September—November 1957 (Table 12). The neighbor synchronous index $\Delta d/\delta t$ (Sec. 7.5) is greater than 0.1 kn.

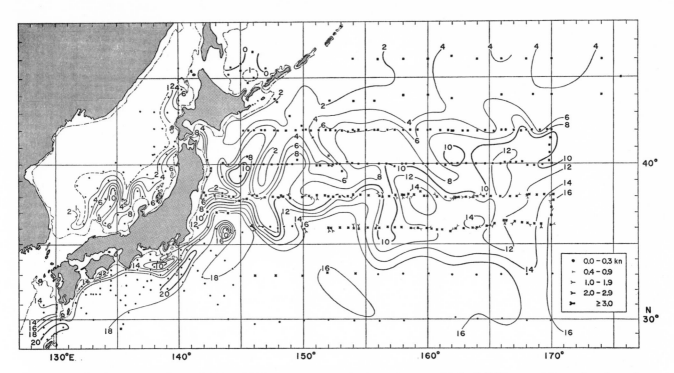

FIG. 42   Surface current from GEK and 200 m temperature in the summer and fall of 1957 around Japan based on data from IGY Poront '57 (Table 12) for the Tohoku Area and its offing and from various surveys principally in November for other seas. Current measurements along 36°N and 38°N are entered only for the values above a speed of 0.7 kn between 155°E and 170°E and above a speed of 1.0 kn between 141°E and 155°E. The neighbor synchronous index $\Delta d/\delta t$ (Sec. 7.5) is greater than 0.1 kn.

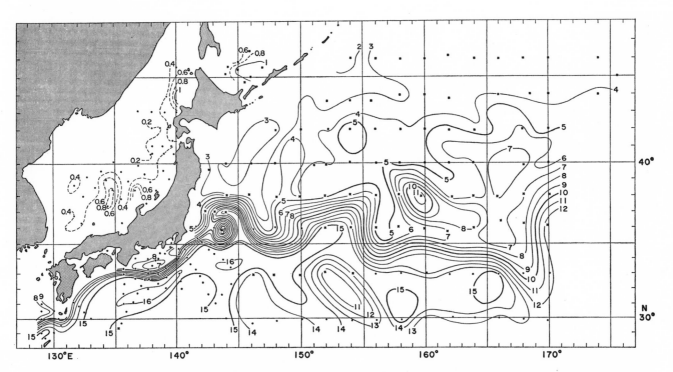

FIG. 43   Temperature at a depth of 400 m in the summer and fall of 1957 around Japan based on data from IGY Poront '57 (Table 12) for the Tohoku Area and its offing and from various surveys principally in November for other seas. The neighbor synchronous index $\Delta d/\delta t$ (Sec. 7.5) is greater than 0.1 kn.

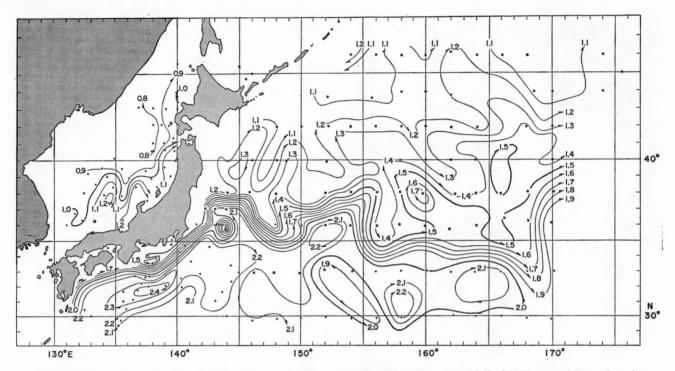

FIG. 44   Dynamic depth anomaly in dynamic meter 0 over 1000 db in the summer and fall of 1957 around Japan based on the same data as in Fig. 43. The values in the Japan Sea are obtained by adding a mean dynamic depth anomaly of 0.360 dynamic meter 500 over 1000 db to the values 0 over 500 db at individual stations. The neighbor synchronous index $\Delta d/\delta t$ (Sec. 7.5) is greater than 0.1 kn.

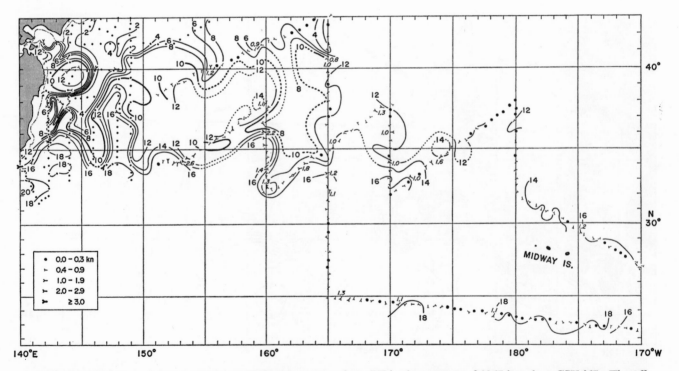

FIG. 45   Surface current from GEK and 200 m temperature from BT in the summer of 1967 based on CSK '67s. The offshore hydrography is based on data from the *Takuyo* of the JHO. The neighbor synchronous index $\Delta d/\delta t$ (Sec. 7.5) is greater than 0.1 kn.

path of the Kuroshio Extension southeast of the
Boso East Coast sometimes shows distorted meanders
convex to the southeast, which he considered as
Bolin's (1950) effect by the Izu-Ogasawara Ridge
(Fig. 1).

Examination of Mann's schematic picture of the
current system southeast of Newfoundland (Mann,
1967, Fig. 15) shows the Gulf Stream fanning out in
the vicinity of the col, southeast of the tail of the
Grand Banks, between the North America Basin and
the Newfoundland Basin. Mann regarded the
branching point as the termination of the Gulf
Stream as an identifiable current, and proposed that
the North Atlantic Current is fed by the Slope Water
Current immediately around the tail of the Grand
Banks and by a branch of the Gulf Stream along the
northern edge of a thermal ridge extending southeast
of the tail.

The hydrography around the termination of the
Kuroshio Extension is not clear, because it is located
far out in the ocean. Only the data listed in Table 12
are available. Figure 41 gives 200 m temperatures
around the Northwest Pacific Rise based on five
expeditions. In August 1940 (Fig. 41b), the axis of
the Kuroshio Extension, as indicated by the 13 or
14°C isotherm (Sec. 6), showed much distorted
meanders and a large loop riding over the top of the
Rise shallower than 3000 m. In August 1939 and 1955
(Figs. 41a and 41c), the axis showed loose meanders
and had detached a big anticyclonic eddy into the
Perturbed Area. The former state seems to be unstable and the latter stable, but it is difficult to allot
the state in May 1962 (Fig. 41d) to either class because the expedition was concentrated on a relatively
narrow area. On the other hand, since the southern
stations in the fall of 1957 (Fig. 41e) were widely
spaced, the contouring was difficult. The isotherms
were able to be contoured to show the Kuroshio
Extension path going through the col of the Northwest Pacific Rise at 35°N, 158°E–160°E, similar to
the isotherms in August 1939 (Fig. 41a), and to show
a countercurrent to the north of the col, similar to
the isotherms in August 1955 (Fig. 41c). Accordingly,
the state in the fall of 1957 seems to be stable.

In the fall of 1957, the stations east of the Rise
were so widely spaced that the feature would be
missed, unless an interpretation analogous to Mann's
(1967) in the Gulf Stream System would be given to
the contouring of isotherms and dynamic isobaths
around the termination of the Kuroshio Extension
(Figs. 41e, 42, 43 and 44). These figures show the
Kuroshio Extension path fanning out into two main
branches, one going to the south, the other to the
northeast, before reaching the Emperor Seamounts.

In July 1967 (Fig. 45), however, the isotherms showed
no similarity to the above by the constraint of GEK
data.

8 SEASONAL VARIATION IN TEMPERA-TURE, SALINITY AND CURRENT SPEED

There are two ways to describe seasonal variation
in such oceanographic elements as temperature,
salinity, thermosteric anomaly and current speed;
one based on values at a fixed spot or area, the other
on values at a floating spot or area with fluctuations
in location of such characteristic hydrographic
structures as stream axis, front and eddy. In any conclusions based on localized data, it must be kept in
mind that a marked change in the oceanographic
elements may be produced by horizontal translation
of the front. The latter needs elaborate analysis of
data, and will be adopted in limited cases in the
following sections.

8.1 Surface temperature

The surface temperature in the Tohoku Area
generally reaches a minimum in March and a maximum in August, while south of Japan the minimum
is in February. This is also shown in the annual
variation of the monthly mean temperature at four
stations described in Sec. 8.2 (Table 13). For this
reason, mean temperatures at the surface in the
Tohoku Area in March and August are shown in
Fig. 46, which is contoured by linear interpolation
along the meridian using the monthly mean temperature by one-degree-square during 1950 to 1957
(Oceanographical Section, JMA, 1958).

According to Wyrtki's chart (1965, Fig. 2) showing
the amplitude of the annual variation of surface
temperature in the North Pacific, the annual variation
in surface temperature off Hokkaido and the northern
Sanriku Coast exceeds a range of 14°C, the largest in
all the North Pacific excluding its adjacent seas (medi-
terranean and marginal seas). In particular, a range as
large as 16°C is localized in a narrow sea south of
Hokkaido. Wyrtki's chart also shows that the range
in the sea off Joban and Boso amounts to a temperature of about 10°C which is still larger than that
in other regions in the North Pacific. These circumstances can be seen by comparing the mean temperatures in March and August (Fig. 46). The sea with
the largest range almost corresponds to the area with
a minimum monthly mean temperature below 1°C in
March. The minimum temperature is brought about
by Oyashio water retaining its low temperature dur-

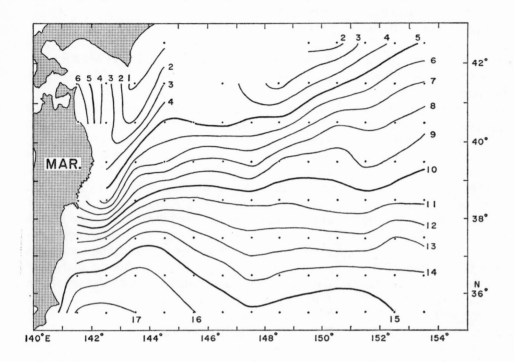

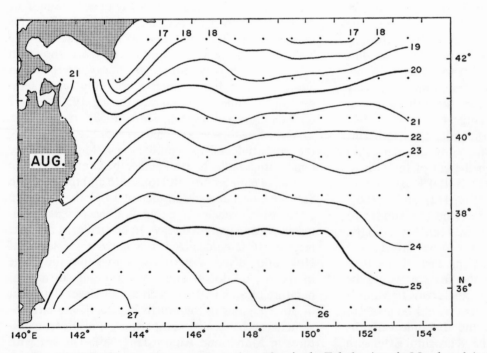

FIG. 46   Monthly mean temperature at the surface in the Tohoku Area in March and August. Contoured on the basis of monthly mean temperature by one-degree-square (Oceanographical Section, JMA, 1958).

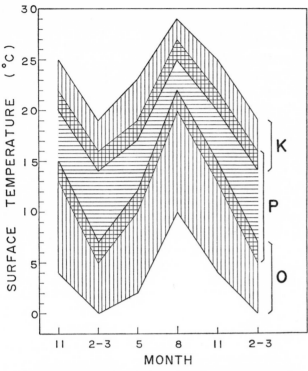

FIG. 47   Seasonal variation of surface temperature in the Kuroshio (*K*), Perturbed (*P*) and Oyashio (*O*) Areas (Kawai, 1965, Fig. 11).

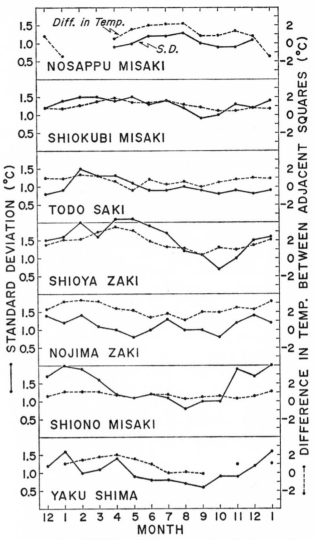

FIG. 48   Solid lines indicate annual variation in the standard deviation of monthly mean surface temperature at seven coastal stations on the Pacific side of Japan (Kawai, 1965, Fig. 7). Dashed lines indicate annual variation in the difference in monthly mean surface temperature between a pair of adjacent one-degree-squares in the vicinity of the seven coastal stations. For the locations of the coastal stations and one-degree-squares see Fig. 49.

ing the course of flowing southwest, close along the southern Kuril Islands and the southeast coast of Hokkaido.

Since a big difference is expected in the mode of seasonal variation in the surface temperature between Kuroshio and Oyashio waters, the variation will be described in the following not by fixed areas but by the Kuroshio, Oyashio and Perturbed Areas, the boundaries of which fluctuate season by season and year by year. Figure 47 shows the seasonal variation in surface temperature in the three areas, obtained from comparisons between distribution of the surface temperature and location of the boundaries of the three areas estimated from subsurface thermal structure for every season of each individual year during 1954–1958. The range of annual variation in the surface temperature is about 10°C in most of the Tohoku Area, but it amounts to 15°C in the vicinity of the Oyashio Front, the boundary between the Perturbed and Oyashio Areas. This result is similar to that obtained from Wyrtki's chart (1965, Fig. 2).

The solid line in Fig. 48 indicates the annual variation in the standard deviation of monthly mean surface temperature at seven coastal stations (Fig. 49) on the Pacific side of Japan. The standard deviation at the upper five stations facing the Tohoku Area was computed by using data through half a century from

1914 to 1963 (at Nosappumisaki, measurement of surface temperature stopped in 1959), and at the lower two stations facing the Kuroshio south of Japan it was computed by Hirano (1957) using data from 1914 to 1955 or 1956. The maximum standard deviation occurs in winter at Yakushima and Shionomisaki, in late winter and spring at Shioyazaki and Todosaki, in spring and summer at Shiokubimisaki, and in summer at Nosappumisaki. The annual variation at Nojimazaki does not show a definite maximum. These occurrences of the maximum seem roughly correlated to the annual excursion of the surface thermal front; a large standard deviation corresponds to a large gradient of surface temperature as inferred from the difference of the monthly mean surface temperature between the two adjacent one-degree-squares (Fig. 49) in the vicinity of the seven coastal stations (Fig. 48, dashed line).

8.2 Thermohaline structure in the water column at four stations

There were two weather ship stations in the Pacific in the neighborhood of Japan. One at 29°N, 135°E in the Kuroshio Area south of Japan is called Station Tango (now active only in summer and fall), and the other at 39°N, 153°E in the Perturbed Area is called Station Extra (stopped in 1953). Originally they were called Tare and X-Ray, and in this study they are referred to as Stations T and X, respectively. Since Station T, located in or near the Kuroshio

Countercurrent 200 nautical miles or more south of the Kuroshio path, does not properly catch water properties directly south of the Kuroshio path, the monthly mean temperature and salinity at standard depths were calculated for a fictional station with the maximum thickness in dynamic height between 0-db and 1000-db surfaces along G-Sec (Fig. 49) south of Shionomisaki. This will be referred to as Station M, because it is located in the area with thickest Subtropical Mode Water (Sec. 4.1). Since there have never been any fixed stations in the Oyashio Area, the bimonthly mean temperature and salinity at standard depths at a spot 41°50′–42°40′N, 143°50′–144°30′E southeast of Hokkaido were computed by using data from the *Yushio Maru* and *Kofu Maru* of the HMO and other ships. This area will be referred to as Station O, because the spot is located within the Oyashio Area for the major part of the

TABLE 13a.   Monthly mean temperature and salinity at Station M (135°45′E).

Depth (m)	Jan.	Feb.	Mar.	Apr.	May	June	July	Aug.	Sept.	Oct.	Nov.	Dec.	Year
						Temperature (°C)							
0	19.80	18.80	19.50	20.03(19.05)	20.55	24.90	27.06	29.08	28.15	26.43	23.40	21.63	23.28
10	20.17	18.73	19.50	20.06(19.03)	20.61	24.48	26.89	28.96	27.85	26.28	23.50	21.92	23.25
25	20.20	18.80	19.44	19.99(19.00)	20.54	23.88	26.33	28.80	27.74	26.24	23.53	21.95	23.12
50	20.20	18.78	19.35	19.65(18.83)	19.94	23.10	23.58	25.75	26.25	26.10	23.50	21.95	22.35
75	20.20	18.78	19.13	19.26(18.80)	19.38	22.70	22.41	23.51	23.34	25.16	23.33	21.97	21.60
100	20.20	18.83	18.90	18.99(18.75)	19.08	21.28	21.52	22.31	21.61	23.45	21.83	21.97	20.83
150	19.53	18.83	18.78	18.77(18.55)	18.75	20.35	20.23	20.56	20.04	20.43	19.75	20.14	19.68
200	19.00	18.80	18.70	18.65(18.48)	18.59	19.30	19.21	19.36	19.19	19.33	19.08	18.98	19.02
300	17.72	18.30	18.31	18.36(18.13)	18.41	17.70	17.49	17.78	18.04	18.11	18.08	17.43	17.98
400	16.37	17.23	17.25	17.29(16.95)	17.33	14.93	15.84	16.21	16.55	16.64	16.65	16.27	16.55
500	14.52	15.50	15.26	15.33(15.10)	15.39	12.38	13.69	14.50	14.75	14.63	15.00	14.30	14.60
600	12.12	13.45	12.93	12.89(12.45)	12.84	10.05	11.23	12.11	12.38	12.19	12.70	12.02	12.24
800	7.72	8.08	8.00	8.01(8.03)	8.01	6.10	6.80	8.33	7.63	7.88	7.85	7.87	7.69
1000	4.70	5.10	5.13	5.22(4.88)	5.30	4.20	4.51	4.89	4.98	4.89	5.25	4.68	4.90
0–200	19.38	18.80	19.04	19.20(18.73)	19.36	21.76	22.23	23.32	22.88	23.25	21.68	21.12	21.04

Depth (m)	Jan.	Feb.	Mar.	Apr.	May	June	July	Aug.	Sept.	Oct.	Nov.	Dec.	Year
						Salinity (‰)							
0	34.71	34.81	34.86	34.83(34.69)	34.79	34.33	34.31	34.26	34.42	34.52	34.57	34.65	34.59
10	34.70	34.79	34.82	34.80(34.70)	34.78	34.36	34.34	34.26	34.38	34.45	34.49	34.63	34.57
25	34.70	34.77	34.81	34.80(34.70)	34.78	34.44	34.40	34.28	34.41	34.47	34.45	34.63	34.58
50	34.70	34.79	34.81	34.82(34.76)	34.82	34.62	34.60	34.66	34.60	34.48	34.49	34.61	34.67
75	34.70	34.79	34.81	34.83(34.76)	34.84	34.78	34.83	34.75	34.75	34.66	34.64	34.63	34.75
100	34.72	34.78	34.82	34.83(34.77)	34.84	34.81	34.86	34.79	34.80	34.83	34.82	34.63	34.79
150	34.82	34.78	34.82	34.82(34.78)	34.82	34.81	34.81	34.81	34.82	34.83	34.79	34.81	34.81
200	34.79	34.78	34.83	34.83(34.78)	34.83	34.77	34.80	34.81	34.80	34.80	34.78	34.81	34.80
300	34.70	34.77	34.79	34.80(34.76)	34.80	34.71	34.76	34.75	34.77	34.76	34.72	34.73	34.76
400	34.64	34.74	34.76	34.76(34.72)	34.75	34.60	34.62	34.67	34.70	34.70	34.68	34.67	34.69
500	34.54	34.65	34.66	34.66(34.61)	34.65	34.47	34.51	34.57	34.56	34.56	34.59	34.57	34.58
600	34.40	34.56	34.50	34.50(34.42)	34.49	34.38	34.37	34.44	34.45	34.38	34.42	34.47	34.45
800	34.24	34.22	34.28	34.28(34.22)	34.28	34.27	34.24	34.29	34.24	34.23	34.17	34.20	34.25
1000	34.30	34.29	34.32	34.32(34.32)	34.31	34.36	34.34	34.30	34.29	34.28	34.26	34.31	34.31
0–200	34.74	34.78	34.82	34.82(34.76)	34.82	34.70	34.71	34.68	34.70	34.69	34.68	34.69	34.74

Reproduced from Kawai & Sakamoto's (unpublished) statistics for the years 1951–53. The values for January have only one observation at the end of the month, and are completed by including one from late December. The values for April are computed as the mean of those in March and May because of the small sample size of the real monthly mean in parentheses. Values on the line 0–200 indicate the mean temperature or salinity in the upper 200 m layer.

TABLE 13b. Monthly mean temperature and salinity at Station T (29°N, 135°E).

Depth (m)	Jan.	Feb.	Mar.	Apr.	May	June	July	Aug.	Sept.	Oct.	Nov.	Dec.	Year
						Temperature (°C)							
0	20.06	19.46	19.51	20.52	22.27	24.30	26.98	28.66	28.18	26.28	24.17	21.54	23.49
10	20.16	19.35	19.48	20.50	22.09	24.11	26.62	28.33	28.03	26.23	24.26	21.66	23.40
25	20.16	19.32	19.47	20.36	21.95	23.57	25.88	27.51	27.86	26.20	24.23	21.66	23.18
50	20.11	19.30	19.40	20.16	21.38	22.07	22.71	23.76	25.76	25.33	24.19	21.66	22.15
75	20.04	19.27	19.32	19.87	20.71	21.23	20.93	21.69	22.87	22.78	23.17	21.56	21.12
100	19.95	19.20	19.15	19.56	20.31	20.33	19.97	20.53	21.20	21.01	21.14	20.78	20.26
150	19.24	18.89	18.70	18.94	19.53	19.40	18.87	19.18	19.35	19.21	19.03	18.66	19.08
200	18.15	18.18	18.01	18.28	18.74	18.61	18.08	18.35	18.45	18.36	18.22	17.62	18.25
300	16.55	16.52	16.48	16.69	17.27	17.10	16.60	16.91	16.98	16.96	16.82	16.01	16.74
400	14.82	14.73	14.63	14.83	15.54	15.35	14.86	15.17	15.23	15.30	14.98	14.12	14.96
500	12.35	12.35	12.17	12.39	13.20	12.91	12.36	12.74	12.74	13.18	12.77	11.68	12.57
600	9.72	9.86	9.54	9.87	10.51	10.15	9.86	10.11	10.38	10.92	10.20	9.01	10.01
800	5.87	5.95	5.86	6.13	6.20	6.24	5.99	6.20	6.43	6.34	6.01	5.64	6.07
1000	4.14	4.11	4.14	4.27	4.37	4.29	4.15	4.26	4.45	4.35	4.20	4.02	4.23
1200	3.30	3.28	3.30	3.37	3.43	3.36	3.29	3.35	3.37	3.40	3.37	3.26	3.34
0–200	19.62	19.04	19.01	19.52	20.43	20.87	21.10	21.86	22.44	21.97	21.46	20.22	20.63

Depth (m)	Jan.	Feb.	Mar.	Apr.	May	June	July	Aug.	Sept.	Oct.	Nov.	Dec.	Year
						Salinity (‰)							
0	34.78	34.85	34.87	34.87	34.79	34.47	34.38	34.36	34.45	34.54	34.61	34.67	34.63
10	34.76	34.81	34.83	34.83	34.78	34.43	34.36	34.34	34.42	34.49	34.56	34.61	34.60
25	34.74	34.79	34.83	34.83	34.79	34.54	34.42	34.38	34.45	34.51	34.56	34.61	34.61
50	34.74	34.79	34.83	34.83	34.81	34.74	34.67	34.63	34.60	34.58	34.58	34.61	34.70
75	34.74	34.79	34.83	34.85	34.85	34.79	34.79	34.78	34.76	34.76	34.63	34.61	34.76
100	34.74	34.79	34.83	34.83	34.83	34.81	34.81	34.81	34.79	34.81	34.78	34.72	34.79
150	34.78	34.78	34.81	34.81	34.83	34.79	34.79	34.81	34.79	34.81	34.81	34.78	34.79
200	34.78	34.76	34.78	34.79	34.81	34.79	34.78	34.78	34.78	34.78	34.78	34.74	34.78
300	34.72	34.70	34.70	34.72	34.74	34.72	34.70	34.72	34.70	34.72	34.72	34.69	34.72
400	34.61	34.60	34.60	34.60	34.65	34.61	34.60	34.61	34.60	34.63	34.61	34.56	34.61
500	34.45	34.43	34.43	34.45	34.49	34.47	34.42	34.45	34.45	34.47	34.47	34.40	34.45
600	34.31	34.31	34.29	34.31	34.33	34.33	34.31	34.33	34.29	34.33	34.34	34.27	34.31
800	34.25	34.22	34.24	34.24	34.24	34.24	34.25	34.22	34.22	34.22	34.22	34.27	34.24
1000	34.36	34.33	34.34	34.33	34.33	34.33	34.34	34.31	34.29	34.33	34.34	34.38	34.33
1200	34.45	34.43	34.45	34.45	34.43	34.43	34.43	34.45	34.43	34.42	34.42	34.45	34.43
0–200	34.76	34.79	34.82	34.82	34.82	34.73	34.70	34.70	34.70	34.71	34.70	34.69	34.74

Reproduced or adapted from Koizumi's (unpublished) statistics for the years 1948–53. Values on the line 0–200 indicate the mean temperature or salinity in the upper 200 m layer.

TABLE 13c. Monthly mean temperature and salinity at Station X (39°N, 153°E).

Depth (m)	Jan.	Feb.	Mar.	Apr.	May	June	July	Aug.	Sept.	Oct.	Nov.	Dec.	Year
						Temperature (°C)							
0	13.37	11.73	10.59	11.82	13.88	17.40	20.77	24.28	23.38	19.93	17.51	15.14	16.65
10	13.48	11.87	10.64	11.71	13.78	17.12	20.49	23.98	23.30	20.01	17.64	15.30	16.61
25	13.44	11.88	10.61	11.62	13.46	16.16	18.49	21.40	22.68	19.86	17.58	15.27	16.04
50	13.34	11.82	10.52	11.40	12.77	14.08	15.07	16.14	16.89	17.95	17.36	15.13	14.37
75	13.23	11.70	10.33	11.06	12.09	12.79	13.43	13.83	14.14	14.99	16.56	14.99	13.26
100	13.08	11.62	10.19	10.78	11.42	11.93	12.31	12.49	12.72	13.38	14.62	14.60	12.43
150	12.60	11.15	9.78	10.18	10.36	10.42	10.76	10.53	10.72	11.28	12.06	12.35	11.02
200	11.44	10.39	9.04	9.27	9.42	9.44	9.31	9.09	9.34	9.63	10.30	10.35	9.76
300	8.87	8.23	7.11	7.15	7.57	7.93	6.86	6.65	6.72	7.17	7.30	7.65	7.44
400	6.74	6.54	5.14	5.42	5.66	5.76	5.50	5.28	5.42	5.39	5.24	6.21	5.69
500	5.24	5.14	4.42	4.54	4.72	4.74	4.65	4.53	4.70	4.58	4.44	5.05	4.73
600	4.48	4.47	4.12	4.19	4.22	4.26	4.22	4.21	4.22	4.10	4.13	4.41	4.25
800	3.73	3.74	3.55	3.58	3.60	3.65	3.58	3.61	3.63	3.61	3.54	3.65	3.62
1000	3.22	3.22	3.07	3.09	3.12	3.14	3.08	3.11	3.16	3.13	3.07	3.13	3.13
1200	2.80	2.80	2.70	2.74	2.74	2.77	2.72	2.72	2.74	2.78	2.70	2.75	2.75
1500	2.41	2.38	2.34	2.37	2.35	2.35	2.32	2.36	2.36	2.38	2.34	2.39	2.36
0–200	12.87	11.43	10.08	10.72	11.57	12.50	13.37	14.10	14.42	14.38	14.58	13.75	12.82

TABLE 13c.   (cont'd)

Depth (m)	Jan.	Feb.	Mar.	Apr.	May	June	July	Aug.	Sept.	Oct.	Nov.	Dec.	Year
						Salinity (‰)							
0	34.43	34.42	34.36	34.43	34.51	34.36	34.24	34.11	34.07	34.18	34.29	34.33	34.31
10	34.42	34.38	34.31	34.40	34.43	34.36	34.22	34.09	34.04	34.13	34.25	34.29	34.27
25	34.40	34.36	34.31	34.38	34.43	34.38	34.33	34.20	34.11	34.15	34.25	34.27	34.29
50	34.40	34.36	34.29	34.38	34.42	34.42	34.45	34.40	34.42	34.29	34.27	34.25	34.36
75	34.38	34.36	34.27	34.34	34.38	34.40	34.43	34.40	34.45	34.43	34.34	34.24	34.36
100	34.34	34.34	34.25	34.33	34.34	34.33	34.38	34.36	34.40	34.40	34.43	34.29	34.34
150	34.34	34.31	34.22	34.25	34.25	34.20	34.27	34.22	34.27	34.29	34.36	34.33	34.27
200	34.31	34.24	34.15	34.18	34.20	34.13	34.15	34.11	34.16	34.16	34.25	34.24	34.18
300	34.13	34.11	34.00	34.02	34.04	34.05	33.96	33.96	33.96	34.02	34.02	34.07	34.02
400	34.04	34.04	33.93	33.96	33.98	33.95	33.96	33.93	33.95	33.93	33.96	34.05	33.96
500	34.02	34.04	34.02	34.04	34.02	33.96	34.02	34.00	34.00	33.98	34.02	34.07	34.00
600	34.09	34.09	34.13	34.13	34.11	34.07	34.13	34.11	34.11	34.11	34.13	34.15	34.11
800	34.25	34.25	34.29	34.29	34.29	34.24	34.27	34.29	34.27	34.27	34.29	34.27	34.27
1000	34.36	34.36	34.40	34.40	34.38	34.34	34.36	34.38	34.38	34.36	34.40	34.38	34.36
1200	34.45	34.45	34.45	34.45	34.45	34.42	34.43	34.43	34.47	34.45	34.47	34.47	34.43
1500	34.52	34.52	34.52	34.54	34.52	34.47	34.51	34.52	34.52	34.52	34.56	34.54	34.52
0–200	34.36	34.33	34.25	34.31	34.33	34.30	34.32	34.26	34.29	34.29	34.33	34.28	34.30

Reproduced or adapted from Koizumi's (1955) statistics for the years 1948–53. Values on the line 0–200 indicate the mean temperature or salinity in the upper 200 m layer.

TABLE 13d.   Bimonthly mean temperature and salinity at Station O (42°15′N, 144°10′E).

Depth (m)	Jan.—Feb.	Mar.—Apr.	May—June	July—Aug.	Sept.—Oct.	Nov.—Dec.	Year
			Temperature (°C)				
0	0.30	0.30	6.83	15.38	14.09	8.91	7.64
10	0.37	0.07	5.69	13.64	13.65	8.99	7.07
25	0.57	0.42	4.23	8.15	12.90	8.98	5.88
50	1.13	0.64	2.50	4.93	8.82	8.24	4.38
75	1.30	1.06	2.04	3.53	5.72	7.00	3.44
100	1.46	1.36	1.64	2.96	4.79	5.21	2.90
150	1.83	1.58	1.54	2.20	3.14	3.41	2.28
200	2.02	1.63	1.57	1.84	2.92	2.56	2.09
300	2.05	1.92	1.93	1.87	2.53	2.03	2.06
400	2.16	2.21	2.30	2.25	2.97	2.24	2.36
600	2.64	2.79	2.83	2.87	3.33	3.17	2.94
800	2.95	2.87	2.88	2.91	3.01	2.98	2.93
1000	2.81	2.79	2.95	2.72	2.86	(2.83)	2.83
0–200	1.38	1.12	2.41	4.44	6.36	5.74	3.57

Depth (m)	Jan.—Feb.	Mar.—Apr.	May—June	July—Aug.	Sept.—Oct.	Nov.—Dec.	Year
			Salinity (‰)				
0	32.69	32.84	32.88	33.08	33.30	33.35	33.02
10	32.69	32.82	32.93	33.12	33.33	33.33	33.04
25	32.76	32.95	33.04	33.26	33.40	33.33	33.12
50	32.94	33.01	33.16	33.32	33.44	33.34	33.20
75	33.05	33.17	33.24	33.32	33.40	33.39	33.26
100	33.13	33.25	33.29	33.36	33.43	33.39	33.31
150	33.24	33.29	33.36	33.39	33.46	33.38	33.35
200	33.32	33.38	33.45	33.44	33.51	33.44	33.42
300	33.43	33.53	33.60	33.58	33.62	33.59	33.56
400	33.60	33.66	33.74	33.73	33.81	33.71	33.71
600	33.92	33.94	33.98	34.05	34.12	34.13	34.02
800	34.17	34.17	34.19	34.29	34.22	34.30	34.22
1000	34.34	34.33	34.32	34.40	34.33	(34.34)	34.34
0–200	33.07	33.17	33.25	33.34	33.43	33.38	33.27

Reproduced from Kawai and Sakamoto's (unpublished) statistics for the years 1948–67. Values on the line 0–200 indicate the mean temperature or salinity in the upper 200 m layer.

year (Fig. 49).

The statistics for Stations M and O by Kawai and Sakamoto (unpublished) are shown in Tables 13a and 13d. The statistics for Stations T and X by Koizumi (unpublished, 1955) are shown in Tables 13b and 13c, where the conversion of chlorinity in his original to salinity have been made. All discussions and figures in this section are based on the monthly or bimonthly temperature and salinity at standard depths at the four stations in Tables 13a–d.

Sounding curves of the annual mean and range

Sounding curves of the annual mean and annual range of temperature, salinity and thermosteric anomaly at the four stations, based on values in Table 13, are shown in Fig. 50. The permanent thermocline is centered at depths of 500–600 m at Stations M and T, and at a depth of about 200 m at Station X. The permanent halocline and stericline are centered at depths similar to that of the center of permanent thermocline at the three stations. At Station O, neither a permanent thermocline nor a permanent stericline is distinct, and a vague and broad halocline is located at depths of 200–600 m, in a manner quite different from that in the central and eastern Pacific subarctic water (Dodimead et al., 1963).

The North Pacific Intermediate Water indicated by a salinity minimum is centered at a depth of 900 m at Station M, 800 m at Station T, and 400 m at Station X, in accord with the depth of a thermosteric anomaly of 125 cl/ton, but no subsurface salinity minimum is seen at Station O because the station is located within the subarctic water. In the upper 50 m layer Station O has the largest range of annual thermal variation of the four stations, at depths of 75–100 m Station M does, at depths of 150–300 m Station X does, and below a depth of 400 m Station M does. These are explained by the difference in the depth of thermocline among the four stations; the larger range of the annual variation at a depth corresponds to the nearer location of the thermocline to this depth. The range of annual variation in thermosteric anomaly is similar to the temperature range, but the salinity range is the largest at Station O at all depths; this can perhaps be attributed to fluctuations of the Oyashio Front, across which the horizontal gradient of salinity is large.

Time sections

By using the values in Table 13, time sections of temperature, salinity and thermosteric anomaly are prepared to describe the seasonal variations in the upper 400 m layer (Figs. 51, 52 and 53), but some of the variations below a depth of about 200 m may be insignificant owing to the small sample size of the monthly or bimonthly means and to the relatively large yearly fluctuations. The depth of the 125-cl/ton isopleth is indicated by the dashed or chain line in the time sections at Stations X and O.

Time sections of temperature (Fig. 51) show that the summer thermocline gets deeper toward fall, reaching a depth of about 80 m at all stations. It appears from the time sections of temperature that the bottom of the upper mixed layer reaches a depth of 100 m or more in winter at all stations, but this is true only for Stations M and T because the upper thermostads (Sec. 4.1) are not homogeneous with respect to salinity at Stations X and O as shown in the time sections of salinity (Fig. 52). The upper mixed layer at Station T is not equivalent to the "18° Water" (Worthington, 1959) in the Sargasso Sea, because the station is located in or near the Kuroshio Countercurrent. A thicker mixed layer reaching a depth of 300 m is found at Station M, which is located at the intersection of the thermal trough of a huge anticyclonic eddy in the Shikoku Basin with the G-Sec line, but its longitude of about 136°E is located a little too far east to pass through the center of the eddy (Sec. 4.1).

Comparison of the time sections of salinity among the four stations (Fig. 52) shows a different situation at Station O; the upper low-salinity water develops in winter at Station O but in summer at the other three stations. The depth of summer low-salinity water at the three stations increases toward fall, while that of winter low-salinity water at Station O decreases toward spring. The difference may be explained as follows. The summer low-salinity water at the surface becomes heavier with cooling and evaporation in fall, and is subject to vertical mixing; the winter low-salinity water at the surface, on the other hand, becomes lighter with heating in spring, and is subject to horizontal mixing. The upper low-salinity water in winter at Station O is the Original Oyashio Water (Sec. 4.3). The subsurface salinity maximum centered at depths of 100–200 m at Station M, at depths of 75–150 m at Station T and at depths of 50–150 m at Station X is only a result of the summer low-salinity water at the surface. The subsurface salinity minimum called the North Pacific Intermediate Water is off the time sections at Stations M and T and only its upper half appears at depths of 300–400 m at Station X, the center of the minimum being in accord with the 125-cl/ton isopleth. At Station O, no salinity minimum is found in the time section as mentioned before, but the subsurface temperature minimum in the time section is centered

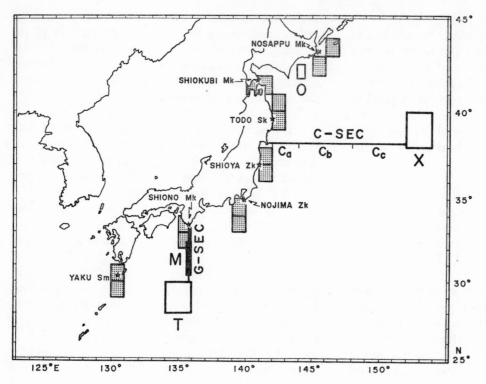

FIG. 49 Locations of seven coastal stations, a pair of one-degree-squares in their vicinity (Fig. 48), four Stations (M, T, X and O), C-Sec and G-sec (for Figs. 50, 51, 52, 53, 54 and 55).

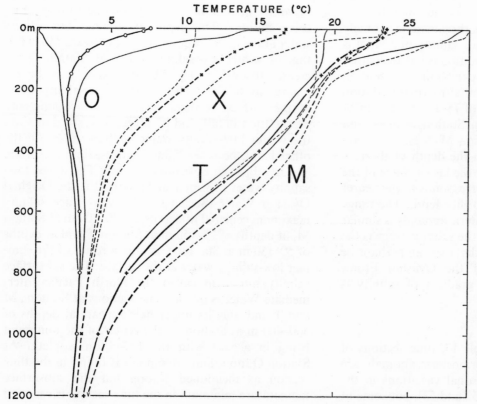

FIG. 50(a) Average temperature soundings at Stations M, T, X and O. Thick lines with small marks indicate the annual mean of temperature, while the thin lines indicate the annual ranges.

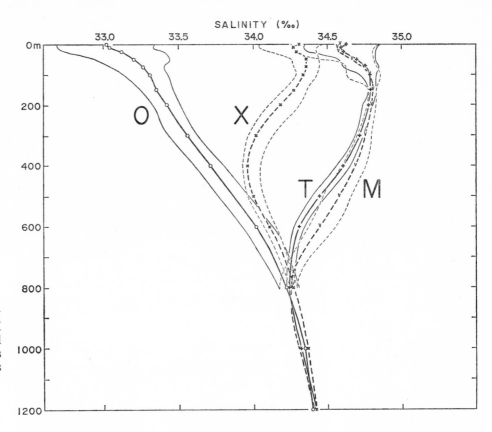

FIG. 50(b)   Average salinity soundings at Stations M, T, X and O. Thick lines with small marks indicate the annual means of salinity, while the thin lines indicate the annual ranges.

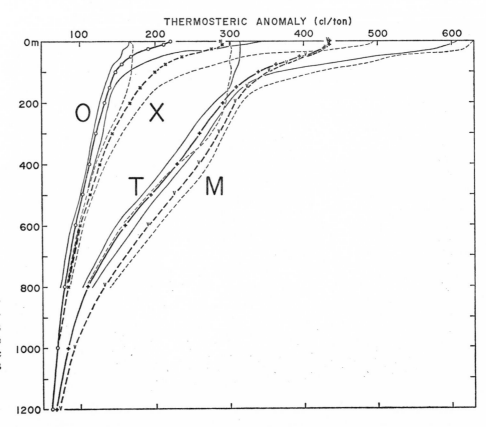

FIG. 50(c)   Average thermosteric anomaly soundings at Stations M, T, X and O. Thick lines with small marks indicate the annual means of thermosteric anomaly, while the thin lines indicate the annual ranges.

roughly at the 125-cl/ton isopleth.

The time sections of thermosteric anomaly (Fig. 53) are similar to those of temperature at Stations M, T and X, but not at Station O. At Stations M, T and X, the low-salinity water spreading over the summer thermocline acts to increase the vertical stability, but

the salinity stratification between the thermocline and the North Pacific Intermediate Water acts in the opposite way, resulting in a marked crowding of *isanosters* (isopleths of thermosteric anomaly) in the upper layer above the salinity maximum, as compared with the lower layer. At Station O, the summer

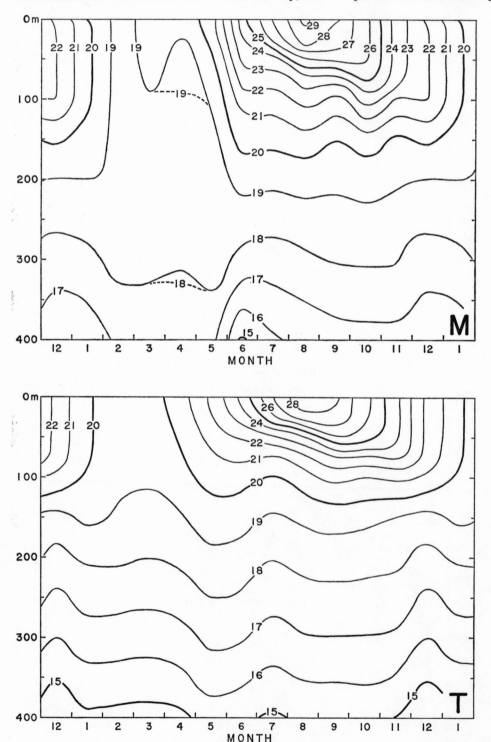

FIG. 51   Time sections of temperature in °C at Stations M, T, X and O. The dashed line indicates the depth of the 125-cl/ton isopleth reproduced from Fig. 53.

surface warm water is well reflected in the time section of thermosteric anomaly, but the winter low-salinity water is not. The lower portion of the time section of thermosteric anomaly at Station O is similar to that of salinity, but is quite different from that of temperature.

T-S annual cycle and the mean values in the upper 200 m layer

The temperature-salinity annual cycle at several depths (Fig. 54) shows that surface salinity at Stations M and T is affected by the rainy season in Japan, especially in June (note that the decrease in

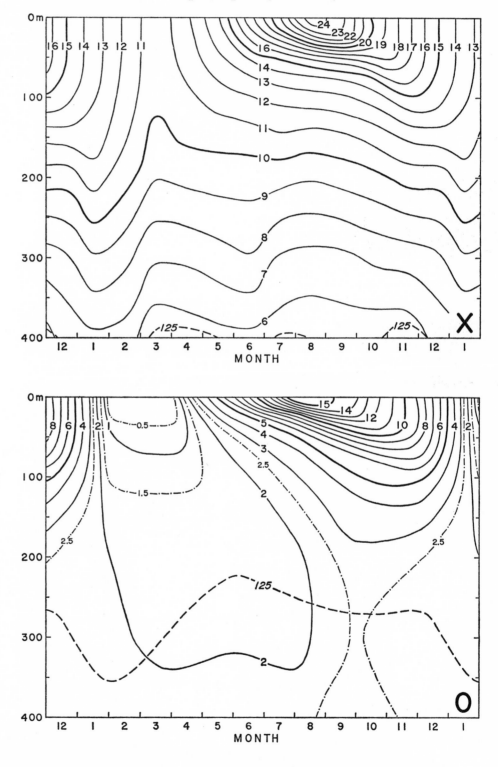

FIG. 51 (cont'd)

salinity from May to June is the steepest of all the seasons). Except for this case, surface water above a depth of 100 m in spring and summer at Stations M, X and T is warmer and more saline than in fall and winter. This may be an indication of northward movement of subtropical water in the upper 100 m layer. The situation at Station O is slightly different.

Using the mean temperature and salinity in the upper 200 m layer at the four stations listed at the bottom of Tables 13a–d and the mean temperature at stations along C-Sec adapted from Masuzawa (1955b, Table 2), the annual variations are shown in

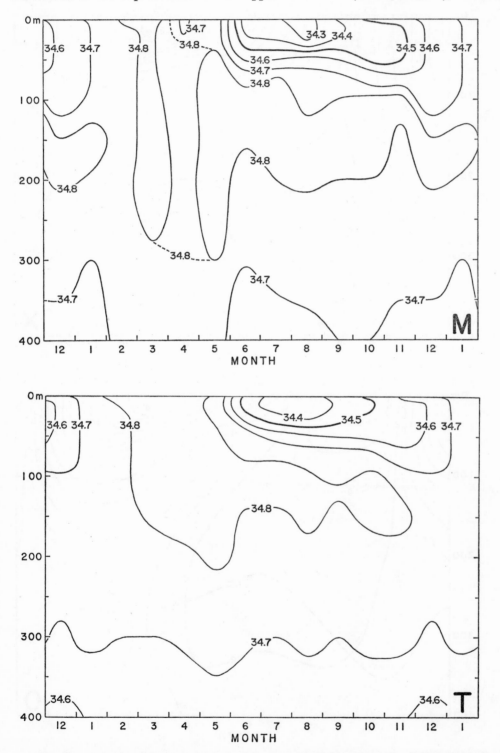

FIG. 52   Time sections of salinity in ‰ at Stations M, T, X and O. The dashed line indicates the depth of the 125-cl/ton isopleth reproduced from Fig. 53.

Figs. 55a and 55b. The annual variations of steric sea-level relative to a pressure of 200 db at the four stations are shown in Fig. 55c. The months in which extreme values occur are listed in Table 14.

In general, the mean temperatures reach their maxima in summer or fall and their minima in winter or spring. The occurrence of the maximum in November or December is especially remarkable at Station X and at stations along C-Sec, showing strong heat transport into the Perturbed Area by northward advection in the fall. The northward advection of warm water in fall has been pointed out by Kawai

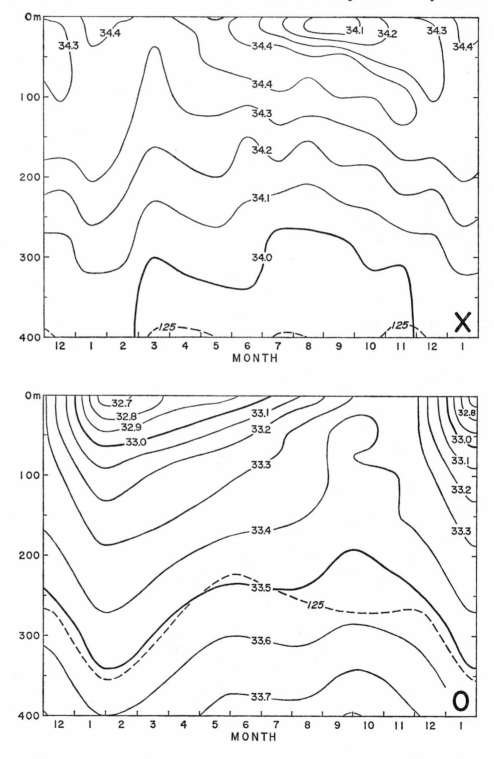

FIG. 52 (cont'd)

(1955a) on the basis of variation in surface isotherms in the five-day-period report of hydrography and fishing conditions in the Tohoku Area (THRFL) and by Koizumi (1956) on the basis of heat exchange between the sea and atmosphere at Station X.

The mean salinity is high in January–June and low in July–December at Stations M and T, but the annual variation is just the reverse at Station O, with low salinity in January–June and high salinity in July–December. The variation at Station X provides a transition from Stations M and T to Station O, characterized by maximal salinity in January–Feb-

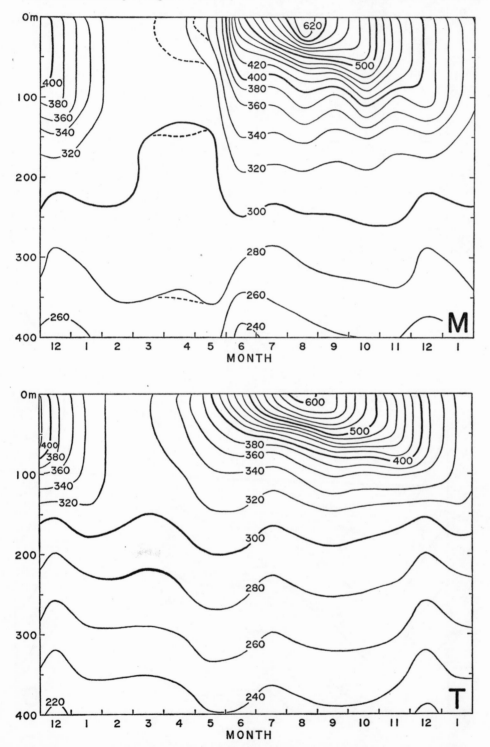

Fig. 53   Time sections of thermosteric anomaly in cl/ton at Stations M, T, X and O.

ruary, May and November and by minimal salinity in March and August. These differences are interpreted to mean that evaporation is surpassed by precipitation in the latter half of the year in the Kuroshio Area, that the southward advection of Oyashio water into the northwestern Tohoku Area is stronger in the

first half of the year, and that the mean salinity in the Perturbed Area is influenced from both sides.

The annual variations in steric sea-level are similar to those of mean temperature. There are, however, remarkable differences in the magnitude of the annual mean. The annual mean of the mean salinity in the

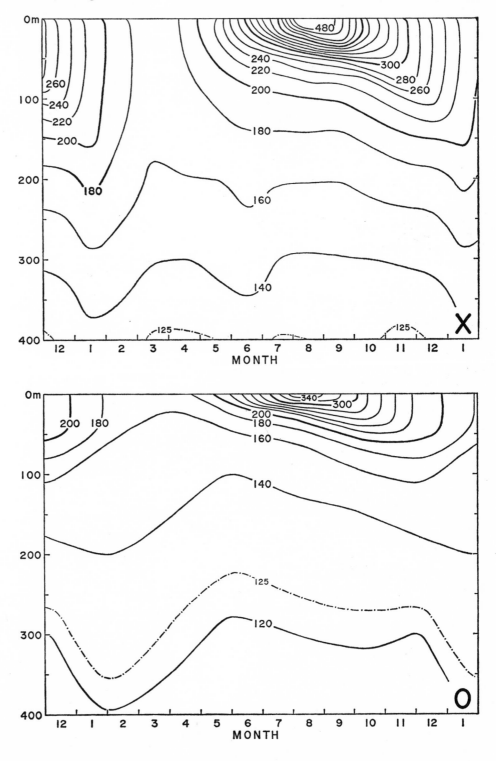

FIG. 53 (cont'd)

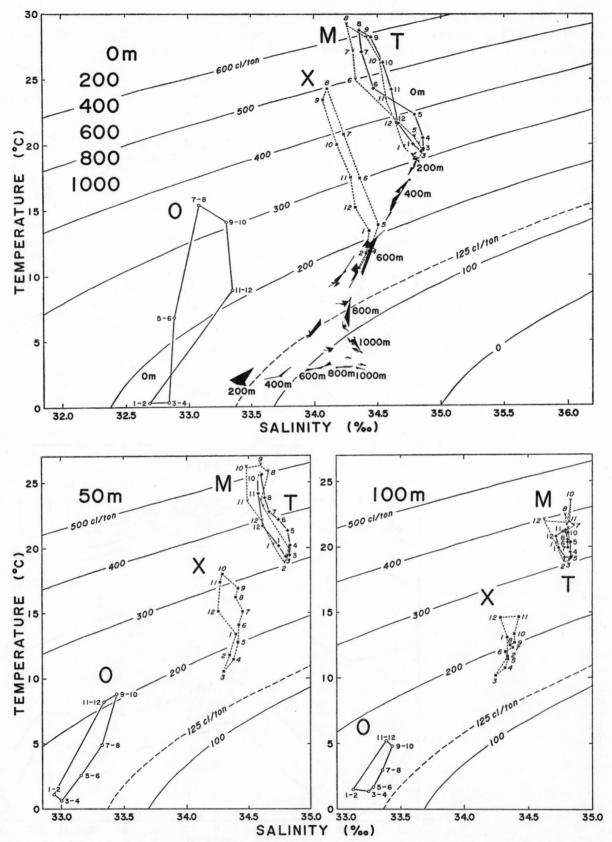

FIG. 54 *T-S* annual cycle at selected depths at Stations M, T, X and O. Small numerals indicate months, omitted for the depths deeper than 100 m because of small ranges of annual fluctuation.

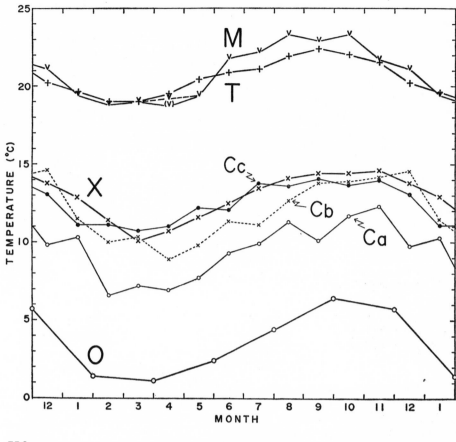

FIG. 55(a) Annual variations in the mean temperature in the upper 200 m layer at four Stations (M, T, X and O) and stations along C-Sec (Table 1).

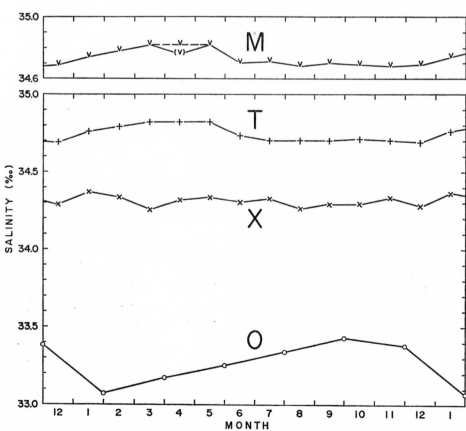

FIG. 55(b) Annual variations in the mean salinity in the upper 200 m layer at four Stations (M, T, X and O).

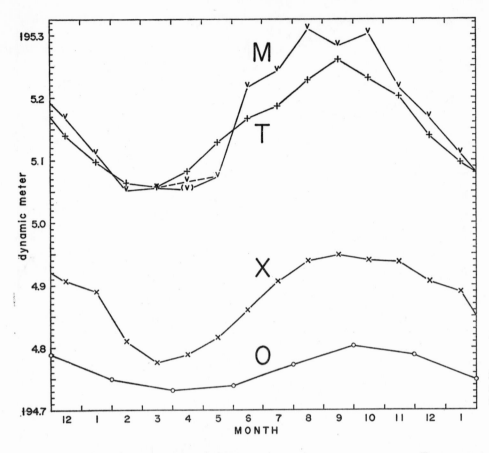

FIG. 55(c) Annual variations in the steric sea level relative to a pressure of 200 db at four Stations (M, T, X and O).

TABLE 14. Months of occurrence of extreme values of the mean temperature ($\overline{T}$) and salinity ($\overline{S}$) in the upper 200 m layer and the steric sea-level ($\overline{D}$) relative to a pressure of 200 db at four Stations (M, T, X and O) and stations along C-Sec.

Stas.	M	T	X	C_c	C_b	C_a	O
$\overline{T}$ max.	Aug.—Oct.	Sept.	Sept.—Nov.	Sept.—Nov.	Dec.	Nov.	Sept.—Oct.
$\overline{T}$ min.	Feb.	Feb.–Mar.	Mar.	Mar.	Apr.	Feb.	Mar.—Apr.
$\overline{S}$ max.	Mar.—May	Mar.—May	Jan.—Feb. May or Nov.	—	—	—	Sept.—Oct.
$\overline{S}$ min.	Aug. or Nov.	Dec. or July—Sept.	Mar. or Aug.	—	—	—	Jan.—Feb.
$\overline{D}$ max.	Aug.—Oct.	Sept.	Aug.—Nov.	—	—	—	Sept.—Oct.
$\overline{D}$ min.	Feb.	Feb.—Mar.	Mar.	—	—	—	Mar.—Apr.

upper 200 m layer at Station X is closer to that at Station M or T than to Station O (Fig. 55b), and the annual mean of the mean temperature in the upper 200 m layer at Station X is almost halfway between those at Station M or T and Station O (Fig. 55a), but the annual mean of the steric sea-level relative to a pressure of 200 db at Station X is closer to that at Station O than to that at Station M or T (Fig. 55c).

8.3 Temperature of the warm core

Figure 56 shows the seasonal variation in 100 m temperatures of the warm core in the Kuroshio Extension at different longitudes. The temperature generally decreases downstream, at a rate of about 2°C per 500 nautical miles measured along the mean path of the meandering Kuroshio Extension. The maximum temperature is found in fall at every longitude, while the minimum is found in winter. The range of annual variation in temperature amounts to 5°C, which is twice the range for 100 m temperatures at Station T and is similar to that at Stations M and X (Fig. 50a).

8.4 Current speed

According to Uda (1964) and Masuzawa (1965), the seasonal variation in the current speed of the Kuroshio in the eastern Shikoku Basin (Fig. 3) shows a maximum in summer and a minimum in fall or

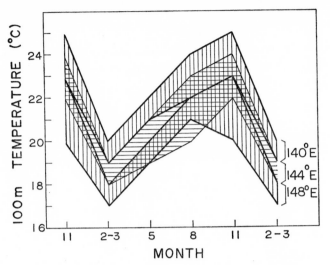

FIG. 56 Seasonal variation in the 100 m temperature of the warm core in the Kuroshio Extension (Kawai, 1965, Fig. 12).

spring. Masuzawa (1960b) and Uda (1964) also suggested a maximum in summer in the sea southeast of Inubozaki. Uda's (1964) study of the seasonal variation in the current speed in the Kuroshio Extension shows an inconsistent result with a minimum in spring between 143°E and 145°E and a maximum in spring between 145°E and 147°E.

The annual variation of surface velocity in the Florida Current shows a pronounced maximum in early summer and a minimum in fall with some semiannual variation (Fuglister, 1951b). Another study (Montgomery, 1938) also shows a semiannual variation in the slope of the sea level across the Florida Current southeast of Charleston, with two maxima in summer and winter and two minima in spring and fall. In the Gulf Stream, however, a different situation is encountered, with a maximum in winter in the sea southwest of the Grand Banks and no significant seasonal variation in the sea south of the Azores Islands (Fuglister, 1951b).

Comparing these results, we may conclude that there is a similar variation of the current speed in both the Kuroshio south of Japan and the Florida Current, showing a maximum in summer and a minimum in fall or spring. It seems difficult to ascribe a definite seasonal variation to either the Kuroshio Extension or the Gulf Stream. This might be attributed to such complicated hydrographic conditions in both that the speed of surface currents is much affected by local conditions of anticyclonic and cyclonic eddies on opposite sides of the two streams.

Though a mere presentation of current speed averaged by season does not necessarily warrant a statistically significant conclusion concerning the seasonal variation in the current speed, a rather good similarity in the seasonal variation between the two stream systems arouses an interest in further inquiry.

9 LONG-PERIOD FLUCTUATIONS

According to Stommel (1965), "there are not sufficient data available to permit us to discuss fluctuations in the Gulf Stream System lasting longer than one year." Japanese oceanographers have a large quantity of data for the study of fluctuations in the Kuroshio path. The change in the flow pattern along the south coast of Japan has grown popular among Japanese fishermen and mariners since Uda (1940) published a paper on the cyclonic eddy south of Enshu Nada. Somewhat inadequate orientation of hydrographic sections in the Tohoku Area, however, had kept us ignorant of the long-period fluctuations in the path of the Kuroshio Extension until recently.

9.1 Long-period fluctuations of the axis in the inshore Tohoku Area

Figure 57 shows long-period fluctuations in the latitude at the north limit of the Kuroshio Extension axis in the inshore Tohoku Area (141°E–146°E), where the current path is convex to the north in almost all cases. Open squares indicate the latitude at the north limit of the axis as based on the position of the 14°C isotherm at 200 m (Sec. 6), which is determined by the exact linear interpolation between two points of temperature measurements, while solid squares indicate the latitude as based on GEK fix. The triangle pointed upward or downward indicates that the north limit is expected to be to the north or south of it, respectively.

Tables 15 and 17 give original data in the vicinity and just at the north limit of the Kuroshio Extension axis. Of course, these were selected from massive data after preparation of working charts of 200 m temperature and GEK current for each individual period, using all published and some unpublished temperature and GEK data from Japanese ships and some temperature data from foreign ships on file in the U.S. National Oceanographic Data Center by the end of 1968.

The following are characteristic features of the fluctuations.

—(i) The long-period fluctuation is more distinct than the seasonal fluctuation, which is hardly detectable. The meridional range of the fluctuations is usually within 200 nautical miles. The north limit in 1942 and 1960 passed 39°N northward and

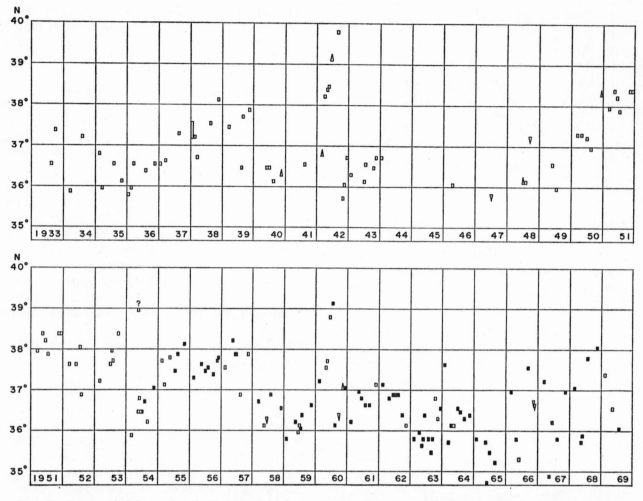

FIG. 57 Long-period fluctuations in the latitude at the north limit of the Kuroshio Extension axis in the inshore Tohoku Area (141°E–146°E). Open squares and triangles are based on 200 m temperature as indicative of the axis (Table 8). Triangles pointed southward and northward indicate the latitude of stations on the north and south sides of the axis, respectively. Solid squares are based on GEK fix.

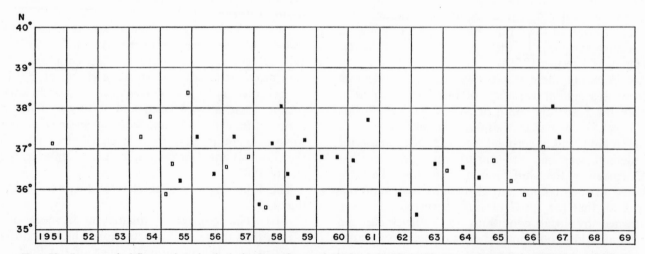

FIG. 58 Long-period fluctuations in the latitude at the north limit of the Kuroshio Extension axis in the offshore Tohoku Area (148°E–152°E). Symbols are the same as in Fig. 57.

that in 1965 and 1967 passed 35°N southward, but these are peculiar cases.

—(ii) The abrupt southward shift of the north limit may be associated with detaching a big anti-cyclonic eddy with a diameter of about 150 nautical miles at the long axis into the inshore Perturbed Area, as observed in the fall of 1942, in the winter of 1953–1954, in the summer of 1960 (Sec. 7.3), in the fall of 1966 and in the spring of 1967. After the abrupt southward shift of the north limit, a northward shift of the limit on a scale smaller than the southward usually follows.

Detachments of such eddies may happen more frequently than those mentioned above, but they cannot be studied in detail owing to lack of suitable data.

—(iii) As the difference in year graduation between the upper and lower panels in Fig. 57 is 18 years, a similar trend of fluctuations in each panel might suggest an 18-year periodicity, the Saros period, but it is statistically insignificant. The lower panel in Fig. 57 shows that the north limit has been shifting south in general trend since 1951.

—(iv) There may be a periodicity of 4.5 years. The north limit of the axis is at the southernmost

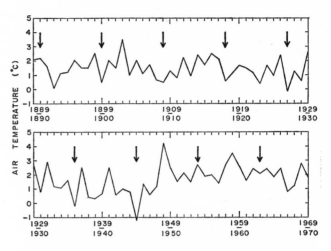

FIG. 59 Long-period fluctuations in air temperature in winter (mean for December–February) at Fukushima (37°45′N, 140°28′E). Arrows indicate very cold winters as assumed from the 9-year periodicity.

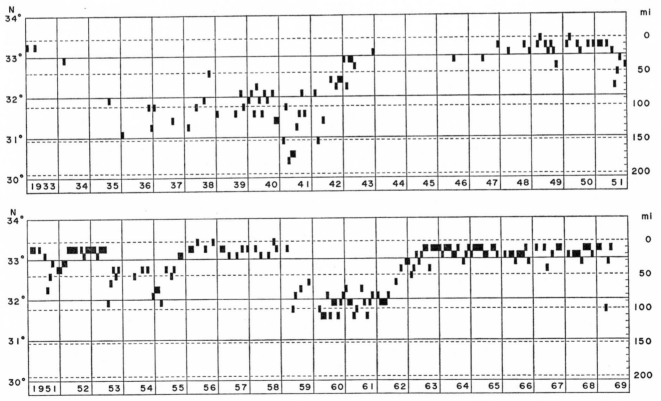

FIG. 60 Long-period fluctuations in the latitude of the Kuroshio axis at 135°40′E–135°50′E south of Shionomisaki. Based on 200 m temperature as indicative of the axis (Table 8).

latitude around the following time with intervals of exactly 4.5 years, though no data are available for 1945 because of World War II.

     Spring   1936, 1945, 1954, 1963
     Fall   1940, 1949, 1958, 1967

A similar periodic variation in surface temperature at some coastal stations facing the Tohoku Area was proposed by Hatanaka (1952), but none of the stations on the south coast of Japan show a 4.5-year periodicity. Neither do the long-period fluctuations in the Kuroshio axis south of Japan show a 4.5-year periodicity (Figs. 60, 61 and 62). According to Tajima (1951), the air temperature in winter at Fukushima (37°45′N, 140°28′E) had been showing extremely low values at intervals of 9 years from the winter of 1899–1900 until that of 1944–1945. Figure 62, prepared for this chapter adding later data, shows that although the 9-year periodicity is not clearly apparent in recent years, the coincidence of the cold winter before 1945 with the spring southernmost

excursion of the north limit of the Kuroshio Extension axis is remarkable. These facts suggest that the 4.5-year periodicity is localized in the Tohoku Area, if it is present at all. Several attempts to relate the periodicity to large-scale meteorological effects have been made, but none of them have proved successful.

9.2 Long-period fluctuations of the axis in the offshore Tohoku Area

Because of a scarcity of data, description of fluctuations in the offshore Tohoku Area 350–550 nautical miles east of Inubozaki (148°E–152°E), where the second crest of meanders convex to the north is often found (Sec. 7.6), is quite uncertain (Fig. 58), and it is difficult to find any relationship in the fluctuations between the inshore and offshore areas. Tables 16 and 18 give original data relating to the north limit of the axis in the offshore Tohoku Area. On the basis of the results given in Sec. 6, the

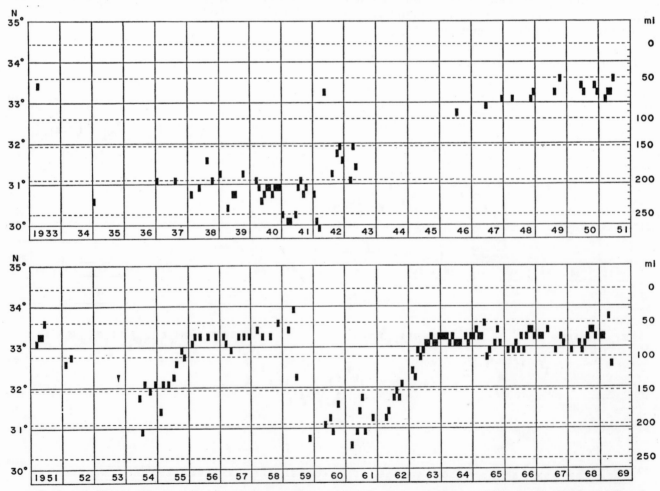

FIG. 61   Long-period fluctuations in the latitude of the Kuroshio axis at 136°40′E–136°50′E south of Daiozaki. Based on 200 m temperature as indicative of the axis (Table 8).

13.5°C isotherm at a depth of 200 m was regarded as indicative of the axis.

9.3   Classification of fluctuations in the north limit of the Kuroshio Extension

Figure 63 shows histograms of latitude at the north limit of the Kuroshio Extension axis in the inshore and offshore Tohoku Areas, the sample size being 164 in the inshore area and 36 in the offshore area. They were easily prepared by counting the squares for each class by 20-minute latitude in Figs. 57 and 58. The histograms in the two areas do not show that the bimodal distribution of the latitude is significant, though there seems to be a mode at about 38°N, which is probably a result of crowded stations along C-Sec (Fig. 49). According to the terminology used in the seasonal forecasts of sea-water and air temperatures by the JMA, seven and five classes are distinguished as shown in Fig. 64. The classifica-

tions are based on the ratio between the deviation from the mean and the standard deviation for sea-water temperature and on probabilities for air temperature, but the alternatives are added to both the classifications in Fig. 64. The seven classes used for description of the southward or northward excursion of the north limit of the Kuroshio Extension axis in the Tohoku Area are shown in the bottom panel of Fig. 64. They are an effective modification of those used for both sea-water and air temperatures.

9.4   Fluctuations in the Kuroshio and Oyashio Areas

The Japanese have long believed that when the power of the Kuroshio was strong, the power of the Oyashio was weak or that when the latter was strong, the former would be weak. Although the terms "Kuroshio power" and "Oyashio power" are vague, they seem to be used somewhat like dimensions of

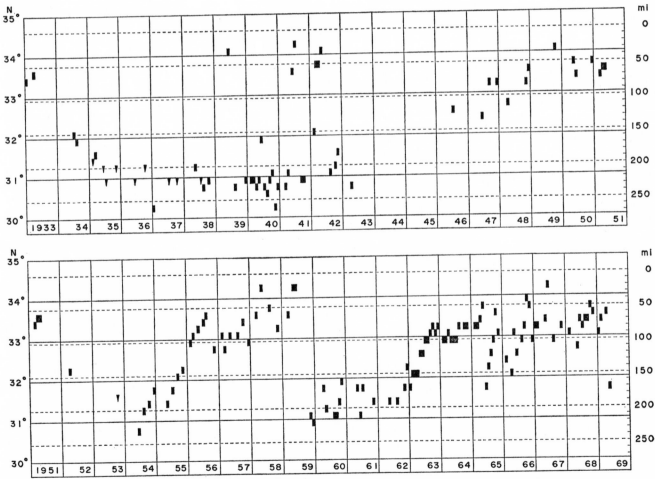

FIG. 62   Long-period fluctuations in the latitude of the Kuroshio axis at 138°10′E–138°20′E south of Omaezaki. Based on 200 m temperature as indicative of the axis (Table 8).

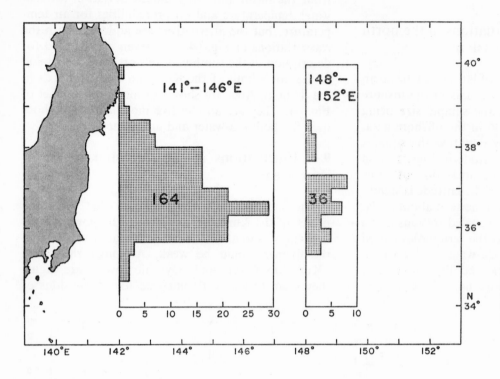

FIG. 63 Histograms of latitude at the north limit of the Kuroshio Extension axis in the inshore (141°E–146°E) and offshore (148°E–152°E) Tohoku Areas.

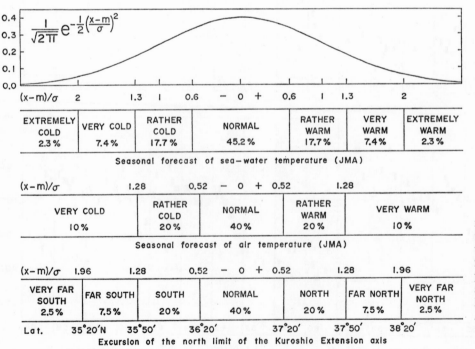

$$\frac{1}{\sqrt{2\pi}}e^{-\frac{1}{2}\left(\frac{x-m}{\sigma}\right)^2}$$

FIG. 64 Top and middle panels show 7 classes of sea-water temperature and 5 classes of air temperature adopted by the JMA for their seasonal forecast. Bottom panel shows 7 classes of latitude at the north limit of the Kuroshio Extension axis in the Tohoku Area.

the Kuroshio and Oyashio Areas. Kawai (1955a) reported that in August there was no immediate relation between fluctuations of dimension of the Kuroshio Area, as determined by the 15°C isotherm at a depth of 100 m [the 14°C isotherm at a depth of 200 m, indicative of the Kuroshio Extension axis, is a far more sensitive indicator, as shown in Sec. 6], and that of the Oyashio Area, as determined by the 5°C isotherm at a depth of 100 m; this was similar to the relation between fluctuations of latitudes of the Kuroshio and Oyashio Fronts at 144°E. He ascribed the relation of fluctuations to the hydrographic structure of the Tohoku Area interlaying the "Transition Area" [Perturbed Area] between the Kuroshio and Oyashio Areas. Following the same method, Hata (1969a) discussed fluctuations of the mean latitudes of the 15°C isotherm at a depth of 100 m in the Tohoku Area between 142°E and 150°E and of the 5°C isotherm between 143°E and 150°E for the summers from 1933 to 1941 and for the summers and winters from 1954 to 1966. Use of the same temperature for winter and summer is improper because of some seasonal variation in the 100 m temperature, but his conclusion that "it is clear that the negative correlation between the mean latitude of 15°C and 5°C exists" seems to dispel the traditional Japanese belief. Inspection of his figure (Hata, 1969a, Fig. 12) shows that as a general trend, the dimensions of both the warm area above a temperature of 15°C and the cold area below a temperature of 5°C have been decreasing since sometime before 1954. This may be related to the fact that the cold core below a temperature of 2°C in the Kuroshio Extension has rarely been observed recently, though this is partly due to a wider spacing of Nansen-bottle stations reaching the deep cold core (Sec. 2.4).

10   CONCLUSION

I would like to conclude with a general introduction to the hydrography in the vicinity of the Kuroshio Extension, omitting, for the sake of brevity, any extensive citations from the voluminous studies on the subject.

In pursuing hydrographic researches or studies on the Kuroshio Extension, one depends much upon synoptic analyses because of variability of the Kuroshio Extension in space and time. The term "synoptic analysis" may be translated into Japanese as an analysis "affording a general view of a whole," which is also the basic meaning of "synoptic" found in Webster's New International Dictionary, 1963 edition. As indicated in the dictionary, however, the term "synoptic" has a qualified meaning in meteorology as well as in oceanography, namely, "relating to or displaying atmospheric and weather conditions as they exist simultaneously over a broad area." Translated into Japanese "synoptic" does not explicitly carry the meaning of "simultaneously." It is perhaps in part due to this lack of explicitness that a synoptic analysis is accepted by some Japanese oceanographers as a broad and vague analysis without due consideration for observation programs or for interpretation of results. This attitude is not confined to Japan, however, as was mentioned in the introduction of "Descriptive Physical Oceanography" by Pickard (1963); the fault lies in the use of the term "synoptic," which in a different context is closely related to "synopsis."

In addition to the synoptic approach as exemlified in Secs. 2, 5 and 7, the climatologic approach is already playing an important role in the hydrographic description. The latter approach is related to the prevailing or average hydrographic conditions of a place as determined by hydrographic elements and their changes over a period of years, and is exemplified in Secs. 1, 4 and 9 and some part of Sec. 8. Since both the approaches are not exclusive, there is an overlapped aspect called synoptic climatology in meteorology as well as in hydrography, as exemplified in Sec. 6 and some part of Sec. 8.

Of course, various approaches are necessary in describing the Kuroshio Extension, but to make those that can really be termed synoptic is not an easy matter.

Acknowledgments

I am greatly indebted to F. C. Fuglister, Henry Stommel, Kozo Yoshida and Jotaro Masuzawa for their useful comments on this manuscript. Thanks are also due to the Data Services Branch, U.S. National Oceanographic Data Center for sending me temperature data from foreign vessels on file in the Center and to Jotaro Masuzawa for showing me various unpublished figures and computed data. Finally, I wish to thank Reiko Watanabe and Noriko Kawashima, who drafted most of the figures for this chapter.

TABLE 15. 200 m temperature data at Nansen-bottle stations or at BT lowerings bracketing the north limit of the Kuroshio Extension axis in the inshore Tohoku Area (141°E–146°E).

Year	Date	Ship	Sta. no.	Lat. (N)	Long. (E)	200m T.	Date	Ship	Sta. no.	Lat. (N)	Long. (E)	200m T.	Lat. (N)	Code	Month
			South side station						North side station				North limit of the axis		
1927	Sept. 19	Manshu	603	34°16'	142°48'	19.0	Aug. 9	Manshu	556	35°53'	143°17'	11.5	35°21	05	8-9
1929	—	—	—	—	—	—	June 27	Carnegie	114	36°38'	143°34'	09.9	▽36°38'	▽20	6-7
1933	Aug. 14	Soyo	023	36°15'	143°10'	17.0	Aug. 13	Soyo	022	36°56'	144°20'	10.5	36°34'	19	8
	Sept. 27	Komahashi	057	37°00'	143°04'	17.8	Oct. 2	Komahashi	075	38°18'	143°03'	05.1	37°23'	29	9-10
1934	Mar. 21	Komahashi	029	35°44'	142°40'	16.1	Mar. 22	Komahashi	031	36°06'	142°48'	05.5	35°52'	11	3
	Aug. 7	FSPFES	011	37°08'	143°37'	14.5	Aug. 7	MGPFES	012	38°14'	144°18'	09.0	37°14'	27	7-8
1935	Feb. 10	Soyo	007	36°36'	144°18'	16.4	Feb. 10	Soyo	008	37°22'	144°19'	06.1	36°47'	22	2
	Mar. 7	Soyo	006	35°33'	144°33'	16.4	Mar. 8	Soyo	007	36°12'	144°46'	12.1	35°55'	12	3
	Aug. 8	Itsukushima	041	36°22'	144°47'	16.6	July 28	Soyo	008	37°02'	144°17'	08.1	36°34'	19	7-8
	Nov. 7	Soyo	007	36°00'	144°32'	15.9	Nov. 7	Soyo	008	36°41'	144°43'	06.8	36°09'	14	10-11
1936	Jan. 9	Soyo	006	35°25'	144°32'	17.1	Jan. 10	Soyo	007	36°22'	144°47'	09.6	35°49'	10	1
	Feb. 10	Soyo	006	35°27'	144°33'	16.0	Feb. 10	Soyo	007	36°10'	144°43'	13.1	35°57'	12	2
	Mar. 11	Soyo	007	36°30'	144°34'	14.9	Mar. 11	Soyo	008	37°12'	144°34'	04.3	36°34'	19	3
	July 31	Soyo	007	36°14'	144°32'	15.5	July 31	Soyo	008	37°09'	144°35'	06.8	36°23'	17	7-8
	Nov. 10	Soyo	007	36°19'	144°26'	17.5	Nov. 10	Soyo	008	37°05'	144°27'	05.6	36°33'	19	11
1937	Jan. 11	Soyo	007	36°16'	144°20'	16.4	Jan. 11	Soyo	008	37°03'	144°19'	09.3	36°32'	19	1
	Mar. 10	Soyo	007	36°23'	144°40'	16.9	Mar. 11	Soyo	008	37°12'	144°40'	07.6	36°38'	20	3
	Aug. 11	Soyo	008	37°05'	143°57'	17.2	Aug. 11	Soyo	009	37°55'	144°00'	05.1	37°18'	28	8
1938	Jan. 12	Soyo	008	37°08'	144°18'	17.9	Jan. 12	Soyo	009	37°58'	144°18'	10.4 (0m)	37°10' / —34'	27-31	1
	Feb. 10	Soyo	008	37°06'	144°57'	15.0	Feb. 11	Soyo	009	37°56'	145°05'	07.4	37°13'	27	2
	Mar. 13	Soyo	007	36°20'	144°13'	16.9	Mar. 13	Soyo	008	37°13'	144°13'	10.1	36°43'	21	3
	Aug. 8	Soyo	008	37°16'	143°58'	16.6	Aug. 8	Soyo	009	37°58'	144°07'	10.6	37°34'	31	8
	Nov. 7	Soyo	009	37°55'	144°20'	16.1	Nov. 8	Soyo	010	38°45'	144°20'	06.0	38°05'	38	11
1939	Mar. 9	Soyo	079	37°22'	142°05'	14.5	Mar. 10	Soyo	078	38°10'	142°23'	06.1	37°25'	30	3
	Aug. 11	Ibaragi	012	35°51'	142°55'	17.7	Aug. 8	Iwaki	011	37°00'	143°04'	10.9	36°29'	18	8
	Aug. 24	Soyo	008	37°05'	144°20'	17.1	Aug. 24	Soyo	009	38°08'	144°22'	11.8	37°42'	33	8-9
	Nov. 10	Soyo	008	37°08'	144°09'	17.5	Nov. 22	MGKSFES	014	38°16'	145°22'	12.3	37°54'	35	11
1940	May 31	#2 Kaiyo	059	36°20'	145°08'	16.2	May 30	#2 Kaiyo	058	37°06'	144°51'	05.1	36°29'	18	5-6
	July 4	#1 Kaiyo	009	36°05'	144°06'	17.2	July 5	#1 Kaiyo	010	36°34'	144°19'	12.9	36°27'	18	6-7
	Aug. 3	Ibaragi	007	36°00'	143°56'	17.4	Aug. 13	Soyo	007	36°14'	144°34'	11.5	36°08'	14	8
	Nov. 12	Soyo	007	36°18'	144°19'	17.3	—	—	—	—	—	—	△36°18'	△16	11
1941	Aug. 12	Soyo	007	36°20'	144°12'	17.0	Aug. 12	Soyo	008	37°07'	144°08'	06.5	36°33'	19	8
1942	Mar. 2	Iwaki	004	36°29'	142°35'	16.8	(Mar. 2)	(Iwaki)	(003)	(37°00')	(142°34')	(11.9)	△36°47'	△22	2-3
	Apr. 3	Yoko	045	38°06'	144°53'	16.3	Apr. 3	Yoko	046	38°32'	145°40'	05.3	38°11'	39	3-4
	Apr. 24	ONG	005	38°18'	142°58'	15.5	Apr. 30	ONG	015	38°37'	143°06'	10.1	38°23'	41	4-5

TABLE 15. (cont'd)   200 m temperature data at Nansen-bottle stations or at BT lowerings bracketing the north limit of the Kuroshio Extension axis in the inshore Tohoku Area (141°E–146°E).

Year	South side station						North side station						North limit of the axis		
	Date	Ship	Sta. no.	Lat. (N)	Long. (E)	200m T.	Date	Ship	Sta. no.	Lat. (N)	Long. (E)	200m T.	Lat. (N)	Code	Month
1942	May 11	ONG	005	38°25'	143°06'	16.1	May 30	ONG	015	38°39'	143°01'	08.9	38°29'	42	5
	June 17	ONG	016	39°08'	143°08'	14.9	—	—	—	—	—	—	△39°08'	△50	6
	Aug. 28	Soyo	040	39°24'	143°58'	17.2	Aug. 21	Tankai	002	40°25'	144°29'	08.8	39°47'	58	8-9
	Oct. 28	Iwaki	021	35°42'	144°47'	14.6	Oct. 28	Iwaki	022	36°06'	144°17'	05.8	35°44'	09	10-11
	Nov. 14	Iwaki	004	35°51'	143°06'	18.3	Nov. 13	Iwaki	003	36°11'	142°25'	11.5	36°04'	13	11
1943	Dec. 18	Iwaki	023	36°32'	143°46'	18.0	Dec. 18	Iwaki	024	36°59'	143°22'	07.8	36°43'	21	12
	Jan. 27	Iwaki	023	36°06'	144°21'	15.8	Jan. 27	Iwaki	024	36°30'	143°44'	12.2	36°18'	16	1-2
	July 7	Iwaki	022	36°04'	144°20'	15.3	July 7	Iwaki	023	36°31'	143°47'	07.1	36°08'	14	6-7
	July 18	Iwaki	022	36°14'	144°20'	17.2	July 19	Iwaki	023	36°42'	143°49'	11.5	36°30'	19	7
	Oct. 13	Iwaki	022	36°10'	144°25'	17.1	Oct. 13	Iwaki	023	36°36'	143°47'	12.2	36°26'	18	10
	Nov. 14	Iwaki	010	36°37'	143°54'	16.5	Nov. 14	Iwaki	011	36°58'	143°22'	08.8	36°44'	21	11
1944	Jan. 9	Iwaki	023	36°33'	143°51'	15.6	Jan. 19	Iwaki	033	37°14'	144°12'	05.6	36°40'	21	1
1946	Apr. 17	Ryofu	003	35°28'	143°17'	17.2	Apr. 21	Ryofu	015	36°49'	145°17'	10.0	36°04'	13	4
1947	—	—	—	—	—	—	July 1	Soyo	007	35°49'	143°53'	07.9	▽35°49'	▽10	6-7
1948	June 30	Shinnan	022	36°06'	143°50'	15.4	July 25	#5 Kaiyo	003	37°05'	143°10'	05.9	△36°06'	△14	6-7
	Aug. 4	Ryofu	014	35°40'	143°30'	18.3	—	—	—	—	—	—	36°09'	14	7-8
	—	—	—	—	—	—	Sept. 12	Ryofu	014	37°14'	144°56'	10.1	▽37°14'	▽27	9
1949	May 26	Ibaragi	015	36°00'	144°00'	16.8	May 25	Fusa	011	36°54'	143°34'	11.9	36°31'	19	5-6
	July 17	#4 Kaiyo	003	35°16'	142°31'	18.0	July 19	#4 Kaiyo	011	36°26'	143°25'	11.0	35°56'	12	7
1950	Mar. 29	Yushio	004	37°08'	143°21'	16.7	Mar. 7	#4 Kaiyo	006	37°30'	143°23'	10.0	37°17'	28	3
	May 4	#4 Kaiyo	011	36°52'	142°44'	17.2	May 8	#4 Kaiyo	019	37°53'	142°47'	09.0	37°16'	28	4-5
	July 2	Yushio	006	37°00'	144°15'	16.3	July 2	Yushio	005	37°34'	144°09'	08.3	37°10'	27	6-7
	[1]Aug. 21	#1 Kaiko	004	36°56'	142°27'	14.0	Aug. 21	#1 Kaiko	004	36°56'	142°27'	14.0	36°56'	24	8
	Dec. 23	Ryofu	001	38°17'	143°56'	14.9	—	—	—	—	—	—	△38°17'	△40	12
1951	Mar. 9	Yushio	329	37°07'	144°12'	17.0	Mar. 6	#1 Tenkai	015	38°00'	144°04'	09.8	△37°29'	36[2]	3
	May 21	Yushio	351	38°00'	144°00'	16.8	May 21	Yushio	349	38°30'	144°00'	13.1	38°23'	41	5
	June 18	Shiga	004	38°05'	144°50'	15.0	June 12	#5 Kaiyo	025	39°08'	144°17'	06.7	38°13'	39	6
	July 20	Soyo	031	37°50'	143°35'	15.0	July 20	Soyo	032	38°08'	143°06'	04.2	37°52'	35	7
	Nov. 11	Shinnan	002	38°19'	144°07'	14.3	Nov. 11	Yushio	030	38°30'	144°00'	13.5	38°23'	41	11
	Dec. 22	Shiga	003	38°14'	144°13'	16.8	Dec. 9	Ikuna	003	38°24'	144°18'	12.5	38°21'	41	12
1952	Mar. 3	Yushio	006	37°27'	144°48'	17.0	Mar. 13	Yushio	011	38°00'	144°00'	07.9	37°38'	32	3
	May 24	Daito	003	37°24'	143°23'	16.3	May 25	Yushio	018	38°12'	144°00'	07.8	37°37'	32	5-6
	July 11	Soyo	012	37°58'	143°40'	15.7	July 17	Chikubu	003	38°16'	143°53'	08.5	38°02'	37	7
	Aug. 6	#5 Kaiyo	042	36°58'	145°00'	17.4	Aug. 5	#5 Kaiyo	039	37°57'	144°57'	03.9	37°13'	23	7-8
1953	Mar. 5	Yushio	664	37°01'	144°54'	16.6	Mar. 5	Yushio	663	38°03'	144°14'	04.1	37°14'	27	2-3
	June 30	Soyo	019	37°27'	145°07'	16.4	June 29	Soyo	018	38°09'	145°02'	07.0	37°38'	32	6-7
	July 16	Soyo	033	37°56'	144°36'	14.6	July 12	Chikubu	004	38°16'	144°55'	03.7	37°57'	36	7
	July 27	Soyo	020	37°30'	144°11'	16.6	July 27	Soyo	019	37°55'	143°56'	11.3	37°42'	33	7-8
	Oct. 4	#4 Kaiyo	014	38°16'	145°00'	15.5	Sept. 29	Baird	077	38°59'	144°51'	07.2	38°24'	41	9-10

TABLE 15. (cont'd) 200 m temperature data at Nansen-bottle stations or at BT lowerings bracketing the north limit of the Kuroshio Extension axis in the inshore Tohoku Area (141°E–146°E).

Year	Date	South side station Ship	Sta. no.	Lat. (N)	Long. (E)	200m T.	Date	North side station Ship	Sta. no.	Lat. (N)	Long. (E)	200m T.	North limit of the axis Lat. (N)	Code	Month
1954	Mar. 1	Ryofu	014	35°46'	144°30'	14.8	Mar. 8	Ryofu	026	37°01'	144°18'	04.2	35°53'	11	2–3
	May 20	Ryofu	061	36°20'	144°06'	14.9	May 20	Ryofu	062	36°28'	144°07'	13.6	36°26'	18	5
	May 22	Yushio	834	38°51'	143°32'	15.0	May 22	Yushio	833	39°13'	142°51'	09.8	38°55'	48?	5
	May 31	Tenyo	021	36°35'	144°17'	17.7	May 31	Tenyo	022	37°18'	144°15'	02.4	36°45'	22	5–6
	June 16	Ryofu	131	36°08'	144°09'	18.3	June 28	#4 Kaiyo	C54	37°00'	143°11'	06.1	36°26'	18	6
	Aug. 25	Tenyo	104	35°50'	144°19'	18.4	Aug. 24	Tenyo	103	36°40'	144°17'	09.3	36°14'	15	8–9
1955	Feb. 18	Ryofu	314	37°40'	144°02'	14.8	Feb. 18	Ryofu	315	37°57'	144°07'	07.8	37°42'	33	2
	Mar. 19	Tenyo	029	36°40'	144°36'	17.0	Mar. 18	Tenyo	028	37°24'	144°10'	12.4	37°09'	26	3
	May 13	#4 Kaiyo	B16	37°42'	144°11'	17.0	May 11	Ryofu	400	38°00'	144°02'	09.3	37°49'	34	5
1957	Feb. 18	Ryofu	026*	37°22'	144°17'	15.8	Feb. 18	Ryofu	027*	37°33'	144°23'	13.3	37°30'	31	2
	Aug. 22	Ryofu	630*	36°43'	142°52'	15.6	Aug. 6	Kaiyo	B05	37°00'	143°52'	13.0	36°53'	23	8
	Nov. 10	Ryofu	864*	37°50'	144°55'	16.0	Nov. 10	Ryofu	863*	37°58'	144°54'	06.0	37°52'	35	11
1958	May 8	Kaiyo	047*	36°04'	144°00'	15.9	May 8	Kaiyo	048*	36°15'	143°04'	10.4	36°08' }	14	5
	May 15	Soyo	018*	36°00'	143°14'	14.9	May 15	Soyo	019*	36°15'	143°14'	11.4	36°04' }		
	June 21	Kitakami	—	—	—	—	June 21	Kitakami	029*	36°18'	143°57'	07.5	▽36°18'	▽16	6
1959[1)]	June 15	Soyo	033*	35°55'	144°04'	14.0	June 15	Soyo	033*	35°55'	144°04'	14.0	35°55' }	12	6
	June 22	Soyo	027*	35°45'	143°23'	16.5	June 22	Soyo	026*	36°00'	143°23'	13.2	35°56' }		
	July 2	Soyo	027*	35°58'	143°28'	16.1	July 2	Soyo	026*	36°20'	143°22'	10.0	36°06'	14	6–7
1960	Apr. 27	Shun'yo	045*	37°31'	143°22'	14.6	Apr. 27	Shun'yo	046*	37°43'	143°19'	12.1	37°34'	31	4–5
	May 12	Shun'yo	040*	37°30'	144°22'	15.2	May 10	Shun'yo	032*	38°01'	144°20'	11.8	37°41'	33	5
	June 19	Kitakami	009*	38°26'	144°35'	16.6	June 11	Kitakami	007*	39°05'	144°52'	12.0	38°48'	46	6
	—	—				—	Oct. 7	Tenyo	009	36°20'	143°33'	10.1	▽36°20'	▽17	9–10
1961	Nov. 8	#2 Chiba	018	37°01'	143°40'	16.0	—	—	—	—	—	—	△37°01'	△25	11
1962	Dec. 1	Wakataka	019*	36°59'	144°00'	16.3	Nov. 28	Wakataka	007*	37°30'	144°10'	09.5	37°09'	26	11–12
	Nov. 15	#2 Chiba	014*	35°59'	144°17'	18.9	Nov. 13	Shun'yo	006*	36°16'	144°09'	13.0 (155m)	36°05' (−36°09')	14	11
1963	Oct. 17	Mito	018*	36°06'	143°35'	16.3	Oct. 22	Mito	029*	37°00'	144°00'	13.1	36°45'	22	10
	Nov. 14	Soyo	035	35°39'	143°17'	16.5	Nov. 17	Soyo	055	36°20'	144°02'	13.9	36°18'	16	11
1964	May 19	Wakataka	009*	36°00'	143°00'	16.3	May 19	Wakataka	008*	36°21'	143°01'	07.3	36°05'	14	5
1966	May 28	Soyo	019*	35°00'	144°08'	16.3	May 27	Soyo	018*	35°30'	144°01'	11.9	35°16'	04	5–6
	Nov. 19	—	—	—	—		Nov. 19	Kofu	465*	36°40'	143°10'	11.5	▽36°40'	▽21	11
	Nov. 24	—	—	—	—		Nov. 24	Kofu	490*	36°39'	143°12'	10.4	▽36°39'	▽20	11–12
1969	Feb. 21	Kofu	060*	37°05'	144°00'	15.7	Feb. 21	Kofu	059*	37°30'	144°00'	13.1	37°21'	29	2
	May 20	Ryofu	042	35°59'	144°03'	18.0	May 20	Ryofu	041	36°58'	144°02'	10.9	36.32'	19	5

* BT 1) The axis is just at the station. 2) Based on the drift record of the *Yushio Maru* in distress.

TABLE 16. 200 m temperature data at Nansen-bottle stations or at BT lowerings bracketing the north limit of the Kuroshio Extension axis in the offshore Tohoku Area (148°E–152°E).

Year	Date	South side station					Date	North side station					North limit of the axis		
		Ship	Sta. no.	Lat. (N)	Long. (E)	200m T.		Ship	Sta. no.	Lat. (N)	Long. (E)	200m T.	Lat. (N)	Code	Month
1951	July 15	Soyo	015	36°27'	149°29'	16.7	July 15	Soyo	013	37°07'	149°56'	13.4	37°06'	26	7
1954	May 18	Ryofu	090	37°04'	149°00'	14.0	May 18	Ryofu	091	37°30'	148°38'	05.0	37°18'	28	5
	Sept. 1	Ryofu	185	37°06'	150°07'	16.0	Sept. 2	Ryofu	186	38°19'	150°00'	11.6	37°47'	34	8-9
1955	Feb. 24	Ryofu	338	35°44'	150°08'	15.8	Feb. 24	Ryofu	339	36°02'	150°06'	09.4	35°50'	11	2-3
	May 19	Ryofu	422	36°31'	150°02'	15.5	May 19	Ryofu	423	37°00'	150°11'	04.5	36°36'	20	5
	Nov. 18	Ryofu	544	38°12'	147°42'	14.7	Nov. 19	Ryofu	546	38°33'	147°56'	12.1	38°22'	41	11
1957	Feb. 21	Ryofu	066*	36°23'	149°50'	16.0	Feb. 21	Ryofu	067*	36°36'	149°45'	12.9	36°33'	19	2
	Nov. 5	Takuyo	013*	36°28'	149°43'	16.6	Nov. 5	Takuyo	012*	37°02'	148°57'	11.5	36°49'	22	10-11
1958	May 11	Ryofu	207*	35°28'	150°01'	14.3	May 11	Ryofu	208*	35°45'	150°10'	09.0	35°31'	07	5
1964	Feb. 12	Ryofu	094*	36°11'	149°55'	15.8	Feb. 14	Ryofu	102*	36°37'	150°52'	11.5	36°25'	18	2
1965	Aug. 2	Zhyemchug	018	36°26'	148°58'	17.2	Aug. 1	Zhyemchug	017	36°56'	148°58'	10.0	36°41'	21	7-8
1966	Feb. 23	Orlick	018	35°57'	148°58'	15.9	Feb. 23	Orlick	017	36°25'	149°00'	11.7	36°13'	15	2
	June 1	Soyo	043*	36°29'	149°01'	17.0	June 1	Soyo	044*	37°00'	149°00'	11.0	36°47'	22	5-6
	1)July 13	Orlick	019	35°54'	149°02'	13.5	July 13	Orlick	019	35°54'	149°02'	13.5	35°54'	11	7
1967	Feb. 8	Orlick	013	37°00'	149°00'	13.6	Feb. 8	Orlick	012	37°30'	149°00'	11.1	37°01'	25	2
1968	July 24	Orlick	020	35°30'	149°02'	16.5	July 24	Orlick	019	36°00'	149°00'	12.5	35°53'	11	7-8

*BT   1) The axis is just at the station.

TABLE 17. GEK data at the north limit of the Kuroshio Extension axis in the inshore Tohoku Area (141°E–146°E).

Date			Ship	Fix no.	Position		Velocity		Lat. code	Month
					Lat. (N)	Long. (E)	(°)	(kn)		
3	July	1954	#4 Kaiyo	063	36°25′	144°06′	111	4.2	18	6–7
5	Aug.	1954	#4 Kaiyo	035	36°42′	144°18′	090	4.3	21	7–8
16	Nov.	1954	#4 Kaiyo	102	37°01′	143°19′	120	3.3	25	11
9	July	1955	Meiyo	027	37°28′	144°06′	099	3.0	30	7
17	Aug.	1955	Ryofu	027	37°53′	145°17′	090	4.5	35	8
2	Nov.	1955	Meiyo	058	38°09′	143°57′	068	3.1	38	10–11
15	Feb.	1956	#4 Kaiyo	045	37°11′	143°56′	103	5.4	27 ⎫ 28	2
18	Feb.	1956	Ryofu	018	37°23′	144°07′	090	3.1	29 ⎭	
17	May	1956	Ryofu	168	37°38′	145°38′	100	2.6	32	5
24	June	1956	Ryofu	280	37°28′	144°56′	110	3.4	30	6–7
3	Aug.	1956	Asachidori	019	37°30′	144°24′	086	4.0	31	7–8
26	Sept.	1956	Ryofu	449	37°10′	144°46′	090	4.9	27 ⎫ 29	9–10
3	Oct.	1956	Miyake	015	37°30′	144°08′	069	3.6	31 ⎭	
9	Nov.	1956	Ryofu	481	37°41′	144°09′	100	4.9	33	11
7	Dec.	1956	Chifuri	024	37°49′	145°03′	084	3.3	34	11–12
9	May	1957	Ryofu	174	38°11′	145°04′	110	4.3	39	5
10	June	1957	Shumpu	020	37°49′	143°31′	080	4.5	34 ⎫	
16	June	1957	Shumpu	042	38°02′	143°37′	090	3.2	37 ⎬ 35	6
23	June	1957	Shumpu	146	37°52′	143°27′	070	3.1	35 ⎭	
28	June	1957	Kaiyo	104	37°48′	143°22′	090	2.8	34 ⎫ 35	6–7
1	July	1957	Ryofu	571	38°00′	142°39′	070	2.3	37 ⎭	
10	Mar.	1958	Ryofu	113	36°40′	145°50′	110	4.3	21	3
3	Aug.	1958	Kaiyo	026	36°51′	143°17′	097	2.5	23	7–8
27	Nov.	1958	Ryofu	752½	36°31′	143°47′	100	3.4	19	11–12
30	Jan.	1959	Ryofu	035	35°46′	144°11′	120	1.8	10	1–2
18	May	1959	Ryofu	210	36°14′	143°24′	120	2.8	15	5
11	July	1959	Takuyo	026	36°04′	143°30′	090	3.2	13	7
3	Aug.	1959	Ryofu	396	36°21′	144°13′	100	3.4	17	7–8
15	Nov.	1959	Ryofu	628	36°39′	143°54′	080	3.3	20	11
19	Feb.	1960	Ryofu	023½	37°13′	143°55′	100	4.0	27	2
10	July	1960	Sagami	057	39°06′	144°32′	095	2.3	50	7
15	Aug.	1960	Ryofu	213	36°08′	143°59′	100	2.8	14	8
19	Dec.	1960	Ojika	047	37°03′	144°56′	129	2.6	25	12
18	Feb.	1961	Ryofu	026	36°12′	144°04′	110	3.9	15	2
15	May	1961	Kuma	014	36°55′	143°44′	055	3.0	24	5
12	June	1961	Chifuri	006	36°47′	143°11′	065	3.4	22	6
6	Aug.	1961	Ryofu	288	36°36′	143°59′	090	3.0	20	7–8
12	Sept.	1961	Takuyo	013	36°37′	142°55′	063	2.4	20	9
8	Feb.	1962	Ryofu	029	37°06′	144°02′	090	2.3	26	2
28	Apr.	1962	Kaiyo	089	36°47′	143°35′	077	2.4	22	4–5
13	June	1962	Meiyo	025	36°51′	143°33′	061	3.7	23	6
13	July	1962	Meiyo	035	36°51′	144°03′	075	4.8	23	7
8	Aug.	1962	Ryofu	403½	36°53′	144°03′	110	4.3	23	8
30	Sept.	1962	Kaiyo	035	36°24′	143°49′	069	2.0	17	9–10
22	Feb.	1963	Ryofu	031	35°45′	144°09′	070	3.8	10	2
23	Apr.	1963	Takuyo	069	35°55′	142°19′	079	3.5	12	4
17	May	1963	Takuyo	034	35°35′	141°55′	083	4.6	08	5
24	May	1963	Shumpu	083	35°45′	143°53′	120	4.4	10	5–6
18	June	1963	Meiyo	075	36°21′	143°07′	104	3.7	17	6
24	July	1963	Kaiyo	074	35°45′	142°07′	063	2.7	10	7–8
31	Aug.	1963	Kaiyo	088	35°26′	141°56′	085	3.9	06	8–9
18	Sept.	1963	Ryofu	327	35°49′	142°01′	090	3.1	10	9
21	Dec.	1963	Meiyo	098	36°34′	142°28′	075	2.7	19	12
4	Feb.	1964	Ryofu	028	37°36′	144°04′	120	1.7	32	1–2
14	Mar.	1964	Soyo	158	35°40′	142°48′	089	2.0	09	3
20	Apr.	1964	Meiyo	067	36°08′	142°58′	082	3.5	14	4
30	June	1964	Kuma	009	36°31′	142°18′	061	2.7	19	6–7
6	Aug.	1964	Ryofu	366	36°28′	144°04′	110	4.1	18	7–8
11	Sept.	1964	Takuyo	025	36°19′	144°05′	090	2.6	16	9
13	Nov.	1964	Soyo	036	36°22′	142°47′	056	2.8	17	11

TABLE 17. (cont'd)   GEK data at the north limit of the Kuroshio Extension axis in the inshore Tohoku Area (141°E–146°E).

Date			Ship	Fix no.	Position		Velocity		Lat. code	Month
					Lat. (N)	Long. (E)	(°)	(kn)		
6	Feb.	1965	*Ryofu*	052	35°45′	143°58′	110	2.1	10	1–2
10	May	1965	*Soyo*	159	35°43′	141°56′	066	3.2	09	5
24	May	1965	*Soyo*	011	34°40′	142°37′	108	3.6	−03	5–6
28	June	1965	*Ryofu*	271	35°27′	142°07′	043	2.1	06	6–7
26	Aug.	1965	*Kaiyo*	106	35°11′	142°22′	065	3.5	03	8–9
3	Mar.	1966	*Kofu*	061	36°57′	143°40′	090	3.3	24	2–3
6	May	1966	*Kofu*	181	35°48′	143°15′	096	2.1	10	4–5
15	Sept.	1966	*Ryofu*	439	37°33′	144°02′	063	2.6	31	9
11	Mar.	1967	*Takuyo*	074	37°12′	142°24′	110	2.3	27	3
18	May	1967	*Kofu*	155	34°53′	144°02′	106	4.3	−01	5
23	June	1967	*Ryofu*	276	36°13′	144°05′	110	2.6	15	6
17	Aug.	1967	*Kofu*	372	35°45′	144°02′	111	2.7	10	8
13	Nov.	1967	*Natori*	337	36°59′	144°27′	098	2.7	24	11
26	Feb.	1968	*Kofu*	075	37°00′	144°07′	099	2.2	25	2–3
16	May	1968	*Kofu*	175	35°42′	142°50′	123	3.0	09	5
24	May	1968	*Ryofu*	149	35°50′	143°09′	103	3.1	11	5–6
24	July	1968	*Kofu*	340	37°46′	144°45′	097	3.0	34	7–8
16	Nov.	1968	*Ojika*	036	38°04′	144°00′	038	2.7	37	11
7	Aug.	1969	*Kofu*	382	36°00′	141°59′	074	3.1	13	7–8

TABLE 18.   GEK data at the north limit of the Kuroshio Extension axis in the offshore Tohoku Area (148°E–152°E).

Date			Ship	Fix no.	Position		Velocity		Lat. code	Month
					Lat. (N)	Long. (E)	(°)	(kn)		
22	Aug.	1955	*Ryofu*	065	36°10′	151°39′	120	4.2	15	8
27	Feb.	1956	*Ryofu*	081	37°16′	148°21′	060	2.1	28	2–3
22	Sept.	1956	*Ryofu*	416	36°23′	150°02′	080	2.4	17	9
11	May	1957	*Ryofu*	200	37°18′	148°23′	110	1.8	28	5
7	Mar.	1958	*Ryofu*	089	35°39′	150°00′	100	3.8	08	2–3
4	Aug.	1958	*Ryofu*	403	37°07′	150°10′	130	2.8	26	7–8
21	Nov.	1958	*Ryofu*	707	38°03′	148°16′	080	1.6	37	11
7	Feb.	1959	*Ryofu*	096	36°24′	150°10′	100	2.3	17	1–2
1	June	1959	*Ryofu*	348	35°46′	151°09′	130	2.5	10	5–6
12	Aug.	1959	*Ryofu*	481	37°12′	152°02′	120	2.7	27	8
28	Feb.	1960	*Ryofu*	095	36°47′	149°58′	080	2.0	22	2–3
26	Aug.	1960	*Ryofu*	281	36°49′	150°09′	100	2.9	22	8–9
1	Mar.	1961	*Ryofu*	085	36°42′	151°02′	110	3.1	21	2–3
13	Aug.	1961	*Ryofu*	346	37°44′	149°56′	110	2.7	33	8
14	Aug.	1962	*Ryofu*	455	35°50′	150°00′	040	3.1	11	8
28	Feb.	1963	*Ryofu*	075	35°22′	150°11′	090	3.5	05	2–3
25	Sept.	1963	*Ryofu*	380	36°39′	148°02′	100	2.1	20	9–10
16	Aug.	1964	*Ryofu*	425	36°32′	148°08′	090	2.9	19	8
12	Feb.	1965	*Ryofu*	098	36°18′	148°17′	080	2.4	16	2
7	June	1967	*Ryofu*	220	38°00′	151°00′	107	2.7	37	5–6
11	Aug.	1967	*Ryofu*	348	37°19′	148°08′	113	2.3	28	8

REFERENCES

Barkley, R. A. (1968): The Kuroshio-Oyashio front as a compound vortex street. *J. Mar. Res.*, **26**: 83–104.

Berghaus, Heinrich (1837): *Physikalischer Atlas*. Erster Band. **2**. Abt. Nr. 4. Gotha (according to bibliog. of Wüst, 1936).

Bogorov, B. G. (1958): Biogeographical regions of the plankton of the North-Western Pacific Ocean and their influence on the deep sea. *Deep-Sea Res.*, **5**: 149–161.

Bolin, Bert (1950): On the influence of the earth's orography on the general character of the westerlies. *Tellus*, **2**: 184–195.

Bratnick, Michael (1970): Convolutions of the surface outcrop of the northern edge of the Gulf Stream. Joint Oceanographic Assembly, Fifteenth General Assembly of International Association for the Physical Sciences of the Ocean. Tokyo, September 1970.

Dodimead, A. J., Felix Favorite and T. Hirano (1963): Review of oceanography of the subarctic Pacific region. In: *Salmon of the North Pacific Ocean*, Part II. *International North Pacific Fisheries Commission, Bull.*, **13**: 195pp.

Ford, W. L., J. R. Longard and R. E. Banks (1952): On the nature, occurrence and origin of cold low salinity water along the edge of the Gulf Stream. *J. Mar. Res.*, **11**: 281–293.

Fuglister, F. C. (1951a): Multiple currents in the Gulf Stream System. *Tellus*, **3**: 230–233.

——— (1951b): Annual variations in current speeds in the Gulf Stream System. *J. Mar. Res.*, **10**: 119–127.

——— (1955): Alternative analyses of current surveys. *Deep-Sea Res.*, **2**: 213–229.

——— (1960): Atlantic Ocean atlas of temperature and salinity profiles and data from the International Geophysical Year of 1957–1958. *Woods Hole Oceanogr. Inst., Atlas Series*, **1**: ii+209pp.

——— (1963): Gulf Stream '60. In: *Progress in Oceanography*, **1**: 265–373.

Fuglister, F. C. and A. D. Voorhis (1965): A new method of tracking the Gulf Stream. *Limnol. Oceanogr.*, **10** (Suppl.): R115–R124.

Fuglister, F. C. and L. V. Worthington (1951): Some results of a multiple ship survey of the Gulf Stream. *Tellus*, **3**: 1–14.

Fujii, Y., T. Ichiye, J. Masuzawa, A. Okubo and R. Marumo (1956): Report of the oceanographic observations in the sea east of Honshu in August–September, 1955 (in Japanese with English abstract). *CMO Kaiyo Hokoku*, **4**(4): 1–11.

Fukuoka, J. (1958a): On the Kuroshio near the Izu Islands. *J. Oceanogr. Soc. Japan*, **14**: 11–14.

——— (1958b): The variations of the Kuroshio Current in the sea south and east of Honshu (Japanese Main Island). *Oceanogr. Mag.*, **10**: 201–213.

Glavnoe Upravlenie Geodezii i Kartografii pri Sovete Ministrov, USSR (1967): *Atlas Sahalinskoi Oblasti*, Moscow, viii+135pp.

Gould, W. J. (1971): Observations of an event in some current measurements in the Bay of Biscay. *Deep-Sea Res.*, **18**: 35–49.

Hansen, D. V. (1970): Gulf Stream meanders between Cape Hatteras and the Grand Banks. *Deep-Sea Res.*, **17**: 495–511.

Hata, K. (1969a): Some problems relating to fluctuation of hydrographic conditions in the sea northeast of Japan (Part I). Relation between the patterns of the Kuroshio and the Oyashio. *J. Oceanogr. Soc. Japan*, **25**: 25–35.

——— (1969b): Some problems relating to fluctuation of hydrographic conditions in the sea northeast of Japan (Part 2). Fluctuation of warm eddy cut off northward from the Kuroshio. *Oceanogr. Mag.*, **21**: 13–29.

Hata, K., S. Hosoda and K. Yamamoto (1964): Report of the detailed oceanographic observations in the Tsugaru Straits from August to September, 1962 (in Japanese with English abstract). In: *Report of a comprehensive survey of the Tsugaru Straits. Bull. Hakodate Mar. Obs.*, Spec. No.: 1–30.

Hatanaka, M. (1952): Studies on the fluctuation of the hydrographic conditions and its effect on the pelagic fisheries resources (in Japanese with English abstract). *Bull. Tohoku Reg. Fish. Res. Lab.*, **1**: 88–119.

Haurwitz, Bernhard and H. A. Panofsky (1950): Stability and meandering of the Gulf Stream. *TAGU*, **31**: 723–731.

Hidaka, K. (1955): *Kairyu* [Ocean Current] (in Japanese). Iwanami Shoten, Tokyo, 291pp., 2 charts.

Hirano, T. (1957): On durability of the surface temperature at hydrographic stations of the Pacific coast II (in Japanese with English abstract). *Bull. Tokai Reg. Fish. Res. Lab.*, **17**: 65–72.

Iselin, C. O'D. (1936): A study of the circulation of the western North Atlantic. *Pap. Phys. Oceanogr. Meteorol.*, **4**(4): 101pp.

Iselin, C. O'D. and F. C. Fuglister (1948): Some recent developments in the study of the Gulf Stream. *J. Mar. Res.*, **7**: 317–329.

Japan Imperial Hydrographic Office (1918): *Nippon Suiroshi* [Japan Sailing Directory] (in Japanese). Vol. 1, Tokyo, 441pp.

Kajiyama, E. (1920): *Nippon Kaiyogaku* [Japanese Oceanography] (in Japanese). Shokabo, Tokyo, iii+286+ii pp.

Kawai, H. (1955a): On the polar frontal zone and its fluctuation in the waters to the northeast of Japan (I) (in Japanese with English abstract). *Bull. Tohoku Reg. Fish. Res. Lab.*, **4**: 1–46.

Kawai, H. (1955b): On the polar frontal zone and its fluctuation in the waters to the northeast of Japan (II) (in Japanese with English abstract). *Bull. Tohoku Reg. Fish. Res. Lab.*, **5**: 1–42.

——— (1956): Notes on the inversion of the upper water temperature in the waters to the northeast of Japan in summer. *Bull. Tohoku Reg. Fish. Res. Lab.*, **6**: 71–80.

——— (1957): On the natural coordinate system and its applications to the Kuroshio System. *Bull. Tohoku Reg. Fish. Res. Lab.*, **10**: 141–171.

——— (1958a): [Methods of data analysis and of preparation of schema of water-mass distribution] (in Japanese). In: *Tohoku Reg. Fish. Res. Lab., Annual report on the fish resources in 1950, Sec. 1. Oceanographic investigation*, 3–4.

——— (1958b): A note on drawing the isotherms at the sea surface (in Japanese with English abstract). *Bull. Tohoku Reg. Fish. Res. Lab.*, **12**: 106–120.

——— (1959): On the polar frontal zone and its fluctuation in the waters to the northeast of Japan (III) (in Japanese with English abstract). *Bull. Tohoku Reg. Fish. Res. Lab.*, **13**: 13–59.

——— (1965): Physiography on the east coast of the mainland of Japan (in Japanese with English abstract). *Bull. Tohoku Reg. Fish. Res. Lab.*, **25**: 105–130.

——— (1969): Statistical estimation of isotherms indicative of the Kuroshio axis. *Deep-Sea Res.*, Suppl. to **16**: 109–115.

——— (1970): Bias, owing to asymmetric distribution, in estimates of values of isotherms indicative of the stream axis. Read before the annual meeting of the Oceanogr. Soc. Japan.

Kawai, H. and H. Sakamoto (1970): A study on convergence and divergence in surface layer of the Kuroshio—II. Direct measurement of convergence and divergence at the top and bottom of surface mixed layer (in Japanese with English abstract). *Bull. Nansei Reg. Fish. Res. Lab.*, **2**: 19–38.

Kawai, H. and M. Sasaki (1961): An example of the short-period fluctuation of the oceanographic condition in the vicinity of the Kuroshio Front. *Bull. Tohoku Reg. Fish. Res. Lab.*, **19**: 119–134.

——— (1962): On the hydrographic conditions accelerating the skipjack's northward movement across the Kuroshio Front (in Japanese with English abstract). *Bull. Tohoku Reg. Fish. Res. Lab.*, **20**: 1–27.

Kawai, H., H. Sakamoto and M. Momota (1969): A study on convergence and divergence in surface layer of the Kuroshio—I. Direct measurement and interpretation of convergence and divergence at the surface (in Japanese with English abstract). *Bull. Nansei Reg. Fish. Res. Lab.*, **1**: 1–41.

Kawasaki, T. (1957): Relation between the live-bait fishery of albacore and the oceanographical conditions in waters adjacent to Japan (I). The fishing grounds south of the Kuroshio Front (in Japanese with English abstract). *Bull. Tohoku Reg. Fish. Res. Lab.*, **9**: 69–109, 2 tables.

Kimura, K. (1949): *Katsuo Gyojo Zushu* [Atlas of the skipjack fishing grounds] (in Japanese). Kuroshio Shobo, Tokyo, 45pp.

Kishindo, S. (1940): Stratification of sea water and the deep layer current of the Pacific (in Japanese). *JHO Hydrographic Bull.*, **19**(11): 351–362.

Koizumi, M. (1955): Researches on the variations of oceanographic conditions in the region of the ocean weather station "Extra" in the North Pacific Ocean (I). Normal values and annual variations of oceanographic elements. *Pap. Meteorol. Geophys.*, **6**: 185–201.

——— (1956): Researches on the variations of oceanographic conditions in the region of the ocean weather station "Extra" in the North Pacific Ocean (III). The variation of hydrographic conditions discussed from the heat balance point of view and the heat exchange between sea and atmosphere. *Pap. Meteorol. Geophys.*, **6**: 273–284.

Konaga, S., K. Shuto, H. Kusano and K. Hori (1967): On the relation between Kuroshio strong currents and the water temperature of 200m depth (in Japanese with English abstract). *Umi to Sora*, **42**: (3/4): 93–97.

Kurashina, S., K. Nishida and S. Nakabayashi (1967): On the open water in the southeastern part of the frozen Okhotsk Sea and the currents through the Kurile Islands (in Japanese with English abstract). *J. Oceanogr. Soc. Japan*, **23**: 57–62.

Kuroda, R. (1959): Notes on the phenomena of "inversion of water temperature" off the Sanriku Coast of Japan (I) (in Japanese with English abstract). *Bull. Tohoku Reg. Fish. Res. Lab.*, **13**: 1–12.

——— (1960): Notes on the phenomena of "inversion of water temperature" off the Sanriku Coast of Japan (II). Results from repeated observations in the fixed point (in Japanese with English abstract). *Bull. Tohoku Reg. Fish. Res. Lab.*, **16**: 65–86.

LaFond, E. C. (1968): Detailed temperature and current data sections in and near the Kuroshio Current. *An Oceanogr. Data Report for CSK*, Marine Environment Division, Naval Undersea Center, San Diego, i+22pp.

Lyman, John (1969): Naviface, oxyty, and epichthon: words versus terms. *J. Mar. Res.*, **27**: 367–368.

Mann, C. R. (1967): The termination of the Gulf Stream and the beginning of the North Atlantic Current. *Deep-Sea Res.*, **14**: 337–359.

Marukawa, H. (1932): *Kaiyogaku* [Oceanography] (in Japanese). Suisangaku Zenshu [Fishery Science Series], 7. Koseikaku, Tokyo, ii+iv+368pp.

Masuzawa, J. (1951): On the intermediate water in the northeastern sea of Japan (First Paper) (in Japanese with English abstract). *CMO Kaiyo Hokoku*, **2**: 5–13.

——— (1952): On the intermediate water in the northeastern sea of Japan (Second Paper) (in Japanese with English abstract). *CMO Kaiyo Hokoku*, **2**: 23–27.

Masuzawa, J. (1954): On the seasonal variation of the Kuroshio east off Cape Kinkazan of Japan proper (Currents and water masses of the Kuroshio System (in Japanese II) with English abstract). *CMO Kaiyo Hokoku*, **3**: 251–255.

—— (1955a): Preliminary report on the Kuroshio in the eastern sea of Japan (Currents and water masses of the Kuroshio System III). *Rec. Oceanogr. Wrs. Japan*, NS, **2**: 132–140.

—— (1955b): An outline of the Kuroshio in the eastern sea of Japan (Currents and water masses of the Kuroshio System IV). *Oceanogr. Mag.*, **7**: 29–48.

—— (1956a): A note on the Kuroshio farther to the east of Japan (Currents and water masses of the Kuroshio System VI). *Oceanogr. Mag.*, **7**: 97–104.

—— (1956b): On the cold belt along the northern edge of the Kuroshio (Currents and water masses of the Kuroshio System VIII). *Oceanogr. Mag.*, **8**: 151–156.

—— (1957a): A contribution to the knowledge on the Kuroshio east of Japan (Currents and water masses of the Kuroshio System IX). *Oceanogr. Mag.*, **9**: 21–34.

—— (1957b): On the fluctuation of the Kuroshio east of Honshu and its forecast (in Japanese). *Proc. of the symposium on the oceanographic forecast near Japan*, Sendai, March, 1957, 25–38, Mimeogr.

—— (1958): A short-period fluctuation of the Kuroshio east of Cape Kinkazan (Currents and water masses of the Kuroshio System X). *Oceanogr. Mag.*, **10**: 1–8.

—— (1960a): Polar front. In: *Kaiyo no Jiten* [Encyclopedia of Oceanography], K. Terada (editor) (in Japanese). Tokyodo, Tokyo, 206–207.

—— (1960b): Statistical characteristics of the Kuroshio Current (Currents and water masses of the Kuroshio System XI). *Oceanogr. Mag.*, **12**: 7–15.

—— (1964): A typical hydrographic section of the Kuroshio Extension (Currents and water masses of the Kuroshio System XII). *Oceanogr. Mag.*, **16**: 21–30.

—— (1965): A note on the seasonal variation of the Kuroshio velocity (in Japanese). *J. Oceanogr. Soc. Japan*, **21**: 117–118.

—— (1968): Cruise report on multi-ship study of short-term fluctuations of the Kuroshio in October to November 1967. *Oceanogr. Mag.*, **20**: 91–96.

—— (1969a): Subtropical Mode Water. *Deep-Sea Res.*, **16**: 463–472.

—— (1969b): A note on the characteristic of the high-velocity belt of the Kuroshio (in Japanese). *J. Oceanogr. Soc. Japan*, **25**: 259–260.

Masuzawa, J. and T. Nakai (1955): Notes on the cross-current structure of the Kuroshio (Currents and water masses of the Kuroshio System V). *Rec. Oceanogr. Wks. Japan*, NS, **2**: 96–101.

Montgomery, R. B. (1938): Fluctuations in monthly sea level on eastern U.S. coast as related to dynamics of western North Atlantic Ocean. *J. Mar. Res.*, **1**: 165–185.

—— (1969): The words naviface and oxyty. *J. Mar. Res.*, **27**: 161–162.

Montgomery, R. B. and E. D. Stroup (1962): Equatorial waters and currents at 150°W in July—August 1952. *Johns Hopkins Oceanographic Studies*, **1**: 68pp.

Morgan, G. W. (1956): On the wind-driven ocean circulation. *Tellus*, **8**: 301–320.

Moriyasu, S. (1961): An example of the conditions at the occurrence of the cold water region. *Oceanogr. Mag.*, **12**: 67–76.

Munk, W. H. (1950): On the wind-driven ocean circulation. *J. Met.*, **7**: 79–93.

Muromtsev, A. M. (1958): *The principal hydrological features of the Pacific Ocean* (Translated from the Russian by A. Birron and Z. S. Cole in 1963). Israel Program for Scientific Translations Ltd., Jerusalem, 417pp.

Nagata, Y. (1967a): Shallow temperature inversions at Ocean Station V. *J. Oceanogr. Soc. Japan*, **23**: 194–200.

—— (1967b): On the structure of shallow temperature inversions. *J. Oceanogr. Soc. Japan*, **23**: 221–230.

—— (1968): Shallow temperature inversions in the sea to the east of Honshu, Japan. *J. Oceanogr. Soc. Japan*, **24**: 103–114.

—— (1970): Detailed temperature cross section of the cold-water belt along the northern edge of the Kuroshio. *J. Mar. Res.*, **28**: 1–14.

Nomitsu, T. (1931): *Kaiyogaku* [Oceanography] (in Japanese). Shizenkagaku Sosho [Natural Science Series], 6. Nippon Hyoronsha, Tokyo, ii + vii + 300pp.

Norpac Committee (1960): *Oceanic observations of the Pacific, 1955: the Norpac Atlas*. Univ. of California Press and Univ. of Tokyo Press, Berkeley and Tokyo, xii + 123 pls.

Nowlin, W. D., Jr., J. M. Hubertz and R. O. Reid (1968): A detached eddy in the Gulf of Mexico. *J. Mar. Res.*, **26**: 185–186.

Oceanographical Section, JMA (1958): Monthly means of sea-surface water temperature in the sea east of Honshu (in Japanese with English abstract). *JMA Kaiyo Hokoku*, **7**(1): 59–65.

Pickard, G. L. (1963): *Descriptive Physical Oceanography*. Pergamon Press, Macmillan Co., New York, 199pp.

Reid, J. L., Jr. (1962): Distribution of dissolved oxygen in the summer thermocline. *J. Mar. Res.*, **20**: 138–148.

—— (1965): Intermediate waters of the Pacific Ocean. *Johns Hopkins Oceanographic Studies*, **2**: 85pp.

—— (1966): Zetes Expedition. *TAGU*, **47**: 555–771.

Rex, D. F. (1950a): Blocking action in the middle troposphere and its effect upon regional climate. I. An aerological study of blocking action. *Tellus*, **2**: 196–211.

—— (1950b): Blocking action in the middle troposphere and its effect upon regional climate. II. The climatology of blocking action. *Tellus*, **2**: 275–301.

Schrenck, L. V. (1873): Strömungsverhältnisse im Ochotskischen und Japanischen Meere und in den zunächst angrenzenden Gewässern. *Mémoires de l'Acad. Imp. d. Sc. Petersbourg.* **7** (according to bibliog. of Wüst, 1936).

Scripps Institution of Oceanography, University of California (1966): *Data report, Boreas Expedition.* SIO Reference, 66–24, xv+ix+164pp.

Seitz, R. C. (1967): Thermostad, the antonym of thermocline. *J. Mar. Res.*, **25**: 203.

Shigematsu, R. (1932): Some oceanographical investigation of the results of oceanic survey carried out by H.I.J.M.S. *Manshu* from April 1925 to March 1928. *Rec. Oceanogr. Wks. Japan*, **4**: 151–170.

Shoji, D. (1965): Description of the Kuroshio (physical aspect). *Proc. of symp. on the Kuroshio*, Tokyo, Oct. 1963, Oceanogr. Soc. of Japan and Unesco, 1–11.

Soule, F. M., P. A. Morrill and A. P. Franceschetti (1961): Physical oceanography of the Grand Banks region and the Labrador Sea in 1960. *U.S. Coast Guard Bull.*, **46**: 31–114.

Stommel, Henry (1963a): Varieties of oceanographic experience. *Science*, **139**(3555): 572–576.

——— (1963b): Some thoughts about planning the Kuroshio survey. *Proc. of symp. on the Kuroshio*, Tokyo, Oct. 1963, Oceanogr. Soc. of Japan and Unesco, 22–23.

——— (1965): *The Gulf Stream—A Physical and Dynamical Description.* Second Edition. University of Cal. Press, Berkeley and Los Angeles, Cambridge Univ. Press, London, xiii+248pp.

Stommel, Henry, W. S. von Arx, D. Parson and W. S. Richardson (1953): Rapid aerial survey of Gulf Stream with camera and radiation thermometer. *Science*, **117** (3049): 639–640.

Suda, K. (1933): *Kaiyo Kagaku* [Ocean Science] (in Japanese). Kokin Shoin, Tokyo, ii+iv+ix+726pp., 1 pl.

——— (1936): On the dissipation of energy in the density currents (2nd paper). *Geophys. Mag.*, **10**: 131–243.

Sugiura, J. (1958): On the Tsugaru Warm Current. *Geophys. Mag.*, **28**, *Takematsu Okada's anniversary vol.*, part 1: 399–409.

Sverdrup, H. U., M. W. Johnson and R. H. Fleming (1942): *The Oceans—Their Physics, Chemistry, and General Biology.* Prentice-Hall, New York, x+1087pp. 7 charts.

Tajima, S. (1951): On the meridional oscillation of anticyclone of horse latitudes in Japan and its effect on the weather. *Geophys. Mag.*, **22**: 109–130.

Takenouchi, Y. (1958): Measurements of subsurface current in the cold-belt along the northern boundary of Kuroshio. *Oceanogr. Mag.*, **10**: 13–17.

Takenouchi, Y., J. Masuzawa and M. Yasui (1957): Report of multiple ship survey in the polar front region from June to July, 1957 (in Japanese with English abstract). *JMA Kaiyo Hokoku*, **6**(4): 181–196.

Takenouchi, Y., T. Nan'niti and M. Yasui (1962): The deep-current in the sea east of Japan. *Oceanogr. Mag.*, **13**: 89–101.

Uda, M. (1929): On the stratification of sea water in the Kuroshio region (in Japanese). *Umi to Sora*, **9**: 175–182.

——— (1931): On the monthly oceanographical charts of the adjacent seas of Japan based on the averages for the thirteen years from 1918 to 1930, with a discussion of the current-system inferred from these charts (Part II: From January to June) (in Japanese with English abstract). *J. Imp. Fish. Exp. Sta.*, **2**: 59–81, 12 pls.

——— (1933): Hydrographical researches on the normal monthly conditions of Oyasiwo and Kurosiwo-area (in Japanese with English abstract). *J. Imp. Fish. Exp. Sta.*, **3**: 79–136.

——— (1934): The results of simultaneous oceanographical investigations in the Japan Sea and its adjacent waters in May and June, 1932 (in Japanese with English abstract). *J. Imp. Fish. Exp. Sta.*, **5**: 57–190.

——— (1935a): The results of simultaneous oceanographical investigations in the North Pacific Ocean adjacent to Japan made in August, 1933 (in Japanese with English abstract). *J. Imp. Fish. Exp. Sta.*, **6**: 1–130.

——— (1935b): On the distribution, formation and movement of the dicho-thermal water in the Tohoku Area (in Japanese). *Umi to Sora*, **15**: 445–452.

——— (1938a): Researches on "Siome" or current rip in the seas and oceans. *Geophys. Mag.*, **11**: 307–372.

——— (1938b): Hydrographical fluctuation in the north-eastern sea-region adjacent to Japan of North Pacific Ocean (A result of the simultaneous oceanographical investigations in 1934–1937) (in Japanese with English abstract). *J. Imp. Fish. Exp. Sta.*, **9**: 1–66.

——— (1940): On the recent anomalous hydrographical conditions of the Kuroshio in the south waters off Japan proper in relation to the fisheries (in Japanese with English abstract). *J. Imp. Fish. Exp. Sta.*, **10**: 231–278.

——— (1941): *Umi no Tankyushi* [History of Investigation of the Sea] (in Japanese). Kagaku Shinsho [New Science Series], 8, Kawade Shobo, Tokyo, v+iv+213pp.

——— (1943): On the structure of the "shiozakai" (oceanic front) (in Japanese). *J. Oceanogr. Soc. Japan*, **2**(4): 9–16.

——— (1949): On the correlated fluctuation of the Kuroshio Current and the cold water mass. *Oceanogr. Mag.*, **1**: 1–12.

——— (1955): *Sekai Kaiyo Tankenshi* [World History of Exploration of the Sea] (in Japanese). Sekai Tanken Kiko Zenshu [World Exploration and Travel Series], Suppl. Kawade Shobo, Tokyo, 436 pp.

Uda, M. (1964): On the nature of the Kuroshio, its origin and meanders. In: *Studies on Oceanography—a Collection of Papers Dedicated to Koji Hidaka,* K. Yoshida (editor). Univ. Tokyo Press, Tokyo, 89–107.

Uda, M. and G. Okamoto (1930): On the monthly oceanographical charts of the adjacent seas of Japan based on the averages for the eleven years from 1918 to 1929, with a discussion of the current-system inferred from these charts (Part I: from July to December) (in Japanese with English abstract). *J. Imp. Fish. Exp. Sta.,* **1**: 39–55, 12 pls.

Uehara, S. (1962): Fishery oceanography around Enshu Nada off the central Pacific coast of Honshu—I. Oceanographic condition for skipjack and shirasu fisheries (in Japanese with English abstract). *Bull. Tokai Reg. Fish. Res. Lab.,* **34**: 55–66.

von Arx, W. S. (1952): Notes on the surface velocity profile and horizontal shear across the width of the Gulf Stream. *Tellus,* **4**: 211–214.

——— (1962): *An Introduction to Physical Oceanography.* Addison-Wesley Pub. Co., Reading, Mass., London, x+422pp.

von Arx, W. S., D. F. Bumpus and W. S. Richardson (1955): On the fine-structure of the Gulf Stream front. *Deep-Sea Res.,* **3**: 46–65.

Warren, B. A. (1963): Topographic influences on the path of the Gulf Stream. *Tellus,* **15**: 167–183.

Worthington, L. V. (1954): Three detailed cross-sections of the Gulf Stream. *Tellus,* **6**: 116–123.

——— (1959): The 18° water in the Sargasso Sea. *Deep-Sea Res.,* **5**: 297–305.

——— (1962): Evidence for a two gyre circulation system in the North Atlantic. *Deep-Sea Res.,* **9**: 51–67.

Worthington, L. V. and W. G. Metcalf (1961): The relationship between potential temperature and salinity in deep Atlantic water. *Rapp. et Proc.-Verb., Cons. int. Explor. Mer,* vol. **149**: 122–128.

Wüst, Georg (1929): Schichtung und Tiefenzirkulation des Pazifischen Ozeans auf Grund zweier Längsschnitte. *Veröff. Institut f. Meereskunde, Berlin Univ.,* N. F., A. Geogr.-naturwiss. Reihe, Heft **20**, 63pp.

——— (1930): Meridionale Schichtung und Tiefenzirkulation in den Westhälften der drei Ozeane. *J. d. Cons.,* **5**: 7–21.

——— (1936): Kuroshio und Golfstrom. *Veröff. Institut f. Meereskunde, Berlin Univ.,* N. F., A. Geogr.-naturwiss. Reihe, Heft **29**, 69pp.

Wyrtki, Klaus (1965): The annual and semiannual variation of sea surface temperature in the North Pacific Ocean. *Limnol. Oceanogr.,* **10**: 307–313.

Yosida, S. (1961): On the variation of Kuroshio and cold water mass off Enshu Nada (part 1) (in Japanese with English abstract). *JHO Hydrographic Bull.,* **67**: 54–57.

Chapter 9

THE TSUSHIMA CURRENT

SHIGEO MORIYASU

Meteorological Research Institute, Tokyo, Japan.

1 INTRODUCTION

The Tsushima Current is defined as follows by Sverdrup et al. (1942): "The warm current that branches off on the left-hand side of the Kuroshio and enters the Japan Sea following the western coast of Japan to the north." This definition is very rough, but expressive of the essential features. According to them, "the Tsushima Current flows into the Japan Sea and carries water of high temperature and high salinity toward the north. Branches of the current flow out through the straits between the northern Japanese islands and part of the water continues along the west side of Sakhalin Island, turns around, and flows south after having been cooled and diluted." Hence, the Tsushima Current is a major feature of the hydrography in the Japan Sea. This is the reason why the Tsushima Current has been studied energetically; especially the cooperative observations carried out by some agencies. A cooperative study was made for five years, from 1953 to 1957, and a comprehensive report was published by the Fisheries Agency in 1958. In the report, the hydrography in the Japan Sea is summarized by Kajiura, Tsuchiya and Hidaka, by Uda and by Miyata.

Kajiura et al. (1958) made a historical review of the past study. Hardly any oceanographic information about the Japan Sea existed before the observations were initiated by the Imperial Marine Observatory (present Kobe Marine Observatory) in the summers from 1928 to 1931. In the summer of 1932 and the autumn of 1933, the Fisheries Experimental Stations carried out cooperative observations covering the whole Japan Sea, and Uda (1934) summarized the results and presented the hydrographical pattern of the Japan Sea for the first time. Thereafter, the Fisheries Experimental Stations (including stations in Korea) around the Japan Sea repeated observations on fixed lines and collected useful data. In addition, cooperative observations by the Fisheries Experimental Stations in May and June of 1941 and observations by the Navy in the summer of 1943 covered the whole area. After World War II, observations were restricted to an area not far from the Japanese Islands and the fluctuation of the Tsushima Current was a main subject of study. This situation was brought about by the restrictions in the area of observation, but it seems to be related to the situation in that the gross hydrographic features in the Japan Sea had been grasped. Recently, the area of observation was extended to the area near Primorskaya and the frequency of observations increased. Thus, the extent of information on the Japan Sea is increasing yearly and more detailed analysis is possible.

After 1958, much effort was directed to studies on volume transport and on the cold and the warm water regions. In addition, meandering was regarded as one of the most important characters of the Tsushima Current and was discussed in relation to the bottom topography. On the other hand, statistical studies were made on temperature, salinity, oxygen content and so on. In this chapter, these results are summarized and analyses are attempted on some problems.

2 TEMPERATURE, SALINITY AND OXYGEN CONTENT IN THE UPPER LAYER

There are some charts showing the normal surface temperature for each month (for example, Tsukuda, 1937; Nakayama, 1951) and for each season (Yasui et al., 1967). They are constructed on the basis of mean value in one degree quadrangles. Nakayama's charts are reproduced here (Fig. 1). The period of the statistics is 30 years (from 1911 to 1940). Though the values are not reliable enough to discuss in detail, the general features are indicated by his charts. The characteristic features are:

(1) The highest and the lowest temperatures are in the Tsushima Strait and near Primorskaya, respectively, for every month.

(2) The isotherms run WSW-ENE for every month.

(3) The gradient running S-N is larger in winter than in summer, and the difference between the highest and the lowest temperatures is from about 6°C for August and September to about 15°C for December.

(4) The annual range is from 15°C in the Tsushima Strait to about 21°C near Primorskaya.

(5) The highest temperature occurs in August and the lowest temperature in February.

The fourth feature is ascribed to the excessive cooling in the northern part in winter and the rising of surface temperatures associated with the development of thermocline in summer (Kajiura et al., 1958). The interpretation is also valid for the third feature. The warm water flowing in through the Tsushima Strait is cooled by cold waters in the Japan Sea during the northward travel. Therefore, the change of surface temperatures is mainly dependent on advection and mixing with the cold water in the northern part.

There is no chart showing normal temperature below the surface for each month. But Yasui et al.

(1967) prepared charts showing normal values of temperature and chlorinity at several depths (including the surface) for each season. Though there is no essential difference in the distribution of surface temperature between Yasui's and Nakayama's charts, marked meandering is found in Yasui's charts. Isotherms at the 100 m depth run in a SW-NE direction, and the location of strong current is inferred from the distribution. In winter, there is little difference in temperature between the surface and the 100 m level because of the development of convection. Temperatures at the 100 m depth are lower in the northern part all through the year, while they are subject to seasonal variation with a range of about 4°C near the Japan coast. Hence, the gradient in a NW-SE direction is larger in summer. This is in marked contrast to conditions at the surface. Temperatures at the surface and the 100 m depth decrease northeastward along the Japan coast and this is an indication that the Tsushima Current carries warm water and that it is cooled by mixing with the cold water in the Japan Sea during the travel. In the distribution of temperature and salinity at the 100 m depth, there is a front at 39–40°N except for coastal regions. The area south (north) of the front is named the warm- (cold-) current region. At the 400 m level, temperatures are nearly homogeneous and no definite

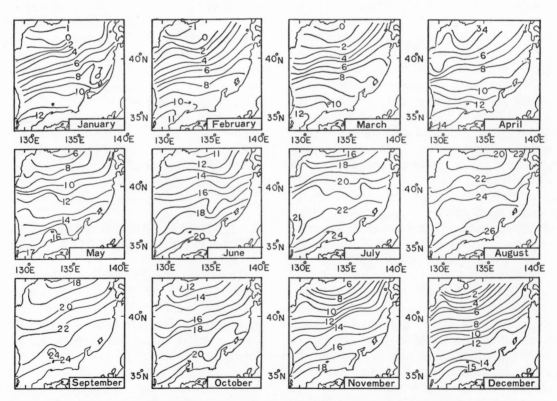

Fig. 1   Mean sea-surface temperatures(°C), according to Nakayama (1951).

characteristic feature is found in the distribution.

The Tsushima Current carries saline surface water to the Japan Sea except in summer, and salinity decreases downstream owing to mixing with less saline water in the Japan Sea. In summer, on the other hand, salinity increase downstream west of the Noto Peninsula, because less saline water in the East China Sea is carried through the Tsushima Strait. Therefore, the distribution of salinity in the upper layer is a good measure of the effect of advection and mixing. The seasonal variation at the surface is discussed by Yamanaka (1951), Kajiura et al. (1958) and Tanioka (1962a). Tanioka's pictures for some areas are reproduced in Fig. 2. West of the Noto Peninsula, an annual variation is predominant and variations of short period are not found, while east of the peninsula, variation is complicated. The variation in the Tsushima Strait is transmitted northeastward, for example, the minimum value of salinity occurs near the Tsugaru Strait two months later than in the Tsushima Strait, where the minimum value occurs in August. The effective current speed of 0.3–0.4 knot is obtained from the phase lag of the annual variation. The value is reasonable. The amplitude of the annual variation decreases downstream from about 1.0‰ to about 0.7‰. East of the Noto Peninsula, the variation is dependent on the cold water of the northern part and on runoff caused by melting snow. The former makes salinity higher in summer and the latter makes it lower in spring. The

influence of melting snow is seen east of Korea as well. West of Hokkaido, salinity is almost constant through the year except for low salinity in October. At the 100 m level, salinity is higher along the coast of Honshu, and the gradient in a NW-SE direction is large in the Tsushima Current. The distribution of salinity at this level signifies the feature of the Tsushima Current, while the distribution at the 400 m level does not (Yasui et al., 1967).

Charts of mean oxygen contents were prepared by Ohwada and Yamamoto (1966). Oxygen contents at the surface are higher in the northeastern part. At the 100 m level, the general features are the same and the gradient in the distribution is large between the cold-current region (6.5–7.2 ml/l) and the warm-current region (less than 6.0 ml/l) in summer and autumn. The seasonal variation occurs in the upper 300 m layer where seasonal variation of temperature and salinity is marked.

3   WATER MASS ANALYSIS

Water mass analysis has been tried by some authors (Akagawa, 1954; Kajiura et al., 1958; Miyata, 1958). Some differences are found in the definitions or the names of water masses, but they are not essential ones. It is not significant to define too many water masses, but four water masses defined by Kajiura et al. are thought to be appropriate in discussing the

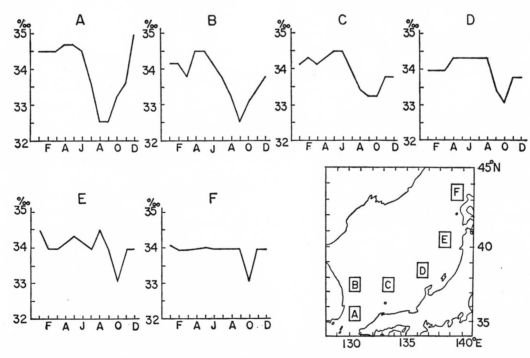

FIG. 2   Annual variations of salinity at the sea surface in selected quadrangles of 1-degree, according to Tanioka (1962a).

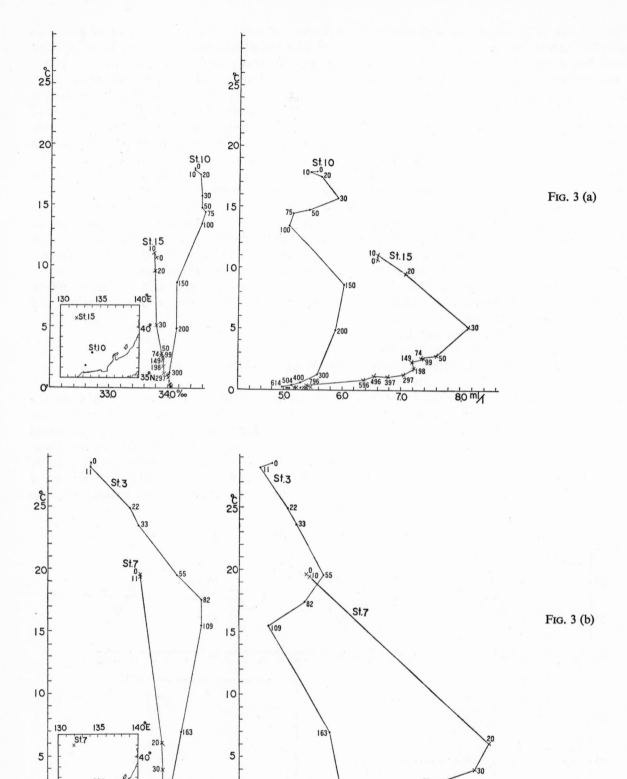

FIG. 3 (a)

FIG. 3 (b)

FIG. 3 (c)

FIG. 3 (d)

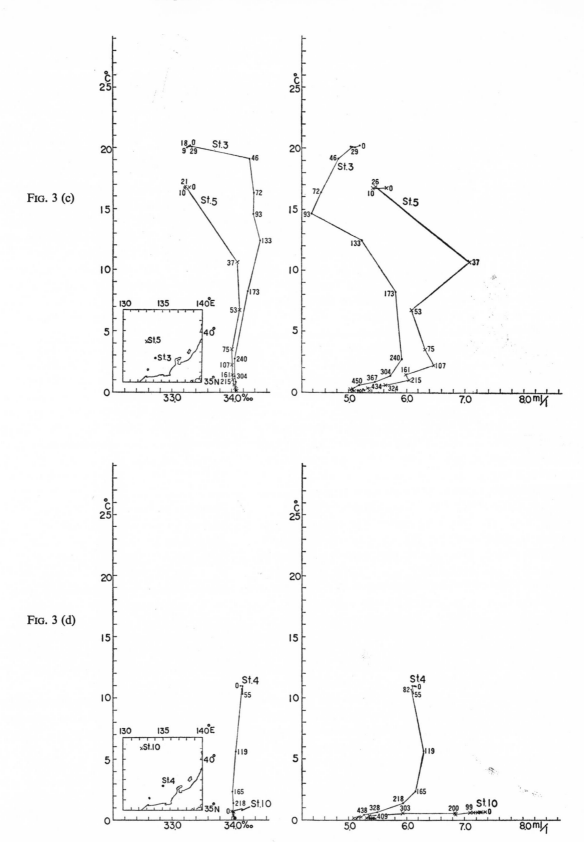

FIG. 3   T-S and T-O$_2$ diagrams based on data obtained during voyages of the *Seifu Maru*. (a) May, 1964; (b) August, 1964; (c) October, 1964; (d) February, 1965.

Numerals in the figures indicate the depth in meter at which the relationship is obtained.

hydrography in the Japan Sea.

Their classification differs from conventional classification and is based on the pattern of distribution and the origin.

(1) Surface water is near the sea surface and is under the marked influence of coastal waters and there is a remarkable thermocline beneath the water in summer. In the warm-current region, waters brought from the East China Sea and runoff from Honshu play an important role, while in the cold-current region, coastal waters off Korea and Primorskaya also play an important role.

(2) Intermediate water is below the surface water and is defined for the warm- and the cold-current regions. The intermediate water in the warm-current region has high temperature and high salinity, and is carried through the Tsushima Strait (see Section 8). The intermediate water in the cold-current region is characterized by high oxygen content.

(3) Proper water in the Japan Sea is nearly homogeneous, that is, temperature is 0.0–0.5°C, salinity 34.0–34.1‰ and oxygen content 5.0–5.5 ml/l. This water is below the intermediate water and is made up of deep water and bottom water. The former is situated above the 1000 m depth and the latter is below the depth. Temperatures increase adiabatically with depth below about 1000 m and oxygen contents are minimal at that depth. These are the reasons why the 1000 m depth is taken as the critical depth. The proper water is distinguished from the deep water in the North Pacific (see Chapter 15).

(4) Another water is situated between the intermediate water and the proper water in the warm-current region. This water is produced at the front by sinking of the low-salinity surface water in the cold-current region and has temperatures of 1–4°C and is characterized by maximum oxygen content.

The characters of the waters mentioned except for the proper water are subject to seasonal variation. Figure 3 shows typical T-S and T-O$_2$ diagrams for the warm- and the cold-current regions. T-S relationships in summer distinguish the intermediate water in the warm-current region, which is located at depths of about 100 m and has high salinity. The intermediate water in the warm-current region is poor in oxygen content, while the water in the cold-current region is rich in oxygen content. The bottom water has nearly constant temperatures, and oxygen content increases with depth. The fourth water in the warm-current region is characterized by low salinity and high oxygen content at a depth of about 200 m. Thus, both T-S and T-O$_2$ relationships are necessary to define waters in the Japan Sea. These T-S and T-O$_2$ relationships are not uniform for other waters than the proper water, because they are under the influence of mixing between waters in the warm- and the cold-current regions.

The surface and the intermediate waters in the warm-current regions are carried northeastward and their relationships are modified by mixing and so on. Therefore, the relationships for those waters near the Tsushima Strait are different from those near the Tsugaru Strait. For spring and autumn, the analysis is possible in the way noted above. For winter, salinity is nearly constant vertically in both current regions and so are temperatures in the cold-current region. Hence, T-O$_2$ relationships are the only measures. But, it is difficult to distinguish individual water on T-O$_2$ diagrams.

The bivariate distributions of volume among potential temperature and chlorinity are discussed for four seasons and throughout the whole year by using Cochrane's method (Yasui et al., 1967). The seasonal variation is prominent in the upper 200 m layer. This is inferred from T-S and T-O$_2$ diagrams. Three modes are found in the bivariate distribution for all throughout the year. The first mode is around 0.5°C and 18.85‰ Cl (34.05‰ S), and corresponding to the proper water occupies about 90% of the entire water of the Japan Sea. The second mode, corresponding to the intermediate water in the warm-current region, is between 8°C and 16°C, and 18.9‰ Cl (34.14‰ S) and 19.2‰ Cl (34.68‰ S). The third mode, corresponding to the surface water in summer, is between 17°C and 20°C, and between 18.4‰ Cl (33.24‰ S) and 18.8‰ Cl (33.96‰ S). According to the analysis of the bivariate distribution, the average T-Cl relationships are represented by a line connecting the third mode with the first mode through the second. The fourth water defined by Kajiura et al. is not found as a prominent mode in the bivariate distribution. The water in the cold-current region does not form a prominent mode, that is, the upper water in the cold-current region hardly contributes to the oceanographic structure of the Japan Sea.

The surface and intermediate waters in the warm-current region are brought through the Tsushima Strait from the East China Sea. The greatest depth in the Tsushima Strait is about 150 m and the water below this depth in the Japan Sea is little influenced by the water in the East China Sea. The proper water is not brought from the East China Sea, but is produced in the Japan Sea; that is, in winter, the water in the northern part is cooled remarkably and increases in density and then sinks to the deep layer. This water spreads horizontally and is found in the deep layer even in the warm-current region. That is the reason why the proper water has high oxygen content. The water is characterized by an approximate horizontal homogeneity. The process was proposed by Suda (1932), Miyazaki (1952, 1953), Ichiye (1954) and Fukuoka (1962 a) and is considered to be reasonable. If the sinking occurs, ascending motion is required somewhere to satisfy the water budget of the Japan Sea. The inflow and the outflow through the straits are restricted in the subsurface layer and are not related to the process. Fukuoka (1965) points out that water temperatures at the depths of 300 m, 400 m and 600 m in the central part are lower in May than in March and suggests the possibility of ascending motion in the central part. This is an approach to solving the mechanism of the production of the proper water.

In the study of water mass, definition and distribution are discussed, but mechanism of production and renewal have not been made clear. These are the problems to be solved in future.

4   THE SEIFU MARU SECTIONS FOR FOUR SEASONS

Vertical sections in a NW-SE direction show the oceanographic structure of warm- and cold-current regions. Here, sections of temperature, salinity, oxygen content and thermosteric anomaly are illustrated (Fig. 4). They are constructed on the basis of data obtained during voyages aboard the *Seifu Maru* of the Maizuru Marine Observatory. The location of the section for February, 1965 deviates a little northeastward from others. The deviation is not important to see the seasonal variation in the section.

The common features of all the sections are that vertical and horizontal gradients are large above the 200–300 m layer, and that the water in the lower layer is approximately homogeneous and that its properties are not subject to seasonal variation. The waters defined in the previous section are seen in the sections except in winter. The intermediate water in the warm-current region is indicated by high salinity and low oxygen content. The depth of its core is 75–100 m. Below this water, water of low salinity and high oxygen content is found in the warm-current region. In the cold-current region, there is the intermediate water characterized by high oxygen content. The characteristic salinity and oxygen content values for each water are variable and the depth of the core also fluctuates with season. In winter, convections develop and the vertical gradients almost vanish so that discrimination of the water is impossible, though the warm- and the cold-current regions can be distinguished from each other.

Near the front, the horizontal gradients are large and high current speeds are obtained according to the geostrophic relationships. The front is the boundary between the warm- and the cold-current regions and its position fluctuates with time. The isobaric surfaces of 200 db, 300 db and 400 db are approximately level when the 500 db surface is taken as the reference level.

5   THE FLOW PATTERN OF THE TSUSHIMA CURRENT

A flow pattern of the Tsushima Current has been accepted since it was proposed by Suda et al. (1932) and Uda (1934). The Tsushima Current is divided into three branches east of the Tsushima Strait; they flow together east of the Noto Peninsula and then the joined currents flow northward. The first branch (the near-shore branch) is the extension of the current which enters the Japan Sea through the eastern channel of the Tsushima Strait. The source of the second branch (the off-shore branch) and the third branch (the Eastern Korean Current) enters the Japan Sea through the western channel of the Tsushima Strait and the two branches are separated from each other near the strait.

Though the features noted above have been accepted for many years, some questions are raised. For example, it is stressed that the characters of the third branch are to be examined (Kajiura et al., 1958) and the marked meandering north of the Oki Islands is pointed out as one of the most characteristic features of the Tsushima Current (Fukuoka, 1957). Figure 5 shows the distribution of temperature at depths of 100 m and 200 m, and the geopotential topography at the surface referred to the 500 db surface in August and September, 1966. The general features of isotherms at the depth of 100 m are nearly the same as those of contours in the geopotential topography, and marked meandering is found. The wave length is about 300 km, which is the same as the value obtained by Fukuoka (1961). In the region of the meandering, the distance between elevations of the bottom is about 300 km and the depth to the bottom is not large. Hence, it is inferred that the meandering is related to the bottom topography and that the wave length is controlled by the topography (Fukuoka, 1957, 1961; Tanioka, 1962b). The amplitude is smaller east of Sado Island. Based on Fig. 5, it would be appropriate to take one meandering major current instead of two non-meandering branches in the model proposed by Suda and Uda. The meandering is slight east of Sado Island, where,

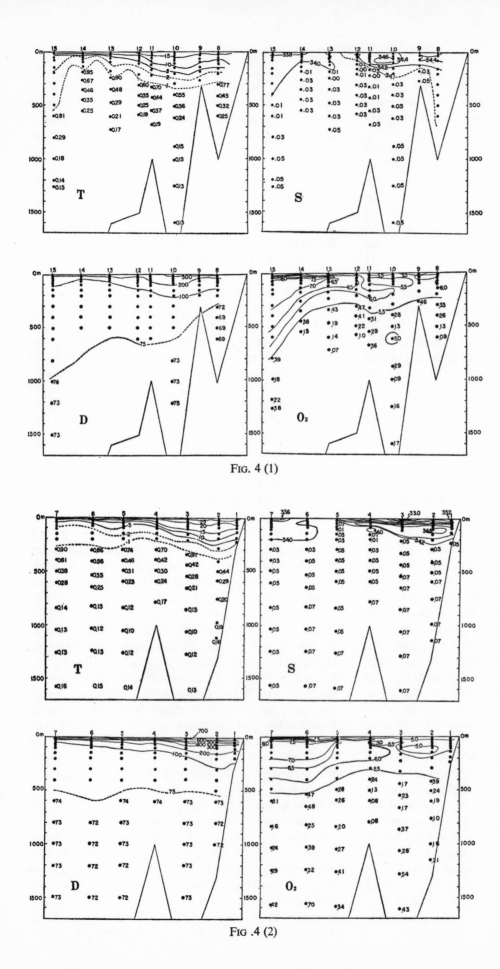

Fig. 4 (1)

Fig. 4 (2)

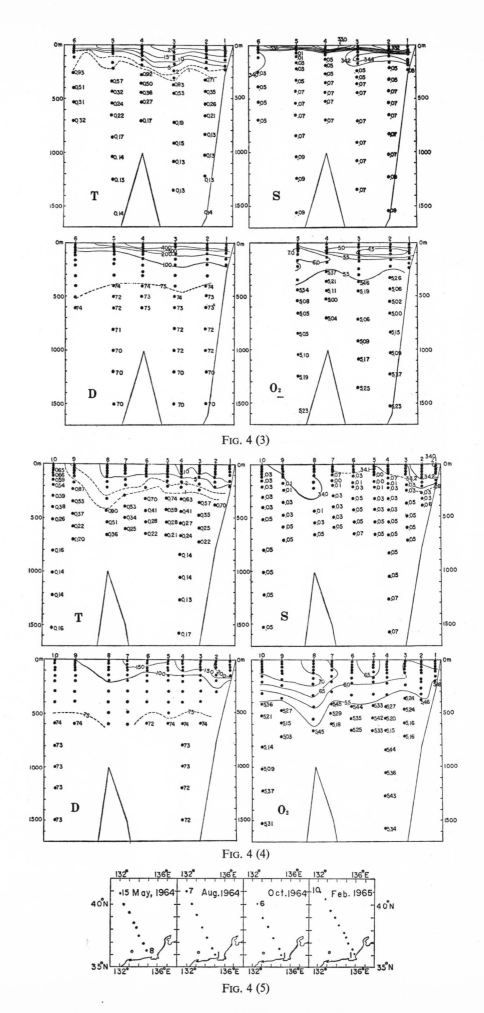

FIG. 4 (3)

FIG. 4 (4)

FIG. 4 *Seifu Maru* Sections.
(1) May, 1964; (2) August, 1964;
(3) October, 1964; (4) February,
1965; (5) Location of sections.
T=temperature in °C; S=salin-
ity in ‰; D=thermosteric
anomaly in cl/t; O₂=oxygen
content in ml/l.

FIG. 4 (5)

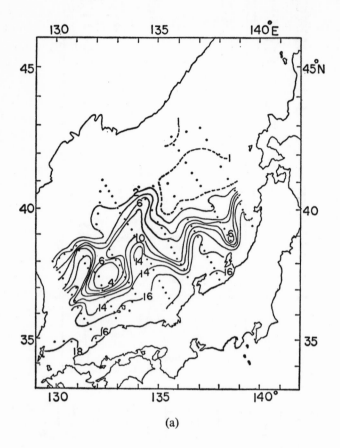

(a)

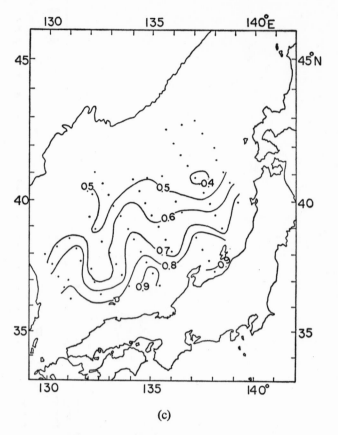

(c)

Fig. 5   Results of observations during a voyage of the *Seifu Maru*, August—September, 1966. (a) temperatures (°C) at a depth of 100 m, (b) temperatures (°C) at a depth of 200 m, (c) geopotential topography (dyn. m.) at the sea surface referred to the 500 db surface.

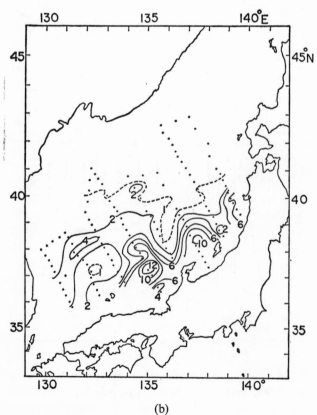

(b)

in their model, the two branches flow together and there is only one major current. It is reported that the Eastern Korean Current near the eastern coast of Korea flows back southward as a counter current just east of the northward current (Tanioka, 1968). This supports the existence of one meandering current.

The cold- and the warm-water regions in Section 7 are taken to be accompanied by the meandering. Since the gross features of the flow pattern do not change with time, the figure is considered to give a representative pattern, though the amplitude of meandering and so the scale of the cold or warm region fluctuate with time.

The first branch has been considered to flow through the channel between the Oki Islands and Honshu. According to Fukuoka's study (1957), the volume transport through the channel is 1/6–1/7 of the volume transport through the Tsushima Strait. Teramoto (1972) tried to obtain the volume transport through the channel from submarine-cable measurements. His measurements do not prove large volume

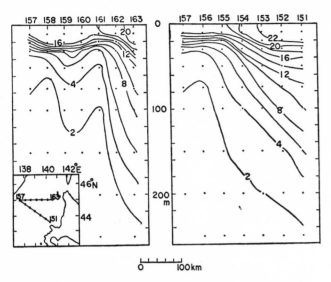

FIG. 6   Vertical sections of temperature (°C) in the sea west of Hokkaido based on data obtained during a voyage of the *Seifu Maru* in August, 1964.

transport, but show the effects caused by wind. Therefore, the current through the channels is not an important factor in studying the hydrography of the Japan Sea, even if the branch exists in the channel. Nevertheless, it is significant to examine the existence of the first branch by means of direct-current measurements.

The main part of the Tsushima Current flows out to the Pacific through the Tsugaru Strait and the rest flows northward in the sea west of Hokkaido. The vertical sections of temperature in Fig. 6 indicate the existence of the current and the current measurements with GEK also support this feature (Oceanographical Section, Maizuru Marine Observatory, 1967).

Though the Tsushima Current is small in scale and influenced by the bottom topography, the hydrographic conditions in the Tsushima Current region are similar to those in the sea east of Honshu where the Kuroshio and the Oyashio waters are in contact with each other; that is, the current is divided into a few branches and eddies are formed.

The outline of the flow pattern of the Tsushima Current is almost steady as noted above, but the details vary greatly. The variation in flow pattern has hardly been discussed. Fukuoka (1962b) suggests that periods of three to seven years are found in the variation though the periods are not obtained through strict examination.

6   VOLUME TRANSPORT OF THE TSUSHIMA CURRENT

The current velocity is determined according to the geostrophic relationships, and the surface-current velocity is measured with GEK. The magnitude of geostrophic-current velocity is dependent on the distance between the hydrographic stations and the depth of the reference level. The distance between the stations is from 20 to 60 nautical miles and the reference level is the 300–500 db surface in general. Though the velocity is computed under various conditions, the velocity at the surface is from 0.5 to 1.0 knot except for the area near the straits. The surface velocity measured with GEK is often higher than 1.0 knot and sometimes shows a distribution which contradicts the dynamic topography. The volume transport is evaluated as the integral of geostrophic-current velocity and may be subject to errors in procedure. Therefore, the magnitude of the volume transport depends on different procedures by different authors, but it probably has a magnitude of $10^6 m^3/$sec.

Since the volume transport through the Tsushima Strait means the amount of inflow into the Japan Sea, it is significant to estimate the volume transport through the strait. According to Yi's computation (1966) referred to the 125 db surface, the northward transport is $1.35 \times 10^6 m^3$/sec on the average. The volume transport at the latitudes near 40°N gives the total volume transport of the Tsushima Current, because three branches join together south of the latitude even if the Tsushima Current is assumed to consist of three branches. Hata's computation (1962) was made for the latitudes of 40–41°N. The northward transport is $2–7 \times 10^6 m^3$/sec and is much greater than that through the Tsushima Strait computed by Yi. Hata's computation is referred to the 400 db surface, and the volume transport through the Tsushima Strait may depend on non-geostrophic current. Nagahara (1965) makes the computation for a section at about 41°N latitude and obtains the value of $2–8 \times 10^6 m^3$/sec, which is compatible with Hata's value. The volume transport across the *Seifu Maru* section in Section 3 is computed referred to the 500 db surface. The value is on the order of $10^6 m^3$/sec. Summarizing these computations, it is inferred from the geostrophic relationships that the volume transport of the Tsushima Current is 10^6 m³/sec.

The water undoubtedly flows into the Japan Sea through the Tsushima Strait. Therefore, the same amount of water must be carried away through other straits to satisfy the continuity. The volume transport through the Tsugaru Strait has indirectly been evaluated by Yasui and Hata (1960) and Hata (1962). According to their studies, 50–100% of the water flowing northward at the latitude of about 41°N is carried away through the strait. The magnitude evaluated by them is $1-4 \times 10^6$ m³/sec which is much larger than the volume transport through the Tsushima Strait obtained by Yi. Miyazaki's value for the Tsugaru Strait is $0.1-1.3 \times 10^6$ m³/sec and is smaller than Yi's value for the Tsushima Strait. The exchange of the water of the Japan Sea is made through the Soya Strait and the Mamiya Strait in addition to the two straits. The Mamiya Strait is very shallow and narrow, and the volume transport through the strait is negligible, while there is evidence that the warm water of the Tsushima Current flows into the Okhotsk Sea through the Soya Strait (Iida, 1962; Maeda, 1968). Therefore, the volume transport through the Tsushima Strait is probably larger than that through the Tsugaru Strait. Here, the excess of precipitation and the run-off from land are not considered to be important. The northward volume transport across the section west of Hokkaido (Fig. 6) is 1.7×10^6 m³/sec, about 40% of that across the section at 41°N which amounts to 4.2×10^6 m³/sec. This evaluation is made on the basis of results obtained in September, 1964 and the reference level is the 500 db surface (Oceanographical Section, Maizuru Marine Obstervatory, 1967). Therefore, about 60% of the volume transport of the Tsushima Current at 41°N is discharged through the Tsugaru Strait. This estimate is made independently of Hata's computation and is in good agreement with his results. Hence, Hata's results are justified and it is possible to estimate the volume transport of the northward current in the sea west of Hokkaido. It is less than 50% of the Tsushima Current and part of the water flows into the Okhotsk Sea through the Soya Strait. Thus, the following model is obtained. The water carried through the Tsushima Strait is added to a cyclonic circulation in the Japan Sea and part of the water is discharged through the Tsugaru Strait and the Soya Strait. The amount of inflow through the Tsushima Strait is equal to the amount of outflow through the Tsugaru Strait and the Soya Strait.

The seasonal variation of volume transport of the Tsushima Current is discussed on the basis of these results. The transport is subject to marked seasonal variation with the maximum value in summer and the minimum value in winter. This feature is also found for the volume transport through the Tsushima Strait (Hidaka and Suzuki, 1950; Miyazaki, 1952; Yi, 1966). Hata suggests that the maximum transport occurs later on the lower reaches. Since the change is

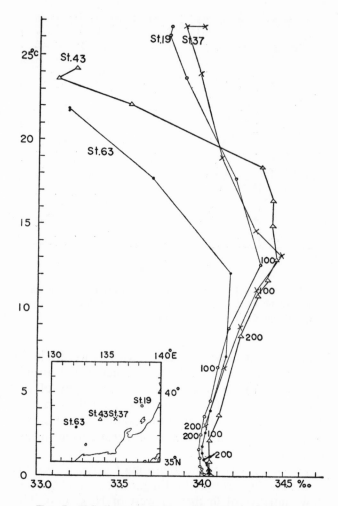

FIG. 7   T-S diagrams in the cold- and the warm-water regions based on data obtained during a voyage of the *Seifu Maru* in August, 1966. Numerals 100 and 200 indicate that the relationship is for the depth of 100 m and 200 m, respectively. Stations 19, 37 and 63 are located in the cold-water region and Station 43 in the warm-water region.

considered to be transferred northward from the Tsushima Strait, this suggestion is probable. The northward increase of the lag, however, is not definite from the data used for the discussion on the seasonal variation, because the sampling interval is too long.

7   COLD- AND WARM-WATER REGIONS

The existence of the cold- and the warm-water regions is one of the most interesting problems relating to the Tsushima Current. The cold- (warm-) water region is called the cold- (warm-) water mass or the low- (high-) temperature region. In the model of the flow pattern proposed in Section 5, the cold or the warm region is formed as accompanied by the meandering of the Tsushima Current. This is the region called the cold- or warm-water region. The cold-water region of this nature is situated in the Tsushima Current to the west of Sado Island. In addition, there is another kind of cold-water region. This is situated just north of the Tsushima Current and is usually formed east at about 138°E by jutting out of the cold water into the warm water (Kajiura et al., 1958) and causes the meandering of the Tsushima Current. This classification of the cold-water region is made on the basis of the distribution of water temperature at the 100 m level, while the warm-water region is distinguished by the distribution of water temperature at the 200 m level. These regions are distinctly shown in Fig. 5.

The waters of the cold- or the warm-water region in the Tsushima Current have common T-S relationships except for the water near the surface (Fig. 7). Stations 19, 37 and 63 are located in the cold-water region and Station 43 is in the warm-water region. Therefore, it is inferred from the water-mass analysis that the water is of the same origin and that the difference in the depth for a certain T-S relationship characterizes these regions. This is the reason why the term cold- or warm-water region is used instead of the cold- or the warm-water mass which is the term I use. The difference in the depth is considered to be caused by vertical displacements (Tanioka, 1962b; Fukuoka, 1961, 1962b). The stronger the upwelling (sinking), the smaller (larger) the depth. The second type of cold-water region is composed of the cold water from the north, and the T-S relationships are quite different from those of the first type of the cold-water region, that is, the T-S relationships of the warm water of the Tsushima Current region (Shimomura and Miyata, 1953).

It is described in Section 5 that the meandering of the Tsushima Current is related to the bottom topography. Since the cold- and the warm-water regions appear to be related to the meandering, their locations are also related to the bottom topography and some places are well-known in studies made before (Kajiura et al., 1958; Miyata, 1958; Tanioka, 1962b; Fukuoka, 1961, 1962b). The name of the place is prefixed to the name of the region, for example, Noto Cold-Water Region, which appears northwest of the Noto Peninsula.

Though the region appears nearly at fixed locations, it moves a little and changes its area corresponding to the fluctuation of the meandering of the Tsushima Current. And, in extreme cases, the meandering and then the region vanish. Such characters differ from one region to another. Some regions are stable and others are variable. The stability is inferred to be related to the bottom topography. The cold-water region formed in elevation of the bottom is stable, while the region in the depression of the bottom is variable (Tanioka, 1962b). The warm-water region in the elevation of the bottom is stable (Fukuoka, 1961, 1962b).

It is accepted that upwelling and sinking contribute to the occurrence of the cold- and warm-water regions. This is reasonable on the basis of the water mass analysis noted above. According to Fukuoka's study, high temperature can be traced down to the depth of 400 m in the warm-water region. This suggests that sinking reaches at least this depth. In addition to vertical motions, horizontal motions are also considered to form the cold- and warm-water regions. Kajiura et al. discuss the warm-water region near the Yamato Rise and suggest a possibility that the region is an extension of the Eastern Korean Current or is composed of warm water flowing northeastward from the Oki Ridge. The cold-water region of the second type is caused by the horizontal motion.

8 HYDROGRAPHY IN THE STRAITS CONNECTING THE JAPAN SEA WITH OTHER SEAS

8.1 The Tsushima Strait

The Tsushima Strait is composed of two channels, the eastern channel and the western channel, which are separated from each other by the Tsushima Islands. Regular monthly observations were carried out for many years and normal conditions were obtained for each month. The Tsushima Current enters the Japan Sea through the strait and the main current is considered to flow through the western channel.

The vertical sections of temperature and salinity are prepared for the eastern channel by Nan'niti and Fujiki (1967) and are constructed for the western channel on the basis of the normal values compiled by the Fisheries Research and Development Agency, the Republic of Korea (1964). In the eastern channel, the distribution is nearly homogeneous in winter and the vertical gradients are large in summer. In the western channel, the lateral gradients are large in winter, while, in summer, they are not large, but the vertical gradients are large in the surface layer (Fig. 8). These features are brought about by strong convection in winter and by the inflow of the coastal water from the East China Sea in summer. Thus, the seasonal variations are marked and the distribution in summer is in striking contrast to that in winter.

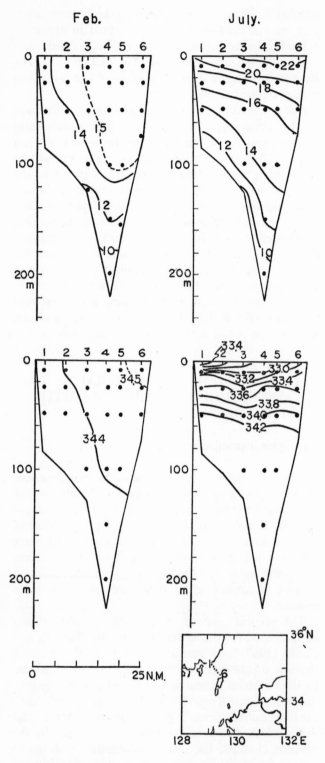

FIG. 8 Vertical sections in the western channel of the Tsushima Strait based on normal values compiled by the Fisheries Research and Development Agency, the Republic of Korea (1964).

Upper temperature in °C; lower salinity in ‰.

The region of high temperatures and high salinity in the eastern part of the western channel is equivalent to the warm core in the Kuroshio, but is obscure in summer because less saline water occupies the surface layer. Though the dimensions of the Tsushima Strait are small and the effects of boundaries are not negligible, the hydrographic structure in the strait is nearly the same as that of the Kuroshio region and the existence of the Tsushima Islands does not disturb the main features. As noted above, the lateral gradients are large in the western channel. This suggests that the main current flows in the western channel. It is inferred that temperature and salinity are higher in the eastern channel. This results from the situation that the strong current flows in the western channel if it is accepted that the hydrographic structure in the strait is similar to that in the Kuroshio region.

Since the Tsushima Current is a branch of the Kuroshio, the characters of the water of the Tsushima Current must be the same as those of the water of the Kuroshio. After the Tsushima Current branches off from the Kuroshio in the East China Sea, the water is modified by mixing with the coastal water and its characters in the Tsushima Strait are different from the original characters. The modification is more marked in the western channel, especially in summer, though in the lower layer, it is not marked and the water has the characters of the Western North Pacific Central Water (Miyazaki and Abe, 1960). The water of high salinity of 34.4–34.5‰ has temperature of about 15°C in T-S diagrams for the strait, and the relation is close to the relation of the intermediate water in the warm-current region shown in Section 4. Neither the inflow of such saline water into the Japan Sea through other straits nor the production of such saline water in the Japan Sea is probable. Therefore, the saline water in the Tsushima

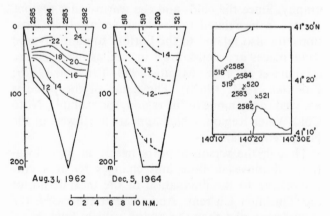

FIG. 9 Vertical sections of temperature (°C) in the Tsugaru Strait. August 31, 1962 voyage of the *Yushio Maru*, December 5, 1964 voyage of the *Kofu Maru*.

Strait is considered to be the source of the intermediate water in the warm-current region of the Japan Sea.

The computation of the volume transport through the Tsushima Strait by Yi (1966) is mentioned in Section 6. According to his results, two branches are found. The main branch flows close to Korea and the secondary branch flows in the eastern part. The volume transport of the secondary branch is about 30% of that of the main branch. The seasonal variation of the volume transport is marked and the maximum value is found around September (Hidaka and Suzuki, 1950; Yi, loc. cit.). The annual range is about 2×10^6 m³/sec.

8.2 The Tsugaru Strait

The Tsushima Current is divided into two branches to the west of the Tsugaru Strait. One of them flows northward along the west coast of Hokkaido and the other flows out to the Pacific through the Tsugaru Strait. The latter is called the Tsugaru Current and its volume transport is estimated at about 75% of that of the Tsushima Current on the average, as noted in Section 6.

Vertical sections of temperature in the Tsugaru Strait are shown in Fig. 9. The features are similar to those in the Tsushima Strait and the pattern is kept undisturbed in spite of long travel, though the values are changed to some degree. Salinity at the surface in the central part of the strait fluctuates with two minimum values (Hakodate Marine Observatory, 1961). The first minimum value occurs in the first ten days of June and is considered to be the effect of melting snow. The second one occurs in the first ten days of August and the decrease in the Tsushima Strait in June reaches this strait in two months. Therefore, this is caused by the influence of the coastal water in the East China Sea (see Section 2).

The volume transport through the Tsugaru Strait is evaluated by Miyazaki (1952) and Hata (1962) and there is much difference between their values (see Section 6). Nevertheless, it is probable that most of the water carried by the Tsushima Current flows into the Tsugaru Strait. The flow pattern at the depth of 5 m is obtained by direct-current measurements (Hikosaka, 1953). Strong current flows in the middle of the strait and the speeds are high in the narrow part. The daily mean speeds amount to 3.5 knots and depend on the slope of the sea level in the direction of S-N. Moreover, periods of a fortnight and a month are found in the fluctuation of the speeds and the range of the fluctuation is more than 1 knot.

8.3 The Soya Strait

The Soya Strait is about 50 km wide and shallow with its greatest depth of about 60 m so that the volume transport through the strait is not large even if a strong current flows there. It is accepted that warm water is carried into the Okhotsk Sea through the Soya Strait (Iida, 1962; Maeda, 1968). Warm water is found along the northeastern coast of Hokkaido in the Okhotsk Sea. Such warm water is neither produced in the Okhotsk Sea nor supplied through passages of the Kuril Islands. The Tsushima Current flows northward in the sea west of Hokkaido (see Section 5) and there is not a southward warm current in the sea west of Sakhalin. These support the existence of an eastward current, branched off from the Tsushima Current, flowing through the Soya Strait. Though it is not justified quantitatively, a vertical section of temperature in the strait may serve as a basis for this discussion. A section from south to north would serve the purpose, but such a section is not available. A vertical section from west to east is shown in Fig. 10, which suggests that the warm current flows northward around Station 461 to the west of the strait and that a warm current flows southward between Station 457 and 458 to the east of the strait. These currents are supposed to be connected with each other though the figure does not provide compelling evidence. The current flowing through the Soya Strait is named the Soya Current.

Surface currents were measured with GEK around the Soya Strait and speeds of 1 knot or more were obtained. But the depth is less than 100 m and the results are not reliable. In addition, the Soya Current flows in the shallow coastal region along the northeastern coast of Hokkaido and evaluation of the volume transport is difficult.

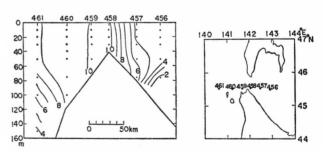

FIG. 10   Vertical section of temperature (°C) around the Soya Strait based on data obtained during a voyage of the *Kofu Maru* in November, 1963.

REFERENCES

Akagawa, M. (1954): On the oceanographical conditions of the north Japan Sea in summer (I). *Journ. Oceanogr. Soc. Japan*, **10**(4): 189–199.

Fisheries Research and Development Agency, the Republic of Korea (1964): *Oceanographic handbook of the neighbouring sea of Korea*, 214 pp.

Fukuoka, J. (1957): On the Tsushima Current. *Journ. Oceanogr. Soc. Japan*, **13**(2): 57–60.

—— (1961): An analysis of the mechanism of the cold and warm water masses in the seas adjacent to Japan. *Rec. Oceanogr. Works in Japan*, **6**(1): 63–100.

—— (1962a): Characteristics of hydrography of the Japan Sea. *Journ. Oceanogr. Soc. Japan, 20th Anniv. Vol.*, 180–188.

—— (1962b): An analysis of hydrographic conditions along the Tsushima Current in the Japan Sea. *Rec. Oceanogr. Works in Japan*, **6**(2): 9–30.

—— (1965): Hydrography of the adjacent sea (1). *Journ. Oceanogr. Soc. Japan*, **21**(3): 95–102.

Hakodate Marine Observatory (1961): Report of the oceanographic observations in the Tsugaru Straits in the period from 1943 to 1958, part I. *Tec. Rep. Japan Meteorological Agency*, No. 9, 62pp (in Japanese).

Hata, K. (1962): Seasonal variation of the volume transport in the northern part of the Japan Sea. *Journ. Oceanogr. Soc. Japan, 20th Anniv. Vol.*, 168–179 (in Japanese).

Hidaka, K. and T. Suzuki (1950): Secular variation of the Tsushima Current (in Japanese). *Journ. Oceanogr. Soc. Japan*, **6**(1): 28–31.

Hikosaka, S. (1953): On the ocean-currents (non-tidal currents) in the Tsugaru Strait (in Japanese). *Hydrogr. Bull.*, **39**: 279–285.

Ichiye, T. (1954): On the distributions of oxygen and their seasonal variations in the adjacent seas of Japan (1). *Oceanogr. Mag.*, **6**(1): 41–66.

Iida, H. (1962): On the water masses in the coastal region of the south-western Okhotsk Sea. *Journ. Oceanogr. Soc. Japan, 20th Anniv. Vol.*, 272–278.

Kajiura, K., M. Tsuchiya and K. Hidaka (1958): The analysis of oceanographical condition in the Japan Sea (in Japanese). *Rep. Develop. Fisher. Resour. in the Tsushima Warm Current*, 1, 158–170.

Maeda, S. (1968): On the cold water belt along the northern coast of Hokkaido in the Okhotsk Sea. *Umi to Sora*, **43**(3): 1–20.

Miyata, K. (1958): Characteristics of the Tsushima Current in the Japan Sea (in Japanese). *Rep. Develop. Fisher. Resour. in the Tsushima Warm Current*, 1: 147–152.

Miyazaki, M. (1952): The heat budget in the Japan Sea (in Japanese). *Rep. Hokkaido Reg. Fisher. Res. Lab.*, 4: 1–45.

—— (1953): On the water masses of Japan Sea (in Japanese). *Bull. Hokkaido Reg. Fisher. Res. Lab.*, **7**.

Miyazaki, M. and S. Abe (1960): On the water masses in the Tsushima Current area (in Japanese). *Journ. Oceanogr. Soc. Japan*, **16**(2): 59–68.

Nagahara, M. (1965): On the seasonal variations of the transport volume and oceanic conditions in the Japan Sea from 1962 to 1964 (in Japanese). *Bull. Japan Sea Reg. Fisher. Lab.*, **14**: 71–79.

Nakayama, I. (1951): On the mean surface temperature of the neighbouring sea of Japan (in Japanese). *Bull. Kobe Marine Observatory*, **159**: 1–16.

Nan'niti, T. and A. Fujiki (1967): Secular variations of hydrographic conditions in the East Tsushima Strait (in Japanese). *Journ. Oceanogr. Soc. Japan*, **23**(4): 201–212.

Oceanographical Section, Maizuru Marine Observatory (1967): Report of the oceanographic observations in the Japan Sea from August to September, 1964 (in Japanese). *The Results of Marine Met. and Oceanogr. Observations*, **36**: 74–86.

Ohwada, M. and K. Yamamoto (1966): Some chemical elements in the Japan Sea. *Oceanogr. Mag.*, **18**(1–2): 31–37.

Shimomura, T. and K. Miyata (1953): On the oceanographic character of the low temperature region off Sado Island (I) (in Japanese). *Bull Japan Soc. Sci. Fisher.*, **19**(4): 424–428.

Suda, K. and K. Hidaka (1932): The results of the oceanographical observations on board R. M. S. "Syunpu Maru" in the southern part of the Japan Sea in the summer of 1929, part 1 (in Japanese). *Journ. Oceanogr. Imp. Mar. Observ.*, **3**(2): 291–375.

Sverdrup, H. U., M. W. Johnson and R. H. Fleming (1962): *The oceans, their physics, chemistry, and general biology*. Prentice Hall, New York, 1087pp.

Tanioka, K. (1962a): The oceanographical conditions of the Japan Sea (I) (in Japanese). *Umi to Sora*, **38**(3): 90–100.

—— (1962b): The oceanographical conditions of the Japan Sea (II) (in Japanese). *Umi to Sora*, **38**(4): 115–128.

—— (1968): On the Eastern Korean Warm Current (Tosen Warm Current). *Oceanogr. Mag.*, **20**(1): 31–38.

Teramoto, T. (1972): Day-to-day to monthly variations in oceanic flows estimated from cross-stream differences in electric potential (to be published).

Tsukuda, K. (1937): On the surface temperature of the neighbouring seas of Japan. *Mem. Imp. Mar. Observ.*, **6**(3): 239–257.

Uda, M. (1934): The results of simultaneous oceanographical investigations in the Japan Sea and its adjacent waters in May and June, 1932 (in Japanese). *Jour. Imp. Fisher. Exp. St.*, **5**: 57–190.

Yamanaka, I. (1951): On the hydrographical condition

of Japan Sea in spring and summer, 1949, part I and part II (in Japanese). *Journ. Oceanogr. Soc. Japan*, **6**(3): 143–156.

Yasui, M., T. Yasuoka, K. Tanioka and O. Shiota (1967): Oceanographic studies of the Japan Sea (I). *Oceanogr. Mag.*, **19**(2): 177–192.

Yasui, Z. and K. Hata (1960): On the seasonal variations of the sea conditions in the Tsugaru Warm Current region. *Mem. Kobe Mar. Observ.*, **14**: 3–12.

Yi, Sok-U (1966): Seasonal and secular variations of the water volume transport across the Korea Strait. *Journ. Oceanogr. Soc. Korea*, **1**(1–2): 7–13.

Chapter 10

COMPARISON BETWEEN DEEP SECTIONS ACROSS THE KUROSHIO AND THE FLORIDA CURRENT AND GULF STREAM

L. V. WORTHINGTON[1])* and HIDEO KAWAI**

* Woods Hole Oceanographic Institution, Woods Hole, USA.
** Department of Fisheries, Kyoto University, Kyoto, Japan.

1 INTRODUCTION

The Kuroshio and the Gulf Stream are two currents which clearly invite comparison. They are both situated on the northwestern edges of their respective oceans and are both characterized by a narrow and rapid flow of water at the surface and in the upper thermocline. In terms of volume transport they are the dominant features of their oceans.

In August and September 1965 *Atlantis II* occupied three deep oceanographic sections across the Kuroshio. The positions of these sections are shown in Fig. 1. A section near the Ryukyu Islands was made during August 1965 by Prof. Henry Stommel of the Massachusetts Institute of Technology and Cdr. William Barbee of the U.S. Coast and Geodetic

Survey who have kindly allowed us to include their data in this study. The sections off Shikoku and off Inubozaki were made in September 1965 by the present authors with the assistance of Mr. Hideo Akamatsu of the Japan Meteorological Agency, Mr. Glenn A. Cannon of the Johns Hopkins University and staff members of the Woods Hole Oceanographic Institution. In the Shikoku and Inubozaki sections direct current measurements were made with neutrally buoyant floats.

In the Atlantic, the Florida Straits section was made from *Atlantis* in June 1955 at the point where the Florida Current emerges from the Straits. The Cape Fear section was also made from *Atlantis* in March 1957 in the course of a cooperative effort with *Discovery II* to determine whether or not there

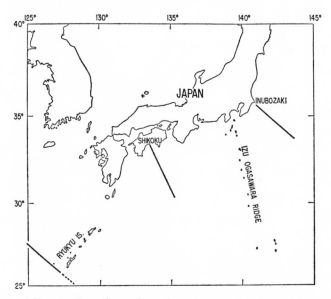

FIG. 1 Location of sections across the Kuroshio.

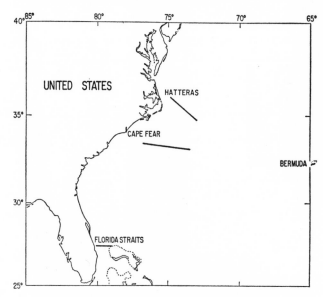

FIG. 2 Location of sections across the Florida Current and Gulf Stream.

[1] Contribution No. 2280 from the Woods Hole Oceanographic Institution. The W.H.O.I. participation in this research was sponsored by the National Science Foundation GP 821 and by the Office of Naval Research 2196 and Co 241.

was a deep countercurrent along the continental slope of the United States (Swallow and Worthington, 1961). Neutrally buoyant floats were tracked by *Discovery II* during this section. The Hatteras section was made in November 1966 from *Atlantis II* and neutrally buoyant floats were tracked immediately after this section.

The correspondence of these three sections is not geographically exact—the Kuroshio has a more marked eastward component of flow than the Florida Current and Gulf Stream—but the sections in each pair lie at roughly the same latitude.

2   THE RYUKYU AND FLORIDA STRAITS SECTIONS

The Ryukyu section (Fig. 3) is compared with the Florida Straits section in terms of temperature, salinity, oxygen and geostrophic current velocity component. The Florida Current is more restricted by the Bahama Banks than the Kuroshio is by the

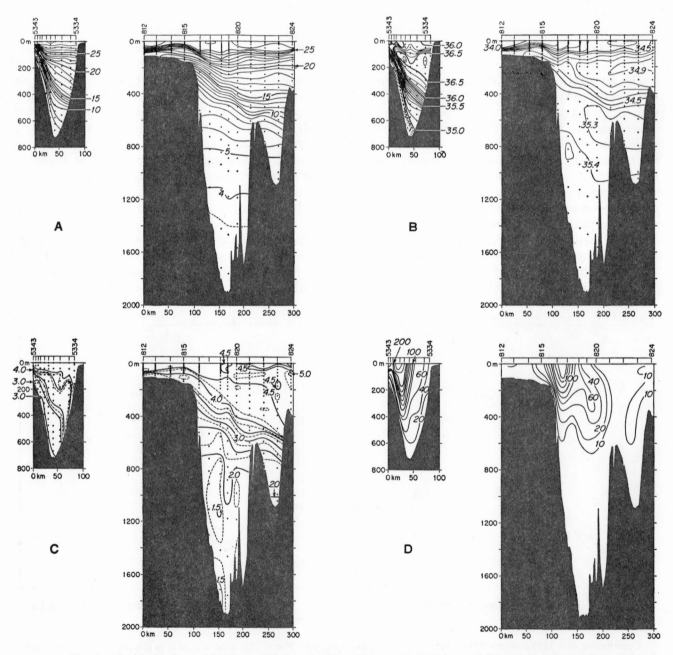

FIG. 3   Florida Straits and Ryukyu sections.   A, Temperature;   B, Salinity;   C, Oxygen;   D, Velocity.

ridge which connects the Ryukyu Islands chain. The effective widths of the currents are about 75 km in the Florida Straits and about 150 km in the Ryukyu section. The temperature sections are quite similar although the slopes of the isotherms are much stronger in the Florida Current than in the Kuroshio. The oxygen and salinity sections are more dissimilar. The oxygen values in the Ryukyu section are far lower at the oxygen minimum and below it, and the salinity values are lower at all depths. This, of course, holds true for all of the sections to be discussed because the Pacific Ocean is generally lower in salinity and oxygen concentration than is the Atlantic Ocean.

The velocity profile for the Ryukyu section is based on a zero velocity reference surface at 800 m, which is approximately the sill depth of the Okinawa Basin which is isolated by the Ryukyu Ridge. The volume transport of the Kuroshio within the ridge is 33 million m³/sec. The profile in the Florida Straits shows higher speeds, the maximum speed is 240 cm/sec, whereas it is 165 cm/sec for the Ryukyu section. No direct subsurface current measurements were made with these two sections but in the case of the Florida Straits section many subsequent direct measurements by Schmitz and Richardson (1968) have shown that the average volume transport through the Straits is 32 million m³/sec. The computed volume transport relative to the bottom in this section was 23 million m³/sec, but this amount has been adjusted to 30 million to agree more closely with Schmitz and Richardson's figure.

The adjustment was made by assuming that the current velocity on the bottom was 18 cm/sec across this section. This figure agrees closely with the near-bottom velocities observed by Schmitz and Richardson. To adjust the transport to the full 32 million m³/sec would require a bottom velocity of 26 cm/sec, which is larger than those observed by Schmitz and Richardson.

In addition to the 33 million m³/sec volume transport flowing northeast within the Ryukyu Ridge there is 26 million m³/sec flowing in the same direction outside the ridge (not shown) according to dynamic computations relative to the bottom, so the maximum flow through the entire Ryukyu section is roughly 59 million m³/sec. No similar flow is found offshore of the Bahama Banks according to Day (1954) who examined a network of oceanographic stations in that area; moreover, a section extending from the Bahama Banks to Bermuda made just prior to the Florida Straits section gave a total of only 4 million m³/sec relative to the bottom.

3 THE SHIKOKU AND CAPE FEAR SECTIONS

The temperature section across the Kuroshio off Shikoku is compared with the corresponding Florida Current section off Cape Fear (Fig. 4). The Cape Fear section was made in March and in consequence the shallow seasonal gradient found in the Shikoku section is lacking, and the warm core is more pronounced. However, the main thermocline is remarkably similar in the Kuroshio and the Florida Current. The Pacific deep water is colder and more uniform in temperature and the adiabatic effect can be noticed in the reappearance of the 1.6°C isotherm at the greatest depths. The potential temperature nevertheless decreases with depth.

One might assume, from the generally steeper slope of the isotherms at all depths in the Cape Fear section, that the calculated velocity and volume transport would be higher in this section than in the Shikoku section but such is not the case. The reason for this (Fig. 5) is the difference in the crosscurrent salinity gradients. While the crosscurrent temperature gradients are stronger in the Florida Current than in the Kuroshio, in the Florida Current there are also strong crosscurrent salinity gradients working in the opposite sense, which tend (relatively) to increase the specific volume on the left-hand side of the current and thus to diminish the dynamic height difference across the current at all depths.

The difference in the salinity values reflects the generally low salinity of the Pacific Ocean compared with that of the Atlantic. The salinity maximum in the Cape Fear section is greater than 36.7‰ at 200 meters in the swiftest part of the current—salinity diminishes to just below 34.9‰ at the bottom. In the Shikoku section the maximum is just above 34.9‰ at 200 meters. The salinity diminishes to a minimum below 34.2‰ at 800 meters and then slowly increases to a value of 34.69‰ at the bottom.

The oxygen profiles of the two sections (Fig. 6) are similar down to the minimum, which is about 1.5 ml/l off Shikoku and about 3.0 ml/l off Cape Fear. In the Shikoku section there is a very gradual increase in oxygen concentration to the bottom where the value is about 3.55 ml/l. In the Cape Fear section there is a much more abrupt increase in oxygen below the minimum. Values greater than 5.75 ml/l are found everywhere below 1600 m. The highest values (greater than 6.0 ml/l) are found along the edge of the continental slope between 1800 and 4000 meters and they roughly distinguish the waters of the deep countercurrent (which is discussed below) from those of the Florida Current.

Dynamic computations of velocity and volume transport are easily performed when the zero-surface is assumed to lie at the ocean bottom. They are also easily performed between pairs of stations where the velocity at a certain depth has been directly measured by a neutrally buoyant float—the computed vertical shear curve relative to the bottom, or some arbitrary depth, is simply adjusted to the measured velocity at the depth of the float. The difficulty lies in assigning a reliable reference surface between pairs of stations where no current measurements were made. In examining the velocity profile of the Shikoku section (Fig. 7) it must be realized that this is only one interpretation and that other guesses could be as good or better. The positions of the neutrally buoyant floats in this section are marked 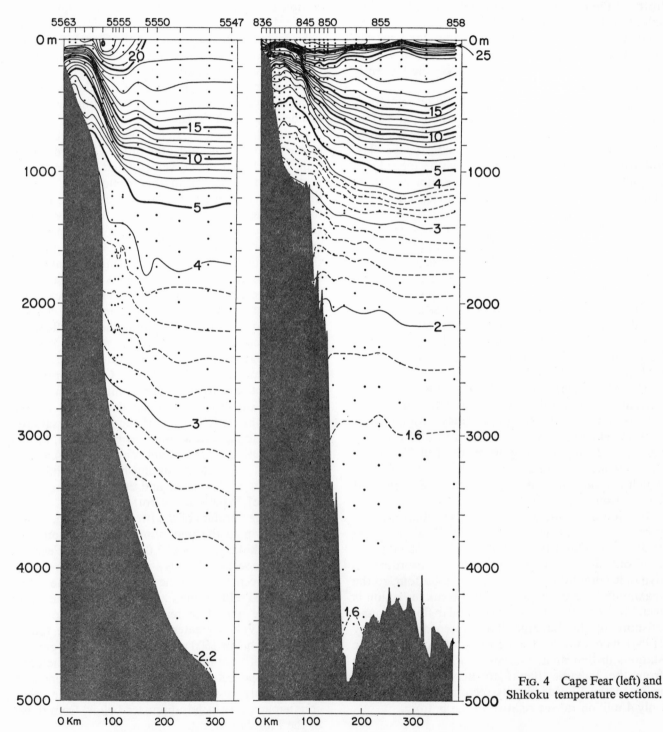. The unshaded areas

Fig. 4   Cape Fear (left) and Shikoku temperature sections.

represent water moving in a northeasterly direction with the main current; the shaded areas represent countercurrent.

The floats shown here are at 1600 m, at 3200 m and at 2250 m. The drift of all the floats used in the Shikoku and Inubozaki sections are summarized in Table 1, with estimates of the possible errors in depth and velocity and the time period for which they were followed. The depth estimates were based on the differences in time of arrival between the direct and bottom reflected signals from the floats. The estimated possible error in the speed and direction of drift is usually based on a Loran error of ±2 microseconds.

The zero-surface on the inshore side of the floats was assumed to be the bottom. This assumption

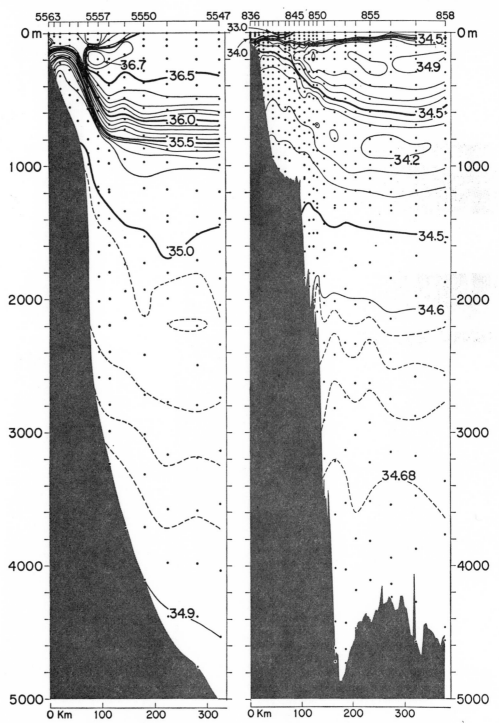

FIG. 5 Cape Fear (left) and Shikoku salinity sections.

caused a second zero level to appear at approximately the same sigma-t surface as that occupied by float No. 10, below which the countercurrent extended from float No. 10 up to the continental slope. The zero level offshore from the floats was also assumed to lie on the bottom. It is indeed arbitrary to use the bottom as a zero-surface in the absence of direct measurements. An alternative would have been to use the nearest computed zero-surface where floats were used. This would have been 1600 m on the inshore side of float No. 10 and 3800 m on the offshore side of float No. 9. In the inshore case the volume of the countercurrent would be reduced somewhat and that of the Kuroshio increased. In the offshore case no appreciable change in transport would result as the vertical shear in the deep water is exceedingly small.

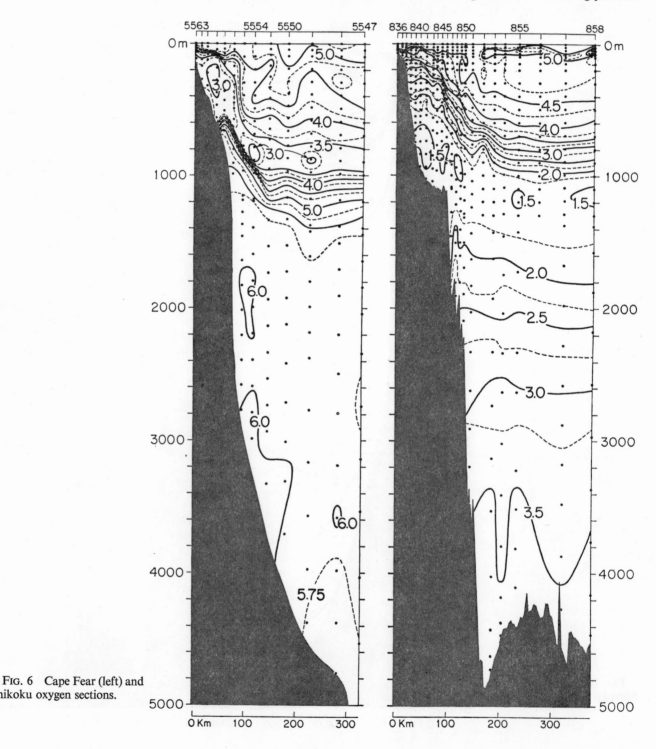

FIG. 6   Cape Fear (left) and Shikoku oxygen sections.

The Shikoku section computed in this manner shows a small countercurrent which transports 1.6 million m³/sec between the Kuroshio and the land. The Kuroshio itself is characterized by very strong surface currents reaching speeds of more than 240 cm/sec. The deep countercurrent lies along the continental slope and has a transport volume of 6.4 million m³/sec. The entire flow between the coastal countercurrent and station 857 amounts to 84 million m³/sec.

The Cape Fear velocity section (Fig. 8, from Swallow and Worthington, 1961) shows a remarkable similarity to the Shikoku section in spite of the differences in the gradients of temperature and salinity which have been noted. A coastal countercurrent is also present and the deep countercurrent carries 6.7 million m³/sec. The total transport through this section, however, is only 57 million m³/sec (excluding the coastal countercurrent). The lower transport is due, in large part, to the strong surface countercurrent found on the offshore side of the swiftest current.

4   THE INUBOZAKI AND HATTERAS SECTIONS

The Inubozaki temperature section and the corresponding Hatteras temperature section are shown in Fig. 9. The Inubozaki section was not as satisfactory as the Shikoku section because it was necessary to break it off for two days in the middle and take refuge from a typhoon in Yokosuka. This probably accounts for the reversal of the slope of the isotherms at station 875; the Kuroshio evidently moved slightly offshore during this two-day period. Observations in this section reached a depth of 7733 m in the Izu-Ogasawara Trench—the depth of water is nearly double that of the Hatteras section. The Hatteras section, however, shows evidence of strong vertical shear all the way to the bottom whereas the Inubozaki section does not.

A striking similarity in the temperature field in these sections is the thermostad at 18°C found south of the Kuroshio and the Gulf Stream. (A thermostad has been defined by Seitz (1967) as a layer in which the vertical gradient of temperature is a relative minimum.) The 18°C water in the Atlantic has been discussed by Worthington (1959) and Istoshin (1961). They agree that it is formed south of the Gulf Stream

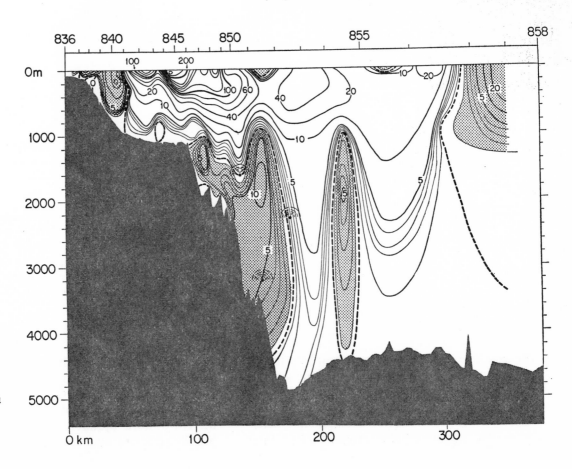

FIG. 7 Shikoku velocity section.

TABLE 1.   Summary of results from neutrally buoyant floats.

Swallow float #	Mean position	Between stations	Depth (meters)	Duration of drift (hours)	Speed (cm/sec)	Direction of drift
10	32°18′N 134°08′E	850 851	1600±130 (11)*	41	0.0±2.0	—
4	32°06′N 134°12′E	851 852	3200±200 (1)*	37	4.9±2.0	225°±10°
9	31°59′N 134°17′E	852 853	2250±85 (5)*	68	1.5±0.5	060°±5°
8	34°24′N 142°56′E	877 878	1200±50 (5)*	34	10.2±1.3	057°±8°
11	34°17′N 142°59′E	878 879	2000±50 (3)*	18	7.4±2.2	072°±5°
7	34°12′N 143°05′E	878 879	3050±25 (6)*	34	4.5±2.0	046°±10°
3	35°16′N 141°16′E	— —	1050±85 (16)*	39	16.1±1.2	235°±5°
13	35°10′N 141°17′E	— —	1000±60 (5)*	24	11.4±1.5	250°±4°

*   Numbers in parentheses indicate the number of depth estimates made for each float.

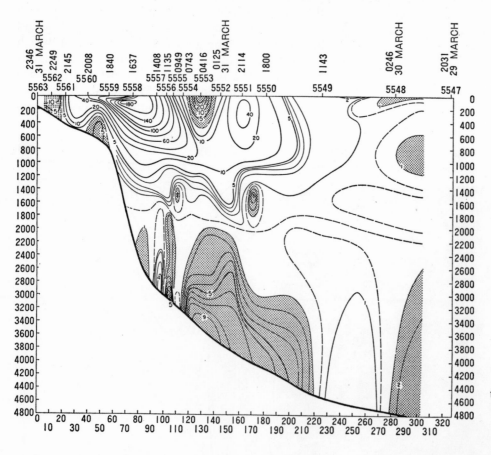

FIG. 8   Cape Fear velocity section (Swallow and Worthington, 1961).

by the excessive heat loss to the atmosphere in late winter and note that it can be found throughout the Sargasso Sea at latitudes as far south as 20°N, which is roughly 1700 miles south of the formation area.

Further investigation is being undertaken to determine to what extent the production of 18°C water affects the depth of the thermocline in the Sargasso Sea and, by inference, to what extent it affects the volume transport of the Gulf Stream. The presence of an identical thermostad in the area to the south of the Kuroshio is probably not coincidental since the heat loss due to evaporation in late winter in this area closely corresponds to that south of the Gulf Stream (Budyko, 1963, plate 28).

The salinity sections (Fig. 10) again show the large difference that exists between the Atlantic and Pacific Oceans and the far larger crosscurrent salinity gradient of the Gulf Stream. The salinity minimum in the Kuroshio in the Inubozaki section is considerably lower than that in the Shikoku section, with values falling below 34.0‰. There is evidently not strict continuity of flow between the Shikoku and the Inubozaki sections; this shows more clearly in the oxygen sections to follow. There is a small area close

to the bottom on the southern side of the Izu-Ogasawara Trench where the salinity rises above 34.7‰. This is possibly due to error in the analysis. From the study of Wooster and Volkmann (1960) on the distribution of properties at 5 km one would, perhaps, not expect to find salinity values this high.

The oxygen profiles for the Inubozaki and Hatteras sections, which are compared in Fig. 11, resemble the Shikoku and Cape Fear profiles and, illustrate the generally lower oxygen values in the deep Pacific Ocean. However, the oxygen-minimum layer in the Inubozaki section is more pronounced

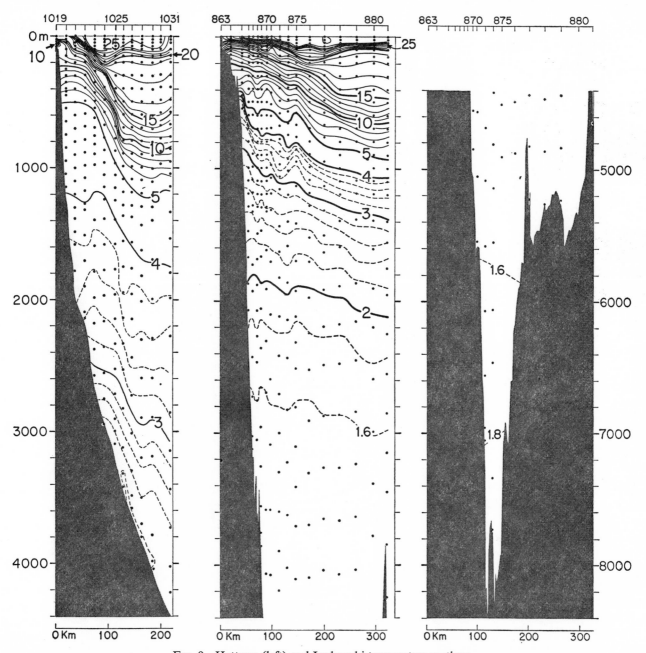

FIG. 9   Hatteras (left) and Inubozaki temperature sections.

than that in the Shikoku section, with some values falling below 1.0 ml/l near the continental slope.

The differences may be more clearly seen on an oxygen/potential density diagram. The method has been used by Kawamoto (1955), who has pointed out differences in the oxygen characteristics in the western North Pacific. An oxygen/potential density diagram for the Shikoku and Inubozaki sections is shown in Fig. 12. At densities greater than 27.3 the oxygen values in the Inubozaki section are consistently lower than in the Shikoku section. This is presumably because the Izu-Ogasawara Ridge, which extends far

to the south between the two sections, acts as a barrier for water denser than 27.3. This density surface is found at a depth of 700 m at the continental slope and descends to 1150 m on the right-hand side of the Kuroshio.

The potential temperature/salinity diagram (Fig. 13) also confirms that there is a clear distinction between the deep water masses on either side of the Izu-Ogasawara Ridge. Values from the Ryukyu section in the Philippine Basin and the Shikoku section in the Shikoku Basin are more saline at all potential temperatures than those in the Inubozaki section in

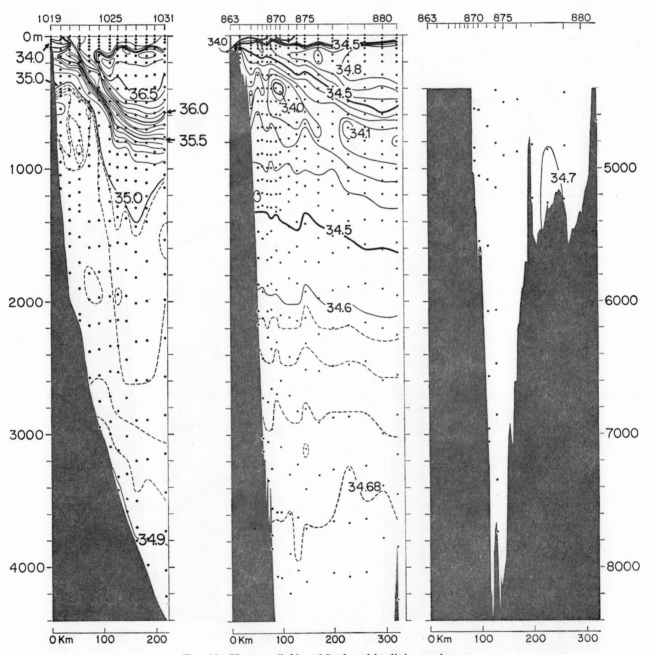

FIG. 10   Hatteras (left) and Inubozaki salinity sections.

the Northwest Pacific Basin. The values within the Okinawa Trough are the most distinct of all, suggesting that the waters in this basin have been cut off from the remainder of the Philippine Basin for some time. A further distinction between the waters of the Philippine and the Shikoku Basins and those outside of them is in the minimum potential temperature—1.19° within the Philippine and the Shikoku Basins and 1.03° without.

The velocity profile (Fig. 14) for the Inubozaki section shows the current bisected by a counter-current over the Izu-Ogasawara Trench. As mentioned in connection with the temperature section this is probably an artifact of the two-day break necessitated by the typhoon.

There is a distinct countercurrent along the continental slope between 500 and 2500 meters, as indicated by neutrally buoyant floats. The floats at 1000 m drifted counter to the Kuroshio at 16 and 11 cm/sec. However, the tracking began six days after the inshore stations were occupied. The zero level used in the dynamic computations for this section was

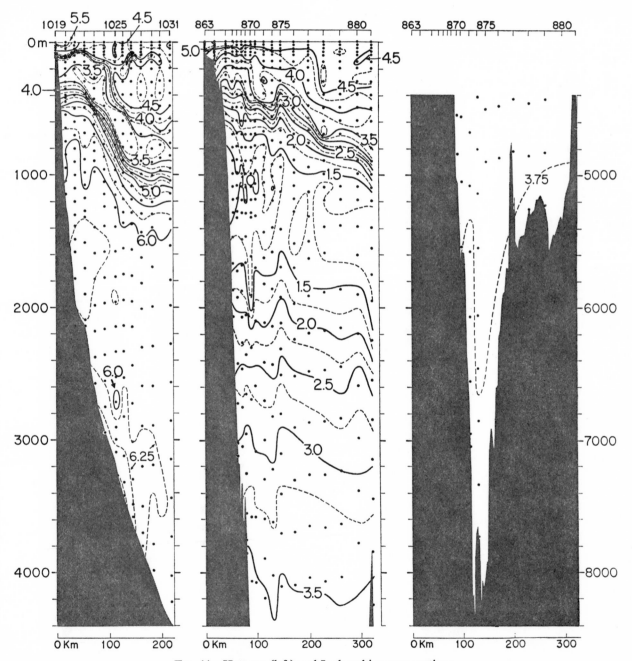

FIG. 11   Hatteras (left) and Inubozaki oxygen sections.

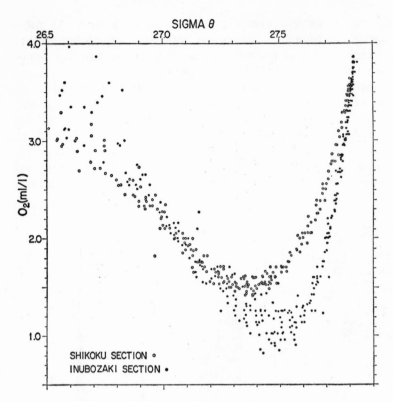

FIG. 12   Oxygen/potential density diagram for Shikoku and Inubozaki sections.

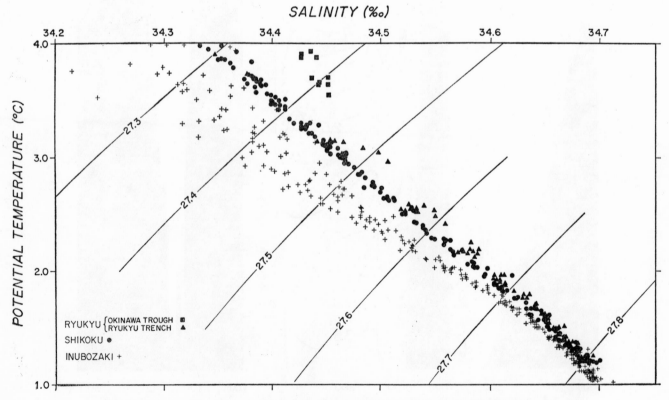

FIG. 13   Potential temperature/salinity diagram for Ryukyu, Shikoku and Inubozaki sections.

based on the drift of the floats and on the vertical shear calculated from three stations made while tracking the floats. The volume transport of the countercurrent was 4.8 million m³/sec.

The remaining neutrally buoyant floats were tracked on the southeast side of the Izu-Ogasawara Trench. Two floats were tracked between stations 878 and 879, one at 2000 meters and one at 3050 meters. Their velocity components at right angles to the section were 6.3 cm/sec at 2000 meters and 4.5 cm/sec at 3050 meters, giving a shear of 1.8 cm/sec. The shear computed from the station data is 3.3 cm/sec. The measured shear was adjusted to agree with the computed value by increasing the 2000 meter velocity from 6.3 to 7.0 cm/sec and reducing the deep value

from 4.5 to 3.7 cm/sec.

The final float was tracked between stations 877 and 878 at 1200 meters. It drifted to the northeast at 10 cm/sec and since the vertical shear in geostrophic velocity was slight below 1200 meters, one would expect the entire deep water column between the stations to move at approximately the same velocity. If this is the case 10 million m³/sec is added to the total transport of the Kuroshio. It will be remembered that the deep salinities in this region seemed extraordinarily high (greater than 34.7‰) so it may be that there is a real flow of saline water along the southeast side of the Izu-Ogasawara Trench. Reid et al. (1968) have detected a deep boundary current along the Tonga-Kermadec Trench in the western

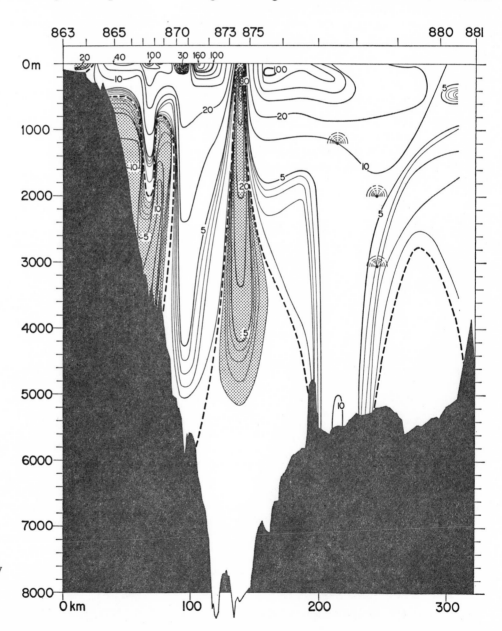

FIG. 14 Inubozaki velocity section.

South Pacific. This current was characterized by salinity values of more than 34.73‰ and had a computed northward volume transport of between 8 and 12 million m³/sec. Consequently, there is undoubtedly a source of deep high-salinity water in the western South Pacific but whether or not it can be continuously traced as far north as the Izu-Ogasawara Trench is problematical.

The velocity profile for the Hatteras section is shown in Fig. 15. At this point the deep countercurrent was directly under the Gulf Stream. On this occasion the deep countercurrent was weak—computed velocity was less than 3 cm/sec and the computed volume transport toward the south below the 2000 m level was only 2 million m³/sec, contrasted with Barrett's (1965) estimate of between 4 and 12 million in approximately the same place. The total volume transport of the Gulf Stream through this section was 82 million m³/sec, although it appears from the relatively high velocities at depth on the offshore end of the section that the whole current was not crossed and that another station should have been made.

For the readers' convenience Table 2 has been prepared showing the transports of the Kuroshio and the Florida Current—Gulf Stream for all these sections.

TABLE 2. Summary of volume transports.

Kuroshio		Florida Current Gulf Stream	
Ryukyu (Total)	59	Florida Straits	30
Shikoku	84	Cape Fear	57
Inubozaki	88	Hatteras	84

The principal difference in transport lies in the large downstream increase in volume transport of the Florida Current and Gulf Stream, although the two currents transport essentially the same volume of water at Inubozaki and Hatteras where they both leave the continental slope. There is evidence that the transport of the Gulf Stream continues to increase after it leaves the continental slope. Warren and Volkmann (1968) have calculated a Gulf Stream transport of 101 million m³/sec in longitude 69°W, 350 km downstream from the Hatteras section and Fuglister (1963) has computed a transport of 147 million 400 km further east at longitude 64°30′W. At present it is not known whether or not the Kuroshio increases in volume transport to the east of Inubozaki; further deep sections made in conjunction with neutrally buoyant floats will be necessary to answer this question.

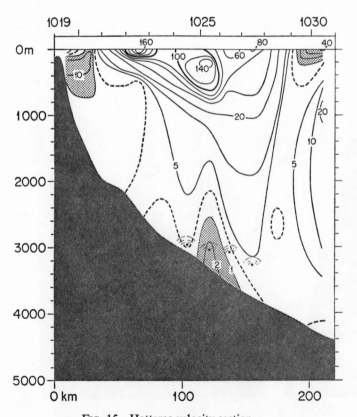

FIG. 15   Hatteras velocity section.

REFERENCES

Barrett, Joseph R. (1965): Subsurface currents off Cape Hatteras. *Deep-Sea Res.*, **12**: 173–184.

Budyko, M. I. (1963): Atlas of the heat balance of the earth sphere. *Joint Geophysical Committee of the Presidium of the Academy of Sciences*, U.S.S.R. Moscow.: pp 1–5, plates 1–69.

Day, C. (1954): A note on the circulation in the region northeast of the Bahama Islands. W.H.O.I. Ref. 54-4, 6pp (unpublished manuscript).

Fuglister, F. C. (1963): Gulf Stream '60. *Progr. in Oceanogr.* **1**: 265–373.

Istoshin, Yu. V. (1961): Formative area of "eighteen-degree" water in the Sargasso Sea. *Okeanologiya* **1**(4): 600–607. (Trans. *Deep-Sea Res.*. **9**: 384–390).

Kawamoto, T. (1955): On the distribution of the dissolved oxygen in the Pacific Ocean. *Umi to Sora.*, **32**(2): 23–37.

Reid, Joseph. Jr., Henry Stommel, E. Dixon Stroup and Bruce A. Warren (1968): Detection of a deep boundary current in the western South Pacific. *Nature*, **217**: 937.

Schmitz, W. J. Jr. and W. S. Richardson (1968): On the transport of the Florida Current. *Deep-Sea Res.*, **15**: 679–693.

Seitz, R. C. (1967): Thermostad, the antonym of thermocline. *Jour. Mar. Res.*, **25**(2): 203.

Swallow, J. C. and L. V. Worthington (1961): An observation of a deep countercurrent in the Western North Atlantic. *Deep-Sea Res.*, **8**: 1–19.

Warren, Bruce A. and G. H. Volkmann (1968): Measurement of volume transport of the Gulf Stream south of New England. *Jour. Mar. Res.*, **26**(2): 110–126.

Wooster, Warren S. and Gordon H. Volkmann (1960): Indications of deep Pacific circulation from the distribution of properties at five kilometers. *Jour. Geophys. Res.*, **65**(4): 1239–1249.

Worthington, L. V. (1959): The 18° water in the Sargasso Sea. *Deep-Sea Res.*, **5**: 297–305.

Chapter 11

DEEP WATERS IN THE WESTERN NORTH PACIFIC

SHIGEO MORIYASU

Meteorological Research Institute, Tokyo, Japan.

1 INTRODUCTION

In this chapter, deep waters are defined as waters below the depth of 1000 m and the area called the western North Pacific is the Northwest Pacific Basin and the Philippine Sea Basin.

Deep waters of the Pacific have been the subject of several past studies and certain main features have been described. It has been accepted (for example, Sverdrup et al., 1942; Knauss, 1962) that there is only one source of deep waters in the Pacific, and that the Pacific is much more nearly homogeneous than the Atlantic. The bottom water, the source of deep waters, derived from the South Pacific is inferred to flow into the North Pacific on the basis of the distribution of some properties. Thus, the modification of the source water is the subject of study.

Ichiye (1960b) says about the hydrographic studies of deep waters in the western North Pacific, ". . . in the western North Pacific stations yielding hydrographic data below 2000 m were only occasionally occupied after World War II. The data available were mostly collected by the Japanese Hydrographic Office vessels from 1930 to 1939. However, these stations are scattered at random in time and space. Also sampling depths at the stations were taken usually at intervals of about 1000 m below 2000 m. Yet some features of deep waters can be derived by adequate interpretation of such data, although they are incomplete from today's situation." During the last ten years, the situation has been improved. The Japanese Expeditions of Deep Sea started in 1959 (Wadati and Terada, 1962) and studies have been made on deep waters in trenches near the Japanese islands. And most recently, the Cooperative Study of the Kuroshio and Adjacent Regions (CSK) has been offering much hydrographic data of deep layers. Most studies on deep waters concern the Northwest Pacific Basin as a main part of the

North Pacific. There are few studies on deep waters in the Philippine Sea Basin, perhaps because oceanographers have believed the Philippine Sea Basin has no relation to the circulation of deep waters in the major part of the Pacific Ocean.

Most studies on the character of deep waters aim to deduce the deep circulation of the entire Pacific Ocean (Kishindo, 1940; Wooster and Volkmann, 1960; Knauss, 1962; Fukuoka, 1962; and so on). These studies are based on the idea that waters flow in the direction of increased temperature, decreased salinity and decreased oxygen content. The changes are considered to be caused by mixing, supply of heat through the ocean floor and consumption of oxygen during travel. Indeed the deduction is reasonable in qualitative treatment, but present knowledge about the factors does not make quantitative discussions possible, and the speed cannot be estimated with reliability.

Studies on circulation of deep waters have also been tried based on geostrophic relationships and other dynamical considerations (Ichiye, 1960a, 1960b, 1962; Masuzawa, 1960). In these studies, many assumptions, the validity of which has not been established yet, are used. Stommel (1958) and Stommel and Arons (1960) propose a model of the circulation of the world's oceans based on theory and experiments. Nan'niti and Akamatsu (1966) propose a model of the circulation in the North Pacific both in the light of Stommel's study and their direct current measurements. Though the results seem to be reasonable, there is not enough evidence to support them.

The study of deep waters is based on data with various accuracies collected by various vessels for long time intervals, and which are considered to be representative values of properties at stations, independently of when they were measured because it has been believed that the values do not change

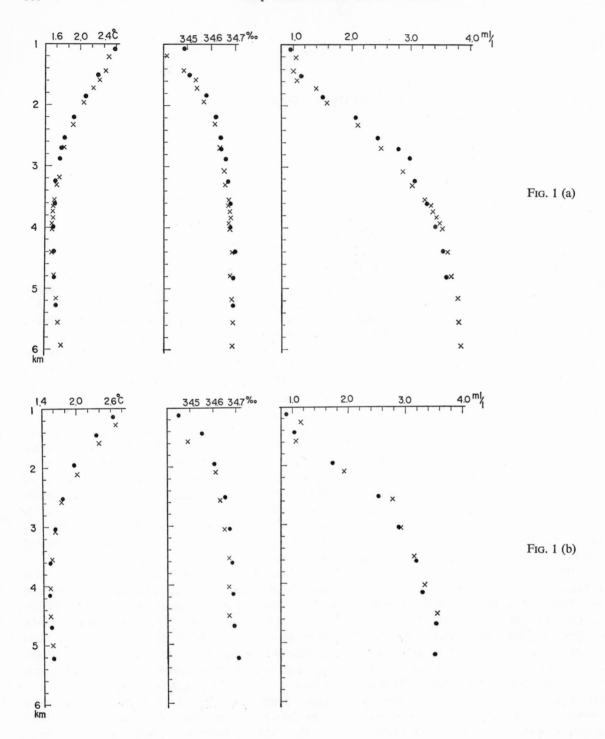

FIG. 1 (a)

FIG. 1 (b)

FIG. 1   Comparison of vertical distributions of temperature, salinity and oxygen content measured at nearly the same position.

(a) ● 43°23.5′N, 149°30.0′E, August 2, 1965, *Ryofu Maru*
     × 43°33′N, 149°51′E, March 22–23, 1966, *Argo*
(b) ● 41°30′N, 145°56′E, July 1, 1966, *Kofu Maru*
     × 41°31′N, 145°58′E, July 23, 1965, *Kofu Maru*
(c) ● 38°00.0′N, 144°01.5′E, July 9, 1965, *Ryofu Maru*
     × 38°01′N, 143°58′E, July 10, 1966, *Kofu Maru*
     ○ 38°00′N, 144°00′E, February 26, 1966, *Ryofu Maru*

(d) ● 32°00.5′N, 144°01.0′E, July 12, 1965, *Ryofu Maru*
     × 32°01′N, 143°59′E, February 8, 1966, *Ryofu Maru*
(e) ● 10°04′N, 126°41′E, July 20, 1965, *Takuyo*
     × 10°01′N, 126°58′E, August 26, 1966, *Takuyo*
(f) ● 22°00′N, 130°00′E, April 18, 1966, *G. Nevelskoy*
     × 22°01′N, 130°01′E, February 11, 1966, *Vitjaz*
     (salinity and oxygen content are not shown)

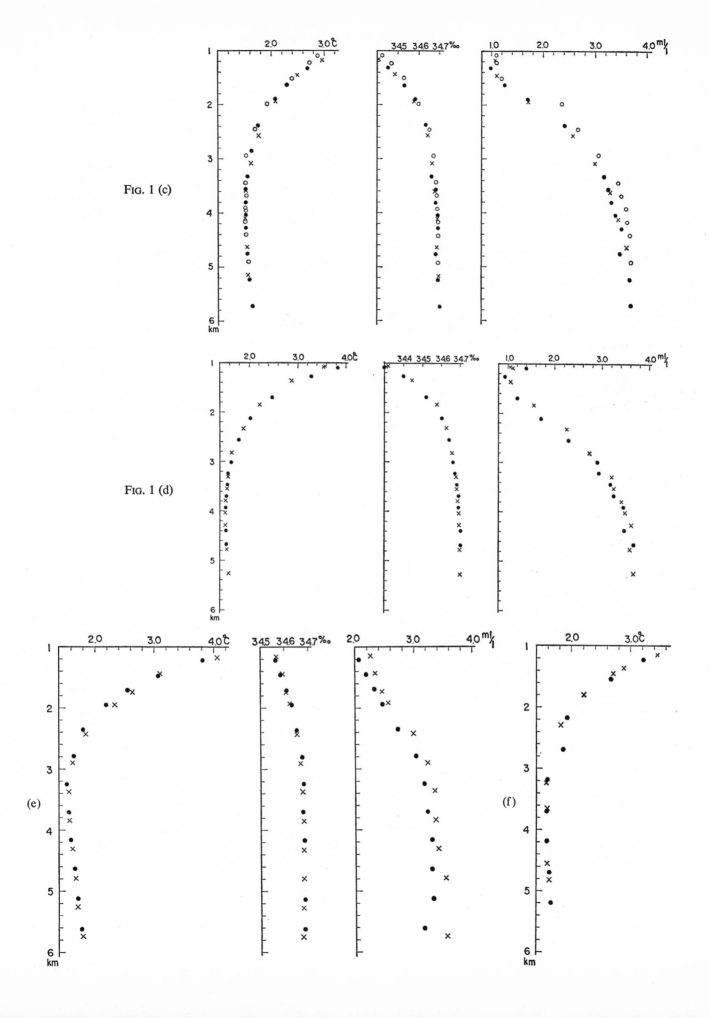

FIG. 1 (c)

FIG. 1 (d)

(e)

(f)

with time in deep waters. Robinson (1960), however, has examined long-term climatic change in deep waters of the Pacific, and although she does not show long-period climatic trends, she finds real temperature and salinity variations at all levels. Some examples are shown in Fig. 1 to illustrate differences of vertical distributions of temperature, salinity and oxygen content obtained in CSK. There are differences beyond the errors of measurements between repeated observations, and changes with time can be found without doubt, most markedly above 2000 m. Neither the cause nor the general magnitude of the changes is determined. As long as these are unknown, a strict analysis cannot be made. In this sense, therefore, the description below is somewhat uncertain. The CSK is carried out with the participation of many vessels and the data are checked and published as "Preliminary Data of CSK" by the Japanese Oceanographic Data Center, and can, therefore, be relied on for accuracy. In this chapter, data collected by vessels listed in Table 1 are used. Aside from this, data collected by the *Takuyo* in the IGY (July 7–15, 1958 and March 8–20, 1959) are used. Horizontal distributions and vertical sections of temperature, salinity and oxygen content are prepared from these data. The locations of four vertical sections are shown in Fig. 2, in which Wooster and Volkmann's arbitrary flow line (1960) and Kishindo's section (1940) are inserted also. Sections I and II were made to find the changes in these properties in the south-north direction and Section III to see the changes in the west-east direction. Section IV was originally made to find the changes along the Japan Trench, but the depth was less than 5000 m at one station.

2 TEMPERATURE

Vertical distributions of temperature in deep waters in the western North Pacific are simple, as described by Masuzawa (1962) and Nitani and Imayoshi (1963). Vertical distributions at five stations are shown in Fig. 3 (a). Station A is located in the Aleutian Trench, Station B in the Kuril-Kamchatka Trench, Station C in the Japan Trench, Station D in the Ryukyu Trench and Station E in the Philippine Trench. Since, at these stations, hydrographic observations were made to depths greater than 6000 m

TABLE 1. Sources of CSK data.

Vessel	Period	KDC Reference No.
Takuyo	June 25—September 7, 1965	49K001
Takuyo	July 1—September 13, 1966	49K024
Takuyo	July 12—August 30, 1967	49K053
Ryofu Maru	July 7—August 3, 1965	49K003
Ryofu Maru	February 4—28, 1966	49K017
Ryofu Maru	September 13—17, 1966	49K033
Ryofu Maru	January 11—February 24, 1967	49K040
Kofu Maru	July 22—28, 1965	49K004
Kofu Maru	June 30—July 10, 1966	49K026
Umitaka Maru	August 7—17, 1965	49K008
Shinyo Maru	July 9—21, 1965	49K009
Shinyo Maru	July 19—August 11, 1966	49K030
Kagoshima Maru	July 28—August 10, 1965	49K010
Kagoshima Maru	August 5—14, 1966	49K031
Oshoro Maru	November 30, 1965—January 25, 1966	49K022
Oshoro Maru	January 15—February 1, 1967	49K045
Tansei Maru	July 30—August 6, 1966	49K419
Atlantis II	August 4—September 23, 1965	31K001
George B. Kelez	February 1—April 6, 1966	31K004
Chautauqua	January 22—27, 1966	31K002
Argo	January 27—April 1, 1966	31K003
Bering Strait	July 8—14, 1966	31K005
Shokalsky	December 12, 1965—March 31, 1966	90K004
G. Nevelskoy	January 27—April 29, 1966	90K005
Vitjaz	December 17, 1965—April 15, 1966	90K007
Yang Ming	August 10—October 13, 1965	21K001

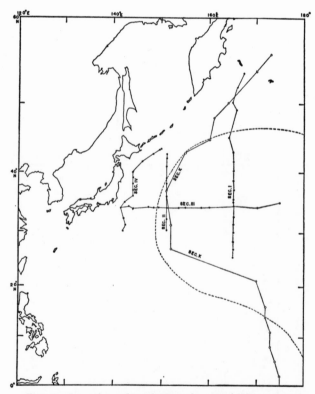

FIG. 2 Location of vertical sections and Wooster and Volkmann's flow line. Sections I, II, III and IV are prepared by the present author. Section K is Kishindo's section. Broken line is Wooster and Volkmann's flow line.

with good accuracy, the vertical distribution is considered to be representative of deep waters along the western boundary of the North Pacific. Temperature decreases with increasing depth to the 3000-m level, but the magnitude of the gradient is much smaller below 2000 m than above. Temperature changes little between 2000 m and 3000 m, and a temperature minimum occurs at around 4000 m, and the temperature is around 1.5°C. Below this level, temperature increases with depth. The distribution divides naturally into two groups, one for Stations A, B and C, and the other for Stations D and E. The temperature-minimum layer is shallower and the minimum temperature is high in the former, but the rate of increase at the 5000–6000-m level is nearly the same (about 0.15°C/1000 m). This causes temperature differences of about 0.15°C between the two groups. Potential temperature is obtained by using the table for salinity 34.85‰ reproduced by Sverdrup et al. (1942). The vertical distribution of potential temperature has nearly the same character above the

3000-m level as that of temperature in situ. Between the 3000-m and 4000-m levels potential temperature decreases with increasing depth and the rate is smaller than that above the 3000-m level. Below the 4000-m level, the decrease is very slight, especially for Stations D and E. In the distribution shown in Fig. 3 (b), potential temperature does not increase with depth, while Nitani and Imayoshi (1963) report a rate of increase of about 0.02°C/1000 m in potential temperature below the 5000–6000-m level for the Kuril-Kamchatka Trench. The discrepancy may result from my rough estimation of potential temperature. Nevertheless, vertical distributions have peculiar features in the Aleutian Trench, the Kuril-Kamchatka Trench and the Japan Trench located in the Northwest Pacific Basin, and in the Ryukyu Trench and the Philippine Trench located in the Philippine Sea Basin.

Depths of the temperature-minimum layer as roughly estimated from CSK data (Fig. 4) distinguish the distribution in the Philippine Sea Basin from that

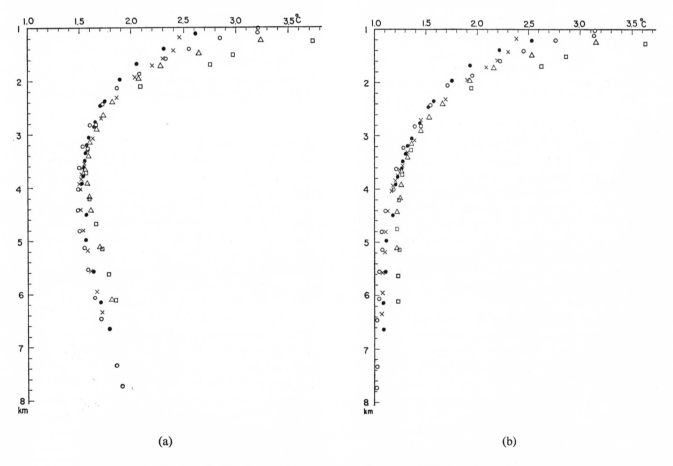

(a)   (b)

Fig. 3 Vertical distributions of (a) temperature in situ and (b) potential temperature.

● Station A 53°43′N, 167°16′E, *Argo*   △ Station D 25°05.5′N, 128°13.5′E, *Atlantis II*
× Station B 43°33′N, 149°51′E, *Argo*   □ Station E 6°59′N, 127°08′E, *Takuyo*
○ Station C 35°02′N, 142°15′E, *Atlantis II*

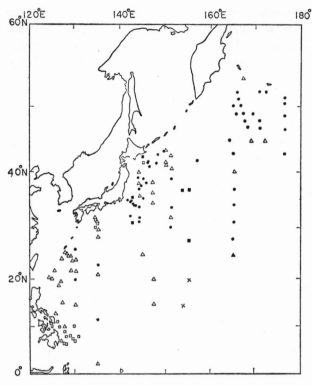

FIG. 4   Depth of minimum temperature layer.

○ ≦ 3000 m   4500 < ■ ≦ 5000 m
3000 < □ ≦ 3500   5000 < ▲ ≦ 5500
3500 < △ ≦ 4000   × not found above 5500 m
4000 < ● ≦ 4500

in the Northwest Pacific Basin. The depth in the Philippine Sea Basin is less than 4000 m in general, and less than 3500 m in the Philippine Trench; whereas in the Northwest Pacific Basin, the depth is usually more than 4000 m. Depths less than 4000 m are seen between the Japan Trench and about 150°E. Depths greater than 4000 m are found from the south to about 40°N to the east of 150°E. The depth is more than 5000 m south of 15°N. At two stations south of 20°N, hydrographic observations were made to depths of more than 5500 m and no temperature minimum was found. Though it is not clear whether a temperature-minimum layer exists below the greatest sample depth, it is inferred that the temperature minimum occurs very near the bottom or not at all. This seems to support Wooster and Volkmann's conclusion (1960) that a temperature minimum is not evident south of 20°N in the Pacific. The depth of the temperature-minimum layer is dependent upon many factors, and various models can be introduced to explain vertical distributions of temperature including the depth of the temperature-minimum layer (e. g., Nitani, 1963). At present, however, there is too little evidence to establish the validity of the models, because the magnitude of the factors has not yet been determined with sufficient accuracy.

Vertical sections of potential temperature in the Northwest Pacific Basin are shown in Fig. 5.

Section I (along 165°E)

Temperature is homogeneous in the north-south direction except at the greatest depths. Low temperature less than 1.0°C is restricted to below the 5000-m level south of 35°N. This cold water is regarded as coming from the south, consistent with bottom water from the South Pacific reaching this latitude. Potential temperatures of 1.0°C or less were also found in the northern basin of the Pacific by Wooster and Volkmann (1960). The isotherm of 1.1°C is deeper around 45°N and around 54°N. The feature is seen in the 1.25°C isotherm as well, though it is not so clear as in the 1.1°C isotherm.

Section II (along 151°E)

Temperature is also homogeneous in the north-south direction. Low temperatures less than 1.0°C are not seen anywhere. The 1.1°C isotherm is deeper north of 40°N and temperatures are higher than 1.1°C down to the bottom at the northern end of the section. The 1.25°C and 1.50°C isotherms between the 2000-m and 3000-m levels are almost level and the isotherm in the layer above 2000 m goes up northward.

Section III (along 34°N)

The distribution is different to the west of the North-west Pacific Rise from that to the east of the rise, that is, temperatures are higher west of the rise in general. This feature is more marked below 3500 m, which is the level of the top of the rise. The 1.1°C isotherm is shallower by about 500 m and low temperatures less than 1.0°C are seen east of the rise. This suggests that the bottom water from the south is less modified east of the rise. The comparison of Section II with Section I supports this conclusion.

Section IV (along the Japan Trench)

Low temperatures less than 1.0°C are seen below the 6000 m level at the southern end of the section alone and the 1.1°C isotherm goes down northward. This means that the bottom water of its initial character from the south reaches about 30°N in the Japan Trench. The isotherm of 3.0°C goes up markedly northward around the 1000 m level and temperatures are almost constant in the north-south direction between 2000 m and 3000 m.

From the distribution in these sections in the Northwest Pacific Basin, the temperature is homogeneous between 2000 m and 3000 m, where the

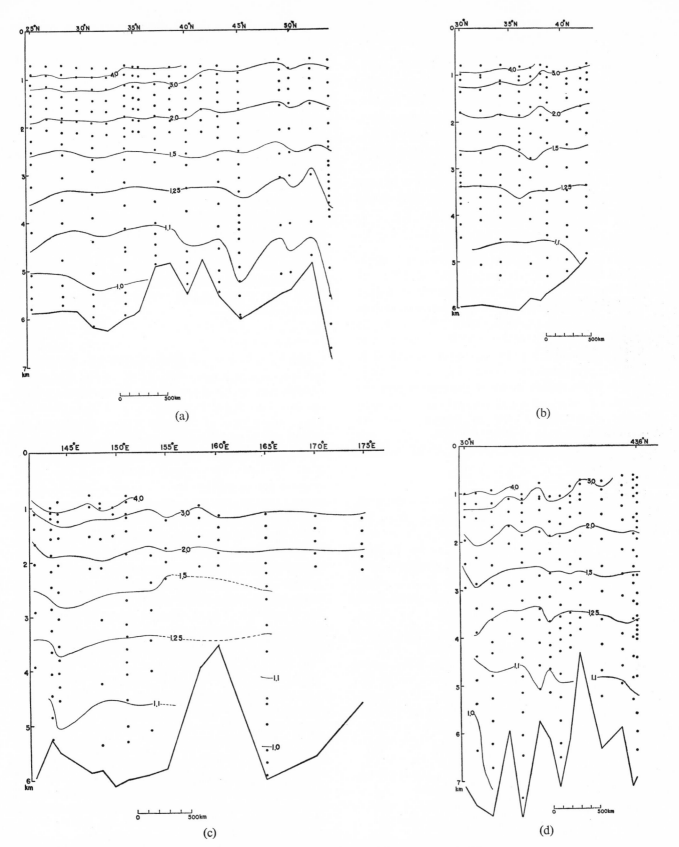

FIG. 5   Vertical sections of potential temperature in °C.
(a) Section I, (b) Section II, (c) Section III, (d) Section IV

effects of upper and lower layers are not marked. In great depths, the bottom water from the south is the most important subject. According to Wooster and Volkmann (1960), the initial potential temperature of the bottom water at the 5000-m level south of New Zealand is less than 0.5°C. Temperatures of the bottom water in our sections are a little lower than 1.0°C and this is caused by the sill effect suggested by them. The Northwest Pacific Rise plays an important part in the extent of the bottom water.

Figure 6 shows the horizontal distribution of temperature in the levels of 2000 m, 3000 m, 4000 m and 5000 m.

2000-m level

Roughly speaking, temperatures decrease northward and the values are between 1.75°C and 2.30°C. In the Philippine Sea Basin, temperatures are higher than 2.0°C for the most part. In the Northwest Pacific Basin, temperatures are higher than 2.0°C (a) in the southeastern part, (b) west of 150°E off Japan, and (c) at 40°N east of 155°E. In the region around 50°N 170°E, temperatures are lower than 1.9°C, while there is a warm region north of the cold region. At this level, temperatures are closely related to those in the subsurface layer above the 1000-m level. As pointed out by Ichiye (1960b), the difference in the temperature between the two basins is not marked.

3000-m level

The pattern of distribution is simple as compared with that at the 2000-m level. Temperatures are higher in the Philippine Sea Basin than in the Northwest Pacific Basin. In the Philippine Sea Basin, temperatures are higher than 1.60°C for the most part and the highest temperature is above 1.70°C. Near the Philippines, temperatures are a little lower than 1.60°C. In the Northwest Pacific Basin, a cold-water belt extends from 20°N, 155°E to a cold region of less than 1.50°C around 50°N, 170°E. Along the western margin of the basin, temperatures are lower than 1.60°C. In the Philippine Sea Basin, temperatures are low along the western margin as well. Though the feature is not ascertained along the northwestern margin north of 40°N, it may be a general feature along the western boundary of each basin where trenches are located. Around 30°N east of the Izu-Ogasawara Ridge, temperatures are higher than 1.60°C and are nearly equal to temperatures to the west of the ridge. This suggests the mixing of waters of the two basins through the gap of the ridge.

4000-m level

The difference between the Philippine Sea Basin and the Northwest Pacific Basin is large and temperatures are nearly homogeneous in each basin. In the Philippine Sea Basin, temperatures are higher than 1.55°C for the eastern part. In the Northwest Pacific Basin, a warm region of more than 1.50°C is seen along the northwestern margin of the basin. A cold belt of less than 1.45°C in the western part reaches around 35°N, 145°E. A cold region of less than 1.45°C is also seen around 50°N, 170°E. At this depth, the basins are completely separated from each other by the Izu-Ogasawara Ridge. Though temperatures in the Northwest Pacific Basin are lower, temperatures at 30°N to the east of the ridge are nearly equal to temperatures to the west of the ridge. This is not caused by horizontal mixing. It might be caused by vertical mixing with the upper water which is mixed with the water of the Philippine Sea Basin or it might be a matter of sloping isotherms connected with geostrophic flow.

5000-m level

In the Philippine Sea Basin, temperatures decrease northward and are a little higher in the Philippine Trench. In the Northwest Pacific Basin, temperatures increase northward from 1.45°C to 1.55°C or more. This is the reverse of the feature at the depth of 2000 m and has been taken as a proof of northward flow of the bottom water in the North Pacific. Knauss' chart (1962) showing the distribution of temperature at the depth of 5000 m has the same feature as noted above. Though the amount of data is not large, the difference in temperature between the two basins is larger than the scatter in values, and this suggests that the water in one basin is completely isolated from the water in the other basin.

From the description of temperature in several levels, a few interesting features are derived. (1) In the Northwest Pacific Basin, there is northward decrease in the 2000 m level, northward increase in the 5000 m level and an intermediate situation between the two levels. (2) A cold region around 50°N, 170°E is seen except for the 5000-m level and is taken as an evidence of upwelling. This is inferred from the vertical section along 165°E (Section I) as well. Knauss (1962) finds upwelling off the Aleutian Islands. The difference in the location of the region of upwelling between Knauss' and the present author's charts is of doubtful significance. (3) Temperature differences between the Philippine Sea Basin and the Northwest Pacific Basin are most marked in the 5000-m level and less marked in the 2000-m level. This is brought about by the change of mixing with depth under the influence of the Izu-Ogasawara Ridge as expressed by Ichiye (1960b).

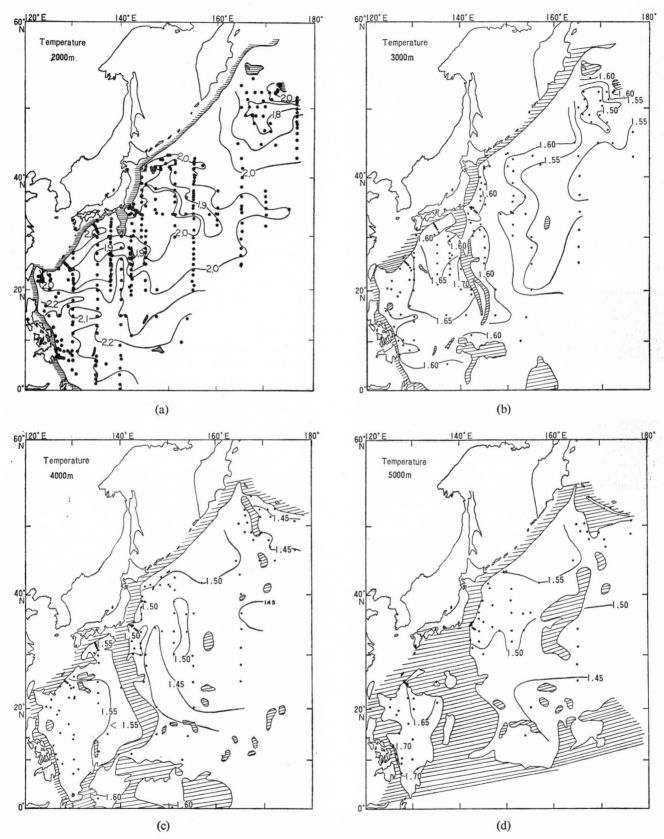

FIG. 6   Horizontal distributions of temperature.
(a) 2000-m level, (b) 3000-m level, (c) 4000-m level, (d) 5000-m level

3 SALINITY

Since salinities at CSK stations were determined mostly by a modern salinometer, the accuracy of measurements has been improved and is thought to be good for the following descriptions. Vertical distributions of salinity in deep waters in the western North Pacific are simpler than those of temperature. Vertical distributions at the five stations selected in Section 2 are shown in Fig. 7. They are considered to be representative of distributions in deep waters along the western boundary of the North Pacific for the reason stated in Section 2. As the salinity-minimum layer is at a depth less than 1000 m south of about 40°N, and does not exist north of the latitude (Sverdrup et al., 1942), salinity increases with depth in deep waters. The rate of increase is smaller between the 2000-m and 3000-m levels than above 2000 m, and very slight below the 3000-m level. This was reported by Masuzawa (1962), who described the distribution in the Kuril-Kamchatka Trench and the Japan Trench. The feature is common to the distribution in the Philippine Sea Basin. Salinity changes from station to station above 2000 m, but the scatter is hardly seen and the value is between 34.65‰ and 34.70‰ below 3000 m. This means that salinity is not a good element for analysis of deep water. Nitani and Imayoshi (1963) point out a salinity-maximum layer around the 5000-m level in the Kuril-Kamchatka Trench, but its existence is not so definite. CSK data also show a salinity-maximum layer at some stations.

Vertical sections of salinity in the Northwest Pacific Basin are shown in Fig. 8.

Section I (along 165°E)

Salinity is homogeneous in the south-north direction except at great depths. Values greater than 34.70‰ are restricted below the 5000-m level south of about 35°N and are not found elsewhere. According to Wooster and Volkmann (1960), salinity is slightly higher than 34.70‰ at 5000 m south of New Zealand. Therefore, it is inferred that the bottom water in the South Pacific reaches about 35°N on the meridian of 165°E with less modification, for the source of this saline water is not found in the North Pacific. The contour of 34.65‰ is between the 2000-m and 3000-m levels and goes down a little northward. The depth of the contour is less at 45–50°N and this may be an indication of upward motion. Knauss (1962) suggests upwelling around the latitude and this feature supports his deduction. There are no definite characteristic features in the distribution of contours of 34.60–34.40‰.

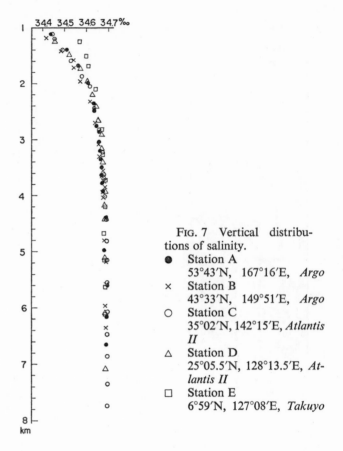

FIG. 7 Vertical distributions of salinity.
● Station A
   53°43′N, 167°16′E, *Argo*
× Station B
   43°33′N, 149°51′E, *Argo*
○ Station C
   35°02′N, 142°15′E, *Atlantis II*
△ Station D
   25°05.5′N, 128°13.5′E, *Atlantis II*
□ Station E
   6°59′N, 127°08′E, *Takuyo*

Section II (along 151°E)

The general features are nearly the same as those for Section I. Saline water of salinity higher than 34.70‰ is found around 5000 m south of about 35°N and the bottom water from the South Pacific reaches the latitude on the meridian of 151°E also. The contour of 34.50‰ goes up northward.

Section III (along 34°N)

This section is not complete because of lack of data. The west-east difference is not marked and the distribution is uniform. Saline water of more than 34.70‰ is found both to the west and to the east of the Northwest Pacific Rise. But the saline water seems to exist in a narrow band around 151°E to the west of the rise and the extent of the saline water to the east of the rise is not inferred. This requires careful selection of vertical sections in the south-north direction to see the extent of the bottom water.

Section IV (along the Japan Trench)

This section shows no marked south-north difference in general. Saline water of more than 34.70‰ is found at some stations. This is dependent on the selection of the section. The saline water seems to

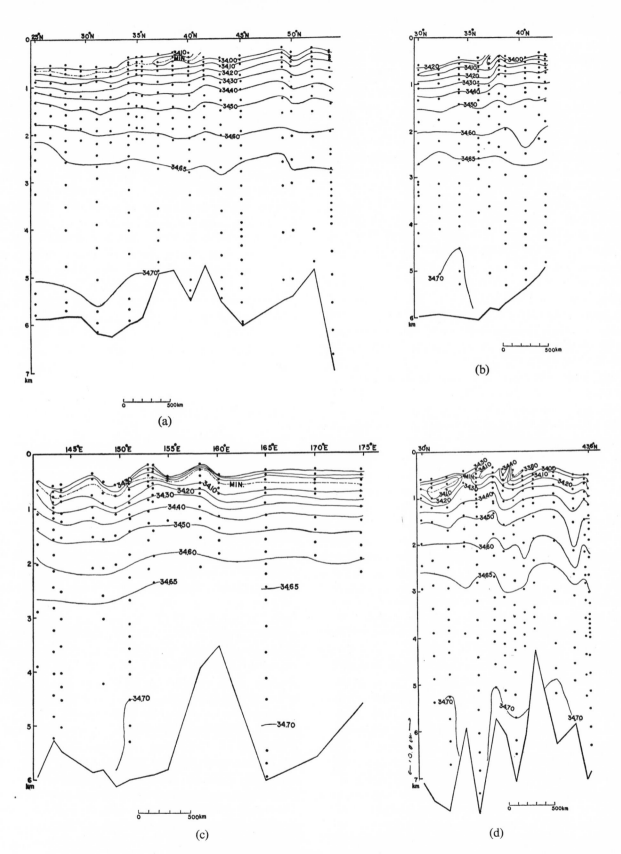

FIG. 8   Vertical sections of salinity in ‰.
(a) Section I, (b) Section II, (c) Section III, (d) Section IV

reach near the northern end of the section below the 5000-m level. At the northern end, however, data at three stations indicate that such saline water does not exist in the great depths. The features in the vertical sections are not interesting, but the existence of saline water of more than 34.70‰ is certain and this supports the view that bottom water from the south extends to the Northwest Pacific Basin. The initial value of salinity south of New Zealand estimated by Wooster and Volkmann is not much changed during the long travel.

As the differences in salinity values are not large, it is difficult to derive characteristic features of the horizontal distribution at standard depths. Though the distributions are not illustrated, some features are inferred below.

2000-m level

Scattering of values is pronounced. This is due to errors of measurement, but most values indicate true salinity. Therefore, the fluctuation is considered to be marked under the influence of the fluctuation in the upper layer. Salinity is higher (34.60–34.68‰) in the Philippine Sea Basin than in the Northwest Pacific Basin, where values less than 34.60‰ are found extensively, but not regularly, and the highest value amounts to 34.65‰.

3000-m level

The range and scattering of values are much smaller than in the 2000-m level. The values are 34.66–34.68‰ in the Philippine Sea Basin and 34.65–34.67‰ in the Northwest Pacific Basin, and the difference between the two basins is much reduced. Around 50°N in the Northwest Pacific Basin, saline water is found and the location is nearly consistent with the location of the cold region. Therefore, the saline water may be an indicator of upwelling.

4000-m level

Salinity is 34.66–34.68‰ in the Philippine Sea Basin and 34.66–34.69‰ in the Northwest Pacific Basin. Therefore, the upper limit is a little higher in the Northwest Pacific Basin. Though it is difficult to derive characteristic features in each basin, there is a saline region around 50°N in the Northwest Pacific Basin. This region is located in the same place as that in the 3000-m level and also suggests the possibility of upwelling at this level.

5000-m level

Salinity is higher in the Northwest Pacific Basin than in the Philippine Sea Basin. In the latter basin, salinity is 34.67–34.69‰ and there are no definite

characteristic features in the distribution. In the former basin, salinity is 34.68–34.70‰ and high values are found in the southern part and in the Aleutian Trench, but the saline region around 50°N does not exist. This means that upwelling does not take place in this level. High salinity in the southern part is an indication that the bottom water comes from the south. If the initial values of the bottom water are slightly higher than 34.70‰ as estimated by Wooster and Volkmann (1960), the decrease during the travel is very slight. The northward decrease in salinity has been taken as a proof that the bottom water goes northward and is found in the 5000-m level, not in the 2000-m, 3000-m and 4000-m levels.

As described above, the distributions in standard depths are not complicated, but some interesting features are inferred. A comparison between the two basins shows the feature that in the 2000-m level, salinity is higher in the Philippine Sea Basin and that in the 5000-m level, it is higher in the Northwest Pacific Basin, though the range of values is very small.

4   OXYGEN CONTENT

Vertical distributions of oxygen content in deep waters are not so uniform as those of salinity. Figure 9 shows the vertical distributions of oxygen content at the five stations used in Sections 2 and 3. The oxygen minimum occurs below 1000 m at Stations B and C, but not at other stations, where it takes place above this level. The oxygen-minimum layer is said to be below the intermediate water, that is, 400–500 m below the salinity-minimum layer (for example, Sverdrup et al., 1942). The depth is less in the southern part (less than 1000 m). In the northern part, the depth is also less than 1000 m, but there are no salinity-minimum layers which form the core of the intermediate water. In other words, the oxygen-minimum layer is deeper in middle latitudes and appears below the 1000-m level at Stations B and C in Fig. 9. Oxygen content increases with depth below the minimum layer, but the rate of increase is not constant. It decreases downward and is very slight below 3000–4000 m. According to Masuzawa (1962), the rate decreases strikingly at the 3000-m level. His estimation is made on the basis of measurements in the Kuril-Kamchatka Trench. Nitani and Imayoshi (1963) find an oxygen-maximum layer at around 6000 m in the Kuril-Kamchatka Trench. The feature is found at some stations in trenches alone, but not all stations in trenches have the feature. This is to be discussed in relation to the occurrence of potential

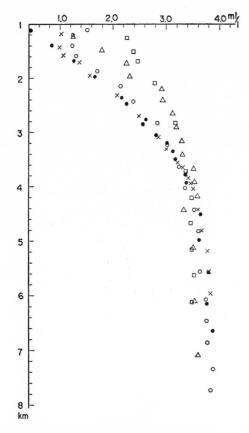

FIG. 9 Vertical distributions of oxygen content.

	Station A	53°43′N, 167°16′E, *Argo*
●	Station A	53°43′N, 167°16′E, *Argo*
×	Station B	43°33′N, 149°51′E, *Argo*
○	Station C	35°02′N, 142°15′E, *Atlantis II*
△	Station D	25°05.5′N, 128°13.5′E, *Atlantis II*
□	Station E	6°59′N, 127°08′E, *Takuyo*

temperature minimum.

Vertical sections of oxygen content in the Northwest Pacific Basin are shown in Fig. 10.

Section I (along 165°E)

Between the 1000-m and 2000-m levels, oxygen content is higher in the southern part. The oxygen-minimum layer occurs above the 1000 m level in the northern part. Around 3000 m, oxygen content is a little higher in the northern part. The feature is more definitely indicated by the contour of 3.5 ml/l. A corresponding feature is pointed out in the 1.1°C and 1.25°C contours in the vertical sections of potential temperature and it is possible to regard this as an indication of upwelling, though it is not a compelling one. The feature is found in nearly the same place as that for temperature. Oxygen content more than 3.75 ml/l is found below the 5000-m level south of about 35°N and in the Aleutian Trench, and not in the middle part. Oxygen content of more than 4.0 ml/l is found below the 5700-m level at the southern

end alone. The high content shows that there is bottom water from the south. Its initial value is 5.0 ml/l according to Wooster and Volkmann (1960) and the content decreases to about four-fifths of the initial content if the initial value is accepted. The water in the Aleutian Trench must come from the south, but it does not go along 165°E, because such a high content is not found in the middle part. The upwelling in the northern part is traced up to 3000 m on the basis of the depth of contours.

Section II (along 151°E)

Oxygen content is homogeneous except for the contours of 2.5 ml/l and 3.0 ml/l, which drop about 500 m in the middle part. Below 5000 m, oxygen content is low as compared with that in Section I. Values more than 3.75 ml/l are hardly found. This is partly due to lack of data in the southern part. Nevertheless, it is certain that the bottom water in this section has less of its initial character. This is in agreement with the deduction on temperature.

Section III (along 34°N)

The location of trough and ridge in the depth of the contours is nearly the same as those in the isotherms, which do not have such marked undulations. If high content is assumed to show the existence of the bottom water, it is inferred that the water goes in narrow bands and that one of them exists west of 145°E. Around the 2000-m level, the distribution is uniform, while between 1000 m and 2000 m, the depth of the oxygen-minimum layer is different from place to place west of the Northwest Pacific Rise.

Section IV (along the Japan Trench)

Above the 3000-m level, the distribution is uniform in general except for the change in the depth of the oxygen-minimum layer. The depth of the contour of 3.5 ml/l changes markedly and, roughly, is small (large) at stations of small (large) depth. This means that the undulations of contours nearly correspond to the undulations of bottom topography. The distribution is much dependent on how the stations are selected which make up the section. Nevertheless, water of more than 4.0 ml/l is found below the 6000-m level at the northern end alone.

Oxygen content is used as a tool for tracing the bottom water. The extreme value in the Northwest Pacific Basin is 4.0 ml/l or a little more and is much smaller than the initial value (about 5.0 ml/l). The reduction is brought about by consumption and diffusion during the long travel. Wooster and Volkmann (1960) concluded that a slight decrease occurs in this basin. In the above description, a flow

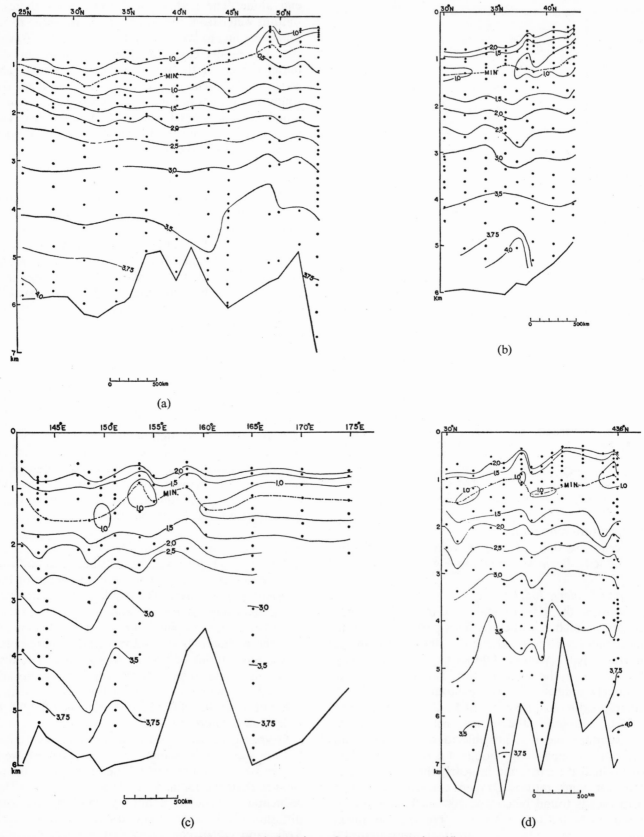

FIG. 10   Vertical sections of oxygen content in ml/l.
(a) Section I, (b) Section II, (c) Section III, (d) Section IV

line is not assumed, but large values of more than 4.0 ml/l in the northern and the southern parts suggest the slight decrease independently of the flow line. The existence of upwelling around 50°N is supported by the vertical section of oxygen content also.

Figure 11 shows horizontal distributions of oxygen content in levels of 2000 m, 3000 m, 4000 m and 5000 m.

2000-m level

Although the distribution is complicated, oxygen content decreases northward in general. Oxygen content is higher than 2.6 ml/l in most of the Philippine Sea Basin and lower than 1.6 ml/l in the cold region around 50°N, 170°E. Oxygen content is higher than 2.0 ml/l in the warm region lying south of the cold region. North of 20°N, the west-east difference is large near the Izu-Ogasawara Ridge. Ichiye (1960b) finds a large west-east difference in content in the oxygen-minimum layer. Those are essentially the same, because the minimum layer is between the 1000-m and 2000-m levels there. He points out that the difference is not found south of 20°N.

3000-m level

Oxygen content is higher in the western parts of the Philippine Sea Basin and of the Northwest Pacific Basin. In the Philippine Sea Basin, there is a region of high oxygen content of more than 3.5 ml/l in the southern part. In the Northwest Pacific Basin, a region of high content extends to about 35°N east of the Izu-Ogasawara Ridge. In the eastern part, oxygen content is almost uniform. This deduction may be partly a product of the small amount of data. Around 50°N, oxygen content is high. The feature is also found in the 2000-m level. Though the separation of the two basins from each other is stronger than at 2000 m, the effect on oxygen content is not so marked. However, oxygen content is higher in the Philippine Sea Basin where values less than 3.0 ml/l are hardly found.

4000-m level

The separation of the two basins is complete, but the difference between them is smaller than the difference in the 2000-m or 3000-m level. In the Northwest Pacific Basin, a belt of high oxygen content (more than 3.5 ml/l) extends from the south to about 40°N. Around 50°N, 170°E, oxygen content is higher than 3.7 ml/l. To the east of the Izu-Ogasawara Ridge, a region of low oxygen content is found.

5000-m level

Oxygen content is lower in the Philippine Sea Basin than in the Northwest Pacific Basin. This is opposite to the feature at 2000 m and 3000 m, as in the distribution of temperature. In the Philippine Sea Basin, the pattern is similar to that at the 4000-m depth. In the Northwest Pacific Basin, lower oxygen content is found at 30–40°N. The Northwest Pacific Rise forms a barrier and oxygen content is higher east of the rise than west of the rise. The region of high content around 50°N, 170°E in the 2000-m, 3000-m and 4000-m levels is obscure in this level. The extreme value, 4.10 ml/l, is found around 20°N, 155°E and is about four-fifths of the initial value, 5.0 ml/l, given by Wooster and Volkmann (1960).

From the above description of the distribution in the Northwest Pacific Basin, it is clear that oxygen content is higher in the northern and southern parts. High content in the southern part has been taken as an indication that the bottom water comes from the south. This is reasonable. Northward decrease in great depths is noted by Kishindo (1940) and Fukuoka (1962) and others. But the results in Fig. 11 show high content in the northern part. This detail cannot be the subject of discussion at present, but the existence of the rich region in the northern part has been reliably demonstrated. The water of high oxygen content must be supplied from the south, because there are no other sources in the Northwest Pacific Basin. The Philippine Sea Basin is completely isolated from the Northwest Pacific Basin below 4000 m. Oxygen content increases downward and this suggests the supply of water of high oxygen content. It is an interesting problem how the bottom water from the south is supplied to the Philippine Sea Basin at great depths.

5   WATER-MASS ANALYSIS

It is supposed that the only source of deep water in the Pacific is the Antarctic, and that deep water in the North Pacific is transferred from the South Pacific, its characteristics being modified during the long travel. Furthermore, it is relatively homogeneous as compared with deep water in other oceans. Cochrane (1958) studied the bivariate distribution of water characteristics in the Pacific and concluded that deep water is homogeneous and that the modal class for the North Pacific is centered at 1.25°C, 34.65‰. In this section, water-mass analysis is made by using T-S and T-O$_2$ relationships. The accuracy of measurements is good and this makes detailed analysis possible.

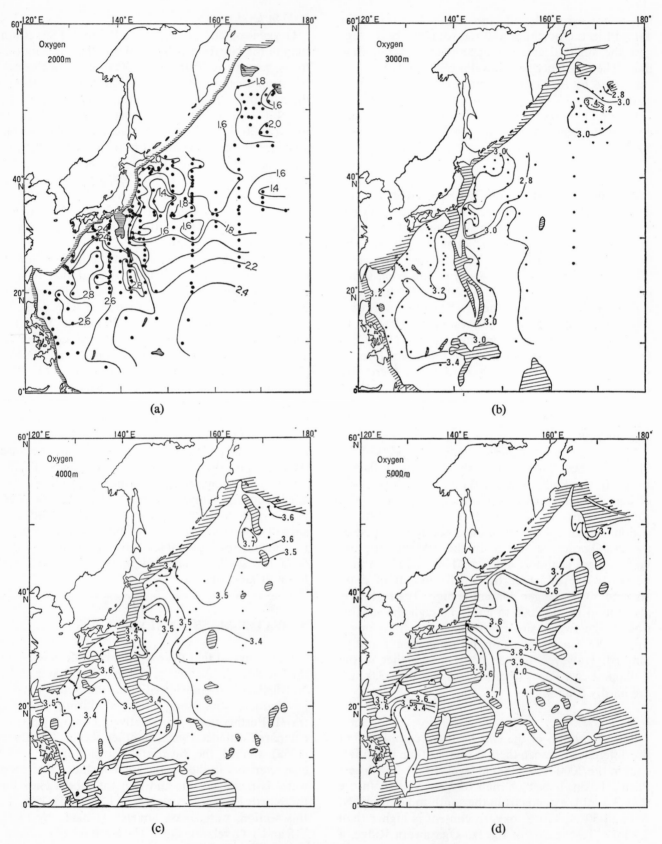

FIG. 11   Horizontal distributions of oxygen content.
(a) 2000-m level, (b) 3000-m level, (c) 4000-m level, (d) 5000-m level

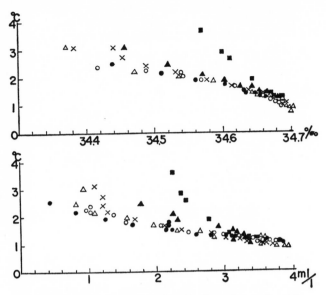

FIG. 12   T-S (upper) and T-O₂ (lower) relationships.
T means potential temperature.

- ● Station A   53°43′N, 167°16′E, *Argo*
- ○ Station B   43°33′N, 149°51′E, *Argo*
- × Station C   35°02′N, 142°15′E, *Atlantis II*
- ▲ Station D   25°05.5′N, 128°13.5′E, *Atlantis II*
- ■ Station E   6°59′N, 127°08′E, *Takuyo*
- △ Station F   31°01′N, 165°02′E, *Takuyo*

TABLE 2. Depths at which T–S curves nearly join the T–S curve at Station F, and potential temperature and salinity at each.

Station	Depth	Potential temperature	Salinity
A	1389(m)	2.21(°C)	34.510(‰)
B	1422	2.30	34.487
C	1400	2.45	34.487
D	1720	2.16	34.570
E	2818	1.45	34.669

The Philippine Sea Basin is a region of the Western North Pacific Central Water and the Northwest Pacific Basin is a region of the Western North Pacific Central Water and the Pacific Subarctic Water (Sverdrup et al., 1942). In the region of the Western North Pacific Central Water, there is intermediate water below the central water, characterized by a salinity-minimum layer lying above 1000 m.

Figure 12 shows T-S relationships below 1000 m at six stations. Here, temperatures are potential temperatures. Five stations (Stations A, B, C, D and E) are used in the previous sections and Station F, taken to make analysis better, is located east of the Northwest Pacific Rise and yields hydrographic data down to a depth greater than 6000 m. T-S curves are different from each other in the upper layer, but they converge at great depths. In the upper layer, differences between Stations A, B, C and F are slight, but the curve at Station E deviates markedly from others. Stations A and B lie in the region of the Pacific Subarctic Water and Stations C and F in the region of Western North Pacific Central Water. Therefore, T-S curves above 1000 m at the former stations are quite different from those at the latter stations. Although the difference almost vanishes below 1000 m, nevertheless, temperatures corresponding to a certain salinity deviate on detailed examination. The deviation is found even at great depths and the temperature is higher at Stations C and F lying in the region of the Western North Pacific Central Water. The curve at Station E located in the Philippine Trench is of another shape and the salinity corresponding to a certain temperature is much higher. This is caused by less influence of the intermediate water of low salinity. The curve at Station D located in the Ryukyu Trench has intermediate shape, but it is close to the curve at Station C. In great depths, the curves at Stations D and E are nearly the same as the curves at Stations C and F.

It is inferred in the previous sections that the bottom water from the south reaches Station F. Though the water at this station is modified, its T-S curve is taken as the reference curve. Then, the curve at Station C is least deviated from the curve at Station F and the curve at Station E is most deviated from it. Those curves at five stations nearly join the curve at Station F at various depths. Table 2 shows the depths and the potential temperature and the salinity at each. In the Northwest Pacific Basin, the depth is around 1400 m, while in the Philippine Sea Basin, the depth is different from place to place. This is attributed to the difference in the effect of the intermediate water. Though the shape of T-S curves is nearly the same below the depth shown in Table 2, there are slight, systematic differences between curves. The differences almost vanish for potential temperature lower than 1.25°C and salinity higher than 34.65‰ and these critical values of temperature and salinity are found at the six stations. Therefore, the modification of the bottom water takes place along the T-S curve independently of place and it is the modified water which exists in the Philippine Sea Basin. The critical values are found around the 4000-m level at Station E.

As stated in Section 4, oxygen content is better for analysis than salinity. T-O₂ relationships below 1000 m at the six stations are shown in Fig. 12. Just as the T-S curves converge at great depths, though they deviate from each other in upper layers, so also do the T-O₂ curves. These have three shapes. The first is a curve representing relationships at Stations A, B, C and F located in the Northwest Pacific Basin, the second is the curve at Station D located in the

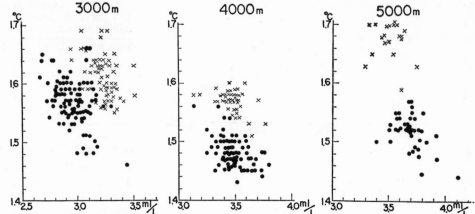

FIG. 13 T-O₂ relationships at the 3000-m, 4000-m and 5000-m levels.
  × Philippine Sea Basin
  ● Northwest Pacific Basin

Ryukyu Trench and the third is the curve at Station E located in the Philippine Trench. The curves at Stations A, B, C and F are different from each other for oxygen content lower than 1.4 ml/l. Though Stations A and B are in the region of the Pacific Subarctic Water and Stations C and F in the region of the Western North Pacific Central Water, T-O₂ relationships are nearly the same. The three shapes of T-O₂ curves correspond well to the three shapes of T-S curves, but the separation of the second from the first is more pronounced in T-O₂ relationships than in T-S relationships and the second curve represents the intermediate relationships between the first and the third. The six curves converge for oxygen content higher than 3.4 ml/l and potential temperature lower than 1.25°C, but potential temperature corresponding to a certain oxygen content is a little higher at Station D. The critical values are found at the six stations and the modification of the bottom water takes place along this curve independently of place; the modified water is found in the Philippine Sea Basin. As seen from the T-S relationships, the critical value takes place around the 4000 m level at Station F.

The Philippine Sea Basin is completely separated from the Northwest Pacific Basin below 4000 m. Therefore, it is interesting to compare T-O₂ relationships below the depth between the two basins. Figure 13 shows T-O₂ relationships at the 3000-m, 4000-m and 5000-m levels. At each level, temperature is higher in the Philippine Sea Basin. Values from 1.54°C to 1.68°C are found in the 3000-m level in both basins, while in the 5000-m level, the value of around 1.58°C separates the temperature in the two basins from each other. Temperatures at 4000 m show an intermediate situation. Oxygen contents are higher at 3000 m and lower at 5000 m in the Philippine Sea Basin than those in the Northwest Pacific Basin. The separation of values is not clear at the 5000-m level. The Izu-

Ogasawara Ridge is a barrier between the two basins and the water from the south is considered to flow northward east of the ridge. Though it is not clear that the deep water in the Philippine Sea Basin is brought over the ridge, the assumption is possible. On such a ground, T-O₂ relationships at 5000 m in the Philippine Sea Basin are compared with those at

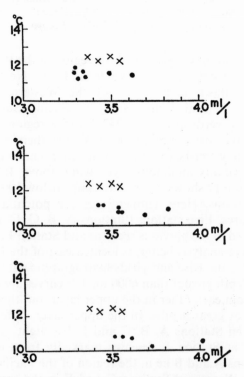

FIG. 14 Comparison of T-O₂ relationships between the east (about 145°E) and the west (about 130°E) of the Izu-Ogasawara Ridge.
  upper   ● at 4000 m east of the ridge
            × at 5000 m west of the ridge
  middle   ● at 4500 m east of the ridge
            × at 5000 m west of the ridge
  lower   ● at 5000 m east of the ridge
            × at 5000 m west of the ridge

4000 m, 4500 m and 5000 m in the Northwest Pacific Basin. Figure 14 shows the comparison of the relationships (temperature means potential temperature) at some stations close to the 5000-m contour both to the east and the west of the ridge. From the comparison, it is inferred that the water in the 5000-m level to the west of the ridge came from the 4000-m level or deeper levels to the east of the ridge. The deep water in the Philippine Sea Basin is not expected to be formed in the basin. And the assumption mentioned above is thought to be possible. If the relationships in levels close to the bottom on the ridge are made clear, the validity of the assumption will be examined.

From the analysis of T-S and T-O$_2$ relationships, the following is inferred.

(1) In the Northwest Pacific Basin, there is only one shape of T-S or T-O$_2$ curve except in upper layers. It does not matter if stations are in the region of the Western North Pacific Central Water or of the Pacific Subarctic Water.

(2) In the Philippine Sea Basin, the relationships above the intermediate depths are quite different in the Ryukyu Trench from those in the Philippine Trench, while in the Northwest Pacific Basin, deviations from the relationships are less marked even though the basin consists of the region of the Western North Pacific Central Water and the region of the Pacific Subarctic Water.

(3) The three curves of different shapes converge at great depths and the relationships at great depths in the Northwest Pacific Basin are maintained in the Philippine Sea Basin, too. The convergence takes place for potential temperature lower than 1.25°C, salinity higher than 34.65‰ and oxygen content higher than 3.4 ml/l.

The depth is different from place to place and is around 4000 m in the Philippine Trench.

(4) The origin of the water at 5000 m in the Philippine Sea Basin is probably the deep water below 4000 m to the east of the Izu-Ogasawara Ridge.

6 DEEP-WATER CIRCULATION

The deep-water circulation of large scale was inferred from the distribution of temperature and oxygen content (for example, Wooster and Volkmann, 1960; Knauss, 1962). This is natural because direct current measurements were very difficult. Such studies led to the conclusion that the deep water in the North Pacific came from the South Pacific, and no one doubts it at present. It is also shown in Stommel's model (Stommel and Arons, 1960), which was obtained from laboratory experiments and theoretical considerations. Though it has not been confirmed by direct current measurements yet, the features of the model are thought to be probable. Recent improvements of measurements make direct current measurements in the deep water possible and a few results are reported. But the data are not enough to derive the features. Besides, inertia motions were found (Nan'niti et al., 1964, 1965) and the removal of the motion is required for the purpose. Thus, it does not necessarily follow that direct current measurements during short period give us accurate informations on the deep circulation. The direct current measurements in the deep water in the North Pacific have just started and only a little knowledge has been obtained (Nan'niti and Akamatsu, 1966). Such a situation forces us to infer the circulation based on the distribution of some oceanographic elements, on the geostrophic relationships or on dynamical considerations.

Here, the second method is taken, though the geostrophic relationships are not strictly examined in the deep water. The choice of the reference surface is important in this method. Examination of the depth has been tried by various methods, but the values are dependent on authors and on the method (Masuzawa, 1962; Nitani and Imayoshi, 1963; Ichiye, 1960b). Geopotential topographic charts of the 3000 db, 4000 db and 5000 db surfaces referring to the 1200 db surface are illustrated in Fig. 15. The reference surface is taken as the 1200 db surface by Ichiye (1960b) in the computation of volume transports. There is no ground other than that noted by him. The charts referring to the 2000 db surface show approximately similar features to those of the charts referred to the 1200 db surface; some characteristic features are described below.

3000 db surface

A clockwise circulation is found along the boundary in the Philippine Sea Basin. Probably there is an eastward current in a narrow break of the Izu-Ogasawara Ridge around 30°N, which connects the basin with the Northwest Pacific Basin. This causes the mixing of the water between the two basins. To the east of the Izu-Ogasawara Ridge, there is a northward current from the south to 30°N. Though the contours do not indicate the northward current close to the ridge to the north of the latitude, the northward current is expected. To the north of 40°N, the current leaves the coast and turns to the east and then to the north again. A southward current is presumed along the northwestern margin of the basin.

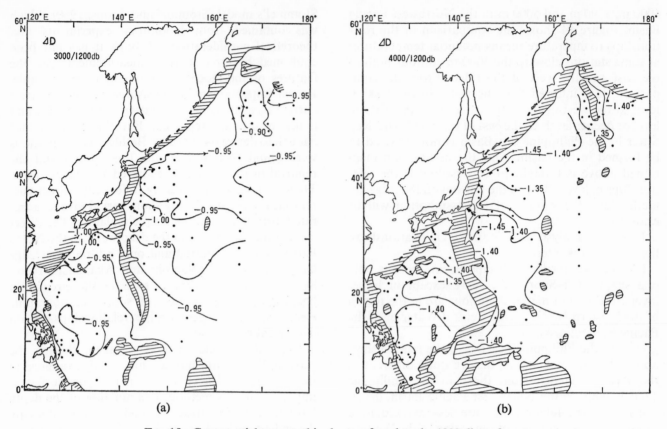

FIG. 15 Geopotential topographic charts referred to the 1200 db surface.
(a) 3000 db surface, (b) 4000 db surface, (c) 5000 db surface

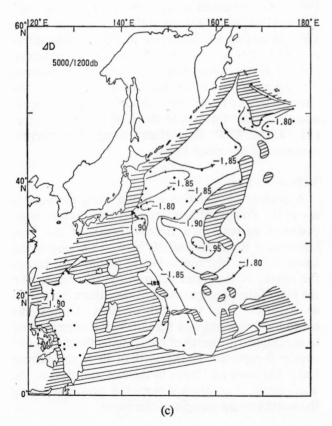

4000 db surface

The Philippine Sea Basin is completely separated from the Northwest Pacific Basin at this surface. A clockwise circulation is found in the northern part of the basin. The northward current to the east of the Izu-Ogasawara Ridge is far from the ridge as compared with the current at the 3000 db surface. The features are nearly the same as those at the 3000 db surface to the north of 30°N.

5000 db surface

The northward current to the east of the Izu-Ogasawara Ridge is close to the ridge as at the 3000 db surface. There is an anticlockwise circulation around the Northwest Pacific Rise. The feature in the northern part is approximately similar to those at the 4000 db surface.

The features noted above correspond to the distribution of temperature and of oxygen content in some areas and do not in other areas. It is not clear if the differences between the three surfaces are significant. But the following is interesting. At the 3000 db surface, the Philippine Sea Basin is open to other basins and the northward current close to the western margin of the basin may be regarded as the western boundary current expected by Stommel. Below the 4000 db surface, where the basin is completely isolated, the western boundary current makes its way to the east of the Izu-Ogasawara Ridge which is the eastern barrier of the basin. According to Stommel and Arons (1960), the northward western boundary current reaches $(30\pm10)°N$, and on our charts, it reaches 30–40°N. This is a good agreement. The existence of the southward current along the northwestern margin of the basin is not confirmed in our charts because of lack of data and should be examined in future. But it is probable in view of continuity. Though Stommel's model does not refer to the dimension of the flow pattern, the region of the anticlockwise circulation occupies a large area of the North Pacific. On the contrary, the circulation is restricted to west of around 170°E and has a much smaller scale on our charts.

In Nan'niti's model (Nan'niti and Akamatsu, 1966), there are remarkable differences in the flow pattern between the 2000–3000-m and the 4000–5000-m levels, while our charts do not show such definite differences. Kishindo (1940) expects the northward bottom water flow and the southward deep-water flow along Section K in Fig. 2. Nan'niti and Akamatsu support his expectation and their model is based on the distribution of temperature, salinity and oxygen content. They summarize the results of direct current measurements carried out up to 1965. The depth of measurements is mostly less than 3000 m and the comparison with our flow patterns is impossible.

As our charts are prepared for the area west of longitude 180°, the circulation in the whole North Pacific is not shown and the existence of a clockwise circulation along the arbitrary flow line (Fig. 2) suggested by Wooster and Volkmann (1960) is not examined here.

According to Ichiye's estimation (1960b) by the water budget method, the western boundary current is northward to the north of 10°N and this northward current vanishes to the north of 40°N. Thus, the northern limit of the northward western boundary current is in agreement with the limit in Stommel's model and on our charts.

Ichiye (1960b) expects the westward current over the Izu-Ogasawara Ridge. The source of the deep water in the Philippine Sea Basin has hardly been discussed and his suggestion is interesting. In the previous section, the deep water is compared with the deep water in the Northwest Pacific Basin by using $T-O_2$ relationships. This is made on the assumption that the deep water in the Northwest Pacific Basin makes its way westward over the Izu-Ogasawara Ridge in part, that is, on the basis of his expectation. It is significant to examine the existence of the westward current over the ridge.

It is believed that the water of the South Pacific enters the North Pacific and that the northward current exists there. Nevertheless, it is unknown where the current exists and how the circulation is constructed. Stommel's model is based on laboratory experiments and on theoretical considerations. Therefore, the model is considered to be a promising clue to the approach. Our charts are not finished, but show the gross features similar to those of Stommel's model. Though there are many ways to study the deep water circulation, it is most important to prove the existence of the western boundary current by direct current measurements in the present situation.

REFERENCES

Cochrane, J. D. (1958): The frequency distribution of water characteristics in the Pacific Ocean. *Deep-Sea Res.*, **5**: 111–127.

Fukuoka, J. (1962): Abyssal circulation in the Atlantic near the poles and abyssal circulation in the Pacific and other oceans in relation to the former. *Journ. Oceanogr. Soc. Japan*, **18**(1): 5–12.

Ichiye, T. (1960a): On water budget in a two-layered ocean. *Oceanogr. Mag.*, **11**(2): 111–126.

———— (1960b): On the deep water in the western North Pacific. *Oceanogr. Mag.*, **11**(2): 99–110.

———— (1962): On formation of the intermediate water in the northern Pacific Ocean. *Geofisica Pura E Applicata*, **51**: 108–119.

Kishindo, S. (1940): On the stratification of sea water and current in the Pacific (in Japanese). *Hydrogr. Bull.*, **19**: 351–362.

Knauss, J. A. (1962): On some aspects of the deep circulation in the Pacific. *Journ. Geophys. Res.*, **67**(10): 3943–3954.

Masuzawa, J. (1960): Western boundary currents and vertical motions in the subarctic North Pacific Ocean. *Journ. Oceanogr. Soc. Japan*, **16**(2): 69–73.

———— (1962): The deep water in the western boundary of the North Pacific. *Journ. Oceanogr. Soc. Japan, 20th Anniv. Vol.*, 279–285.

Nan'niti, T., H. Akamatsu and T. Nakai (1964): A further observation of a deep current in the East-North-East Sea of Torishima. *Oceanogr. Mag.*, **16**(1–2): 11–19.

———— (1965): A deep current measurement in the Honshu Nankai, the sea south of Honshu, Japan. *Oceanogr. Mag.*, **17**(1–2): 77–86.

Nan'niti, T. and H. Akamatsu (1966): Deep current

observations in the Pacific Ocean near the Japan Trench. *Journ. Oceanogr. Soc. Japan*, **22**(4): 154–160.

Nitani, H. and B. Imayoshi (1963): On the analysis of the deep sea observations in the Kurile-Kamchatka Trench. *Journ. Oceanogr. Soc. Japan*, **19**(2): 75–81.

Nitani, H. (1963): On the analysis of the deep sea in the region of the Kurile-Kamchatka, Japanese and Izu-Bonin Trench. *Journ. Oceanogr. Soc. Japan*, **19**(2): 82–92.

Robinson, M. K. (1960): Statistical evidence indicating no long-term climatic change in the deep waters of the North and South Pacific Oceans. *Journ. Geophys. Res.*, **65**(7): 2097–2116.

Stommel, H. (1958): The abyssal circulation. *Deep-Sea Res.*, **5**: 80–82.

Stommel, H. and A. B. Arons (1960): On the abyssal circulation of the world ocean—II. *Deep-Sea Res.*, **6**: 217–233.

Sverdrup, H. U., M. W. Johnson and R. H. Fleming (1942): *The oceans–their physics, chemistry and general biology*. Prentice-Hall, New York, 1087pp.

Wadati, K. and K. Terada (1962): Deep-sea research in Japan. *Journ. Oceanogr. Soc. Japan, 20th Anniv. Vol.*, 1–3.

Wooster, W. S. and G. H. Volkmann (1960): Indications of deep Pacific circulation from the distribution of properties at five kilometers. *Journ. Geophys. Res.*, **65**(4): 1239–1249.

Chapter 12

DIRECT CURRENT MEASUREMENTS

TOSHIHIKO TERAMOTO

Ocean Research Institute, University of Tokyo, Tokyo, Japan.

1 INTRODUCTION

Our present knowledge of the velocity field in the Kuroshio and adjacent regions is derived mainly from dynamic computations of hydrographic observations except for the surface, where additional information has been obtained from direct measurements of current, using towed electrodes and other methods. The direct measurement of current implies velocity determination through the detection of signals which are induced in direct association with the movement of water.

Observations and measurements made in the past indicated qualitatively that the velocity field varied irregularly in time and space. The irregularly varying velocity field is composed of harmonic constituents of a large variety of temporal and spatial frequencies. The temporal and spatial constituents are associated with local and inertial accelerations, which give rise to departure of the velocity field from geostrophy. Consequently, it is clear that dynamic computations of hydrographic observations cannot reveal correct spectral structures of the velocity-field variations. Direct current measurements made in the past also failed to reveal the correct structures, because there were no instruments reliable enough to give correct data on the actual velocity on the one hand, and because the measurements were not made with the definite purpose of exploring the spectral structures on the other. Thus, the structures of the oceanic velocity field have remained unclear. Clarification of these structures is essential to understanding the processes that generate and maintain the velocity field and its variations. Direct current measurements made in the past, though inadequate, may still show some aspects of the structures and mechanisms of the velocity field and its variations in the Kuroshio and adjacent regions. In this chapter a brief review of these direct current measurements will be made.

2 MEASUREMENT OF VELOCITY AND CURRENT PATTERN AT THE SURFACE

2.1 Crude tracing of surface current patterns by the use of drifting objects

Measurement with drift bottles

In the preliminary stages of current investigation, where synoptic features of surface current patterns are still totally unknown, an economical, effective method for a crude determination of the current path is the use of drift bottles or cards. Wada (1894a, 1894b, 1895) and his collaborator Kumata (1922) successfully used this method and first revealed a synoptic pattern in the surface current system in the region adjacent to Japan. The Hydrographic Office of the Japanese Navy adopted the method for a more extensive region starting in 1908 and showed a synoptic current pattern in the regions east and south of Japan (Konishi, 1921a, 1921b, 1922a, 1922b, 1923). The same method was recently adopted by Huzii and Kimura (1961) to examine the branching of the Kuroshio southwest of Kyushu into the main current and the Tsushima Warm Current. Recent measurements using this method have been chiefly conducted in inner bays or nearly enclosed seas, where a high percentage of bottle recovery could be expected (Kimura 1950; Nakamiya, 1953; Tamiya, 1960; Kawakami, 1957, 1959). Normally, however, information on the pick-up point in reference to the point of release is available for only a very low percentage of the bottles released, and Lagrangian tracking of actual drift paths cannot be carried out.

Determination by dead reckoning

Estimates of surface velocity from differences in ship position determined by astronomical observation and by dead reckoning were made by the Hydrographic Office in the early days of surveying. Starting in 1881 all merchant ships and warships collaborated in determining surface velocity distributions by providing data from their log books. This method of velocity determination is easy and economical, but due to the influence of wind drift it is not generally accurate. Another shortcoming is the lack of homogeneity in distribution of available data, which are concentrated along the regular merchant-ship routes. The measurements taken in the North Pacific from 1924 to 1943 have been summarized and published by the Hydrographic Office (Chart No. 6031 A-D).

Measurements with drift buoys

Measurement of surface velocity by means of drift-buoy tracking was used preferentially in oceanographic surveys by the Hydrographic Office of the Japanese Navy. In particular, current measurements with this method were made extensively in the sea south of Japan from 1928 to about 1938. In these measurements 20 to 30 buoys were released along a straight line at intervals of 1 to 2 miles and were tracked for 24 hours to explore the synoptic pattern of the current in the region. The simple construction of the buoys is illustrated in Fig. 1. The measurements revealed small eddies, divergences and convergences of the weak current in the subtropical region.

Measurement of horizontal divergence

Similar surface buoys were used recently by Kawai et al. (1969) and by Kawai and Sakamoto (1970a, 1970b) to measure the horizontal divergence of velocity field in the surface mixed layer of the strong current region of the Kuroshio south of Shikoku. At the vertices of a quadrilateral with sides about 1 mile in length four buoys were released. Temporal changes in the area of the quadrilateral were measured by the ship's radar for two sets of buoys at depths of 2 to 3 m and of 30 m, and by aerial photography for a set of buoys at the surface. The measurements showed, as illustrated in Fig. 2, that the strongest convergence occurred at the surface, an intermediate convergence at a depth of 2 to 3 m and a weak convergence or divergence at a depth of about 30 m. The importance of divergence in treating the distribution of suspended matter in the surface layer as well as diffusion was emphasized. Although the appropriateness of using quadrilateral scales for this particular study and

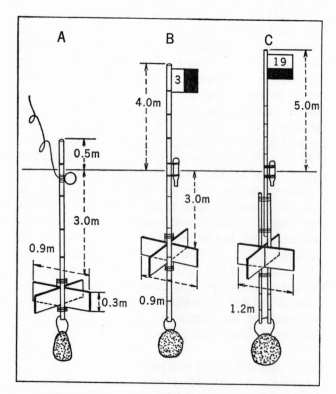

FIG. 1 Construction of three types of drift buoy used by the Hydrographic Office of the Japanese Navy for tracking surface currents.

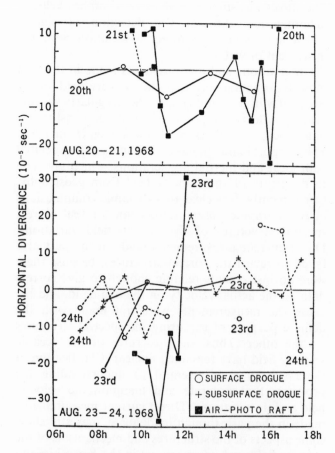

FIG. 2 Measurements of horizontal divergence in the surface layer by the use of a set of four buoys. (after Kawai)

the influence of internal waves upon the movement of the buoys should be examined further, the measurements are useful in suggesting one approach to making direct measurements of horizontal divergence.

2.2   Measurement of surface velocity by means of towed electrodes

Examination of measurements with towed electrodes
(1)   Comparison of surface velocities from measurements with towed electrodes to those from hydrographic observations and from direct current measurements

Soon after the method of measuring surface water velocities with electrodes towed from a moving ship was established by von Arx (1950), the method was introduced to Japan and has since become one of the most common means for making rough but rapid measurements during routine monitoring of ocean currents.

Two methods for determining velocity were widely used in the past: dynamic computation of hydrographic observations and direct measurement with two propeller-type current meters at two layers simultaneously. With the first method geostrophic, baroclinic velocities relative to the velocity at a temporarily specified reference level are obtainable. With the second method, actually existing velocities relative to the velocity at a temporarily specified depth for the reference current meter are obtained

irrespective of the hydrodynamic mechanisms and structures of the current. In order to examine the response characteristics of the new method of velocity determination to actually existing velocity fields of various mechanisms and structures, a comparison of velocities obtained by the new method with those determined by the two existing methods should be made for regions with different mechanisms and structures of current. In addition, further examination is necessary for regions of the sea bed composed of different materials, because surface electric-current densities, which are measured by the new method to give a direct indication of surface velocities, depend upon sea-bed conductivity. The influence of the sea-bed conductivity will be treated later.

Prior to routine use, Shoji (1955) examined the method with reference to hydrographic observations in the Kuroshio region south of Japan, the Oyashio region and the polar front region east of Japan, where water depths reach more than a few thousand meters. No comparison with velocities from measurements using current meters was made. The velocity component perpendicular to each hydrographic section was examined. The reference level selected for the dynamic computations was 800 db. As Shoji pointed out, surface velocities from measurements with towed electrodes accorded well qualitatively with those from hydrodynamic computations in any region (Figs. 3 (a), (b) and (c)).

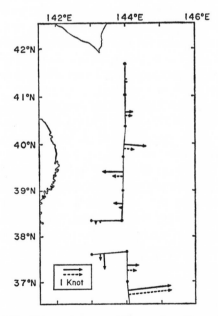

FIG. 3(a)   Dynamic computations of surface velocity from hydrographic observations and surface velocities determined by means of towed electrodes at the sections in the Kuroshio region south of Japan. (after Shoji)

FIG. 3(b)   Dynamic computations of surface velocity from hydrographic observations and surface velocities determined by means of towed electrodes at the sections in the Oyashio region east of Japan. (after Shoji)

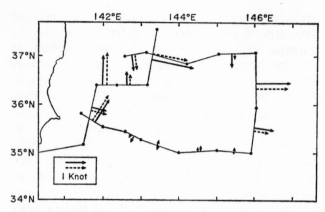

FIG. 3(c)   Dynamic computations of surface velocity from hydrographic observations and surface velocities determined by means of towed electrodes at the sections in the polar front region east of Japan. (after Shoji)

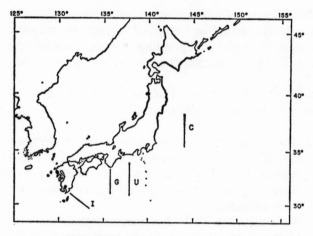

FIG. 4   Locations of hydrographic sections I, G, U and C. (after Nakai et al.)

The slight discrepancy between the velocities determined by these two methods was mainly attributed to the difference in response of the two methods to the dynamical structures and mechanisms of the current. That is, measurements with towed electrodes may include any baroclinic currents, irrespective of whether the currents are geostrophic or non-geostrophic, while hydrodynamic computations include only geostrophic, baroclinic currents but not non-geostrophic currents even if they are baroclinic. Shoji showed from the figures that for the Oyashio region surface velocities from measurements with towed electrodes were systematically smaller than those from hydrographic observations, whereas for the other regions the former velocities were usually a little larger. His points were interesting and of great practical importance, but there was too little data available to form a conclusive opinion.

Similar examinations were recently carried out by Nakai et al. (1972) using hydrographic observations conducted in parallel with measurements made by towed electrodes in the same regions in the past ten years (Fig. 4). As illustrated in Figs. 5 (a)–(d), cross-stream distributions of surface-velocity components determined by the two methods are qualitatively in good accordance. Surface velocities estimated from hydrodynamic computations represent the cross-stream components of geostrophic velocity averaged usually over at least 20 to 30 miles, while velocities measured with towed electrodes indicate roughly the mean velocities over an area less than 1 mile square. Hence, any comparison between these velocities should be made with respect to distributions of determined velocities on a scale far larger than that used for averaging in these methods.

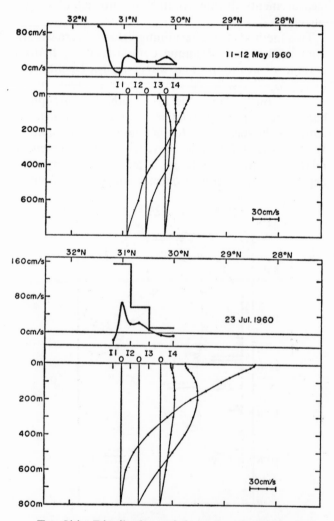

FIG. 5(a)   Distributions of dynamic computations of surface velocity from hydrographic section I and distributions of surface velocities determined by means of towed electrodes at that section. (after Nakai et al.)

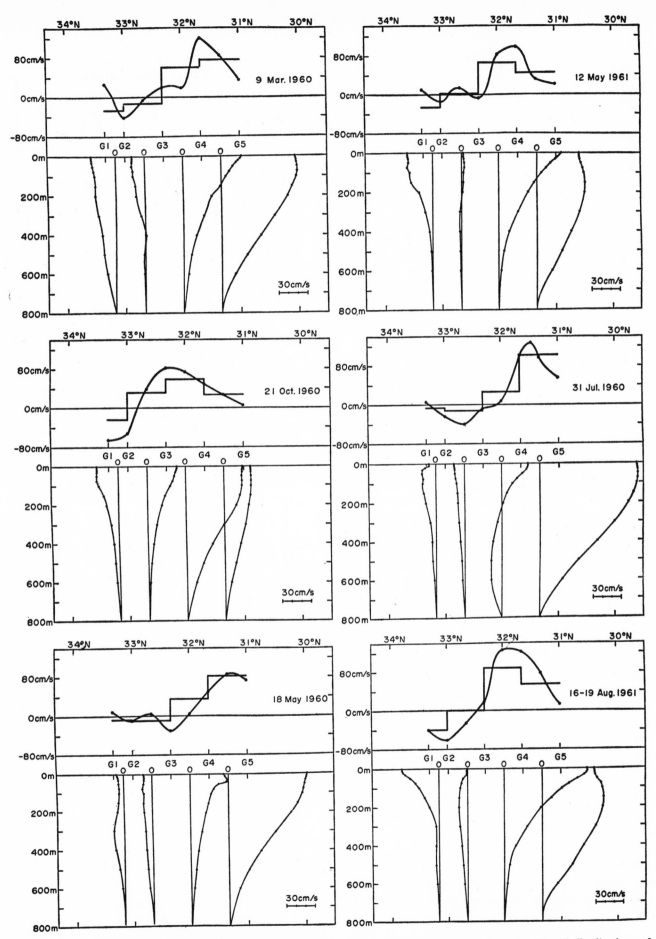

FIG. 5(b) Distributions of dynamic computations of surface velocity from hydrographic section G and distributions of surface velocities determined by means of towed electrodes at that section. (after Nakai et al.)

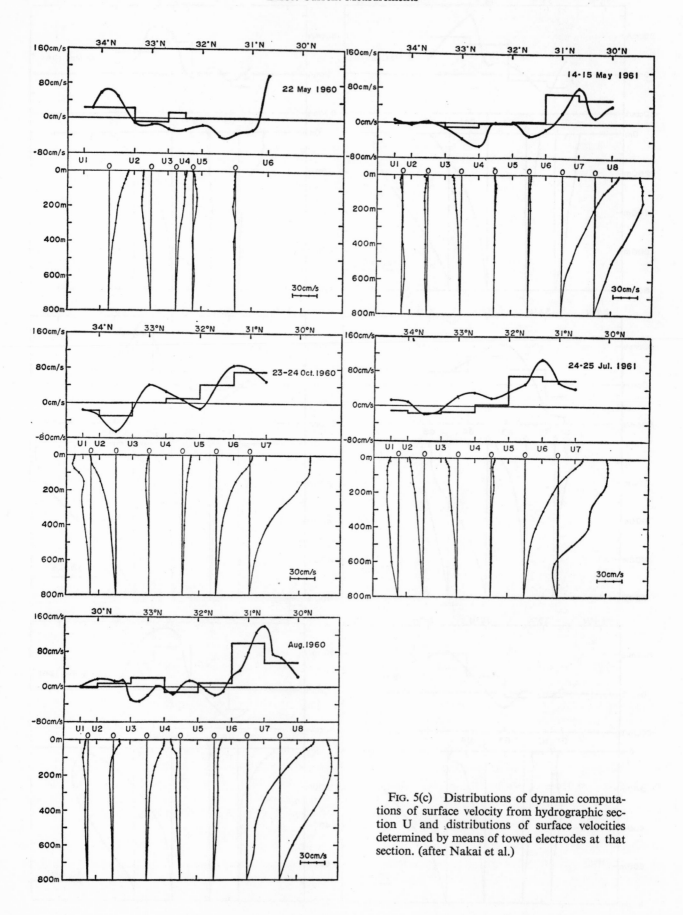

FIG. 5(c) Distributions of dynamic computations of surface velocity from hydrographic section U and distributions of surface velocities determined by means of towed electrodes at that section. (after Nakai et al.)

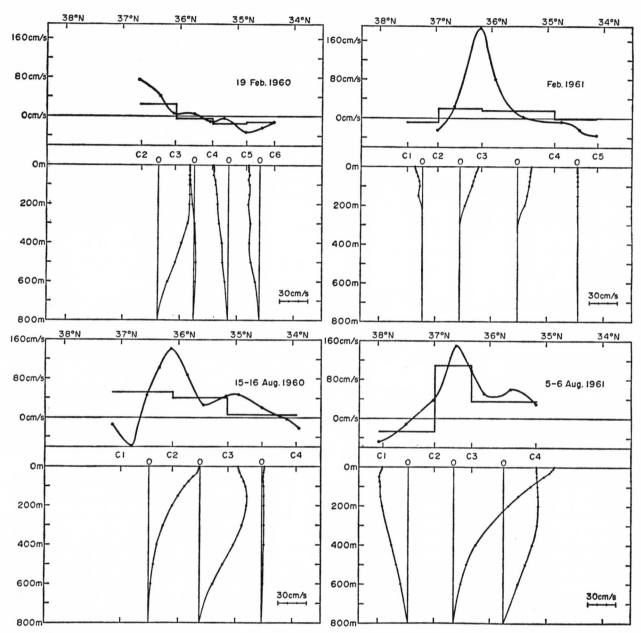

FIG. 5(d)   Distributions of dynamic computations of surface velocity from hydrographic section C and distributions of surface velocities determined by means of towed electrodes at that section. (after Nakai et al.)

Nakai et al. estimated ratios of averaged surface velocities by these two methods for the four sections. Results are shown in Fig. 6. Ratios for sections G and U, where water depths are about 3500 to 4000 m, are approximately unity as an average. However, ratio for section I, where water depth is about 2000 m, is a little less than unity as an average, and on the contrary ratio for section C, where water depth is about 5000 to 6000 m, is a little larger than unity as an average. For any section the cases in which ratios are greatly deviated from unity are found to correspond always to periods in which the Kuroshio made short-

term fluctuations in the vicinity of the section. This fact may indicate that the Kuroshio fluctuations were not in the geostrophic balance. From relations between the averaged velocity ratios and water depths for sections I, G and U, it can be concluded that for short-circuiting of an electromotive force induced by the Kuroshio water depth of 2000 m is not enough and that of 3500 to 4000 m is necessary.

With respect to surface velocities measured with towed electrodes, Ichiye tried to determine the depth of reference level used in hydrodynamic computations for the Kuroshio region. He estimated statistic-

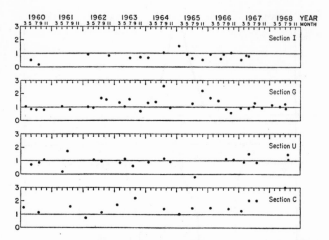

FIG. 6 Ratios of surface velocities measured with towed electrodes to surface velocities determined by dynamic computation of hydrographic observations for the four sections, I, G, U and C, respectively. The velocities used are those averaged over each of these sections.

ally that the level was 1000 to 1200 db. Due to insufficient data, however, his estimate cannot be confirmed. Further comparisons should be made between surface velocities determined by the two methods, taking velocity profiles into account.

Shoji also examined surface velocities measured with towed electrodes with reference to those measured with propeller-type current meters in the Straits of Tsugaru, where the maximum water depth is as little as 200 m. No comparison with hydrodynamic computations could be made because of the steep bottom topography in cross section of the very shallow Straits. Measurements using towed electrodes were carried out from a ship cruising along the perimeter of a rectangle, inside which fixed stations, E_1 and S_1, were specified for measurements with an Ekman current meter and a self-recording current meter, respectively (Fig. 7).

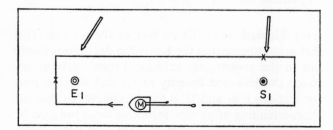

FIG. 7 Locations of stations E and S for current measurements with an Ekman current meter and a self-recording current meter, respectively from anchored ships, and the course of a ship from which a measurement with towed electrodes was made in the Straits of Tsugaru. (after Shoji)

Current meters were suspended from anchored ships at these fixed stations. The comparison was made with respect to velocity components in a direction perpendicular to the course of the cruising ship. Prior to the comparison, data processing was made as follows. Measurement readings were made once every hour with the Ekman current meter, so that velocities averaged over an hour were obtained. Measurements from a self-recording current meter were averaged over 20 minutes. Velocity measurement with towed electrodes takes several minutes and the determined velocity is supposed to be roughly the mean velocity for that period. Insofar as we treat a steady component of velocity and superimposed tidal components, which were both predominant in the records, the data processing would not have deformed the records seriously. Taking into consideration the fact that the straits are very shallow and especially that the tidal components would extend down near to the bottom without being remarkably reduced, the surface velocity determined from measurements with towed electrodes would depart considerably from those determined from measurements with current meters. But actually the surface velocity obtained by the first method is in good agreement with those by the latter two methods as shown in Figs. 8(a) and (b). The discrepancy between the obtained and the theoretical results will be discussed next.

(2) The influence of sea-bed conductivity upon measurements with towed electrodes

The role of sea-bed conductivity in the induction of a surface electric field, $(rj_x)_{z=0}$, which was induced by crossing the water flow with the geomagnetic field and which was measured by means of towed electrodes to give a surface velocity, $V_{z=0}$, was studied by Teramoto (1971) on the basis of a model of a uniform stream in a long, straight channel of rectangular cross-section. When the x-, y- and z-axes of the cartesian coordinates are taken cross-stream, downstream and vertically upwards in the channel, the electric field is estimated by the following equation:

$$(rj_x)_{z=0} \sim V_{z=0} \, H_z \left(1 - \frac{\bar{V}}{V_{z=0}} \cdot \frac{1}{1 + \frac{\bar{r}D}{R_e h}} \right), \quad (1)$$

where H_z is the vertical component of the geomagnetic field, $\bar{V}$ is the vertically averaged speed of water, $\bar{r}$ is the vertically averaged electric resistivity of water, R_e is the effective electric resistivity of the sea bed and h is the depth of the channel. The effective resistivity mentioned here is the sea bed resistivity averaged with depth from the surface to a depth comparable to the channel breadth, D. The quantity in parentheses

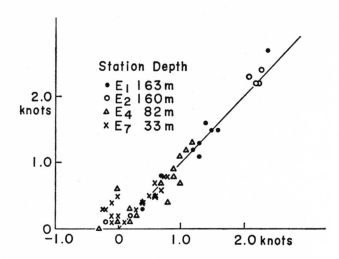

FIG. 8(a)   Relation of surface velocites measured with towed electrodes to surface velocities with an Ekman current meter. (after Shoji)

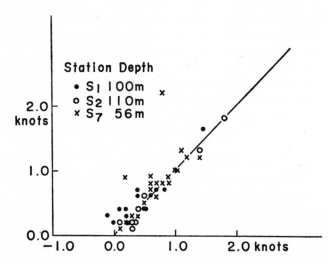

FIG. 8(b)   Relation of surface velocities measured with towed electrodes to surface velocities with a self-recording current meter. (after Shoji)

on the right-hand side of the equation is a reduction factor used in determining $V_{z=o}$ from $(rj_x)_{z=o}$.

Since R_e for the Kuroshio region is deduced to be approximately $4.5 \times 10^3 \Omega cm$, as was estimated for the Izu Island region, the reduction factor is estimated to be about $(1-0.7\ \bar{V}/V_{z=o})$ on taking $\bar{r}$ to be $2.5 \times 10^2 \Omega cm$, D to be 10^7 cm, and h to be 10^5 cm. Therefore, the reduction factor is $0.8{\sim}0.9$ provided $\bar{V}/V_{z=o}$ is $1/3.5{\sim}1/7$.

For the Straits of Tsugaru R_e is $1.6 \times 10^3 \Omega cm$, so that $1/(1+\bar{r}D/R_e h)$ is estimated to be about $1/10$ by taking $\bar{r}$ to be $2.6 \times 10^2 \Omega cm$, D to be 6.0×10^6 cm and

h to be 1.3×10^4 cm. Hence, the reduction factor is about 0.95 when $\bar{V}/V_{z=o}$ is $1/2$, which does not differ greatly from the actual data. This provides a reasonable solution to the question of the fairly good agreement between surface velocities measured with towed electrodes and with current meters in the shallow straits.

(3)   Examination of noise from electromagnetic induction associated with geomagnetic fluctuations

One more question remains with respect to velocity determination from measurements with towed electrodes, that is, contamination of a signal with noise from electromagnetic induction of an electric field associated with fluctuations of a geomagnetic field. Cox et al. (1964) treated the coherent relations between fluctuations of periods from 1 to 8 hours in the telluric electric and the geomagnetic fields of the sea. In regions far from shore, the theoretical ratio of the horizontal electric field to the horizontal magnetic field in a direction perpendicular to the former field was estimated to be about 0.043 mv/km/γ. But in regions not far from shore, the ratio was considered to be far larger owing to the shoreline effect. For the Izu Island region, for example, the ratio was estimated to be 0.2 to 0.3 mv/km/γ from measurements. When the fluctuation amplitude of the geomagnetic field is as much as 50γ, the amplitude of noise induced between towed electrodes reaches 0.6 to 0.9 mv, provided the spacing of the electrodes is 60 m. This magnitude of noise can be incorrectly interpreted as an ocean current of 0.6 to 0.9 knot. In determining the surface velocity with towed electrodes, a great departure from the velocity measured by other methods is sometimes encountered, particularly in regions not far from shore. This difference could be attributed to contamination due to the described noise.

Availability of the method

Although the validity of surface velocity measurements with towed electrodes has not yet been examined sufficiently in the Kuroshio and adjacent regions, the measurements are of practical significance because of their ease of application. Moreover, the fact that the measurements give absolute velocities, i. e. velocities relative to the earth, should be emphasized as a merit of the method.

Results from routine measurements by the method are useful in drawing a rough pattern of current at the surface. A figure showing the current pattern in regions around Japan has been published regularly by the Hydrographic Office every two months since 1960. The figures have been utilized, for example, by

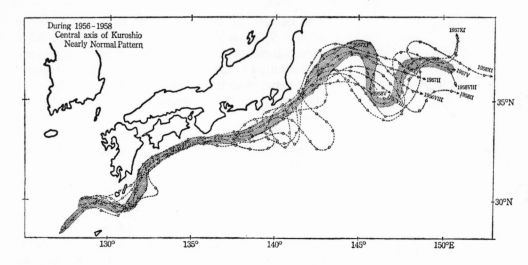

FIG. 9(a) Fluctuations of the Kuroshio path in the region south of Japan inferred from measurements with towed electrodes during 1956 to 1958. (after Uda)

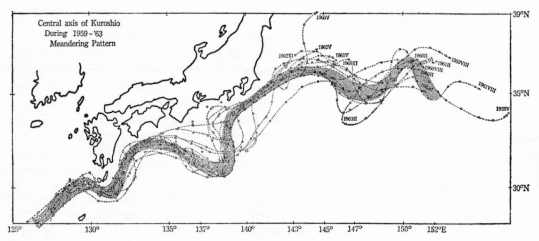

FIG. 9(b) Fluctuations of the Kuroshio path in the region south of Japan inferred from measurements with towed electrodes during 1959 to 1963. (after Uda)

Uda (1964) to make clear long-term fluctuations such as meanders of the Kuroshio (Figs. 9(a) and (b)). Masuzawa (1965) statistically treated seasonal variations in velocity of the Kuroshio off Enshunada on the basis of measurements with towed electrodes in the period from 1955 to 1964, and indicated that the current became stronger in summer and weaker in autumn.

Measurements with towed electrodes were used by Shoji and Nitani (1966, 1967) and by Masuzawa (1968) in special expeditions for examining short-term variations of the Kuroshio south of the Kii Peninsula. The results, which are illustrated in Figs. 10(a) and (b) and Fig. 11, show that diurnal or semi-diurnal variations prevailed in the time series of surface velocity components in a cross-stream direction. Results suggested that the density field varied in response to the variations in velocity field in such a way as to approach a geostrophic balance of force.

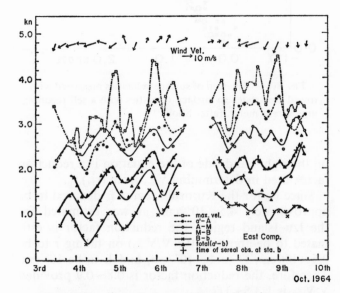

FIG. 10(a) Short-term variations of surface velocity measured with towed electrodes in the Kuroshio region south of the Kii Peninsula during 3 to 10 October 1964. (after Shoji and Nitani)

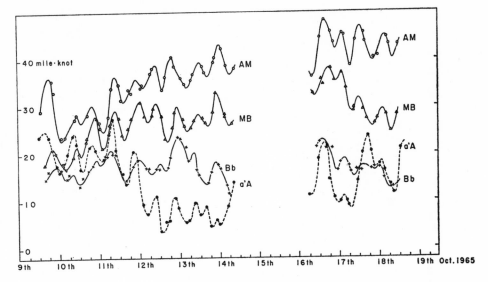

FIG. 10(b) Short-term variations in distance integral of surface velocity from measurements with towed electrodes in the region south of the Kii Peninsula during 9 to 19 October 1965. (after Shoji and Nitani)

3 MEASUREMENT OF CURRENT IN SUBSURFACE LAYERS

3.1 Measurement by Lagrangian methods

Measurement with parachute drogues

One Lagrangian method of current measurement was recently developed, in which a tracking was made by means of parachute drogues submerged to a definite layer. The velocity of subsurface water is determined by tracking a small surface buoy to which a parachute drogue is attached by a piano wire down to a desired depth. Consequently, the accuracy of velocity determination depends upon the accuracy of position determination. The method was introduced by Takenouti (1958) to observe subsurface water movement in the cold water belt along the northern boundary of the Kuroshio Extension. The cold water belt was a long belt or a succession of filaments of cold, low-salinity water lying in contact with the northern boundary of the Kuroshio Extension in the surface layer down to about 200 m. The belt was expected to play an important role in the formation of isolated, anticyclonic warm-eddies north of the Kuroshio Extension as well as in the mixing of water in the region. Four drogues, submerged at 50, 100, 250 and 600 m, respectively, were tracked by ship's radar. In the first measurement, carried out from 19 to 22 June 1957, three drogues at 50, 200 and 600 m were tracked for about 48 hours; in the second measurement, carried out from 25 to 26

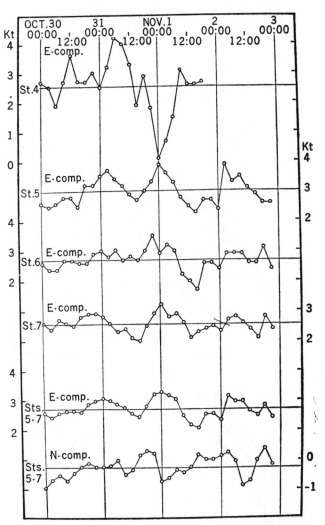

FIG. 11 Short-term variations of surface velocity measured with towed electrodes in the Kuroshio region south of the Kii Peninsula during 30 October to 2 November 1967. (after Masuzawa)

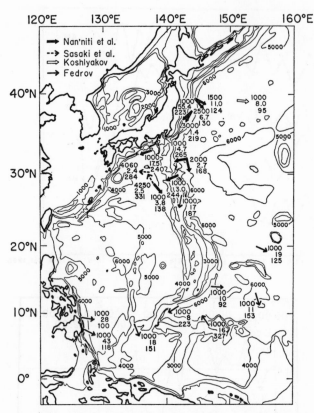

FIG. 12 Distribution of deep-current measurements in the adjacent region of Japan. (after Nan'niti and Akamatsu)

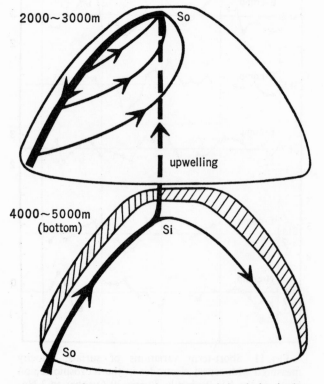

FIG. 13 Schematic picture of abyssal circulation in the North Pacific Ocean. (after Nan'niti and Akamatsu)

June 1957, three drogues at 50, 100 and 200 m were tracked for about 24 hours. From analysis of the drogue movement, especially of the drogue at 200 m, Takenouti inferred, on the basis of isotherm patterns in the region, that the cold water belt was formed by horizontal translation of the cold water mass of the Oyashio and that a mixing of this cold water with the Kuroshio water took place at the boundary of the Kuroshio.

Measurement with neutrally buoyant floats

Measurements of subsurface water movement with neutrally buoyant floats have been carried out mainly by Nan'niti and his collaborator at the Meteorological Agency since 1960 (Takenouti, 1961; Takenouti et al., 1962; Nan'niti et al., 1963, 1964a, 1964b, 1965, 1966). Except for the measurement made in the Japan Sea in 1966, all the measurements were made in the adjacent region of the Japan Trench, where the water depth exceeded 4000 m. The depths of submerged floats in the nine trackings were 1500, 2500, 1000, 3000, 990, 1000, 1000, 2000 and 1000 m, respectively and the periods of trackings were 19 hours, 21 hours, 13 hours, 60 hours, 21 hours, 85 hours, 74 hours, 78 hours and 64 hours, respectively. In most of the measurements, floats made irregular translations accompanied by circular motions of the inertia period at the latitude of measurement. For revealing a deep current, periods of measurement shorter than a day were inadequate and even periods of four to five days were not long enough considering the results of longer-period measurements in the Atlantic. Although the available data were insufficient, especially at layers deeper than 2000 m, Nan'niti and Akamatsu (1966) speculated on the deep circulation in the North Pacific (Fig. 12) on the basis of these measurements and those made by Fedrov (1960) and by Koshlyakov (1963), mainly at 1000 m, and by Sasaki et al. (1965) near the bottom (Fig. 13) as well as on Stommel's model (1958, 1960a, 1960b).

Measurements with floats were also made from the *Atlantis II* of the Woods Hole Oceanographic Institution in the Kuroshio region in 1965 to explore the subsurface current in the Kuroshio region.

Measurement of bottom current with sea-bed drifters

Using sea-bed drifters developed by Bumpus (1965), Kawai et al. (1970b) made measurements in the East China Sea. Sea-bed drifters, with a density slightly more than that of sea water at the bottom, are released at the desired positions. Retrieval of the drifters is made during fishing by trawl nets or dragnets. The synoptic features of the bottom cur-

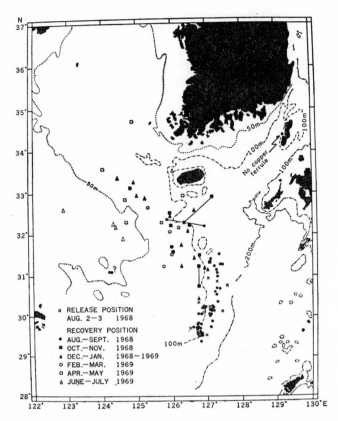

FIG. 14(a) Positions of release and recovery of seabed drifters. (after Kawai)

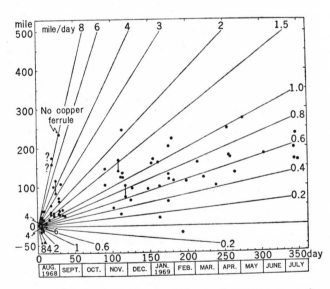

FIG. 14(b) Relation of drifters' displacements and periods of drift. (after Kawai)

rent are inferred from displacement of the drifters and from the periods of displacement. The principles for determining current features are essentially the same as for surface current determination by means of

drift bottles. The release by Kawai was made at nine points in a region southwest of Kyushu from 2 to 3 August 1968. The total number of drifters was 1788, 93 of which had been retrieved by the end of July 1969. The positions of release and pickup are shown in Fig. 14(a), and the relation of the drifters' displacements to periods of drift is shown in Fig. 14(b). The mean of the retrieved drifter's speeds, which would be slightly less than the speed of the bottom current due to friction between the drifters and the bottom, is therefore estimated to be between 1~3 cm/s in the region.

3.2 Measurement by Eulerian methods

Measurement with bottom current meters

Measurements of the bottom current made by photographing the displacement of pingpong balls suspended with a thin wire from a stable stand were carried out by Sasaki et al. (1965, 1967). The measurements were accurate enough for a synoptic survey of the bottom current, and efforts are being made to establish the technique for long-period recording.

Measurement with a moored array of current meters

Recently, Takano et al. initiated measurement of water velocity at subsurface layers using a moored array of Geodyne Current Meters (Hara and Takano, 1969; Hara et al., 1969). The meters, already widely used in various oceanic regions, have proved to be a good means of long-term recording of subsurface-water velocity. Besides measurements, Takano has maintained an interest in general problems encountered in making deep-sea moorings (Kubouchi et al., 1970).

4 MEASUREMENT OF VERTICAL DISTRIBUTION OF HORIZONTAL VELOCITY

4.1 Measurement by simultaneous use of two current meters

Measurements of velocity with two current meters from a drifting ship, one at the layer of measurement and the other at a specified reference layer, were initiated by Nishida et al. of the Korea Fisheries Experimental Station in 1931. The method was later used extensively by the Hydrographic Office of the Japanese Navy to obtain vertical profiles of horizontal velocity in the sea south of Japan. Numerous measurements were made, but they were conducted with-

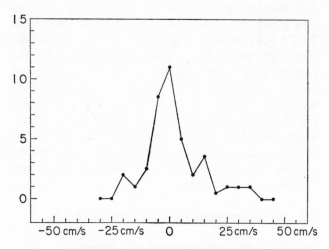

FIG. 15(a) Frequency distribution of differences in the N-S component of velocity at 10 m between successive measurements by the simultaneous use of two current meters in the region south of Japan.

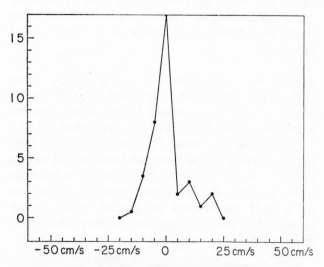

FIG. 15(b) Frequency distribution of differences in the E-W component of velocity at 10 m between successive measurements by the simultaneous use of two current meters in the region south of Japan.

out any specific purpose and so have not contributed greatly to the study of the velocity field in the region. In many instances the velocity at the 10 m layer was measured twice, perhaps to check whether the meter was operating correctly. A frequency distribution of the difference between these two measurements was plotted in Figs. 15(a) and (b). The standard deviations of the distributions with respect to N-S and E-W components of velocity were estimated as 9 cm/s and 7 cm/s, respectively, which are considered to give a measure of validity in interpreting the measurements. In order to examine the availability of the measure-

ments in surveying a synoptic current pattern, velocities measured at 10 m in the region south of Japan, mainly in the summer of 1938, 1940, 1941 and 1942, are summarized in Figs. 16(a)–(d), respectively, together with distributions of geopotential at the surface. Meanders of the Kuroshio path which were supposed to have existed during 1934 to 1943 (Yosida, 1961) are clearly indicated. Unfortunately, measurements for the deeper layers were rare and not even the synoptic conditions of the Kuroshio meanders and the subtropical eastward current in the deeper layer could be clarified.

4.2 Measurement with a current-profile recorder

Measurements of the vertical profile of horizontal velocity were carried out by Nan'niti (1959) using a current-profile recorder from the *Ryofu-Maru* in the region east of Japan (Fig. 17). The profile recorder was a combination of a propeller-type current meter and a bourdon-tube type depth-gauge. Vertical distributions of magnitudes of the water velocity are illustrated in Fig. 18 together with those of water temperature. Nan'niti pointed out that in the Kuroshio region east of Japan, the water speed was frequently as much as 1 knot even at a depth of 500 m. He also pointed out that in some cases the speed reached a maximum at a depth of about 200 m, while in other cases it varied discontinuously in a layer at 100 to 200 m where the temperature gradient became maximum. The orientation of the velocity vectors changed clockwise with depth in some cases and anticlockwise in other cases, as shown in Fig. 19. What he indicated is of interest; however, samples of the velocity profile which he presented are too few. In measurements of velocity profiles by the simultaneous use of two current meters mentioned earlier and with a sonic current meter to be mentioned later, such a regular spiraling in the distribution of the velocity vector is not generally found.

The current-profile recorder can be compared to the recording bathypitotmeter developed by Malkus (1953), which was used in measurements of the Gulf Stream. For reference, a vertical profile of velocity measured in the Gulf Stream by Malkus with this meter is illustrated in Fig. 20.

4.3 Measurement with a sonic current meter

Measurements of the vertical profile were made by Teramoto (1972b) using a sonic sing-around current-meter from the *Hakuho-Maru* in October 1967. Six stations were made across the Kuroshio at an interval

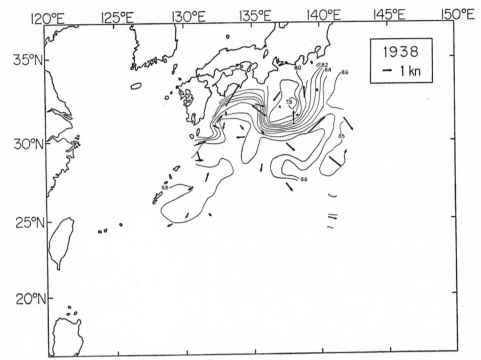

FIG. 16(a) Distribution of velocity vectors at 10 m from measurements by the simultaneous use of two current meters in the region south of Japan during 1938 and iso-geopotential lines at the sea surface in the same period. Numerical values in the figure are surface geopotential anomalies from 770 dm which are estimated in reference to 800 db layer.

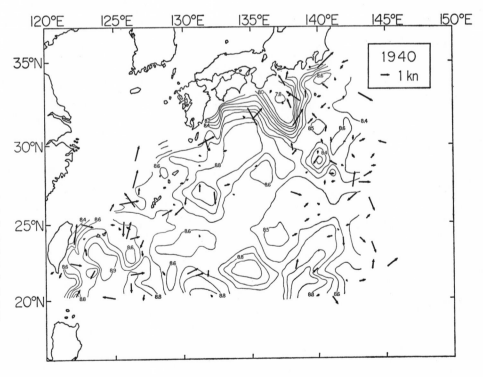

FIG. 16(b) Distribution of velocity vectors at 10 m from measurements by the simultaneous use of two current meters in the region south of Japan during 1940 and iso-geopotential lines at the sea surface in the same period. Numerical values in the figure are surface geopotential anomalies from 770 dm which are estimated in reference to 800 db layer.

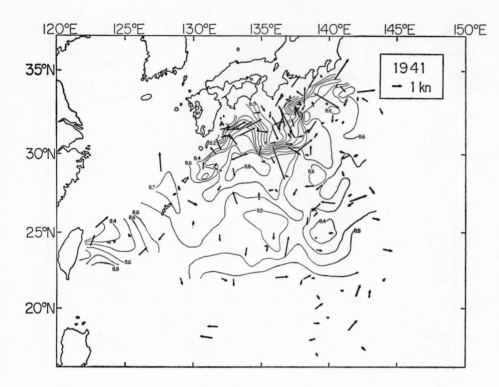

FIG. 16(c) Distribution of velocity vectors at 10 m from measurements by the simultaneous use of two current meters in the region south of Japan during 1941 and iso-geopotential lines at the sea surface in the same period. Numerical values in the figure are surface geopotential anomalies from 770 dm which are estimated in reference to 800 db layer.

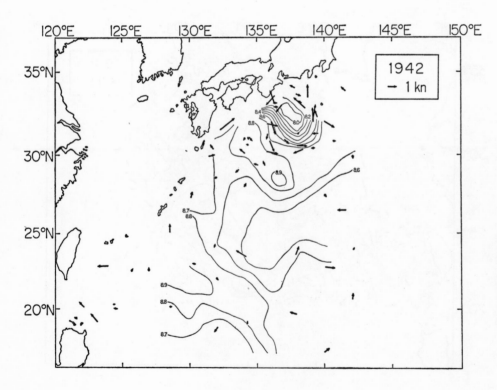

FIG. 16(d) Distribution of velocity vectors at 10 m from measurements by the simultaneous use of two current meters in the region south of Japan during 1942 and iso-geopotential lines at the sea surface in the same period. Numerical values in the figure are surface geopotential anomalies from 770 dm which are estimated in reference to 800 db layer.

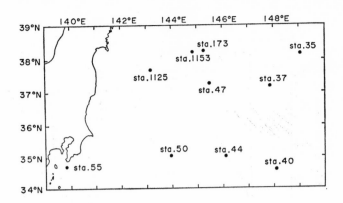

FIG. 17 Locations of stations for current measurements with current-profile recorder. (after Nan'niti and Iwamiya)

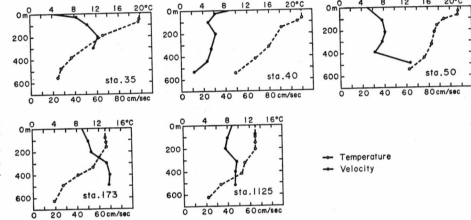

FIG. 18 Vertical distributions of magnitudes of water velocity measured with current-profile recorder along with the distributions of water temperature. (after Nan'niti and Iwamiya)

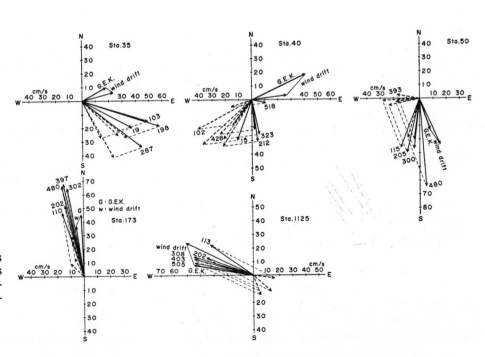

FIG. 19 Vertical distributions of orientations of velocity vectors measured with current-profile recorder. (after Nan'niti and Iwamiya)

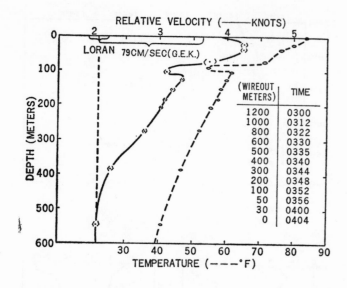

FIG. 20 Vertical profile of velocity measured with recording bathypitotmeter and of temperature at the northern edge of the Gulf Stream. (after Malkus)

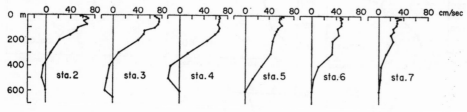

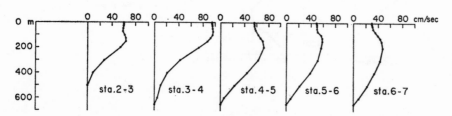

FIG. 21 Cross-stream distributions of time-averaged vertical profiles of velocity from measurements with a sonic current meter (apper panel) and from dynamic computations of hydrographic observations (lower panel) along 135°20′E.

of 10′ along 135°20′E once a day for five successive days. Velocity profiles relative to that at 600 m were averaged over the period and are illustrated in Fig. 21. Velocity profiles obtained from dynamic computation of hydrographic observations made at the stations just after every velocity measurement were averaged over the period and are illustrated in the same figure. The difference in profiles indicates a departure of the Kuroshio from geostrophy.

5 MEASUREMENT OF WATER TRANSPORT BY MEANS OF THE CROSS-STREAM ELECTRIC-POTENTIAL METHOD

The cross-stream electric-potential difference induced in sea water flowing across the geomagnetic field was shown to be closely associated with transport of the stream (Malkus and Stern, 1952; Longuet-Higgins et al. 1954). Continuous measurement of the

difference is not difficult, provided a submarine cable across the stream is available.

In regions of the Kuroshio and its branch currents, several submarine cables were laid across the currents: the telephone cable from Kawazuhama (in Shizuoka Prefecture) to Oshima and the cable from Oshima to Miyakejima through Niijima in the Izu Islands region, the cable from Chikumi (in Shimane Prefecture) to Dogoshima in the Oki Islands region, and the cable from Tobetsu (in Hokkaido) to Ishizaki (in Aomori Prefecture) in the Straits of Tsugaru (Fig. 22). A branch current of the Kuroshio is expected to flow across the first cable, a part of the main current of the Kuroshio across the second, the Tsushima Warm Current across the third and the Tsugaru Warm Current across the fourth. The Tsushima Warm Current is the branch current of the Kuroshio flowing into the Japan Sea through the Straits of Korea and of Tsushima and the Tsugaru Warm Current is its extension flowing out from the Japan

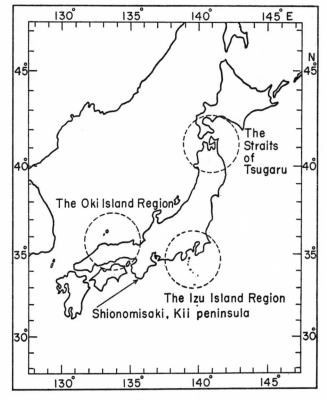

FIG. 22   Locations of cross-stream potential measurements in the adjacent regions of Japan.

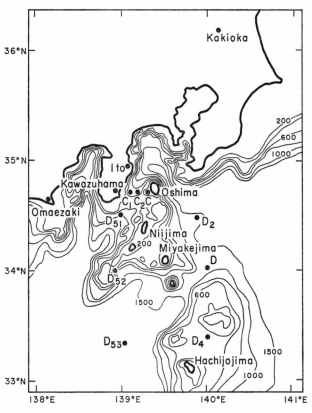

FIG. 23   Locations of stations for electric-potential measurement in the Izu Island region.

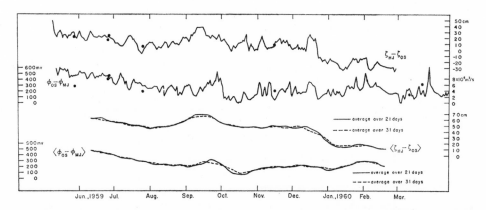

FIG. 24   Record of the time-averaged cross-stream electric-potential difference between Oshima and Miyakejima, $\langle \phi_{OS} - \phi_{MJ} \rangle$ (scale at lower left), of the time-averaged cross-stream sea-level difference between Hachijojima and Oshima, $\langle \zeta_{HJ} - \zeta_{OS} \rangle$ (scale at lower right), prefiltered record of the cross-stream electric-potential difference, $\phi_{OS} - \phi_{MJ} \equiv (\Delta\phi)_X$ (scale at upper left) together with the transport estimated from the geopotential difference between stas. D_2 and D_3, Q_{GP} (scale at intermediate right) and that of the cross-stream sea-level difference between Hachijojima and Oshima, $\zeta_{HJ} - \zeta_{OS} \equiv (\Delta\zeta)_X$, together with the cross-stream steric sea-level difference between D_2 and D_3, ΔD, (scale of both at the upper right).

Sea through the Straits of Tsugaru. With these cables measurements were carried out by Teramoto (1972a). Measurements with the second cable were also carried out by Nitani et al. (1959) and by Kubota and Iwasa (1961).

5.1 Relations between cross-stream differences in electric potential, geopotential and sea level at the Oshima-Miyakejima-Hachijojima section

Teramoto compared the real transport estimated from the cross-stream electric-potential difference between Oshima and Miyakejima, $\phi_{OS}-\phi_{MJ}$, with transports estimated from vertically integrated cross-stream geopotential-differences between station D_2 (or D_{51}) and station D_3 (or D_{52}). These stations were selected in the vicinity of Oshima and Miyakejima, respectively (Fig. 23), and were occupied seven times during the period of potential measurement (Fig. 24).

The comparison of a real sea-level difference between Hachijojima and Oshima, $\zeta_{HJ}-\zeta_{OS}$, to a steric sea-level difference between station D_4 (or D_{53}) and station D_2 (or D_{51}) was also carried out (Fig. 23).

The figure suggests that through vicinal averaging over 21 to 31 days the real transport estimated from $\phi_{OS}-\phi_{MJ}$ and real sea-level difference estimated from $\zeta_{HJ}-\zeta_{OS}$ approach the geostrophic transport and steric sea-level difference, respectively. Running averages of $\phi_{OS}-\phi_{MJ}$ and $\zeta_{HJ}-\zeta_{OS}$ over periods of 21 days and 31 days indicate close coherency between longer-term components of these quantities, as is also shown in Fig. 24.

5.2 Day-to-day to monthly variations of cross-stream electric-potential differences compared in reference to variations of physical quantities such as cross-stream and downstream differences in sea-level and atmospheric pressure

Teramoto made cross-spectral analysis of cross-stream electric-potential records together with records of the reference quantities for Oshima-Miyakejima section, Kawazuhama-Oshima section, Tobetsu-Ishizaki section and Dogoshima-Chikumi section. The records analysed are taken for 10 to 12 months in the period from June 1959 to May 1961

TABLE 1. Coherence Table for Oshima-Miyakejima section. Coherences between any two of data on cross-stream electric-potential difference and on reference quantities are plotted against frequency in a corresponding small frame of the table.

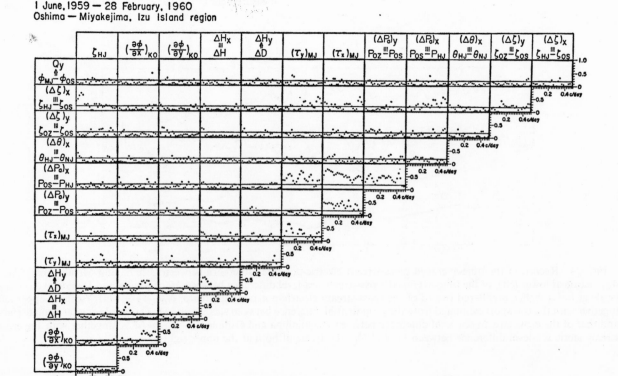

TABLE 2.   Coherence Table for Kawazuhama-Oshima section. Coherences between any two of data on cross-stream electric-potential difference and on reference quantities are plotted against frequency in a corresponding small frame of the table.

21 January, 1960 — 15 August, 1961
Kawazuhama — Oshima, Izu Island region

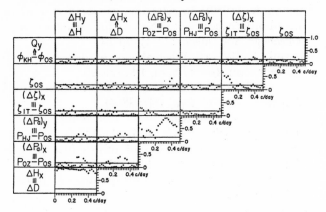

TABLE 3.   Coherence Table for Tobetsu-Ishizaki section Coherences between any two of data on cross-stream electric-potential difference and on reference quantities are plotted against frequency in a corresponding small frame of the table.

20 October, 1959 — 29 November, 1962
Tobetsu — Ishizaki, The Straits of Tsugaru

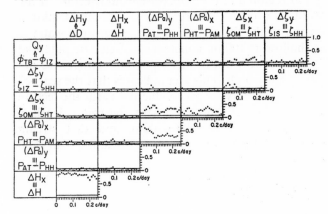

TABLE 4.   Coherence Table for Dogoshima-Chikumi section. Coherences between any two of data on cross-stream electric-potential difference and on reference quantities are plotted against frequency in a corresponding small frame of the table.

9 September, 1960 — 27 May, 1961
Dogoshima — Chikumi, Oki Island region

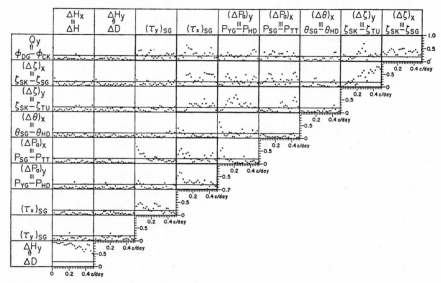

The coherencies were estimated at 50 frequencies in the range from 0 to 1 c/day or ½ c/day and are presented in Tables 1, 2, 3 and 4. As is supposed from the previously described fact that the induced electromotive forces are almost short-circuited through still deep-water and underlying sea bed, coherencies of the electric potential differences with any reference quantities are not high for Oshima-Miyakejima section and Kawazuhama-Oshima section in the Izu Island region and for Tobetsu-Ishizaki section in the Straits of Tsugaru. Whereas, for Dogoshima-Chikumi section in the Oki Island region, where the water is as shallow as 100 m, coherencies of electric potential difference with atmospheric pressure differences are significantly high. This fact indicates that the sea bed for this section is not so conductive as for the section in the Straits of Tsugaru where the water depth is also as shallow as 100~200m. and that stream fluctuation are mainly caused in association with wind variations in that region.

REFERENCES

Bumpus, D. F. (1965): Residual drift along the bottom on the continental shelf in the Middle Atlantic Bight area. *Limnol. Oceanogr.*, Suppl. to **10**: 50–53.

Cox, C. S., T. Teramoto and J. Filloux (1964): On coherent electric and magnetic fluctuations in the sea. *Studies on Oceanography. Hidaka Jubilee Committee.* University of Tokyo Press, 449–457.

Fedrov, K. N. (1960): On the evaluation of the deep-currents in the West Pacific Ocean (in Russian). Works of the Institute of Oceanology, Academy of Science. Vol. XL, 162–166.

Hara, H. and K. Takano (1969): Mesure du courant près du fond (in Japanese). *La mer*, **7**(1): 27–28.

Hara, H., S. Nagumo, H. Kobayashi, K. Takano and Y. Sugimori (1969): A measurement of bottom current on a shoulder of a sea mount. *Jour. Oceanogr. Soc. Japan*, **25**(6): 307–309.

Huzii, M. and M. Kimura (1961): Drifting of 20,000 current bottles, released in the sea south-west of Kyusyu, July, 1960 (in Japanese). *Hydrogr. Bull.*, New Ser. **67**: 58–62.

Kawai, H., H. Sakamoto and M. Momota (1969): A study on convergence and divergence in surface layer of the Kuroshio-I (in Japanese). *Bull. of the Nansei Regio. Fish. Res. Labor.*, **1**: 1–14.

Kawai, H. and H. Sakamoto (1970a): A study on convergence and divergence in surface layer of the Kuroshio II. *Bull. of the Nansei Regio. Fish. Res. Labor.*, **2**: 19–38.

—— (1970b): On bottom currents in the East China Sea measured with sea-bed drifters-I (in Japanese). *Bull. of the Nansei Regio. Fish. Res. Labor.*, **2**: 39–48.

Kawakami, K. (1957): The report of the drift-bottle experiments at the west entrance of Tsugaru Straits, Hokkaido, Japan (1). *Jour. Oceanogr. Soc. Japan*, **13**(4): 131–137.

—— (1959): The report of the drift-bottle experiments at the west entrance of Tsugaru Straits, Hokkaido, Japan (2). *Jour. Oceanogr. Soc. Japan*, **15**(1): 5–10.

Kimura, K. (1950): Investigation of ocean current by drift-bottle experiments (No. 1) (in Japanese). *Jour. Oceanogr. Soc. Japan*, **5**(2–4): 70–84.

Konishi, T. (1921a): Whereabouts of drift-bottles (I) (in Japanese). *Umi to Sora*, **1**(8): 86–90.

—— (1921b): Whereabouts of drift-bottles (II) (in Japanese). *Umi to Sora*, **1**(4): 38–41.

—— (1922a): Whereabouts of drift-bottles (IV) (in Japanese). *Hydrogr. Bull.*, **1**(3): 111–122.

—— (1922b): Whereabouts of drift-bottles (III) (in Japanese). *Hydrogr. Bull.*, **1**(2): 59–66.

—— (1923): Whereabouts of drift-bottles (V) (in Japanese). *Hydrogr. Bull.*, **2**(5): 31–41.

Koshlyakov, M. N. (1963): Vertical circulation of waters in the region of the Kuroshio. *Deep-Sea Res.*, **10**: 284–292.

Kubota, T. and K. Iwasa (1961): On the currents in Tugaru Strait (in Japanese). *Hydrogr. Bull.*, **65**: 19–26.

Kubouchi, Y., H. Shimamura and K. Takano (1970): A Simple Tensionmeter (in Japanese). *La mer.*, **8**(1): 13–17.

Kumata, T. (1922): Contributions from investigations of ocean currents around Japan during 1913 to 1917. *Osaka Mainichi Press.*

Longuet-Higgins, M., M. E. Stern and H. Stommel (1954): The electrical field induced by ocean current and waves, with applications to the method of towed electrodes. *Pap. Phys. Oceanogr. Meteor.*, **13**(1): 1–37.

Malkus, W. V. R. (1953): A recording bathypitotmeter. *Jour. Mar. Res.*, **12**(1): 51–59.

Malkus, W. V. R. and M. E. Stern (1952): Determination of ocean transports and velocities by electromagnetic effects. *Jour. Mar. Res.*, **11**(2): 97–105.

Masuzawa, J. (1965): Materials on seasonal variations of the Kuroshio velocity (in Japanese). *Jour. Oceanogr. Soc. Japan*, **21**(3): 23–24.

—— (1968): Cruise report on multi-ship study of short-term fluctuations of the Kuroshio in October to November 1967. *Oceanogr. Mag.*, **20**(1): 91–96.

Nakai, T., H. Otobe and K. Hasumoto (1972): Comparison between surface velocities from measurements with towed electrodes and from hydrographic observations (in preparation).

Nakamiya, T. (1953): On the ocean current in the Japan Sea inferred from results of drift-bottle experiments (in Japanese). *Hydrogr. Bull.*, Special **10**: 1–9.

Nan'niti, T. (1959): On the structure of ocean currents (III)—A new Nan'niti-Iwamiya current meter and the current profile observed by it—. *Papers in Meteorology and Geophysics*, **10**(2): 124–134.

Nan'niti, T., A. Watanabe, H. Akamatsu and T. Nakai (1963): Deep current measurements in the cold water mass and vicinity, the Enshu-Nada Sea. *Oceanogr. Mag.*, **14**(2): 135–139.

Nan'niti, T., H. Akamatsu, T. Nakai and K. Fujii (1964a): An observation of a deep current in the southern east sea of Torishima. *Oceanogr. Mag.*, **15**(2): 113–122.

Nan'niti, T., H. Akamatsu and T. Nakai (1964b): A further observation of a deep current in the East-North-East sea of Torishima. *Oceanogr. Mag.*, **16**(1–2): 11–19.

——— (1965): A deep current measurement in the Honshu Nankai, the sea south of Honshu, Japan. *Oceanogr. Mag.*, **17**(1–2): 77–86.

Nan'niti, T., H. Akamatsu and T. Yasuoka (1966): A deep current measurement in the Japan Sea. *Oceanogr. Mag.*, **18**(1–2): 63–71.

Nan'niti, T. and H. Akamatsu (1966): Deep current observations in the Pacific Ocean near the Japan Trench. *Jour. Oceanogr. Soc. Japan*, **22**(4): 154–160.

Nitani, H., K. Iwasa and W. Inada (1959): On the oceanic and tidal current observation in the channel by making use of induced electric potential (in Japanese). *Hydrogr. Bull.*, **61**: 14–24.

Nitani, H. and D. Shoji (1970): On the variability of the velocity of the Kuroshio-II. In *The Kuroshio—A Symposium on the Japan Current* (edited by J. C. Marr), East-West Center Press, Honolulu, 107–116.

Sasaki, T., S. Watanabe and G. Oshiba (1965): New current meters for great depths. *Deep-Sea Res.*, **12**: 815–824.

——— (1967): Current measurements on the bottom in the deep waters in the western Pacific. *Deep-Sea Res.*, **14**: 159–167.

Shoji, D. (1955): Measurement of the Currents in the Sea with Geomagnetic Electro-Kinetograph (in Japanese). *Hydrogr. Bull.*, Special **17**: 28–32.

Shoji, D. and H. Nitani (1966): On the variability of the velocity of the Kuroshio-I. *Jour. Oceanogr. Soc. Japan*, **22**(5): 10–14.

Stommel, H. (1958): The abyssal circulation. *Deep-Sea Res.*, **5**: 80–82.

Stommel, H. and A. B. Arons (1960a): On the abyssal circulation of the world ocean-I. *Deep-Sea Res.*, **6**: 140–154.

——— (1960b): On the abyssal circulation of the world ocean-II. *Deep-Sea. Res.*, **6**: 217–233.

Takenouti, Y. (1958): Measurements of subsurface current in the cold-belt along the northern boundary of Kuroshio. *Oceanogr. Mag.*, **10**(1): 13–17.

——— (1961): Deep-current measurements by means of the neutrally buoyant float (in Japanese). *Oki Review*, **28**(2).

Takenouti, Y., T. Nan'niti and M. Yasui (1962): The deep-current in the sea east of Japan. *Oceanogr. Mag.*, **13**(2): 89–101.

Tamiya, Y. (1955): Observation of the current in Japan Sea with drift bottles (in Japanese). *Hydrogr. Bull.*, Special **17**: 13–21.

Teramoto, T. (1971): Estimation of sea-bed conductivity and its influence upon velocity measurements with towed electrodes. *Jour. Oceanogr. Soc. Japan*, **27**(1): 7–19.

——— (1972a): Day-to-day to monthly variations in oceanic flows estimated from cross-stream differences in electric potential (in preparation).

——— (1972b): Vertical profiles of horizontal velocity in the Kuroshio region off the Kii Peninsula. Preliminary report of the *Hakuho Maru* Cruise KH- 67–4, Ocean Research Institute (in preparation).

Uda, M. (1964): On the nature of the Kuroshio, its origin and meanders. *Studies on Oceanography. Hidaka Jubilee Committee*. University of Tokyo Press, 89–107.

von Arx, S. (1950): An electromagnetic method for measuring the velocities of ocean currents from a ship underway. *Pap. Phys. Oceanogr. Meteor.*, **11**(3): 1–62.

Wada, Y. (1894a): A study on the Japan Current (in Japanese). *Tokyo Butsuri Gakko Zasshi*, **3**(33): 241–243.

——— (1894b): Investigation on the ocean currents in the east of Japan. Part 1 (in Japanese). *Suisan Chose Hokoku*, **2**(2): 1–28.

——— (1895): Investigation on the ocean currents in the east of Japan. Part 2 (in Japanese). *Suisan Chose Hokoku*, **3**(3): 1–34.

Watanabe, S., T. Sasaki and M. Okazaki (1970): Nouvel essai de la mesure de courant marin (in Japanese). *La mer*, **8**(1): 41–47.

Yosida, S. (1961): On the variation of Kuroshio and Cold Water Mass off Ensyū Nada (Part 1) (in Japanese). *Hydrogr. Bull.*, New Ser. **67**: 54–57.

Chapter 13

SOME ASPECTS OF THEORETICAL STUDIES ON THE KUROSHIO ——A REVIEW

KOZO YOSHIDA

Geophysical Institute, University of Tokyo, Tokyo, Japan.

1 KUROSHIO AS A WESTERN BOUNDARY CURRENT: DYNAMICAL EXPLANATION OF ITS EXISTENCE

There was no physical theory explaining the existence of the Kuroshio until Stommel (1948) first demonstrated dynamically that the latitudinal variation of the Coriolis parameter is responsible for the westward intensification of the ocean circulation. This discovery by Stommel led to our understanding that the Kuroshio is the counterpart of the Gulf Stream in the North Pacific. The Kuroshio and the Gulf Stream were thus interpreted essentially as the same dynamical phenomena; their mutual resemblance had for many years been recognized from observations (Bent, 1857; Wüst, 1936). Soon afterward, essentially employing Stommel's idea of the importance of the Coriolis parameter variation, Munk (1950) successfully depicted the Kuroshio for the first time in his theoretical model of the Pacific circulation. Of course, these theories were not complete, and during the following twenty years further investigations have been made in order to resolve such ambiguities as frictional forces, inertial effects, thermohaline circulation and vertical structures. However, those earliest achievements by Stommel and Munk are still of great value in many aspects such as in showing that local topographic conditions are essentially unimportant to the major features of the phenomena.

Development of nonlinear theories of western boundary currents was initiated by Charney (1955) and Morgan (1956). The theories of inertial (nonlinear) boundary currents have been developed further during these last fifteen years, but the general problem is essentially unsolved especially for the decay regions where the current leaves the coast to join the interior circulation. The papers by Fofonoff (1954), Carrier and Robinson (1962) and Moore (1963) have provided various stimulating results though the theories are still unsatisfactory in many aspects. More recently, Bryan and Cox (1967, 1968 a,b) carried out numerical experiments of the ocean circulation obtaining somewhat more realistic features of the Gulf Stream in its velocity and density fields. Although the developments in this direction have largely been focused on the Gulf Stream, they should be valid for the Kuroshio as well, as far as the gross features of the current are concerned.

Much less understood are physical processes responsible for the general feature of vertical distributions of temperatures and current velocities. One of the difficulties in solving the problem is that one should include nonlinear and diffusion terms in the heat equation even for the oceans far from the coasts. It is true that the model of the oceanic thermocline and the associated thermohaline circulation has been much advanced during these last fifteen years. However, these models do not apply to the boundary current regions which are of immediate interest to the dynamics of the Kuroshio.

2 DETAILS OF THE STRUCTURE AND TIME VARIATIONS: NEED FOR OBSERVATIONS

Further details of the structures of the Kuroshio and the Gulf Stream are not all similar to each other, although some of the features still present a dynamical resemblance. First of all, even the best observations have been inadequate for examining the details of time and spatial scales. Especially after World War II, various new observational techniques progressively have been devised and employed in exploring the Gulf Stream, but these methods have only gradually been introduced in the observations of the Kuroshio.

Thus recent observational studies of the Kuroshio have been very different from those of the Gulf Stream. Even Gulf Stream measurements have not resolved most of the detailed features of the fine structures and the time fluctuations. Various small-scale features have been observed in the Gulf Stream, such as meanders, eddies, multiple current structure, and time variations of various scales. There is no theory yet to account for these small-scale features, however.

Owing to an extreme difficulty in obtaining adequate programs of sampling, our knowledge of time variations and fine structures in the Kuroshio still remains exceedingly limited. A greater part of the descriptions reviewed in the present volume should not be considered as final. A great deal of future work will be required to establish their validity. Because of the limitations of the observations a theoretical analysis could hardly be developed to explain the observed features. First of all, present knowledge is very incomplete as to the time variable responses of the stratified ocean with complex spatial variabilities. Although there have been various attempts to account for those features of the Kuroshio (Ichiye, 1954a, b; 1955, 1957, 1958), these theories are still largely hypothetical and cannot be tested carefully by observations. As described in various chapters of the present volume, time variations over scales from hours to years have been indicated from observations. It is desirable that in the future theoretical models be constructed after more adequate measurements have been accumulated to reveal more realistic features of the current. Prior to 1950, there was no dynamical theory that could be used to study the time variable behavior of the Kuroshio. Veronis and Stommel (1956) developed a simple theory of the responses of a two-layer ocean to variable wind stresses. This model excludes, however, coastal boundaries and regions of strong currents, so that its immediate application to the western boundary currents like the Kuroshio is difficult. This problem remains to be solved.

One of the features of the Kuroshio, which varies over periods of many years and has been a subject of the most intensive study, is the meander of the current path to the south of Honshu between 136°E and 140°E. A survey is given below of the Japanese investigations of this phenomenon during the last forty years in the hope that it will stimulate a theoretical study of this controversial problem.

3   A SURVEY OF THE JAPANESE INVESTIGATIONS OF THE COLD WATER MASS SOUTH OF HONSHU

Although many small-scale features of the Kuroshio are likely to resemble those present in the Gulf Stream, there are some other features of the Kuroshio which appear to be entirely missing in the Gulf Stream. One such feature is the large semi-permanent cold eddy occurring off the south coast of Japan, associated with the large-scale meandering of the Kuroshio. This is one of the most outstanding phenomena of the Kuroshio which has no counterpart in the Gulf Stream. Also in the downstream regions of the Kuroshio after it leaves the Japanese coast, complex structures are observed in the form of meanders, eddies, fronts and other fine structures. These meanders, which show greater variability than those south of Japan, appear to be rather similar to those occurring in the Gulf Stream. Descriptions of these features are given by Kawai in this book (Chapter 8) and will not be considered here.

During these forty years, the large Cold Water Mass has occurred three times: 1935–1944, 1953–1955 and 1959–1963. Its remarkable persistency should be noted. Owing to the intriguing nature of this phenomenon itself, as well as to its practical importance for fisheries and climate, Japanese oceanographic surveys have largely been accelerated each time one of these large meanders appeared. Thus the observational effort has been intensified at three principal stages starting in 1936, 1954 and 1960, respectively. This can be seen in the illustration presented in this book by Stommel and Winterfeld (Chapter 3). Although a large number of Japanese oceanographers have proposed various hypotheses in their attempts to find a plausible physical explanation, it is fair to state that the dynamical mechanism of the processes is entirely unknown as yet. This notable phenomenon in the Kuroshio seems to have drawn surprisingly little attention from the Gulf Stream oceanographers and has not been a subject of any particular interest to foreign investigators.

3.1   A historical review

The first scientific documentation of the Cold Water Mass and the meander of the Kuroshio south of Honshu appeared in 1937 in a preliminary article written by Uda. Soon afterwards, during the period from 1938 to 1939, the first systematic observations were undertaken in this region jointly by the Fisheries Experimental Station and the Japanese Hydrographic Office. Detailed descriptions of the results of these surveys were published by Uda in 1940, and this early literature not only revealed a great number of outstanding features hitherto unknown but also pointed out many problems to be solved in the future. Somewhat systematic surveys which were initiated in 1936 immediately after this discovery were unfortunately interrupted by World War II. Following the end of the War, a period with no conspicuous meander extended for almost ten years, and little advance was made in the study of the meander, until the first post-war meander appeared in 1953, leading to a considerable increase in the research activities after 1954. Prior to 1950, there were practically no observations which were designed to reveal the details of the current in these regions. During the period from 1950 to 1953, a section was repeatedly taken by ocean weather ships of the Central Meteorological Observatiory (present Japan Meteorological Agency) at an approximate interval of three weeks (Masuzawa, 1954). During this survey period a remarkable change in the position of the current axis took place in July 1953. Fortunately in 1954 the BT and GEK became widely available for rapid and detailed surveys of the Kuroshio region, and in November of this year a more systematic survey program was initiated with the cooperation of the Japanese Hydrographic Office and the Japan Meteorological Agency. This observational program, with surveys regularly repeated four times a year, had continued till the end of 1959. The meander which was present from 1953 to 1955 was observed during this program. A summary description using the data from this six-year program was given by Masuzawa (1960, 1961); particular attention was given to the fluctuations in the path of the Kuroshio.

Since 1960, immediately after the second post-war meander reappeared, the Hydrographic Office has intensified its observational activities in the region south of Honshu by taking surveys as frequently as twice a month (Shoji, 1964). This Cold Water Mass, whose appearance was recognized during May 1959, decayed after July, 1962, during the period of this program. Following the finding of the cold eddy south of Shikoku during a cruise of the Kobe Marine Observatory in May, 1959, the behavior of the eddy was described from the results of a number of surveys conducted by various organizations including the Japan Meteorological Agency and the Japanese Hydrographic Office. On the basis of these results, Moriyasu (1961b; 1963) pointed out an eastward propagation of the meander pattern.

In particular, the detailed survey during June to July 1959 by the Japanese Hydrographic Office had originally been planned to detect short-term varia-

tions in this part of the Kuroshio and not specifically to trace the meandering pattern. This special survey was able to track successfully the variations in the path taking place over this earlier period. S. Yosida estimated the approximate speed of eastward travel of the meander pattern to be 4 miles per day (S. Yosida, 1961a). A transient feature during the decay period of the meander from 1960 to 1963 was summarized in a review paper by Shoji, in which he indicated that the Cold Water region moved also eastward but at a much lower speed (Shoji, personal communication).

Obviously, a study of such long-period fluctuations requires sufficiently long records of systematic observations. Until the early part of the 1950's, scientific information on this subject had by no means been sufficient.

It is not surprising that one might completely change his earlier views simply because the length of the historical record was increased, or a better coverage of reliable oceanographic observations over longer periods of years became available. In fact, there have been a number of Japanese references, especially among earlier ones, which regarded the Cold Water Mass as an anomalous or abnormal phenomenon, whereas recent investigators with the advantage of longer historical records tend to refer to it rather as stable, long-term variations. It may be worth noting at this point, that Uda suggested in his oldest document (Uda, 1940) a possibility of this phenomenon being one phase of long-period oscillation instead of an anomalous feature, and that he stressed the vital importance of long-continued, as well as systematic, observations for testing such a possibility. On the other hand, the same author has so frequently employed the terminology "abnormal" or "anomalous" in his subsequent papers that some confusing impressions may have been left with the readers. What is needed now are observations as well as dynamical models of the Cold Water Mass; some of these are suggested in various chapters of this volume. In this connection, it is of importance to realize that the observations should also be made in the period of no meander for fuller understanding of the processes involved. In reality, however, the distribution of oceanographic observation in Japan has been considerably influenced by this "abnormal" event, while the "normal" period has been comparatively ignored.

Most of the questions raised by Uda in his early papers (1940 and others) still remain unanswered. Although it has long been desired that we make a specific survey to observe as much detailed structure as possible of the meander at each stage of its life history, no satisfactory attempt has ever been made in this aspect. However, sea-level studies were developed under a long-range program of the Japanese Hydrographic Office, under the direction of Shoji and his colleagues (Shoji, 1951, 1954, 1957a,b; 1961, 1964), and they will prove to be useful if they are extended into the future. In 1964, a number of tide-gauge stations were newly established at various oceanic islands across the path of the Kuroshio. Future development of buoy technology should provide us with means of revealing further details of the phenomenon.

The lack of long-term investigation has been felt also in the field of descriptive work on the Kuroshio. In spite of the long history of our oceanographic explorations in this region, most of the papers have dealt only with fragmentary parts of those observational records. The recognition of these shortcomings in the past studies was indeed one of the essential motives for bringing together comprehensive reviews on the Kuroshio, and publishing them in this treatise.

3.2 General features hitherto revealed

As mentioned before, the large Cold Water Mass has occurred during three distinct periods over these forty years. The first one persisted over about ten years from 1935 to 1944, although the time of its appearance is somewhat uncertain owing to the lack of systematic observations during this early period; also the time of its disappearance is uncertain owing to the complete absence of observations during the War. The second occurrence took place during the period from 1953 to 1955. The third, the latest one, lasted from 1959 to 1963. In most cases, it was inferred that the meander formed in a period as short as a few months, while the decay period was much more prolonged. As has been suspected by many authors, it now appears quite certain that two different characteristic paths existed for the Kuroshio, both of which were stable over a period of at least 3 years (Taft, Chapter 6). This remarkable feature is convincingly shown in the marked "bimodal" distributions of the cold water centers (Stommel and Winterfeld, Chapter 3) and of the position of the Kuroshio axis (Taft, Chapter 6). Similar characteristic distributions were also shown by S. Yosida (1964).

Many of the features noted by Uda in 1940 have been substantiated by later investigations. The physical interpretation of the features in most cases is not known, however. General features requiring explanation may be summarized as follows:

(i) The large meander of the current and the large cold water mass are closely associated with each other.

(ii) Some of these phenomena persist quite stably over the period of 3 to 10 years.

(iii) Transient processes of generation and decay of the meander may take place rather abruptly. Decay of the meander appears to be a much slower process than generation.

(iv) There exist short-period transient meanders which appear to travel eastward.

(v) The persistent Cold Water Masses which appeared continuously during three different periods (1935–44, 1953–55 and 1959–63) over the last forty years, suggests that these phenomena are not to be considered simply abnormal.

(vi) The regions of the persistent Cold Water Masses are somewhat fixed. The locality is usually to the west of the Izu-Ogasawara Ridge, and this strongly suggests that bottom topography is important.

(vii) The Cold Water Mass is very likely maintained by upwelling and this upwelling water may originate from the intermediate layer.

(viii) Fluctuations in speeds of the Kuroshio and the development of the meander appear to be closely correlated.

(ix) Fluctuations in the current path seem to be related to large-scale atmospheric and oceanic circulations.

3.3 Topographic effects on the meander

As mentioned in item (vi) above, it seems reasonable to consider that certain topographic effects arising from the bottom topography and the coastal boundaries are responsible for the occurrence of this persistent meander of the Kuroshio path and the Cold Water Mass at certain fixed localities. Various investigators suggested the possible importance of such effects on the large bending of the current path or on the upwelling of cold waters, but none have developed quantitative dynamical models. Uda (1937a,b; 1940, 1949), Saito (1960) and Nan'niti (1958, 1959) speculated on possible influences of topographic features on upwelling. Hayami et al. (1955) and Hayami (1957) suggested a possible role of changes in water depth on the deflection of the current; the current is assumed to extend to the very bottom. The earlier thoughts had been based on nothing more than Ekman's classical model for a very wide current. Although Fukuoka (1957, 1958a, b,c; 1960a,b) and Ichiye (1960) gave some different views on the dynamics of this aspect of the flow, their discussions did not go beyond applying to the problem a very simplified barotropic model (Bolin, 1950) of the effect of a mountain on a very narrow atmospheric current. They suggested that the deflection of the stream path would be greater in inverse proportion to the speed of jet. The comparison of these computations with the actual behavior of the current was not justified because so far we are completely ignorant of the deep motions beneath the Kuroshio. In connection with the study of the mechanism of the meander, it is indeed crucial to gain some reliable knowledge about the flow deep below the high speed regions of the Kuroshio. As described elsewhere in the present volume (Teramoto, Moriyasu, Taft), there are no direct measurements of velocity beneath the surface Kuroshio Current. Although some direct measurements of deep currents have been made in this region since 1959, the results are not yet very useful for the present problem.

Warren's work in 1963 was the first successful attempt to account quantitatively for the meanders of the Gulf Stream. The validity of this theory was essentially based on the deep measurements on Gulf Stream '60 (Fuglister, 1963). There are no deep data in the Kuroshio which are comparable with those obtained in the Gulf Stream region.

"The path equation" for the Kuroshio meanders east of Honshu has been discussed by Ichiye (1957). He followed Rossby's work on stationary Rossby waves which excludes bottom topographic effects (Rossby, 1940). Warren's more generalized model including topographic waves seems to be more applicable to the Kuroshio meander. Robinson and Niiler (1967) and Niiler and Robinson (1967) developed Warren's theory further to construct a more refined model of free inertial currents. Although the large Cold Water Mass associated with the persistent meander to the south of Honshu appeared during the period between 1959 and 1962, no Japanese workers have developed a dynamical model of the meander. Most of the literature published since 1963 is restricted to review papers (Moriyasu, 1963; Uda, 1964; Shoji, 1964; Masuzawa, 1965; Ichiye, 1965).

Despite a large number of reports and papers dealing with the subject, previous descriptions of the current path of the Kuroshio have been rather incomplete. For instance, descriptions of the shifts in the path in relation to the underlying bottom topography have been insufficient. The first detailed description of this subject has only recently been made available by Taft as presented in this treatise. On the basis of nine years of data, including the periods of presence and absence of the meander, Taft pre-

sents a full analysis. This work may possibly lead to the construction of the first theoretical model of topography-controlled meanders in the Kuroshio.*

3.4 Upwelling

It has quite generally been believed that the Cold Water Mass in this southern region is associated with upwelling of cold water from the intermediate layer (Shoji, 1951; Shoji and Suda, 1956; Uda, 1937a,b; 1940, 1949; Saito, 1960; Ichiye, 1955; Koenuma, 1937; Moriyasu, 1955, 1961a,b; and Masuzawa, 1965). Uda further asserts that the upwelling cold water originates in the Oyashio region. However, this is not well documented. Although many authors have indicated the indirect evidences of upwelling based on the water mass analysis for this region, more careful analysis with better data is needed to derive a result. If upwelling is really the mechanism for the Cold Water Mass, such a grand upwelling of a persistent nature would indeed be a remarkable phenomenon. No satisfactory theoretical work has been attempted on this kind of upwelling (Hikosaka, 1954; Ichiye, 1954a; 1955), which possibly differs from that in other oceanic regions. A construction of a dynamical model is desirable for this marked upwelling which may be associated with the topographic steering of a strong current.

3.5 Possible relations to fluctuations in the large-scale circulation

The correlation between development of the meander and changes in speed of the Kuroshio has been pointed out by various authors. In particular, Shoji (1964) finds a fairly good correlation between the weakening of the current and the growth of the meander. Taft indicates a similar correlation based on a more detailed analysis (Taft, chapter 6). As Taft remarks, this important result could be incorporated with topographic effects (Robinson and Niiler, 1967) and also with a more recent finding of Namias (1970) of the positive correlation between sea-surface temperature in the subtropics in the western Pacific and the maximum speed of the Kuroshio south of Honshu. A close relationship between the meander and the oceanic and atmospheric circulations in the western Pacific is thus suggested. In particular, the systematic study of the whole Subtropical Gyre may be fruitful.

---

* Recently Robinson and Taft (personal communication) applied the theory of Robinson and Niiler (1967) to the dynamics of the Kuroshio, and provided a possible explanation for the bimodal behavior of the Kuroshio.

4 THE SOUTHERN HALF OF THE SUB-TROPICAL GYRE

The recent result obtained by Namias, as mentioned above, suggests the importance of thorough examination of the entire Subtropical Gyre, particularly its southern half, in relation to the long-period variations of the Kuroshio. This has been realized for many years but has not gone beyond mere speculation. In closing the present article, I would like to give a brief summary of a feature newly discovered in this southern Subtropical Gyre, which might eventually be linked to the variations in the path of the Kuroshio.

The large area in the western Pacific between latitudes 30°N and 20°N had remained unsurveyed to a great extent before the CSK Program (The Co-operative Study of the Kuroshio and its Adjacent Regions) began in the summer of 1965. More or less systematic surveys of this area repeated during the CSK period between 1965 and 1968 have provided useful data to examine general features of this oceanic area. The preliminary analysis of the data has already revealed a new feature in the southern half of the Subtropical Gyre. The feature is described in the papers by Yoshida and Kidokoro (1967a,b; 1968, 1969, 1970) and by Uda and Hasunuma (1969). The Subtropical Countercurrent, flowing eastward between 20°N and 25°N in the west and somewhat further north in the mid-Pacific, had first been suggested by the computations of Sverdrup transport from the statistical data of seasonal winds (Yoshida and Kidokoro, 1967a). The computations of seasonal transport charts revealed pictures which were similar to the annual average charts produced by various authors before, except for one remarkable departure— the eastward transport at these lower subtropical latitudes during winter and spring. Computations with annual mean winds had not shown this transport. The average wind stresses during these seasons are quite different from those during summer and fall. They indicate a minimum of negative wind stress curl (a trough zone) very close to the boundary between the Westerlies and the Trades. This trough zone is an apparent consequence of the nonlinear form of the wind stress function, which strongly diminishes wind stress values at low values of wind speeds. Transport computations from monthly average wind stresses for individually years made earlier by Fofonoff et al. (1963) did reveal more distinct eastward transports well separated from the main portion of the Kuroshio. The temperature distributions at 50 m to 200 m levels do show a zone of marked concentration in isotherms along 20°N to 25°N. Some previously published charts have been found to show similar eastward flows near these latitudes.

The large amount of data from the CSK cruises have been analyzed to test this earlier hypothesis. Many of CSK observations of standard type have been taken mostly with USSR ships. Although the area of survey has been limited to west of 155°E, convincing evidence of the existence of a quite stable eastward flow is obtained between 20°N and 24°N. However, the flows to the north of 24°N are by no means simply westward, but appear rather complicated with a few eastward flowing bands. These preliminary indications are not yet considered to be confirmative, but it has become at least certain that the circulation in the southern half of the Gyre is much more complex than was believed before. More systematic measurements are needed to examine the detailed structure of the currents, probably requiring more careful re-examination of the entire Subtropical Gyre. Theoretical models should perhaps be reconsidered; perhaps the explanation of the basic features may not necessarily require any singularity in the wind stress distribution. Our present knowledge of wind stress values is very inadequate. It would appear to be desirable to recompute the stresses so that critical tests of theoretical models may be carried out.

Acknowledgments

I am indebted to Prof. B. A. Taft for his valuable comments and advices. Thanks are also due to Dr. Jiro Fukuoka for his various suggestions and assistance.

REFERENCES

Bent, S. (1857): The Japanese Gulf Stream. *Bull. Amer. Geog. Statis. Soc.*, **2**: 203–213.

Bolin, Bert (1950): On the influence of the earth's orography on the general character of the westerlies. *Tellus*, **2**(3): 184–195.

Bryan, Kirk and M. D. Cox (1967): A numerical investigation of the oceanic general circulation. *Tellus*, **19**(1): 54–80.

———— (1968a): A nonlinear model of an ocean driven by wind and differential heating; Part I. Description of the three-dimensional velocity and density fields. *J. Atmos. Sci.*, **25**(6): 945–967.

———— (1968b): A nonlinear model of an ocean driven by wind and differential heating; Part II. An analysis of the heat, vorticity and energy balance. *J. Atmos. Sci.*, **25**(6): 968–978.

Carrier, G. F. and A. R. Robinson (1962): On the

theory of the wind-driven ocean circulation. *J. Fluid Mech.*, **12**(1): 49–80.

Charney, J. G. (1955): The Gulf Stream as an inertial boundary layer. *Proc. Nat'l. Acad. Sci. Wash.*, **41**: 731–740.

Fofonoff, N. P. (1954): Steady flow in a frictionless homogeneous ocean. *J. Mar Res.*, **13**(3): 254–262.

Fofonoff, N. P. and F. W. Dobson (1963): Transport computations for the North Pacific Ocean, 1950–1959. *Fish. Res. Bd. Canada, Ms. Reports, Oceanogr. and Limnol. Series*, No. **162**: 150pp.

Fuglister, F. C. (1963): Gulf Stream '60. In *Progr. Oceanogr.*, M. Sears (ed.): 265–373.

Fukuoka, Jiro (1957): On the variation of oceanic conditions in the sea adjacent to Japan. *Oceanogr. Mag.*, **9**(1): 95–106.

—— (1958a): On the Kuroshio near the Izu Islands. *J. Oceanogr. Soc. Japan*, **14**(1): 11–14.

—— (1958b): The variations of the Kuroshio current in the sea south and east of Honshu (Japan Main Island) *Oceanogr. Mag.*, **10**(2): 201–213.

—— (1958c): On the variation of the Kuroshio current south of Honshu (Japanese Main Island). *Geophys. Mag.*, **28**(3): 343–355.

—— (1960a): An analysis on the mechanism of the cold water mass appearance in the Enshu-nada. *Oceanogr. Mag.*, **11**(2): 127–143.

—— (1960b): An analysis of the mechanism of cold and warm water masses in the seas adjacent to Japan. (Part 1). *Umi to Sora*, **36**(1): 1–12.

Hayami, Shoitiro (1957): On the dynamics of Kuroshio off the southern coast of Japan. (abstract). *Proc. UNESCO Symp. Phys. Oceanogr.*: 139.

Hayami, Shoitiro, Hideo Kawai and Masao Ouchi (1955): On the theorem of Helland-Hansen and Ekman and some of its applications. *Rec. Oceanogr. Wks. Japan*, New Ser., **2**(2): 56–67.

Hikosaka, Shigeo (1954): Movement of sea water under a circular wind system (1) *Hydrogr. Bull.*, Spec. No. **15**: 20–23.

Ichiye, Takashi (1954a): On the variation of oceanic circulation (VI). *Geophys. Mag.*, **25**(3–4): 185–217.

—— (1954b): On the variation of oceanic circulation (VII). *Oceanogr. Mag.*, **6**(1): 1–14.

—— (1955): On the variation of oceanic circulation (V). *Geophys. Mag.*, **26**(4): 283–342.

—— (1957): On the variation of oceanic circulation in the adjacent seas of Japan. *Proc. UNESCO Symp. Phys. Oceanogr.*: 116–129.

—— (1958): On the decay of a cyclonic vortex in the southern sea of Honshu (Hydrography of the Kuroshio in relation to its productivity, II). *Rec. Oceanogr. Wks. Japan*, New Ser., Spe. No. **2**: 90–100.

—— (1960): On the deep water in the western North Pacific. *Oceanogr. Mag.*, **11**(2): 99–110.

—— (1965): The Kuroshio System. *Oceanus*, **12**(1): 18–21.

Koenuma, Kan'ichi (1938): On the recent variations in the Kuroshio. *Umi to Sora*, **18**(5): 185–187.

Masuzawa, Jotaro (1954): On the Kuroshio south off Shiono-Misaki of Japan (Currents and water masses of the Kuroshio System I). *Oceanogr. Mag.*, **6**(1): 25–33.

—— (1960): Statistical characteristics of the Kuroshio Current. *Oceanogr. Mag.*, **12**(1): 7–15.

—— (1961): Recent research on the Kuroshio Current. *Geographical studies presented to Prof. Tsujimura in honour of his 70th birthday*: 354–364.

—— (1965): Meanders of the Kuroshio—The cold water mass south of Honshu—. *Kagaku*, **35**(11): 588–593.

Moore, D. W. (1963): Rossby waves in oceanic circulation. *Deep-Sea Res.*, **10**: 735–748.

Morgan, G. W. (1956): On the wind-driven ocean circulation. *Tellus*, **8**(3): 301–320.

Moriyasu, Shigeo (1955): An example of the variation of the oceanographical condition of Enshu-nada. *J. Oceanogr. Soc. Japan*, **11**(2): 43–46.

—— (1961a): On the influence of the monsoon on the oceanographic conditions. *J. Oceanogr. Soc. Japan*, **17**(2): 74–79.

—— (1961b): An example of the conditions at the occurrence of the cold water region. *Oceanogr. Mag.*, **12**(2): 67–76.

—— (1963): The fluctuation of hydrographic conditions in the sea south of Honshū, Japan (Review). *Oceanogr. Mag.*, **15**(1): 11–29.

Munk, W. H. (1950): On the wind-driven ocean circulation. *J. Meteor.*, **7**: 79–93.

Namias, Jerome (1970): Macroscale variations in sea-surface temperatures in the North Pacific. *J. Geophys. Res.*, **75**(3): 565–582.

Nan'niti, Tosio (1958): A theory of the mechanism of the generation of the cold water region in the offing of Enshūnada. *Pap. Met. Geophys.*, **8**(4): 317–331.

—— (1959): A supplementary note to the previous paper "A theory of the mechanism of the generation of the cold water region in the offing of Enshunada." *Pap. Met. Geophys.*, **10**(1): 51–53.

Niiler, P. P. and A. R. Robinson (1967): The theory of free inertial jets. II. A numerical experiment for the path of the Gulf Stream. *Tellus*, **19**(4): 601–619.

Robinson, A. R. and P. P. Niiler (1967): The theory of free inertial currents. I. Path and structure. *Tellus*, **19**(2): 269–291.

Rossby, C. -G. (1940): Planetary flow patterns in the atmosphere. *Quart. J. Roy. Met. Soc.*, **66**, Supplement: 68–87.

Saito, Yukimasa (1960): The cold water-mass off the south coast of central Japan (Part I). *Memo. Kobe Mar. Obs.*, **14**: 35–43.

Shoji, Daitaro (1951): The variations of oceanic currents

in the adjacent seas of Japan. *Hydrogr. Bull.*, New Ser., **25**: 230–244.

Shoji, Daitaro (1954): On the variation of daily mean sea level and the oceanographic condition (I) *Hydrogr. Bull.*, Spec. No. **14**: 17–25.

—— (1957a): Kuroshio during 1955. *Rec. Oceanogr. Wks. Japan*, Spec. No. **1**: 21.26.

—— (1957b): On the variations of daily mean sea levels and the Kuroshio from 1954 to 1955. *Proc. UNESCO Symp. Phys. Oceanogr.*: 130–136.

—— (1961): On the variations of the daily mean sea levels along the Japanese Islands. *J. Oceanogr. Soc. Japan*, **17**(3): 141–152.

—— (1964): A note on the anomalous cold water regions along the Japanese coasts and their relation to the fluctuations in the current speeds of the Kuroshio *Bull. Japan. Soc. Fish. Oceanogr.*, **4**: 31–40.

Shoji, Daitaro and Kanji Suda (1956): On the variation of the Kuroshio near the Japan Islands. *Proc. Eighth Pacific Science Congress*, III: 619–636.

Stommel, Henry (1948): The westward intensification of wind-driven ocean currents. *Trans. Amer. Geophys. Union*, **29**: 202–206.

Uda, Michitaka (1937a): On the recent abnormal condition of the Kuroshio to the south of Kii Peninsula. *Kagaku*, **7**(9): 360–361.

—— (1937b): On the recent abnormal condition of the Kuroshio to the south of Kii Peninsula (Supplement). *Kagaku*, **7**(10): 403–404.

—— (1940): On the recent anomalous hydrographical conditions of the Kuroshio in the south waters off Japan proper in relation to the fisheries. *J. Fish. Exp. Sta. Tokyo*, **10**: 231–278.

—— (1949): On the correlated fluctuation of the Kuroshio Current and the cold water mass. *Oceanogr. Mag.*, **1**(1): 1–12.

—— (1964): On the nature of the Kuroshio, its origin and meanders. In *Stud. Oceanogr.*, Kozo Yoshida (ed.), Univ. Tokyo Press: 89–107.

Uda, Michitaka and Keiichi Hasunuma (1969): The eastward Subtropical Countercurrent in the western North Pacific Ocean. *J. Oceanogr. Soc. Japan*, **25**(4): 201–210.

Veronis, George and Henry Stommel (1956): The action of variable wind stresses on a stratified ocean. *J. Mar. Res.*, **15**(1): 43–75.

Warren, B. A. (1963): Topographic influences on the path of the Gulf Stream. *Tellus*, **15**(2): 167–183.

Wüst, George (1936): Kuroshio und Golfstrom. Eine vergleichende Hydro-dynamischen Untersuchung *Veröff. des Inst. f. Meereskn.* Neufolg A. Hf. **29**: 1–69.

Yoshida, Kozo (1968): Subtropical Countercurrent (Abstract). *Notes on the 1968 Summer Study Program in Geophysical Fluid Dynamics at the Woods Hole Oceanographic Institution*, Vo. **1**: 190–191.

—— (1969): Subtropical Countercurrents—a preliminary note on further observational evidences—. *Rec. Oceanogr. Wks. Japan*, **10**(1): 123–124.

—— (1970): Subtropical Countercurrents—Band structures revealed from CSK data—. In *The Kuroshio —A Symposium on the Japan Current* (edited by J. C. Marr), East-West Center Press, Honolulu, 197–204.

Yoshida, Kozo and Toshiko Kidokoro (1967a): A Subtropical Countercurrent in the North Pacific—an eastward flow near the Subtropical Convergence. *J. Oceanogr. Soc. Japan*, **23**(2): 88–91.

—— (1967b): A Subtropical Countercurrent (II)— A prediction of eastward flows at lower subtropical latitudes. *J. Oceanogr. Soc. Japan*, **23**(5): 231–246.

Yosida, Shozo (1961): On the short period variation of the Kuroshio in the adjacent sea of Izu Islands *Hydrogr. Bull.*, New Ser., **65**: 1–18.

—— (1964): A note on the variations of the Kuroshio during recent years. *Bull. Japan Soc. Fish. Oceanogr.*, **5**: 66–69.

Chapter 14

BIBLIOGRAPHY OF STUDIES ON KUROSHIO

KOZO YOSHIDA and YOKO SHIMIZU

Geophysical Institute, University of Tokyo, Tokyo, Japan.

INTRODUCTION

This chapter is devoted to a bibliography containing some 800 publications on the oceanography of the Kuroshio. In making this collection, our attention has been focused mainly on the physical aspects of the Kuroshio; as a consequence, a greater number of papers and reports appearing in this list deal with physical oceanography of the Kuroshio rather than its chemistry, geology, geophysics or fisheries. For inclusion in this bibliography, materials have also been confined to those which were published by the year 1968. It is true some of the 1969 issues have been added, but not to the extent that all the important publications of that year are included. In particular, only a part has been taken from the international CSK program in spite of a large number of recent publications produced by the project. Furthermore, the authors acknowledge that publications, dealing with the Gulf Stream as well as the larger areas of the Pacific and in this sense indirectly related to the Kuroshio, have been greatly limited in number in this list.

In checking up foreign publications, some difficulties arose. Although in recent years studies on the Kuroshio have been remarkably active in such countries as the USA and USSR, it has not been an easy task to identify all the foreign issues, especially those of the USSR, mainly due to the lack of information on our side.

In spite of these unsatisfactory factors, the present survey is believed to contain a greater number of publications prepared with more accuracy than similar lists issued before.* We hope that this new bibliography will serve as a useful guide to future students as well as researchers of the Kuroshio.

The publications are arranged in the alphabetical order of the authors' names. This arrangement, however, may not always be convenient. For instance, many researchers may find it more helpful if the materials are grouped by various subjects rather than authors. In order to compensate for such a disadvantage, classification of each reference by area and subject was attempted (Fig. 1, Table 1) and the corresponding numbers are added to each reference. In addition to the papers written by individual authors, data reports published by various oceanographic institutes are listed at the end of this bibliography.

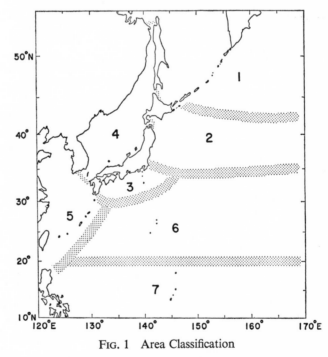

FIG. 1   Area Classification

---

* e.g.(1) Oceanographic Papers in Japan (Annotated Bibliography) Vols. I (1957), II (1958) and III (1961) published by Jap. Nat. Comm. UNESCO.

(2) Bibliography of Japanese Study on Kuroshio (Physical Oceanography); Information paper distributed at Symposium on the Kuroshio, October 1963, Tokyo.

We are indebted to all who have contributed to the compilation of this bibliography, which took some five years to complete. Especially we wish to acknowledge the great deal of assistance offered by Dr. Mizuki Tsuchiya, who made a major part of the classification work and acquainted us with a number of Russian literature. Thanks are also due to Mrs. Hiroko Nakanishi and Miss Mikiko Uzaki, who assisted during the earlier stage of this work, and to Messrs. Keiichi Hasunuma and Nobuo Suginohara, who assisted in preparing the data sources.

TABLE 1. Classification by area and subject.

Area	Subject
1. Subarctic	1. Data report, cruise report
2. East of Tohoku, south of Hokkaido	2. Current field
3. South of Honshu from Kyushu to Chosi	3. Distribution of physical and chemical properties— climatological averages
4. Japan Sea	4. Distribution of physical and chemical properties— individual data
5. Philippines to Kyushu, Yellow Sea, East China Sea	5. Time variations in currents and properties
6. Subtropical	6. Meander of the Kuroshio cold and warm eddies
7. North Equatorial Current	7. Dynamics, theory, hypothesis, speculation
8. No specific area (e.g. general circulation theories)	8. Miscellaneous

BIBLIOGRAPHY

Original forms of the abbreviations used for various publications are given on p. 501 and 502.

		Area	Subject	Chapter
AKAGAWA, Masaomi				
1954	On the oceanographical conditions of the North Japan Sea (west off the Tsugaru-Straits) in summer (Part 1). (In Japanese with English abstract). *JOSJ*, **10**(4): 189–199.	4	4	9
1955	On the oceanographical conditions of the North Japan Sea (west of the Tsugaru-Straits) in summer (Part 2). (In Japanese with English abstract). *JOSJ*, **11**(1): 5–11.	2,4	2,4	
AKAMATSU, Hideo				
1964	Preliminary report of the Japanese Expedition of Deep-Sea, the sixth cruise, 1963 (JEDS-6). (In English). *OM*, **15**(2): 123–125.	2	1	
1965	(with Tsutomu AKIYAMA and Tsutomu SAWARA) Preliminary report of the Japanese Expedition of Deep-Sea, the tenth cruise, 1965 (JEDS-10). (In English). *OM*, **17**(1–2): 49–68.	3,6	1	
1966	Measurements of the deep currents. (In Japanese with French abstract). *La mer*, **4**(1): 40–45.	3,6	2	
1966	(with Tsutomu SAWARA) Cruise report of CSK survey by JMA in 1965. (In English). *OM*, **18**(1–2): 53–56.	1–7	1	
1967	(with Tsutomu SAWARA) Cruise report of CSK survey by JMA in 1966. (In English). *OM*, **19**(2): 157–162.	1–7	1	
1968	(with Tsutomu SAWARA) Cruise report of CSK survey by JMA in 1967. (In English). *OM*, **20**(1): 97–104.	1–7	1	
1969	The preliminary report of the third cruise for CSK, January to March 1969. (In English). *OM*, **21**(1): 83–96.	3,5,6,7	1,2,4	

	Area	Subject	Chapter
AOKI, Takeo			
1918 Drifting of drift bottles. (In Japanese). *Taiwan Suisan Zasshi*, **31**: 13–18.	3,5	2	
ARAKAWA, Hidetoshi			
1935 On a method for predicting the distributions of temperature and salinity in the sea. (In Japanese). *US*, **15**(7): 230–233.	2	5	
ASANO, Hikotaro			
1915 Oceanographical investigation in the area between O-shima in Izu Province and Shiono-misaki in Kii Province (Un'yo-maru). (In Japanese). *Gyogyo Kihon Chosa Hokoku*, **4**: 57–97.	3	1	
1919 Oceanographical investigation in the Tsushima Strait and the Tokara-gun-to (carried out by the Urrupu-maru from 1915–1916). (In Japanese). *Gyogyo Kihon Chosa Hokoku*, **7**: 1–10.	4,5	1	
ASAOKA, Osamu			
1966 (with Shigeo MORIYASU) On the circulation in the East China Sea and the Yellow Sea in winter (Preliminary Report). (In English). *OM*, **18**(1–2): 73–81.	5	3,7	5
BARKLEY, R. A.			
1968 The Kuroshio-Oyashio front as a compound vortex street. (In English). *JMR*, **26**(2): 83–104.	2,3	6,7	8
1968 *Oceanographic Atlas of the Pacific Ocean*. (In English). Univ. of Hawaii Press, Honolulu: 156 figs.	1–7	3	4
BATALIN, A. M.			
1960 The state of the Kuroshio and fisheries problems. (In Russian). *Trudy soveshch. ikhtiol. komiss. Akad. Nauk, SSSR*, **10**: 198–204.	8	7	
1961 The meandering of the Kuroshio. (In Russian). *Okeanologiya, Akad. Nauk, SSSR*, **1**(6): 961–975.	2,3	6	
BEKLEMISHEV, C. W.			
1958 (with V. A. BURKOV) The relationship between the plankton and the water masses in the frontal region of the North-West Pacific. (In Russian). *Trudy Inst. Okeanol. Akad. Nauk, Sci. SSSR*, **27**: 55–65.	2	4	
BENT, S.			
1856 Report made to Commodore M. C. Perry upon the Kuroshiwo or Gulf Stream of the North Pacific Ocean. (In English). *Narrative of the expedition of an American squadron to the China Seas and Japan 1852–1854*, **2**: 363–370.	3	5	
1857 The Japanese Gulf Stream. (In English). *Bull. Amer. Geog. Statis. Soc.*, **2**: 203–213.	8	2,3	13

		Area	Subject	Chapter

| 1868 | The Japan Current Kuro-siwo, or the Gulf Stream of the North Pacific Ocean.*
(In English). *Mercantile Mar. Mag.*, **15**: 161–168. | — | — | |

BOGDANOV, M. A.

| 1965 | On the circulation changes in the atmosphere and the hydrosphere. —Investigations in line with the programme of the International Geophysical Year.
(In Russian). *Trudy Vses. Nauchno-Issled. Inst. Morsk. Ribn. Choz. Okeanogr.* (VNIRO), **57**: 43–52. | 1–7 | 5 | |

BUBNOV, V. A.

| 1960 | Water dynamics in the frontal zone of the Kuroshio and Oyashio currents.
(In Russian). *Trudy Morsk. Gidrofiz. Inst. Akad. Nauk*, **22**: 15–25. | 2 | 2,4 | |

BULGAKOV, N. P.

| 1967 | The main features of the structure and position of the subarctic front in the northwestern Pacific.
(In Russian with English abstract). *Okeanologiya, Akad. Nauk, SSSR*, **7**(5): 879–888.
(In English): 680–690. | 2 | 4 | |

BURKOV, V. A.

| 1958 | (with V. S. ARSENJEV)
An attempt to classify the water-masses in the contact zone of the Kuroshio and the Kuril Current.
(In Russian). *Trudy Inst. Okeanol., Akad. Nauk, SSSR*, **27**: 5–11. | 1 | 4 | |

| 1960 | (with I. M. OVCHINNIKOV)
Structure of zonal streams and meridional circulation in the central Pacific during the northern hemisphere winter.
(In Russian). *Trudy Inst. Okeanol. Akad. Nauk, SSSR*, **40**: 93–107. | 1,2,6,7 | 1,2,4 | |

| 1963 | (with Yu. V. PAVLOVA)
Geostrophic circulation on the surface of the northern part of the Pacific in summer.
(In Russian). *Result at y Issledovaii po Programme Mezhdunarodnogo Geofizicheskogo Goda, Okeanologicheskie Issledo vaniya, Sbornik Statey, X Razdel Programmy Mgg, Izdatelstvo Akademii Nauk, SSSR*, **9**: 21–31. | 1–7 | 2,3 | 4 |

| 1963 | Water circulation in the North Pacific.
(In Russian). *Okeanologiya, Akad. Nauk, SSSR*, **3**(5): 761–776. | 1–7 | 2 | |

| 1966 | The structure of currents in the Pacific Ocean and their nomenclature.
(In Russian). *Okeanologiya, Akad. Nauk, SSSR*, **6**(1): 3–14.
(In English): 1–11. | 1–3,6,7 | 2 | |

| 1967 | Determination of absolute geostrophic current velocities in the oceans.
(In Russian). *Okeanologiya, Akad. Nauk, SSSR*, **7**(1): 41–50.
(In English): 31–38. | 1–7 | 2,3,7 | |

| 1969 | A model of steady-state water transport in a nonhomogeneous ocean. | 1–7 | 2,3,7 | |

* Area and subject unclassificable due to the unavailable material.

		Area	Subject	Chapter
	(In Russian). *Okeanologiya, Akad. Nauk, SSSR,* **9**(1): 15–17. (In English): 9–12.			
1969	The bottom circulation of the Pacific Ocean. (In Russian). *Okeanologiya, Akad. Nauk, SSSR,* **9**(2): 223–234. (In English): 179–188.	1–7	2,3,7	

CANNON, G. A.

| 1966 | Tropical waters in the western Pacific Ocean, August to September 1957. (In English). *Deep-Sea Res.,* **13**: 1139–1148. | 7,8 | 2,4 | 4,5 |

CHAEN, Masaaki

| 1960 | Mixing stages of water masses in the western Equatorial Pacific. (In Japanese). *Mem. Fac. Fish., Kagoshima Univ.,* **9**: 37–47. | 7 | 3 | |

COCHRANE, J. D.

| 1958 | The frequency distribution of water characteristics in the Pacific Ocean. (In English). *Deep-Sea Res.,* **5**: 111–127. | 1–7 | 3 | 11 |

COX, C. S.

| 1964 | (with Toshihiko TERAMOTO and Jean FILLOUX) On coherent electric and magnetic fluctuations in the sea. (In English). *SO* (dedicated to Prof. Hidaka in commemoration of his sixtieth birthday): 449–457. | 3 | 5 | 12 |

DOBROVOL'SKIY, A. D.

1949	Concerning the position of the zero surface for dynamic calculations in the North Pacific. (In Russian). *Trudy in-ta okeanol. Akad. Nauk, SSSR,* **4**: 3–26.	1–7	3	
1958	Important problems of physical oceanology in the northwestern Pacific. (In Russian). *Tr. Okeanogr. kommiss. Akad. Nauk, SSSR,* **3**: 24–27.	8	7	
1962	(with V. V. LEONT'YEVA and M. A. RADZIKHOVSKAYA) Deep-water hydrological research in the Pacific Ocean. (In Russian). *Trudy Inst. Okeanol. Akad. Nauk, SSSR,* **60**: 130–141.	1–7	8	

DODIMEAD, A. J.

| 1963 | (with Felix FAVORITE and Toshiyuki HIRANO) Salmon of the North Pacific Ocean, Part II. —Review of oceanography of the Subarctic Pacific region. (In English). *Bull. Int. North Pacific Fish. Comm.,* **13**: 195pp. | 1 | 2,4 | 1,4,8 |

ENDO, Hiroshi

| 1961 | On the correlation between the surface water temperature and current axis in the Kuroshio region. (In Japanese with English abstract). *HB, NS,* **65**: 42–47. | 3 | 2,5 | 6 |

FAVORITE, Felix

| 1964 | Drift bottle experiment in the northern North Pacific Ocean, 1962–1964. (In English). *JOSJ,* **20**(4): 160–167. | 1 | 2 | |

		Area	Subject	Chapter

1967   The Alaskan Stream.   1   2,4   4
        (In English). *Bull. Int. North Pacific Fish. Comm.*, **21**: 1–20.

FEDROV, K. N.

1960   On the evaluation of the deep-currents in the West Pacific Ocean.   6,7   2   12
        (In Russian). *Trudy Inst. Okeanol. Akad. Nauk, SSSR*, **40**: 162–166.

FISHERIES RESEARCH and DEVELOPMENT AGENCY, the REPUBLIC of KOREA

1964   *Oceanographic handbook of the neighbouring seas of Korea.*   4,5   3,5   9
        (In Korean with English foreword and legends): 214pp.

FUGLISTER, F. C.

1955   Alternative analyses of current surveys.   1–3,6   2,4,6,7   8
        (In English). *Deep-Sea Res.*, **2**: 213–229.

FUJII, Masayuki

1954   The ocean current in the south-western sea area of Kyūsyū in   5   2
        January, 1954.
        (In Japanese with English abstract). *HB, NS*, **42**: 138.

1960   (with Minoru KIMURA)   5   5   5
        Concerning the relation between the variation of Kuroshio in Osumi
        Gunto Area and that of the water temperature in Yakushima coastal
        area.
        (In Japanese with English abstract). *JOSJ*, **16**(2): 55–58.

1961   (with Minoru KIMURA)   3–5   2   5,12
        Drifting of 20,000 current bottles, released in the sea south-west of
        Kyūsyū, July, 1960.
        (In Japanese with English abstract). *HB, NS*, **67**: 58–62.

FUJII, Yoshiyuki

1955   (with Jotaro MASUZAWA, Ryuzo MARUMO and Kantaro WATANABE)   2   1
        On the current, water masses and plankton in the Kuroshio region east
        off Honshu in the summer of 1954.
        (In Japanese). *KSH*: 39–46.

FUJIMORI, Saburo

1912   On the ocean currents in the vicinity of the Bonin Islands.   6   2
        (In Japanese). *Chigaku Zasshi*, **24**(286): 718–726.

FUJIMORI, Takashi

1964   Preliminary report on the oceanographic anomaly in the early half of   2,3   4–6   7
        1963 in the waters adjacent to Kanto District.
        (In English). *TKRFL*, **38**: 77–98.

1967   (with Toshiyuki HIRANO and Susumu UEHARA)   2,3   5,6
        The fluctuations and its characteristics of current systems in the sea
        adjacent to Kanto District.
        (In Japanese). *Rep. of Fisheries Resources Investigations by the Scientists
        of the Fisheries Agency*, **7**: 9–18.

FUKUDA, Kiyoshi

1942   On the relation between the summer air temperature of the Tohoku   2   5

		Area	Subject	Chapter
	District and the sea surface temperature off the district. (In Japanese). *JMSJ*, *II*, **20**(11): 432–433.			
1943	On the secular variation of sea water temperature. (In Japanese). *JMSJ*, *II*, **21**(1): 36–39.	1–5	5	

Fukuoka, Jiro

		Area	Subject	Chapter
1950	Note on the dichoterm layer off Sanriku-District. (In Japanese with English abstract). *JOSJ*, **6**(2): 23–28.	2	4	
1950	The oceanographical conditions at the "Fixed Point," Long. 153°E, Lat. 39°N. (In Japanese with English abstract). *KH*, **1**(2): 104–110.	2	5	
1950	On the "Oyasio." (In Japanese with English abstract). *KH*, **1**(3): 124–128.	2	4	
1950	A method of forecasting oceanographic conditions. (In Japanese with English abstract). *KH*, **1**(3): 144–152.	2	5	
1951	The sea water motion in the low-latitude districts of the North Pacific Ocean. (In English). *OM*, **3**(3): 97–101.	7	4	
1952	(with Tsuyoshi Yusa) An example of the variation of oceanic condition in the adjacent sea east of the Tohoku District. (In Japanese with English abstract). *JOSJ*, **8**(1): 1–5.	2	5	
1952	(with Tsuyoshi Yusa) The variation of the oceanic condition in the sea adjacent to Tohoku District. (In English). *OM*, **4**(2): 57–65.	2	5	
1953	(with Iwao Tsuiki) On the variation of the oceanographic condition of the sea near the fixed point "Extra." (In English). *ROWJ*, *NS*, **1**(1): 23–27.	2	5	
1953	(with Iwao Tsuiki) The rising of the coastal sea water temperature in winter. (In English). *ROWJ*, *NS*, **1**(2): 1–6.	2,3	4,5,7	
1954	(with Iwao Tsuiki) On the variation of chlorinity in the Kuroshio Area. (In English). *OM*, **6**(1): 15–23.	3,6	5	1
1955	The variation of the polar front in the sea adjacent to Japan. (In English). *OM*, **6**(4): 181–195.	2	5	1
1955	Oceanic condition in the North-western Pacific Ocean (Preliminary Report). (In English). *ROWJ*, *NS*, **2**(2): 102–107.	1–3,6,7	4	
1957	Variation of the oceanic conditions in the North Western Pacific. (In English with French abstract). *Proc. UNESCO Symp. Phys. Oceanogr., Tokyo, 1955*: 76–86.	2,3	5	
1957	A note on the westward intensification of ocean current. (In English). *ROWJ*, *NS*, **4**(1): 7–13.	8	8	
1957	On the variation of oceanic conditions in the sea adjacent to Japan. (In English). *OM*, **9**(1): 95–106.	2–5	3,5	13

		Area	Subject	Chapter
1957	On the Tsushima Current. (In English). *JOSJ*, **13**(2): 57–60.	4	4	9
1958	On the Kuroshio near the Izu Islands. (In English). *JOSJ*, **14**(1): 11–14.	3	7	7,8,13
1958	The variations of the Kuroshio Current in the sea south and east of Honshu (Japanese Main Island). (In English). *OM*, **10**(2): 201–213.	2,3	5,7	6,8,13
1958	On the variation of the Kuroshio current south of Honshu (Japanese Main Island). (In English). *GM*, **28**(3): 343–355.	3	6	13
1959	On the periodicity of the variations of the oceanic conditions. (In Japanese with English abstract). *US*, **35**(1): 13–20.	2–4	5	
1960	An analysis of the mechanism of cold and warm water masses in the seas adjacent to Japan (Part 1). (In Japanese with English abstract). *US*, **36**(1): 1–12.	3,4	6	13
1960	An analysis of the mechanism of cold and warm water masses in the sea adjacent to Japan (Part 2). (In Japanese). *US*, **36**(2): 29–36.	3,4	6	
1960	An analysis on the mechanism of the cold water mass appearance in the Enshu-nada. (In English). *OM*, **11**(2): 127–143.	3	6	13
1961	An analysis of the mechanism of the cold and warm water masses in the seas adjacent to Japan. (In English). *ROWJ*, *NS*, **6**(1): 63–100.	3,4	6	9
1962	Abyssal circulation in the Atlantic near the poles and abyssal circulation in the Pacific and other oceans in relation to the former. (In English). *JOSJ*, **18**(1): 5–12.	1–3, 5–7	4	11
1962	Characteristics of hydrography of the Japan Sea—In comparison with hydrography of the North Pacific. (In Japanese with English abstract). *JOSJ*, *20th Anniv. Vol.*: 180–188.	1–6	2,4,5	9
1962	An analysis of hydrographical condition along the Tsushima Warm Current in the Japan Sea. (In English). *ROWJ*, *NS*, **6**(2): 9–30.	4	4,5	9
1965	Hydrography of the adjacent sea (1). The circulation in the Japan Sea. (In English). *JOSJ*, **21**(3): 95–102.	4	4	9

FUKUSHIMA, Shinichi

| 1958 | Relation between the fishing of the Pacific saury and the oceanographical conditions in the northeastern sea area along the Pacific coast of Japan. (In Japanese with English abstract). *THRFL*, **12**: 1–27. | 2,3 | 4,5 | |
| 1962 | On the relation between the pattern of the Kuroshio Current in spring and summer and the saury fishing conditions in fall. (In Japanese with English abstract). *THRFL*, **21**: 21–37. | 2,3 | 4,5 | |

GAMBO, Kanzaburo

| 1963 | The role of sensible and latent heats in the baroclinic atmosphere. (In English with Japanese abstract). *JMSJ*, *II*, **41**(4): 233–246. | 4 | 7 | |

	Area	Subject	Chapter

GRIGORKINA, R. G.

1969 (with A. N. MICHURIN, P. P. PROVOTOROB and B. R. FUKC) — Area 1–3, Subject 5
The short-period variability of oceanic conditions in productive region of the Kuroshio waters system (Part 1).
(In Russian). *Izvestija, Tichookeanskogo nautchnoiszle-dovatelskogo instituta ribnogo chozyaystva i okeangrafii*, **68**: 45–66.

GUILCHER, Andre

1955 Hydrology of the equatorial and Northwest Pacific. — Area 1–3,5–7, Subject 3,5,6
(In French). *Annales de Geographie*, Paris, **64**(342): 81–96.

GURIKOVA, Z. F.

1966 Calculation of surface and deep ocean currents in the North Pacific in summer. — Area 1–7, Subject 2,7
(In Russian). *Okeanologiya, Akad. Nauk, SSSR*, **6**(4): 615–631.
(In English): 504–519.

HAKODATE MARINE OBSERVATORY

1958 The charts of the mean surface water temperatures in the neighbouring sea of Japan (1911–1940). — Area 1–6, Subject 3
(In Japanese with English abstract). *KH*, **7**(1): 67–74.

1961 Report of the oceanographic observations in the Tsugaru Straits in the period from 1943 to 1958. Part 1. — Area 2, Subject 1, Chapter 9
(In Japanese). *Tec. Rep. Japan Meteor. Agency*, **9**: 62pp.

HANZAWA, Masao

1949 Oceanographical considerations on the bad rice crop in the Tohoku District of Japan (Synthetic Review). — Area 1–3,5, Subject 7,8
(In English). *OM*, **1**(4): 194–203.

1950 On the annual variation of evaporation from the sea-surface in the North Pacific Ocean. — Area 8, Subject 8, Chapter 4
(In English). *OM*, **2**(2): 77–82.

1952 (with Tamio INOUE) Relation between oceanic and atmospheric states in the North Pacific Ocean for the year of bad and good rice crop. — Area 1–7, Subject 3,5,7
(In Japanese with English abstract). *OM*, **2**(3): 283–288.

1952 On some examples of abrupt change of oceanographical conditions at the ocean weather station Extra adjacent to the oceanic polar front. — Area 2, Subject 5
(In English). *OM*, **4**(3): 67–79.

1953 On the eddy diffusion of pumices ejected from *Myojin*-Reef in the southern sea of Japan. — Area 3,6, Subject 1,2,4,7
(In English). *OM*, **4**(4): 143–148. *ROWJ, NS*, **1**(1): 18–22.

1956 Note on the variation of the polar front in the sea adjacent to Japan. — Area 2, Subject 5
(In English). *OM*, **7**(2): 95–96.

1956 Preliminary report on a close relationship between surface water temperatures at the two separated ocean stations, Papa and Tango (Studies on the interrelationship between sea and atmosphere. Part 1). — Area 1–3, 6, Subject 5
(In English). *OM*, **8**(2): 157–160.

1957 Studies on the inter-relationship between the sea and the atmosphere (Part 2). — Area 1,2, Subject 5,7
(In English). *OM*, **9**(1): 87–93.

		Area 1–7	Subject 5	Chapter
1958	Studies on the inter-relationship between the sea and the atmosphere (Part 3). (In English). *OM*, **10**(1): 91–96.	1–7	5	
1958	Studies on the inter-relationship between the sea and the atmosphere (Part 4). (In English). *OM*, **10**(2): 215–226.	1–3,5,6	5,7	
1962	Studies on the inter-relationship between the sea and the atmosphere (Part 5). Sea surface temperature anomalies in connection with the incoming incident radiation anomalies. (In English). *OM*, **14**(1): 1–14.	1–3,5,6	3,5	
1964	Preliminary report on the abnormal oceanic conditions in the seas adjacent to Japan in the winter of 1963. (In English). *SO* (dedicated to Prof. Hidaka in commemoration of his sixtieth birthday): 59–67.	1–6	5	7
1964	A statistical relationship between sea surface temperature and geostrophic wind in the Kuroshio waters. (In English). *ROWJ, NS*, **7**(2): 1–9.	1–3	3,5	
1966	Verification of the Fournier approximation of the space-wise interpolation of BT data in the Pacific Ocean. (In English). *OM*, **18**(1–2): 113–138.	3,6	3	

HARA, Hisako

1969	(with Kenzo TAKANO) Test measurements of bottom currents. (In Japanese). *La mer*, **7**(1): 27–28.	3	1,5	12
1969	(with Shozaburo NAGUMO, Heihachiro KOBAYASHI, Kenzo TAKANO and Yasuhiro SUGIMORI) A measurement of bottom current on a shoulder of a sea mount. (In English). *JOSJ*, **25**(6): 307–309.	2	1,5	12

HASHIGUCHI, Yukio

1968	(with Ryoki SARUWATARI) Taking part in the Cooperative Study of Kuroshio (in pursuit of the origin and extension of Kuroshio). (In Japanese). *HB, NS*, **85**: 13–22.	2,5,6	1	1

HATA, Katsumi

1961	Relation between hydrographical condition in the sea area of Tsugaru Straits and the area off Sanriku. (In Japanese). *Tec. Rep. Japan Meteor. Agency*, **9**: 52–57.	2	3	
1962	Seasonal variation of the volume transport in the northern part of the Japan Sea. (In Japanese with English abstract). *JOSJ, 20th Anniv. Vol.*: 168–179.	4	3	9
1963	The report of drift bottles released in the North Pacific Ocean. (In Japanese with English abstract). *JOSJ*, **19**(1): 6–15.	1,2	2	
1964	(with Sadao HOSODA and Koji YAMAMOTO) Report of the detailed oceanographic observations in the Tsugaru Straits from August to September, 1962. (In Japanese with English abstract). *Bull. Hakodate Marine Observatory* (Special Number): 1–30.	4	2,3	8

		Area	Subject	Chapter
1965	The state of the sea off Tohoku District in 1963. (In Japanese). *JOSJ*, **21**(3): 118–120.	2	8	
1965	Seasonal variation of the volume transport in the Oyashio area. (In Japanese with English abstract). *JOSJ*, **21**(5): 193–201.	2	5	
1968	Fluctuation of the warm eddy in the sea of south of Hokkaido —Relation to the Oyashio. (In Japanese). *Bull. Japan. Soc. Fish. Oceanogr.*, **12**: 47–57.	2	6	
1969	Some problems relating to fluctuation of hydrographic conditions in the sea northeast of Japan (Part 1)—Relation between the patterns of the Kuroshio and the Oyashio. (In English). *JOSJ*, **25**(1): 25–35.	2,3	2,4,5	8
1969	Some problems relating to fluctuation of hydrographic conditions in the sea northeast of Japan (Part 2)—Fluctuation of the warm eddy cut off northward from the Kuroshio. (In English). *OM*, **21**(1): 13–29.	2,3	2,4–6	8

HATANAKA, Masayoshi

		Area	Subject	Chapter
1948	On the secular variation of coastal water temperature in the North-Eastern Sea region of Japan. (In Japanese). *JSSFB*, **13**(1): 41–43.	2	5	
1948	On the variation of the temperature and salinity of sea water at Enoshima Islands, Miyagi Prefecture. (I) Long period variation. (In Japanese with English abstract). *JSSFB*, **13**(4): 161–163.	2	5	
1948	Yearly variations of temperature and salinity of sea water at Enoshima Islands, Miyagi Prefecture. (II) Cyclic relation. (In Japanese with English abstract). *JSSFB*, **13**(5): 213–215.	2	5	
1952	Studies on the fluctuation of the hydrographic conditions and its effect on the pelagic fisheries resources. (In Japanese with English abstract). *THRFL*, **1**: 88–119.	2	5	8

HAYAMI, Shoitiro

		Area	Subject	Chapter
1955	(with Hideo KAWAI and Masao OUCHI) On the theorem of Helland-Hansen and Ekman and some of its applications. (In English). *ROWJ, NS*, **2**(2): 56–67.	8	2, 7	13
1957	On the dynamics of Kuroshio off the southern coast of Japan. (In English with French abstract). *Proc. UNESCO Symp. Phys. Oceanogr., Tokyo, 1955*: 139.	3	7	13

HAYASHI, Shigeaki

		Area	Subject	Chapter
1968	(with Ikihiko MIYASHITA and Igoro FUJITA) Maritime meteorological summary of western North Pacific Ocean in winter. (In English). *OM*, **20**(2): 105–120.	3,5–7	4	

HIDAKA, Koji

		Area	Subject	Chapter
1927	Experimental studies on the North Pacific ocean currents near Japan (1st Paper). (In Japanese). *KKI*, **13**: 1–7. *US*, **7**(3): 37–43. (In English). *GM*, **1**(3): 68–75.	2–6	2	

		Area	Subject	Chapter
1927*	Experimental studies on the North Pacific ocean currents near Japan (2nd Paper). (In Japanese). *KKI*, **13**: 8–12. *US*, **7**(4): 51–55.	2–6	2	
1928**	Experimental studies on the North Pacific Ocean currents near Japan (2nd Paper). (In English). *GM*, **1**(4): 203–210.	2–6	2	
1933	(with Yasuo MATUDAIRA, Hideo KAWASAKI, Tsutomu TAKAHATA, Tokio KUBO and Zen'iti YASUI) Report of the oceanographical observations on board the R.M.S. "Syunpu Maru" in Tugaru Strait. (In Japanese with English abstract). *JO*, **4**(2): 341–357.	2,4	1,2,4	
1934	(with Yasuo MATUDAIRA, Matuichi MIZUUTI, Tokio KUBO, Zen'iti YASUI, Hidejiro KURASIGE, Tadazane YANAGISAWA and Takeo KATO) The results of oceanographical expedition of R.M.S. "Syunpu Maru" to the northern part of the Japan Sea in the summer of 1931 and 1932. (In Japanese with English abstract). *JO*, **6**(1): 1–101.	4	1,2,4	
1940	The Kuroshio current in the adjoining seas of Kyushu and the Nansei Shoto. (In Japanese). *KKI*, *Extra No.*: 1–20.	5	1	
1949	Mass transport in ocean currents and lateral mixing. (In English). *JMR*, **8**(2): 132–136.	8	7	
1949	(with Tamiro YAMAGIWA) On the absolute velocity of the subarctic intermediate current to the south of Japan. (In English). *OM*, **1**(2): 99–102. *GN*, **2**(5): 1–4.	3	2, 4	
1950	(with Tadasu SUZUKI) Secular variation of the Tsushima Current. (In Japanese with English abstract). *JOSJ*, **6**(1): 28–31.	4	2, 5	9
1966	Kuroshio Current. (In English). In *Encyclopedia of Oceanography*, Editor, R. W. Fairbridge, Reinhold Publishing Corp., New York: 433–437.	2, 3, 5	2, 4, 6	
1969	Deep-sea current research in Japan. (In Russian). *Okeanologiya, Akad. Nauk, SSSR*, **9**(3): 430–434. (In English): 351–354.	8	8	

HIGANO, Ryozi

| 1967 | (with Yoshiro SETO, Ryoki SARUWATARI and Yosiyuki IWANAGA) Distribution of chemical constituents of sea water in the eastern sea off the Philippines and in the Luzon Straits. (In Japanese with English abstract). *Rep. Hydrogr. Res.*, **3**: 33–47. | 5, 6, 7 | 4 | |

HIKOSAKA, Shigeo

1953	On the ocean-currents (non-tidal currents) in the Tugaru Strait. (In Japanese with English abstract). *HB*, *NS*, **39**: 279–285.	4	2	9
1954	On the upwellings in the cold water mass. (In Japanese with English abstract). *HB*, *NS*, **43**: 181–184.	2, 3	6	
1955	(with Ryuzo WATANABE) Observations of the Kuroshio region off Shionomisaki and Enshu-nada	3	4	

** English translation of the paper marked by * .

		Area	Subject	Chapter
	from 1942 to 1943. (In Japanese with English abstract). *HB, Special No.* **17**: 63–69.			
1957	(with Ryuzo WATANABE) Areas of divergence and convergence of surface currents in the North-Western Pacific. (In English with French abstract). *Proc. UNESCO Symp. Phys. Oceanogr., Tokyo, 1955*: 101–103.	1–3, 5–7	2	
1959	(with Ryoji HIGANO) Water temperature fluctuations off Muroto, Shikoku. (In English). *ROWJ, NS, Special No.* **3**: 97–101.	3	5	
1964	Differences between observed and predicted water levels. (In English). *ROWJ, NS,* **7**(2): 11–15.	3	5	

HIRANO, Toshiyuki

1953	Horizontal mixing and flow in a boundary area of the Kuroshio off Nojima-Zaki. (In English). *JOSJ,* **8**(3–4): 105–112.	2, 3	4, 7	
1955	(with Nobuo WATANABE) An attempt to predict the surface temperature of the North-Eastern Sea adjacent to Japan for the summer, 1955. (In English). *JOSJ,* **11**(2): 47–55.	2	5	
1957	The oceanographic study on the Subarctic Region of the north-western Pacific Ocean—I. On the water systems in the Subarctic Region. (Based upon the oceanographic survey by the R. V. Tenyo-Maru in summer, 1954.) (In English). *TKRFL,* **15**: 39–55.	1	1	1
1957	The oceanographic study on the Subarctic Region of the north-western Pacific Ocean—II. On the formation of the Subarctic Water systems. (In English). *TKRFL,* **15**: 57–69.	1	4, 7	1
1957	(with Chieko BOSHU) On durability of the surface temperature at hydrographic stations of the Pacific coast—I. (In Japanese with English abstract). *TKRFL,* **16**: 23–38.	1–3, 5	5	
1957	On durability of the surface temperature at hydrographic stations of the Pacific coast—II. (In Japanese with English abstract). *TKRFL,* **17**: 65–72.	1–3, 5	5	8
1958	The oceanographic study on the Subarctic Region of the North-western Pacific Ocean—III. On general condition of the sea in spring and summer. (In English). *TKRFL,* **20**: 23–46.	1	4	1
1961	The oceanographic study on the Subarctic Region of the North-western Pacific Ocean—IV. (In English). *TKRFL,* **29**: 11–39.	1	2, 4	1

HISHIDA, Kozo

| 1950 | On the coastal water and its relation to the oceanic water.
(First Report: Studies on the specific gravity of coastal sea waters).
(In English). *OM,* **2**(3): 123–128. | 2–5 | 3,5 | |

		Area	Subject	Chapter
1953	Annual variation of sea surface temperature in the current region. (In English). *ROWJ, NS,* **1**(2): 7–12.	3, 4	5	
1955	Annual variation of sea-surface water temperature and air temperature in the western parts of North Pacific Ocean. (In English). *ROWJ, NS,* **2**(3): 79–84.	1–7	3, 5	
1955	On the coastal water and its relation to the oceanic water (2). (Second report: Statistical study on the specific gravity of sea waters in the monthly interval.) (In English). *OM,* **7**(1): 1–4.	3–5	3, 5	
1955	On the coastal water and its relation to the oceanic water (3). (Third report: Analytical study on the annual variation of coastal surface water temperature.) (In English). *OM,* **7**(1): 5–10.	3–5	3, 5	
1957	On the forecasting of water temperature in the Japan Sea. (In English). *OM,* **9**(1): 51–54.	4	3,5	
1959	On the coastal water and its relation to the oceanic water (4). (On the standard deviation and the fluctuation of the surface water temperature.) (In English). *KMOM,* **13**(2): 1–11.	1–5	5	

HOLOPAINEN, E. O.

		Area	Subject	Chapter
1967	A determination of the wind-driven ocean circulation from the vorticity budget of the atmosphere. (In English). *Pure Appl. Geophys.,* **67**(2): 156–165.	1–7	7	

HORIBE, Yoshio

		Area	Subject	Chapter
1968	(with Nobuko OGURA) Deuterium content as a parameter of water mass in the ocean. (In English). *JGR,* **73**(4): 1239–1249.	2, 3, 6, 7	4	

ICHIYE, Takashi

		Area	Subject	Chapter
1950	The theory of ocean currents (I). (In English). *KMOM,* **8**: 31–34.	8	7	
1950	On the fluctuations of ocean currents. (In Japanese with English abstract). *US,* **28**(2): 16–20.	2	6, 7	
1951	On the annual variations of water temperatures in the Kii Suido (1). (In Japanese with English abstract). *KH,* **2**(2): 119–131.	3	5	
1951	On the annual variations of water temperatures in the Kii Suido (2). (In Japanese with English abstract). *KH,* **2**(2): 133–139.	3	5	
1951	On the variation of the oceanic circulation (II). (In English). *OM,* **3**(3): 89–96.	3, 5	5, 7	
1952	On the hydrography off Shionomisaki and Enshu-Nada (1951). (In Japanese with English abstract). *KH,* **2**(3): 231–240.	3	4	
1952	On the hydrography of the Kii-Suido (1951). (In Japanese with English abstract). *KH,* **2**(3): 253–278.	3	4, 7	
1952	On the annual variations of water temperature. (In English). *KMOM,* **10**: 11–24.	3	5, 7	
1952	On the use of T-S diagrams in shallow water. (In English). *KMOM,* **10**: 25–45.	3	4, 7	

Year	Reference	Area	Subject	Chapter
1952	On the variation of oceanic circulation (III). (In English). *OM*, **4**(2): 37–47.	1–3, 6, 7	5, 7	
1953	On the hydrographical conditions in the Kuroshio region (1952). (Southern Area of Honshu.) (In Japanese with English abstract). *KKI*, **163**: 1–30.	3	4, 5	
1953	On the variation of oceanic circulation (IV). (In English). *OM*, **5**(1): 23–44.	2	6, 7	1
1954	On the variation of oceanic circulation (VI). (In English). *GM*, **25**(3–4): 185–217.	3	6, 7	6, 7, 13
1954	On the variation of oceanic circulation (VII). (In English). *OM*, **6**(1): 1–14.	3	5, 6, 7	1, 6, 7, 13
1954	On the distributions of oxygen and their seasonal variations in the adjacent seas of Japan (I). (In English). *OM*, **6**(2): 41–66.	2–4	4, 7	9
1954	On the distributions of oxygen and their seasonal variations in the adjacent seas of Japan (II). (In English). *OM*, **6**(2): 67–100.	2–4	4, 7	
1954	On the distributions of oxygen and their seasonal variations in the adjacent seas of Japan (III). (In English). *OM*, **6**(3): 101–131.	2,3	4, 5, 7	
1955	On the possible origin of the intermediate water in the Kuroshio. (In English) *ROWJ*, *NS*, **2**(2): 82–89.	2,3	2, 4, 7	4
1955	On the variation of oceanic circulation (V). (In English). *GM*, **26**(4): 283–342.	3	5–7	13
1956	On the annual variation of chlorinities at the upper layer of the Kuroshio. (In English). *OM*, **7**(2): 87–93.	3	5, 7	
1956	On the behaviour of the vortex in the polar front region (Hydrography of the polar front region, I). (In English). *OM*, **7**(2): 115–132.	2	6, 7	
1956	On the distributions of oxygen and their seasonal variations in the adjacent seas of Japan (IV). (In English). *OM*, **8**(1): 1–27.	2, 3	2, 4, 7	
1956	On the movement of the upper water in the polar front region (Hydrography of the polar front region, II). (In English). *OM*, **8**(1): 29–41.	2	6, 7	
1956	On the mechanism of a cold water domain on the northern boundary of the Kuroshio (Hydrography of the polar front region, III). (In English). *OM*, **8**(1): 43–51.	2	6, 7	
1956	On the properties of the isolated cold and fresh water along the edge of the Kuroshio (Hydrography of the polar front region, IV). (In English). *OM*, **8**(1): 53–64.	2	6, 7	
1957	On the variation of oceanic circulation in the adjacent seas of Japan. (In English with French abstract). *Proc. UNESCO Symp. Phys. Oceanogr., Tokyo, 1955*: 116–129.	2, 3	5, 7	13
1957	A note on the horizontal eddy viscosity in the Kuroshio. (In English). *ROWJ*, *NS*, **3**(1): 16–25.	2, 3	7	

		Area	Subject	Chapter
1958	On the decay of a cyclonic vortex in the southern sea of Honshu (Hydrography of the Kuroshio in relation to its productivity, II). (In English). *ROWJ, NS, Special No.* **2**: 90–100.	3	6, 7	13
1958	The response of a stratified, bounded ocean to variable wind stresses. (In English). *OM*, **10**(1): 19–63.	3	7	
1960	On critical regimes and horizontal concentration of momentum in ocean currents with two-layered system. (In English). *Tellus*, **12**(2): 149–158.	8	7	
1960	On the deep water in the western North Pacific. (In English). *OM*, **11**(2): 99–110.	2–3, 6, 7	2, 4	1, 11, 13
1960	On water budget in a two-layered ocean. (In English). *OM*, **11**(2): 111–126.	1–3, 5–7	2, 7	11
1962	On formation of the intermediate water in the northern Pacific Ocean. (In English). *Pure Appl. Geophys.*, **51**: 108–119.	1–3, 6, 7	4, 7	11
1963	(with J. R. PETERSEN) The anomalous rainfall of the 1957–58 winter in the Equatorial Central Pacific arid area. (In English with Japanese abstract). *JMSJ, II*, **41**(3): 172–182.	6, 7	4, 5	
1965	Geostrophic eddies in the ocean. Part I. (In English). *LGO Tech. Rep. No. CU-21-65 to AEC No. CU-4-65 to NSF and No. CU-14-65 to ONR*: 27pp.	2	6, 7	
1965	Dynamics of Gulf Stream meanders. (In English). *LGO Tech. Rep. No. CU-1806-3 to NSF No. CU-266(48)-13 to ONR and No. CU-2663-20 to AEC*: 90pp.	2, 3	6, 7	
1965	The Kuroshio System. (In English). *Oceanus*, **12**(1): 18–21.	2, 3	2, 4, 6	13
1966	Oyashio Current. (In English). In *Encyclopedia of Oceanography*, Editor, R. W. Fairbridge, Reinhold Publishing Corporation, New York: 651–652.	1, 2	2, 4	

IIDA, Hayato

		Area	Subject	Chapter
1960	On the dynamical structure of the Tsugaru Warm Current. (In English). *KMOM*, **14**: 13–18.	2	4, 7	
1962	On the water masses in the coastal region of the south-western Okhotsk Sea. (In English). *JOSJ, 20th Anniv. Vol.*: 272–278.	1	4	9

IMPERIAL FISHERIES INSTITUTE

		Area	Subject	Chapter
1918	Oceanographic observations. (In Japanese). *KY*, **1**: 36.	1–5	1	

IMPERIAL JAPANESE NAVY (Hydrographic Department)

		Area	Subject	Chapter
1891	The northern boundary of the Kuro-Shio and the counter-current observed in the area between O-Shima and Miko-moto-Jima. (In Japanese). *Kampo, No.* 2387: 203.	3	2	
1925	Statistics of the surface density values of sea water in the North Pacific Ocean. (In Japanese). *HB*, **4**(2): 600–601.	1–7	3	

		Area	Subject	Chapter
1926	Oceanographic observations on board the survey vessels. (In Japanese). *HB*, **5**(1): 13–14.	3, 6, 7	4	
1926	On the ocean currents off the coast of Tosa. (In Japanese). *HB*, **5**(4): 101–102.	3	2	
1927	Oceanographical investigations in the Kuroshio region. (In Japanese). *HB*, **6**(1): 18–19.	2–3, 5–7	3	
1928	On the ocean and tidal currents in the vicinity of Shionomisaki in winter. (In Japanese). *HB*, **7**(7): 269–281.	3	2	
1929	On the ocean currents off the coast of Tosa in summer. (In Japanese). *HB*, **8**(7): 303.	3	2	
1933	The report of oceanic survey in the western part of the North Pacific Ocean carried out by H.I.J.M.S. Manshu from April 1925 to March 1928. (In Japanese with English abstract). *HDB*, **6**: 1–496.	1–3, 5–7	1	
1935	On the ocean current from the east of Formosa to Shionomisaki. (In Japanese). *HB*, **14**(3): 85–88.	3, 5	5	
1936	On the abnormal ocean current in the Kuroshio region off the south coast of Japan. (In Japanese). *HB*, **15**(6): 235–237.	3	2	
1936	On the Kuroshio observed off the south coast of Honshu. (In Japanese). *HB*, **15**(12): 491–494.	3	2	
1937	On the abnormal condition of the Kuroshio. (In Japanese). *HB*, **16**(6): 223–224.	3	2	
1938	On the oceanic currents in the west Pacific south of Honshu. (In Japanese). *KI*, **6**: 1–4.	3, 6	2, 3, 6	
1939	Marine conditions of the Taiwan-Kaikyo. (In Japanese). *KI*, **10**: 1–15.	5	1	
1941	Hydrographical data of the southern seas. (In Japanese). *Nanpo Suiro Shiryo*: 68pp.	2, 3, 5–7	1	

INOUE, Motoo

| 1959 | Studies on movements of albacore fishing grounds in the Northwest Pacific Ocean—(II). Influence of fluctuations of the oceanographical conditions upon the migration and distribution (pattern) of albacore in the winter-summer period and its fishing grounds in the southern waters off Japan. (In Japanese with English abstract). *JSSFB*, **25**(6): 424–434. | 3, 6 | 4 | |

ISELIN, C. O'D.

| 1936 | A study of the circulation of the western North Atlantic. (In English). *Pap. Phys. Oceanogr. Meteor.*, **4**: 101pp. | 8 | 2, 4 | 4, 8 |
| 1939 | The influence of vertical and lateral turbulence on the characteristics of the waters at mid-depths. (In English). *Trans. Amer. Geophys. Union*: 414–417. | 8 | 3 | 4 |

ISHINO, Makoto

| 1955 | Results of the oceanographical observations by Umitaka-Maru in 1954. (In Japanese). *KSH*: 65–78. | 5, 6, 7 | 1 | |

		Area	Subject	Chapter
1955	Hydrographic survey in the Equatorial Pacific Ocean, the South China Sea and the Formosa-Satsunan Kuroshio Region. (In English). *ROWJ, NS,* **2**(1): 125–131.	5, 6, 7	4	
1967	(with Kazuyuki OTSUKA, Akihiro SETOGUCHI and Keinosuke MOTO-HASHI) Fishery oceanographical studies for oceanic current system in the adjacent seas of Japan; Part I. Coastal "Kyucho" current in summer of 1966 caused by inflow of the Kuroshio Water. (In Japanese with French abstract). *La mer,* **5**(4): 244–250.	3	5	7
1967	(with Kazuyuki OTSUKA) Fishery oceanographical studies for oceanic current system in the adjacent seas of Japan; Part II. On the branch current of the Kuroshio in the region of Zunan-Boso in summer of 1967. (In Japanese with French abstract). *La mer,* **5**(4): 251–260.	3	2, 5	

ISOZAKI, Ichiro

| 1969 | An investigation on the variations of sea level due to meteorological disturbances on the coast of Japanese Islands (III). On the variation of daily mean sea level. (In English). *JOSJ,* **25**(2): 91–102. | 1–4 | 8 | 7 |

ISTOSHIN, Yu. V.

1961	Formative area of 'eighteen-degree' water in the Sargasso Sea. (In Russian). *Okeanologiya, Akad. Nauk, SSSR,* **1**: 600–607. (In English). *Deep-Sea Res.,* **9**: 384–390.	8	3	4, 10
1965	The research vessel Yu. M. Shokal'skiy expedition to Pacific Ocean in 1963–64. (In Russian). *Sovetskaya entsiklopediya Yezhegodnik BSE,* No. **9**: 509.	3	1	
1968	(with Ye. M. Sauskan) Countercurrents of the Kuroshio. (In Russian). *Okeanologiya, Akad. Nauk, SSSR,* **8**(6): 949–959. (In English): 755–762.	2, 3, 5, 6	2, 4, 5	6

IWASA, Kinji

| 1968 | On the establishment of continual measuring instrument of water temperature with submarine cable. (In Japanese). *HB, NS,* **85**: 27–28. | 5 | 8 | 1 |

JAPAN METEOROLOGICAL AGENCY (Oceanographical Section)

1956	On the long-range forecasting of oceanic condition. (In Japanese). *KH,* **5**(Special Issue): 1–4.	2, 3	5	
1956	Current measurements in the Kuroshio with the parachute current drogue. (In Japanese). *KH,* **5**(3): 157–169.	2, 3	2	
1957	Report of multiple ship survey in the polar front region from June to July, 1957. (In Japanese with English abstract). *KH,* **6**(4): 181–196.	2	1	
1958	Monthly means of sea-surface water temperature in the sea east of Honshū. (In Japanese with English abstract). *KH,* **7**(1): 59–65.	2	3, 5	

		Area	Subject	Chapter

——(Maritime Meteorology and Oceanographical Section)

| 1958 | Report of the oceanographic observations made on board the ocean weather ships on the way to the Ocean Weather Station Tango in 1957. (In Japanese). *KH*, **7**(1): 15–20. | 3, 6 | 1 | |

KAJIURA, Kinjiro

1949	On the hydrography of the Okhotsk Sea in summer. (In Japanese with English abstract). *JOSJ*, **5**(1): 19–27.	1	4	
1953	On the influence of bottom topography on ocean currents. (In English). *JOSJ*, **9**(1): 1–14.	8	7	
1954	Seasonal variation in surface water of the Japan Sea. (In Japanese). *TKH*, **1**: 33–36.	4	5	
1958	(with Mizuki TSUCHIYA and Koji HIDAKA) The analysis of oceanographical condition in the Japan Sea. (In Japanese). *TKH*, **1**: 158–170.	4	4	9
1965	Physical processes associated with the Kuroshio. (In English). *Proc. Symp. on the Kuroshio*: 15–20.	8	7	

KARAVAYEVA, V. I.

| 1965 | (with M. A. RADZIKOVSKAYA) The volumes of the main water masses in the northern part of the Pacific Ocean. (In Russian). *Okeanologiya, Akad. Nauk, SSSR*, **5**(2): 230–234. (In English): 32–37. | 1–3, 6, 7 | 3 | 4 |

KATADA, Seiichi

| 1939 | (with Tadasu KAMEDA and Humio TAMAKI) Notes on the correlations between the summer temperature in the Tohoku District and the wind velocity at Tateno, and the surface water temperature of the western part of the northern Pacific. (In Japanese with English abstract). *JMSJ*, *II*, **17**(7): 282–294. | 2 | 5 | |

KATO, Takeo

| 1958 | Analysis of distribution and correlation of water masses in the East China Sea in summers, 1954–1956. (In Japanese). *Jour. Met. Res.*,: **10**(8) 693–697. | 5 | 4 | 5 |

KAWAI, Hideo

1955	On the polar frontal zone and its fluctuation in the waters to the northeast of Japan. (I). (In Japanese with English abstract). *THRFL*, **4**: 1–46.	2	4	1, 4, 8
1955	On the polar frontal zone and its fluctuation in the waters to the northeast of Japan. (II). (In Japanese with English abstract). *THRFL*, **5**: 1–42.	2	4, 7	1, 8
1956	Notes on the inversion of the upper water temperature in the waters to the northeast of Japan in summer. (In English with Japanese abstract). *THRFL*, **6**: 71–80.	2	4	8
1957	On the natural coordinate system and its applications to the Kuroshio System. (In English with Japanese abstract). *THRFL*, **10**: 141–171.	8	7	1, 8
1958	A note on drawing the isotherms at the sea surface. (In Japanese with English abstract). *THRFL*, **12**: 106–120.	8	8	8

		Area	Subject	Chapter
1959	On the polar frontal zone and its fluctuation in the waters to the northeast of Japan (III). Fluctuation of the water mass distribution during the period 1946–1950 and hydrographic conditions in the fishing grounds of skipjack and albacore. (In Japanese with English abstract). *THRFL*, **13**: 13–59.	2	4	1, 8
1961	(with Minoru SASAKI) An example of the short-period fluctuation of the oceanographic condition in the vicinity of the Kuroshio Front. (In English with Japanese abstract). *THRFL*, **19**: 119–134.	2	3, 5	8
1962	(with Minoru SASAKI) On the hydrographic condition accelerating the skipjack's northward movement across the Kuroshio Front. (In Japanese with English abstract). *THRFL*, **20**: 1–27.	2	4	8
1965	Physiography on the east coast of the Mainland of Japan. (In Japanese with English abstract). *THRFL*, **25**: 105–130.	2	3	8
1969	Statistical estimation of isotherms indicative of the Kuroshio axis. (In English). *Deep-Sea Res., Suppl.* to **16**: 109–115.	2, 3	2, 3, 6	6, 8
1969	(with Hisao SAKAMOTO and Masako MOMOTA) A study on convergence and divergence in surface layer of the Kuroshio —I. Direct measurement and interpretation of convergence and divergence at the surface. (In Japanese with English abstract). *Bull. Nansei Reg. Fish. Lab.*, **1**: 1–14.	3	1, 2, 8	1, 8

KAWAKAMI, Kiyoshi

1957	The report of the drift-bottle experiments at the west entrance of Tsugaru Straits, Hokkaido, Japan (1). (In English). *JOSJ*, **13**(4): 131–137.	1, 2, 4	1, 2	12
1959	The report of the drift-bottle experiments at the west entrance of Tsugaru Straits, Hokkaido, Japan (2). (In English). *JOSJ*, **15**(1): 5–10.	1, 2, 4	1, 2	12

KAWAMOTO, Takeo

1955	On the distribution of the dissolved oxygen in the Pacific Ocean (Part 1. On the δ_t–O_2 diagram in the western North Pacific Ocean). (In Japanese with English abstract). *US*, **32**(2): 23–37.	6, 7	4	4, 10
1956	On the distribution of the dissolved oxygen in the Pacific Ocean (Part 2. On the δ_t–O_2 diagram in the equatorial and the eastern region of the North Pacific Ocean). (In Japanese with English abstract). *US*, **32**(5–6): 92–98.	1–3, 6, 7	4	4
1957	On the distribution of the dissolved oxygen in the Pacific Ocean (Part 3. On the δ_t–O_2 diagram in the South Pacific Ocean and the general summary). (In Japanese with English abstract). *US*, **33**(1–2): 28–33.	1–3	4	

KAWANA, Takeshi

1935	(with Yoshitaro NAKAJIMA) Results of drift-bottle experiments from 1921 to 1926. (In Japanese). *SCHH*, **38**: 103–108.	1–7	2	

	Area	Subject	Chapter
KHARCHENKO, A. M.			
1968 Currents and water masses in the East China Sea. (In Russian with English abstract). *Okeanologiya, Akad. Nauk, SSSR,* **8**(1): 38–48. (In English): 28–36.	5	2, 3	
KIKUCHI, Shigeo			
1958 Researches on the sea surface temperature of the East China Sea (1st Report)—Characters on the normal conditions. (In Japanese with English abstract). *Jour. Met. Res.,* **11**(3): 169–183.	1, 5	3, 5	5
1964 On the abnormal low sea temperature in the East China Sea due to the cold wave. (In Japanese with English abstract). *Jour. Met. Res.,* **16**(2): 84–95.	1, 5	3, 5	5
KIMURA, Kinosuke			
1936 Approach of the Kuroshio to the coast and the growth and decay of the energy of the current (Abstract). (In Japanese). *JSSFB,* **5**(2): 131.	3	2, 4	
1940 Hydrography and fisheries of yellowtail in Sagami Bay. (In Japanese with English abstract). *JIFES,* **10**: 38–230.	3	4	
1942 On the coastal stormy current "Daikyutyo." (In Japanese). *CI,* **19**(1): 1–85.	3	2,4,5	
1950 Investigation of ocean current by drift-bottle experiments (No. 1). (In Japanese). *JOSJ,* **5**(2–4): 70–84.	2, 3	1, 2	12
KISHINDO, Saburo			
1928 Chart showing the drift of bottles. (In English), *ROWJ,* **1**(1): Plate I.	7	2	
1930 On the vortical currents along the south side of the Kuroshio. (In English). *4th Pacific Sci. Cong. Proc. 2B*: 1059.	3, 5, 6	2	
1931 On the method of ocean current obervation now used by the Hydrographic Department and some results obtained. Part 1. (In Japanese). *HB,* **10**(9): 349–355.	3, 5	1, 2	
1931 On the method of ocean current observation now used by the Hydrographic Department and some results obtained. Part 2. (In Japanese). *HB,* **10**(10): 393–404.	3	1, 2	
1931 On the method of ocean current observation now used by the Hydrographic Department and some results obtained. Part 3. (In Japanese). *HB,* **10**(12): 483–503.	5	1, 2, 4	5
1932 On the method of ocean current observation now used by the Hydrographic Department and some results obtained. Part 4. (In Japanese). *HB,* **11**(2): 60–63.	3, 5	1, 2	5
1940 On the bottom temperature and the bottom current in the Pacific. (In Japanese). *HB,* **19**(8): 259–270.	1–7	4	1
1940 Stratification of sea water and the deep layer current of the Pacific. (In Japanese). *HB,* **19**(11): 351–362.	1, 2, 6, 7	4	1, 8, 11
KISI, Seiichi			
1967 Operation of the oceanographic survey by the Hydrographic Office under the Imperial Navy (Part 2). (In Japanese). *HB, NS,* **82**: 39–54.	8	8	1

		Area	Subject	Chapter

KITAHARA, Tasaku

| 1912 | On the report by the captain of the Kinkazan Maru, in 1910. (In Japanese). *Gyogyo Kihon Chosa Hokoku*, **1**: 34–37. | 2, 3 | 2, 4 | |
| 1912 | (with Mitsuhiko SHIMAMURA) On the reports by the Fisheries Experimental Stations in 1911. (In Japanese). *Gyogyo Kihon Chosa Hokoku*, **2**: 1–36. | 2–4 | 3 | |

KITAMURA, Hiroyuki

1955	The study of horizontal σ_t–O_2 relation to water mass separation. (In Japanese with English abstract). *US*, **32**(3): 51–55.	2, 3	4	
1957	On the phosphate distribution of the Western North Pacific. Chemical oceanography of the Western North Pacific (I). (In Japanese with English abstract). *US*, **33**(4–5): 65–69.	1–3, 5–7	4	
1958	Oxygen distribution in the North Pacific Ocean. Chemical oceanography of the North Pacific (IV). (In Japanese with English abstract). *US*, **34**(4): 71–75.	1–3, 6, 7	4	
1959	A short-term variation of the dissolved oxygen. Chemical oceanography of the North Pacific (V). (In Japanese with English abstract). *US*, **35**(1): 7–12.	3	4, 5	
1959	Chemical oceanography in the North Pacific (VI). Phosphate in the Equatorial Pacific Ocean. (In Japanese with English abstract). *JOSJ*, **15**(3): 131–135.	7	4	
1964	(with Takeshi SAGI) On the chemical elements in the sea south of Honshū, Japan. (In Japanese with English abstract). *KKI*, **172**: 6–54.	3	3, 5	
1965	(with Takeshi SAGI) On the chemical elements in the sea south of Honshū, Japan (Part II). (In Japanese with English abstract). *KKI*, **174**: 39–55.	3	3, 5	

KITAMURA, Naoharu

| 1962 | On the relationship between formation of typhoons and sea surface temperatures in the North Pacific Ocean. (In Japanese with English abstract). *US*, **38**(3): 75–83. | 6, 7 | 8 | |

KITANI, Kozo

| 1969 | (with Michitaka UDA) Variability of the deep cold water in Japan Sea—Particularly on the abnormal cooling in 1963. (In English with Japanese abstract). *JOSJ*, **25**(1): 10–20. | 4 | 3, 5 | |

KITANO, Kiyomitsu

1953	On the formation of the skipjack, *Katsuwonus Pelamis* (LINNAEUS), fishery ground off Kinkazan in the North-Eastern sea area along the Pacific coast of Japan. (In Japanese with English abstract). *THRFL*, **2**: 1–10.	2	4	
1956	Note on the eddies generated around the turbulent jet flow. (In Japanese with English abstract). *JOSJ*, **12**(4): 121–124.	2	2, 6	
1958	A possible interpretation of the warm core and the cold low-salinity layer. (In English). *HRFL*, **17**: 103–110.	2	6	

		Area	Subject	Chapter
1958	On the characteristics of the water mass stratifications over the frontal zone of "Tohoku-Kaiku." (In Japanese with English abstract). *HRFL*, **17**: 111–124.	2	4, 5	
1958	Oceanograhic structure of the Bering Sea and the Aleutian waters. Part 1. Based on the oceanographic observations by R. V. "Tenyo-maru" of 1957. (In English). *HRFL*, **19**: 1–9.	1	2, 4	
1958	Oceanographic structure of the Bering Sea and the Aleutian waters. Part 2. Based on the oceanographic observations by the Oshoro-maru, Komabashi, Iwate-maru and Soyo-maru during the 4 years 1956, 1936, 1935 and 1934. (In English). *HRFL*, **19**: 10–15.	1	3, 5	
1958	Observations of surface phenomena in ocean and coastal waters near Japan. (In English). *JMR*, **16**(3): 175–179.	2	4	
1967	On the Alaskan Stream. (In Japanese). *JOSJ*, **23**(6): 306–307.	1	8	4
1967	Oceanographic structure near the western terminus of the Alaskan Stream. (In English). *HRFL*, **32**: 23–40.	1	4	
1967	A note on the tendency of the long-term fluctuation of the vertical distribution of temperature over the western subarctic waters. (In English). *HRFL*, **33**: 85–93.	1	5	

Knauss, J. A.

		Area	Subject	Chapter
1962	On some aspects of the deep circulation of the Pacific. (In English). *JGR*, **67**(10): 3943–3954.	1–7	2, 3	4, 11

Kobe Marine Observatory (Oceanographic Section)

		Area	Subject	Chapter
1956	On the forecasting of oceanic conditions in the seas south of Honshū. (In Japanese). *KH*, **5**(*Special Issue*): 11–14.	3	5	
1957	Report of the oceanographic observations in the sea south of Honshū in August, 1957. (In Japanese). *KH*, **6**(4): 217–224.	3	1	
1958	Report of the oceanographic observations in the sea south of Honshū from November to December, 1957. (In Japanese with English abstract). *KH*, **7**(1): 21–28.	2	1	

Koenuma, Kan'ichi

		Area	Subject	Chapter
1933	Some oceanographical conditions of the Kuroshio area. (In Japanese with English abstract). *JO*, **5**(2): 469–476.	3, 5	4	1
1936	On the hydrography of the south-western part of the North Pacific. (In Japanese). *US*, **16**(4): 133–155. *KKI*, **93**: 1–23.	3, 5–7	4, 5	
1937	On the hydrography of the south-western part of the North Pacific and the Kurosio. Part 1: General oceanographical features of the region. (In English with Japanese abstract). *KMOM*, **6**(1): 279–331.	2–3, 5–7	4, 5	1
1937	On the water masses in the south-western North Pacific and the mixing of them. (In Japanese). *US*, **17**(4): 155–184.	3, 5–7	4	

		Area	Subject	Chapter
1937	On the water masses and their mixed states in the south-western part of the North Pacific Ocean. (In Japanese). *KKI*, **101**: 1–34.	3, 5–7	4	
1938	On the hydrography of the south-western part of the North Pacific and the Kurosio. Part II: Characteristic water masses which are related to this region, and their mixtures, especially the water of the Kurosio. (In English with Japanese abstract). *KMOM*, **6**(4): 349–414.	2–3, 5–7	4, 5	1, 4, 5
1938	On the recent variations in the Kuroshio. (In Japanese). *US*, **18**(5): 185–187.	2, 3, 5–7	6	13
1939	On the hydrography of the south-western part of the North Pacific and the Kurosio. Part III: Oceanographical investigations of the Kurosio area and its outer region; development of ocean currents in the North Pacific. (In English with Japanese abstract). *KMOM*, **7**(1): 41–114.	3, 5–7	2, 4, 5	1, 4, 5
1939	Study on the Kuroshio. (In Japanese). *US*, **19**(9): 247–291. *KKI*, **127**: 1–45.	3, 5–7	4, 5	
1940	On the energy of the Kuroshio. (In Japanese). *KKI*, **134**: 1–18. *US*, **20**(9): 219–236.	3, 5	5, 7	
1941	A story of Kuroshio. (In Japanese). *CG*, **9**(1): 33–39.	1–7	8	
1941	On the hydrography of the south-western part of the North Pacific and the Kuroshio. Part IV: Energy of the Kuroshio. Part V: Concluding remarks. (In English). *KMOM*, **7**(4): 399–435.	3	5, 7	1
1943	The unusual conditions of the Kuroshio. (In Japanese). *Tenki to Kiko*, **10**(10): 379–383.	3	6	

KOIZUMI, Masami

		Area	Subject	Chapter
1950	(with Osamu ASAOKA) The oceanographical conditions of the sea off Tohoku District. (In Japanese with English abstract). *KH*, **1**(2): 74–85.	2	4	
1951	On the oceanographical conditions around the fixed point "X-Ray" (Lat. 39°N, Long. 153°E) (1948–1949). (In Japanese with English abstract). *KH*, **2**(2): 163–169.	2	3, 5	
1952	The report of regular oceanographical observations in Sagami Bay in 1951 and 1952. (In Japanese). *KH*, **2**(4): 323–332.	3	4	
1953	On the annual variation in oceanographical elements at a fixed point (39°N, 153°E) in the Pacific Ocean. (In English). *ROWJ*, NS, **1**(1): 36–43.	2	5	
1953	On the oceanographic and meteorological conditions at the fixed station "Extra" in the North Pacific Ocean (Part 1). (In English). *ROWJ*, NS, **1**(2): 13–17.	2	5	
1955	Researches on the variations of oceanographic conditions in the region of the ocean weather station "Extra" in the North Pacific Ocean (I)—"Normal" values and annual variations of oceanographic elements. (In English). *MGP*, **6**(2): 185–201.	2, 6	5	4, 8

		Area	Subject	Chapter
1956	Researches on the variations of oceanographic conditions in the region of the ocean weather station "Extra" in the North Pacific Ocean (II)— Air temperature, sea-surface temperature and wind. (In English). *MGP*, **6**(3–4): 261–272.	2	5	
1956	Researches on the variations of oceanographic conditions in the region of the ocean weather station "Extra" in the North Pacific Ocean (III)—The variation of hydrographic conditions discussed from the heat balance point of view and the heat exchange between sea and atmosphere. (In English). *MGP*, **6**(3–4): 273–284.	2	5	8
1956	Researches on the variations of oceanographic conditions in the region of the ocean weather station "Extra" in the North Pacific Ocean (IV)—On the diurnal variations in air and sea-surface temperatures. (In English). *MGP*, **7**(2): 144–154.	2	5	
1956	Researches on the variations of oceanographic conditions in the region of the ocean weather station "Extra" in the North Pacific Ocean (V)—On a short-periodic variation of sea-surface temperature. (In English). *MGP*, **7**(2): 155–160.	2	5	
1956	Researches on the variations of oceanographic conditions in the region of the ocean weather station "Extra" in the North Pacific Ocean (VI)—A note on the diurnal variation of air temperature on the open sea. (In English). *MGP*, **7**(3): 322–326.	2, 6	5	
1957	On the forecast of oceanographic conditions in the seas west of Japan. (In English). *OM*, **9**(1): 43–49.	5	5	
1962	Seasonal variation of surface temperature of the East China Sea. (In English). *JOSJ, 20th Anniv. Vol.*: 321–329.	5	5	5, 7
1964	On the standard deviation of the surface temperature of the East China Sea. (In Japanese with English abstract). *SO* (dedicated to Prof. Hidaka in commemoration of his sixtieth birthday): 140–144.	5	3, 5	5, 7
1964	(with Taro Shinjo) Summary of oceanographic conditions of the sea west of Japan in 1963. (In Japanese). *JOSJ*, **20**(6): 286.	5	2, 4	

Konaga, Shunji

		Area	Subject	Chapter
1957	Kuroshio in Enshūnada (I). On "Shiome." (In Japanese with English abstract). *US*, **33**(4–5): 61–64.	3	2, 4	
1958	Kuroshio in Enshunada (II). On the T-σ_t relation. (In Japanese). *US*, **34**(1): 20–24. (In English). *KMOM*, **12**(2): 27–31.	3	2, 4	
1959	The water temperature at the sea surface. (In Japanese with English abstract). *US*, **35**(2–3): 44–50.	8	8	
1960	The water temperature at the sea surface (III). An influence of the weather condition (1). (In Japanese with English abstract). *JOSJ*, **16**(3): 128–133.	3, 5	5, 7	
1961	Kuroshio in Enshu-nada (III)—Variations of the depth of isotherms. (In English). *OM*, **13**(1): 31–40.	2	4, 5	
1961	The water temperature at the sea surface (IV). The influence of weather (2). (In Japanese with English abstract). *JOSJ*, **17**(2): 68–73.	3	5	

		Area	Subject	Chapter
1962	Water temperature at the sea surface (V)—Influences of the internal wave. (In Japanese with English abstract). *US*, **38**(4): 107–114.	3	4, 5	
1964	On the current velocity measured with geomagnetic electro-kinetograph. (In Japanese). *JOSJ*, **20**(1): 1–6.	3	2, 4	
1965	The observation of the internal waves. (In English). *OM*, **17**(1–2): 141–174.	3	2, 4, 5	
1966	Water temperature at the sea surface (VII)—A calculation of energy budget near the sea surface. (In Japanese with English abstract). *US*, **41**(3–4): 148–154.	3	5	
1967	(with Kenzo SHUTO, Hiromitsu KUSANO and Kimisuke HORI) On the relation between Kuroshio strong currents and the water temperature of 200 m depth. (In Japanese with English abstract). *US*, **42**(3–4): 93–97.	3	2, 4	8
1967	(with Kenzo SHUTO, Hiromitsu KUSANO and Kimisuke HORI) On the relation between Kuroshio strong currents and the water temperature of 200 m depth (II). (In Japanese with English abstract). *US*, **43**(2): 48–53.	3	2	
1968	Variations of the oceanographic condition south of Japan in relation to the mean sea level at Kushimoto and Uragami tidal station, Japan. No. 1. (In Japanese with English abstract). *US*, **43**(4): 125–134.	3	5	
1968	Variations of the oceanographic condition south of Japan in relation to the mean sea level at Kushimoto and Uragami tidal stations, Japan. No. 2. (In Japanese with English abstract). *US*, **44**(1): 1–12.	3	5	

KONDO, Masato

		Area	Subject	Chapter
1968	Short-period fluctuation of the oceanographic condition of the Tsushima warm current (I): Fluctuations of temperature and chlorinity. (In Japanese with English abstract). *SFLB*, **36**: 1–20.	5	5	1

KONISHI, Tatehiko

		Area	Subject	Chapter
1921	Whereabouts of drift-bottles (I). (In Japanese). *US*, **1**(8): 86–90.	5–7	1, 2	1, 12
1921	Whereabouts of drift-bottles (II). (In Japanese). *US*, **1**(4): 38–41.	3, 5, 6	1, 2	1, 12
1922	Whereabouts of drift-bottles (III). (In Japanese). *HB*, **1**(2): 59–66.	1–7	1, 2	1, 12
1922	Whereabouts of drift-bottles (IV). (In Japanese). *HB*, **1**(3): 111–122.	1–7	1, 2	1, 12
1923	Whereabouts of drift-bottles (V). (In Japanese). *HB*, **2**(5): 31–41.	1–7	1, 2	1, 12

KORT, V. G.

		Area	Subject	Chapter
1958	Investigations of the far eastern seas and the Pacific Ocean, conducted by the Institute of Oceanology. (In Russian). *Trudy Okeanogr. Komissii, Akad. Nauk, SSSR*, **3**: 3–12.	1–4	8	

		Area	Subject	Chapter

1966 The 38th cruise of the R/V "Vityaz" (main scientific results). — Area 6, 7; Subject 1
(In Russian). *Okeanologiya, Akad. Nauk, SSSR*, **6**(6): 1099–1107.
(In English): 886–894.

Koshlyakov, M. N.

1961 Vertical circulation of waters in the region of the Kuroshio. — Area 2, 6; Subject 2, 7; Chapter 12
(In Russian). *Okeanologiya, Akad. Nauk, SSSR*, **1**(5): 805–814.
(In English). Deep-Sea Res., **10**: 284–292.

1961 Problems of the water dynamics of the north-western Pacific. — Area 1–3, 6; Subject 2, 7
(In Russian). *Trudy Inst. Okeanol., Akad. Nauk, SSSR*, **38**: 31–55.

1965 (with V. G. Neiman) — Area 7; Subject 2
Some results of measurements and calculations of zonal currents in the equatorial part of the Pacific Ocean.
(In Russian). *Okeanologiya, Akad. Nauk, SSSR*, **5**(2): 235–249.
(In English): 37–49.

Kosugi, Akira

1966 The state of the sea off Tohoku District in 1964 (Notes). — Area 2; Subject 4
(In Japanese). *JOSJ*, **22**(4): 161–163.

Kubota, Terumi

1961 (with Kinji Iwasa) — Area 2; Subject 5; Chapter 12
On the currents in Tugaru Strait.
(In Japanese with English abstract). *HB, NS*, **65**: 19–26.

Kuksa, V. I.

1959 Hydrological features of the North Kuril waters. — Area 1; Subject 4
(In Russian). *Trudy Inst. Okeanol., Akad. Nauk, SSSR*, **36**: 191–214.

1962 On the formation of the layer of low salinity in the northern part of the Pacific Ocean. — Area 2; Subject 3; Chapter 4
(In Russian). *Okeanologiya, Akad. Nauk, SSSR*, **2**: 769–782.

1963 Basic regularity in formation and distribution of intermediate waters in the northern part of the Pacific Ocean. — Area 2; Subject 3; Chapter 4
(In Russian). *Okeanologiya, Akad. Nauk, SSSR*, **3**: 30–43.

Kun, M. S.

1969 (with G. N. Gladkikh, E. P. Karedin, V. P. Pavlychev, V. I. Rachkov and E. G. Starodubtsev) — Area 1–3; Subject 4
Hydrological conditions and biological characteristic of the Kuroshio waters.
(In Russian). *Izvestija, Tichookeanskogo nautchno-issledovatelskogo instituta ribnogo chozyaystva i okeangrafii*, **68**: 3–14.

Kurashina, Shoji

1952 Result of oceanographic observations in the Eastern China Sea in April and May, 1952. — Area 5; Subject 1
(In Japanese with English abstract). *HB, NS*, **33**: 224–225.

1967 (with Koji Nishida and Shunji Nakabayashi) — Area 1; Subject 2, 4, 8; Chapter 8
On the open water in the southeastern part of the frozen Okhotsk Sea and the currents through the Kurile Islands.
(In Japanese with English abstract). *JOSJ*, **23**(2): 57–62.

		Area	Subject	Chapter

KURODA, Ryuya

1958	Notes on the "inversion of water density" in the vertical distributions of sea water density. (In Japanese with English abstract). *THRFL*, **11**: 82–87.	2	4	1
1959	Notes on the phenomena of "inversion of water temperature" off the Sanriku Coast of Japan (I). (In Japanese with English abstract). *THRFL*, **13**: 1–12.	2	4	1, 8
1960	Notes on the phenomena of "inversion of water temperature" off the Sanriku Coast of Japan (II). Results from repeated observations in the fixed point. (In Japanese with English abstract). *THRFL*, **16**: 65–86.	2	4	1, 8
1962	On the states of appearances of "Shiome" in the waters to the northeast of Japan. (In Japanese with English abstract). *THRFL*, **22**: 45–115.	2	8	1

KURUME, Shinemon

| 1918 | On the oceanic current in the sea adjoining Taiwan. (In Japanese). *Taiwan Suisan Zasshi*, **36**: 47–49. | 5, 6 | 2 | |

LaFOND, E. C.

| 1968 | Detailed temperature and current data sections in and near the Kuroshio Current. (In English). *An Oceanogr. Data Report for CSK, Marine Environment Division, Naval Undersea Center, San Diego*: 1+22pp. | 3, 6 | 1, 4 | 8 |

LEONTYEVA, V. V.

1959	(with A. E. GAMUTILOV) The influence of Pacific Ocean water on the hydrological conditions in Kronotski Bay (according to the survey in the spring of 1955). (In Russian). *Trudy Inst. Okeanol., Akad. Nauk, SSSR*, **36**: 59–72.	1	4	
1960	Latest results of hydrological study of the Kuroshio Current from the viewpoint of fishery oceanology. (In Russian). *Trudy Soveshchanii Ikhtiol. Komissii Akad. Nauk, SSSR*, **10**: 205–209.	8	8	
1961	Waters of Kuroshio in the northwestern Pacific in summer 1953 and 1954. (In Russian). *Trudy Inst. Okeanol., Akad. Nauk, SSSR*, **38**: 3–30.	1–3	4	
1963	(with M. A. RADZIKHOVSKAYA) Identification of hydrological structures and water masses in the ocean. (In Russian). *Trudy In-ta. Okeanol. Akad. Nauk, SSSR*, **66**: 79–90.	8	8	

LISITZIN, Eugenie

| 1967 | Sea level variation in the Sea of Japan. (In English). *Int. Hydrogr. Rev.*, **44**(2): 11–22. | 4 | 5 | |

LYNN, R. J.

| 1968 | (with J. L. REID, Jr.) Characteristics and circulation of deep and abyssal waters. (In English). *Deep-Sea Res.*, **15**: 577–598. | 1–7 | 3 | |

		Area	Subject	Chapter

MAEDA, Akio

1965   On the variation of the vertical thermal structure. — Area 6, Subject 5
(In English). *JOSJ*, **20**(6): 255–263.

MAEDA, Sonosuke

1968   On the cold water belt along the northern coast of Hokkaido in the — Area 1, Subject 4, Chapter 9
Okhotsk Sea (Hydrography of the Okhotsk Sea, Part I).
(In English with Japanese abstract). *US*, **43**(3): 1–20.

MANABE, Shukuro

1957   On the modification of air-mass over the Japan Sea when the out- — Area 4, Subject 8
burst of cold air predominates.
(In English). *JMSJ, II*, **35**(6): 311–326.

1958   On the estimation of energy exchange between the Japan Sea and the — Area 4, Subject 8
atmosphere during winter based upon the energy budget of both the
atmosphere and the sea.
(In English with Japanese abstract). *JMSJ, II*, **36**(4): 123–134.

MAO, H. L.

1955   (with Kozo YOSHIDA)
Physical oceanography in the Marshall Islands area. — Area 6, 7, Subject 2–4, Chapter 5
(In English). *Geol. Surv. Prof. Pap.*, 260-R: 645–684.

MARITIME SAFETY AGENCY (Hydrographic Dept.)

1948   Hydrography of Izu Shoto and its adjacent region in 1947. — Area 3, Subject 2, 4, 6
(In Japanese). *HB, NS*, **8**: 31–32.

1948   Hydrography of the seas west of Kyushu and south of Japan in Mar.– — Area 3, 5, Subject 2, 4
Apr., 1948.
(In Japanese). *HB, NS*, **9**: 65–67.

1949   Oceanographic condition of the adjacent sea of the Izu-Shoto in — Area 3, Subject 1
December 1946.
(In Japanese). *KJ*, **4**: 1–8.

1950   Oceanographic results in the area east and south of Honsyu, May, — Area 2, 3, Subject 2, 4
1950.
(In Japanese). *HB, NS*, **20**: 210–211.

1950   Oceanographic results in the area south of Honsyu, Jun.–Jul., 1950. — Area 3, Subject 2, 4
(In Japanese). *HB, NS*, **20**: 210–212.

MARUKAWA, Hisatoshi

1926   On the Ogasawara Current. — Area 6, Subject 2
(In Japanese). *SSK*, **523**: 17–18.

MASUZAWA, Jotaro

1950   The oceanographical conditions off Ensyu in December, 1948. — Area 3, Subject 4
(In Japanese with English abstract). *KH*, **1**(2): 85–92.

1950   Some examples of typhoons as a motive of variations of oceanographic — Area 6, Subject 4, 5
conditions.
(In Japanese with English abstract). *KH*, **1**(3): 118–123.

1950   On the intermediate water in the Southern Sea of Japan. — Area 1–3, 6, Subject 4, Chapter 1
(In English). *OM*, **2**(4): 137–144.

		Area	Subject	Chapter
1951	On the intermediate water in the Northeastern Sea of Japan (first paper). (In Japanese with English abstract). *KH*, **2**(1): 5–13.	2	4	8
1951	On the transport of Kuroshio Current south off Shionomisaki. (In Japanese with English abstract). *KH*, **2**(2): 115–118.	3	2, 5	
1951	Diurnal variations of the surface water temperature in the open sea. (In Japanese with English abstract). *JOSJ*, **6**(3): 165–167.	6	5	
1952	On the heat exchange between sea and atmosphere in the Southern Sea of Japan. (In English). *OM*, **4**(2): 49–55.	3	8	4
1952	On the intermediate water in the Northeastern Sea of Japan (Second Paper). (In Japanese with English abstract). *KH*, **2**(3): 241–245.	2	4	8
1953	(with Toshisuke NAKAI) On the fluctuation of the Kuroshio Current off Shionomisaki from 1950 to 1952. (In English). *ROWJ, NS*, **1**(2): 25–32.	3	2, 4, 5	1
1954	On the seasonal variation of the Kuroshio east of Cape Kinkazan of Japan Proper. (In Japanese with English abstract). *KH*, **3**(4): 251–255.	2	4, 5	1, 8
1954	On the Kuroshio south off Shiono-Misaki of Japan (Currents and water masses of the Kuroshio System I). (In English). *OM*, **6**(1): 25–33.	3, 6	2, 4, 5	1, 4, 6, 7, 13
1955	Preliminary report on the Kuroshio in the Eastern Sea of Japan (Currents and water masses of the Kuroshio System III). (In English). *ROWJ, NS*, **2**(1): 132–140.	2	1, 2, 4	8
1955	An outline of the Kuroshio in the Eastern Sea of Japan (Currents and water masses of the Kuroshio System IV). (In English). *OM*, **7**(1): 29–48.	2	2, 4, 5	8
1955	(with Toshisuke NAKAI) Notes on the cross-current structure of the Kuroshio (Currents and water masses of the Kuroshio System V). (In English). *ROWJ, NS*, **2**(2): 96–101.	2, 3	2, 4	6
1956	A note on the Kuroshio farther to the east of Japan (Currents and water masses of the Kuroshio System VI). (In English). *OM*, **7**(2): 97–104.	2	4	4, 8
1956	On the cold belt along the northern edge of the Kuroshio (Currents and water masses of the Kuroshio System VIII). (In English). *OM*, **8**(2): 151–156.	2	4	1, 8
1957	An example of cold eddies south of the Kuroshio (Currents and water masses of the Kuroshio System VII). (In English). *ROWJ, NS*, **3**(1): 1–7.	2	6	6
1957	A contribution to the knowledge on the Kuroshio east of Japan. (In English). *OM*, **9**(1): 21–34.	2, 3	2, 4, 6	4, 8
1958	A short-period fluctuation of the Kuroshio east of Cape Kinkazan. (In English). *OM*, **10**(1): 1–8.	2	2, 4, 5	1, 7, 8
1960	Western boundary currents and vertical motions in the subarctic North Pacific Ocean. (In English). *JOSJ*, **16**(2): 69–73.	1	2	4, 11

		Area	Subject	Chapter
1960	Statistical characteristics of the Kuroshio current. (In English). *OM*, **12**(1): 7–15.	2, 3	2, 5	6, 8, 13
1961	Recent research on the Kuroshio current. (In Japanese with English abstract). *Geographical studies presented to Prof. Tsujimura in honor of his 70th birthday*: 354–364.	2, 3, 5	2	13
1962	The deep water in the western boundary of the North Pacific. (In English). *JOSJ, 20th Anniv. Vol.*: 279–285.	2, 3	4	11
1962	(with Takeo TSUCHIDA and Tamio INOUE) The monthly mean sea-surface temperature in the north-western Pacific. (In English). *OM*, **13**(2): 77–87.	1–3, 5, 6	3, 5	4
1964	Flux and water characteristics of the Pacific North Equatorial Current. (In English). *SO* (dedicated to Prof. Hidaka in commemoration of his sixtieth birthday): 121–128.	7	2, 4	4, 5
1964	A typical hydrographic section of the Kuroshio Extension. (In English). *OM*, **16**(1–2): 21–30.	2	4	1, 4, 8
1965	Water characteristics of the Kuroshio. (In English). *OM*, **17**(1–2): 37–47.	2, 3	2, 4	4, 5, 6
1965	A proposal for the standard sections of the Kuroshio System (Shorter contribution). (In Japanese). *JOSJ*, **21**(2): 68.	1–7	8	
1965	A note on the seasonal variation of the Kuroshio velocity. (In Japanese). *JOSJ*, **21**(3): 117–118.	3	2, 5	4, 6, 8
1965	Meanders of the Kuroshio—The cold water mass south of Honshu. (In Japanese). *KG*, **35**(11): 588–593.	3	6	6, 7, 8, 13
1967	Ryofu-Maru's survey for CSK in January to March 1967. (In English). *OM*, **19**(1): 1–5.	5–7	1	
1967	An oceanographic section from Japan to New Guinea at 137°E in January 1967. (In English). *OM*, **19**(2): 95–118.	5–7	4	1, 4, 5
1968	Cruise report on multi-ship study of short-term fluctuations of the Kuroshio in October to November 1967. (In English). *OM*, **20**(1): 91–96.	3	1	5, 6, 7, 8, 12
1968	Second cruise for CSK, Ryofu-maru, January to March 1968. (In English). *OM*, **20**(2): 173–185.	5–7	1	1, 4
1969	The Mindanao Current. (In Japanese with English abstract). *Uda's Anniv. Vol.*: 99–104.	5	2, 4	4
1969	Subtropical Mode Water. (In English). *Deep-Sea Res.*, **16**: 463–472.	5	2, 4	4, 8
1969	A short note on the Kuroshio stream axis. (In Japanese). *JOSJ*, **25**(5): 259–260.	1, 2, 4	3	6

MATSUDAIRA, Yasuo

		Area	Subject	Chapter
1941	On the hydrography of the North Pacific Ocean and its influence upon the climate of Japan. (In Japanese). *KKI*, **136**: 1–10.	3, 5–7	4	
1941	Hydrographic conditions in the North Pacific Ocean and the climate of Japan (Preliminary report). (In Japanese). *US*, **21**(10): 261–270.	3, 5–7	4	

		Area	Subject	Chapter
1964	The meaning of the high salinity region in the ocean. (In Japanese with English abstract). *SO* (dedicated to Prof. Hidaka in commemoration of his sixtieth birthday): 85–88.	3, 6, 7	4	5

MATSUMOTO, Sadayuki

		Area	Subject	Chapter
1962	On the periodical fluctuation of the seal level in the western parts of Japan. (In Japanese with English abstract). *US*, **38**(2): 48–52.	2, 3	3, 5	

McGARY, J. W.

		Area	Subject	Chapter
1961	(with J. J. GRAHAM and Tamio OTSU) Oceanography and North Pacific albacore. (In English). *Calif. Cooper. Ocean. Fish. Inv.*, **8**: 45–53.	1, 2	4	

MIYAKE, Yasuo

		Area	Subject	Chapter
1939	Chemical studies of the western Pacific Ocean, Part 5—The vertical variation of minor constituents in the Kuroshio region. (In English). *CSJB*, **14**: 461–466.	3	4	
1939	Chemical studies of the western Pacific Ocean, Part 6—The vertical variation of minor constituents in the Kuroshio region. (In English). *CSJB*, **14**: 467–471.	3	4	
1946	Kuroshio and Oyashio. (In Japanese). *Agronomy*, **2**: 425–427.	2, 3	3	

MIYATA, Kazuo

		Area	Subject	Chapter
1958	Characteristics of the Tsushima Current in the Japan Sea. (In Japanese). *TKH*, **1**: 147–152.	4	4	9

MIYAZAKI, Masamori

		Area	Subject	Chapter
1955	Seasonal variations of the sea level along the Japanese coasts. (In English). *ROWJ, NS*, **2**(3): 1–8.	1–4	5	

MIYAZAKI, Michio

		Area	Subject	Chapter
1952	The heat budget of the Japan Sea. (In Japanese with English abstract). *HRFL*, **4**: 1–54.	4	4	9
1953	On the water masses of the Japan Sea. (In Japanese with English abstract). *HRFL*, **7**:1–65.	4	4	9
1960	(with Shigeo ABE) On the water masses in the Tsushima Current area. (In Japanese with English abstract). *JOSJ*, **16**(2): 59–68.	4, 5	4	9

MIZUKOSHI, Masayoshi

		Area	Subject	Chapter
1897	On the Kuroshio near Izu Islands. (In Japanese). *JG*, **9**: 431–433, 476–479.	3	2	

MOISEYEV, L. K.

		Area	Subject	Chapter
1967	Approximate calculation of the depth of the wind-mixed layer in the ocean. (In Russian). *Okeanologiya, Adad. Nauk, SSSR*, **7**(1): 51–63. (In English): 39–50.	1–7	3, 7	

		Area	Subject	Chapter
MORIYASU, Shigeo				
1953	On the hydrographical condition in the Kuroshio region (1952) (Kii Suido). (In Japanese with English abstract). *KKI*, **163**: 31–40.	3	4	
1955	(with Yasuo MATSUDAIRA, Takeo KAWAMOTO, Fukuzo UENO and Kenzo FURUHASHI) On the Kuroshio to the south of Honshu. (In Japanese). *KSH*: 47–63.	3	1	
1955	An example of the variation of the oceanographical condition of Enshu-nada. (In English). *JOSJ*, **11**(2): 43–46.	3	5, 6	13
1956	On the oceanographical conditions south off Honshu in 1954. (In English). *OM*, **7**(2): 105–113.	3	4	
1956	On the fluctuation of the Kuroshio south of Honshū (1) (Statistical treatment of the displacement of the stream axis). (In English). *OM*, **8**(2): 143–149.	3	5	1
1957	An attempt to forecast oceanographic conditions south of Honshū. (In English). *OM*, **9**(1): 35–42.	3	5	
1958	On the fluctuation of the Kuroshio south of Honshū (2): Seasonal variations of water temperature and chlorinity in the upper layer. (In English). *KMOM*, **12**(2): 1–18.	3	5	1,7
1958	On the fluctuation of the Kuroshio south of Honshū (3) (A short note on the computation of the relative volume transport). (In English). *OM*, **10**(1): 81–89.	3	2, 5	1
1958	An attempt to estimate the dynamic depth anomaly. (In Japanese with English abstract). *US*, **34**(2): 40–44.	3	8	
1958	On the fluctuation of the Kuroshio south of Honshū (4)—(The influence of the oceanographic conditions upon the monthly mean sea level). (In English). *JOSJ*, **14**(4): 137–144.	3	5	1
1959	On the fluctuation of the Kuroshio south of Honshū (5): The horizontal distribution of density in a circular eddy. (In English). *KMOM*, **13**(2): 12–21.	3	6, 7	1
1959	Supplementary note on the dynamical property of the cold water region. (In English). *OM*, **11**(1): 13–19.	3	6	7
1960	The thickness of the upper homogeneous layer. (In English). *ROWJ, NS*, **5**(2): 44–51.	3	4	
1960	On the monthly sea level on the south coast of Japan. (In English). *KMOM*, **14**: 19–31.	3	5	1
1961	On the influence of the monsoon on the oceanographic conditions. (In English). *JOSJ*, **17**(2): 74–79.	3	5, 7	13
1961	On the difference in the monthly sea level between Kushimoto and Uragami, Japan. (In English). *JOSJ*, **17**(4): 197–200.	3	5	1
1961	An example of the conditions at the occurrence of the cold water region. (In English). *OM*, **12**(2): 67–76.	3	6	1, 6, 7, 8, 13

		Area	Subject	Chapter
1963	The fluctuation of hydrographic conditions in the sea south of Honshū, Japan (Review). (In English). *OM*, **15**(1): 11–29.	3	5, 6	1, 6, 7, 12, 13
1964	Preliminary note on the water temperature in the surface layer. (In English). *SO* (dedicated to Prof. Hidaka in commemoration of his sixtieth birthday): 150–155.	3	5	
1966	The water temperature in the surface layer south of Honshu, Japan (Data). (In Japanese with English abstract). *KKI*, **176**: 34–42.	3	3, 5	
1967	On the anomaly of the sea surface temperature in the East China Sea (I). (In English). *OM*, **19**(2): 201–220.	5	5	
1968	On the anomaly of the sea surface temperature in the East China Sea (II). (In English). *OM*, **20**(2): 121–132.	5	5	

MOROCHKINE, K. V.

		Area	Subject	Chapter
1955	Water masses of the north-western Pacific and the Kurile-Kamchatka Trench region. (In Russian). *Trudy Inst. Okeanol., Akad. Nauk, SSSR*, **12**: 155–160.	1	4	
1958	Les masses d'eau de la partie mod-ouest de l'Ocean Pacifique dans la region de la fosse Kourilo-Kamtchat-kienne. (In French). *Bull. d'Info., C.C.O.E.C.*, **10**(7): 407–412.	1	4	

MOROZ, I. F.

		Area	Subject	Chapter
1969	General characteristic of the subarctic front of the Pacific Ocean. (In Russian). *Izvestija, Tichookeanskogo nautchno-issle-dovatelskogo instituta ribnogo chozyaystva i okeangrafii*, **68**: 15–32.	1, 3	3, 4	

MUROMTSEV, A. M.

		Area	Subject	Chapter
1958	The basic features of Pacific hydrology. (In Russian). *Gidrometeoizkat*: 631pp.	1–7	3, 5	
1963	Atlas of temperature, salinity and density of water in the Pacific Ocean. (In English). *Mezhduvedomstvennuy Geofizicheskii Komitet pri Prezidiume Akad. Nauk, SSSR, Moskba*: 120pp.	1–7	2, 3, 5	4

NAGAHARA, Masanobu

		Area	Subject	Chapter
1965	On the seasonal variations of the transport volume and oceanic conditions in the Japan Sea from 1962 to 1964. (In Japanese with English abstract). *JSRFL*, **14**: 71–79.	4	5	9

NAGASAKI MARINE OBSERVATORY (Oceanographical Section)

		Area	Subject	Chapter
1955	Report of the oceanographical observations in the Eastern China Sea in winter, 1954. (In Japanese with English abstract). *KH*, **4**(1): 70–73.	5	3, 5	
1955	Report of the oceanographical observations in the Eastern China Sea in summer, 1954. (In Japanese with English abstract). *KH*, **4**(1): 74–84.	5	3,5	
1956	Forecasting of summer oceanic conditions in 1956. (In Japanese). *KH*, **5**(Special Issue): 15–17.	5	5	

	Area	Subject	Chapter

1958 The sea surface temperature of every ten days of the East China Sea in 1953 to 1957.
(In Japanese). *KH*, **7**(Special Issue): 100pp. — Area 5, Subject 3–5

Nagata, Yutaka

1967 Shallow temperature inversions at Ocean Station V.
(In English with Japanese abstract). *JOSJ*, **23**(4): 194–200. — Area 2, 3; Subject 3, 5; Chapter 8

1967 On the structure of shallow temperature inversions.
(In English with Japanese abstract). *JOSJ*, **23**(5): 221–230. — Area 2, 3; Subject 4; Chapter 8

1968 Shallow temperature inversions in the sea to the east of Honshu, Japan.
(In English with Japanese abstract). *JOSJ*, **24**(3): 103–114. — Area 2, 3; Subject 3; Chapter 8

Nagayama, Moriyoshi

1957 The heat budget of the Eastern China Sea.
(In Japanese with English abstract). *J. Met. Res.*, **9**(2): 67–75. — Area 5; Subject 3

Nakai, Jinjiro

1955 (with Nobuo Watanabe, Shigekazu Hattori, Yasushi Honjo, Susumu Uehara and Hansuke Ota)
Report on the investigation of the Kuroshio by Soyo Maru in July, 1954.
(In Japanese). *KSH*: 83–110. — Area 3; Subject 1

1967 (with Shigemasa Hattori, Koji Honjo and Sigeiti Hayasi)
Fluctuations in the fish populations related to the environmental changes. Taking changes of the sardine, anchovy, mackerel, and jack mackerel caused by an oceanographic anomaly in the early 1963 as an example.
(In English). *J. Coll. Mar. Sci. Tech., Tokai Univ.*, **2**: 115–130. — Area 3; Subject 8

Nakamiya, Terutoshi

1949 Results of oceanographic observations in the latter part of 1948.
(In Japanese). *HB, NS*, **11**: 3–7. — Area 2–4; Subject 1

1953 On the ocean current in the Japan Sea inferred from results of drift-bottle experiments.
(In Japanese with English abstract). *HB, Special No.* **10**: 1–9. — Area 4; Subject 1; Chapter 1, 12

Nakamura, Akikazu

1958 Report on recoveries of drift-bottles in the western Pacific Ocean, 1956.
(In English). *Mem. Osaka Univ. of Liberal Arts & Education, B. Nat. Sci.*, No. **7**: 125–162. — Area 3, 5–7; Subject 2, 5

1962 On the surface oceanographic observations across the North Pacific Ocean in winter, 1959–1960, and in summer and autumn, 1960.
(In English). *ROWJ, NS*, **6**(2): 31–57. — Area 1–3; Subject 4

Nakamura, Hiroshi

1959 (with Hajime Yamanaka)
Relation between the distribution of tunas and the ocean structure.
(In Japanese with English abstract). *JOSJ*, 15(3): 143–149. — Area 1–3, 5–7; Subject 4

	Area	Subject	Chapter
NAKAYAMA, Ichizō			
1951 On the mean surface temperature of the neighbouring sea of Japan. (In Japanese). *KKI*, **159**: 1–16.	2–6	3, 5	9
NAN'NITI, Tosio			
1950 On the oceanographical conditions of the sea off Tokaido. (In Japanese with English abstract). *KH*, **1**(3): 129–136.	3	4	
1951 On the variation of the oceanographical condition along the so-called "C" line (38°N, from 141°E to 153°E) from August 1948 to December 1949. (In English). *OM*, **3**(1): 27–48.	2	4, 5	1
1951 On the Austausch coefficient in the sea. (In English). *OM*, **3**(1): 49–52.	2, 3	7	
1951 On the fluctuation of the Kuroshiwo and the Oyashiwo. (In English). *MGP*, **2**(1): 102–111.	2	5	
1952 On the fluctuation of the Kuroshio and the wind. (In English). *JOSJ*, **8**(1): 23–29.	2, 3, 5	5	
1957 A preliminary study for numerical prediction of oceanic conditions. (In English). *OM*, **9**(1): 107–120.	3	7	
1958 A theory of the mechanism of the generation of the cold water region in the offing of Enshūnada. (In English with Japanese abstract). *MGP*, **8**(4): 317–331.	3	6, 7	7, 13
1958 Relation between the velocity and the location of the front of the Kuroshio off the Tohoku district. (In English). *OM*, **10**(2): 185–192.	2	4, 5	
1958 On the boundaries of ocean currents (Correspondences). (In Japanese). *JOSJ*, **14**(2): 65.	8	8	
1959 A new simple current profile recorder. (In Japanese with English abstract). *JOSJ*, **15**(2): 57–60.	8	8	
1959 Time variation of vorticity in the sea. (In English with Japanese abstract). *MGP*, **10**(1): 38–43.	2	7	
1959 A supplementary note to the previous paper, "A theory of the mechanism of the generation of the cold water region in the offing of Enshūnada." (In English with Japanese abstract). *MGP*, **10**(1): 51–53.	3	6, 7	7, 13
1959 On the structure of ocean currents (III). A new Nan'niti-Iwamiya current meter and the current profile observed by it. (In English with Japanese abstract). *MGP*, **10**(2): 124–134.	2	2	1, 12
1960 Long-period fluctuations in the Kuroshio. (In English with Japanese abstract). *MGP*, **11**(2–4): 339–347.	3, 5	5	7
1962 Deep-sea current measurements. (In English). *JOSJ*, **18**(2): 73–77.	3	2	
1962 New oceanographic routine observations necessary for Japan (Correspondences). (In Japanese). *JOSJ*, **18**(2): 98.	3, 6, 7	8	
1962 (with Masashi YASUI) Deep-sea current measurement during the JEDS-4. (In English). *OM*, **13**(2): 133–136.	2	2	

		Area	Subject	Chapter
1962	A study of the temperature of the surface layer in the Pacific Ocean near Japan—Mean values and annual variations of temperature. (In English with Japanese abstract). *MGP*, **13**(2): 196–205.	2, 3	3, 5	
1963	(with Akira WATANABE, Hideo AKAMATSU and Toshisuke NAKAI) Deep current measurements in the cold water mass and vicinity, the Enshū Nada Sea. (In English). *OM*, **14**(2): 135–139.	3	2	12
1964	Some observed results of oceanic turbulence. (In English). *SO* (dedicated to Prof. Hidaka in commemoration of his sixtieth birthday): 211–215.	3	5, 7	
1964	(with Hideo AKAMATSU, Toshisuke NAKAI and Kazutoshi FUJII) An observation of a deep current in the southern east sea of Tori Shima. (In English). *OM*, **15**(2): 113–122.	6	2	12
1964	(with Hideo AKAMATSU and Toshisuke NAKAI) A further observation of a deep current in the East-North-East Sea of Torishima. (In English). *OM*, **16**(1–2): 11–19.	3	2	11, 12
1964	Oceanic turbulence. (In English). *OM*, **16**(1–2): 35–45.	8	7	
1965	(with Hideo AKAMATSU and Toshisuke NAKAI) A deep current measurement in the Honshu Nankai, the sea south of Honshu, Japan. (In English). *OM*, **17**(1–2): 77–86.	6	2	11, 12
1966	(with Hideo AKAMATSU) Deep current observations in the Pacific Ocean near the Japan Trench. (In English with Japanese abstract). *JOSJ*, **22**(4): 154–160.	2, 3, 5–7	2	1, 6, 11 12
1966	(with Hideo AKAMATSU and Takeo YASUOKA) A deep current measurement in the Japan Sea. (In English). *OM*, **18**(1–2): 63–71.	4	2, 4, 5	12
1967	(with Akimitsu FUJIKI) Secular variation of hydrographic conditions in the East Tsushima Strait. (In Japanese with English abstract). *JOSJ*, **23**(4): 201–212.	4	3, 5	9

NITANI, Hideo

		Area	Subject	Chapter
1959	(with Kinji IWASA and Wataru INADA) On the oceanic and tidal current observation in the channel by making use of induced electric potential. (In English). *HB*, *NS*, **61**: 14–24.	2–4	2, 5	1, 12
1961	On the general oceanographic conditions at the western boundary region of the North Pacific Ocean. (In Japanese with English abstract). *HB*, *NS*, **65**: 27–35.	5, 7	2, 4	1, 5
1963	(with Bunkichi IMAYOSHI) On the analysis of the deep-sea observations in the Kurile-Kamchatka Trench. (In English). *JOSJ*, **19**(2): 75–81.	1	4	11
1963	On the analysis of deep sea in the region of the Kurile-Kamchatka, Japanese and Izu-Bonin Trench. (In English). *JOSJ*, **19**(2): 82–92.	2, 3, 6	2, 4, 7	1, 11

		Area	Subject	Chapter
1969	On the variability of the Kuroshio in recent several years. (In Japanese). *Bull. Japan. Soc. Fish. Oceanogr.*, **14**: 13–18.	3	2, 5, 6	7

NOVOZHILOV, V. N.

1961	Hydrological conditions in the regions of the Komandorski Islands and Kamchatka Peninsula in the Pacific Ocean in summer 1956. (In Russian). *Trudy Inst. Okeanol., Aka. Nauk, SSSR*, **38**: 56–60.	1	2, 4	

OHKAWA, Aifusa

1964	(with Naruji SUZUKI) Summary of oceanographic conditions of the sea west of Japan in 1962 (Note). (In Japanese). *JOSJ*, **20**(2): 95.	5	4	

OHWADA, Mamoru

1966	(with Katsumi YAMAMOTO) Some chemical elements in the Japan Sea. (In English). *OM*, **18**(1–2): 31–37.	4	3	9

OKADA, Takematsu

1922	On the possibility of forecasting the summer temperature and the approximate yield of rice-crop for northern Japan. (In English). *2nd Pap. Mem. Imp. Mar. Obs.*, **1**: 18–26.	1–5	2, 3, 5–7	1
1936	Note on the correlative oscillations of the surface water temperature of the Oyasio and the Kurosio. (In English). *KMOM*, **6**: 97–103.	2, 3, 5	5	
1936	On the relation between the sea water temperature of the Kuroshio and the "Kyorei" of Tohoku District. (In Japanese). *Tenki to Kiko*, **3**(3): 100–102.	2, 3, 5	5	

OKUBO, Akira

1956	(with Kaname MATSUMURA, Hirozo YOSHIMURA, Itsuro TAKEDA and Shiro HASEGAWA) On nutrient substances in the sea area east of Honshu, Japan (1)—Phosphate–P. (In Japanese with English abstract). *J. Met. Res.*, **8**(11): 707–714.	2	4	
1958	The distribution of dissolved oxygen in the north-western part of the North Pacific Ocean in the aspect of physical oceanography. Part 1. General features of the oxygen distribution (I). (In English). *OM*, **10**(1): 137–156.	1–3, 5, 6	4	4

OMORI, Makoto

1967	*Calanus cristatus* and submergence of the Oyashio water. (In English). *Deep-Sea Res.*, **14**: 525–532.	2, 3	2, 6	
1967	(with Otohiko TANAKA) Distribution of some cold-water species of copepods in the Pacific water off east-central Honshu, Japan. (In English). *JOSJ*, **23**(2): 63–73.	3	8	

OTSUKA, Kazuyuki

1964	On the unusual low temperature in the East China Sea in winter of 1963 discussed from the heat balance.	5	5	

	Area	Subject	Chapter

(In Japanese with English abstract). *J. Met. Res.*, **16**(6): 326–333.

Ōuti, Masao

1965	On the long-term variations in sea level(I) —Japanese waters under the influence of Kuroshio. (In Japanese with English abstract). *Bull. Kyoto Gakugei Univ.*, Ser. B, No. **27**: 41–58.	3	5	
1966	Long-term variations of the atmosphere and oceanic conditions. (In English). *Sp. Contributions, Geophys. Inst., Kyoto Univ.*, **6**: 69–78.	2, 3, 5	3, 5, 7	7
1967	On the long-term variations in sea level (II)—The waters along the coast of Japan Sea. (In Japanese with English abstract). *Bull. Kyoto Kyoiku Univ.*, Ser. B, No. **31**: 7–21.	4	5, 7	

Panfilova, S. G.

1964	Surface water temperature anomalies in the North Pacific. (In Russian). *Okeanologiya, Akad. Nauk, SSSR*, **4**(4): 617–620.	1–7	5	
1965	Latitude-mean values of water temperature and salinity in the Pacific. (In Russian). *Okeanologiya, Akad. Nauk, SSSR*, **5**(1): 84–88. (In English): 60–63.	1–7	3	
1967	Bottom water temperature and salinity in the Pacific. (In Russian). *Okeanologiya, Akad. Nauk, SSSR.* **7**(5): 889–893. (In English): 690–694.	1–3, 5–7	3	
1968	Seasonal variations of sea surface temperature of the Pacific. (In Russian). *Okeanologiya, Akad, Nauk, SSSR*, **8**(5): 801–806. (In English): 639–644.	1–7	3, 5	

Parlychev, V. P.

| 1969 | Some features of the chemistry and thermal structure of the waters of interaction region of the Kuroshio and Oyashio currents. (In Russian). *Izvestija, Tichookeanskogo nautchnoissle-dovatelskogo instituta ribnogo chozyaystva i okeangrafii*, **68**: 45–66. | 1–3 | 4, 5, 6 | |

Pavlova, Yu. V.

| 1964 | The seasonal variation of the Kuroshio Current. (In Russian). *Okeanologiya, Akad. Nauk, SSSR*, **4**(4): 625–640. | 1–3, 5, 6 | 2, 5 | 6 |

Petelin, V. P.

| 1959 | Oceanographic investigations in the North-West Pacific, April–June 1955. (In Russian). *Trudy Inst. Okeanol., Akad. Nauk, SSSR*, **16**: 98–132. | 1, 2 | 4 | |
| 1966 | The 37th cruise of "Vityaz" in the central part of the Pacific Ocean. (In Russian). *Okeanologiya, Akad. Nauk, SSSR*, **6**(1): 172–175. (In English): 140–143. | 2, 6, 7 | 1 | |

Polyushkin, V. A.

| 1955 | (with S. I. Ushakov) Ekspiditsionno-issledovatil'skoye sudno "Vityaz." (In Russian). *Morskoy Flot*, **15**: 19–21. | 1–7 | 8 | |

		Area	Subject	Chapter

REID, J. L., Jr.

1961	On the geostrophic flow at the surface of the Pacific Ocean with respect to the 1,000-decibar surface. (In English). *Tellus*, **13**(4): 489–502.	1–7	2, 3	4, 5
1965	Intermediate waters of the Pacific Ocean. (In English). *Johns Hopkins Oceanogr. Studies*, **2**: 85pp.	1–7	2, 3	4, 5
1966	Zetes Expedition. (In English). *Trans. Amer. Geophys. Union*, **47**(4): 555–561.	1, 2	1, 2, 4	4
1969	Sea-surface temperature, salinity and density of the Pacific Ocean in summer and in winter. (In English). *Deep-Sea Res., Suppl. to* **16**: 215–224.	1–7	3	

ROBINSON, M. K.

| 1960 | Statistical evidence indicating no long-term climatic change in the deep waters of the North and South Pacific Oceans. (In English). *JGR.*, **65**(7): 2097–2116. | 1–7 | 3, 5 | 11 |

RODEN, G. I.

| 1961 | (with J. L. REID, Jr.) Sea surface temperature, radiation, and wind anomalies in the North Pacific Ocean. (In English). *ROWJ, NS*, **6**(1): 36–52. | 1–7 | 4 | |
| 1964 | Spectral analysis of Japanese sea level records. (In English). *SO* (dedicated to Prof. Hidaka in commemoration of his sixtieth birthday): 166–180. | 2–5 | 5 | |

SAITO, Yukimasa

| 1960 | The cold water-mass off the south coast of central Japan (Part I). (In English). *KMOM*, **14**: 35–43. | 2, 3, 5, 6 | 4, 6 | 13 |

SAKAMOTO, Ichitaro

| 1966 | (with Ichiro FUJII) Meanders of the Kuroshio and coastal temperature in the Kumano-Nada. (In Japanese). *Rep. on the effects of the warm drainage of atomic-electric generation on the fishery of Kumanonada coast. Committee of the Fishery of Kumanonada coast, Mie Pref.*: 23–31. | 3 | 6 | |
| 1966 | (with Ichiro FUJII) Coastal currents in the Kumano-Nada. (In Japanese). *Rep. on the effects of the warm drainage of atomic-electric generation on the fishery of Kumanonada coast. Committee of the Fishery of Kumanonada coast, Mie Pref.*: 33–34. | 3 | 2 | |

SASAKI, Tadayoshi

| 1955 | (with Noboru OKAMI, Seiichi WATANABE and Gohachiro ŌSHIBA) Optical properties of the water in the Kuroshio Current. (In English). *ROWJ, NS*, **2**(2): 133–140. | 3 | 4 | |
| 1957 | (with Noboru OKAMI, Seiichi WATANABE and Gohachiro ŌSHIBA) Optical properties of the water in the Kuroshio Current (II). (In English). *ROWJ, NS*, **3**(1): 92–103. | 3 | 4 | |

		Area	Subject	Chapter
1958	(with Noboru OKAMI, Gohachiro ŌSHIBA and Seiichi WATANABE) Spectral energy distribution of submarine daylight off Kii Peninsula. (In English). *ROWJ, NS, Special No.*, **2**: 120–127.	3	4	
1962	(with Noboru OKAMI, Gohachiro ŌSHIBA and Seiichi WATANABE) Studies on suspended particles in deep sea water. (In English). *Sci. Pap. Inst. Phys. Chem. Res.*, **56**(1): 77–83.	2	4	
1965	(with Seiichi WATANABE and Gohachiro ŌSHIBA) New current meters for great depths. (In English). *Deep-Sea Res.*, **12**: 815–824.	3	1, 2	12
1967	(with Seiichi WATANABE and Gohachiro ŌSHIBA) Current measurements on the bottom in the deep waters in the western Pacific. (In English). *Deep-Sea Res.*, **14**: 159–167.	3	2	12
1968	(with Noboru OKAMI, Motoaki KISHINO and Gohachiro ŌSHIBA) Optical properties of the water in adjacent regions of the Kuroshio. (In English). *JOSJ*, **24**(2): 45–50.	3	4	
1968	(with Noboru OKAMI and Satsuki MATSUMURA) Scattering functions for deep sea water of the Kuroshio. (In English with Japanese abstract). *La mer*, **6**(3): 165–176.	3	4	

SCHOTT, Gerhard

| 1906 | (with P. PERLEWITZ) Lotungen I.N.M.S. "Edi" und des Kabeldampfers "Stephen" im westlichen Stillen Ozeans. (In German). *Arch. Deutschen Sweewarte*, **29**(2): 1–38. | 5–7 | 4, 8 | |

SECKEL, G. R.

| 1968 | A time-sequence oceanographic investigation in the North Pacific trade-wind zone. (In English). *Trans. Amer. Geophys. Union*, **49**(1): 377–387. | 6, 7 | 2, 4, 5 | |

SETTE, O. E.

| 1955 | Consideration of midocean fish production as related to oceanic circulatory systems. (In English). *JMR*, **14**(4): 398–414. | 6, 7 | 4 | |

SHIGEMATSU, Ryoichi

1923	Oceanographical investigation on the results of surveys made by the Matsue and Manshu, survey ships. (In Japanese). *HB*, **2**(7): 95–111.	3, 5–7	1, 4	
1925	On the results of oceanographic observations carried out on board the HIMS Yamato in the Japan Sea in the summer of 1924. (In Japanese). *HB*, **4**(6): 296–304.	4	1, 4	
1932	On the dynamical investigation of the ocean current in Japan Sea. (In Japanese). *HB*, **11**(1): 27–28.	4	2	
1932	Some oceanographical investigation of the results of oceanic survey carried out by H. I. J. M. S. Mansyû from April 1925 to March 1928. (In English). *ROWJ*, **4**(1): 151–170.	2, 3, 5–7	2, 4	5, 8
1933	On the current in the sea of Okhotsk and the origin of Oyasio. (In Japanese). *HB*, **12**(8): 325–328.	1	1, 2, 4	

		Area	Subject	Chapter
1935	On the ocean current from the east coast of Taiwan to the Cape of Shiono-misaki. (In Japanese). *HB*, **14**(3): 85–88.	3, 5	2	

SHIMOMURA, Toshimasa

1950	(with Masami KOIZUMI) Report on the oceanographical observations from Torishima to the sea off Sanriku District. (In Japanese with English abstract). *KH*, **1**(2): 55–74.	2, 3	4	
1951	(with Jiro SUGIURA and Masamori MIYAZAKI) Report on the plankton and hydrographical conditions off Miyako during the period July 29-August 2, 1946. (In Japanese with English abstract). *KH*, **2**(1): 49–57.	2	1	
1953	(with Kazuo MIYATA) On the oceanographic character of the low temperature region off Sado Islands. (1). (In Japanese with English abstract). *JSSFB*, **19**(4): 424–428.	4	4	9
1953	Water-mass analysis by use of microplankton. (In Japanese). *KG*, **23**: 639–642.	3	4	
1957	(with Kazuo MIYATA) The oceanographical conditions of the Japan Sea and its water systems, laying stress on the summer of 1955. (In Japanese with English abstract). *JSRFL*, **6**: 23–119.	4	4	
1968	On the influences on the anchovy fisheries, by warm water drainage from the atomic power plant. (In Japanese). *Sendai Genshiryoku-hatsudensho Ricchijoken Chosa Hokoku-sho*: 95–101.	5	8	

SHINJI, Fukutaro

| 1962 | On the relation between the sea surface water temperature in the sea east of Japan and 500mb height difference. (In Japanese with English abstract). *JOSJ*, **18**(3): 111–114. | 3 | 5 | |
| 1965 | Note on the occurrence of the cold water mass off Enshu-nada. (In Japanese). *Bull. on Coastal Oceanography*, **4**(1): 51. | 3 | 6 | |

SHISHKOV, Yu. A.

| 1966 | Sea surface temperature anomalies in the northern part of the Pacific. (In Russian with English abstract). *Okeanologiya, Akad. Nauk, SSSR*, **6**(3): 416–429. (In English): 342–353. | 1–3, 6, 7 | 3, 5 | |

SHOJI, Daitaro

1950	The oceanographical results, southern area of Honsyu, September and October, 1949. (In Japanese). *HB, NS*, **16**: 277–278.	3	1	
1951	The variations of oceanic currents in the adjacent seas of Japan. (In Japanese with English abstract). *HB, NS*, **25**: 230–244.	3	5	1, 7, 13
1954	On the variation of daily mean sea level and the oceanographic condition (I). (In Japanese with English abstract). *HB, Special No.* **14**: 17–25.	3	5	1, 7, 13

		Area	Subject	Chapter
1955	Measurements of the currents in the sea with geomagnetic electro-kinetograph. (In Japanese). *HB, Special No.* **17**: 28–32.	2, 3	1, 2	12
1955	(with Terumi KUBOTA and Shozo YOSIDA) The Kuroshio current in the waters near the Izu Islands. (In Japanese). *KSH*: 3–24.	3	4–6	
1956	(with Kanji SUDA) On the variation of the Kuroshio near the Japan Islands. (In English). *Proc. Eighth Pacific Sci. Congress, III*: 619–636.	3	5, 6	13
1957	On the variations of daily mean sea levels and the Kuroshio from 1954 to 1955. (In English with French abstract). *Proc. UNESCO Symp. Phys. Oceanogr., Tokyo, 1955*: 130–136.	2, 3, 5	5	Pre., 13
1957	Kuroshio during 1955. (In English). *ROWJ, NS, Special No.* **1**: 21–26.	2, 3	2, 4	Pre., 13
1958	(with Ryuzo WATANABE, Naruji SUZUKI and Katsumi HASUIKE) On the "Shiome" at the boundary zone of the Kuroshio and the coastal waters off Shionomisaki. (In English). *ROWJ, NS, Special No.* **2**: 78–84.	3	2, 4	6, 7
1961	On the variations of the daily mean sea levels along the Japanese Islands. (In English). *JOSJ*, **17**(3): 141–152.	2–5	5	Pre., 7, 13
1964	A note on the anomalous cold water regions along the Japanese coasts and their relation to the fluctuations in the current speeds of the Kuroshio. (In Japanese). *Bull. Japan. Soc. Fish. Oceanogr.*, **4**: 31–40.	3	5, 6	7, 13
1965	Description of the Kuroshio (Physical Aspect). (In English). *Proc. Symp. on the Kuroshio*: 1–11.	2, 3, 5–7	2, 4–6	6, 8
1966	(with Hideo NITANI) On the variability of the velocity of the Kuroshio–I. (In English). *JOSJ*, **22**(5): 192–196.	3	5	5, 6, 7, 12

SMETANIN, D. A.

		Area	Subject	Chapter
1958	Hydrochemistry in the region of the Kurile-Kamchatka Trench. I. Certain questions of the hydrology and chemistry of the lower subarctic water in the region of the Kurile-Kamchatka Trench. (In Russian). *Trudy Inst. Okeanol., Akad. Nauk, SSSR*, **27**: 22–54.	1, 2	2, 4	
1959	Hydrochemistry of the Kurile-Kamchatka Trench. (In Russian). *Trudy Inst. Okeanol., Akad. Nauk, SSSR*, **33**: 43–86.	1	2, 4	

SMETANINA, N. S.

		Area	Subject	Chapter
1962	Charts of vorticity of the tangential wind stress over the Pacific Ocean. (In Russian). *Trudy Inst. Okeanol., Akad. Nauk, SSSR*, **57**: 133–155.	1–7	3, 5	

STOCKMANN, W. B.

		Area	Subject	Chapter
1945	An attempt of a qualitative analysis of temperature conditions in the region of Kuroshio. (In English). *Dokl. Akad. Nauk, SSSR*, **46**(2): 56–59.	3	5, 7	
1946	On the thermal regime of the Kuro-Shio. (In Russian with English abstract). *Trudy Inst. Okeanol., Akad. Nauk, SSSR*, **1**: 74–98.	3	3, 5, 7	

			Area	Subject	Chapter

STOMMEL, Henry

1958	The abyssal circulation. (In English). *Deep-Sea Res.*, **5**: 80–82.	1–7	2, 3, 7	11, 12
1958	*The Gulf Stream*: *A physical and dynamical description*. 1st Edition. (In English). Univ. of Calif. Press, Berkeley and Los Angeles: 202pp.	8	2–8	5
1960	(with A. B. ARONS) On the abyssal circulation of the world ocean, II. (In English). *Deep-Sea Res.*, **6**: 217–233.	1–7	2, 3, 7	4, 11, 12
1965	Comparison on Kuroshio and Gulf Stream. (Abstract). (In English). *Proc. Symp. on the Kuroshio*: 21.	2, 3	8	
1965	Some thoughts about planning the Kuroshio survey. (In English). *Proc. Symp. on the Kuroshio*: 22–23.	2, 3, 5	8	
1965	*The Gulf Stream*: *A physical and dynamical description*. 2nd Edition. (In English). Univ. of Calif. Press, Berkeley and Los Angeles, Cambridge Univ. Press, London: 248pp.	8	2–8	6, 8

SUDA, Kanji

1929	(with Kazuo SEKI) Report on the mean surface temperature in the adjacent seas of Japan. (In Japanese). *JO*, **1**(1): 57–58.	2–6	3, 5	
1929	(with Kazuo SEKI) Report on the mean temperatures of the neighbouring seas of Japan. (In Japanese). *JO*, **1**(2): 132–133.	2–6	3, 5	
1930	(with Kazuo SEKI) Report on the mean surface temperatures of the neighbouring seas of Japan. (In Japanese). *JO*, **1**(3): 379–382.	2–6	3, 5	
1930	(with Kazuo SEKI) Report on the mean surface temperatures of the neighbouring seas of Japan. (In Japanese). *JO*, **2**(1): 74–77.	2–6	3, 5	
1930	(with Kazuo SEKI) Report on the mean surface temperatures of the neighbouring seas of Japan. (In Japanese). *JO*, **2**(2): 266–267.	2–6	3, 5	
1930	(with Kazuo SEKI) Report on the mean surface temperatures of the neighbouring seas of Japan. (In Japanese). *JO*, **2**(3): 447–448.	2–6	3, 5	
1930	On the systems of the Kuroshio water and the state of its turbulent flow (Part I). (In Japanese). *JO*, **2**(3): 483–499.	2–6	4	
1931	(with Kazuo SEKI) Report on the mean surface temperature of the neighbouring seas of Japan. (In Japanese). *JO*, **2**(4): 644–646.	2–6	3, 5	
1931	On the dissipation of energy in the density currents. (In Japanese). *US*, **11**(10): 229–239.	3	7	
1931	On the dissipation of the energies in the density current. (In Japanese). *KKI*, **43**: 1–11.	3	7	

		Area	Subject	Chapter
1931	On the systems of the Kuroshio water and the state of its turbulent flow (Part II). (In Japanese). *JO*, **3**(1): 201–209.	3	5, 7	
1932	The results of the oceanographical observations on board R.M.S. Shumpu Maru in the Japan Sea during the summer of 1929 (Part I). (In Japanese with English abstract). *JO*, **3**(2): 291–375.	4	1	9
1932	On the dissipations of energy in the density current. (In English). *GM*, **6**: 297–314.	3, 4	4, 7	
1932	On the dissipation of energy in the density current. (In Japanese). *US*, **12**(1): 1–10.	4	4, 7	
1932	On the bottom-water of the Japan Sea. (In Japanese). *JO*, **4**(1): 221–240.	4	4, 7	
1932	Further investigation on the dissipation of the energy in the density currents. (In Japanese). *US*, **12**(1): 1–10, *KKI*, **44**: 1–10.	4	4, 7	
1932	(with Koji HIDAKA, Yasuo MATSUDAIRA, Hideo KAWASAKI, Hidejiro KURASHIGE, and Tokio KUBO) The results of oceanographical observations on board H.M.S. "Syunpu Maru" in the principal part of the Japan Sea in the summer of 1930. (In Japanese with English abstract). *JO*, **4**(1): 1–173.	4	1	
1935	On the turbulence in the ocean. (In Japanese). *US*, **15**(1): 21–26.	3, 4	7	
1936	On the dissipation of energy in the density currents. (2nd paper). (In English). *GM*, **10**: 131–243.	2–4	7	8
1936	A practical study on the variations of hydrographic conditions. (In Japanese). *US*, **16**(12): 417–426.	3	4	
1938	On the variations of the oceanographical state of the Kuroshio in the original region (Part I). (In English). *GM*, **11**: 373–410.	5	4, 5	4
1941	The Kuroshio current at the time of Commodore Perry's visit to Japan. (In Japanese). *Kaiyo no Kagaku*, **1**(4): 242–244.	3	2	
1941	Surface water temperature and air temperature in the North Pacific Ocean. (In Japanese). *CG*, **9**(1): 1–24.	2–5, 7	1, 2, 5–7	

SUDO, Hideo

		Area	Subject	Chapter
1960	On the distribution of divergence and convergence of surface drift vectors in the Western Pacific Oceans. (In English). *ROWJ, NS*, **5**(2): 25–43.	1–3, 5–7	2	
1969	An attempt to estimate the vertical component of the current velocity in the south off the main island of Japan on the basis of heat conservation. (In English). *ROWJ, NS*, **10**(1): 1–11.	3	5, 6	

SUETOV, S. W.

		Area	Subject	Chapter
1959	Oceanographic investigations in the Bering Sea, October–December, 1953. (In Russian). *Trudy Inst. Okeanol., Akad. Nauk, SSSR*, **16**: 47–69.	1	4	

		Area	Subject	Chapter

SUGIMURA, Yukio

		Area	Subject	Chapter
1965	Distribution of the dissolved oxygen in the deep waters of the northwestern Pacific Ocean. (In Japanese with French abstract). *La mer*, **2**(2): 111–115.	1–3, 5, 6	4	

SUGIURA, Jiro

		Area	Subject	Chapter
1954	On the transport in the sea area off Sanriku from March to August 1950. (In Japanese with English abstract). *KH*, **3**(3): 182–186.	2	2, 6	
1955	On the transport in the eastern sea of Honshu (Part 1. Conditions during 1949–1950). (In English). *OM*, **6**(4): 153–163.	2	2, 5	
1955	On the transport in the eastern Sea of Honshu (Part 2. Conditions during 1946–1948). (In English). *OM*, **7**(1): 21–28.	2	2	
1957	Forecast of water temperature in the sea south of Hokkaidō. (In English). *OM*, **9**(1): 13–19.	2	5	
1957	Variations of sea conditions in the sea south off Hokkaidō. (In English). *OM*, **9**(1): 121–131.	2	4, 5	
1957	On the seasonal variation of the water mass transport in the sea south off Hokkaidō. (In Japanese with English abstract). *J. Met. Res.*, **9**(3): 162–166.	2	5	
1958	On the Tsugaru Warm Current. (In English). *GM*, **28**(3): 399–409.	2	2, 4, 5	8
1958	Oceanographic conditions in the northwestern North Pacific based upon the data obtained on board the "Komahashi" from 1934 to 1936. (In Japanese with English abstract). *JOSJ*, **14**(3): 81–85.	1, 2	4	
1959	On the currents south off Hokkaidō in the western North Pacific. (In English). *OM*, **11**(1): 1–11.	2	2	

SUGIURA, Yoshio

		Area	Subject	Chapter
1965	Distribution of reserved (preformed) phosphate in the subarctic Pacific region. (In English with Japanese abstract). *MGP*, **15**(3–4): 208–215.	1	4	
1967	(with Hirozo YOSHIMURA) The significance of the difference in conductometric chlorinity minus titrimetric chlorinity. (In English). *ROWJ, NS*, **9**(1): 55–64.	8	7	
1967	Total carbon dioxide and its bearing on the dissolved oxygen in the Oyashio and in the frontal region of the Kuroshio. (In English). *Geochem. Jour.*, **1**: 125–130.	1, 2	3, 4	
1968	Distribution of AOU in the surface water around Japan and its significance—Delay of gas exchange between ocean and atmosphere. (In Japanese with French abstract). *La mer*, **6**(1): 9–16.	1–3, 6	7	
1968	Distribution of pH of seawater around Japan and its significance. (In Japanese with French abstract). *La mer*, **6**(2): 115–119.	1, 2	7	
1968	(with Katsumi YAMAMOTO) Distribution of iron and aluminum in the seawater of the Japan Sea	4	3, 7	

		Area	Subject	Chapter
and its oceanographical significance. (In Japanese with French abstract). *La mer*, **6**(3): 177–189.				
Sverdrup, H. U.				
1942	(with M. W. Johnson and R. H. Fleming) *The oceans: their physics, chemistry and general biology.* (In English). Prentice-Hall, New York: 1087pp.	1–7	2–7	4, 5, 6, 8 9, 11
Taguchi, Ryuzo				
1962	(with Takeo Ueno) On the variation of the mean sea levels in the eastern parts of Japan. (In Japanese with English abstract). *US*, **38**(2): 53–58.	2–4	3, 5	
Takahashi, Tadao				
1959	Hydrographical researches in the western equatorial Pacific. (In English). *Mem. Fac. Fish., Kagoshima Univ.*, **7**: 141–147.	5–7	1, 2, 4, 6	5
1962	Temperature and humidity profiles over the sea. (In English). *JOSJ, 20th Anniv. Vol.*: 257–264.	5	8	
1967	(with Masaaki Chaen) Oceanic conditions near the Ryukyu Islands in summer of 1965. (In English). *Mem. Fac. Fish., Kagoshima Univ.*, **16**: 63–75.	5	2, 4	5
Takano, Kenzo				
1961	Circulation générale permanente dans les océans. (In French). *JOSJ*, **17**(3): 121–131.	8	7	
1961	Circulation générale permanente dans les océans—Deuxième Partie. (In French). *JOSJ*, **17**(3): 132–140.	8	7	
1961	Circulation générale permanente dans les océans—Deuxième Partie (suite et fin). (In French with English abstract). *JOSJ*, **17**(4): 179–189.	8	7	
1961	Distribution de densité à la surface d'un océan de longueur indéfinie en fonction de la latitude. (In French). *JOSJ*, **17**(4): 190–196.	8	7	
1962	Circulation générale permanente dans un océan. (In French). *ROWJ, NS*, **6**(2): 59–155.	8	7	
1962	Circulation générale permanente dans les océans—Un calcul numérique complémentaire. (In French with English abstract). *JOSJ*, **18**(2): 59–67.	8	7	
1962	Circulation générale permanente dans les océans—Un calcul numérique complémentaire (suite et fin). (In French with English abstract). *JOSJ, 20th Anniv. Vol.*: 200–212.	8	7	
1964	Thermohaline circulation in the oceans. (In Japanese). *JOSJ*, **20**(3): 135–147.	8	7	
1964	Variation annuelle de la circulation générale dans les océans. (In French). *La mer*, **1**(2): 51–61.	8	5, 7	
1964	Variation annuelle de la circulation générale dans les océans (suite et fin). (In French). *La mer*, **2**(1): 1–21.	8	5, 7	

		Area	Subject	Chapter
1965	Courants marins induits par le vent et la non-uniformité de la densité de l'eau superficielle dans un océan. (In French with Japanese abstract). *La mer*, **2**(2): 81–86.	8	7	
1965	Un exemple numérique des courants marins induits par le vent et la non-uniformité de la densité de l'eau superficielle dans un océan. (In French with Japanese abstract). *La mer*, **3**(2): 57–65.	8	7	
1965	General circulation due to the horizontal variation of water density with the longitude, maintained at the surface of an ocean. (In English). *ROWJ, NS*, **8**(1): 1–11.	8	7	
1965	Periodic variation of the barotropic components of a wind-driven circulation in an ocean. (In English). *JOSJ*, **21**(1): 1–5.	8	5, 7	
1966	A possible effect of the bottom topography on the general circulation in an ocean. (In English with Japanese abstract). *JOSJ*, **22**(6): 264–273.	8	7	

TAKENOUTI, Yositada

		Area	Subject	Chapter
1957	(with Katsumi HATA and Mitsugu TORII) On the forecast of surface water temperature for the frontal zone of western north Pacific. (In English with French abstract). *Proc. UNESCO Symp. Phys. Oceanogr., Tokyo, 1955*: 96–100.	2	5	
1957	Long-range oceanographic forecast, historical review and future problem. (In English). *OM*, **9**(1): 1–12.	2–5	5	
1958	Measurements of subsurface current in the cold-belt along the northern boundary of Kuroshio. (In English). *OM*, **10**(1): 13–17.	2	2	1, 8, 12
1960	(with Masashi YASUI) On the short period variations of oceanographical conditions in the Kuroshio region (Preliminary Report). (In English). *ROWJ, NS, Special No.* **4**: 55–57.	5, 6	5	
1960	The 1957–1958 oceanographic changes in the western Pacific. (In English). *Calif. Cooper. Ocean. Fish. Inv., Rep.* **7**: 67–76.	1–3, 5, 6	4, 5	
1962	(with Tosio NAN'NITI and Masashi YASUI) The deep-current in the sea east of Japan. (In English). *OM*, **13**(2): 89–101.	2	3	1, 8, 12

TAMIYA, Yoshimi

		Area	Subject	Chapter
1949	The oceanographic phenomena, south eastern area of Honsyū in June 1949. (In Japanese). *HB, NS*, **15**: 187–200.	2, 3	1	
1950	A simple method to infer the current from the distribution of water temperature. (In Japanese). *HB, Special No.* **5**: 82–83.	3	8	
1950	Current charts in the area southward and eastward of Honshu. (In Japanese). *HB, NS*, **19**: 136–140.	2, 3	2, 4	
1955	Observation of the current in Japan Sea with drift bottles. (In Japanese). *HB, Special No.* **17**: 13–21.	2, 4	1, 2	12

		Area	Subject	Chapter

Tanioka, Katsumi

1962	The oceanographical conditions of the Japan Sea (1). (In Japanese with English abstract). *US*, **38**(3): 90–100.	4	3, 5	9
1962	The oceanographical conditions of the Japan Sea (II). (In Japanese with English abstract). *US*, **38**(4): 115–128.	4	4, 5	9
1968	On the East Korean Warm Current (Tōsen Warm Current). (In English). *OM*, **20**(1): 31–38.	4	2	9

Terada, Kazuhiko

1953	(with Koichiro Osawa) On the energy exchange between sea and atmosphere in the adjacent seas of Japan. (In English). *GM*, **24**(3): 155–170.	1–4	8	4
1962	Recent activities of the marine division and its related branches of the Japan Meteorological Agency. (In English). *GM*, **31**(1): 127–145.	1–7	8	

Terada, Torahiko

1922	(with J. T. Liu and Seiti Yamaguti) Influences of wind, air temperature and solar radiation upon sea water temperature. (In English). *JJAG*, **1**(2): 46–48.	2	3	

Tezuka, Takio

1963	The state of sea off Tohoku District in 1961 (Notes). (In Japanese). *JOSJ*, **19**(2): 112–113.	2	4	

Tsuchiya, Mizuki

1961	An oceanographical description of the equatorial current system of the western Pacific. (In English). *OM*, **13**(1): 1–30.	5–7	2, 3	1, 4
1968	Upper waters of the intertropical Pacific Ocean. (In English). *Johns Hopkins Oceanogr. Studies*, **4**: 50pp.	8	2, 3	4

Tsukuda, Kazuyoshi

1931	(with Taizo Yoshikawa) On the distributions of atmospheric pressure, cloudiness and surface water temperature over the North Pacific Ocean. (In Japanese). *KKI*, **37**: 1–10.	1–7	3, 5	
1935	(with Taizo Yoshikawa) Surface water temperature in the seas adjacent to Japan and its variation. (In Japanese). *US*, **15**(8): 257–275. *KKI*, **84**: 1–19.	1–7	3, 5	
1937	On the surface temperature of the neighboring seas of Japan. (In English with Japanese abstract). *KMOM*, **6**(3): 239–257.	2–7	3, 5	9

Tsumura, Kensiro

1963	Investigation of the mean sea level and its variation along the coast of Japan (Part I). Regional distribution of sea level variation. (In Japanese with English abstract). *J. Geodetic Soc. Japan*, **9**(2): 49–90.	2–4	8	4

		Area	Subject	Chapter
1964	Japanese contributions to the study of mean sea level (A Review). (In English). *J. Geodetic Soc. Japan*, **10**(3–4): 192–202.	2–5	5	

TULLY, J. P.

		Area	Subject	Chapter
1964	Oceanographic regions and processes in the seasonal zone of the North Pacific Ocean. (In English). *SO* (dedicated to Prof. Hidaka in commemoration of his sixtieth birthday): 68–84.	1–3, 5–7	4, 5	

UDA, Michitaka

		Area	Subject	Chapter
1927	Relation between the daily catch of fish and the meteorological elements. I. Statistical studies on the influences of the motion of cyclone upon the fishing. (In English). *J. Imp. Fish. Inst.*, **23**(3): 80–88.	5	5	
1929	On the stratification of sea water in the Kuroshio region. (In Japanese). *US*, **9**(11): 175–182.	1–6	4	8
1930	On some oceanographical researches of the sea water of Kurosiwo. (In English). *ROWJ*, **2**(2): 59–81.	1–5	3, 4	
1930	On the distributions of color of the sea and transparency in the neighboring seas of Japan and their annual variations. (In Japanese). *US*, **10**(8): 173–180.	2–5	4	
1930	(with Gorozo OKAMOTO) On the monthly oceanographical charts of the adjacent seas of Japan based on the averages for the eleven years from 1918 to 1929, with a discussion of the current-system inferred from these charts (Part I: from July to December). (In Japanese with English abstract). *JIFES*, **1**: 39–55, 12 pls.	1–6	3, 5	8
1931	On the monthly oceanographical charts of the adjacent seas of Japan based on the averages for the thirteen years from 1918 to 1930, with a discussion of the current-system inferred from these charts (Part II: from January to June). (In Japanese with English abstract). *JIFES*, **2**: 59–81, 12 pls.	1–6	3, 5	8
1932	Hydrographical investigations in the seas adjacent to Wakasa Bay. (In English). *ROWJ*, **4**(1): 1–29.	4	2, 4	
1932	Oceanography for fisheries conditions. (In Japanese). *KG*, **2**: 62–65.	2–6	8	
1933	Hydrographical researches on the monthly conditions of Oyashio and Kuroshio area. (In Japanese with English abstract). *JIFES*, **3**: 79–136.	2, 3, 5	3, 5	
1933	(with Noboru WATANABE) Hydrographical researches on the normal monthly conditions of the Seto-Naikai. (In Japanese with English abstract). *JIFES*, **3**: 137–164.	3	4	
1934	Hydrographical studies based on simultaneous oceanographical surveys made in the Japan Sea and its adjacent waters during May and June, 1932. (In English). *ROWJ*, **6**(1): 19–107.	4	4	
1934	Dense fog and other miscellaneous in the north-eastern sea of Japan in August, 1934. (In Japanese). *Tenki to Kiko*, **1**: 386–389.	2	8	

		Area	Subject	Chapter
1934	Study of the ocean currents by use of drift-nets. (In Japanese). *KG*, **4**: 51–52.	2	2	
1934	The results of simultaneous oceanographical investigations in the Japan Sea and its adjacent waters in May and June, 1932. (In Japanese with English abstract). *JIFES*, **5**: 57–190.	4, 5	1	1, 9
1935	Hydrographical researches on the normal monthly conditions in the Japan Sea, the Yellow Sea and the Okhotsk Sea. (In Japanese with English abstract). *JIFES*, **5**: 191–236.	1, 4, 5	4	
1935	The results of simultaneous oceanographical investigations in the North Pacific Ocean adjacent to Japan made in August, 1933. (In Japanese with English abstract). *JIFES*, **6**: 1–130.	1–7	4	1, 4, 8
1935	Oceanic drift-nets and long-lines in fishing. (In Japanese with English abstract). *JSSFB*, **3**(2): 93–95.	2–6	2	
1935	Low temperature of the sea in the poor harvest years and its forecasting. (In Japanese). *KG*, **5**: 7–8.	2	5	
1935	On the distribution, formation and movement of the dicho-thermal water off the northeast of Japan. (In Japanese). *US*, **15**(2): 445–452.	2	4	4
1935	Skipjack shoals concentrating near the oceanic front. (In Japanese). *KG*, **5**: 503–504.	2	4	
1936	Oceanic circulation and its variation. (In Japanese). *KG*, **6**(10): 449–453.	1–7	2, 5	
1936	(with Gorozo OKAMOTO) Effect of oceanographic conditions on "Iwasi" (Sardine) fishing in the Japan Sea. (In Japanese with English abstract). *JIFES*, **7**: 19–49.	4	4	
1936	Distribution of drifting bodies in the ocean current. (In English). *JSSFB*, **4**(5): 289–293.	4	8	5
1936	Results of simultaneous oceanographic investigations in the Japan Sea and its adjacent waters during October and November, 1933. (In Japanese with English abstract). *JIFES*, **7**: 91–151.	4, 5	4	1
1936	Locality of fishing center and shoals of "Katuwo," *Euthynnus vagans* (Lesson), correlated with the contact zone of cold and warm currents. (In Japanese with English abstract). *JSSFB*, **4**(6): 385–390.	2	4	
1937	(with Eimatu TOKUNAGA) Fishing of *Germo germo* (Lacépède) in relation to the hydrography in the North Pacific waters (Report I). (In Japanese with English abstract). *JSSFB*, **5**(5): 295–300.	2, 3, 6	8	
1937	Results of hydrographical investigations in the Sagami Bay in connection with the "Buri" (*Seriola Quinqueradiata, T & S*) fishing. (In Japanese with English abstract). *JIFES*, **8**: 1–50.	3	4	
1937	On the recent abnormal condition of the Kuroshio to the south of Kii Peninsula. (In Japanese). *KG*, **7**(9): 360–361.	3	6	1, 6, 7 13

		Area	Subject	Chapter
1937	On the recent abnormal condition of the Kuroshio to the south of Kii Peninsula (Supplement). (In Japanese). *KG*, **7**(10): 403–404.	3	6	1, 6, 13
1938	Researches on "Siome" or current rip in the seas and oceans. (In English). *GM*, **11**(4): 307–372.	1–6	2, 4	8
1938	Hydrographical fluctuation in the north-eastern sea-region adjacent to Japan in North Pacific Ocean (A result of the simulataneous oceanographic investigations in 1934–1937). (In Japanese with English abstract). *JIFES*, **9**: 1–66.	2	4, 5	1, 8
1938	On the names of the Kuroshio and the Oyashio. (In Japanese). *KG*, **8**(10): 400–401.	2, 3	2, 4	
1938	Secular variations of the water temperature along the coast of Japan and her neighborhood. (In Japanese). *US*, **18**(9): 343–346.	2–5	5	
1938	(with Nobuo WATANABE) Autumnal fishing of skipper and bonito influenced by the rapid hydrographic change after the pass of cyclones. (In Japanese with English abstract). *JSSFB*, **6**(5): 240–242.	2	4, 5	
1939	Recent progress of hydrographic observations and experimental studies in physical oceanography in Japan. (In Japanese). *PMSP*, **13**: 133–150.	2, 3	2, 4	
1940	Relation between the heavy rainfall, dry spell, and oceanographic conditions. (In Japanese). *CI*, **16**: 183–192.	3	4, 5	
1940	On the recent anomalous hydrographical conditions of the Kuroshio in the south waters off Japan proper in relation to the fisheries. (In Japanese with English abstract). *JIFES*, **10**: 231–278.	3	6	6, 8, 13
1941	The results of the hydrographical surveys in the China Sea in summer 1939. (In Japanese with English abstract). *JIFES*, **11**: 39–97.	5	1	
1942	On the fluctuation of oceanic current 1. (In Japanese). *JOSJ*, **2**(1): 22–26.	4	4, 5	1
1943	On the cyclogenesis and the passage of cyclone in relation to sea conditions. (In Japanese). *US*, **23**(6): 219–228.	2–6	4, 5	
1943	On the structure of oceanic front. (In Japanese). *JOSJ*, **2**(4): 9–16.	2, 3, 5	2, 4	1, 8
1949	On the correlated fluctuation of the Kuroshio Current and the cold water mass. (In English). *OM*, **1**(1): 1–12.	3	5, 6	1, 6, 7 8, 13
1950	On the fluctuation of oceanic current (2nd paper). (In Japanese with English abstract). *JOSJ*, **5**(2–4): 55–69.	4. 5	2	1
1950	On the variation of water temperature in the East China Sea 1. (In Japanese). *SFLB*, **1**: 1–10.	5	5	
1951	On the fluctuation of the main stream axis and its boundary line of Kuroshio. (In English). *JOSJ*, **6**(4): 181–189.	3	2, 5	6
1952	On the characteristic feature of the yearly variation of coastal water temperature. (In Japanese). *US*, **30**(1–2): 5–7.	1–3, 5	5	

		Area	Subject	Chapter
1952	On the relation between the variation of the important fisheries conditions and the oceanographical conditions in the adjacent waters of Japan 1. (In English). *TUFJ*, **38**(3): 363–389.	1–7	4, 5	
1952	On the hydrographical fluctuation in the Japan Sea (Preliminary Report) (Appendix: The extraordinary abundant catch of "*Euphausia pacifica* HANSEN" in winter and spring of 1948 along the coast of the Japan Sea side). (In Japanese with English abstract). *JSRFL, 3rd Anniv. Vol.*: 291–300.	4, 5	4	
1953	On the convergence and divergence in the NW Pacific in relation to the fishing grounds and productivity. (In English). *JSSFB*, **19**(4): 435–438.	1–7	4	
1953	On the stormy current ("Kyūtyō") and its prediction in the Sagami Bay. I. (In Japanese with English abstract). *JOSJ*, **9**(1): 15–22.	3	2, 4, 5	7
1953	The Kuroshio and its branch currents in the seas adjacent to Hachijo Island in relation to fisheries (Report I). (In English). *ROWJ, NS*, **1**(1): 1–10.	3	4	
1954	Studies of the relation between the whaling grounds and the hydrographical conditions (I). (In English). *SRW*, **9**: 179–187.	1–6	4	
1955	(with Yoshimi MORITA) On the Kuroshio in Zunan area in the summer of 1954. (In Japanese). *KSH*: 25–37.	3	1	
1955	On the relation between Kuroshio and fishing grounds near Hachijo Island (3). (In Japanese). *KSKH*, **I**: 9–13.	3	4	
1955	On the Subtropical Convergence and the currents in the Northeastern Pacific. (In English). *ROWJ, NS*, **2**(1): 141–150.	6	4	5
1955	Researches on the fluctuation of the North Pacific circulation (I. The fluctuation of Oyasiwo current in relation to the atmospheric circulation and to the distribution of the dichothermal waters in the North Pacific Ocean). (In English). *ROWJ, NS*, **2**(2): 43–55.	1	4, 5	
1955	Correlation of hydrographic and meteorological features with the fluctuations in certain fisheries in the waters off Japan. (In English). *Proc. Indo-Pacific Fish. Council*, Section *II*: 1–2.	1–4	5	
1955	On the variation of water temperature due to the passage of typhoon. (In English). *Assoc. Oceanogr. Phys. Proc.-Verb.*, **6**: 297–298.	1–5	2, 3	
1956	(with Nobuo WATANABE and Makoto ISHINO) General results of the oceanographic surveys (1925–1955) on the fishing grounds in relation to the scattering layer. (In English). *TUFJ*, **42**(2): 169–207.	3	4	
1957	(with Atsushi DAIROKUNO) Studies of the relation between the whaling grounds and the hydrographic conditions II. A study of the relation between the whaling grounds off Kinkazan and the boundary of water masses. (In English). *SRW*, **12**: 209–224.	2	3	

		Area	Subject	Chapter

1957   A consideration on the long years trend of the fisheries fluctuation in   1–5   8
relation to sea conditions.
(In English). *JSSFB*, **23**(7–8): 368–372.

1957   A note on the modification of air masses over the seas adjacent to   2–7   8
Japan.
(In English). *JMSJ, 75th Anniv. Vol.*: 372–377.

1957   (with Yoshimi MORITA and Makoto ISHINO)   2, 3, 6   4
Results from the oceanographic observations in the North Pacific
(1955–56) with Umitaka Maru and Shinyo Maru.
(In English). *ROWJ, NS, Special No.* **1**: 1–20.

1957   Research on the fluctuation of the North Pacific circulation (1).   1, 2, 4   5
(In English with French abstract). *Proc. UNESCO Symp. Phys.
Oceanogr., Toyko, 1955*: 112–113.

1958   On the abyssal circulation in the Northwest Pacific Area.   1, 2, 4, 5   2
(In English). *GM*, **28**(3): 411–416.

1958   A note on the hydrographical conditions of Kurosiwo and its northern   2   4, 6
boundary in 1956.
(In English). *ROWJ, NS, Special No.* **2**: 68–77.

1958   (with Makoto ISHINO)   1–5   2, 3
Enrichment pattern resulting from eddy systems in relation to fishing
grounds.
(In English). *TUFJ*, **44**(1–2): 105–129.

1958   (partly with Hisayasu OTSUBO)
Fluctuations of hydrographic and fisheries conditions in the Japan   4   4, 5
Sea and oceanic fronts in the East China Sea.
(In Japanese). *TKH*, **1**: 501–539.

1959   Oceanographic seminars.   1–7   2–6
(In English). *MS Rep. Series (Oceanogr. and Limno.), Fish. Res. Bd.
Canada*, **51**: 110pp.

1959   The fisheries of Japan.   1–7   2–7
(In English). *MS Rep. Series (Biol.), Fish. Res. Bd. Canada*, **686**: 96pp.

1960   On the fluctuation and prediction of ocean temperature in the North   1, 2   3, 5
Pacific.
(In English). *KMOM*, **14**: 53–68.

1960   (with Makoto ISHINO)   2, 3   1, 2, 4
Researches on the currents of Kuroshio.
(In English). *ROWJ, NS, Special No.* **4**: 59–72.

1961   Fisheries oceanography in Japan, especially on the principles of fish   1–7   8
distribution, concentration, dispersal and fluctuation.
(In English). *Calif. Cooper. Oceanogr. Fish. Inv.*, **8**: 25–31.

1961   Cyclical fluctuation of the Pacific tuna fisheries in response to cold   1–3   5
and warm water intrusions.
(In English). *Rep. Pacific Tuna Biology Conference, Honolulu*, **39**: V-7.

1962   Subarctic oceanography in relation to whaling and salmon fisheries.   1   2, 4
(In English). *Sc. Rep. Whale Res. Inst.*, **16**: 105–119.

1962   Cyclic, correlated occurrence of world-wide anomalous oceanographic   1–3, 5–7   5   7
phenomena and fisheries conditions.
(In English). *JOSJ, 20th Anniv. Vol.*: 368–376.

		Area	Subject	Chapter
1963	Oceanography of the Subarctic Pacific Ocean. (In English). *J. Fish. Res. Bd. Canada*, **20**(1): 119–179.	1	2–7	4
1964	Dissolved oxygen in the Pacific Ocean and adjacent seas as an important element to study the circulation and structure in the changing marine environment. (In English). *Recent Researches in the Fields of Hydrosphere, Atmosphere and Nuclear Geochemistry, Prof. Sugawara's Festival Vol.*: 349–356.	1–3, 6, 7	4	
1964	Structure of Kuroshio—The behavior and problems of changing Kuroshio. (In Japanese). *KG*, **34**(8): 413–418.	2, 3, 5	4–6	
1964	On the nature of the Kuroshio, its origin and meanders. (In English). *SO* (dedicated to Prof. Hidaka in commemoration of his sixtieth birthday): 89–107.	2, 3, 5	2–6	5, 6, 8, 12, 13
1965	Some problems on the deep-sea circulation. (In Japanese with French abstract). *La mer*, **2**(2): 125–132.	1–3, 5–7	2, 4	
1966	The accomplishment of fisheries oceanography in Japan and its future prospect. (In English). *Advances in Fish. Oceanography*, **1**: 13–18.	1–6	8	
1966	Synoptic studies of ocean climate. (In Japanese). *JOSJ*, **22**(5): 231–235.	1–7	8	
1969	(with Keiichi Hasunuma) The eastward Subtropical Countercurrent in the western North Pacific Ocean. (In English with Japanese abstract). *JOSJ*, **25**(4): 201–210.	3, 5–7	2, 4	4, 5, 13

Udintsev, G. B.

		Area	Subject	Chapter
1962	Geomorphology of the northwest part of the Pacific Ocean. (In Russian). *Sbornik Dokl. II Plen., Komissii Ribochoz, Issledov. Zapadnoi Chasti Tichogo Okeana.*: 40–52.	2, 3	8	
1965	The 36th cruise of R/V Vityaz. (In Russian). *Okeanologiya, Akad. Nauk, SSSR*, **5**(6): 1113–1119. (In English): 142–148.	6, 7	1	

Uehara, Susumu

		Area	Subject	Chapter
1962	Fishery oceanography around Enshu Nada off the central Pacific coast of Honshu—I. Oceanographic condition for skipjack and shirasu fisheries. (In Japanese with English abstract). *TKRFL*, **34**: 55–66.	3	2, 4–6	8

Wada, Yuji

		Area	Subject	Chapter
1894	A study on the Japan Current. (In Japanese). *Tokyo Butsuri Gakko Zasshi*, **3**(33): 241–243.	3, 5, 6	2	12
1894–95	Investigation on the ocean currents in the east of Japan. (In Japanese). Part 1, *SCHH*, **2**(2): 1–28. Part 2, *SCHH*, **3**(3): 1–34.	2	2	12

Wadati, Kiyoo

		Area	Subject	Chapter
1962	(with Kazuhiko Terada) Deep-sea research in Japan. (In English). *JOSJ, 20th Anniv. Vol.*: 1–3.	1–3, 5	8	11

		Area	Subject	Chapter

WATANABE, Kantaro

1965   The state of the sea off Tohoku District in 1962.   2   4
       (In Japanese). *JOSJ*, **20**(6): 287–288.

1969   (with Matsuyuki Jo)   3   2, 3, 6
       Mean pattern of the Kuroshio axis south off Honshu (Main Island of
       Japan).
       (In Japanese with English abstract). *KKI*, **181**: 7–21.

WATANABE, Nobuo

1955   Hydrographic conditions of the north-western Pacific. Part 1. On the   1   5
       temperature change in the upper layer in summer.
       (In English). *JOSJ*, **11**(3): 111–122.

1957   A preliminary report on the oceanographic survey in the "Kuroshio"   3   1
       area, south of Honshu, June–July, 1955.
       (In English). *ROWJ, NS, Special No.* **1**: 197–208.

1958   On the mechanism of variation in water temperature and chlorinity   5   4, 7
       of upper layer in the seas southwest off Kyushu and around Nansei-
       shoto.
       (In Japanese with English abstract). *TKRFL*, **21**: 15–24.

1966   (with Hideo INABA)   3   4
       Oceanographic conditions in the Kuroshio area south of Honshu in
       summer and autumn 1963.
       (In Japanese with English abstract). *J. Fac. Oceanogr., Tokai Univ.*, **1**: 11–26.

WATANABE, Ryuzo

1954   Results of the oceanographic observations in the eastern sea area of   2   1
       Honsyū, at the beginning of October, 1953.
       (In Japanese with English abstract). *HB, NS*, **41**: 67–68.

WOOSTER, W. S.

1960   (with G. H. VOLKMANN)   1–3, 5–7   2–4   10, 11
       Indications of deep Pacific circulation from the distribution of pro-
       perties at five kilometers.
       (In English). *JGR*, **65**(4): 1239–1249.

WRIGHT, Redwood

1969   Temperature structure across the Kuroshio before and after Typhoon   3   5
       Shirley.
       (In English with Russian abstract). *Tellus*, **21**(3): 409–413.

WÜST, Georg

1929   Schichtung und Tiefenzirkulation des Pazifischen Ozeans auf Grund   1–7   3, 4   8
       zweier Längsschnitte.
       (In German). *Veröff. Institut f. Meereskunde, Berlin Univ.*, N.F., A.
       Geogr.-naturwiss. Reihe, Heft **20**: 63pp.

1930   Meridionale Schichtung und Tiefenzirkulation in den Westhälften der   1–7   3, 4   8
       drei Ozeane.
       (In German). *J. d. Cons.*, **5**: 7–21.

		Area	Subject	Chapter
1936	Kuroshio und Golfstrom. Eine vergleichende hydrodynamische Untersuchung. (In German). *Veröff. Institut f. Meereskunde, Berlin Univ.*, N.F., A. Georg.-naturwiss. Reihe, Heft **29**: 69pp.	1–7	2, 3, 6	5, 6, 8, 13

WYRTKI, Klaus

		Area	Subject	Chapter
1961	Physical oceanography of the Southeast Asian Waters. Scientific results of marine investigations of the South China Sea and Gulf of Thailand, 1959–1961. (In English). *Naga Report*, **2**: 195pp.	7, 8	2, 3, 5	4, 5, 6
1965	The average annual heat balance of the North Pacific Ocean and its relation to ocean circulation. (In English). *JGR*, **70**(18): 4547–4559.	8	8	4
1965	The annual and semiannual variation of sea surface temperature in the North Pacific Ocean. (In English). *Limnology and Oceanography*, **10**(3): 307–313.	8	8	4, 8

YAGURA, Masami

		Area	Subject	Chapter
1961	(with Taneyasu GOTO) On the effect of the sea conditions upon the daily mean tide level differences between Kusimoto and Uragami. (In Japanese). *US*, **36**(4): 93–96.	3	5	

YAMAGATA, Tadakazu

		Area	Subject	Chapter
1961	Strong current belt of the Kuroshio and cyclones. (In Japanese). *Ship and Maritime Met.*, **5**: 1–2, 8–9.	2	8	

YAMAGIWA, Tamio

		Area	Subject	Chapter
1951	Variation of the flow in the intermediate water at the fixed point 153°E, 39°N. (In Japanese with English abstract). *JOSJ*, **6**(3): 160–164.	2	4, 5	

YAMAGUTI, Seiti

		Area	Subject	Chapter
1961	Changes in the heights of mean sea-levels at Kusimoto and Urakami in Kii Peninsula, and their difference. (In English with Japanese abstract). *J. Geodetic Soc. Japan*, **7**(1): 30–33.	3	5	
1968	On the changes in the heights of yearly mean sea-levels in recent years. (In English). *Bull. Earthq. Res. Inst.*, **46**, Part 4: 901–906.	2–4	8	
1968	On the changes in the heights of yearly mean sea-levels preceding the Great Earthquakes. (In English with Japanese abstract). *Bull. Earthq. Res. Inst.*, **46**, Part 6: 1269–1273.	2–4	8	

YAMANAKA, Hajime

		Area	Subject	Chapter
1962	Tunas and oceanic conditions. (In Japanese with English abstract). *JOSJ, 20th Anniv. Vol.*: 663–678.	3, 5–7	4	
1965	(with Noboru ANRAKU and Jiro MORITA) Seasonal and long-term variations in oceanographic conditions in the western North Pacific Ocean. (In Japanese with English abstract). *Rep. Nankai Reg. Fish. Lab.*, **22**: 35–70.	7, 8	2, 3	5

	Area	Subject	Chapter

1968   Variations in the equatorial currents in the western equatorial   **Area 7**   **Subject 2, 5**
Pacific.
(In Japanese). *Rep. C.S.K. (Fisheries Aspects) in 1967*: 99–101.

YAMANAKA, Ichiro

1951   On the hydrographical condition of Japan Sea in spring and summer   **4**   **4, 5**   **9**
1949. Part I.
(In Japanese with English abstract). *JOSJ*, **6**(3): 143–149.

1951   On the hydrographical condition of Japan Sea in spring and summer   **4**   **4**   **9**
1949. Part II.
(In Japanese with English abstract). *JOSJ*, **6**(3): 150–156.

1953   On the hydrographic conditions of the Sea of Japan in spring and   **4**   **1, 4, 5**
summer 1949.
(In English). *ROWJ, NS*, **1**(1): 71–80.

YAMASHITA, Kaoru

1943   On the anomaly of sea water temperature.   **7**   **5**
(In Japanese). *HB*, **21** *Special No.—Kenkyu Chosa Shiryo*, **1**: 77–80.

YAMASHITA, Yukinari

1958   The oceanographical conditions of the sea east of Honshu, Japan, in   **2**   **4**
1956 (Correspondences).
(In Japanese). *JOSJ*, **14**(1): 31.

1967   (with Shizuo YAMAUCHI)   **2, 3, 5**   **1–6**   **7**
On the states of the Kuroshio in recent years.
(In Japanese). *HB, NS*, **82**: 9–22.

YASUI, Masashi

1955   On the rapid determination of the dynamic depth anomaly in the   **3, 6**   **8**
Kuroshio Area.
(In English). *ROWJ, NS*, **2**(2): 90–95.

1958   Internal waves in the open ocean.   **2**   **5, 7**
(In English). *OM*, **10**(2): 227–234.

1961   Internal waves in the open ocean—An example of internal waves pro-   **2**   **5**   **7**
gressing along the oceanic frontal zone.
(In English). *OM*, **12**(2): 157–183.

1961   Internal waves in the open ocean—Internal waves of long periods in   **6**   **5**
the ocean weather station "Tango".
(In English). *OM*, **12**(2): 185–205.

1963   Note of the Japanese Expedition of Deep-Sea the Fifth Cruise, 1962   **2, 3, 6**   **1**
(JEDS-5).
(In English). *OM*, **14**(2): 131–133.

1967   (with Takeo YASUOKA, Katsumi TANIOKA and Osami SHIOTA)   **4**   **3**   **9**
Oceanographic studies of the Japan Sea (1)—Water characteristics.
(In English). *OM*, **19**(2): 177–192.

YASUI, Zen'ichi

1941   (with Kiyoshi KOSAKA)   **2, 3, 5–7**   **5**
Annual variations of temperature and density of coastal waters along

		Area	Subject	Chapter
	the Pacific coast of Japan. (In Japanese). *US*, **21**(2): 59–64.			
1953	(with Masaharu MATUNO) On the circulation of currents in the China Eastern Sea. (In English). *ROWJ, NS*, **1**(1): 28–35.	5	2, 4, 5	
1960	(with Katsumi HATA) On the seasonal variations of the sea conditions in the Tsugaru Warm Current region. (In English). *KMOM*, **14**: 3–12.	4	3	9

YI, S.-U.

| 1966 | Seasonal and secular variations of the water volume transport across the Korea Strait. (In English). *JOSK*, **1**(1–2): 7–13. | 4 | 2, 5 | 9 |

YONEZAWA, Hisashi

| 1950 | The oceanographic result, in the area east of Honshu, July and August 1949. (In Japanese). *HB, NS*, **16**: 245–256. | 2 | 1 | |

YOSHIDA, Kozo

1957	(with Mizuki TSUCHIYA) Note on the thermosteric anomaly. (In English). *JOSJ*, **13**(4): 127–129.	1–3, 6, 7	8	
1965	A theoretical model on wind-induced density field in the oceans. I. (In English). *JOSJ*, **21**(4 : 154–173.	1–3, 5–7	7	
1967	(with Toshiko KIDOKORO) A Subtropical Counter-Current in the North Pacific—An eastward flow near the Subtropical Convergence. (In English). *JOSJ*, **23**(2): 88–91.	6	7	4, 5, 13
1967	(with Toshiko KIDOKORO) A Subtropical Countercurrent (II)—A prediction of eastward flows at lower subtropical latitudes. (In English). *JOSJ*, **23**(5): 231–246.	3,5–7	2,7	5,13
1969	Subtropical Countercurrents—A preliminary note on further observational evidences. (In English). *ROWJ, NS*, **10**(1): 123–124.	5,6	2,4	13

YOSIDA, Shozo

1950	Oceanographic result in the area east of Honsyū, August 1950. (In Japanese). *HB, NS*, **21**: 240–241.	2	1	
1951	Oceanographic result in the area south of Honsyū, June and July, 1950. (In Japanese). *HB, ŃS*, **22**: 16–18.	3	1	
1951	Oceanographic result in the area south of Honsyū, December, 1950. (In Japanese). *HB, ŃS*, **23**: 72–74.	3,5	1	
1951	Oceanographic result in the area east of Honsyū, August and September, 1950. (In Japanese). *HB, ŃS*, **24**: 124–125.	2	1	

		Area	Subject	Chapter
1951	Oceanographic result in the area south of Honsyū, December, 1950 and February and March, 1951. (In Japanese). *HB*, *ÑS*, **24**: 124–126.	3,5	1	
1951	Oceanographic result in the area south and east of Honsyū, February and March, 1951. (In Japanese). *HB*, *NS*, **26**: 328–330.	2,3	1	
1951	Oceanographic result in the area south of Honsyū, May and June, 1951. (In Japanese). *HB*, *ÑS*, **27**: 376–377.	3,5	1	
1952	Oceanographic result in the area east of Honsyū, June 1951. (In Japanese). *HB*, *NS*, **29**: 22–23.	2	1	
1952	Current and other phenomena in the area south of Honsyū in February–May, 1952. (In Japanese). *HB*, *NS*, **32**: 172–175.	3	1	
1953	Results of oceanographic observations in the area east of Honsyū in July and August, 1952. (In Japanese with English abstract). *HB*, *ÑS*, **35**: 56–59.	2,3	1	
1959	(with Hideo NITANI and Naruji SUZUKI) Report of multiple ship survey in the equatorial region (I.G.Y.) Jan.–Feb., 1958. (In English). *HB*, *NS*, **59**: 1–30.	7	1,2,4	1
1961	On the short period variation of the Kurosio in the adjacent sea of Izu Islands. (In Japanese with English abstract). *HB*, *ÑS*, **65**: 1–18.	3	5	Pre., 1 6, 7, 13
1961	On the variation of Kurosio and cold water mass off Ensyū Nada (Part 1). (In Japanese with English abstract). *HB*, *NS*, **67**: 54–57.	3	5,6	6, 7, 12
1964	A note on the variations of the Kuroshio during recent years. (In Japanese). *Bull. Japan. Soc. Fish. Oceanogr.*, **5**: 66–69.	3	5,6	13

YOSHIZAWA, Hiroshi

| 1957 | On the variations of the monthly mean sea levels in the western parts of Japan. (In Japanese with English abstract). *US*, **33**(4–5): 82–86. | 3 | 3,5 | |

ZAVERNIN, J. P.

| 1969 | Some features of the interaction between atmosphere and hydrosphere in the northwestern Pacific Ocean. (In Russian). *Izvestija, Tichookeanskogo nautchnoissle-dovatelskogo instituta ribnogo chozyastva i okeanografii*, **68**: 67–77. | 1–3 | 3 | |

ZENKEVICH, L. A.

| 1957 | Oceanographic research conducted by the USSR in the north-west Pacific. (In English with French abstract). *Proc. UNESCO Symp. Phys. Oceanogr., Tokyo, 1955*: 251–252. | 1–4 | 8 | |

LIST OF ABBREVIATIONS

CG   Geography (Chirigaku)
          Geographical Society of Japan, Tokyo

CI   Chuokishodai Iho (Memoirs of the Central
          Meteorological Observatory)
          Central Meteorological Observatory, Tokyo

CSJB   Bulletin of Chemical Society of Japan
          Chemical Society of Japan, Tokyo

GM   Geophysical Magazine
          Japan Meteorological Agency, Tokyo

GN   Geophysical Notes
          Geophysical Institute, Faculty of Science,
          University of Tokyo, Tokyo

HB   Hydrographic Bulletin (Suiro Yoho)
          Imperial Japanese Navy

HB, NS   Hydrographic Bulletin (Suiro Yoho), New
          Series
          Maritime Safety Agency, Tokyo

HDB   Bulletin of the Hydrographic Department
          Imperial Japanese Navy

HRFL   Bulletin of Hokkaido Regional Fisheries Re-
          search Laboratory
          Hokkaido Regional Fisheries Research Labor-
          atory, Yoichi, Hokkaido

JG   Journal of Geography (Chigaku Zasshi)
          Geographical Society of Japan, Tokyo

JJAG   Japanese Journal of Astronomy & Geophysics
          National Research Council of Japan, Tokyo

JIFES   Journal of Imperial Fishery Experimental
          Station (Suisan Shikenjo Hokoku)
          Imperial Fishery Experimental Station

JMR   Journal of Marine Research
          Bingham Oceanographic Laboratory, Yale
          University

JMSJ   Journal of Meteorological Society of Japan
          (Kisho Shushi)
          Meteorological Society of Japan, Tokyo

JO   Journal of Oceanography (Kaiyo Jiho)
          Kobe Marine Observatory, Kobe

JOSJ   Journal of Oceanographical Society of Japan
          Oceanographical Society of Japan, Tokyo

JOSK   Journal of Oceanographical Society of Korea
          Oceanological Society of Korea, Seoul

JSRFL   Bulletin of Japan Sea Regional Fisheries Re-
          search Laboratory
          Japan Sea Regional Fisheries Research Labor-
          atory, Niigata

JSSFB   Bulletin of the Japanese Society of Scientific
          Fisheries
          Japanese Society of Scientific Fisheries, Tokyo

KG   Kagaku (Science)
          Iwanami Shoten, Tokyo

KH   Kaiyo Hokoku (The Oceanographical Report
          of Japan Meteorological Agency)
          Japan Meteorological Agency, Tokyo

KI   Kaisho Iho
          Imperial Japanese Navy

KI, NS   Kaisho Iho (Oceanographic Bulletin), New
          Series
          Maritime Safety Agency, Tokyo

KJ   Kaisho Jiho (Oceanographic Bulletin)
          Maritime Safety Agency, Tokyo

KKI   Kobe Kaiyo Kishodai Iho (Bulletin of the
          Kobe Marine Observatory)
          Kobe Marine Observatory, Kobe

KMOM   Memoirs of the Kobe Marine Observatory
          Kobe Marine Observatory, Kobe

KSH   Kaiyo Shigen Kaihatsu Chosa Hokoku
          Japanese National Commission for UNESCO
          Tokyo

KSKH   Kaiyo Shigen Kaihatsu Chosa Kenkyu
          Hokoku
          The Japan Society for the Promotion of Science
          Tokyo

KY   Kaiyo Chosa Yoho (Oceanographic Observa-
          tion Report)
          Imperial Fisheries Experimental Station

MGP   Papers in Meteorology and Geophysics
          Meteorological Research Institute, Tokyo

OM   Oceanographical Magazine
          Japan Meteorological Agency, Tokyo

PMSP   Proceedings of Physico-Mathematical Society
          of Japan (Nihon Sugaku Butsuri Gakkai Kiji)
          Physico-Mathematical Society of Japan, Tokyo

ROWJ   Records of Oceanographic Works in Japan
          Science Council of Japan, Tokyo

ROWJ,   Records of Oceanographic Works in Japan,
NS   New Series
           Science Council of Japan, Tokyo

SCHH   Suisan Chosa Hokoku
           Fisheries Agency, Tokyo

SFLB   Bulletin of the Seikai Regional Fisheries
           Research Laboratory
           Fisheries Agency, Tokyo

SO   Studies on Oceanography
           University of Tokyo Press, Tokyo

SRW   The Scientific Reports of Whales Research
           The Whales Research Institute, Tokyo

SSK   Suisan-Kai (The Fisheries World)
           Japan Fisheries Association, Tokyo

THRFL   Bulletin of Tohoku Regional Fisheries Re-

search Laboratory
Tohoku Regional Fisheries Research Laboratory, Sendai

TKH   Tsushima Danryu Kaihatsu Chosa Kenkyu
           Hokoku
           Fisheries Agency, Tokyo

TKRFL   Bulletin of Tokai Regional Fisheries Research
           Laboratory
           Tokai Regional Fisheries Research Laboratory
           Tokyo

TUFJ   Journal of the Tokyo University of Fisheries
           Tokyo University of Fisheries, Tokyo

US   Umi to Sora (Sea and Sky)
           Marine Meteorological Society, Kobe Marine
           Observatory, Kobe

APPENDIX: *Data Sources*

FISHERIES AGENCY

Publ. Year	Period	Total Page
1966	The results of fisheries oceanographical observations. January–December 1963.	1026.
1967	The results of fisheries oceanographical observations. January–December 1964.	1643.
1968	The results of fisheries oceanographical observations. January–December 1965.	1578.
1969	The results of fisheries oceanographical observations. January–December 1951.	144.
1969	The results of fisheries oceanographical observations. January–December 1966.	1671.
1970	The results of fisheries oceanographical observations. January–December 1952.	141.
1970	The results of fisheries oceanographical observations. January–December 1967.	1702.

TOKAI REGIONAL FISHERIES RESEARCH LABORATORY

(IMPERIAL FISHERIES INSTITUTE)

1918—   Oceanographic Investigation (Quarterly Report).
         (In Japanese with English abstract).

Publ. Year		Period	Total Page
1918a	No. 1	April–June 1918.	53.
1918b	No. 2	July–September 1918.	38.
1919a	No. 3	October–December 1918.	64.
1919b	No. 4	January–March 1919.	12.
1919c	No. 5	April–June 1919.	16.
1919d	No. 6	July–September 1919.	14.

Publ. Year		Period	Total Page
1920a	No. 7	October–December 1919.	179.
1920b	No. 8	January–March 1920.	50.
1920c	No. 9	April–June 1920.	49.
1920d	No. 10	July–September 1920.	53.
1921a	No. 11	October–December 1920.	89.
1921b	No. 12	January–March 1921.	82.
1921c	No. 13	April–June 1921.	85.
1921d	No. 14	July–September 1921.	86.
1922a	No. 15	October–December 1921.	85.
1922b	No. 16	January–March 1922.	70.
1922c	No. 17	April–June 1922.	79.
1922d	No. 18	July–September 1922.	84.
1923a	No. 19	October–December 1922	81.
1923b	No. 20	January–March 1923.	13.
1923c	No. 21	April–June 1923.	15.
1923d	No. 22	July–September 1923.	13.
1924a	No. 23	October–December 1923.	258.
1924b	No. 24	January–March 1924.	64.
1924c	No. 25	April–June 1924.	77.
1924d	No. 26	July–September 1924.	84.
1925a	No. 27	October–December 1924.	74.
1925b	No. 28	January–March 1925.	11.
1925c	No. 29	April–June 1925.	16.
1925d	No. 30	July–September 1925.	14.
1926a	No. 31	October–December 1925.	179.
1926b	No. 32	January–March 1926.	45.
1926c	No. 33	April–June 1926.	52.
1926d	No. 34	July–September 1926.	58.
1927a	No. 35	October–December 1926.	51.
1927b	No. 36	January–March 1927.	46.
1927c	No. 37	April–June 1927.	64.
1927d	No. 38	July–September 1927.	70.
1928a	No. 39	October–December 1927.	56.
1928b	No. 40	January–March 1928.	44.
1928c	No. 41	April–June 1928.	63.
1929a	No. 42	July–September 1928.	72.
1929b	No. 43	October–December 1928.	55.

(THE IMPERIAL FISHERIES EXPERIMENTAL STATION)
(Semi-annual Report)

1930	No. 44	January–June 1929.	168.
1930	No. 45	July–December 1929.	185.
1930	No. 46	January–June 1930.	276.
1931	No. 47	July–December 1930.	284.
1931	No. 48	January–June 1931.	235.
1932	No. 49	July–December 1931.	236.
1932	No. 50	January–June 1932.	406.
1933	No. 51	July–December 1932.	283.
1933	No. 52	January–June 1933.	242.
1934	No. 53	July–December 1933.	394.
1934	No. 54	January–June 1934.	239.
1935	No. 55	July–December 1934.	342.
1935	No. 56	January–June 1935.	297.
1936	No. 57	July–December 1935.	336.
1936	No. 58	January–June 1936.	205.
1937	No. 59	July–December 1936.	246.
1937	No. 60	January–June 1937.	265.

Publ. Year		Period	Total Page
1938	No. 61	July–December 1937.	218.
1938	No. 62	January–June 1938.	181.
1939	No. 63	July–December 1938.	163.
1940	No. 64	January–June 1939.	197.
1940	No. 65	July–December 1939.	147.
1941	No. 66	January–June 1940.	151.
1941	No. 67	July–December 1940.	126.
1942	No. 68	January–June 1941.	182.
1942	No. 69	July–December 1941.	116.
1943	No. 70	January–June 1942.	107.
1943	No. 71	July–December 1942.	70.

(TOKAI REGIONAL FISHERIES RESEARCH LABORATORY)

1951	No. 72	January 1943–December 1944.	105.
1952	No. 74	January–December 1950.	143.
1967	No. 73	January 1945–December 1949.	186.
1968	Suppl., No. 73 and 74	January 1949–December 1950.	121.

TOHOKU REGIONAL FISHERIES RESEARCH LABORATORY

Annual Report on the Fish Resources, Oceanographic Investigation

Publ. Year	Period	Total Page
1958	1947 and 1948	38.
1952	1949	53.
1958	1950	90.
1960	1951	123.
1961	1952	96.
1962 (March)	1953	90.
1962 (November)	1961	90.

Annual Report on the Fish Resources, Oceanographic Investigation
(Preliminary Report)

1959 October	1958 (May–Nov.)	171.
1960 March	1959 (May–Aug.)	228.
1960 October	1959 (Aug.–Nov.)	113.
1961 March	1960 (Feb.–Aug.)	138.
1961 October	1960 (Aug.–Nov.)	88.

HYDROGRAPHIC OFFICE, MARITIME SAFETY AGENCY

1948—   Results of Oceanographic Observation in the north-western Pacific Ocean.
           (In English).

Publ. Year		Period	Reference: Total Page
1948	No. 1	April 1923–March 1929.	KJ, 3: 224.
1949	No. 2	April 1930–March 1931.	KJ, 6: 44.
1950	No. 3	April 1931–March 1935.	HB, NS, Special 6: 142.
1951	No. 4	May 1935–February 1938.	HB, NS, Special 8: 146.
1952	No. 5	February 1938–November 1941.	HB, NS, Special 9: 199.
1954	No. 6	May 1938–September 1939.	HB, NS, Special 13: 178.
1955	No. 7	June 1939–March 1941.	HB, NS, Special 16: 156.
1962	No. 8	August 1931–January 1941.	HB, NS, 69: 245.
1962	No. 9	January–December 1941.	HB, NS, 71: 218.
1963	No. 10	January–December 1942.	HB, NS, 74: 230.
1966	No. 11	January–December 1943.	DHO, SO, 2: 202.
1968	No. 12	January–December 1944.	DHO, SO, 6: 194.

1947—   Tables of Results from Oceanographic Observation.
         (In English).

Publ. Year	Period	Reference: Total Page
1947	1946	KI, NS, 1: 89.
1949	1946	KJ, 4: 25.
1949	1949	HB, NS, 14: 166–176.
1949	1949	HB, NS, 15: 201–213.
1949	1947	KJ, 5: 10
1950	1949	HB, NS, 16: 257–276.
1950	1948	HB, NS, Special 5: 52.
1950	1949	HB, NS, Special 7: 123–169.
1953	1946–1948	HB, NS, Special 10: 110.
1954	1950	HB, NS, Special 14: 26–164.
1954	1951	HB, NS, Special 15: 133.
1956	1952–1953	HB, NS, 51: 171.
1959	1954–1955	HB, NS, 58: 139.
1959	1956	HB, NS, 62: 91.
1960	1957	HB, NS, 64: 103.
1961	1958	HB, NS, 66: 153.
1961	1959	HB, NS, 68: 112.
1964	1960	HB, NS, 75: 86.
1964	1961	HB, NS, 77: 82.
1965	1962	DHO, SO, 1: 77.
1966	1963	DHO, SO, 3: 85.
1967	1964	DHO, SO, 4: 98.
1967	1965	DHO, SO, 5: 125.
1968	1966	DHO, SO, 7: 117.
1970	1967	DHO, SO, 8: 91.

1955—   State of the adjacent seas of Japan (Quarterly).
         Surface current observed with G.E.K.
         Temperature at 0m, 100m and 200m
         28°N–46°N, 122°E–155°E

1951   Current charts of the adjacent seas of Nippon (Vol. 1).   24 figs.

1971   Current charts of the adjacent seas of Nippon (Vol. 2).   40 figs.
         May 1955–October to November 1964

JAPANESE OCEANOGRAPHICAL DATA CENTER

1966—   Preliminary data reports of CSK.
         (In English).

Publ. Year

1966   No. 1, 2, 3, 4, 5, 6, 7, 8, 9, 10, 11, 12, 13, 14, 15, 16, 17, 18, 19, 20, 21, 22, 23, 24, 25, 26, 27,
         28, 30, 31, 32, 33, 37, 38.

1967   No. 29, 34, 35, 36, 39, 40, 41, 42, 43, 44, 45, 46, 47, 48, 49, 50, 51, 52, 53, 54, 55, 56, 57, 58,
         59, 60, 61, 62, 63, 66, 67, 68, 69, 70, 71, 72, 74, 76, 77, 78, 79, 80, 81, 83, 84, 85, 87, 88, 99,
         100, 104, 113, Supplement to No. 24, Supplement (Volume of errata).

1968   No. 64, 65, 82, 86, 89, 90, 92, 96, 97, 103, 105, 106, 108, 109, 114, 145.

1968—   Data report of CSK.
         (In English).

1968   No. 73, 75, 91, 94, 98, 102, 110, 111, 115, 116, 117, 119, 124, 125, 126, 127, 128, 130, 133, 134,
         136, 137, 138, 139, 142, 154, 155, 157, 158, 161, 162, 163.

Publ. Year
1969   No. 93, 95, 101, 107, 112, 121, 122, 123, 131, 132, 140, 141, 143, 144, 150, 156, 159, 160, 165,
166, 167, 168, 169, 171, 172, 173, 174, 176, 177, 179, 180, 181, 182, 183, 184, 185, 186, 187,
188, 189, 191, 198, 199, 200, 202, 203, 204, 205, 206, 210, 213, 214, 215, 217.

1970   No. 135, 148, 175, 190, 193, 194, 195, 196, 197, 201, 209, 211, 216, 218, 219, 220, 221, 222, 223,
227, 228, 229, 231, 232, 233, 234, 235, 236, 238, 239, 240, 257, 264, 266, 267, 268, 270, 271, 272.

1967—   CSK Atlas

Publ. Year		Period	Total Page
1967	CSK Atlas	vol. 1, summer 1965	32.
1968		vol. 2, winter 1965–66	44.
1969		vol. 3, summer 1966	31.
1970		vol. 4, winter 1967	32.

JAPAN METEOROLOGICAL AGENCY

1946—   The ten-day marine report (In Japanese).
Mean sea surface temperature (°C) averaged for 1° quadrangle 10°N–53°N, 110°E–180°E.

1950—   The results of marine meteorological and oceanographical observations.
(In Japanese).

(CENTRAL METEOROLOGICAL OBSERVATORY)

Publ. Year		Period		Total Page
1950	No. 1	January–June 1947.		113.
1950	No. 2	July–December 1947.		244.
1951	No. 3	January–June 1948.		256.
1951	No. 4	July–December 1948.		414.
1951	No. 5	January–June 1949.		337.
1951	No. 6	July–December 1949.		423.
1952	No. 7	January–June 1950.		220.
1952	No. 8	July–December 1950.		299.
1952	No. 9	January–June 1951.		177.
1952	No. 10	July–December 1951.		310.
1953	No. 11	January–June 1952.		362.
1954	No. 12	part 1: Oceanography	July–December 1952.	138.
1954	No. 13	part 1: Oceanography	January–December 1953.	210.
1955	No. 14	part 1: Oceanography	January–June 1954.	93.
1955	No. 15	part 1: Oceanography	July–December 1954.	134.
1955	No. 16	part 1: Oceanography	January–June 1955.	120.
1956	No. 17	NORPAC Expedition (Special Number) July–September 1955.		131.

(JAPAN METEOROLOGICAL AGENCY)

Publ. Year		Period		Total Page
1956	No. 18	part 1: Oceanography	July–December 1955.	90.
1957	No. 19	January–June 1956.		184.
1957	No. 20	July–December 1956.		191.
1958	No. 21	January–June 1957.		168.
1958	No. 22	July–December 1957		183.
1959	No. 23	January–June 1958.		240.
1959	No. 24	July–December 1958.		289.
1960	Supplement			149.
1960	No. 25	January–June 1959.		258.
1960	No. 26	July–December 1959.		256.
1961	No. 27	January–June 1960 (including the results of JEDS*-2 & 3).		257.
1962	No. 28	July–December 1960.		304.
1962	No. 29	January–June 1961.		284.
1962	No. 30	July–December 1961 (including the results of JEDS*-4).		326.

Publ. Year		Period	Total Page
1964	No. 31	January–June 1962.	313.
1964	No. 32	July–December 1962 (including the results of JEDS*-5).	328.
1964	No. 33	January–June 1963 (including the results of JEDS*-6).	289.
1965	No. 34	July–December 1963.	360.
1966	No. 35	January–June 1964 (including the results of JEDS*-8).	328.
1967	No. 36	July–December 1964.	367.
1968	No. 37	January–June 1965 (including the results of JEDS*-10).	385.
1968	No. 38	July–December 1965.	404.
1969	No. 39	January–June 1966.	349.
1970	No. 40	July–December 1966.	336.
1970	No. 41	January–June 1967 (including the results of JEDS*-11).	332.
1970	No. 42	July–December 1967.	273.
1970	No. 43	January–June 1968.	289.

1962—   Marine climatological tables of the North Pacific Ocean.

Publ. Year	Period	Reference	Total Page
1962	1942–1960 (Part 1)	Tech. Report, No. 17	321.
1963	1942–1960 (Part 2)	Tech. Report, No. 23	297.
1966	1961	Tech. Report, No. 51	176.
1967	1962		178.
1967	1963		179.
1968	1964		184.
1970	1965		184.

*The Japanese Expedition of the Deep Sea

KOBE MARINE OBSERVATORY

1921—   Journal of Oceanography (Kaiyo Jiho): The temperature and densities of the sea-water along the Japanese coasts. (In Japanese).

Publ. Year & No.	Period	Page
1929: 1(1)	January–February 1929.	46–49.
1929: 1(2)	March–May 1929.	118–123.
1930: 1(3)	June–September 1929.	368–376.
1930: 2(1)	October–December 1929.	78–85.
	Japan Sea, July–October 1928.	1–73(data: 45–73).
1930: 2(2)	January–April 1930.	268–275.
	Japan Sea, July–October 1928.	1–264(data: 200–264).
1930: 2(3)	May–August 1930.	449–462.
1930: 2(4)	September–December 1930.	647–664.
1931: 3(1)	January–March 1931.	154–165.
1932: 3(2)	April–August 1931.	376–398.
	Japan Sea, July–September 1929.	291–375(data: 344–375).
1932: 3(3)	September–December 1931.	620–637.
	Japan Sea, July–September 1929.	545–619(data: 585–619).
1932: 4(1)	January–April 1932.	175–190.
	Japan Sea, July–September 1930.	1–173(data: 70–173).
1933: 4(2)	May–August 1932.	445–462.
	Tsugaru Strait, August 1930.	341–370(data: 358–370).
1933: 5(1)	September–December 1932.	211–229.
1933: 5(2)	January–April 1933.	397–416.
1933: 6(1)	May–August 1933.	113–132.
	Japan Sea, {June–September 1931. / July–October 1932.	1–106(data: 34–101).
1934: 6(2)	September–December 1933.	247–270.
	Tsugaru Strait, August–September 1932.	177–207(data: 191–207).

Publ. Year & No.	Period	Page
1934: 7(1)	January–April 1934.	367–394.
1935: 7(2)	May–August 1934.	461–488.
1935: 8(1)	September–December 1934.	65–92.
1935: 8(2)	January–April 1935.	213–240.
1936: 8(3)	May–December 1935.	333–388.
1936: 9(1)	January–April 1936.	19–48.
1936: 9(2)	May–August 1936.	271–304.
1937: 10(1)	September–December 1936.	77–109.
1937: 10(2)	January–April 1937.	191–220.
1938: 11(1)	May–August 1937.	129–160.
	Kuroshio region, summer 1933–35.	1–128(data: 48–128).
1938: 11(2)	September–December 1937.	363–396.
1938: 11(3)	January–April 1938.	631–652.
1939: 12(1)	May–December 1938.	89–128.
1940: 12(2)	January–August 1939.	259–300.
1941: 13(2)	September 1939–April 1940.	368–407.
	Off Sanriku, May–June 1940.	330–367(data: 331–367).
	Off Kishu, April–May 1940.	318–329(data: 319–329).
1942: 13(3)	May–October 1940.	676–707.
	Off Sanriku, October 1940–May 1941.	653–675(data: 653–675).
1943: 13(4)	November 1940–October 1941.	1002–1075.
	Off Sanriku, July 1941–February 1942.	978–1001(data: 978–1001).
1943: 14(1)	November 1941–September 1942.	145–225.
1944: 14(2)	October 1942–July 1943.	425–512.

The mean atmospheric pressure, cloudiness, air temperature and sea surface temperature of the North Pacific Ocean and the neighbouring seas.

Publ. Year	Period	Total Page
1927	For the year, 1926	122.
1928	For the year, 1927	171.
1929	For the year, 1928	176.
1930	For the year, 1929	173.
1931	For the year, 1930	176.
1932	For the year, 1931	167.
1933	For the year, 1932	167.
1934	For the year, 1933	168.
1935	For the year, 1934	168.
1936	For the year, 1935	168.
1938	For the year, 1936	195.
1939	For the year, 1937	195.
1939	For the year, 1938	195.
1940	For the year, 1939	195.
1941	For the year, 1940	195.
1943	For the year, 1941	186.
1925	For the lustrum 1916–1920	621.
1928	For the lustrum 1911–1915	581.
1929	For the lustrum 1921–1925	623.
1932	For the lustrum 1926–1930	234.
1935	During the years, 1911–1930	173.
1937	During the years, 1911–1935	197.
1943	During the years, 1911–1940	157.

HAKODATE MARINE OBSERVATORY
The Journal of Oceanography of the Hakodate Marine Observatory

1944	No. 1	114.
1945	No. 2	226.

UNIVERSITY OF CALIFORNIA AND UNIVERSITY OF TOKYO PRESS

Publ. Year	Reference & Period	Total Page
1960	Oceanic observations of the Pacific. 1955. (The NORPAC Atlas). (In English).	123 (pls.).
1960	Oceanic observations of the Pacific. 1955. (The NORPAC Data). (In English).	532.

UNIVERSITY OF CALIFORNIA

1961	Oceanic observations of the Pacific. Pre–1949.	349.
1962	Oceanic observations of the Pacific. 1955. Operation Troll.	477.
1965	Oceanic observations of the Pacific. 1953. Expedition Transpac.	576.

SCRIPPS INSTITUTION OF OCEANOGRAPHY, UNIVERSITY OF CALIFORNIA

1966	Data report: Boreas Expedition. 27 January–1 April 1966.	164.
1970	Data report: Zetes Expedition, Leg 1. 11–24 January 1966.	67.

OCEAN RESEARCH INSTITUTE, UNIVERSITY OF TOKYO

	Reference	Page
1968	Preliminary report of the Hakuho Maru cruise KH-68-3.	111–116.

UNITED STATES DEPARTMENT OF THE INTERIOR
U.S. FISH AND WILDLIFE SERVICE

1968	Monthly mean charts, sea surface temperature, North Pacific Ocean 1949–1962 (U.S. Fish and Wildlife Service: Circular 258).	168 figs.

IGY WORLD DATA CENTER A, A & M COLLEGE OF TEXAS

		Total Page
1961	IGY Oceanography Report No. 3 Oceanographic Observations in the Intertropical Region of the World Ocean during IGY and IGC. 1956–1959. Part IIa: Pacific Ocean.	225.

FACULTY OF FISHERIES, UNIVERSITY OF HOKKAIDO

1957—   Data record of oceanographic observations and exploratory fishing

Reference No.	Publ. Year	Period	Total Page
No. 1	1957	1953–1956	247.
No. 2	1958	1957	199.
No. 3	1959	1957–1959	296.
No. 4	1960	1959	221.
No. 5	1961	1935–1939 1959–1960	392.
No. 6	1962	1961	283.
No. 7	1963	1962	262.
No. 8	1964	1963	303.
No. 9	1965	1964	343.
No. 10	1966	1965	388.
No. 11	1967	1965–1966	383.
No. 12	1968	1966–1967	421.
No. 13	1969	1968	406.
No. 14	1970	1969	529.

MARITIME SELF DEFENCE FORCE, DEFENCE AGENCY, JAPAN

Publ. Year	Reference	Period	Total Page
1962	Annual oceanographic observation report.	1961	231.
1963	//	1962	308.
1964	//	1963	301.
1965	//	1964	363.
1966	//	1965	431.
1967	//	1966	527.
1968	//	1967	517.
1969	//	1968	507.
1970	//	1969	555.

(BT Observation only)

Author Index

Heavy type indicates the page numbers at which the authors' own chapters begin.
The names of the authors appearing in Chapter 14(Bibliography) are not included here.

Subject Index